Polyhedral Graphical Statics

Polyhedral Graphic Statics is a geometry-based structural design method based on polyhedral reciprocal diagrams first proposed by Rankine and Maxwell 150 years ago. Providing a unique design guide for structural form finding of efficient spatial structures, this book offers up-to-date insights into the field of geometry-based structural design.

The author introduces a different design domain, the force domain, and explains the relationship between the geometry of efficient structural forms and the geometric equilibrium of forces called the force diagram. Beginning with step-by-step geometric methods to construct the reciprocal force diagram for simple structural forms, the book then advances to more articulated systems. These include spatial compression-only forms and systems with mixed tensile and compressive forces. Presenting computational applications, the text provides an algebraic formulation for form finding in 3D.

Discover the applications of this pioneering method in Material Science, Mechanical Engineering, Aerospace, and the design of carbon-absorbing structures.

Masoud Akbarzadeh is an Associate Professor of Architecture in Structures and Building Technologies and the Director of the Polyhedral Structures Laboratory. In 2020, he received the National Science Foundation CAREER Award for extending the methods of 3D/Polyhedral Graphic Statics for Education, Design, and Optimization of High-Performance Structures.

An indispensable guide for students, engineers, and scientists, providing a beautifully illustrated, step-by-step guide to the polyhedral geometry that bridges forms and forces. The author combines modern computational geometry with classical graphic statics to create revolutionary spatial structures and cutting-edge cellular materials.

Tomohiro Tachi, The University of Tokyo

A joy to read! This book provides clear insight and original design concepts based on the author's immense knowledge of form and structure. An inspiration for us to explore, create, and invent new approaches in design.

Paul Kassabian, SGH

Polyhedral Graphical Statics

For Funicular Structural Form Finding

MASOUD AKBARZADEH
University of Pennsylvania

CAMBRIDGE
UNIVERSITY PRESS

CAMBRIDGE
UNIVERSITY PRESS

Shaftesbury Road, Cambridge CB2 8EA, United Kingdom

One Liberty Plaza, 20th Floor, New York, NY 10006, USA

477 Williamstown Road, Port Melbourne, VIC 3207, Australia

314–321, 3rd Floor, Plot 3, Splendor Forum, Jasola District Centre,
New Delhi – 110025, India

103 Penang Road, #05–06/07, Visioncrest Commercial, Singapore 238467

Cambridge University Press is part of Cambridge University Press & Assessment,
a department of the University of Cambridge.

We share the University's mission to contribute to society through the pursuit of
education, learning and research at the highest international levels of excellence.

www.cambridge.org
Information on this title: www.cambridge.org/9781108494731

DOI: 10.1017/9781108859769

When citing this work, please include a reference to the DOI 10.1017/9781108859769

First published 2025

A catalogue record for this publication is available from the British Library

A Cataloging-in-Publication data record for this book is available from the Library of Congress

ISBN 978-1-108-49473-1 Hardback

To Yasaman, Zarrin Taj, and Mohsen

Contents

Contributors

Mostafa Akbari
PhD Candidate at the Weitzman School, Polyhedral Structures Laboratory, University of Pennsylvania

Hamid Akbarzadeh
Associate Professor and the director of AM^3, Department of Mechanical Engineering, McGill University

Georgios Athanasopoulos
Senior BIM Modeler and CORE Studio Representative, Thornton Tomasetti

William F. Baker
Consulting Partner, Skidmore Owings & Merrill

Mathias Bernhard
Postdoctoral Researcher, Media and Design Laboratory, The Swiss Federal Institute of Technology in Lausanne

Philippe Block
Professor of Architecture and Structures at the Institute of Technology in Architecture, ETH Zurich

Mohammas Bolhassani
Assistant Professor of Architecture, City College, New York

Hua Chai
PhD Student at the Weitzman School, Polyhedral Structures Laboratory, University of Pennsylvania

Marton Hablicsek
Assistant Professor at Leiden University

Zakaria Hsain
Postdoctoral Candidate at the University of Pennsylvania

Marina Konstantatou
Design Systems Analyst, Forster and Partners Office

Juney Lee
Assistant Professor of Architecture at The Carnegie Mellon University

Yao Lu
PhD Candidate at the Weitzman School, Polyhedral Structures Laboratory, University of Pennsylvania

Allan McRobie
Professor of Structural Engineering, Cambridge University

Armin Mirabolghasemi
Associate Professor and the director of AM^3, Department of Mechanical Engineering, McGill University

Hossein Mofatteh
PhD Student at AM^3, Department of Mechanical Engineering, McGill University

Salma Mozaffari
Postdoctoral Research Associate, Princeton University

Maximilian E. Ororbia
Postdoctoral Fellow at the University of Pennsylvania

James Pikul
Associate Professor, Department of Mechanical Engineering, University of Wisconsin at Madison

Jens Schneider
Professor Structural Mechanics and Design at TU Darmstadt

Shu Yang
Chair of the Department of Materials Science and Engineering at the University of Pennsylvania

Paria Yavartanoo
Postdoctoral Candidate at The ABC Lab, City College, New York

Joseph Robert Yost
Professor of Structural Engineering, Villanova University

Kun-Hao Yu
Postdoctoral Candidate at the University of Pennsylvania

Hao Zheng
Assistant Professor of Architecture, City University of Hong Kong

Yefan Zhi
PhD Student at the University of Pennsylvania

Foreword by Dr. William F. Baker

Geometry is key.

The geometry of the struts and cables of a bridge or the geometry of the atoms in a molecule sets the limits of the efficiency and behavior of a structure or a material. The more the designer, be it an architect, an engineer, or a material scientist, understands the effects of geometry on behavior, the more effective and creative they can become.

As I matured in my career as a structural engineer, the more I came to understand the importance of geometry. It became central to the structures I was designing, such as Exchange House in London or the Burj Khalifa in Dubai. The geometric arrangement of the structural elements has an outsized effect on the efficiency of a structure. Furthermore, if the geometry is such that the forces of gravity and wind can be resisted by uniform axial stresses with little or no bending, the material quantities are greatly reduced, and the stiffness of the structure increases.

Along the way, I learned many things about structural geometry that I was not taught in school. At a conference in Belgium, I heard about the *Michell truss* (a geometry that is the stiffest form for a truss with uniform stress).[1,2] I was intrigued so I started studying A. G. M. Michell's 1904 paper, and I found that his work was based on a generally forgotten energy theorem by James Clerk Maxwell.[2] I discovered that this energy theorem, now named the *Maxwell Load Path Theorem*,[3] was a powerful tool for designing efficient structures. What else did Maxwell do? From Timoshenko's *History of Strength of Materials*,[4] I read that Maxwell was responsible for graphic statics. I had heard of graphic statics but did not know what it was. I had one of the engineers in the studio read about graphic statics; he then showed me and others how to use it.

Graphic statics methods were incredible! I could see the geometry of the structure and the forces at the same time. I could design the forces directly. I could manipulate the geometry and forces while using Maxwell's Load Path Theorem to guide the process of creating minimum load path structures. What else did Maxwell do? I bought his collected works and attempted to read them. What about his contemporaries? I started looking closely at the incredible creativity in structural theory that was happening in the mid nineteenth century.

It was a time when the invention of railroads led to the need for bridges that could span the rivers and streams that crossed the routes of these heavy trains. Although modern structural theory dates back to Galileo in the early seventeenth century, much theoretical work was left to be done as the nineteenth century engineers designed and built bridges to support these heavy loads but with a limited theoretical basis. This caught the attention of the era's great thinkers, particularly in Britain.

Science in the nineteenth century was rapidly progressing with fundamental discoveries in thermodynamics, optics, fluids, electricity, and magnetism. The boundaries between science and engineering were porous. British researchers such as Thompson, Tait, Jennings, Airy, and, particularly, Rankine and Maxwell made important contributions to engineering mechanics and structural engineering.

In January 1864, William John Macquorn Rankine, a Scottish engineer/scientist from the University of Glasgow, published the paper "Principle of the Equilibrium of Polyhedral Frames."[5] The paper consisted of only three paragraphs and four sentences. This succinct publication was like a match that lit a fire in the brain of Maxwell. Just three months later, Maxwell published "On Reciprocal Figures and Diagrams of Forces." Maxwell's remarkable paper led to two-dimensional (2D) graphic statics and major contributions to graph theory.

Rankine's and Maxwell's papers developed a theoretical basis for the geometry of equilibrium for frames with axial forces between the joints, what today we would call trusses. Rankine's paper was on three-dimensional (3D) polygonal trusses, and Maxwell's was on 2D and 3D trusses.

The nineteenth century was ripe for creating a theoretical basis for 3D trusses but much too early for the application to design. The complexity of 3D geometry was beyond the computational power of the time. That power came roughly 150 years later. The theoretical basis for 2D trusses was embraced by the world of engineering, and the resulting method of graphics statics was the technology used to design structures from bridges and roof trusses to the Eiffel Tower.[6]

Graphic statics was about the geometry of 2D equilibrium. The architect or engineer could perform design and analysis at the drafting table with only a straight edge and a ruler. The feedback loop between creating a geometry for a structure and the effect on the forces was immediate. The designer could easily see how to manipulate the geometry to adjust the forces in the members. Better yet, the designer could design the forces and then derive the corresponding geometry of the structure. Over time, graphical methods were replaced by analytical methods. Equations replaced the straight edge and ruler. For flexural systems such as moment frames, *Hardy Cross* and moment distribution was the method of choice.[7]

For trusses, the numerical precision of calculations won over the measurement of a force with a ruler. This accelerated with the advent of the computer that could calculate the forces to many decimals and the forces in structures with redundant members through solving simultaneous equations. Engineers shifted from using graphical methods to design the geometry and forces together to guessing a shape and then calculating the forces. For trusses with redundant members, the possibility of locked-in forces from lack-of-fit is ignored as linearly elastic analysis ruled the day. The computer said: If there are no external forces, there are no internal forces. This contrasts with the era of iron structures, where engineers knew there were forces locked in from residual stresses, and they were only calculating the additional forces. They knew the fundamental difference between a statically determinate three-legged stool and a statically indeterminate four-legged stool. These concerns were evident when the engineer was graphically addressing the interrelated issues of geometry, redundancy, and equilibrium fell by the wayside. The focus on designing for equilibrium was replaced with a focus on running a computer analysis.

It should also be recognized that graphical analysis with a pencil and a straight edge was difficult and often impossible for 3D structures. So, the shift to computers for complex 3D structures was understandable and often necessary. So, what about the theories of Rankine and Maxwell for the equilibrium of 3D structures? By the twenty-first century, computational and graphical tools had developed to the extent that what was impossible in the 1860s and 1870s was possible in the 2010s and 2020s. But someone needed to connect current tools with ignored theories.

Professors such as Edward Allen and Waclaw Zalewski at Massachusetts Institute of Technology (MIT) returned their students' focus to equilibrium as revealed by 2D graphic statics. John Ochsendorf and Caitlin Meuller followed in their footsteps at MIT. Ochsendorf and his student Philippe Block studied how to treat 2D equilibrium as a projection of the equilibrium of a 3D structure and how to create the 3D equilibrium geometry.[8] The focus of these people is now clearly on equilibrium. Block finished his studies at MIT in 2009, took a position at ETH Zurich, and formed the Block Research Group (BRG) with a focus on creating structures and design methodologies where the geometry of equilibrium is central.

But in all this, Rankine's article of four sentences contained in three paragraphs, the article that inspired Maxwell and led to 2D graphics statics, is still mostly unexplored. I, among others, had tried to unpack this very dense package with little description to show for our efforts.

With knowledge of today's sophisticated computational and graphical tools, Masoud Akbarzadeh joined the BRG in 2012 and unpacked Rankine leading to his doctoral thesis.[9] The thesis is just the beginning of a process where he developed the concepts, design methods, and tools presented in this book.

Dr. Akbarzadeh was well prepared for this role with a background in both structural engineering and architecture. Allan McRobie at the University of Cambridge and Juney Lee at Carnegie Mellon University have both made important contributions.[10,11] However, Dr. Akbarzadeh's Polyhedral Structures Laboratory at the University of Pennsylvania has a special focus on the design potential of Rankine's reciprocal polyhedra.

This book is a wonderful guide to this elegant world of 3D equilibrium, with exceptional graphics and clear descriptions. Dr. Akbarzadeh leads the reader through the concepts and processes needed to understand and use this world of geometry, architecture, structural engineering, and equilibrium. On the route to 3D graphic statics, the book provides a strong background in 2D graphic statics and the many applications of graphic-based structural design.

Enjoy the journey!

William F. Baker
Consulting Partner, Skidmore Owings & Merrill

Foreword by Dr. Philippe Block

For the past 15 years, the Block Research Group (BRG) at ETH Zurich has dedicated much of its efforts to demonstrating the power of gaining strength through geometry rather than through material quantities or strength. This geometry is, of course, structural, which we understand as a form that follows the flow of forces or that is cleverly equilibrated to work predominantly through axial forces.[12] Our motivation for bringing back structures that follow first principles is inspired by the Gothic master builders of the past who followed masonry principles, no longer known or used in today's practice, to create spectacular vaulted structures. Over the years, our fascination with structural geometry has evolved into guiding design principles that permeate all our research activities.

Our dedication has deepened in the context of the pressing challenges posed by climate change and the undeniable negative impact of the building and construction industry on the environment. This awareness has strengthened our conviction in the significance of our motto, "Strength through Geometry." Such an approach not only allows for a significant reduction in mass but, crucially, because of the alignment of structural elements with the natural flow of forces, also the reduction of stresses, steering away from the common practice of designing for peak stresses in bending systems. Efficiently utilized sections enable the safe use of lower-strength materials, which often have a significantly lower carbon footprint.[13] Combining this with discrete, masonry-like logic, structural geometry can also offer all kinds of exciting circularity opportunities; for example, low stresses enable longevity, dry-assembly reuse, and separation of materials recycling.[14]

The exploration of structural geometry is essential. We have long argued that graphic statics stands out as the most powerful framework to explore, discover, and steer structural form.[15] As a technique, it beautifully relates geometry and the equilibrium of the forces in it through a graphical, intuitive, and interactive manner. We have argued that such a white-box approach to structural design is essential to fully discover the exciting design space of equilibrium. Unlike a black-box approach, where there is a disconnect between input and output, white-box methods give insight and allow an intuition to emerge. They become increasingly vital in a context where we aim to do better, design better, and build better. Graphic statics, in particular, provides a very intuitive and approachable way to work with high-level structural concepts. Despite trends favoring machine learning and artificial intelligence and the belief that the computer or the algorithm will solve it all without us needing to fully understand the problem, I am convinced that white-box approaches will become increasingly important. Tools such as graphic statics help discover, explore, and exploit equilibrium and structural forms. They not only allow human creativity and intellect but also experience and unique insights to be combined.

It is with excitement and pride that I write this short foreword for Masoud, our first PhD student who has extended graphic statics from 2D over 2.5D (the topic of my own PhD thesis with John Ochsendorf at MIT[8]) to fully spatial explorations. Polyhedral graphic statics, compared to other structural design approaches, provides visual feedback of spatial equilibrium and an effective control mechanism in 3D structures. Masoud's work delves into strategies, manipulations, and algorithms that vividly demonstrate how polyhedral graphic statics in 3D offers unparalleled design exploration and sophisticated constrained form finding.

Enjoy discovering the beauty and power of equilibrium through the reciprocal polyhedral form and force diagrams, an exciting framework that was first described by Rankine and Maxwell but expertly implemented as a generative design tool by Akbarzadeh!

Philippe Block Professor of Architecture and Structures
Co-lead of the Block Research Group
Institute of Technology in Architecture
ETH Zurich

Preface

This book is the product of academic research, experimentation, and studies spanned over three continents involving elite universities. The journey enabled the discovery of true personal passion coupled with the hope to excite those who read it.

The Path toward Architecture

I was born in a small mountainous town with very cold winters in the northwest of Iran into a family involved in construction. My early fascination with building design led to my dreams of becoming an architect. A degree in civil engineering and a master's in seismic engineering and dynamic structures firmly grounded me in building techniques. Separately, my creativity skills in illustration, drawing, and conceptual art yielded national awards. These credentials were fused to gain acceptance to the MIT professional architecture program in the United States. My fascination with designing breathtaking architectural structures revealed the path to pursue a Master of Science in Design Computation simultaneous to the Master of Architecture degree at MIT. My architecture thesis was distinguished with an award of Skidmore, Owings, and Merrill fellowship to travel and explore international infrastructure projects of varying scales. After obtaining my architecture degrees, the delight of exploring highly efficient forms through rigorous analysis spawned a clear career path in architecture, fusing my design talents with my technical abilities.

MIT–ETH Zurich

Time at MIT presented me with the eye-opening possibilities of optimizing and generating architectural structures using geometric diagrams. John Ochsendorf introduced me to the methods of graphic statics, and the book by Allen and Zalewski, *Form and Forces: Designing Efficient, Expressive Structures*,[16] changed my world about structural design. Discovering graphic statics was my threshold to enter the advanced world of 3D structures. It serves as a vehicle to generate efficient architectural forms. However, its analytical methodology and digital tools needed to be expanded, advanced, and refined.

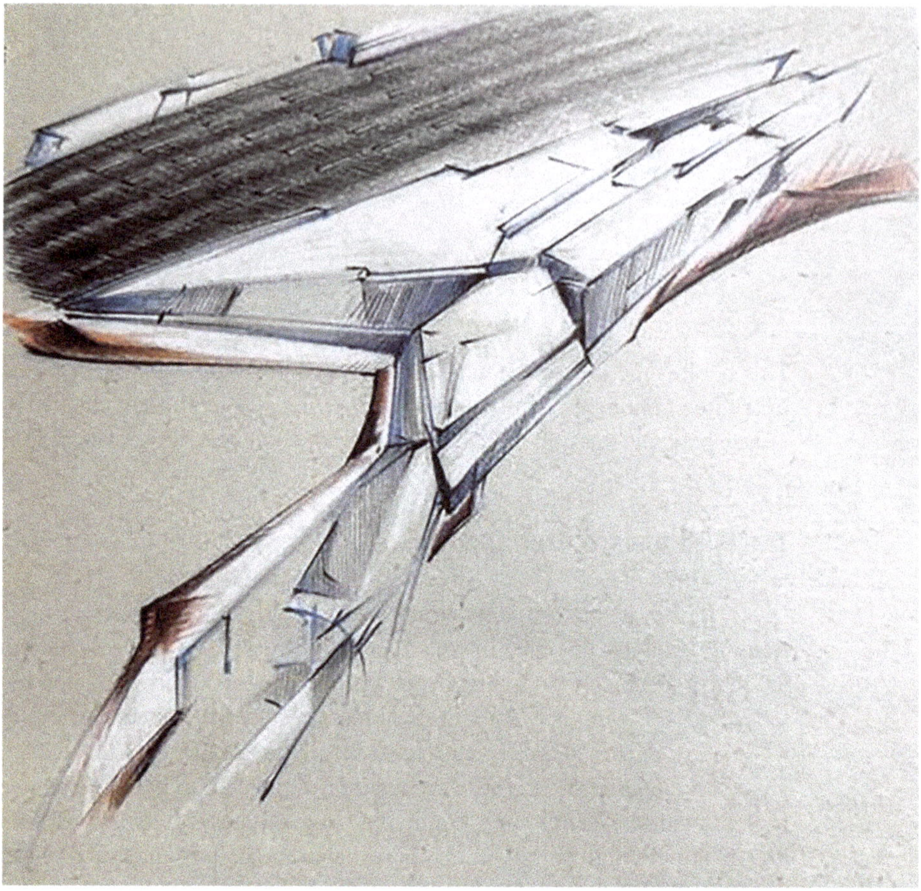

A free-hand drawing by the author before entering the architecture program at MIT, from speculative structural forms sketch series, December 2005.

At that time, Philippe Block, a PhD student of John, was working on "*Thrust Network Analysis*"[8] to design free-form compression-only vaults by applying the methods of graphic statics. Through my work and associations at MIT, a path became evident for me to pursue a PhD at ETH Zurich in Switzerland, where Philipe started his professorship and established the Block Research Group. Since then I focused on geometry-based structural design methods and digital fabrication, always maintaining the ultimate goal of creating fascinating architecture.

In my doctoral thesis at ETH, I developed initial recipes for the methods of graphic statics in three dimensions. The thesis is entitled "3D Graphic Statics using Reciprocal Polyhedral Diagrams,"[9] the genesis stems from a publication by William John Macquorn Rankine in 1864.[5] In less than a half-page article, he proposes the relationship between the equilibrium of forces in three dimensions and closed polyhedral diagrams. Yet he did not offer any illustrations or proof to support his proposal, and my life's work substantiates the proposal and reveals a revolutionary approach to design.

University of Pennsylvania

In 2016, I joined the faculty of the Weitzman School of Design and soon after established the Polyhedral Structures Laboratory, which continues the theoretical research on methods of graphic statics and its application to the design and construction of innovative structures. My collaborators include my colleagues in the Architecture, Material Science, Engineering, and Mathematics Departments, and my research encompasses the application of graphic statics in various topics and the potential use in the future of manufacturing and carbon-absorbing 3D printed structures.

Why Writing a Book

The more I work with these methods, the more I become aware of their limitations. However, I have accepted that any method or tool comes with a limitation, and in the end, it is upon the user and his/her creativity to work with or around the limitations. This book is the culmination of the past 10 years of my life, aiming to learn more about and explore the geometry-based methods of graphic statics and their application in architectural engineering and beyond.

The 2D methods of graphic statics were established around 1864 and continue to be practiced and researched until now. Undoubtedly, there are still many unknown topics related to the 3D extensions of these methods as polyhedral graphic statics. In this book, I do not intend to cover all topics related to polyhedral graphic statics. Quite the opposite, I decided to write this book because I wanted to share what my team and I have learned and developed based on these methods for structural design purposes.

The methods of graphic statics, in general, can be used for structural *design* and *analysis* problems. By "structural design," I do not refer to the dimensioning process, which is quite common in engineering procedures. I refer to *structural design* as a process in which the efficient geometry of the structure is created. It was not known before the design process and was developed using the methods of graphic statics given the specific boundary conditions and the applied loads. Thus, this book is about the design process and not analysis!

I do not enter into the analysis topics for clear reasons: (i) to analyze a structure, one needs to have the geometry. If the geometry is given, then the design is mainly about studying the stability or sizing of the members. Many tools and approaches exist that can provide such services. (ii) It is simply challenging to analyze structural geometries that have not been developed using polyhedral methods. Because polyhedral graphic statics is a subset of all possible equilibrium configurations in three dimensions, these methods cannot be generalized to all possible equilibrium configurations. James Clark Maxwell also pronounced this observation in his ground-breaking paper about reciprocal polyhedral diagrams.*

* "It is manifest that the mechanical problems may be solved, though the reciprocal figure cannot be constructed owning to the condition of all the sides of the face lying in a plane not being fulfilled,

It is worth mentioning that it is possible to show the equilibrium of nonpolyhedral systems using the "Disjointed Force Polyhedra" method developed by Juney Lee[18] that breaks the reciprocity between the form and force but gives nonpolyhedral equilibrium solutions. Allen McRobie and his students Marina Konstantatou and Georgios Athanasopoulos at Cambridge University take 3D graphic statics design and analysis to another level. Marina links the structural design and analysis by linking graphical methods to Airy Stress Functions.[19] Allen and the team developed *graphic kinematics*, which is the extension of graphic statics for dynamic loadings.[20] This book would not cover these topics, but I highly recommend his related publications and many other exciting aspects of polyhedral graphic statics explored by Allen and his team.

Who Should Read This Book

The book has been written such that everyone interested in geometry-based methods should be able to use it. It starts with simple examples in 2D and extends the same concept with similar examples in 3D. Although the book covers content related to 3D/polyhedral systems, 2D examples are incorporated in almost every section and related to every main topic of the book. The simple 2D examples, in conjunction with their 3D equivalent, are quite beneficial in conveying the message effectively to the reader and seamlessly linking the world of 2D graphic statics with this 3D version. The book's primary emphasis is to instigate an intuitive understanding of force flow in three dimensions and geometric procedures to control and understand the equilibrium of force using polyhedral reciprocal diagrams. The primary intent has been to make the topics accessible to everyone interested in geometry. Hence, the examples are designed to be reproduced by following the procedures provided in this book. Even for complicated examples, a matrix barcode beside the drawing links the reader to an online platform where the interactive instance of the structural geometry is stored.

Changing the Convention

However, there is a main difference in presenting the examples of 2D graphic statics in the context of this book. In the past 150 years, the drawings of form and force diagrams of 2D graphic statics followed the parallelity in the construction process. That is, every edge in the structure's geometry has a corresponding parallel edge in the force diagram. It is unclear to me where this parallel convention was adopted in the history of the methods, but given the drawing tools at that time, it made great sense to do so. However, in the reciprocal relationship between the form and force diagrams proposed by Maxwell and Rankine, the members of these diagrams are perpendicular

or owning to a face belonging to more than two cells. Hence, the mechanical interest of reciprocal figures in space rapidly diminished with the complexity ..." from the paper "On Reciprocal Figures, Frames, and Diagrams of Forces."[17]

instead of parallel. Whether parallel or perpendicular, the relationship between the form and force and the design approach do not change as it has been practiced as parallel diagrams for the past 150 years.

In this book, for the first time in the history of 2D graphic statics, I took the liberty to present 2D examples based on the perpendicular convention for the first time. This might seem counterintuitive for those familiar with conventional drawings of 2D graphic statics. However, presenting the 2D examples using a perpendicular convention makes a clear connection to the examples in 3D. In polyhedral graphic statics, the members of the form and force diagrams are perpendicular, which helps significantly in understanding similar topics in 3D. Moreover, the main emphasis of this book is on the topological and dual relationships between the form and force diagrams of graphic statics, which is quite understandable in the perpendicular convention of drawing.

Before Starting

Anyone interested in polyhedral graphic statics should know that polyhedral equilibrium is only a subset of all possible spatial equilibrium. That is, explaining and exploring the equilibrium for some nonpolyhedral systems is quite challenging using this method. Nevertheless, this book will open the door to an exciting world of efficient spatial structures and teach geometric rules to design such systems and underlying equilibrium principles that result in such compelling and expressive geometries.

I have divided the content of this book into nine main chapters, including Introduction, which is a brief introduction to the history of graphical methods and their application in structural design. I have also been greatly inspired by how William S. Wolfe wrote his "*Graphical Analysis: A Text Book on Graphic Statics*."[21] Nevertheless, as mentioned earlier, this book emphasizes design rather than analysis.

Chapter 1 starts with the equilibrium of a single node in both 2D and 3D and explains the geometric rules for constructing a force diagram for a single node in equilibrium. The main outcome of the chapter is how some structural systems can be abstracted into a single node and how knowing the equilibrium of a single node can help us design various simple structural systems.

Chapter 2 introduces the concept of equilibrium for a group of forces in 3D, explains the relationship between the group of polyhedral forces, and provides recipes to find one force or a group of balancing forces to be added to the system. The main outcome of the chapter is to show how one might be able to design sculptures similar to the works by Alexander Calder using graphical methods both in 2D and 3D.

Chapter 3 opens the door to the structural form finding of compression or tension-only structures and provides various examples in both 2D and 3D. It revisits some famous structural design examples, such as the Salginatobel Bridge by Robert Millart. It provides recipes for designing such structures using polyhedral graphic statics and highlights the differences between the 2D and 3D versions.

Chapter 4 goes deeper into the design techniques for compression-only systems and shows how a force diagram can be designed and subdivided to generate articulated features in the geometry of the structural form. The concepts of curvatures are explained with respect to the geometry of the form and force diagrams. In the end, the novel subdivision technique to translate cellular systems into shellular systems is discussed.

Chapter 5 expands on systems with combined tensile and compressive members and provides design techniques that result in various topologies of tension and compression members.

Chapter 6 discusses the geometric degrees of freedom of the form and force diagram and shows how, in particular cases, the magnitude of the applied loads or reaction forces in the supports can be controlled to change control the internal force distribution in the system. For instance, how one can control the flow of forces in spatial branching structures or what other forms of equilibrium exist for a single force diagram.

Chapter 7 provides an algebraic approach to solving the methods of graphic statics in 3D and presents a mathematical formulation of the problem. This formulation opens the door to various exciting equilibrium possibilities that could be otherwise left unexplored.

Finally, Chapter 8 shows the application of this method in other fields of science, including cellular solids and mechanical and aerospace engineering, and advocates for more sustainable construction methods for the future of manufacturing. We will also show the equilibrium of nonpolyhedral systems developed by Juney Lee in his PhD thesis.

I hope this book provides a compendium for the graphical statics methods in the long term and becomes a reference for those who would like to gain a more in-depth understanding of the methods of graphic statics and would be able to appreciate the extraordinary elegance of this method in design and equilibrium of efficient structural forms.

Acknowledgments

I am extremely grateful to the following individuals, without whom this book could have never come to life: the members of the team of Polyhedral Structures Laboratory at the Weitzman School of Design who contributed to the content of the book, Mostafa Akbari, Yao Lu, Hua Chai, Yefan Zhi, Hao Zheng, Teng Teng, and Dr. Mathias Bernhard; individual researchers and practitioners who contributed to the content and the broader impacts of the work, William F. Baker, Dr. Juney Lee, Hossein Mofatteh, Dr. Hamid Akbarzadeh, Armin Mirabolghasemi, Dr. Zakaria Hsain, Dr. James Pikul, Dr. Shu Yang, Dr. Kun-Hao Yu, Dr. Mohammad Bolhassani, and Dr. Paria Yavartanoo.

This book contains over 460 drawings, and the visualization team played a significant role in facilitating understanding of the concepts in this huge project: Jasmine Liu, Rongxuan (Roxanne) Zhou, Simone Yang, Tian Ouyang, Chenyang Yu, Michael Ting, Hadi Elkebbi, Hua Chai, Yefan Zhi, Mostafa Akbari, and Yao Lu.

I am extremely thankful to Richard Farley for his support through my academic career at Weitzman, his remarks about this book, and his remarkable friendship and wisdom. Thanks to Dr. Edmond Saliklis and Dr. Faisal Mukhtar for their great discussions about the book's topics and content.

I am thankful to Dr. Márton Hablicsek for his fantastic contribution to the algebraic formulation of the methods, which opened another door to the world of research in the algebraic methods of polyhedral graphic statics.

I would also like to thank Dr. Andrei Nejur for his efforts in implementing the methods into an efficient computational tool for education and research. Such computational implementation is essential in using and applying this method in practice and education.

I especially thank Dr. Damon (Mohammad) Bolhassani for his invaluable contribution to the numerical analysis of funicular polyhedral structures and his continued collaboration on this topic.

I am also grateful to Hua Chai, Wenxi Chen, and Yao Lu for developing an online platform, including supporting materials and examples for the book's content.

I am deeply indebted to Dr. John Ochsendorf for introducing me to the world of graphic statics. I also extend my deepest gratitude to Dr. Philippe Block for guiding me during my doctoral research, which was the starting point of my research in the polyhedral graphic statics methods and flourished into the content of this book. I am also grateful to Dr. Tom Van Mele for his fantastic insights and excellent discussions on graph theory and topology.

I cannot begin to express my thanks to William F. Baker, who supported me both in my PhD and the preparation of this book and always inspired me with his professional insights and the approach he takes in applying geometry-based methods in practice.

I am also grateful to my mentor, Dr. William Braham, who guided me through my academic path at the Weitzman School. I would also like to thank Dr. Franca Trubiano for her wonderful comments on the editorial aspects of the book.

This book would not have been possible without the support and nurturing of my family: my beloved wife Yasaman, my mother Zarrin Taj, Shohreh, Morteza, Maryam, Roya, Saeed, Mohammad, Ali, Yasamin, Zahra, Golnaz, Shahryar, Sima, and Malileh.

I thank the US National Science Foundation for providing the NSF CAREER-CMMI 1944691 fund to develop the book.

Special thanks to Dr. Fredrick R. Steiner, the dean of Weitzman School of Design, University of Pennsylvania, and his office for the generous and continuous support of my research and its related publications.

Lastly, I would like to express my gratitude to Steven Elliot and Cambridge University Press for the opportunity to make the content of this book accessible to the public via their excellent services and distribution.

Introduction: A Brief History

Masoud Akbarzadeh

Science must become sensual and poesy scientific.

Friedrich Gilly. Essays zur Architektur 1796–1799.[22]

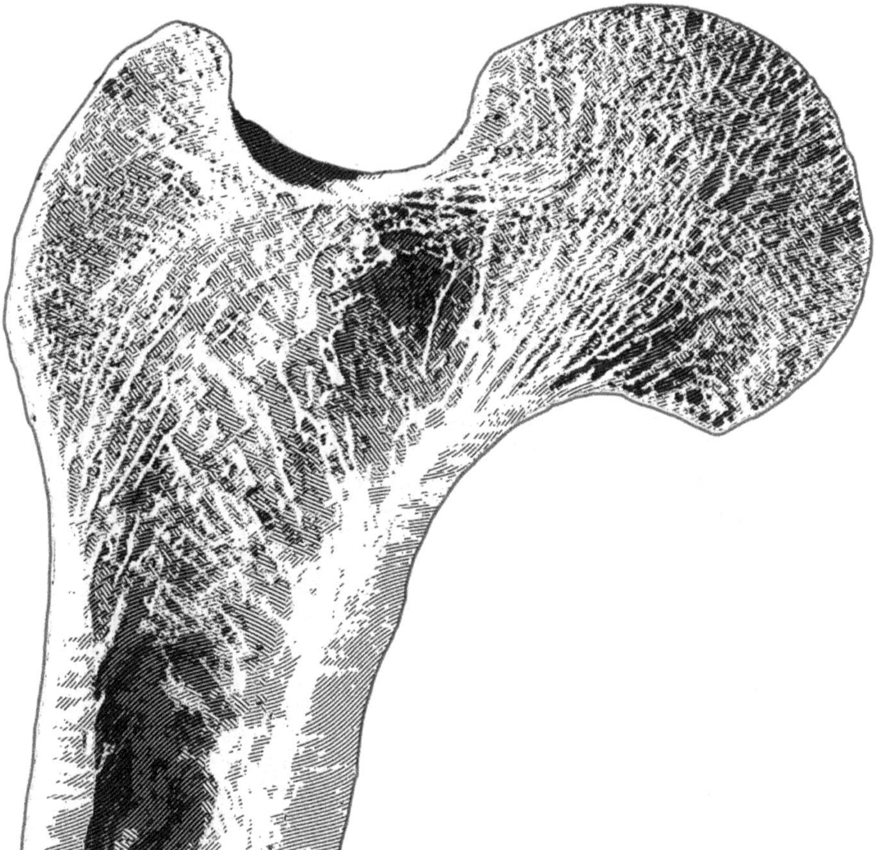

Figure I.1 Cross-section of a human bone showing the trabecular bone, also called cancellous bone, aligned with the stress direction within the bone structure to minimize mass and maximize performance.[23]

I.1 Structural Efficiency

Before introducing the methods of graphic statics and their application in finding efficient structural forms, perhaps it would be best to define *efficiency* in structural geometry and force transfer and visit some historical techniques developed to design efficient structures. Here, efficiency refers to the minimum amount of material needed to effectively transfer internal and external forces throughout the structural system. The structural system might be natural, like a bone, or man-made, like a piece of furniture, a vehicle, a home, a bridge, or a skyscraper. Hence, in all these cases, the ability to transfer forces with the minimum use of material while effectively maintaining desired performance can be referred to as structural efficiency (Figure I.1).

Minimizing construction material in the built environment is receiving much attention. Minimum mass is equivalent to reduced embodied energy, the energy used to produce the material, thus fewer carbon emissions. Given that the mass of (material of) structural systems used for various infrastructure projects, including buildings and bridges, will remain throughout and beyond the system's life cycle, its pre-reduction will contribute to fewer adverse environmental impacts. Therefore, finding efficient structural systems for various projects is quite essential.

Robert Hooke, in 1675, proposed an intuitive method for finding an efficient structural form. At that time, the range of construction materials was quite limited.[24] Masonry construction was very common due to the accessibility of the material. It worked well in compression but could not carry tensile forces. Thus, understanding force flow was quite necessary for architects and builders to guarantee the stability of their structures.

Hooke implicitly explained that an efficient structure is a system where the construction materials are placed where internal forces flow, also referred to as the load path in the structure. He showed this concept using a simple example of a hanging chain: if a chain is hung by holding its two ends, it will form a geometry in which the direction of the chain follows the direction of the force flow within the system (Figure I.2). In fact, at any point along the geometry of the hung chain, the cross-section of the chain is normal to the direction of internal force or the tension in the system, or the material of the structure precisely follows the direction of a force flow. More importantly, he showed that if the chain form is rigidified, its reflection will represent a perfect form in compression. He mentioned this in an acronym deciphered years after his death: "*As hangs the flexible line, so but inverted will stand the rigid arch.*"[25] This efficient geometry was later referred to as *catenary* geometry and is used for structural systems that are formed based on the self-weight of the system. In general, systems that carry the forces in pure tension or pure compression in their members are referred to as *funicular* forms, a term that is used extensively throughout this book. Hooke's studies on hanging chains are very important not only because they provide an intuitive and quick method to find efficient structural forms but also clearly demonstrate that the *geometry* of an efficient structure is placed along the direction of the internal force flow.

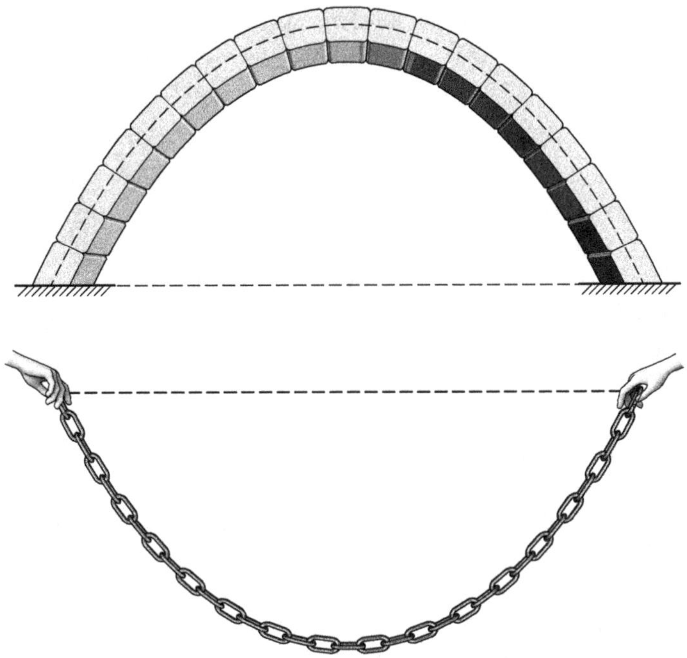

Figure I.2 Hanging a chain under its self-weight results in a form in pure tension. The reversed form gives a perfect arch in compression.

The methods established by Hooke were adapted for hanging chains or other physical scale models made of ropes and sandbags, networks of tensile elements, and so on to predict the structural behavior of materials and find efficient structural forms in many instances throughout the history of architecture and structural design. Antoni Gaudí, Heinz Isler, Frei Otto, and many other master builders used similar physical models to derive efficient structural forms for their designs.[26]

Physical form-finding techniques are straightforward and do not require the need for an extensive amount of structural design experience or knowledge to be readily used. A simple system of hanging chains or cloth can represent an optimal form of a shell under gravitational loads.[27] However, this technique has the following limitations:

- Translating the geometric properties of the resulting form to the building scale might be quite tedious.
- Building models for complex architectural concepts may take a long time to finish, mainly because the weight distribution needs to be updated for each design iteration.[28]
- Controlling the design features of the resulting form is extremely difficult because local changes have global effects.

- Measuring the internal forces in the elements of the structure and designing them accordingly is infeasible unless sophisticated measuring devices are at hand.
- The behavior of the materials cannot be scaled for all structural and material types.[29]

While physical models are quite powerful in providing the initial step into the entire structural design process and in conveying the concept of efficient structural forms, they do not provide the designer with a desirable level of control in design. In the meantime, other approaches, such as geometry-based structural form finding, have been developed to make the form-finding process quite intuitive, explicit, and less cumbersome.

I.2 A Brief History of Graphical Statics

Geometric methods of structural design represent a group of techniques that investigate the static equilibrium of structures using geometric constructions. The use of geometry in finding the equilibrium of forces has a history as early as 1586. Stevin[30] in his book *The Principles of the Art of Weighing** proposed that the forces on two sides of inclined planes are in equilibrium if the magnitude of the force on each side is proportional to the lengths of the side planes that are connected by a horizontal line as shown in Figure I.3. This approach related the lengths of the sides to the magnitude of forces.

Around 1725, Varignon[31] was probably the first person who laid the foundation for the use of geometry in the field of structural design by introducing the concept of

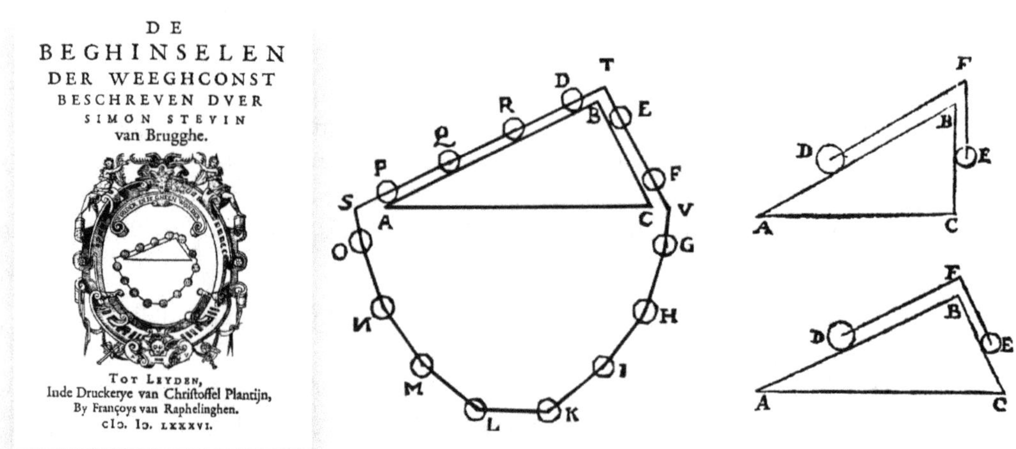

Figure I.3 Drawings by Simon Stevin representing the equilibrium of forces on the sides of inclined planes where the magnitude of the forces is proportional to the length of each side.[30]

* De Beghinselen der Weeghconst.[30]

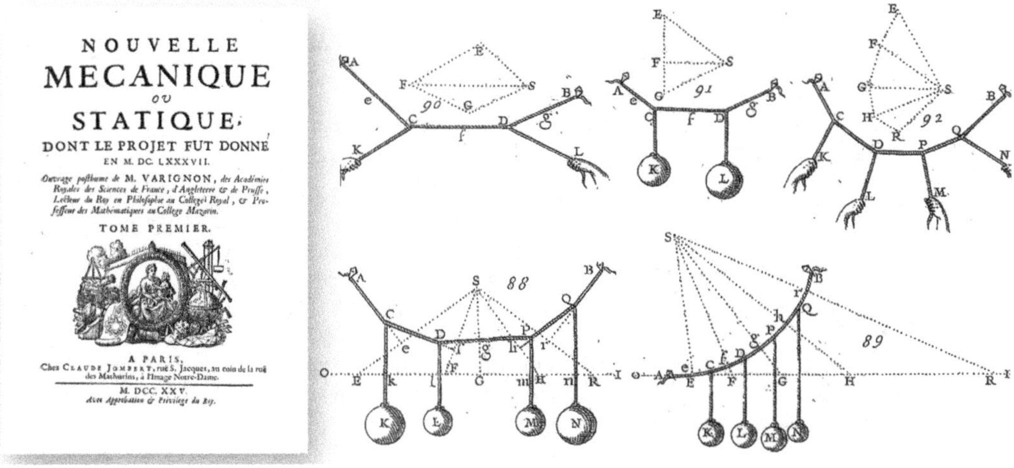

Figure I.4 Drawings by Varignon representing the equilibrium/funicular form of an inelastic rope loaded in various directions and the force polygon (dashed) representing the magnitude of the forces in equilibrium.[31]

the *funicular polygon* and the *polygon of forces* in his book *Nouvelle Mécanique ou Statique* (Figure I.4).

According to Varignon, the funicular polygon is the equilibrium geometry of an inelastic rope suspended from two points and subjected to applied loads. This is very similar to the funicular geometry introduced by Hooke using a hanging chain example. Figure I.4 shows a funicular polygon *ACDPQB* of a suspended rope at *A* and *B* and loaded by *K*, *L*, *M*, and *N* weights. In addition, there is a secondary drawing as the *force polygon* that includes force triangles representing the equilibrium of each node in the funicular polygon. The dashed line in the figure shows the force polygon as a group of polygons *perpendicular* to the rope segments of the funicular form and in the bottom *parallel* to the segments. For instance, triangle *SEF* represents the equilibrium of the forces in the node *C* of the funicular form. The length of each edge of the triangle represents the magnitude of the forces in the members connected to the node *C*.

Poncelet[32] subsequently used the funicular polygon to determine the center of gravity in his lectures at the artillery and military engineering school in Metz.[33,34] Later, Lamè and Clapeyron[35] applied Varignon's funicular polygon and polygon of forces for the first time in construction engineering and used it to analyze the dome of the St. Isaac Cathedral.

In Britain, the funicular polygon was used by Rankine[36] to determine the internal forces of a statically determinate truss system in 1858. He applied this concept in developing the Theory of Equilibrium of Polygonal Frames in his *Manual of Applied Mechanics*.

Nevertheless, James Clark Maxwell[17,37] was the first person to consider the relationship between the form and force diagrams as *reciprocal* and provide geometric recipes to construct them. Maxwell used the theory of *duality* to define the properties of the reciprocal diagrams. The duality was originally discovered by Möbius.[38]

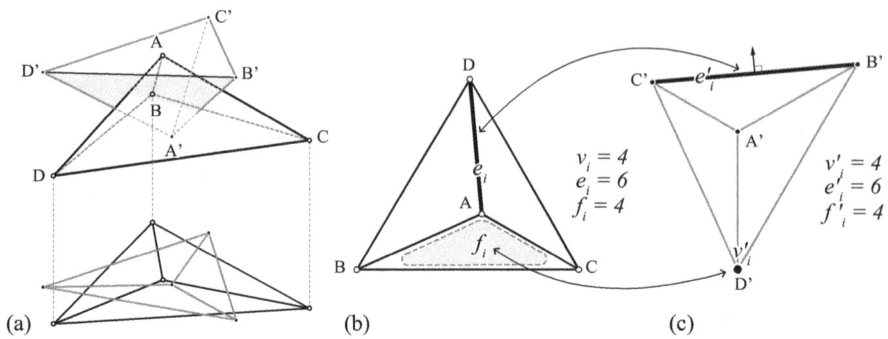

Figure I.5 (a) Möbius's reciprocal tetrahedrons in three dimensions and their orthographic projection; (a) and (b) Maxwell's reciprocal figures, which are similar to the orthographic projection of the Möbius's reciprocal tetrahedrons.

Möbius, in his theory of duality, stated the following relationship between two tetrahedrons in three-dimensional (3D) space:

> If the faces of a tetrahedron, e.g., $B'C'D'$ of Figure I.5, pass respectively through the vertices of another tetrahedron, e.g., A, and if three faces of the latter, e.g., ADB, DAC, ABC pass through three vertices of the former, e.g., C', B', D', then the fourth face of the second tetrahedron DBC will pass through the fourth vertex of the first A'. As a result, these tetrahedrons are reciprocal.

This theory is extendable to two polyhedrons, and the orthographic projection of these reciprocal polyhedrons appears in Maxwell's "reciprocal figures," where each edge of one diagram is perpendicular to an edge of the other diagram, and each closed polygon with n edges in one diagram corresponds to a vertex where n edges meet in the other diagram (Figure I.5). These reciprocal diagrams, obtained from the orthographic projection of two dual polyhedrons, are incorporated directly into the traditional methods of the two-dimensional (2D) graphical statics we use today.

In Germany, Culmann[39] published the methods of graphical statics for the first time in his monograph *Die Graphische Statik* in 1866 and invested his time to implement the methods of graphical statics in the curriculum of the Swiss Federal Institute of Technology. Culmann discovered that the funicular polygon and the polygon for the concurrent system of forces are interchangeable and used Maxwell's term to call the diagrams *reciprocal*.[34,40]

In 1872, Cremona[41] extended the work by Maxwell and combined it with Culmann's methods of graphical statics. He generalized Maxwell's theory of reciprocal diagrams and developed a methodology to draw a diagram of forces for structural systems with nonconcurrent, externally applied loads[40] (Figure I.6). Cremona's methods encouraged engineers to use graphical methods to analyze the forces in a given structural form, such as trusses.

Later in 1873, Bow[42] in his monograph *The Economics of Construction in Relation to Framed Structures* presented a catalog of various types of trusses and divided them into four categories with their reciprocal force diagrams and simplified the use of graphical methods of analysis for engineers in the design of trussed frameworks (Figure I.7). Bow was probably one of the first designers who explicitly visualized

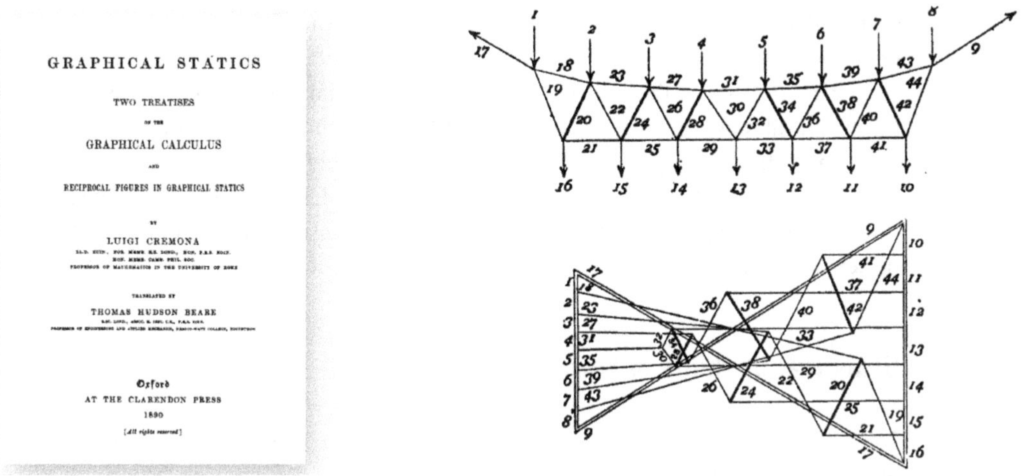

(a) (b)

Figure I.6 Cremona diagrams representing the form of a truss (a) and its reciprocal force diagram (b).[41]

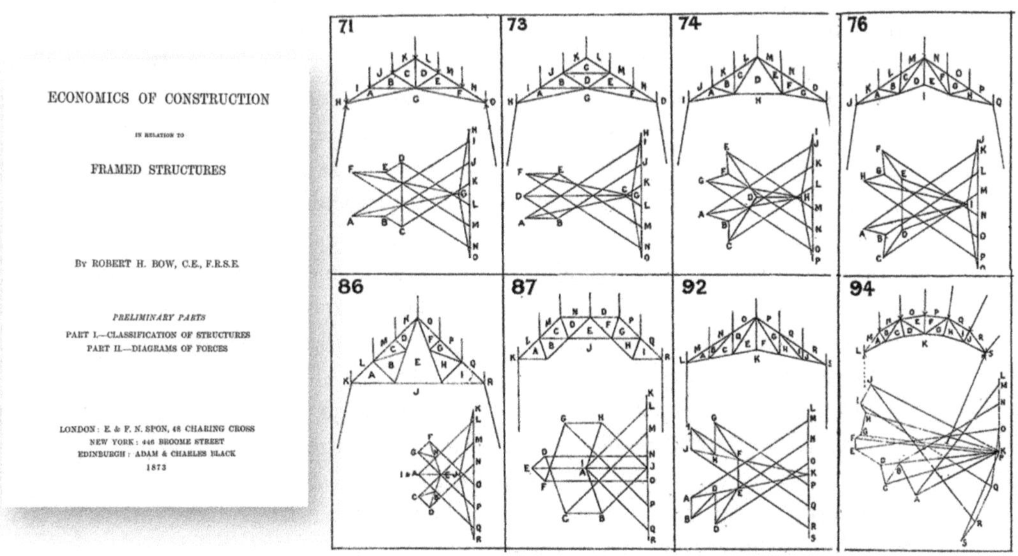

Figure I.7 Bow's drawings of various truss systems with their corresponding force diagrams highlight the power of graphical methods to control forces in truss design.[42]

the effect of geometry on the magnitude of the internal force flow for truss structures. For instance, in his drawings, as illustrated in Figure I.7, one can tell which truss carries more force by comparing the lengths of lines in the force diagram of trusses. The longer the line is, the more force is carried in the member, and thus a thicker cross-section should be used for the structure, which results in a heavier and more expensive system. In comparison, the same span can be covered by a relatively similar truss with different angles of members but with less internal force and more economical construction.

During the end of the nineteenth century, numerical computational methods based on linear algebra defeated graphical analysis methods in dealing with complex problems; therefore, engineers at that time lost interest in using graphical analysis. Nevertheless, the *form-finding* potential of graphical methods has preserved its importance among architects and engineers until today. It has been extended to three dimensions recently, which is the main emphasis of this book.

I.3 Graphical Methods of Form Finding

Despite their limitations in analyzing complex structural systems, graphical methods are a powerful technique for finding efficient structural forms. As briefly discussed, in graphical statics, the structural equilibrium and force distribution are controlled by the geometric construction of *form* and *force* diagrams. Moreover, these geometric constructions are reciprocally dependent. That is, one diagram can be constructed from the other. This property makes the graphical method a powerful tool for finding efficient structural forms. For instance, by visiting the example of a funicular form drawn by Varignon in Figure I.4, we learn that the geometry of the force diagram consists of a group of closed triangles. This is an important observation because one can start from the geometry of the force and easily construct or *find* the geometry of the funicular form. This approach is at the core of the form-finding methods using a graphical approach. The force diagram is often much easier to understand and draw, and its closeness guarantees equilibrium in the structural form. Thus, a designer can start by aggregating closed triangles or polygons to make a system of force that is in equilibrium and extract an efficient structural geometry based on the starting force distribution (Figure I.8). Dealing with boundary conditions, changing the force distribution, and many other details relevant to a design case should also be learned, which is the main intention behind preparing this book.

Advances in computer science and engineering have allowed engineers and architects to develop numerical techniques to find efficient structural forms in a computer-aided environment. Some of the developed techniques simulate the physical transformation of materials and find an optimal structural geometry under the given loading conditions. Particle-spring systems,[44,45] physics-based simulation engines,[46] force density methods,[47,48] and dynamic relaxation[49] are among them.

Numerical methods are powerful substitutes for tedious physical form-finding techniques. In these techniques, the final geometry results from the computational processes, which are not straightforward and explicit to control for beginners and sometimes even professionals. Unlike numerical methods, the relationship between the optimal geometry and its force diagram is exceptionally transparent in the graphical method. In this regard, the designer can easily observe the effects of a change in the form diagram due to a modification in the force or vice versa. Therefore, graphical statics can intuitively explain structural concepts to architecture and engineering students.

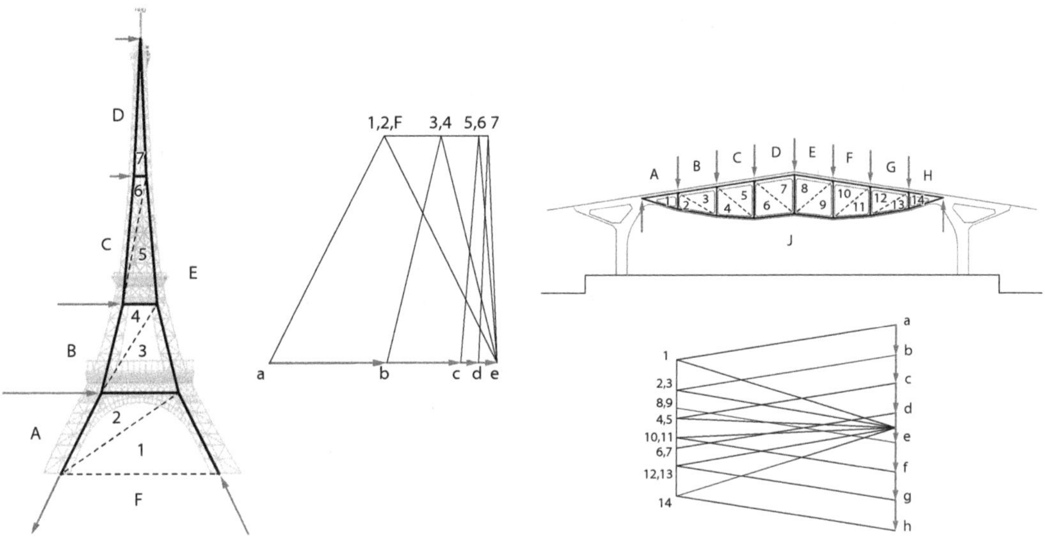

Figure I.8 (a) Graphical form finding of the Eiffel Tower by Maurice Koechlin; (b) form and force diagram of Robert Maillart's Magazzini Generali Shed in Chiasso.[43]

I.4 2D Limitations of Graphical Statics

Like other methods of structural form finding, graphical statics has limitations: until recently, it existed as a 2D method. Thus, only 2D abstractions of 3D structures could be designed. Since the nineteenth century, several methods have been developed to extend graphical statics to 3D. One of these 3D extensions is the result of the work of Föppl,[50] who used projective geometry to analyze 3D trusses. This method is a method of analysis and cannot be used as a form-finding method. Moreover, at a certain level of analysis, the process becomes extremely counterintuitive due to the complicated representation of forces in the third dimension.*

This historical method has been revitalized by 3D graphic statics, enabling its application to the design of complex spatial structures. Amidst the different approaches using reciprocal diagrams, vector-based[52] and polyhedron-based[9,11] strategies have been actively investigated in the past few years. In the force diagram of the vector-based approach, equilibrium is guaranteed by closed cycles of force vectors, where each force vector is parallel to the corresponding edge in the form diagram, and the vector length indicates its force magnitude. In contrast, the polyhedron-based approach uses closed polyhedrons to represent equilibrium. This book is precisely prepared to introduce the latter approach and its application in the design of structural systems.

* See Schrems and Kotnik,[51] who suggested using the force-pair technique to design 3D structural forms based on Föppl's approach.

I.5 Polyhedral Reciprocal Diagrams

In 1864, Rankine[5] proposed a reciprocal relationship between the form and forces in 3D in a half-page publication in Philosophical Magazine. Figure I.9 shows the entire publication. He called it the *Principle of the Equilibrium of Polyhedral Frames*. In this publication, he proposed that the equilibrium of a spatial node with four or more members can be described by a closed force polyhedron such that each face of the polyhedron is normal to an edge of the node. In that case, the area of each face shows the magnitude of the force in the members, and all forces will be in equilibrium.

In the same year, in response to Rankine's proposition,[17] provided a geometric procedure to construct 3D reciprocal diagrams for a specific case. However, in the same publication, he acknowledged the complexity of solving these problems in the third dimension and the subsequent lack of interest in continuing the topic. Therefore, this topic has been left intact since 1864.

In 2016, I was among the first people to develop an extension of graphic statics in three dimensions based on a 150-year-old publication by Rankine[5] in Philosophical Magazine.[9,11,53] Rankine's principle was limited to a short paragraph without any proof or illustration, and the highly ambiguous language of the proposition and technological limitations at that time made it very well hidden among other historic scientific publications.

In simple terms, if in 2D graphical form finding, the force diagram is represented by closed force polygons, in Polyhedral Graphic Statics, this equilibrium is shown by closed force polyhedron, as shown in Figure I.10. This simple principle can open

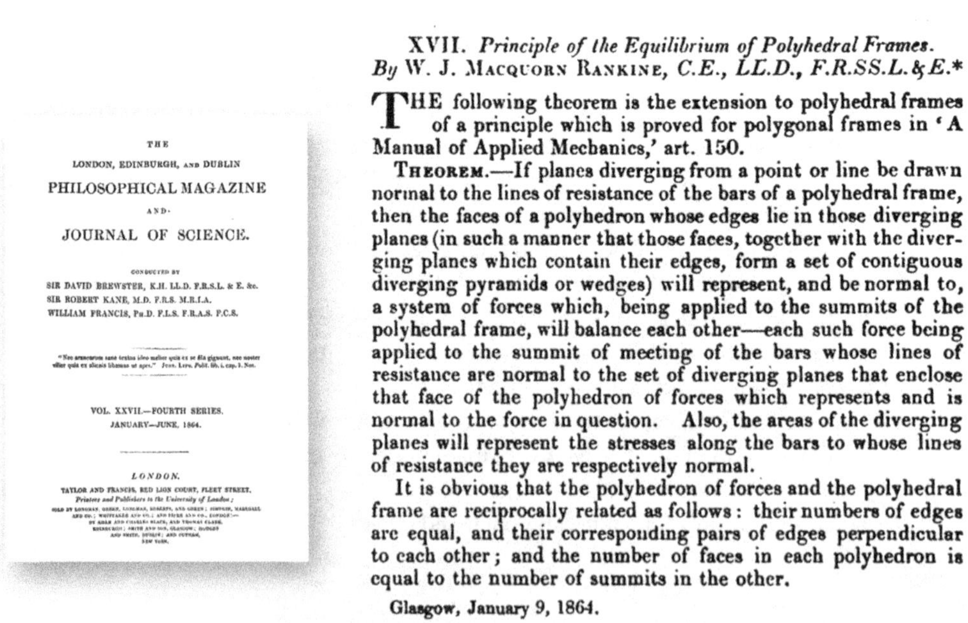

THE

LONDON, EDINBURGH, AND DUBLIN

PHILOSOPHICAL MAGAZINE

AND

JOURNAL OF SCIENCE.

CONDUCTED BY

SIR DAVID BREWSTER, K.H. LL.D. F.R.S.L. & E. &c.

SIR ROBERT KANE, M.D. F.R.S. M.R.I.A.

WILLIAM FRANCIS, PH.D. F.L.S. F.R.A.S. F.C.S.

VOL. XXVII.—FOURTH SERIES.

JANUARY—JUNE, 1864.

LONDON.

TAYLOR AND FRANCIS, RED LION COURT, FLEET STREET.

XVII. *Principle of the Equilibrium of Polyhedral Frames.* *By* W. J. Macquorn Rankine, *C.E., LL.D., F.R.SS.L.&E.**

THE following theorem is the extension to polyhedral frames of a principle which is proved for polygonal frames in 'A Manual of Applied Mechanics,' art. 150.

Theorem.—If planes diverging from a point or line be drawn normal to the lines of resistance of the bars of a polyhedral frame, then the faces of a polyhedron whose edges lie in those diverging planes (in such a manner that those faces, together with the diverging planes which contain their edges, form a set of contiguous diverging pyramids or wedges) will represent, and be normal to, a system of forces which, being applied to the summits of the polyhedral frame, will balance each other—each such force being applied to the summit of meeting of the bars whose lines of resistance are normal to the set of diverging planes that enclose that face of the polyhedron of forces which represents and is normal to the force in question. Also, the areas of the diverging planes will represent the stresses along the bars to whose lines of resistance they are respectively normal.

It is obvious that the polyhedron of forces and the polyhedral frame are reciprocally related as follows: their numbers of edges are equal, and their corresponding pairs of edges perpendicular to each other; and the number of faces in each polyhedron is equal to the number of summits in the other.

Glasgow, January 9, 1864.

Figure I.9 Rankine's half-page publication on the equilibrium of polyhedral frames and 3D reciprocal diagrams in Philosophical Magazine.[5]

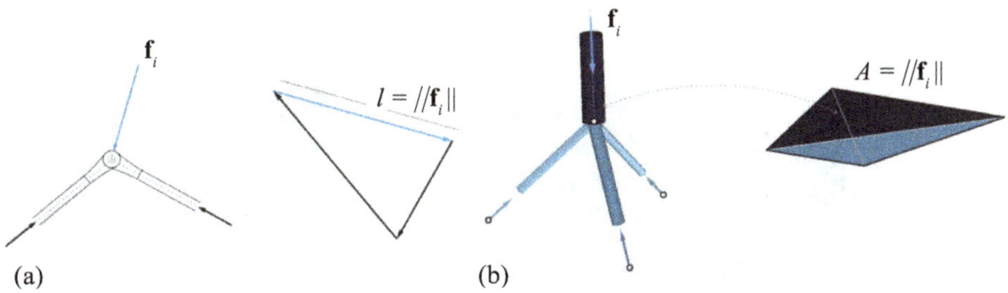

Figure I.10 (a) A single 2D node and applied load and its reciprocal force polygon and (b) a single node in 3D and its reciprocal force polyhedron.

a design door to various spatial funicular forms. Since a closed cell represents the equilibrium of a single spatial node, aggregating closed polyhedral cells should represent force equilibrium in a spatial system consisting of multiple nodes. With a bit of creativity in design, aggregating and designing closed polyhedral cells as the force diagram can result in various interesting spatial equilibrated structures, as shown in Figure I.11. This might be enough on the surface, but many exciting design techniques can be learned beyond this simple aggregation of polyhedral cells. That is the primary intention behind preparing this manuscript for those aspiring to explore this structural design method.

I.6 Material Limitations of Graphical Methods

In geometry-based form-finding methods, it is assumed that the stiffness of the construction material is almost infinite. That is, the construction material has a very large modulus of elasticity and a very small Poisson's ratio; therefore, the deflection in the system is negligible. That assumption is, in fact, the main reason that such techniques were quite powerful in the design of masonry structures.[54] Moreover, the joints in the structural form are also rotation-free with no moment resistance.

In reality, the modulus of elasticity of common construction materials such as steel or concrete is not infinite. Thus, the self-weight might cause deformation in the system, particularly in the structural forms designed for applied loads much larger than the self-weight of the system. In such scenarios, if the self-weight is not considered in the form-finding process, the global deformation of the system under the self-weight should be measured using finite element analysis to avoid excessive internal stress. Adding moment-resistance stiffness at the location of the joints for the frame structures could also help to resist an internal bending moment caused by deflection in the system.

Regardless of the mentioned restrictions, graphical statics is still considered an intuitive structural design method among architects and engineers. The explicit relationship, including the geometric dependencies between the form and force diagrams, can explain complex structural concepts in simple terms and give designers a rigorous control level in design. Therefore, since 1864, graphical statics-based approaches

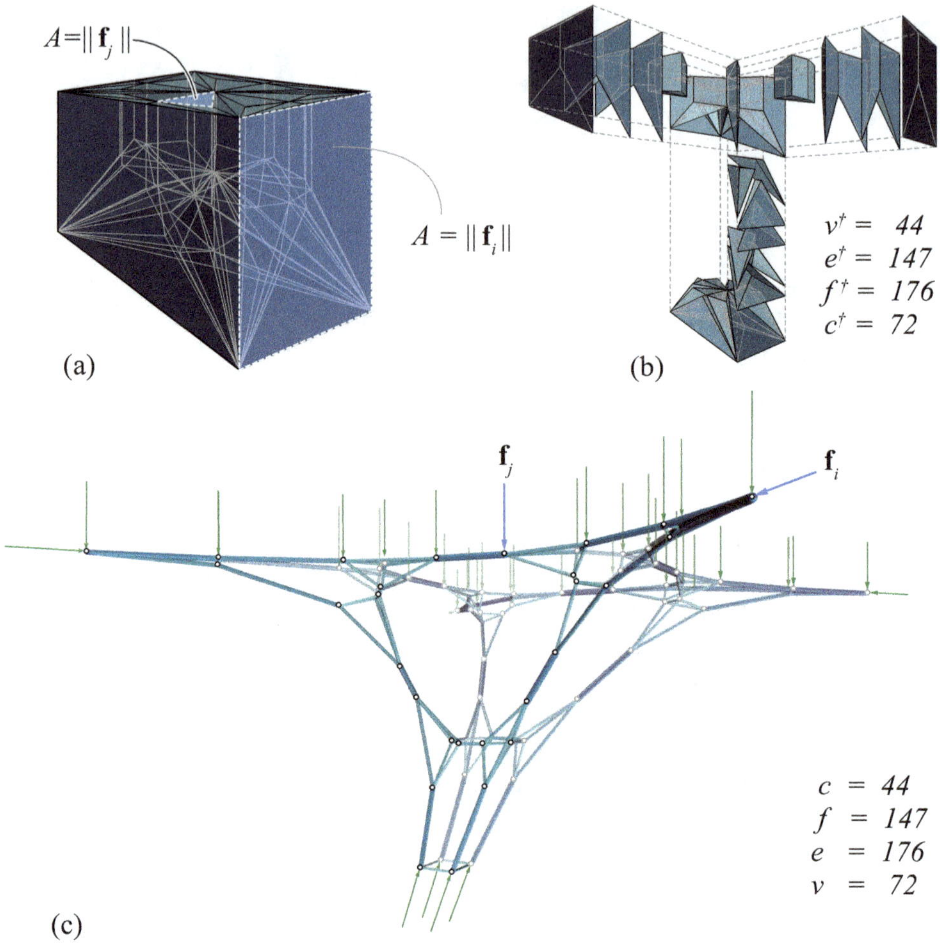

Figure I.11 (a) A force diagram constructed by aggregating closed convex polyhedral cells; (b) the exploded axon of the closed cells of the force diagram; and (c) the reciprocal spatial funicular form extracted from the designed force diagram.

have been used and further developed as extremely powerful methods of structural form finding for professional and educational purposes.[15,21,43,55–58]

I.7 Why Using Polyhedral Graphic Statics

Advances in computer science and engineering have allowed the development of techniques such as physics simulation engines,[46] particle-spring systems,[44,45] force density methods,[47,48] and dynamic relaxation[49] to simulate the physical transformation of materials and find an optimal geometry of the structure under the given loading conditions and substitute tedious physical form-finding techniques.[59] In addition to form-finding methods, over the past twenty years, extensive research has been carried out on developing methodologies to automatically generate these load paths in a continuum field following the actual flow of stresses. These studies are either based on the various formulations of finite element analysis,[60–65] or discrete (ground truss) and continuum topology optimization algorithms.[66–74]

In all these techniques, a designer starts with an arbitrary geometry as an input and receives an optimized result as an output. The final geometry is the result of a *black-box* computational process where the designer's contribution is minimal. The lack of an explicit relationship between the derived form and the internal and external forces makes it difficult for a designer to recognize the effective parameters in the form-finding process. Moreover, these methods cannot be used as an intuitive pedagogical tool to explain structural concepts for educational purposes. The mathematical concepts underlying their calculations are not trivial for most architects and traditional engineers.

In contrast to these black-box numerical methods, graphic statics is a powerful and intuitive structural design method based on pure geometry. A designer can explicitly control the magnitude of the forces in a structural system simply by geometrically modifying its force diagram. The structures designed by graphic statics-based methods are among the best examples of innovative use of material and efficiency. Many eminent engineers and designers, such as Guastavino, Maillart, Eiffel, Nervi, and Dieste, constantly used graphic statics to design their masterpieces.[16,43]

Intuitive Design, Explicit Optimization, and Efficient Fabrication

Unlike any other structural design technique, a designer can design the geometry of the *force* diagram that guarantees the equilibrium of the resulting structural form. For instance, a designer can aggregate force polyhedrons and explore a variety of nonconventional spatial structural forms with minimum bending moments in their members.[75] Besides, since the areas of the polyhedral faces represent the magnitude of forces, the optimization criteria become entirely explicit and understandable. Further, the inherent planarity constraints between the form and force diagrams dictate that the resulting concepts include planar facets, making their construction much easier. Thus, polyhedral graphic statics can combine structural form-finding and fabrication optimization in a single method. Therefore, it opens a new horizon to an unexplored realm of efficient spatial structural solutions for prefab, lightweight, high-performance systems.

A Valuable Teaching Tool

In an interactive environment, the reciprocal relationships between the form and force diagrams help students understand structural concepts more intuitively than ever. For instance, a designer can change the magnitude of the applied forces and observe the resulting change in the form of the structure and its internal forces. This property can teach students what parameters control their design and how they can deliberately

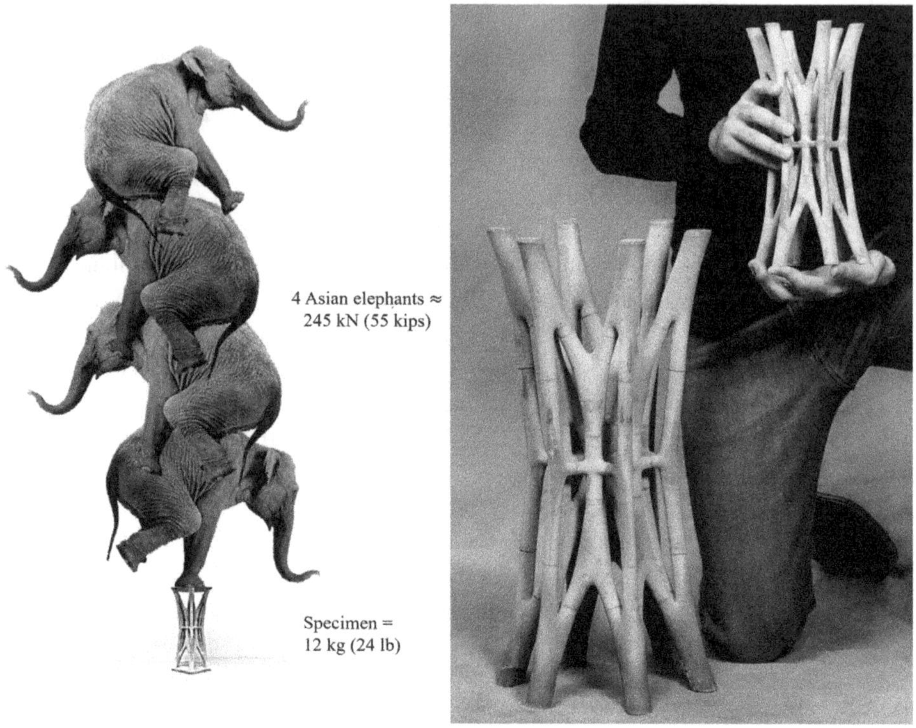

Figure I.12 The 12 kg (26.45 lbs) specimen could take 240 kN load (2,000 times more than its self-weight).

modify those parameters to achieve specific design criteria, a feature that a few other existing structural design methods have.

Attesting Structural Efficiency

In the funicular forms resulting from polyhedral graphic statics, if the stiffness is not infinite, then the efficiency of the structural system becomes a function of material property and load-bearing capacity. The structural efficiency of funicular polyhedral structures has recently been investigated in multiple studies.[76] For instance, a small-scale prototype with dimensions proportional to the standard concrete cylindrical compression test was constructed using in-situ high-performance, self-consolidating concrete. The 12 kg (26.45 lbs) specimen could take 240 kN load (2,000 times more than its self-weight) (Figure I.12). The experimental results revealed the *resiliency* of the system as the local buckling did not immediately cause the global failure of the system due to the internal indeterminacy (states of self-stress) of the funicular geometry. The structural efficiency of the system was also verified as all remaining members of the specimen simultaneously collapsed under their maximum strength[76] (Figure I.13).

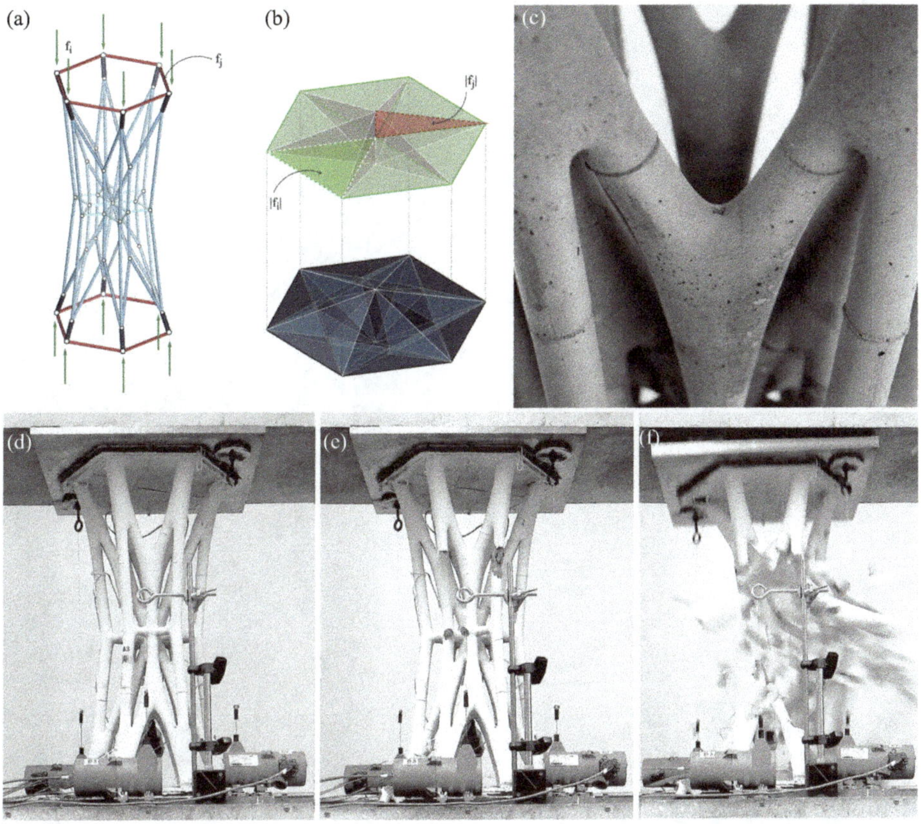

Figure I.13 (a) to (c) form and force diagram for a cylindrical sample designed by using polyhedral graphic statics; and (d) to (f) the final compression-only prototype tested to failure.

Impact on the Construction of High-Performance Structures

Using geometry-based methods in designing efficient structures can significantly reduce the use of the material in construction without compromising the structure's load-bearing capacity and thus challenge contemporary and traditional building construction techniques. In 2017, we designed and constructed a prefabricated concrete polyhedral structure using polyhedral graphic statics. This structure consists of 2.1 m^3 of fiber-reinforced concrete distributed in 50 m^3 of space with a capacity to transfer 1,016 kN of external loads to the supports.[77,78] Taking the same volume of material and designing a conventional reinforced concrete structure result in a system that can only take a maximum load of 82 kN, and to reach the same load-bearing capacity, 300% more concrete and 20% more rebars are needed, which testifies to the efficiency of the systems designed using polyhedral graphic statics.

Impact on the Construction with Sheet Materials

One important feature of polyhedral graphic static is the intrinsic planarity of the polyhedral geometries, which can be exploited for efficient fabrication and construction using flat sheet materials. Flat sheet materials have certain advantages over other material types in building complex spatial forms due to their accessibility, processibility, relatively low cost, and applicability on large scales. More importantly, as a structural material, flat-sheet-based systems provide more load paths and increase stability under a wider range of loading cases than the traditional bar-node spatial framework.

We have recently explored the structural efficiency and behavior of a lightweight, ultra-transparent glass pedestrian bridge constructed out of planar glass sheets in compression to take advantage of the planar geometry of the reciprocal polyhedral diagrams in construction. We designed and built a compression-dominant glass pedestrian bridge constructed mainly of flat sheets of glass and revisited conventional structural glass detailing to extend existing techniques for the future of efficient construction with glass.[79–81] This project showed how construction materials can be significantly reduced without compromising the structural performance of the system.

Impact on Material Science, Biomedical Engineering, and Aerospace

The cellular structures of polyhedral graphic statics[82] have applications in the design of cellular solids with micro-architectures tuned for particular mechanical behavior. The cellular structures of polyhedral graphic statics can be used in biomedical research where the porosity reduces the weight and increases the biodegradability of an implant. We have discovered that subdividing the cells of the force diagram following certain axes/labyrinths can translate buckling-prone cellular polyhedral structures to the spatial shell, which is called *shellular*.* These systems have at least three times more load-bearing capacity than cellular ones. Thus, using polyhedral graphic statics, we can design funicular systems with a very low density of the material, which can be used to design highly efficient structural systems on a large scale and metamaterials on smaller scales[83,84] (Figure I.14). We have even applied graphic statics in designing resilient wings for airplanes,[85] which we will discuss in Chapter 8.

Multi-layered, Falt-Sheet Construction

Multi-layer spatial structures usually take considerable external loads with minimal material usage at all scales. Polyhedral graphic statics can be used to design multi-layered spatial funicular networks usually materialized as space frames. The intrinsic planarity of the polyhedral geometries can be harnessed for efficient fabrication and construction processes using flat-sheet materials. Sheet-based structures are advantageous over conventional space frame systems because sheets can provide more load

* The term Shellular is used for the spatial shell geometries, see[118] for the definition of the term.

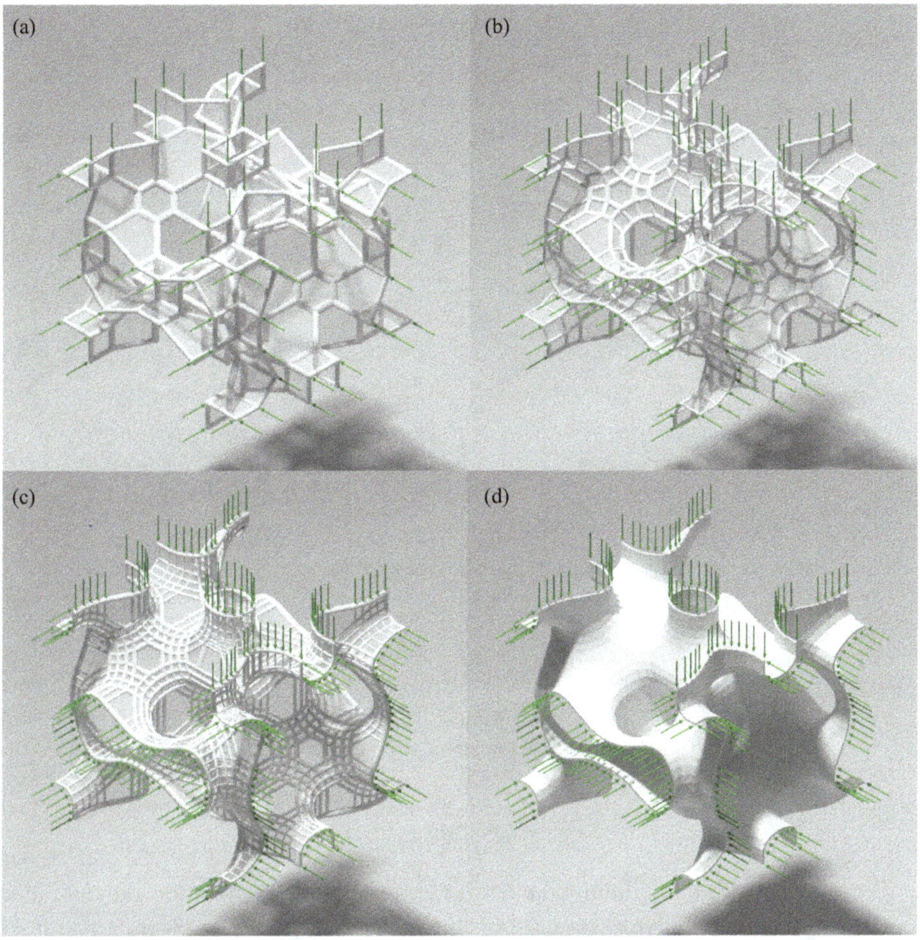

Figure I.14 (a) to (d) a cellular compression-only force flow (form diagram) with its applied loads on the boundaries can be translated to a shellular funicular system with a fine network of axially loaded members by subdividing the reciprocal force diagram (not shown here).

paths and take a broader range of loads than space frames. Moreover, sheet materials can be fabricated into complex shapes using computer numerical control (CNC) milling, laser cutting, water jet cutting, and CNC bending techniques. In contrast, the spatial frames required the careful production of complex spatial nodes.

Sheet-based structures designed using polyhedral graphic statics can be optimized further to reduce construction material. Indeed, several faces can be removed from the system without compromising the structural performance. We have shown that we can achieve higher material efficiency in sheet-based construction by combining the matrix analysis method with Polyhedral Graphic Statics. We have developed algorithms for the system made of sheet-based materials and predict the formation of mechanisms under the applied loads. The algorithm in an iterative approach reduces the number of systems, ensuring that no mechanism happens in the structure. As a result, we will end

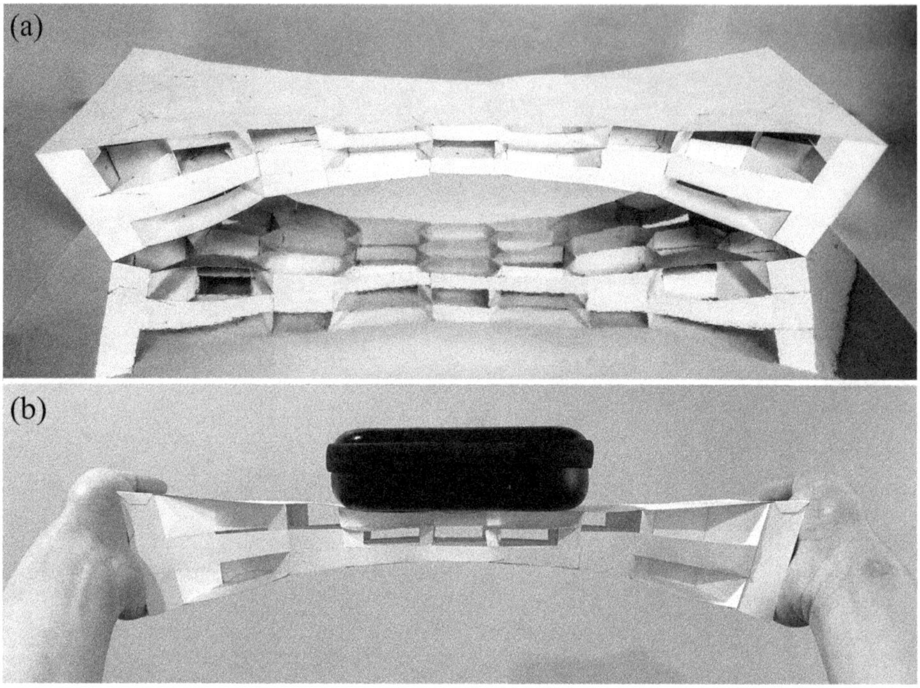

Figure I.15 (a) A small-scale physical model made of Bristol paper; (b) a simple physical load test on the small-scale model made of Bristol paper (0.006 in, 0.15 mm), the span is 50 cm and the weight is 2.3 kg

up with a minimum number of faces needed to restrain the structure kinematically.[86] A comparative study used finite element analysis to verify the structural performance and compare the sheet-based versus polyhedral frame. The result shows that, with the same material usage, the sheet-based system outperforms spatial frames and resists a wide range of loading scenarios (Figure I.15).

Reducing Carbon Emissions

Building construction contributes to a disproportional amount of landfill waste and annual carbon emissions. Using the methods of polyhedral graphic statics, we can optimize the design; thus, less material will be required overall, making the buildings much lighter, resulting in fewer carbon emissions. Besides, prefabrication methods following the inherent planarity of the funicular polyhedral systems can waste fewer materials than in-situ construction on a project site. Moreover, the amount of material to be disposed of after the structure's lifecycle will also be reduced. Further research is needed to quantifiably argue the application of graphic statics in reducing carbon emissions. However, we have shown in various works and research that at least a 40% reduction in the use of material can be achieved due to optimization that decreases the embodied carbon and significantly reduces adverse environmental effects.

I.8 Limitations of Polyhedral Graphic Statics

This structural design method's main limitation is its "polyhedral" nature. The fact that both the structural form and its force diagram consist of planar faces attests that the equilibrium states represented by this method are only a subset of all possible equilibrium states that do not fall into the "polyhedral" version. Therefore, this method may not capture all possible scenarios one might face during the design stage.

Nevertheless, we should not forget that its predecessor, 2D graphic statics, is used with the incorporation of lots of assumptions by experienced practitioners. Experienced designers and practitioners like Millart used the 2D version and created structures still considered wonders of structural engineering. Thus, this extension to 3D can still provide a valuable tool for learning complex equilibrium concepts in 3D and opens the door to various design possibilities that were not fully explored.

With this brief introduction, I would invite you to gain a deeper understanding of the methods of polyhedral graphic statics in Chapters 1 to 8.

1 Equilibrium of a Single Node

Masoud Akbarzadeh and Juney Lee

1.1 Introduction

Efficient structural forms carry the applied loads in the form of pure tension and/or compression forces. In such systems, each member of the structure carries an internal force axially, that is, the member is aligned with the direction of internal force, and its cross-section is normal to the direction of the force. Carrying the force in pure tension/compression will limit the type of stress to axial only and minimize other forms of stress, such as bending, shear, and torsion. It has been proven that structures can withstand larger forces if axial stress is the dominant form of stress in their members. Also, less material is needed to carry axial forces; thus, less material is used to build such structures, making them quite efficient.

These structural systems are often referred to as *funicular* structures. The geometry of the funicular structure and its members precisely matches the direction of their internal flow of forces. Tensile structures are excellent examples of the funicular system because the structure's geometry is a true representative of the direction of the force flow within the system.

In this book, we show that graphic statics is a powerful tool for designing funicular structures. The first step in this journey is understanding their equilibrium under given applied loads and boundary conditions. If the structure is in static equilibrium, then every node is also in equilibrium. Thus, in this chapter, we concentrate on the equilibrium of a single node in two-dimensional (2D) and three-dimensional (3D) and learn to find this equilibrium using geometric methods of graphic statics. Chapters 2, 3 and 4 will extend this concept to designing more complex funicular structures with compressive and tensile forces.

1.1.1 Chapter Objectives

Figure 1.1 illustrated an example of a funicular bridge consisting of a single node in 2D that is connected to three cables and carries loads of a bridge from the midpoint of the deck. This is a classic example by Edward Allen and Waclaw Zalewski in their famous book *Form and Forces: Designing Efficient, Expressive Structures*,[16] which was the author's first encounter with the methods of graphic statics. A similar structure in 3D consists of two similar nodes attached by a shared edge. Each node in 3D is connected to four cables. Let's assume that the bridge's dead load and live load are

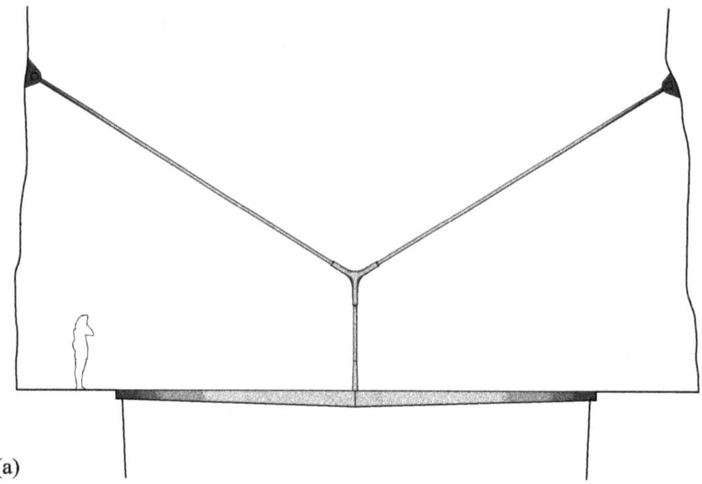

(a)

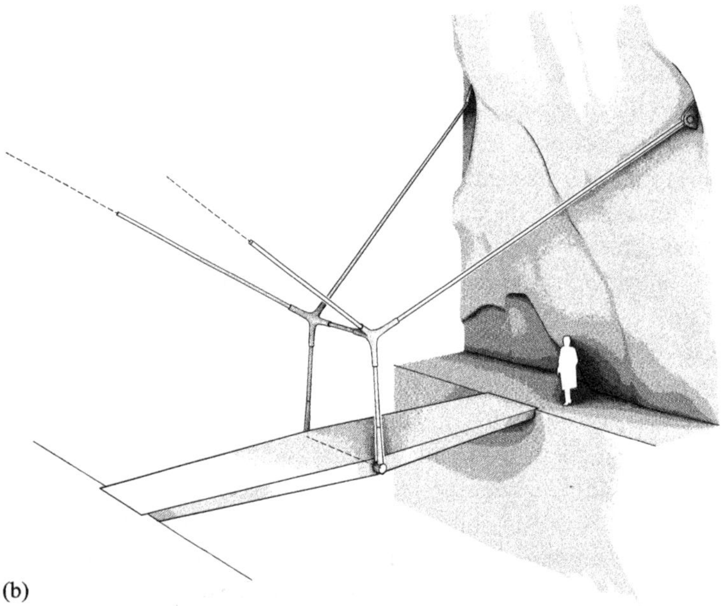

(b)

Figure 1.1 (a) A 2D structure of a bridge consisting of a node in pure tension; (b) a 3D structure of a bridge with two attached nodes in pure tension, after.[43]

known, and we are interested in finding the magnitude of the internal force in each cable connected to the node of the tensile structure. Moreover, it is also crucial to learn how the change in the angles of the cables will affect the internal flow of forces in the structure using pure geometric constructions. Therefore, the main objective of this chapter is to provide the necessary means and methods to:

- verify the equilibrium of a single node of a structure in 2D and 3D using the methods of graphic statics;

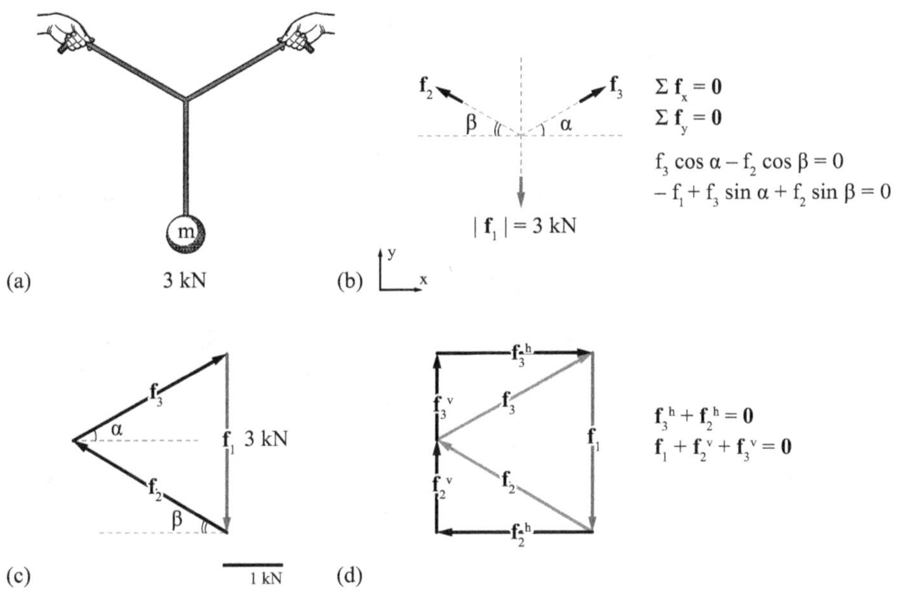

Figure 1.2 (a) The geometry of a node connected to three ropes in tension under the load 3 kN; (b) the numerical equilibrium of the node; (c) the closed polygon of forces to assure the equilibrium; and (d) the decomposition of the force vectors satisfying the equilibrium in *x*- and *y*-directions, or say, in horizontal and vertical directions.

- explains the geometric and topological relations between the form and the force diagrams of a single node in 2D and 3D; and
- learn how the change in the geometry of one diagram informs the geometry of the other and use that technique to design simple structures.

1.1.2 Assumptions

The primary assumption in this chapter is that the structure either consists of a single node similar to the example of Figure 1.1 or a node is part of a larger structure in global equilibrium. Thus, the node is connected to multiple internal members and usually receives an externally applied load with known direction and magnitude. For instance, the 2D and 3D nodes of the structure of Figure 1.1 carry the dead load and live load of the bridge that is approximately 3 kN. Thus the applied load is given, and the magnitude of the internal forces in each cable/member connected to the node in the structure will be calculated using graphic statics.

1.1.3 Numerical Approach

Let us begin with a very simple example of Figure 1.2 to describe certain properties related to an equilibrated geometry. Figure 1.2 illustrates a system of three ropes and a knot. Two ropes are pulled from the sides to keep a hanging weight in equilibrium. The weight is represented by *m*, and the internal forces in each rope are in question.

Based on the numerical definition of the equilibrium, the sum of the forces acting on a node, including external and internal, should be zero, that is:

$$\sum \mathbf{f}_x = \mathbf{0}, \quad \sum \mathbf{f}_y = \mathbf{0}. \tag{1.1}$$

The applied load is shown by $\mathbf{f}_1$, and the magnitude of $\mathbf{f}_2$ and $\mathbf{f}_3$ can be found numerically by solving the equations in x- and y- direction, as shown in Figure 1.2b.

1.1.4 Graphical/Geometric Approach

The same equilibrium can be represented geometrically. We can draw a closed force polygon such that each edge of the polygon is parallel to the direction of the rope and the applied load. If the internal and external forces are in equilibrium, then the force vectors must make a closed polygon such that they are connected tip-to-tail, as illustrated in Figure 1.2c. Furthermore, decomposing the vectors in x- and y- directions shows how the force components cancel each other and make the sums of the forces zero (Figure 1.2d).

As shown briefly, the geometric representation of equilibrium is quite intuitive. However, to learn and apply this method for various configurations of nodes and extend the technique to the design and analysis of structures, one may need to understand the underlying rules and guidelines pertinent to the *form* of a structure and its *diagram of forces* in 2D and 3D. These rules establish the methods of graphic statics, and the rest of the chapters will expand on them.

1.2 Equilibrium of a Tensile/Compressive Node in 2D

This section explains how the geometry of a single node and its external and internal forces in 2D can be abstracted into a graph referred to as the *form diagram*. Chapters 2 to 4 will describe the relationship between the elements of this graph and will show how a second graph can be extracted based on the information of the first one. The second graph is related to the equilibrium of forces and will eventually become what would be referred to as the *force diagram* for the node. In the end, it will explain how these two graphs are related both geometrically and topologically.

1.2.1 A Graph Representation

Let us start with the simple node of Figure 1.3a with its three edges in tension. The node and its members can be abstracted into a graph consisting of one vertex and three edges (Figure 1.3b). We will call this graph that represents the geometry of a node the form diagram. The vertex and its edges of the form diagram divide the page into three *subspaces* named by capital letters A, B, and C. In this example, the subspaces are adjacent to each other. For instance, A is adjacent to B and C, and B is adjacent to C and A. Each subspace is connected to its adjacent subspace via an edge (Figure 1.3b).

We can make another graph and describe the connectivity between these subspaces. Pick an arbitrary point in each subspace and connect it to that of the adjacent subspace

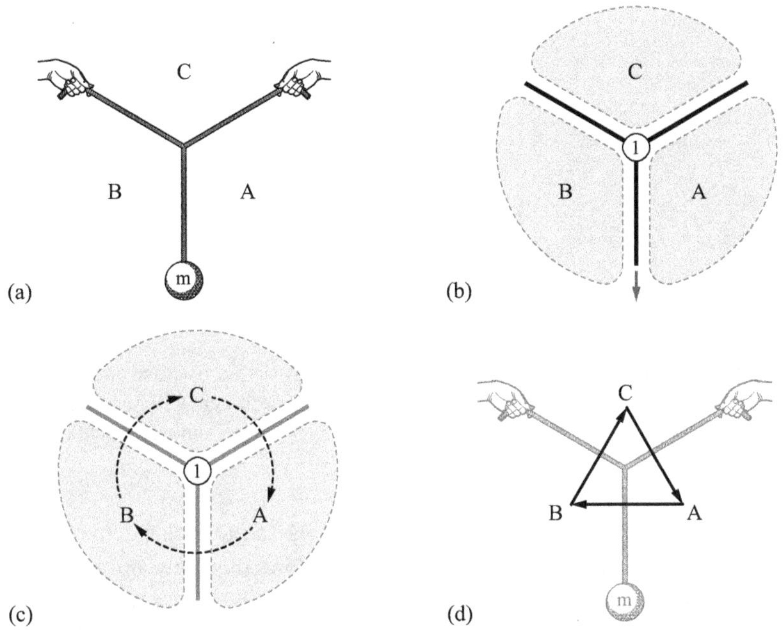

Figure 1.3 (a) A node in tension as a result of hanging a weight; (b) the graph representation of the structural node with its division of subspaces; (c) the connectivity of the subspaces as a secondary graph; and (d) the edges of the second graph drawn normal to the edges of the form establishing the force polygon for the node.

by drawing a line crossing over the shared edge of the form diagram (Figure 1.3c). Each subspace is connected to its adjacent subspace by one and only one edge. Let's use a clockwise direction to give the new graph edges a consistent direction.

Suppose every edge of the new graph becomes perpendicular to the edges of the form diagram. In that case, there will be a closed polygon with edge vectors perpendicular to the edges of the initial graph. The sum of the vectors for this polygon is zero, and each edge of the new graph can be related to its perpendicular edge. Therefore, the new graph shows the equilibrium of the node and the magnitude of the force in each member of the node. This polygon is similar to the closed polygon of the force of Figure 1.2c that describes the equilibrium of the node; only its edges are perpendicular to the edges of the structure instead of parallel.

The edge vectors of this polygon are vectors connected tip-to-tale, and the polygon has a consistent direction to ensure that the sum of the forces is zero. This polygon is called the force diagram, and the cycling direction around the node defines the direction of the vectors around the polygon (Figure 1.3d). There is a rule about choosing a consistent direction, which will be discussed in Section 1.2.2.

1.2.2 The Cycling Direction

The cycling direction is crucial in giving a consistent direction to edge vectors in the force diagram. The direction of cycling can be defined by the direction of the applied

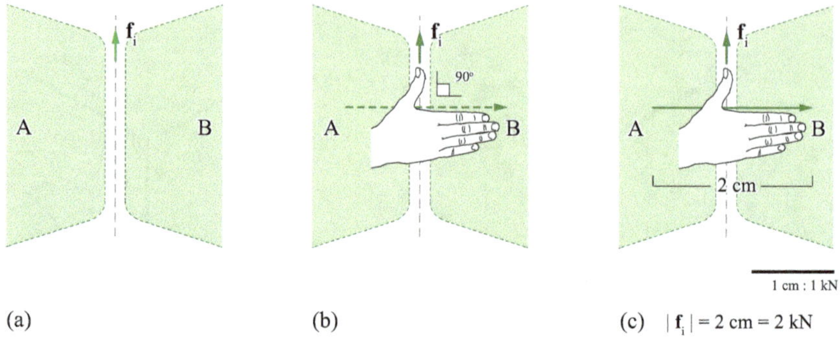

Figure 1.4 (a) The line of action of an applied load is shared between two subspaces; (b) putting the thumb in the direction of the applied load, the four fingers will be a point to the direction of travel from space *A* to *B* with (c) a vector whose length is 2 cm equivalent to 2 kN force magnitude.

load given in the system. In this book, we use a particular convention based on the *right-hand rule*, and the cycling direction follows this convention.

Consider an applied load $\mathbf{f}_i$ of Figure 1.4a, as an edge of a graph of the form diagram in a 2D space. The edge of the applied load divides the 2D plane into two subspaces. Connecting one subspace to the next by crossing the applied load makes an edge of the force polygon. Nevertheless, the edge is a vector and should have a proper geometric direction and a length.

A particular right-hand rule convention defines the direction of the traversing in this book. We put the thumb in the direction of the applied load while your palm is on the paper (Figure 1.4b). The stretch of the four fingers represents the direction of the force vector. In this drawing, the length of the travel defines the length of the force vector; for instance, if 1 cm is equivalent to 1 kN force, then a 2 cm vector is needed to represent 2 kN applied load, as illustrated in Figure 1.4c.

1.2.3 Constructing a Closed Force Polygon

Example 1.1 provides a step-by-step procedure to complete a closed force polygon for a tensile node of Figure 1.4.

Example 1.1 Construct a closed force polygon for the node of Figure 1.4a and report the magnitude of the internal force if the applied load $\mathbf{f}_1$ is 3 kN.

Construction steps:

1 Illustrate the node by a graph of three edges and a vertex including three subspaces made by the edges of the graph (Figure 1.5a).*

* This book annotates the subspaces of the form with capital letters in the same manner as Henry Bow did in his influential book *Economics of Construction in Relation to Framed Structures*.[42]

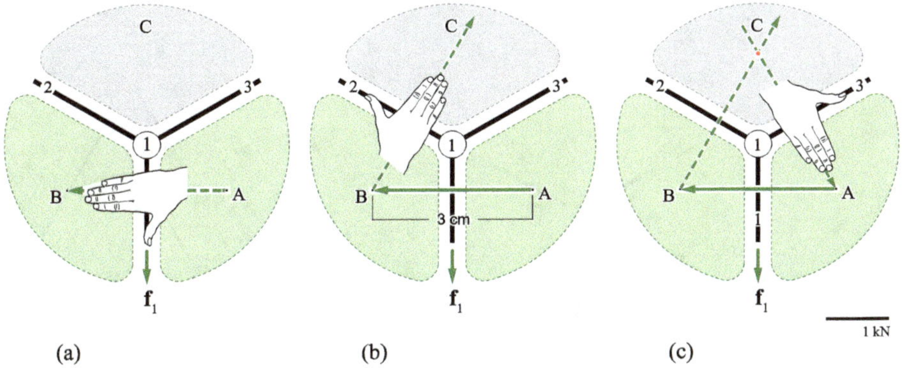

Figure 1.5 (a) to (c) Moving from space *A* to *B*, and *C* and back to space *A* while keeping the perpendicular direction to the ropes establishes the geometry of the force diagram.

2 Start from the applied load $\mathbf{f}_i$ shared by subspaces *A* and *B*.
3 Put the thumb of the right hand in the direction of the applied load $\mathbf{f}_i$ while your palm is on the paper. Then the four fingers are crossing over the edge 1 and will show the direction to travel from *A* to *B*, *B* to *C*, and *C* to *A*.
4 Start from an arbitrary point in space *A*, draw a vector in the direction of four fingers perpendicular to the $\mathbf{f}_1$ from subspace *A* to *B* and a 3-cm length representing 2 kN of force (Figure 1.5b).
5 Locate point *B* at the end point of the vector *AB* and put the four fingers normal to the edge 2 shared by subspaces *C* and *B*.
6 Draw a line normal to the edge 2 from point *B*. Note that the length of the vector perpendicular to the edge 2 is not known to locate point *C*. However, at some point during the process, the loop needs to be closed to represent equilibrium, and the chain of vectors needs to return to the starting point *A*.
7 To locate point *C*, put the four fingers normal to the edge 3 while your palm is on the paper and draw a normal line from point *A*.
8 The intersection of the lines drawn from *B* and *A* locates point *C*.
9 By finding point *C*, the force polygon is completed, and the lengths of the vectors normal to the edges of the form diagram can give the magnitude of the forces.

□

Remark 1.1. Note that the location of the construction of the force does not have to start from the subspaces of the form. We used that starting point to relate the geometry of the form and force diagrams. Later in this chapter and Chapters 2, 3, and 4, the form and force will be constructed separately while following the direction given by the right-hand rule. As a result, the vertex of the form will not necessarily be inside the force polygon.

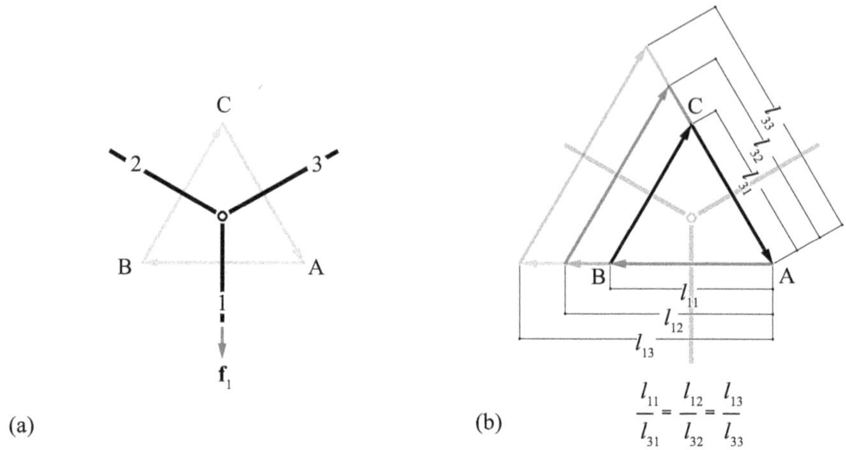

Figure 1.6 (a) The form of the node with its edges and subspaces; and (b) the completed force polygon scaled by choosing a different length for the applied load vector *AB*.

1.2.4 Scaling the Force Diagram

Figure 1.6 shows a completed, closed force polygon for the given configuration of the ropes and hanging mass. Note that choosing different lengths for the vector *AB* results in proportional force triangles, meaning that the proportion of the edges of the triangles does not change regardless of the scale. This property highlights a vital aspect of the graphical approach: The magnitude of the internal forces in the edges of the node does not change by scaling the force diagram because the change ratio stays constant (Figure 1.6a,b).

1.2.5 The Direction of Internal Forces

Once the force diagram is constructed, the magnitude and the direction of the internal forces in the node can be identified using the same right-hand rule. As explained, the length of each vector in the force diagram represents the magnitude of the force in the corresponding member of the form diagram. To find the type of force, either tension or compression, in the member, we need both the geometry of the form and the force diagram. The following procedure can be used to find the direction of the internal force in the members of the form diagram:

1. In the force diagram, put the four fingers of the right hand on the edge vector such that the tip of the fingers match the tip of the edge vector (Figure 1.7a).
2. Move the right hand and the thumb from the edge vector of the force to the edge in the form diagram.
3. If the thumb points the direction toward the node, then the internal force is compression, and otherwise, it is tension as in Figure 1.7b.

Remark 1.2. Note that in this particular example, we already knew the direction of the internal force in the members of the node. Nevertheless, the same procedure applies

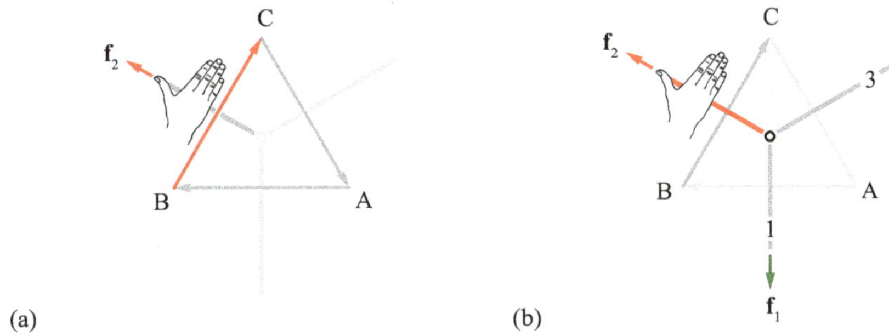

Figure 1.7 Finding the direction of the internal force (tensile) in the members by putting four fingers in the direction of the force vector in the force diagram (a) and moving the right hand to the related edge of the form diagram (b).

to identify the direction of the internal force in any other configuration of the node, including the tension and compression system that will be discussed in Sections 1.5 and 1.6.

1.2.6 Controlling the Magnitude of the Internal Forces

Once the form and force diagrams are constructed, their geometric dependencies can be used to explore a variety of formal configurations with explicit control over the force magnitudes in the members of the structure. Since the edges of the form diagram are perpendicular to the edges of the force diagram, the shallower angles in the ropes result in bigger forces in the force diagram. For instance, in the example of Figure 1.3, changing the angle of the ropes holding the same weight results in different internal forces within the system. Figure 1.8 illustrates multiple configurations with the same scale of the force diagram to emphasize the change in the magnitude of the internal forces. As shown in Figure 1.8, the configurations with shallower angles of the ropes carry larger internal forces within the ropes.

1.2.7 Topological Relationship between Form and Force in 2D

Each element of the form diagram is related to an element in the force diagram. For instance, the number of edges in both diagrams matches, and the number of vertices in the form diagram equals the number of faces in the other diagram. This unique mathematical property is known as *duality*.* The form and force diagrams are the *topological dual* of each other; that is, per member of one diagram exists a member in the other. Thus, the topological information of one diagram can lead to the topological properties of the other.

* The concept of duality was proposed by Möbius and later adopted by many mathematicians and was also used in graphic statics.[38]

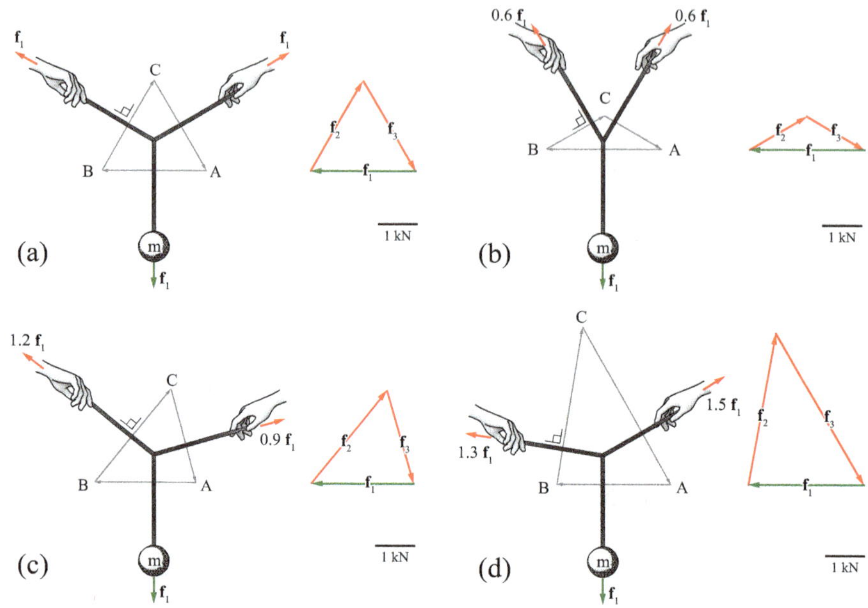

Figure 1.8 Multiple angles of the ropes under the same applied load and the change in the magnitude of the internal forces. (a) Starting configuration where the angles of the ropes 2, and 3 are the same, (b) making the angle between the ropes 2 and 3 more obtuse results in shallower force diagram and less internal force in the members; (c) and (d) variable angles for the ropes results in unequal internal force in the members.

Also, each edge of the form diagram is perpendicular to an edge in the force diagram. The geometric dependencies of the form and force and their duality establish the *reciprocity* between these diagrams. Various people in the history of graphic statics have enunciated this relationship. Rankine, in his *Manual of Applied Mechanics*, emphasizes this relationship,[36] and Maxwell formulates this relationship mathematically in his groundbreaking paper, *On Reciprocal Figures and Diagrams of Forces*, which became the basis of the methods of graphic statics.[17,37]

If we denote the form diagram by a Greek letter Γ and its force diagram by $\Gamma^\dagger$, the reciprocity between the two can be summarized as follows:

- the number of internal vertices v in the form diagram Γ is equal to the number of faces $f^\dagger$ in the force diagram $\Gamma^\dagger$;
- the number of edges e in the form and edges $e^\dagger$ of the force are equal; and
- the number of faces f (open/closed) in the form diagram is equal to the number of vertices $v\dagger$ in the force diagram (Figure 1.9).

The methods presented in Section 1.2.6 make it quite clear how to calculate the forces in the tension-only node of the bridge of Figure 1.10. In addition, it provides a design mechanism to control the edges of the force polygon to measure or minimize the internal forces in the cables that hold the bridge from the cliffs.

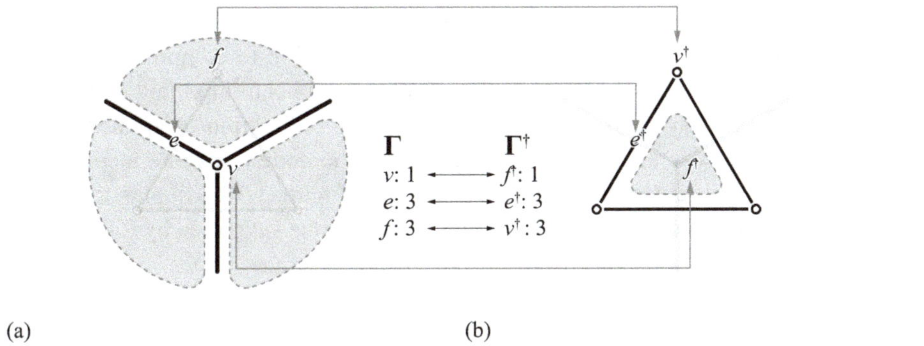

Figure 1.9 The topological relationship between (a) the form diagram and (b) the force diagram, including the number of vertices, edges, and faces.

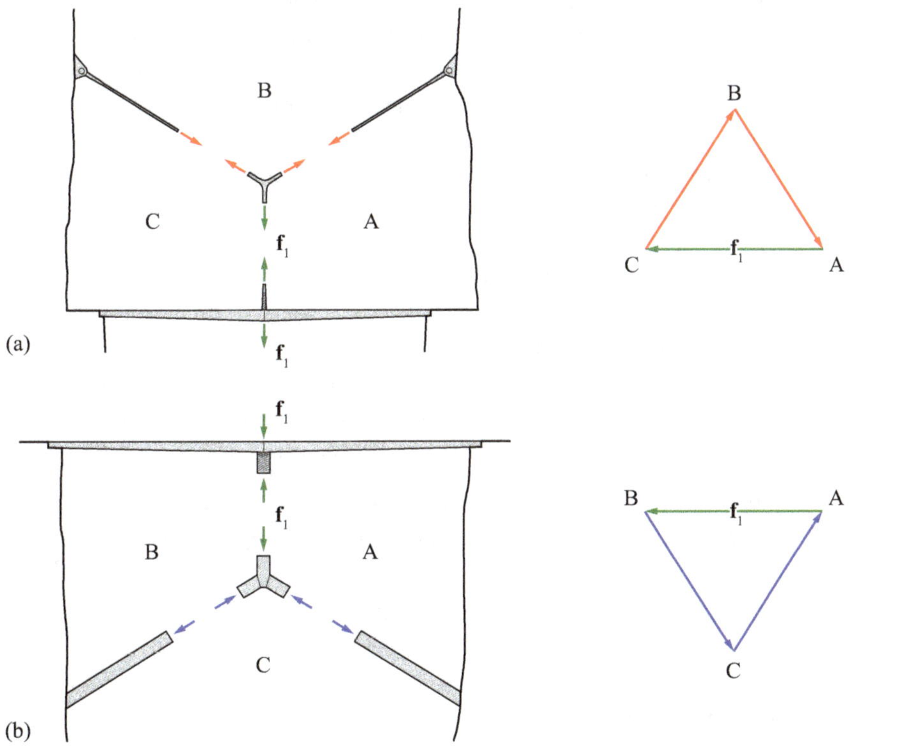

Figure 1.10 (a) The form and the force diagram of a 2D bridge with a tension-only node; (b) the flipped geometry of the node as a compression-only system for the same applied load of the deck.

1.2.8 A Compression-Only Node in 2D

The procedures explained for a tension-only node can be simply extended to a compression-only node as well. In fact, there is an underlying relationship between the equilibrium of a tension-only node and a compression-only node: for the *same geometry of the node*, if the direction of the applied load is reversed, then the direction of the internal forces will flip from tension to compression. This property will be extensively used in the other chapters and is an important feature in the design of funicular systems. In addition, for the same direction of applied load, if the node's geometry is mirrored around the x-axis, the node turns into a compression-only system. This example has been illustrated in Figure 1.10. Flipping a tension-only system to get a compression-only system is a technique suggested by Robert Hooke[24] and will be discussed in Chapter 3.

1.3 A Closed Polyhedron and Its Area-Weighted Normals

There are many examples of polyhedrons in natural and artificial structures.[87] The faces of a polyhedron are planar or flat, and they can be convex or non-convex and even self-intersecting. In a single polyhedron, every two adjacent faces share a straight edge. Finally, each edge connects two vertices, namely its endpoints. A convex polyhedron, as shown in Figure 1.11a, is a polyhedron whose faces are convex polygons (Figure 1.11b).

Any closed polyhedron can have two orientations that are as follows: for each face, a consistent choice of directions of its edges is given by attaching the edges to each other tip to tail. This defines a normal unit vector of the face using the right-hand rule. Once we require that the directions of an edge shared by two adjacent faces are opposite in those two faces, a consistent choice of normal vectors for the faces can be found (Figure 1.11c,d).

Let us denote the consistent choice of the unit normal vector of a face f by $\mathbf{n}_f$. The *Divergence Theorem* states that the weighted average of the consistent choice of unit normal vectors by the area of the faces is the zero vector.*

Theorem 1.1 (Divergence Theorem). *Given a closed polyhedron with faces $f_1, f_2, \ldots, f_k$ and a consistent choice of unit normal vectors $\mathbf{n}_i$ (where $1 \leq i \leq k$), we have that*

$$\sum_{i=1}^{k} A_i \mathbf{n}_i = \mathit{0}$$

where A_i denotes the area of the face f_i.

We apply the theorem earlier as follows. Consider the consistent choice of unit normal vectors of the faces and apply them to a single node. In order to have all these forces in equilibrium in any closed polyhedron, the normal vectors need to be weighted by some chosen lengths. If the chosen lengths of the normals are proportional to the area of the faces of the polyhedron, then the weighted sum is zero. Thus, the node will be in equilibrium or the sum of the forces will be zero.

* See, for instance, Chapter 10.[88]

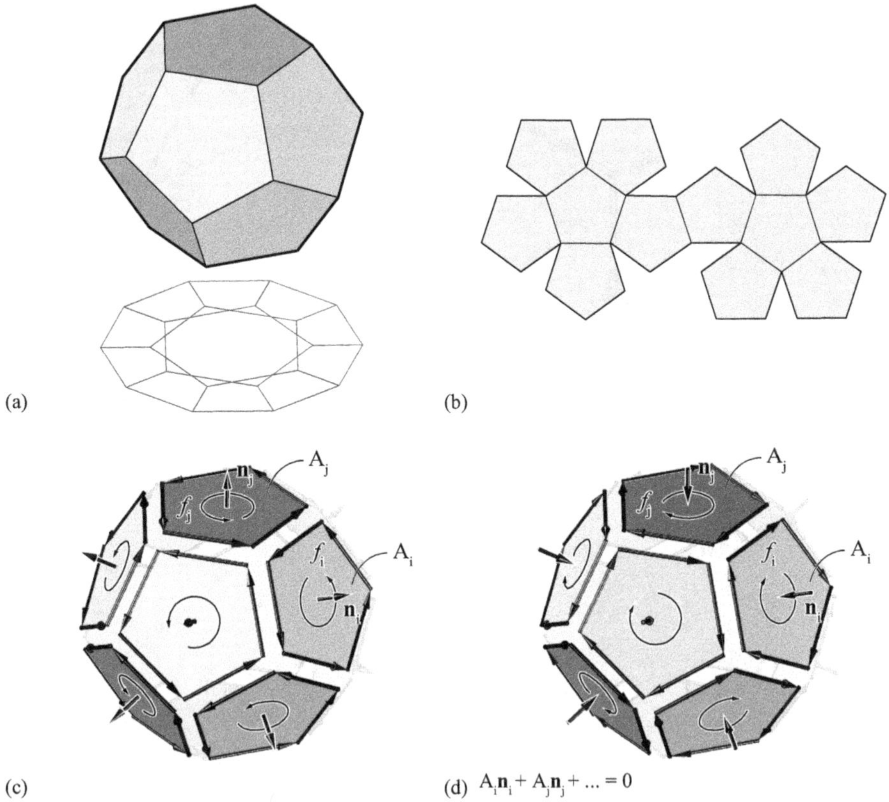

Figure 1.11 (a) A regular dodecahedron; (b) a net of the dodecahedron showing its convex polygonal faces; (c) outward and (d) inward orientations of the faces.

1.4 Equilibrium of a Tensile/Compressive Node in 3D

Like the 2D process, a geometric diagram can represent the equilibrium of forces in a 3D node. To understand this relationship, let us revisit a historic proposition by Rankine published in the Philosophical Magazine of 1864.[5]

1.4.1 The Principle of the Equilibrium of Polyhedral Frames

Rankine in The Principle of the Equilibrium of Polyhedral Frames proposes that a closed polyhedron with planar faces can represent the equilibrium of forces in a spatial node. This polyhedron can be constructed by putting planes normal to the edges connected to the node. Consequently, the intersection of the planes identifies the edges of the polyhedron. If so, each face of this polyhedron is normal to an edge, or the line of action of an applied load in the node, and its area represents the magnitude of the applied load or the internal force in the members of a node in equilibrium.*

* This proposition can be mathematically proven using Stokes's theorem.[75,89]

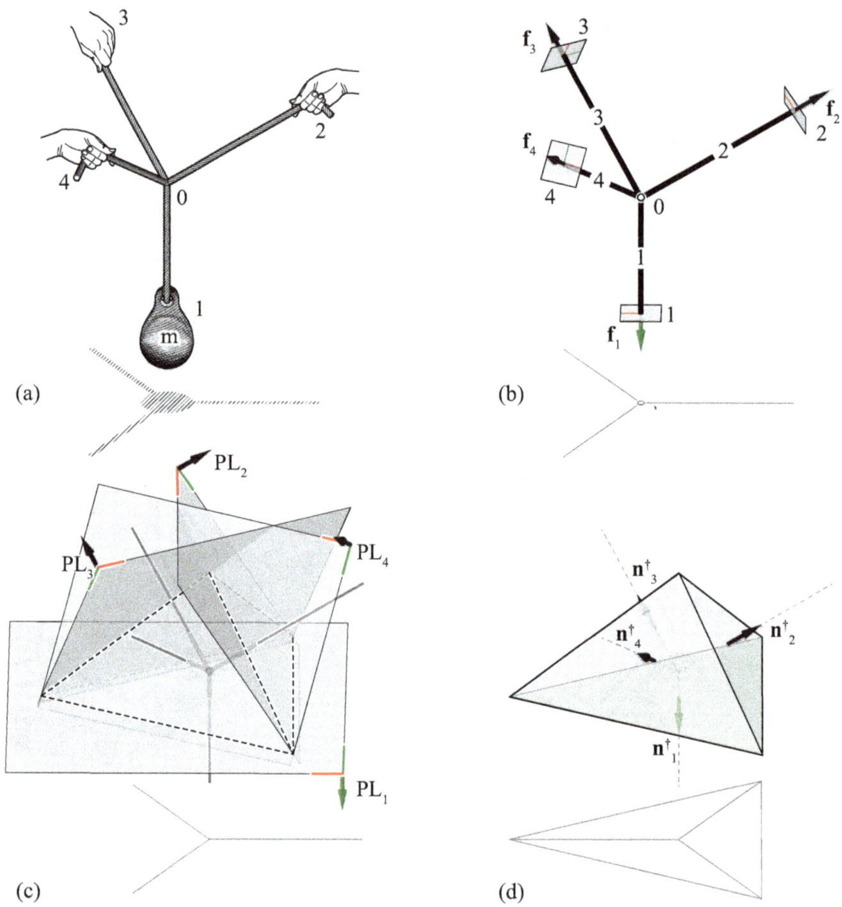

Figure 1.12 The equilibrium of a node in 3D consisting of three ropes and a hanging load can be represented by a closed force polyhedron with planes perpendicular to the members and force meeting at the node.

Following this proposition, we can construct a force diagram for the spatial node of Figure 1.12 held by three ropes that carry a weight of *m*. By putting perpendicular planes at an arbitrary distance from the node, the four planes intersect each other and establish the edges of a closed polyhedron shown in Figure 1.12b–d. According to Rankine, this closed force polyhedron represents the equilibrium of the spatial node, and the area of each face corresponds to the magnitude of the force in the ropes connected to the node. There are some geometric properties of closed force polyhedrons proposed by Rankine's method that will be visited next.

1.4.2 Properties of a Closed Force Polyhedron

A closed force polyhedron, similar to the one representing the equilibrium of a node of Figure 1.12a, has its face normals consistently pointing either toward *inside* or *outside*

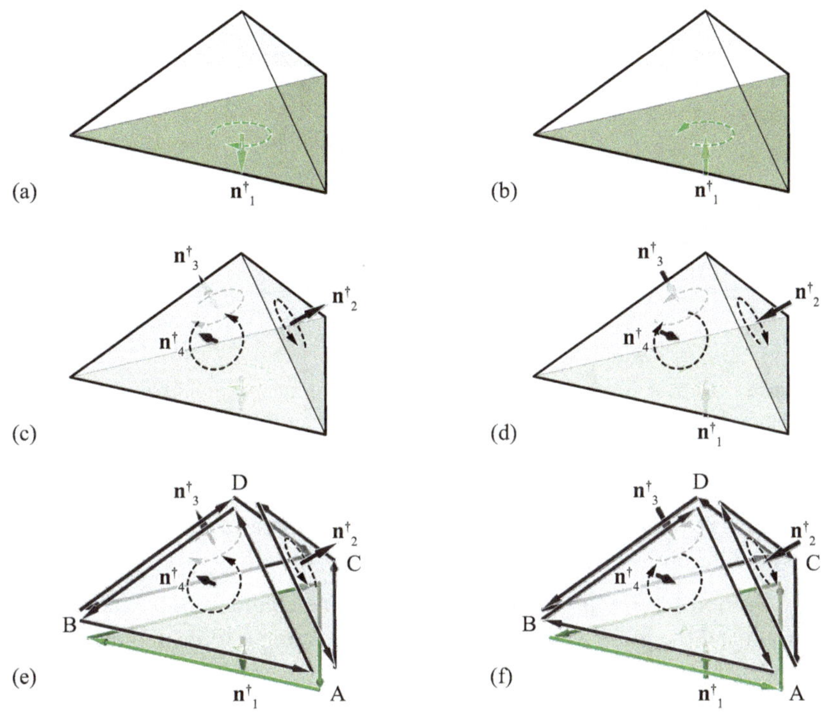

Figure 1.13 The direction of a face normal in a force polyhedron, (a), (b), defines the direction of the rest of the faces in (c) and (d); the direction of the edge vectors around each face follows the direction of the normal using the right-hand rule and is opposite to the direction of the adjacent edge vectors (e) and (f).

of the polyhedron. If the direction of a single face is known, the direction of the rest can be found, knowing this consistency. In Polyhedral Graphic Statics, the direction of the applied load and the normal of its corresponding face in the force diagram are often given. If the direction of an applied load is given, it can be used to find the direction of the rest of the polyhedron faces. For instance, Figure 1.13a shows a closed polyhedron with a green face representing the face of an applied load. From the face of the applied load, we can deduce the direction of the rest of the faces in the force polyhedron, as illustrated in Figure 1.13b–d for two polyhedrons with opposite directions of applied loads.

Moreover, each face of the polyhedron has *edge vectors* that form a closed polygon of vectors around each face. These vectors are connected tip-to-tail. The direction of the edge vectors follows the curl of the right hand if the thumb is pointing to the direction of the face normal. In addition, each of the two *adjacent faces* shares an edge vector with equal and opposite direction (Figure 1.13e,f).

Remark 1.3. Although putting planes perpendicular to the edges of a spatial node can result in a closed force polyhedron for the system (Figure 1.12), it does not provide us with more insights into the relationship between individual elements in both form and force diagrams in 3D. Moreover, a procedural method is needed similar to the

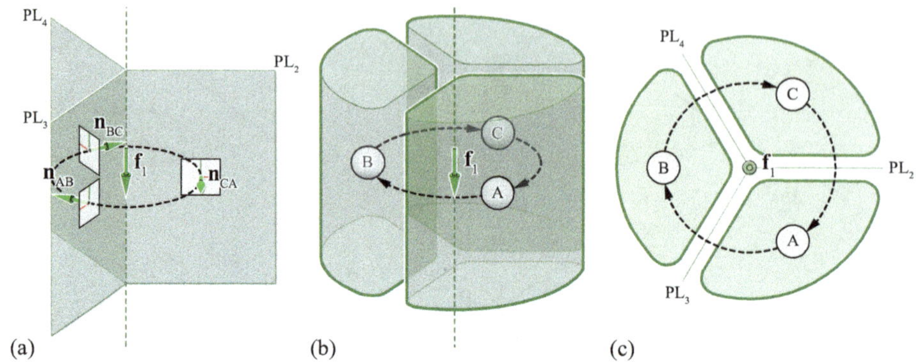

Figure 1.14 (a) A line of action in 3D is the place of intersection of at least three nonco-planar planes (b) and (c), which divide the space into three subspaces.

methods explained for 2D Graphic Statics to generalize the technique to various nodes with different configurations.

Section 1.4.3 will expand on the underlying elements of a 3D node as the form diagram and the procedural methods for constructing a force diagram for various spatial nodes in general.

1.4.3 Applied Load as Intersection of Planes

It might sound counterintuitive, but one of the critical properties of the form diagram in Polyhedral Graphic Statics is that the line of action of a force or an edge connected to the node is the line of intersection of *at least three planes*. Consider Figure 1.14 as an instance. The planes intersecting at the line of action of $\mathbf{f}_1$ divide the space into subspaces *A*, *B*, and *C*. Each subspace shares a face plane PL_i with its adjacent subspaces (Figure 1.14b). These face planes have normal directions shown in the drawings. The subspaces are also referred to as open cells. Cycling these subspaces is necessary to construct a face reciprocal to the applied load.

1.4.4 Construction of the Applied Load Face

The view perpendicular to the line of action of force $\mathbf{f}_i$ on Figure 1.14c shows similar space partitioning as described for a 2D case of Figure 1.5. If we place the thumb of the right hand in the direction of the applied load, then the curl of the fingers shows the direction of cycling subspaces around the load. In this regard, the normals of the dividing planes are aligned with the cycling direction. They are indexed based on the start and endpoint of the travel from one subspace to the next. For instance, $\mathbf{n}_{AB}$ is pointing from subspace *A* to *B* and is indexed respectively.

The geometry of the face reciprocal to the applied load can be made by cycling the subspaces around the load. We can start from an arbitrary point *A* and draw a vector with the direction $\mathbf{n}_{AB}$ and a chosen length, for instance, 3 cm, to locate the point *B*.

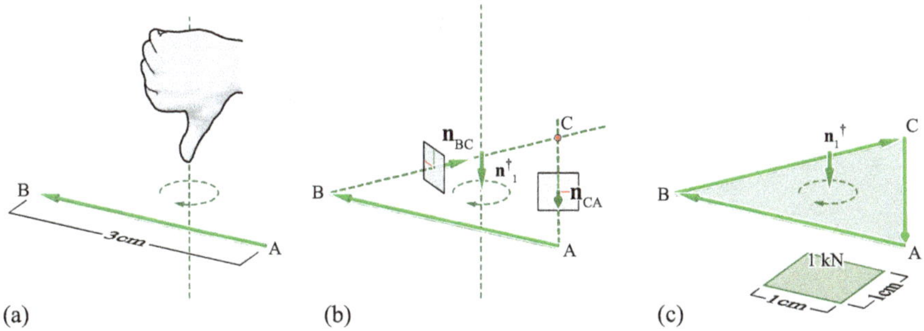

Figure 1.15 Putting the thumb in the direction of the applied load, curling up the fingers points out to the direction through which we can travel from subspace *A* to *C* by using the normal vectors of the planes.

The new vector is the first edge vector of the face corresponding to the force $\mathbf{f}_i$ (Figure 1.15). Note that the length of this vector needs to be chosen based on the scale of the drawing, which is very similar to the case in 2D (Figure 1.15a). The rest of the edge vectors need to be constructed to complete the face. By defining the length of the vector *AB*, the location of *C* can be found by intersecting the lines at both ends of the vector *AB* with the directions of $\mathbf{n}_{BC}$ and $\mathbf{n}_{CA}$ (Figure 1.15b). These lines are normal to the planes separating subspace *C* from *A* and *B*. Once *C* is found, the face is completed, and its area represents the magnitude of the force $\mathbf{f}_i$ (Figure 1.15c). Obviously, choosing a different edge length as a scaling parameter for the vector *AB* results in a smaller or larger triangle, but the proportion of the edges would not change as explained in the 2D example of Figure 1.5.

Remark 1.4. The minimum number of faces/planes around an edge or an applied load of the form is three because the smallest reciprocal face that can be made for the force diagram is a triangle.

1.4.5 Faces of the Form Diagram

Let us revisit the same node of Figure 1.12, knowing that lines of action of forces or the edges in a spatial node are the lines of intersection of at least three planes. Rope 1 where the mass is hanged from the node should make three faces with its adjacent ropes, as illustrated in Figure 1.16b. These faces satisfy the minimum number of planes that meet at the line of action of a load in 3D, as discussed in Section 1.3, and are shown by green color around the node. Besides, other ropes or edges share similar faces, as illustrated in Figure 1.16c.

1.4.6 Subspaces around the Node

The most important step in constructing a force diagram for a spatial node is to identify the subspaces around the node. Every edge in the form diagram is at the intersection

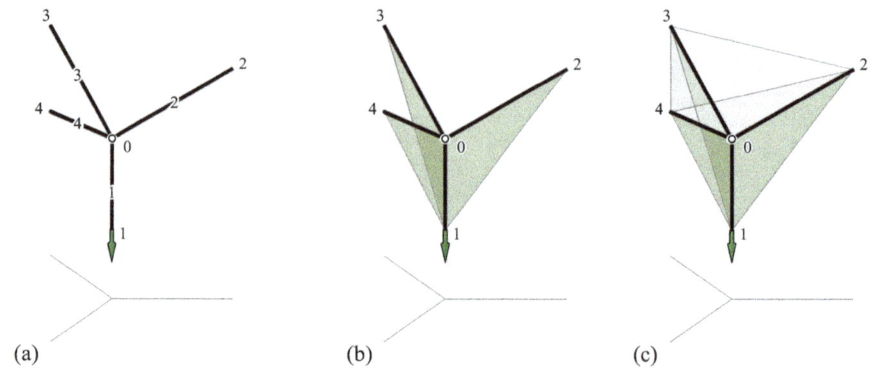

Figure 1.16 The node in 3D space and the planes shared by the edges/ropes meeting at the node dividing the space into subspaces.

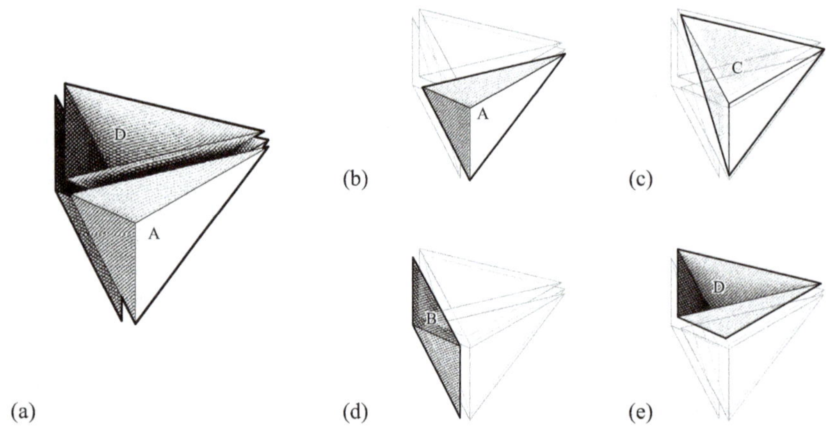

Figure 1.17 (a) Every three faces in the form diagram make an open polyhedral cell, which results in four open cells *A*, *B*, *C*, and *D* (b)–(e).

of three planes, and each of the two edges shares a single and only single plane with each other. In 3D, the subspaces are formed by planes shared between the edges. Every three planes connected to a node establish an open polyhedral cell dividing the space into subspaces, as illustrated in Figure 1.17. The subspaces of the form diagram are labeled with capital letters in the context of this book.

In the case of Figure 1.16, there are six triangular faces shared by the edges of the node. Every three faces make an open polyhedron around the node. There are four open polyhedrons around the node (Figure 1.17a–e). These open cells are the subspaces used for the construction of the force diagram. The number of subspaces in the form corresponds to the number of vertices of the force diagram, as it was explained in the 2D example of Figure 1.9.

1.4.7 Subspaces around an Edge

We know that the number of faces of the force polyhedron must equal the number of edges connected to a 3D node. Also, each face of the force diagram is normal to an edge in the node. The subspaces around each edge of the node give the reciprocal face vertices in the force diagram. If so, in addition to identifying subspaces around the node, each edge can be described with its surrounding subspaces. For instance, edge 1 is surrounded by subspaces *A*, *B*, and *C* and edge 2 is surrounded by subspaces *D*, *B*, *A*, and so on. This definition is the last information needed to complete the force polyhedron for the node.

1.4.8 Constructing the Closed Force Polyhedron

Example 1.2 will provide step-by-step procedures to complete a force diagram for a node that is in pure tension.

Example 1.2 Complete a force diagram for the node of Figure 1.12 and report the magnitude of the internal forces based on the applied load of 3 kN.

Construction steps:

1. Illustrate the system of ropes of Figure 1.12 with a diagrammatic representation of vertices and edges as we did in Figure 1.16a.
2. Start from the applied load and identify the subspaces around the line of action of force $\mathbf{f}_1$ as *A*, *B*, and *C*.
3. Place the thumb in the direction of the applied load, the curl of the figures shows a full cycle around it starting from subspace *A* to *B*, then to *C*, and back to *A* (Figure 1.18a).
4. Start from an arbitrary point *A* and draw a vector in the direction of $\mathbf{n}_{AB}$ with the length of *l*.
5. Name the tip of the vector *AB* point *B*.
6. The intersection of the normals $\mathbf{n}_{BC}$ and $\mathbf{n}_{CA}$ locates point *C* and completes the edge vectors of the face *ABC* with a normal $\mathbf{n}_1^\dagger$ (Figure 1.18b).
7. $\mathbf{n}_1^\dagger$ is parallel to the applied load $\mathbf{f}_1$ and the area of the triangle is proportional to the magnitude of the applied load, which is 3 kN for this example (Figure 1.18c).
8. Let us construct the face *BAD*. Using the direction of edge vector *AB*, draw edge *BA*. Note that the edge vectors of two adjacent faces in the force polyhedron are equal in length but opposite in direction. Also, the direction of an edge vector in a face leads to the direction of the entire face using the right-hand rule (Figure 1.18d).
9. Find an edge in the form diagram that includes subspaces *A* and *B* aside from the applied load.

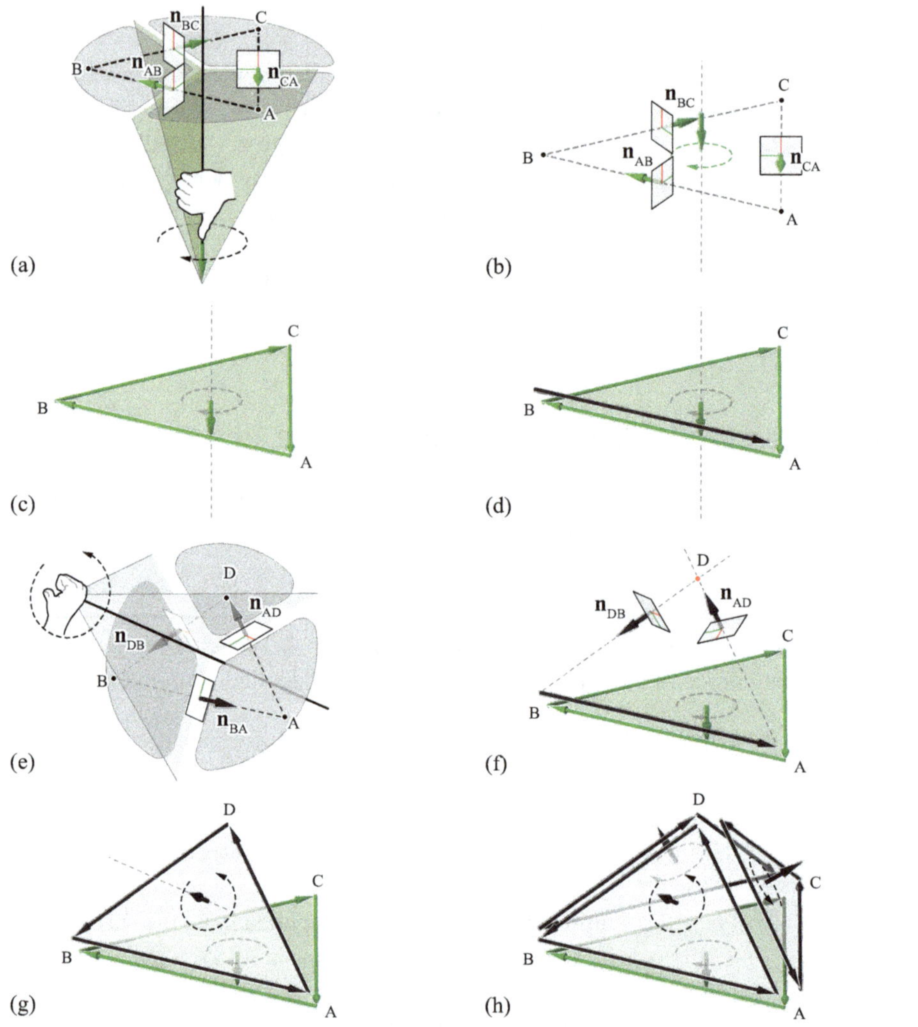

Figure 1.18 (a) Starting from the applied load and identifying its subspaces and the direction of the cycle; (b), (c) using the normal vectors of the faces in the form, we can find the location of the vertices of the reciprocal face *ABC*; (d), (e), (f) the edge vector *AB* of the face *ABC* provides for the direction and the length of the edge vector *BA* of the face *BAD*; (g) the completed face *BAD*; and (h) the completed force polyhedron.

10 Edge 4 is surrounded by subspaces *A*, *B*, and *D* and thus gives normals $\mathbf{n}_{AD}$ and $\mathbf{n}_{DB}$ to complete the face *BAD* (Figure 1.18e,f).

11 Once the location of the point *D* is found, the rest of the faces can easily be constructed by connecting the edges *AC* and *CB* to *D* (Figure 1.18g,h).

12 Measure the areas of *ADB*, *CDA*, and *BDC* to find the internal forces in their corresponding ropes of the node (Figure 1.18h).

□

1.4.9 The Direction of Internal Forces

The completed force diagram with the direction of the faces includes information related to the magnitude and direction of the internal forces in the ropes. To identify the direction of the internal force in the ropes, we can use the right-hand rule similar to the 2D example. Curling the fingers around the face in the force diagram and repeating the same rotation around the reciprocal edge in the form diagram without changing the direction or angle of the hand shows the direction of the internal force in the member (Figure 1.19). The direction of the thumb shows whether the force is tensile, moving out from the node, or compressive, pushing toward the node. In this example, all the ropes are in tension, and the internal force vector pulls the node from points 1, 2, 3, and 4 (Figure 1.19).

1.4.10 Controlling the Internal Forces in 3D by Force Polyhedron

Once the geometry of the force diagram is completed with all the faces perpendicular to the direction of the ropes in the system, we can explicitly see the effect of changing the geometry on the magnitude of the internal forces in the structure. Similar to the 2D example, if the ropes are connected to the node with shallower angles, more force needs to be applied from the hands to keep the weight hanging in the 3D space (Figure 1.20). In contrast, if the angles get closer to the vertical configuration, the magnitude of the force decreases in the system. In all the examples illustrated in Figure 1.20, the area of the face corresponding to the applied load is the same.

1.4.11 Topological Relationships between Form and Force in 3D

Having the geometry of the form and the force diagram with all involving components, we can highlight their topological relationships. Each diagram consists of vertices,

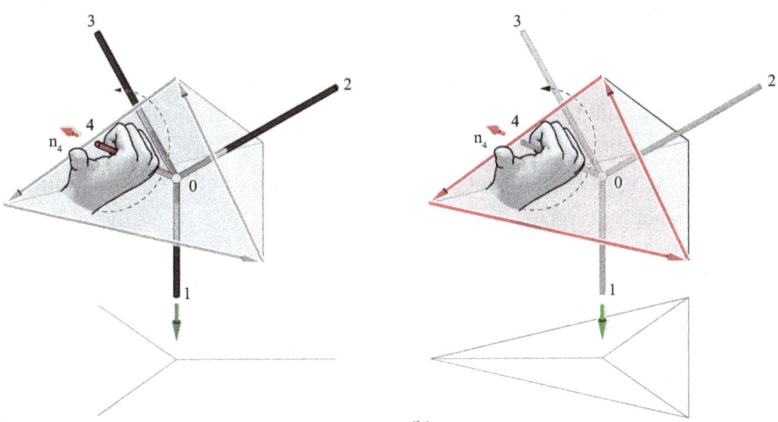

(a) (b)

Figure 1.19 Matching the direction of the face (a) with the direction of the internal force in the member of the structure (b) using the right-hand rule.

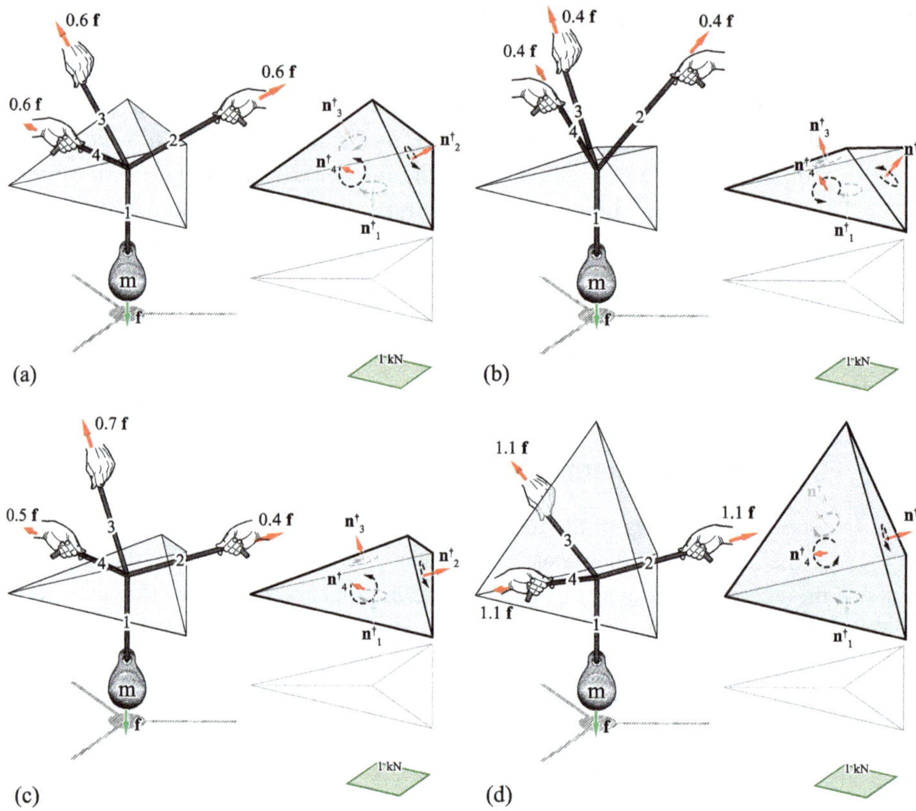

Figure 1.20 Various angles for the ropes and their effects on the geometry of the force diagram and the magnitude of the internal forces in the ropes.

edges, faces, and cells. We can annotate the form diagram by Γ and the force diagram by $\Gamma^\dagger$. Figure 1.21 summarizes the topological relationship including the number of vertices, faces, edges, and cells of the form diagram and their corresponding elements in the force diagram. The components of the form diagram Γ are represented by v, e, f, and c, and the ones of the force diagram $\Gamma^\dagger$ are shown with a superscript "†." The following dual/reciprocal relationships exist between the components of the two diagrams:

- The number of vertices v in the form diagram is equal to the number of cells $c^\dagger$ in the force diagram.
- The number of edges e in the form diagram and the number of faces $f^\dagger$ in the force are equal.
- The number of faces f of the form corresponds to the number of edges $e^\dagger$ in the force diagram.
- The number of open/closed cells c in the form is equal to the number of vertices $v^\dagger$ in the force.

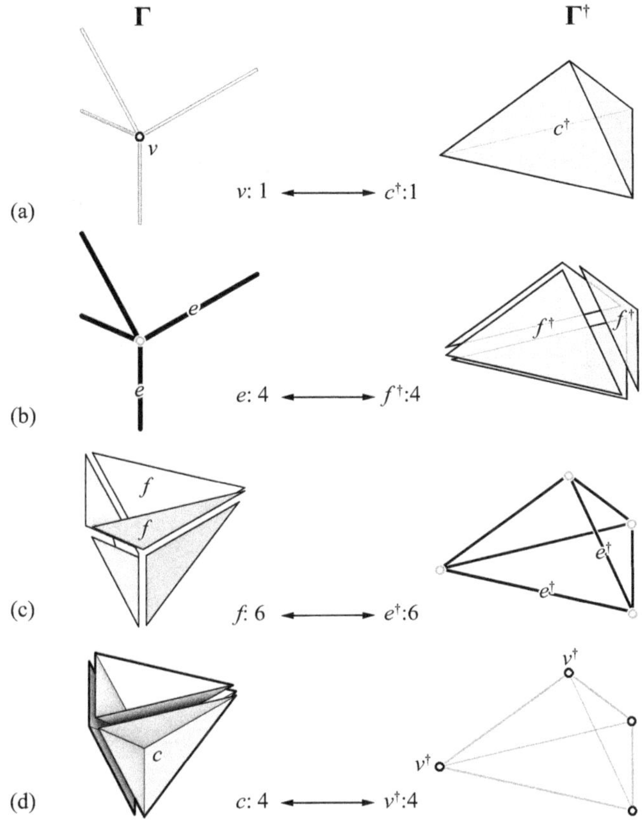

Figure 1.21 The topological relationship between the elements of the form and the force diagram.

In addition to the duality of the components of the diagrams, what establishes reciprocity is the geometric dependencies of the diagrams where the edges of each diagram are perpendicular to the faces of the other.

1.4.12 Example of a Compression-Only Node in 3D

Similar to the 2D case, without changing the geometry of the node, if the direction of the applied load is changed, the type of internal forces in the member switches from tension to compression. Figure 1.22a illustrates a node with an applied load on its top. The process for constructing the force diagram in 3D for this node is exactly the same as in the previous example. The construction process can be summarized as follows:

- In the form diagram, identify all the faces shared between the edges of the node.
- Annotate the subspaces formed by the faces of the form diagram.
- Develop the topology of the force by connecting the subspaces to each other.
- In the force diagram, construct the face corresponding to the applied load.
- Complete the rest of the faces by using the edge vectors of the face of the applied load.

Example 1.3 Complete a force diagram for a node of Figure 1.22 and report the magnitude of the internal forces based on the applied load.

Construction steps:

1 Start by identifying the faces of the form diagram. The edge corresponding to the applied load shares a face with every other edge connected to the node and is shown in green (Figure 1.22a).

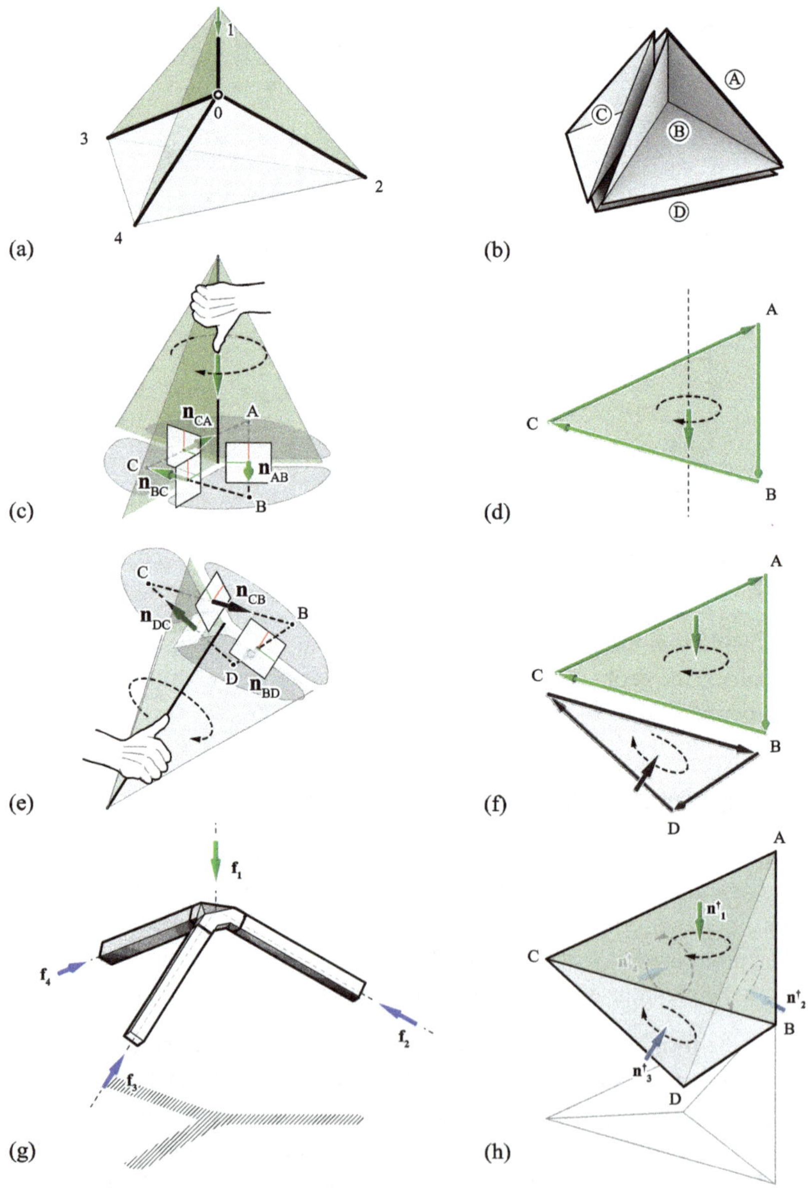

Figure 1.22 A compression-only node and the process of construction of the force diagram to find the magnitude of the force.

2 In the form diagram, every three faces make an open cell labeled *A*, *B*, *C*, and *D* and illustrated in Figure 1.22b.
3 Connect subspaces around each edge to get the topology of the corresponding face as well as the entire force diagram.
4 In the form diagram, put the thumb in the direction of the applied load, and the curl of the fingers gives the direction of cycling around the subspaces of the load (Figure 1.22c).
5 Use the direction of cycling around the load and complete the force face by normals $\mathbf{n}_{AB}$, $\mathbf{n}_{BC}$, and $\mathbf{n}_{CB}$ direction and magnitude and complete the force face diagram, as shown in Figure 1.22d.
6 Use the opposite edge of the force face, for example, *CB* in the force diagram, and find another edge in the form diagram to round (Figure 1.22e).
7 The reciprocal face *BDC* is completed using the normals $\mathbf{n}_{BD}$ and $\mathbf{n}_{DC}$ and the location of point *D* is found (Figure 1.22f).
8 The rest of the faces of the force diagram can be completed by using the location of *D*.
9 The direction of the internal force in the members of the form can be found by moving the right-hand thumb from the face of the force diagram to the reciprocal member in the form (Figure 1.22g,h).
10 The force diagram is completed, and the magnitude of the forces can be found by measuring the area of the polyhedron's faces formed perpendicular to the force directions.

□

Using the earlier example, the design of a tension-only bridge of Figure 1.1 can be changed into a design where two adjacent compressive nodes support the bridge, as shown in Figure 1.23.

1.5 Combined Tensile-Compressive Nodal Equilibrium in 2D

In both 2D and 3D examples, the ropes' internal force was either all tensile or compression. Let us move to some examples of nodes with both tensile and compression internal forces. Figure 1.24a shows a 2D configuration where the node has both compression and tension internal forces. But how can we find the force in the member? In the first step, we identify the spaces between the edges and applied load in the node; starting from the applied load, putting the thumb in the direction of the load, the four fingers show the direction to connect the subspace *A* to *B*, and eventually complete the cycle by passing through the *C* and returning to the space *A*.

Following the cycle sequence from *A* to *B*, you need to rotate your four fingers 180 degrees to close the polygon. Otherwise, the force polygon does not get closed (Figure 1.24b,c). Did you notice the difference between the cycle direction around the node and the force polygon? Although the sequential order of passing through the subspaces *A*, *B*, and *C* in the form and the force polygon is the same, the geometric

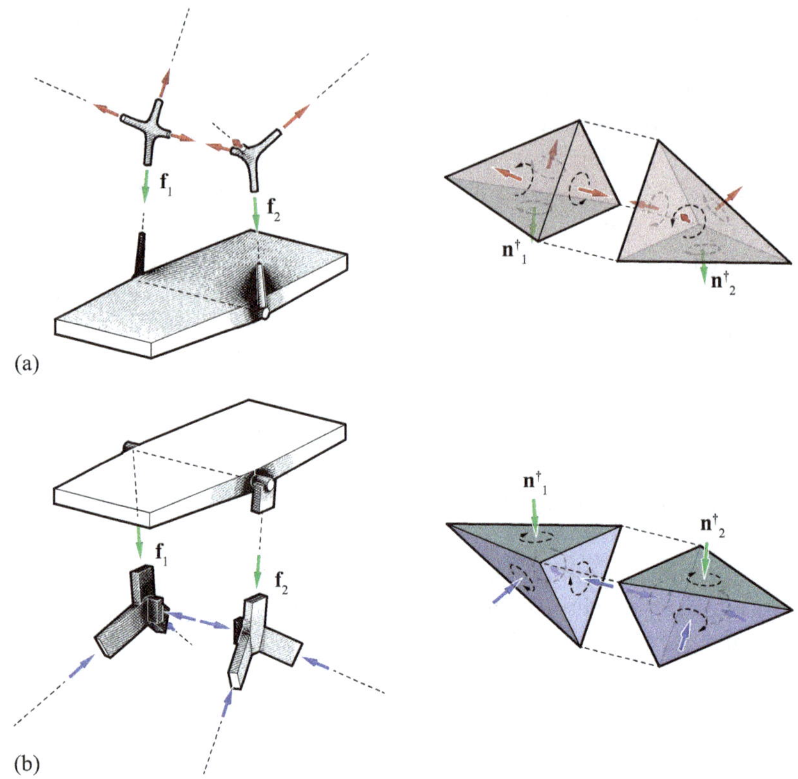

Figure 1.23 (a) The form and the force diagram of a 3D bridge with two tension-only nodes; (b) the flipped geometry of the nodes as a compression-only system for the same applied load of the deck.

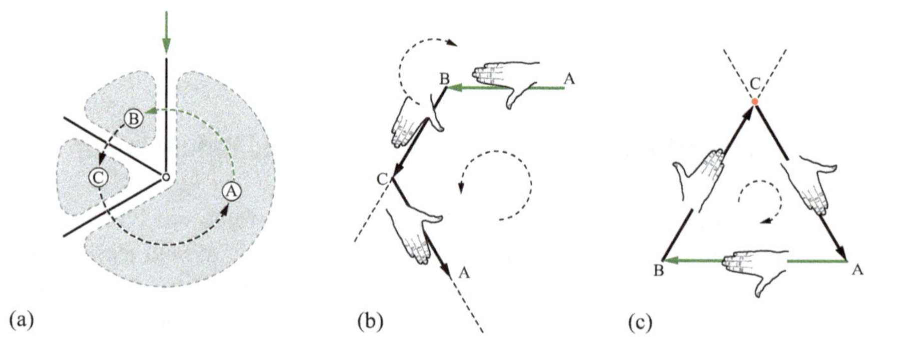

Figure 1.24 The construction of a force polygon for a node with both tension and compression forces; (a) the subspaces around the node; and (b) an open polygon of force that can get closed by revering the vector *BC* in the construction process (c).

direction of the cycle in the form is the opposite of the direction of the force polygon. Once the force diagram is closed, by matching the direction of the thumb in the form and force, we can identify the type of internal forces, which is compression in the edge e_2 and tension in the edge e_1 (Figure 1.25).

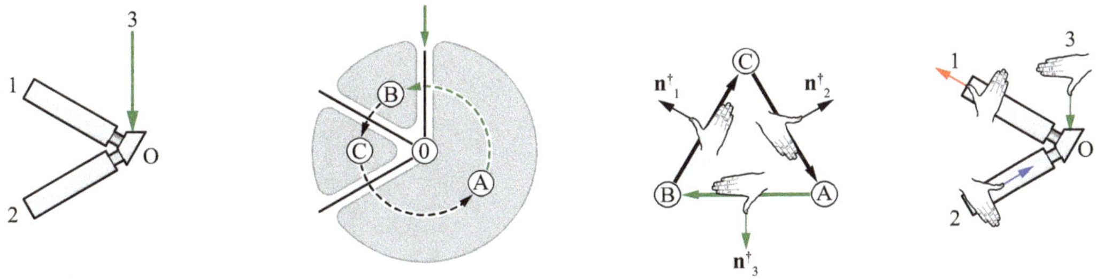

Figure 1.25 A 2D joint with compression and tension combined forces.

1.6 Combined Tensile-Compressive Nodal Equilibrium in 3D

The process of finding the force diagram for a node with compression and tension in 3D is also similar to the previous cases and includes the following steps: (1) identifying the faces of the form diagram; (2) designating the subspaces and finding the topology of the force diagram by relating them to their neighbors; (3) constructing the face of the applied load in the force diagram; (4) completing the rest of the faces; and (5) finding the direction and the magnitudes of the internal forces.

1.6.1 Identifying the Faces of the Form

Consider Figure 1.26a as a given configuration of edges meeting at node 0 with an applied load $\mathbf{f}_1$. The first step is constructing the faces between the applied load and the members meeting at the node (Figure 1.26a). The figure highlights these faces in green to show that they are all attached at one side to the edge corresponding to the applied load (Figure 1.26a). The rest of the edges should also share faces with their adjacent edges, resulting in the triangular faces connecting vertices 0, 2, 3, and 4.

1.6.2 Identifying the Subspaces of the Form

Once the faces of the form are identified, we can start by recognizing the subspaces or cells formed by the faces. Note that each cell is separated from its neighbors by a face. In other words, every two adjacent cells share a face. These cells establish the subspaces of the form diagram that are topologically related to the force diagram's vertices. In this example, four open cells are formed by the faces of the form annotated by *A*, *B*, *C*, and *D*. To better understand the domain of these subspaces, we can intersect these open cells with a sphere centered at the vertex v_0 with a unit radius (Figure 1.26c). This intersection results in multiple solids, each representing the space formed by the open cells. Cutting these solids with a plane perpendicular to the edge e_1 explicitly shows their surrounding subspaces, as illustrated in Figure 1.26d.

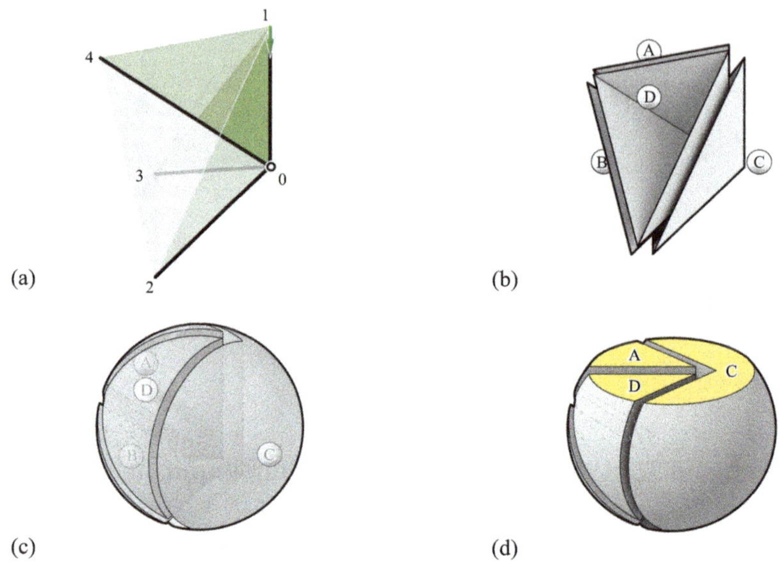

Figure 1.26 (a) A 3D node with an applied load and the faces shared by the edges; (b) open cells in the form and their resulting subspaces; (c) intersecting a sphere with the open cells to represent the subspaces physically; and (d) a section cut perpendicular to the applied load revealing the configuration of the subspaces around the edge corresponding to the applied load.

1.6.3 Subspaces around the Edges

Each subspace or open cell shares one and only one face with every adjacent cell/subspace. Every cell shares an edge with its adjacent cells in the form diagram. Looking through the edge e_1 of Figure 1.26 shows that it is surrounded by subspaces A, C, and D. Since each cell of the form diagram corresponds to a vertex in the force diagram, every face shared between two cells corresponds to an edge connecting two vertices in the force diagram. This topological information is enough to construct the force diagram.

1.6.4 Constructing the Face Reciprocal to the Applied Load

Similar to the 2D case, in 3D, the construction of the force diagram should start from the face corresponding to the magnitude and direction of the applied load $\mathbf{f}_i$. We can construct the face reciprocal to the applied load by putting the thumb in the direction of the applied load and passing through the cells/subspaces surrounding the line of action of the load (Figure 1.26a). As illustrated in Figures 1.26d and 1.27a, the subspaces around the applied load can be cycled using the right-hand rule.

Putting the thumb in the direction of the applied load gives a clockwise cycling direction for subspaces. In this regard, the normals of the shared faces between the subspaces are named in the direction of the cycle as $\mathbf{n}_{AC}$, $\mathbf{n}_{CD}$, and $\mathbf{n}_{DA}$. Following the direction of the cycle and the normals, start from point C with the direction of $\mathbf{n}_{CD}$

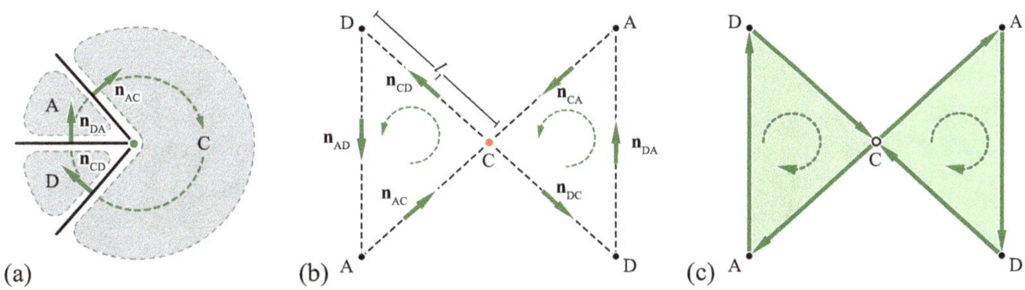

Figure 1.27 Looking perpendicular to the direction of the applied load of Figure 1.26: (a) the subspaces around the applied load; (b) the vertices of the force face corresponding to the applied load; and (c) the completed force face and its edge vectors with point symmetry.

and draw a line with a length of l_1 to locate point D and use the same approach to find point A (Figure 1.27b). You may notice that we have constructed a face with the same cycle connectivity, but the direction of the edge vectors and the normal of the face are opposite to the direction of the applied load $\mathbf{f}_1$. Even if we flip the normal directions and use $\mathbf{n}_{AD}$ instead of $\mathbf{n}_{DA}$, we will end up with a face that has a normal opposite to the applied load.

This result clearly indicates that the system is experiencing tension and compression from internal forces. In such scenarios, it is sufficient to flip the directions of the edge vectors after the construction. In such systems, the length of the starting edge vector should be negative $-l_1$ with the correct normal direction to get the correct face normal, and we explain this phenomenon in detail in Chapter 7. Also, the force triangle has a *point symmetry* because each triangle can turn into the other by a 180-degree rotation.

Remark 1.5. In systems with both tension and compressive internal forces, the direction of the cycling edges in the form diagram matches the direction of the edge vectors in the force diagram, but the order of the cycle in the force diagram is reversed, that is, $A \rightarrow C \rightarrow D$ in the form changes to $D \rightarrow C \rightarrow D$ in the force diagram.

1.6.5 Completing the Force Polyhedron

Once the face of the applied load is completed, it provides us with the information on edge vectors for the rest of the faces to be completed (Figure 1.29a,b). First, we know that the edge vectors of the adjacent faces of a single force polyhedron are equal and opposite in direction. Moreover, the direction of the edge vector defines the direction of the normal and the direction of the curl. We can complete the face corresponding to each edge of the form by cycling the subspaces in the corresponding edge of the form diagram. Consider edge $e_{(0,4)}$ of the node. The face corresponding to this edge should connect spaces B, D, and A in the form and the same vertices in the force diagram (Figure 1.28a,b). We can annotate the edge vector connecting A to D by $\mathbf{e}^{\dagger}_{AD}$. We may know that the edge $\mathbf{e}^{\dagger}_{DA}$ belongs to the face DAB and is opposite

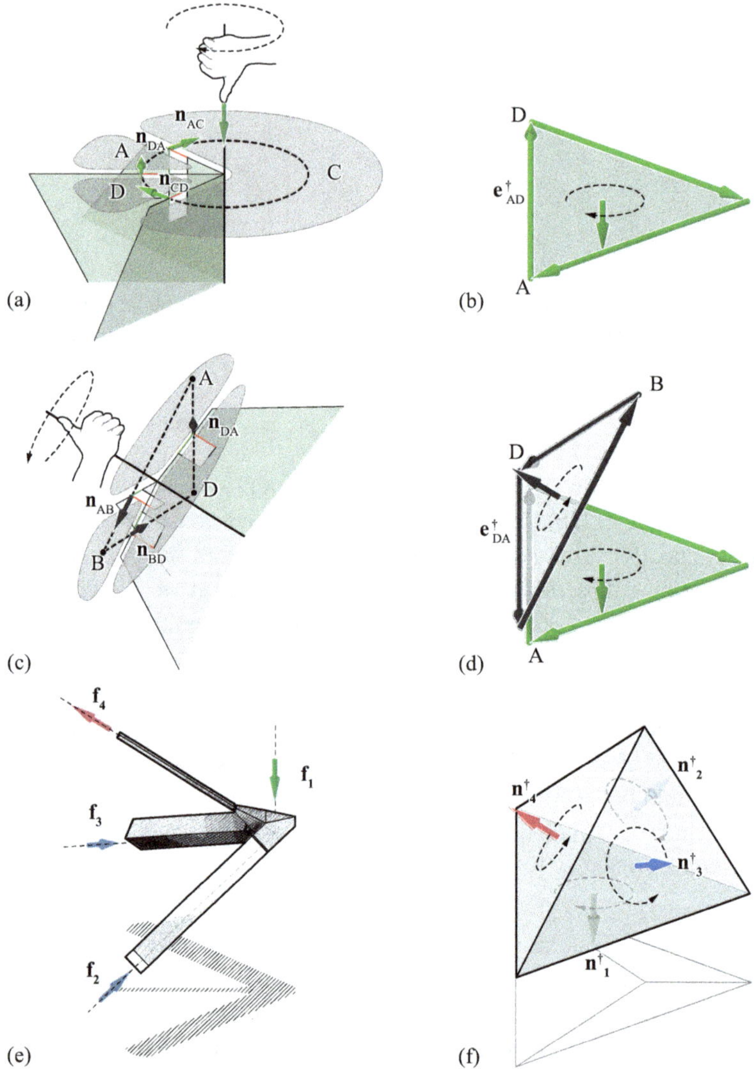

Figure 1.28 (a) Aligning the thumb in the direction of the applied load passing through spaces *A*, *C*, and *D*; (b) completing the face *ADC* by choosing a scale factor; (c), (d) taking the direction of the face *BDA* and cycling through the spaces around the edge $e_{(0,4)}$ in the form resulting in the location of *B*; (e), (f) finding the direction of forces in the members by matching the normal of the face and the geometry of the member in the node.

in the direction to the edge $\mathbf{e}^{\dagger}_{AD}$. Also, the face *BDA* corresponds to the form's edge $e_{(0,4)}$. Thus, knowing the direction of one edge vector gives us the cycling direction around the spaces surrounding the edge $e_{(0,4)}$ (Figure 1.28c,d). The location of vertex *B* can be found by intersecting normals $\mathbf{n}_B D$ and $\mathbf{n}_A B$, which completes the face *DAB*. The rest of the faces can be constructed by simply connecting the vertex *C* to the vertex *B*.

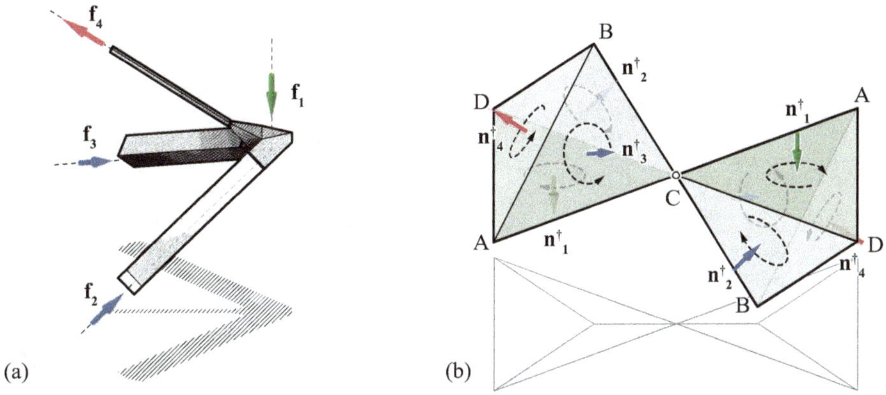

Figure 1.29 (a) The node with both compression and tension forces and (b) the point symmetry of its force diagram according to Figure 1.27.

1.6.6 Direction of Internal Forces

The direction of the internal forces in the members can be found by using the right-hand rule, similar to all previously shown examples in 3D. Aligning the thumb in the direction of the normal of a face in the force diagram and matching it with the geometry of the corresponding member in the form leads to the direction of the force in the node (Figure 1.29e,f).

1.6.7 The Point Symmetry

As explained in Figure 1.27a,b, the face corresponding to the force diagram has a point symmetry, and there are two possibilities for its construction while preserving the direction of the normal. Regardless of the chosen configuration for the force face, the resulting polyhedron can very well represent the equilibrium of the node with both tension and compression forces.

1.6.8 Multiple Configurations of Tension and Compression

Using the same procedures explained in Section 1.6.7, one can easily find the force equilibrium for all the configurations of the members intersecting at the nodes of Figure 1.30. Let us assume all the edges connected to node 0 in the top of the column of Figure 1.30a have positive edge lengths. Other node configurations can be generated by simply assuming negative edge lengths for the members. For instance, the edge $e_{(0,4)}$ has a negative length in the second row of the column. Using the same approach, one can devise all configurations in the left column (Figure 1.30a). Finding the force diagram for all the configurations is followed by (1) finding the cells of the

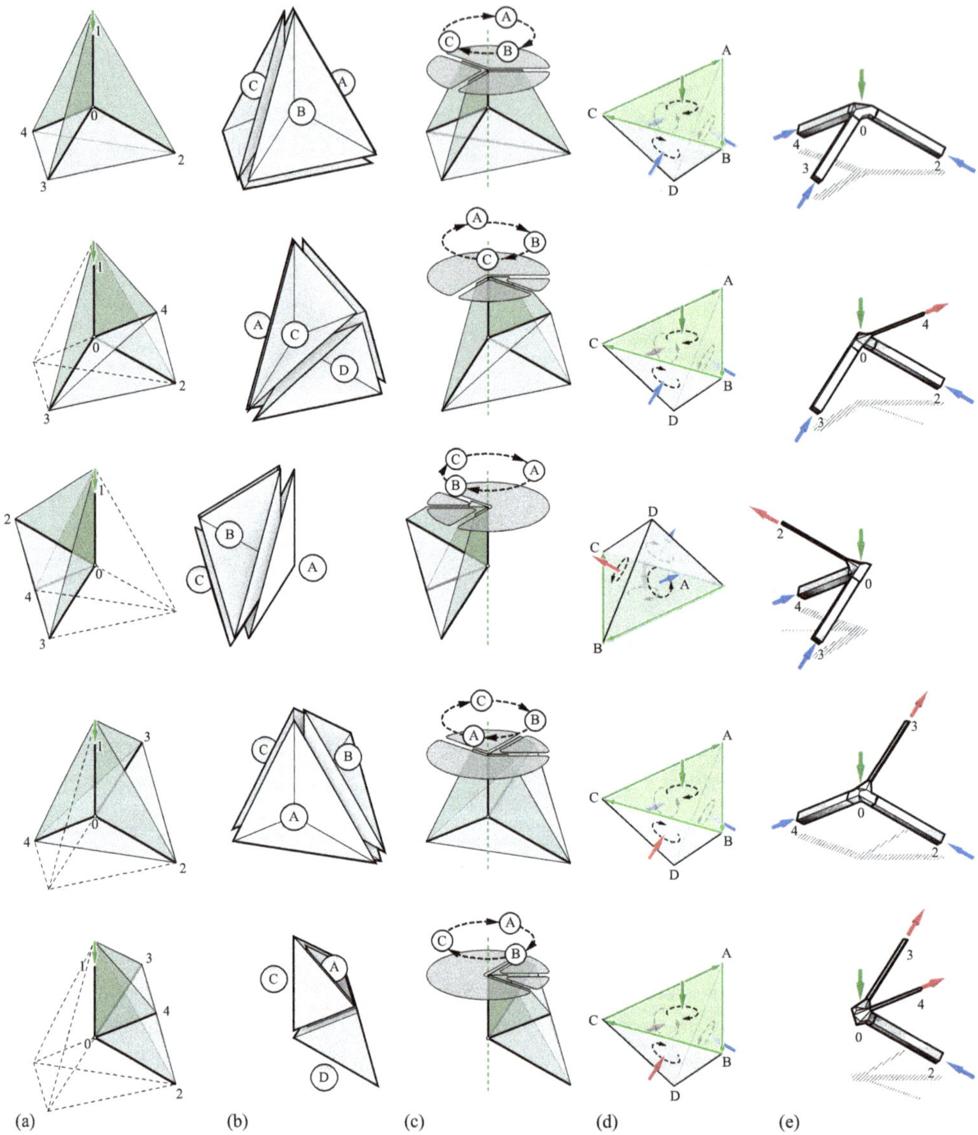

Figure 1.30 Multiple 3D configurations of members meeting at a node (a) and the process of constructing the force diagram for the node by (b) finding the cells of the form; (c) cycling around the spaces of the applied load; (d) constructing the force diagram; and (e) finding the direction of the internal forces in the node.

form (Figure 1.30b); (2) cycling the spaces around the applied load (Figure 1.30c); (3) constructing the applied force face and the rest of the force diagram (Figure 1.30d); and (4) identifying the direction of the internal forces in the members (Figure 1.30e). This figure summarizes the steps to construct the reciprocal force diagram for a given node, as explained in this chapter.[25]

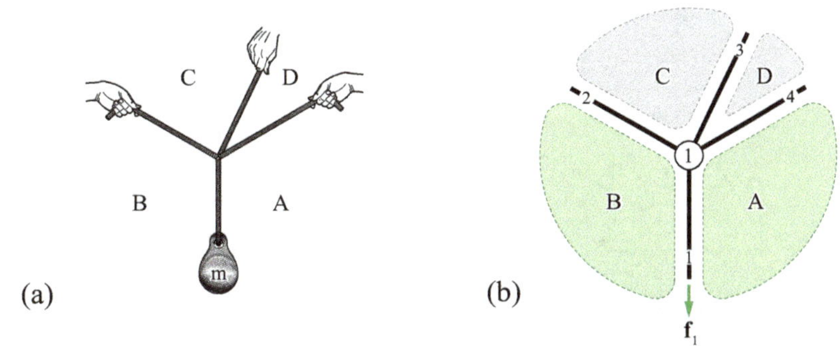

Figure 1.31 (a) An example of a system where three ropes keep the node in equilibrium and (b) the annotated spaces between the edges connected to the node.

1.7 Determinacy of the Force System

In all previously presented examples, choosing one particular scale for the force diagram resulted in a single force distribution for the system. Note that in the example of Figure 1.3, choosing the length (magnitude) and the direction of the applied load was enough to construct the force triangle and define the magnitude of the internal force in the members. Thus, the scale of the force triangle does not affect the internal force distribution because the proportion of the triangle's edges does not change by scaling it up or down (Figure 1.6b). In other words, the force distribution for these nodes was *determinate*. In contrast to the determinate force distribution, the *indeterminate* equilibrium suggests that the force distribution is not unique in a system. The difference between the determinate and indeterminate equilibrium can be explicitly described by the geometry of the force diagram in both 2D and 3D.

Let's consider a 2D example of ropes where three ropes keep the weight from falling, as illustrated in Figure 1.31a. Using the methods described in this chapter, we can identify and annotate the spaces between the ropes, as shown in Figure 1.31b. Constructing the force diagram for this node requires two edge lengths to be defined, as shown in Figure 1.32a–d. One edge length can be chosen according to the magnitude of the applied load. In addition, we need to arbitrarily guess the internal force magnitude in one of the ropes to close the force polygon and find the other missing internal forces to complete the equilibrium. As shown in Figure 1.32, there is not a single force polygon that can describe the equilibrium of internal forces in the system. In other words, the force distribution is indeterminate. This property can also be described using the geometric properties of the force polygon.

1.7.1 Geometric Degrees of Freedom

As illustrated for the polygon *ABCD*, two edge lengths must be defined to construct a closed polygon. Comparing this polygon with the force triangle in Figure 1.5 can reveal the relationship between the number of edges and a determinate force distribution. For a polygon with e number of edges, the length of $e-2$ edges should be

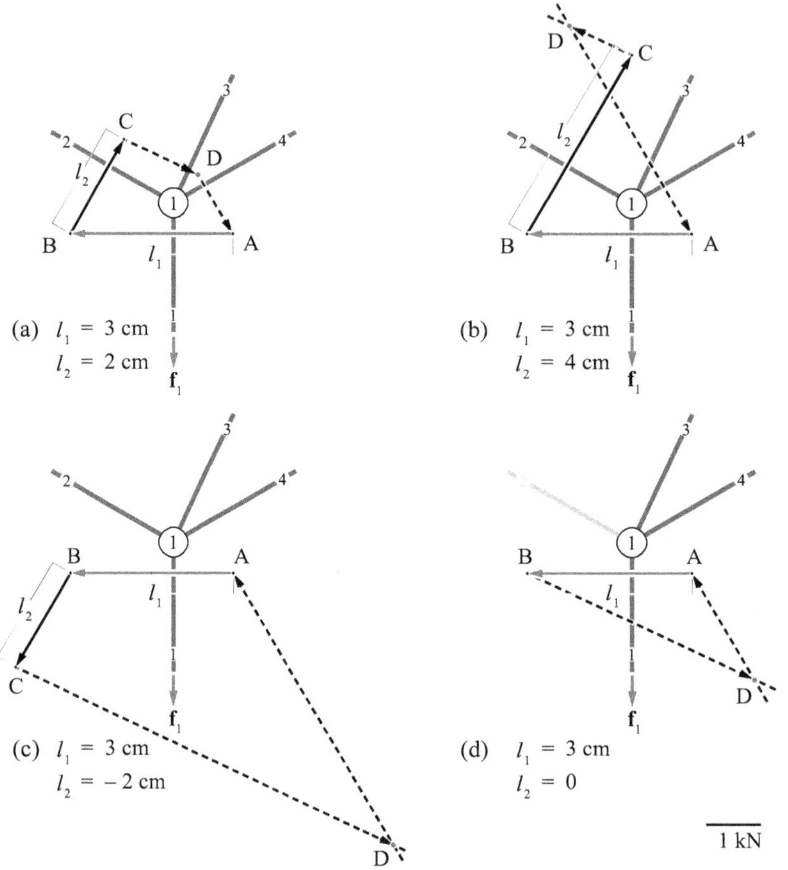

Figure 1.32 Multiple force distribution for an indeterminate system in 2D including compression and tension internal forces.

chosen to have a determinate force distribution. For instance, for a triangular force polygon, the length of one edge must be defined to complete the triangle ($e - 2 = 1$). Whereas for a polygon with four edges, two edge lengths must be chosen to complete the geometry of the force polygon. This property is referred to as the *Geometric Degrees of Freedom (GDoF)* of the force polygon, and it also means that there are $e-2$ edges that we can freely choose to determine the force distribution in the system.

1.7.2 Degrees of Statics Indeterminacy

Since the force diagram describes the static equilibrium of the system, the GDoF of the force diagrams correspond to the *degrees of static indeterminacy* of the system. For instance, if the GDoF of a force diagram is one, then the system is statically determinate under one given external force. That is, depending on the scale chosen for the force diagram, the distribution of the internal forces is always the same. Whereas, if the GDoF of a force diagram is more than one, then that system is statically indeterminate

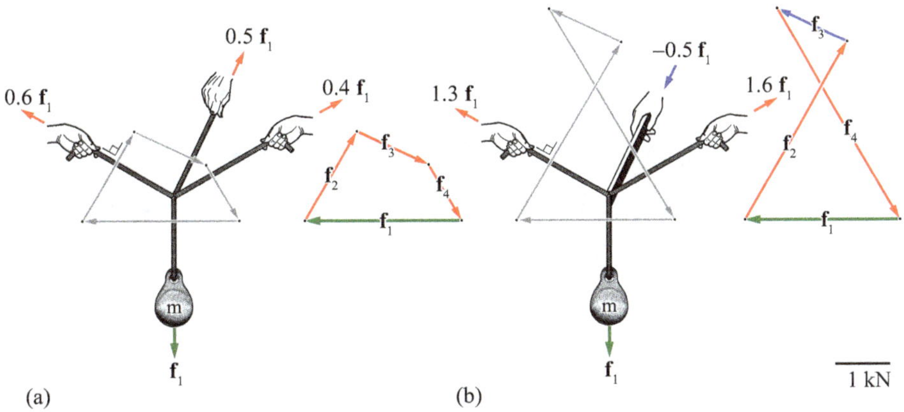

Figure 1.33 (a) A tension-only equilibrium of forces in the node versus (b) a system with both tension and compression.

under a given external force. According to this explanation, the degree of static indeterminacy is two for the indeterminate example of Figure 1.33. It means that, besides the chosen value for the applied load, one of the forces in the four ropes can also be controlled. As a result, multiple combinations of forces can keep the weight from falling *without* changing the angles of the ropes. For instance, if we pull rope 2 by a 2 kN force, that is, $l_2 = 2$ cm in the force diagram, the forces in the rest of the ropes will be tensile, as shown in Figure 1.33a. Whereas if the force in the rope 2 increases to 4 kN, then the rope 3 must be replaced by a compressive member to keep the node in equilibrium (Figure 1.33b), which clearly suggests that the force distribution is not unique in the system.

1.7.3 Indeterminate Node in 3D

As mentioned in Section 1.4.3, the line of action of a single force can be defined by the intersection of three planes. But it can also be defined by intersecting more than three planes. Figure 1.34 illustrates a configuration where four ropes are pulled from the sides to keep the weight from falling. Constructing the geometry of the force polyhedron starts by identifying the faces shared by the applied load and the rest of the ropes (Figure 1.34b). These faces are illustrated by green color. The face corresponding to the applied load can be constructed in the force diagram using the right-hand rule and aligning the thumb in the direction of the load, as shown in Figure 1.35a,b. Note that in this case, the force polygon is a quadrilateral and has two GDoF. That is, two edge lengths of the quadrilateral can be chosen freely to complete the edge vectors of the face (Figure 1.35a,b). Figure 1.35b shows a quadrilateral with two edge lengths l_{11} and l_{21} and another quadrilateral with two edge lengths l_{12} and l_{22}, which can represent two different quadrilaterals with different edge lengths but the same area. There are many different polygons with various edge lengths that can represent the applied load in this case, and the geometry of the force polyhedron won't be unique. This geometric property best describes the static indeterminacy of the system.

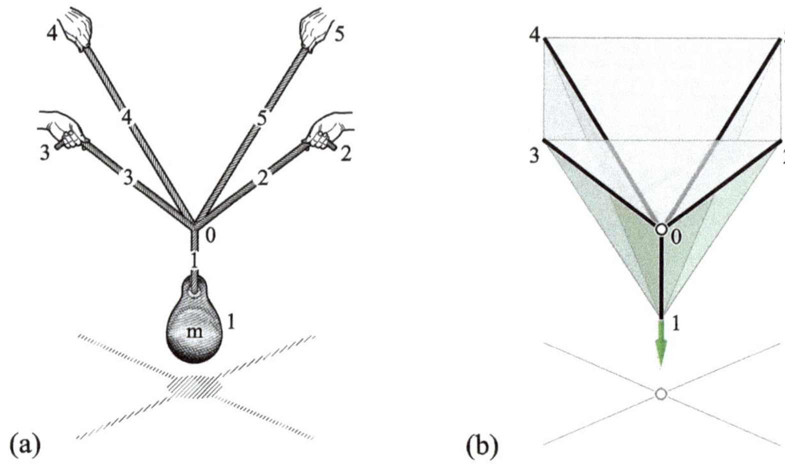

Figure 1.34 (a) A system of four ropes pulled from the sides holding a weight in 3D; (b) additional face (highlighted) shared between the edges across from each other.

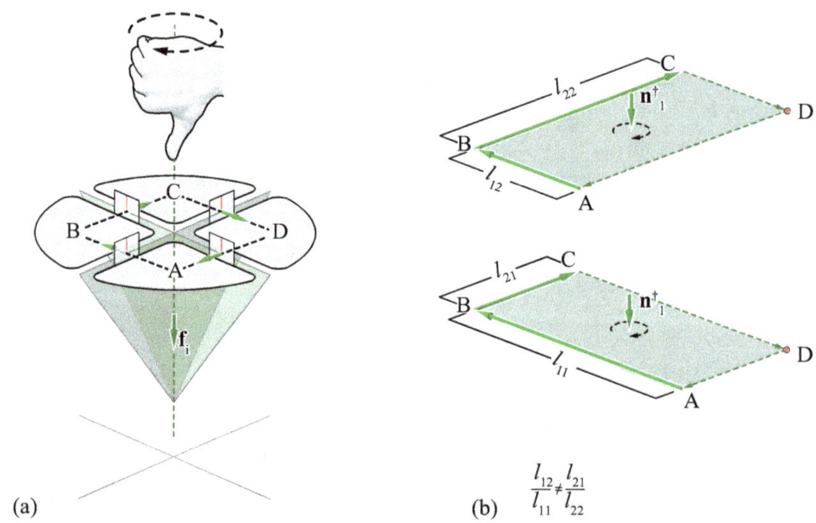

Figure 1.35 (a) Using the thumb and the right-hand rule to cycle the subspaces around the applied load, and (b) the force face corresponding to the applied load with two geometric degrees of freedom.

Closing the Force Polyhedron

Continuing with the rest of the faces to complete the force polyhedron for the force face of the indeterminate node of Figure 1.34 reveals more interesting features about the force diagram for this configuration. Cycling around the edges of the form diagram results in an open force polyhedron, as shown in Figure 1.36a,b. The top vertex of the side faces should either collapse to a single point, which is impossible geometrically, or be connected by an edge, which induces a topological mismatch between the two diagrams.

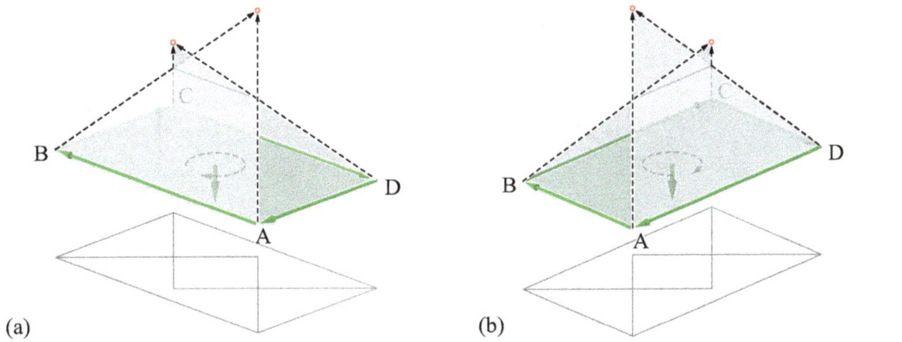

Figure 1.36 (a) and (b) Completing the construction of the force polyhedron for the node of Figure 1.34 results in two open force polyhedrons.

Revisiting the form diagram and the shared faces between the edges might solve this problem. Besides the line of action of the applied load, the members/ropes of the node can also be the intersection of more than three planes. Figure 1.37a,b shows two different configurations where the edges across from each other share a face. Both of these configurations are correct, and both can result in a closed force polyhedron describing the equilibrium of forces in the system. The force polyhedron for each configuration includes an additional edge $e_i^{\dagger}$ corresponding to the added face f_i in the form diagram. Since there is not a single force distribution for the indeterminate systems, both configurations and their force diagrams are correct and can be used to design the system.

Revisiting the topological relationship between the form and force diagrams corroborates the correctness of the construction. The open cells and faces in the form diagram correspond to the vertices and edges of the force diagram, as shown in Figure 1.21a,b. While the force diagram always consists of closed polyhedral faces, the existence of open faces in the form diagram is quite counterintuitive. So, how do we know the number of open faces for a node with multiple members or applied loads to construct its force diagram? Section 1.8 will answer this question using the extended gaussian image (EGI).

1.8 Indeterminate Force Polyhedrons

The procedural construction method presented in Section 1.7 is not sufficient to determine the correct topology of an indeterminate force polyhedron for a specific distribution of forces. Iterative or algorithmic computation of polyhedral geometry, more commonly known as *polyhedral reconstruction*, is a well-researched topic for various applications in many disciplines, such as computer vision, computational geometry, and combinatorics. In most applications, the objective is to reconstruct the polyhedral geometry from partial information about the polyhedra (i.e., from projected images, vertex locations, edge lengths, face geometries, face normals, face areas, dihedral

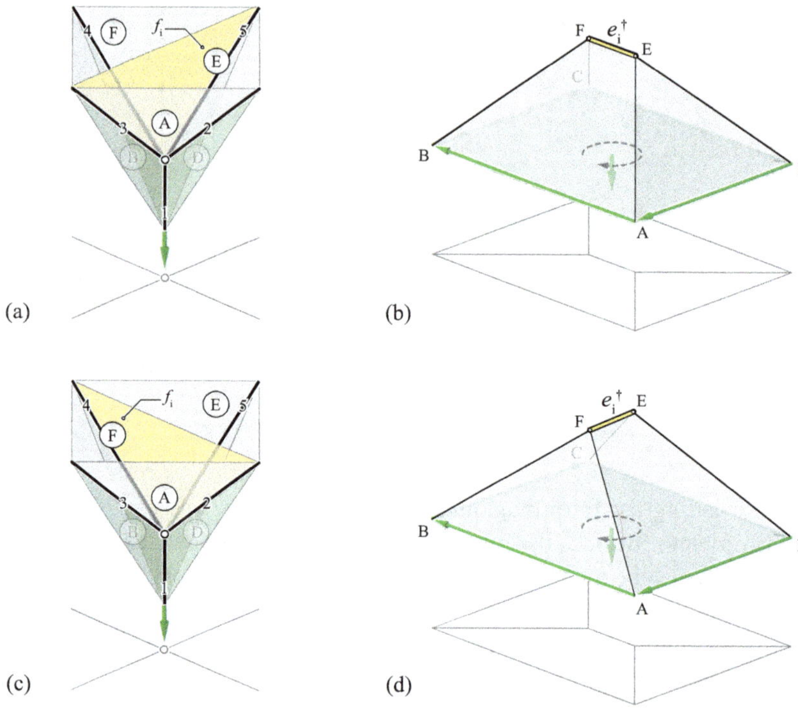

Figure 1.37 (a) A form configuration with additional face f_i shared by the edges e_3 and e_5; (b) its force diagram with an additional edge $e_i^\dagger$; (c) another configuration with the face f_i shared between edges e_4 and e_2; and (d) its corresponding force diagram.

angles, etc.).[90] The reconstruction method most relevant to polyhedral graphic statics is based on face normals (force directions) and areas (force magnitudes).

1.8.1 Polyhedral Reconstruction from Force Vectors

For a determinate system of forces, the *invisible* faces shared between each pair of force vectors can be used to define the topology of the corresponding faces of the force polyhedron. To find these invisible faces, we borrow a concept from computer science called the *convex hull.* A convex hull is the smallest convex polyhedron constructed from a given set of points, with all its faces being triangles.[91] In the case of a node with multiple applied loads, for instance, in the node of Figure 1.38a, a convex hull is the triangulation of the tails of the load vectors with unitized lengths (Figure 1.38b). The triangulation provides primary information about how these forces can be related in 3D space, as the edges of these triangles reveal the adjacencies of the corresponding faces of the force polyhedron. Rounding each force vector with all its attached faces and face normals gives the edges of the corresponding face in the force polyhedron (Figure 1.38c,d).

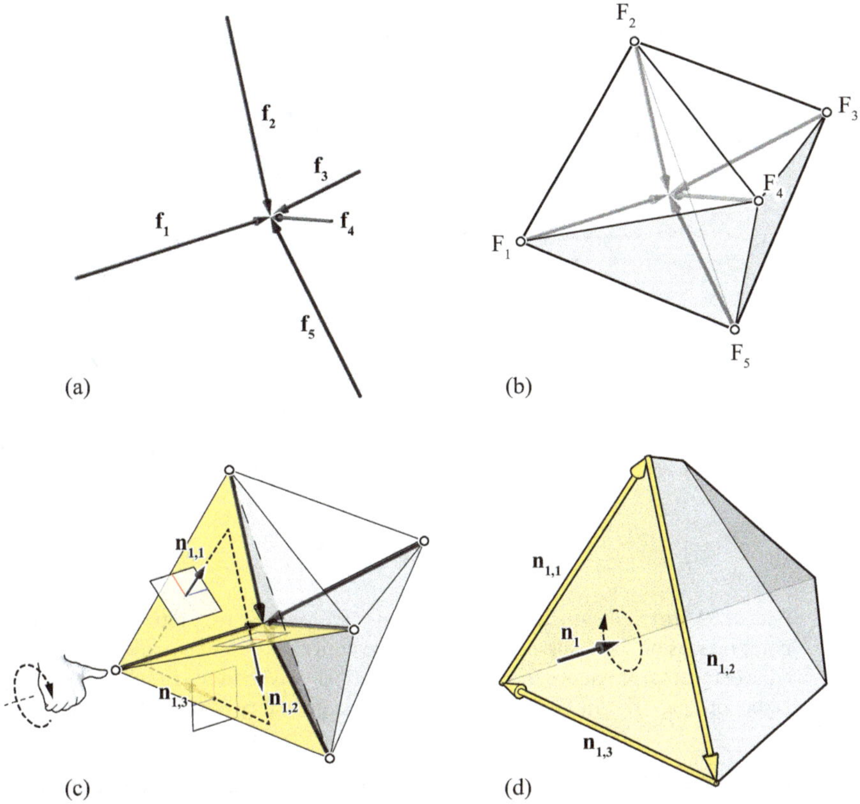

Figure 1.38 (a) A system of five forces in equilibrium; (b) the convex hull of the endpoints (tails) of the unitized force vectors; (c) the internal faces of the convex hull are the invisible shared faces between the applied forces; and (d) rounding each force vector with its face normals gives the geometry of the corresponding face in the reciprocal force polyhedron.

1.8.2 Indeterminate Force Polyhedrons from Extended Gaussian Images

For an indeterminate system of forces, the topology of the force polyhedron depends on the distribution of forces (Figure 1.37). Because the topology of the force polyhedron can be different for the same configuration of forces, the reconstruction method using the convex hull of the tails of the unitized force vectors is not sufficient.

The topology of an indeterminate force polyhedron can be explained using EGIs.[18] EGI represents surfaces or solid objects on a unit, a *Gaussian sphere* using its face orientations and areas.[92] If the heads of the unitized normals of the faces are placed at the center of the Gaussian sphere, then the tails of the unitized normals lie on the surface of the Gaussian sphere. The face areas, whether already known or input as targets, are placed as *point masses* at the heads of the corresponding normals on the Gaussian sphere (Figure 1.39b).

Based on the duality principle of projective geometry, which maps faces into points, points into faces, and edges to edges, the EGI is then a spherical *dual image* or topological dual of a convex polyhedron[93] (Figure 1.39a–d); a polyhedron with *n* number

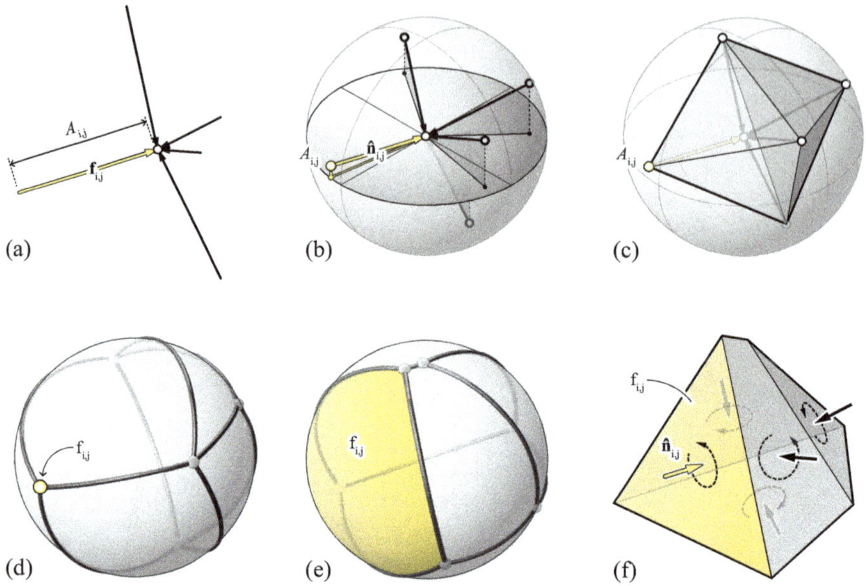

Figure 1.39 (a) Spike model of equilibrated force vectors; (b) unitized force vectors and the point masses on the Gaussian sphere; (c) the convex hull of the point masses; (d) EGI with adjacency arcs, or equivalently the dual spherical polyhedron; (e) spherical polyhedron, which is the topological primal of the EGI; and (f) the geometry of the polyhedron.

of faces will have n number of point masses on the Gaussian sphere. Since any two faces of a polyhedron are adjacent if they share a common edge, an edge of a polyhedron can be represented on the EGI as a connection between two point masses as an *arc*. This arc is defined as the minor arc of the great circle containing any pair of points on the EGI,[94] and equivalent to the corresponding edge of the convex hull (Figure 1.39c) of the point masses projected onto the Gaussian sphere. It is possible for two antipodal points to be connected, for example, in the case of a dihedron or a flattened polyhedron. However, such instances are not considered in this book, and any two diametrically opposite points are assumed to be not adjacent.

However, the face adjacency information of a polyhedron or a cell is not directly recoverable from the location of the point masses and arcs alone. Consider the three cells in Figure 1.40. These three cells have the same face orientations but different face areas. Based on the definition of an arc stated earlier, the face normals and areas of the three cells result in the same convex hull (middle row of Figure 1.40). However, the actual EGIs of the cells (bottom row of Figure 1.40) are different, as the different face area distributions result in different face adjacencies. Depending on the face area distribution, various face adjacencies can occur: faces 1 to 4 are all adjacent with one another at a vertex; faces 1 and 3 are adjacent; or faces 2 and 4 are adjacent.

This indeterminacy of face adjacencies occurs when an arc crosses another arc on the EGI. Moni[94] defines these arc intersections as *cross-adjacencies*, where various face adjacencies could occur depending on different face area distributions. By adding a fictitious, *zero-point mass* at these arc intersections, and subsequently a face with

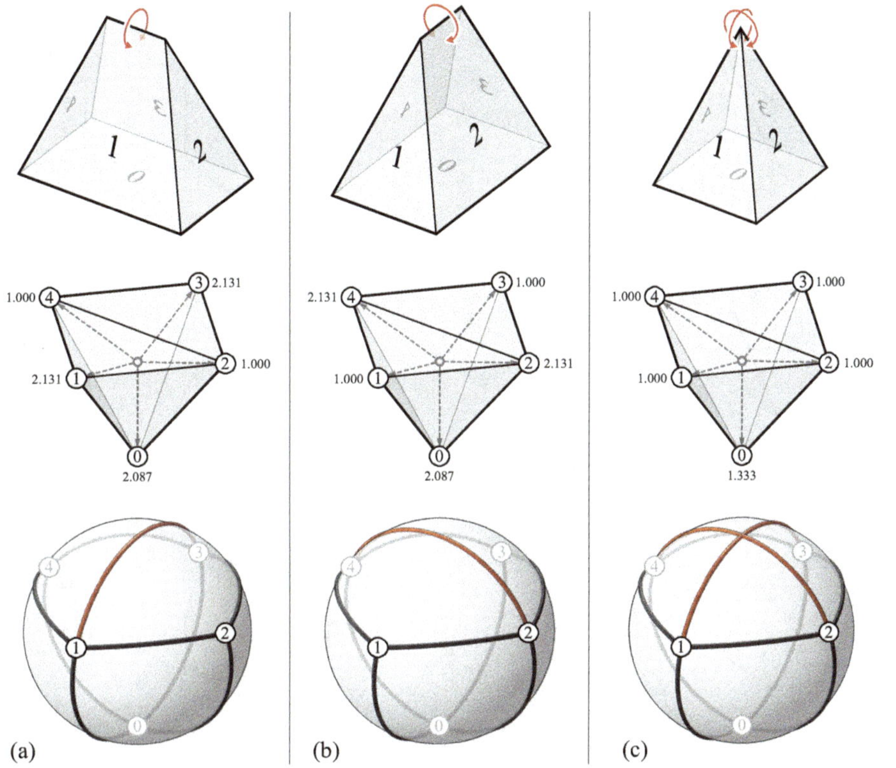

Figure 1.40 Three cells with the same face orientations but with different areas: (a) faces 1 through 4 are adjacent to one another at a single vertex; (b) faces 1 and 3 are adjacent along an edge; and (c) faces 2 and 4 are adjacent along an edge.

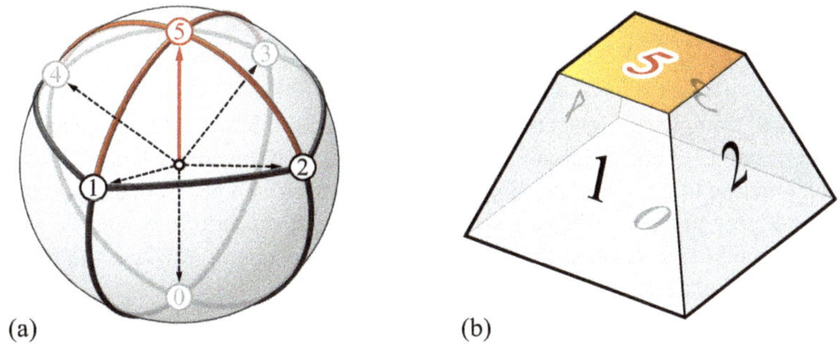

Figure 1.41 (a) The EGI with zero-point mass 5 added; and (b) the corresponding unit cell with an added zero face.

zero area, or a *zero face*, all possible face adjacencies can be embedded and represented in a single EGI (Figure 1.41a). Once all zero-point masses have been added, the topology of the EGI as the dual spherical polyhedron is complete, and the *unit polyhedron* can now be constructed, which is simply the topological primal of the dual spherical polyhedron (Figure 1.41b).

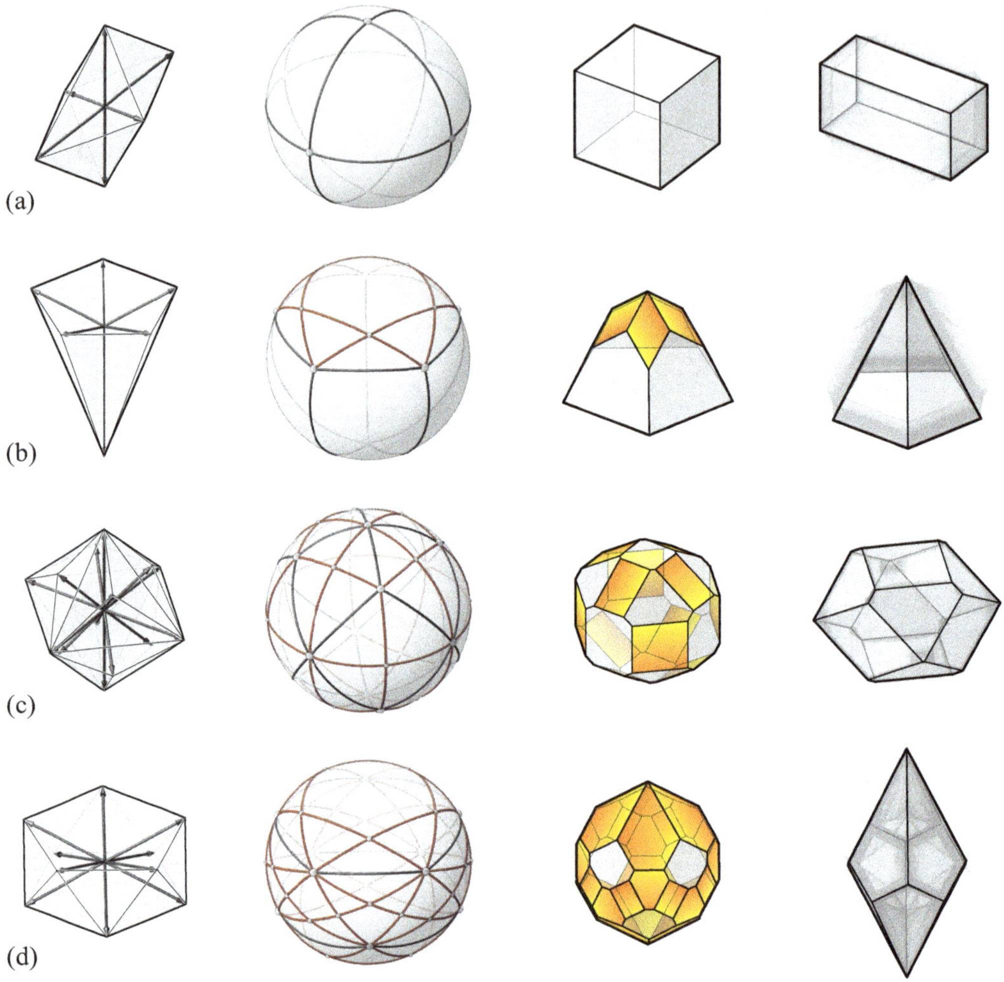

Figure 1.42 Simple examples of EGI-based reconstruction of force polyhedrons (force vectors, the EGI, unit cell, and the final geometry of the force polyhedron, from left to right): (a) reconstruction of a rectangular box; (b) reconstruction of a pentagonal pyramid, where the zero faces collapse to a point; (c) reconstruction of a tetradecahedron, where the zero faces collapse to edges; and (d) reconstruction of a pentagonal trapezohedron, where the zero faces collapse to points and edges.

1.8.3 Iterative Resizing of Faces

Finally, the faces of the unit cell can be resized through iterative methods such that each face ends up with an area equivalent to the magnitude of the corresponding force vector.[18] The zero faces collapse to either a point or an edge depending on the force distributions. Figure 1.42 shows four simple examples of the EGI-based reconstruction.

1.8.4 Indeterminacy and Self-Intersecting Faces

The indeterminate force polyhedron has some very interesting geometric properties. For instance, changing the length of the edge vectors of the face corresponding to the applied load can alter the convex geometry of the side faces and turn them into self-intersecting or *complex* face. Consider the face *FBAE* of the force diagram of Figure 1.43a. Changing the length of the vector *AB* from l_1 to l_2 results in the relocation of vertex *F* and *B*, as illustrated in Figure 1.43b. In other words, the vector *FB* is moved along the direction of *BA* with no change in its length. Moving the vertex *F* to its new location results in the complex polygon of *FBAE* of Figure 1.43b. A complex face has a continuous vector direction along its edges and includes regions with opposite normal directions. If we consider the direction of the normal $\mathbf{n}_3$ of a convex face *FEBA* of Figure 1.43a positive, then the complex face of Figure 1.43b has two normal directions $\mathbf{n}_{31}$ and $\mathbf{n}_{32}$, which are negative and positive, respectively. In addition, the magnitude of the force in a complex face is the sum of the signed areas of the regions of the face. In the example of Figure 1.43b, the length of the vector *BA* is precisely chosen as half of the original length minus the length of the edge vector *EF*. This results in two identical triangular regions on the self-intersecting face with the same area and opposite normal direction. As a result, the total area of a face adds up to zero, which corresponds to a zero-force member in the form diagram. Section 1.9 provides more examples of the force distribution for the indeterminate case of Figure 1.34.

Multiple Examples of an Indeterminate Force Face

Having the geometry of the force diagram is quite useful in exploring various equilibrium states for an indeterminate node. Figure 1.44a illustrates an equilibrium state where all the ropes are pulled from sides with a force polyhedron comprised of convex-only faces. In the force diagram of Figure 1.44b, the length of the edge has been reduced such that a self-intersecting face is created with positive (red) and negative (blue) areas. In this example, the area belonging to the positive region is larger than the area of the negative region, and thus, the total sum is positive. This equilibrium state results in a tension force in the corresponding rope. If the area of the negative region becomes larger than the positive region, the force in the rope flips from tension to compression (Figure 1.44c). This property of graphic statics is quite fascinating since it shows another state of equilibrium where the combination of tension and compression for the same node geometry can keep the weight from falling.

Pushing this to the other extreme results in having zero-area faces where the negative and positive regions are equal. Having zero-area faces in the force diagram translates into zero-force members in the form diagram, as shown in Figure 1.44d. In this case, only two ropes are sufficient to keep the node in equilibrium.

All these configurations can be used to design the system using the *master safe theorem of plasticity* as explained by:[54]

> If any equilibrium state can be found, that is, one for which a set of internal forces is in equilibrium with the external loads, and further, for which every internal portion of the structure satisfies a strength criterion, then the structure is safe.[54]

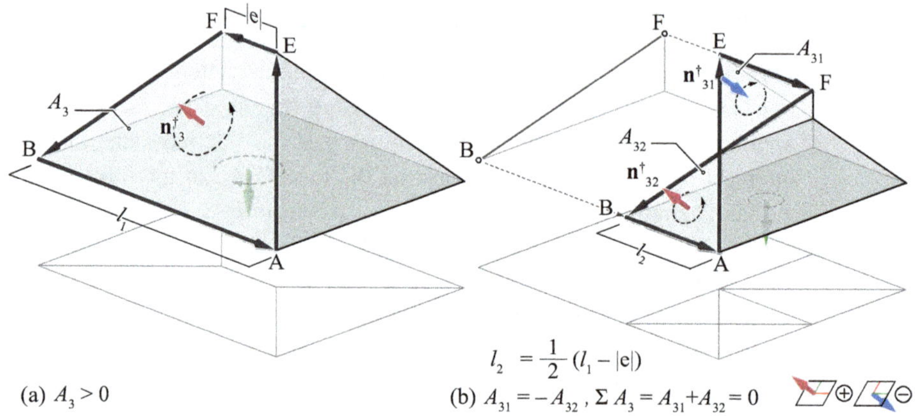

Figure 1.43 (a) An indeterminate force diagram with all convex faces; and (b) the same force diagram with self-intersecting faces.

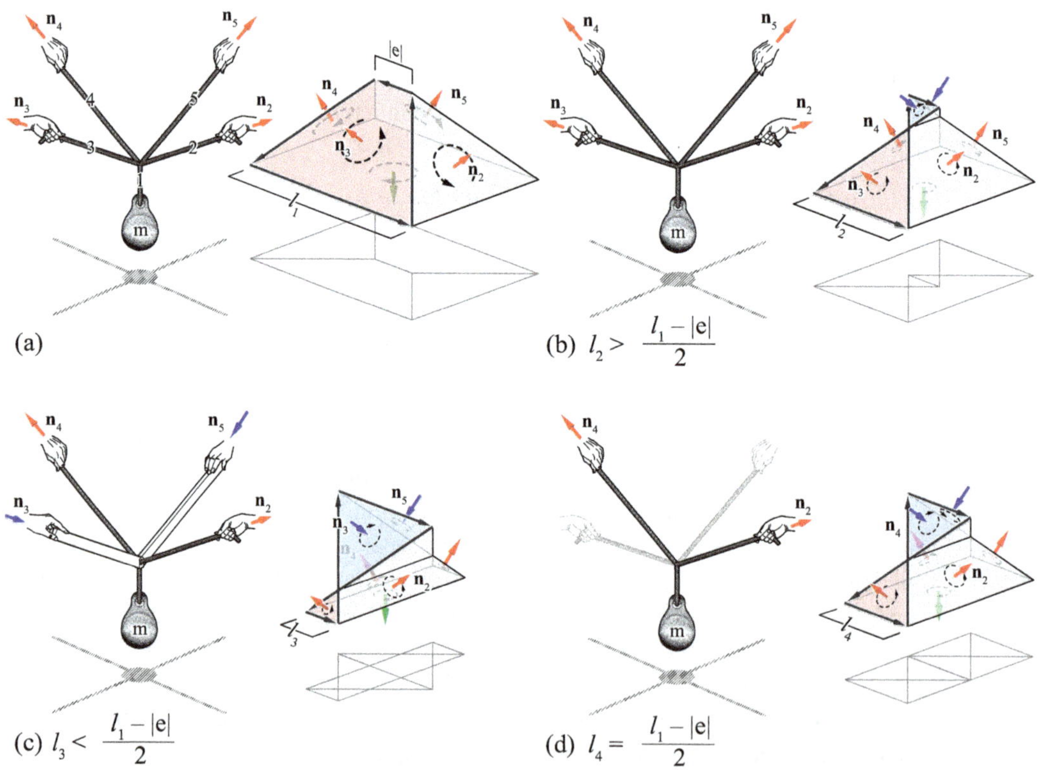

Figure 1.44 Multiple equilibrium states for the same form by controlling the geometry of an indeterminate force diagram: l_1 is the initial length of the bottom edge of face f_3, and $|e|$ is the length of its top edge. Form has (a) all tension members and a convex force polyhedron; (b) all tension members with complex force polyhedron; (c) combined tension and compression and complex force polyhedron; and (d) only two tension members and zero area faces.

Remark 1.6. Note that the *strength criterion* is related to material properties and member sizing to withstand the internal forces. The main limitation of the methods of graphic statics is that they do not include material properties. Thus, to realize structures in real-world applications, the forces calculated by the graphic methods should be used to properly size the members of the structure according to material properties.

The GDoF of force diagrams are quite useful in controlling the boundary conditions and redistribution of internal force in spatial funicular forms. This feature will be explained and used extensively in Chapter 6 of this book for the design of funicular spatial structures with both tension and compression forces.

1.9 Designing Nodal Equilibrium

All simple nodes discussed in this chapter can be used to design structural systems. Figure 1.45 shows multiple supporting structures for a bridge deck based on a compression-only node or compression and tension-combined nodes. In all the designs shown in Figure 1.45, a single member carries the load of a speculative bridge deck using compression-only members or a combination of compressive and tensile members. These nodes can be deliberately combined to make a complete structure by simplifying connecting the nodes that carry the same force in their shared members. The following designs of Figures 1.46 and 1.47 show how this simple principle can help us design structures.

The design of the bridge of Figure 1.46 started with two nodes with tension and compression members that are symmetrically placed away from each other to carry

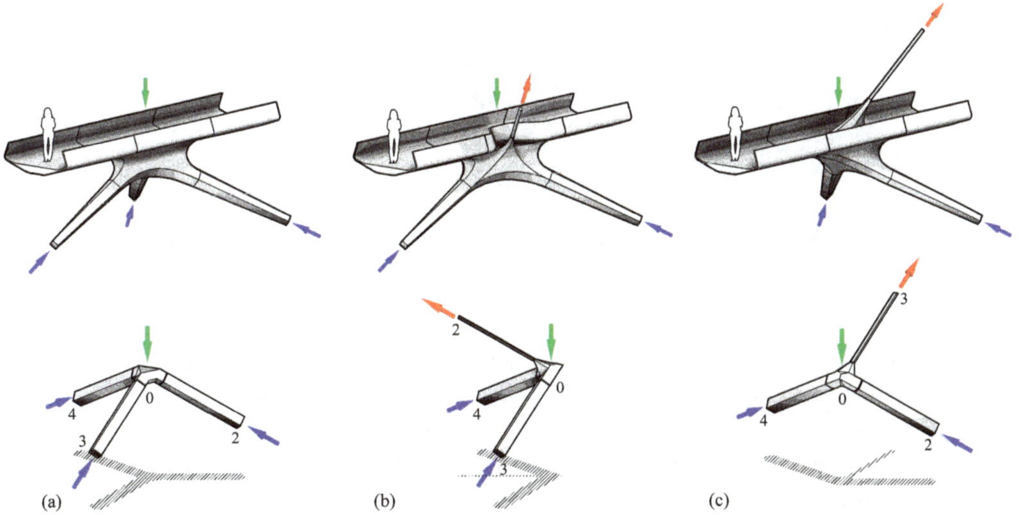

Figure 1.45 Multiple configurations of a single node and its application in carrying a load of a conceptual bridge deck by their (a) compression-only members and (b) and (c) compression and tension combined members.

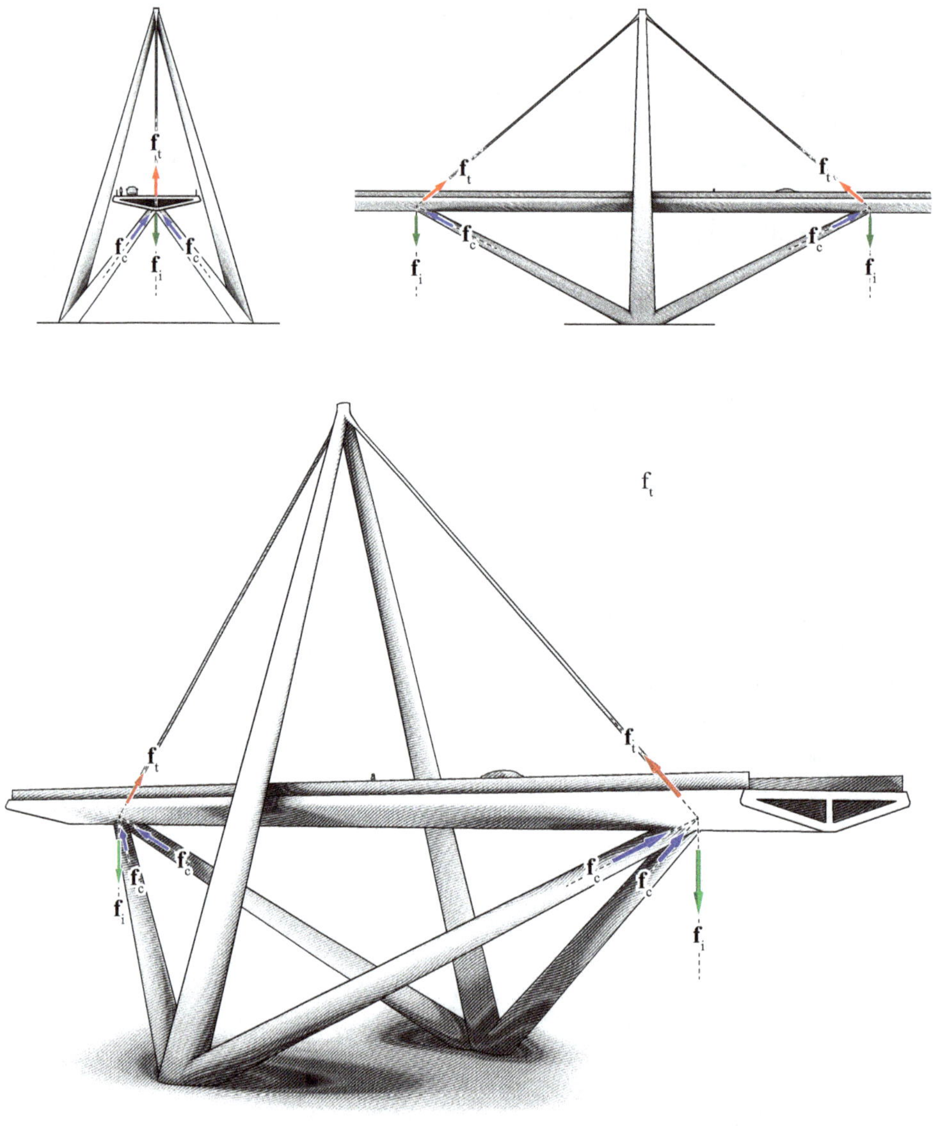

Figure 1.46 A conceptual bridge design consists of three simple nodes with combined tension and compression forces. The nodes with the applied load are highlighted.

the load of a bridge deck. These nodes are then connected to a single node in the middle, establishing the bridge's pylon and the main support.

The second design in Figure 1.46 started with a similar idea, but it required two nodes on the top to connect all members properly to carry the load of the deck. Can you construct the force polyhedron for each node of the bridges of Figures 1.46 and 1.47 and see how they would be connected? Use the example of Figure 1.23 as a hint. Consider each node in its compression-only configuration and flip members of the form to turn them into tensile members. If you still cannot complete this task,

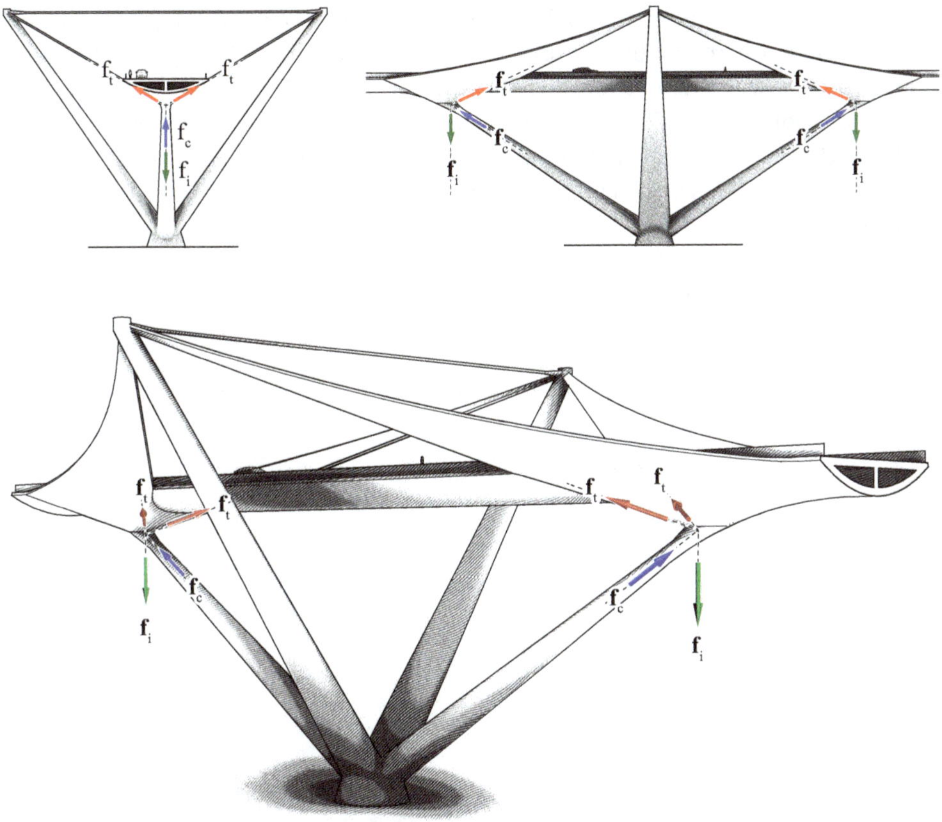

Figure 1.47 A conceptual bridge design of four nodal equilibriums with tensile and compressive forces. Note that only the nodes with an applied load (green) are highlighted in this figure.

you may need to follow the rest of the chapters and learn more about the relationship between multiple force polyhedrons within a force diagram of a structure.

Summary

This chapter introduced the geometric principles of equilibrium for concurrent systems of forces both in 2D and polyhedral graphic statics. It starts with a simple example of a tensile or a compression-only node and explains constructing a closed force diagram for the given geometry in 2D and 3D. Suppose the magnitude of the applied load is known. In that case, the closeness of the force diagram ensures the node's equilibrium, and the measurements in the force diagram, including the length of the edges in 2D and the area of the faces in 3D, correspond to the magnitude of the applied load and internal forces in the members connected to the node.

The chapter explained the geometrical and topological properties of a single force polygon in 2D and a polyhedron in 3D and its relationship with the single node as the form diagram. In addition, the static determinacy and indeterminacy of forces are explained by the number of independent edges in the force diagram, referred to as the GDoF. Furthermore, the construction of an indeterminate force polyhedron for a general node with multiple members or applied loads is discussed by explaining the EGI of a polyhedron. The final examples of this chapter show how the knowledge of this chapter can be applied in the conceptual design of simple bridges by simply combining multiple nodes that are in equilibrium with both tension and compression members.

2 Equilibrium of a System of Forces

Masoud Akbarzadeh

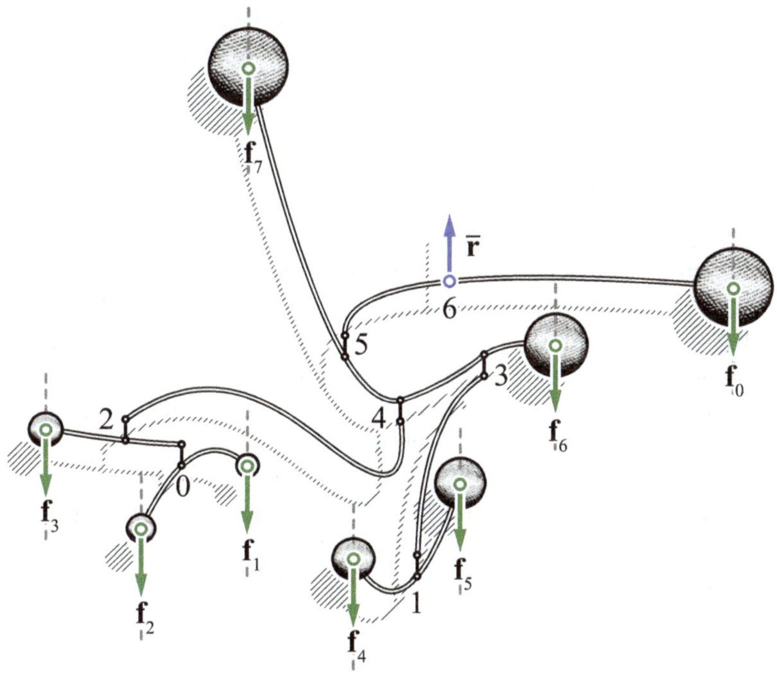

Figure 2.1 A design inspired by the works of Calder showing the equilibrium of multiple circular objects held by a single force $\bar{\mathbf{r}}$.[95]

2.1 Introduction to Graphical Equilibrium

In Chapter 1, we showed how the equilibrium of several forces around a single node can be checked and represented using graphic statics. Assuming that the system is in equilibrium, the internal force in the members connected to a single node was calculated by the geometric construction of the force diagram, where an applied load's magnitude and direction were given. What if we have a system consisting of multiple loads in three-dimensional (3D) space that are not applied to a single node, and we intend to find a force or forces to keep this system in equilibrium? What are different scenarios for which we can find equilibrating forces?

2.1.1 Chapter Objectives

In this chapter, we look into an isolated system of forces free from any structure or form and provide procedures to induce equilibrium for the given loads. To understand the concept, let's begin with a balancing sculpture of Figure 2.1 inspired by the works of the renowned artist Alexander Calder.[95] This structure consists of multiple small and large parts with their associated weights held in balance in a particular configuration. You will be surprised that there is only *one location* on the entire sculpture from which the system can be hung to maintain equilibrium for the form you see as it is. This point is where the resultant force $\bar{\mathbf{r}}$ is applied. If you hang the sculpture from any other point, it will rotate and disturb the arrangement of the weights for which it was designed. In this chapter, you will learn what it takes to find an equilibrating force for a system and where to put it to keep it from rotating or being imbalanced using methods of graphic statics.

2.1.2 Single Force Equilibrium in 2D

According to Newton's Third Law, all forces in the universe occur in pairs: for every force, there is a reaction force of equal magnitude and opposite direction. Consider a weight hanging from a string, as illustrated in Figure 2.2a. In the form diagram, we depict the force of the weight as a downward green vector. If the mass is at rest, an internal force in the rope is equal in length but opposite in direction, which keeps the weight in equilibrium. As described in more detail in Chapter 1, the forces can be described by demarcating the subspaces in the form diagram and creating a force diagram by drawing vectors perpendicular to the applied load.

In the example illustrated in Figure 2.2, the line of action of the force **f** divides the two-dimensional (2D) space into two subspaces, *A* and *B*. If the right-hand thumb aligns with the direction of the given load, four fingers show the direction of a path

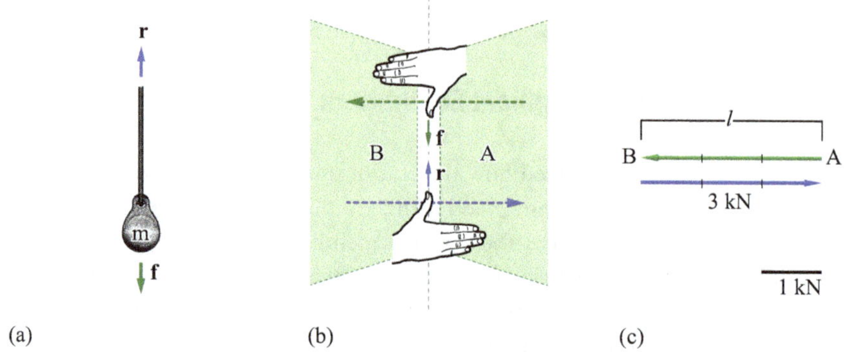

Figure 2.2 (a) A weight hanging from a rope in 2D; (b) its form diagram; and (c) its closed force polygon.

from subspace A to B to construct the geometry of the force diagram. The force diagram can then be drawn on a particular scale proportional to the scale of the force **f**; for instance, 1 cm is equivalent to 1 kN, as shown by vector AB in Figure 2.2c.

To ensure equilibrium, the force diagram must be closed. The force **r** closes the force polygon by joining the start and ends of the applied load in tip-to-tail order. In this case, there is only one applied load, so the resultant is identical but in the opposite direction. As illustrated in Figure 2.2c, **r** equal and opposite to vector AB must exist as an internal force to keep **f** in equilibrium. Vector BA closes the force diagram by connecting the start and endpoints of the vector AB and ensures the equilibrium of the system.

2.1.3 Single Force Equilibrium in 3D

The same operation may also be performed in 3D, with some changes in the process to accommodate the extra dimension. Note that in polyehdral graphic statics, each load in the form diagram is at the intersection of at least three planes. Consider the 3D equivalent of the mass hung from a string, as illustrated in Figure 2.3a,b. First, with 3D constructions, the subspace boundaries are defined by planes instead of lines. Let us assume that in this example, the line of action of the force **f** is at the intersection of at least three planes dividing the space into three subspaces A, B, and C. The configuration of the planes and their dihedral angles in 3D are chosen arbitrarily.

Second, an applied force vector in 2D becomes an applied force face in 3D, with its magnitude indicated by the face's area instead of the vector's length. The force face ABC can be constructed using the right-hand rule as examined in Chapter 1. The area of this triangular face represents the magnitude of **f**, its normal $\mathbf{n}^{\dagger}$ parallel to **f**, and the edge vectors perpendicular to the planes intersecting in the form diagram of Figure 2.3b.

Third, a reaction force of equal magnitude and opposite direction should be found to balance the force in 3D. The magnitude of the resultant **r** in 3D translates into the

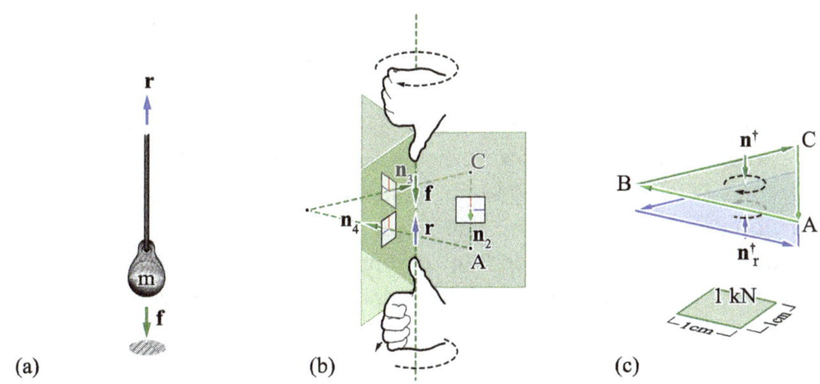

Figure 2.3 (a) A single force **f** in 3D; (b) the minimum number of planes intersecting at the line of action of the force; and (c) its closed force diagram, which includes the face related to the reaction force **r**.

area of a face in the force diagram. Thus, the face corresponding to the reaction force should have the same area as the face *ABC* but in the opposite normal direction.

Last, the force polyhedron needs to be closed to guarantee equilibrium. Thus, the edge vectors of the face corresponding to the force **r** should be identical but opposite to the edge vectors of the triangle *ABC*. The face *CBA*, reciprocal to force **r**, closes the force diagram and results in a *flat* polyhedron consisting of two coplanar faces (Figure 2.3c).

2.2 Concurrent Force Systems

The system of forces may involve more than one force in both 2D and 3D and take various geometric configurations. In general, the force in a system can either be *concurrent* or *nonconcurrent*. If the lines of action of all the forces in a system intersect at a single point, then the system is concurrent; otherwise, it is nonconcurrent. If the sum of the forces in both *x*- and *y*- directions is zero, then a system of concurrent forces in 2D space is in equilibrium. The latter statement can be written in numerical terms as follows:

$$\Sigma \mathbf{f}_x = 0$$

$$\Sigma \mathbf{f}_y = 0$$

But how these numerical equations can be translated into geometric diagrams?

2.2.1 Closed Force Polygon in 2D

Consider a Y-shaped column in which two members direct loads downward and into a point of intersection, where a third member resists movement by pushing up against the intersection. This is a system of two applied loads and one resultant force, with all lines of action intersecting at a single point, and is a concurrent system. Figure 2.4a illustrates this concept with two forces intersecting at a point in the 2D space. These two forces divide the space into three subspaces, *A*, *B*, and *C*, which results in the force diagram made by the vectors *AB* and *BC*.

After dividing the subspaces, the force diagram can be drawn. The two force vectors *AB* and *BC* establish an open polygon, where the forces placed tip-to-tail are not in a closed loop and, therefore, not in equilibrium. An unknown third force should be added to the system to keep it in equilibrium. A new vector, **r**, can be drawn to close the force polygon. Connecting the endpoint *C* of the polygon to the starting point *A* will close the polygon and define the magnitude and direction of the equilibrating force.

In other words, if the force polygon/diagram is closed, the sum of the forces in *x*- and *y*- is zero, which is the geometric translation of the static equilibrium equations. Vector **r** is the resultant of the forces in the system. Drawing **r** in the form diagram will

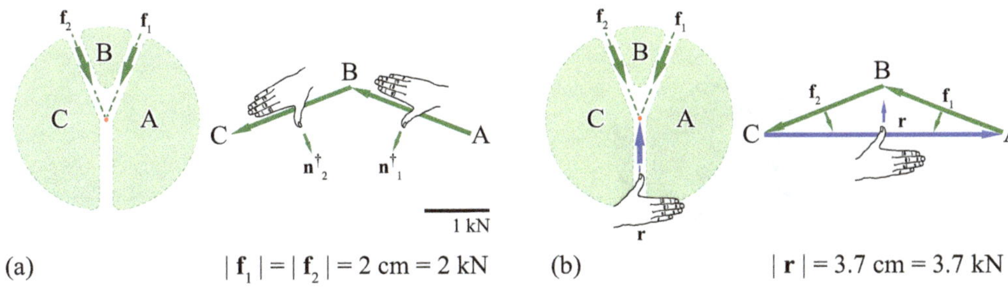

Figure 2.4 (a) A system of two concurrent forces in 2D and its open force polygon; and (b) the closing edge of the force polygon defines the magnitude and direction of the reaction force **r**, which keeps the system in equilibrium.

keep the system in equilibrium, as shown in Figure 2.4b. The system of two applied loads can be easily extended to more loads, and the procedure stays the same. Note that all loads intersect at a point, and the system has no moment or rotation. Thus, the applied loads create more subspaces in the form diagram, and a single resultant applied at the point of concurrence can still ensure equilibrium.

2.2.2 Closed Force Polyhedron in 3D

The 3D equivalent of the concurrent forces may take various configurations. Figure 2.5 illustrates a column module of the Sagrada Familia by Antoni Gaudi.[96] These are 3D columns that Gaudi designed by developing meticulous physical structural models made of ropes and sandbags. To demonstrate how the force diagram can manipulate the reactions required to maintain equilibrium, we will provide an example with two concurrent applied forces and an example with three concurrent applied forces.

Two Concurrent Forces in 3D

Consider a V-shaped column in 3D with two members that direct loads downward and into a point of intersection and the ground. Figure 2.6a illustrates this V-shaped column as two concurrent forces $\mathbf{f}_1$ and $\mathbf{f}_2$ in 3D. Note that in polyhedral graphic statics, two neighboring forces share a face. Since there are only two forces, they inherently share a plane. Each force in the form diagram has a line of action at the intersection of at least three planes and is reciprocal to a triangular face in the force diagram as shown in Figure 2.6b. The direction of the edge vectors is defined using the right-hand rule, and the length of one edge chooses the scale of the triangles.

The areas of the triangles define the magnitude of the forces. The free edges of these triangles can take any angle or length except for the shared edge between the triangles. The change in the length and angle of the free edges will also alter the area and magnitude of the applied load. Since two forces in the form diagram share a plane, their reciprocal faces in the force diagram share an edge. The edge vectors *CA* and *AC* are equal in length and are opposite in direction (see Figure 2.6b).

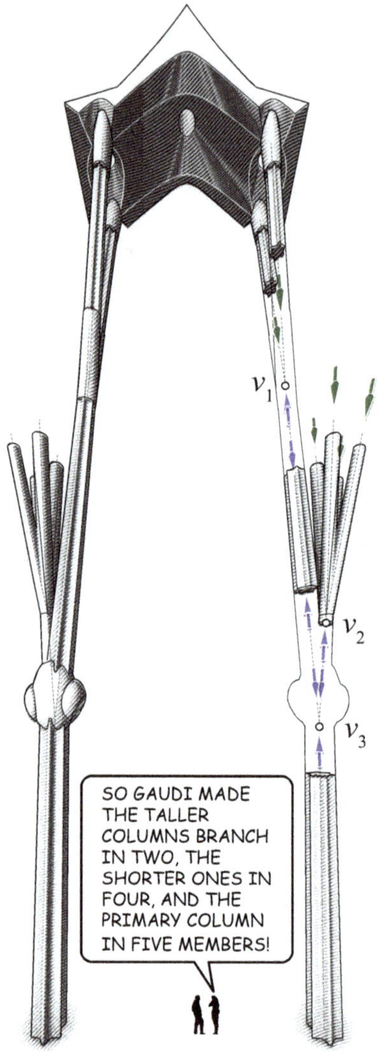

Figure 2.5 A schematic drawing of branching columns of the Sagrada Familia church by Antoni Gaudi where v_1 and v_3 branch in two and v_2 branches in four members, after.[96]

The two adjacent triangles of the force diagram have four open edges, establishing a nonplanar polygon of *DCBA*. This force diagram, that is, the open force polyhedron, should get closed to maintain the equilibrium condition for this system. Two triangular faces, *ABD* and *DBC*, can close the nonplanar polygon of this force diagram, as shown in Figure 2.6d. These two additional faces correspond to two forces in the form that can keep the forces $\mathbf{f}_1$ and $\mathbf{f}_2$ in equilibrium.

Since $\mathbf{f}_1$ and $\mathbf{f}_2$ are concurrent, forces $\mathbf{r}_1$ and $\mathbf{r}_2$ should also be applied to the point of concurrency in the direction parallel to the normals $\mathbf{n}_1$ and $\mathbf{n}_2$ and with the magnitude proportional to the area of the triangles *ABD* and *DBC* (see Figure 2.6c). Figure 2.6d

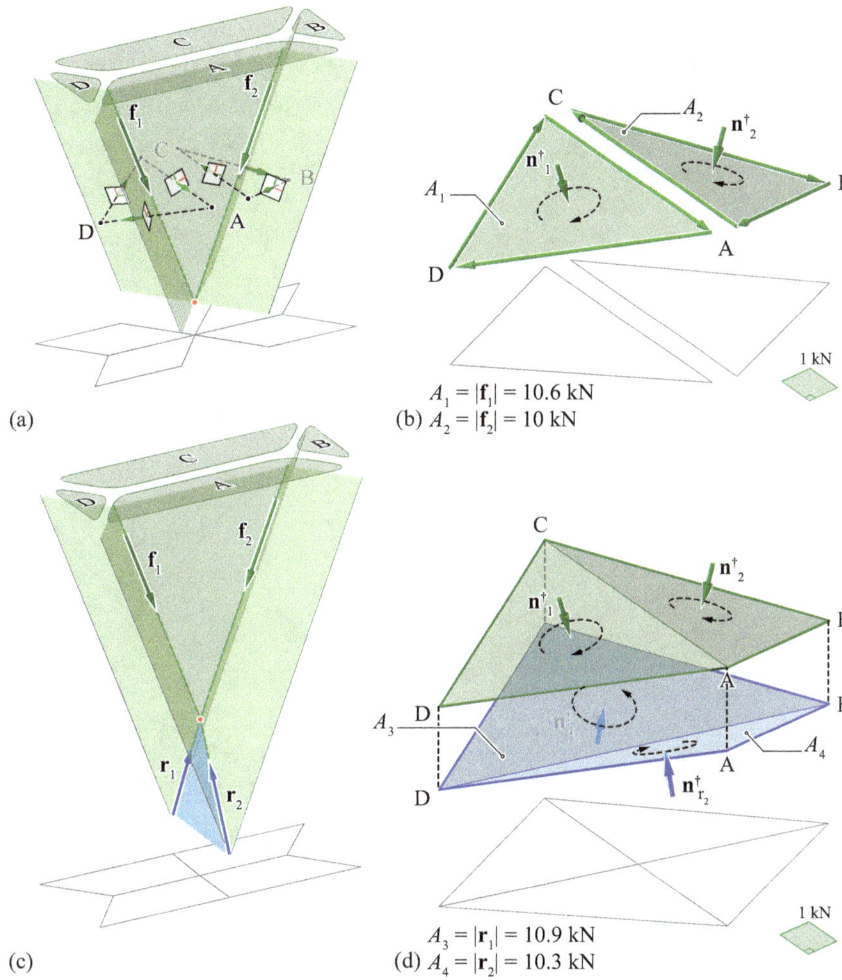

Figure 2.6 (a) Two concurrent forces $\mathbf{f}_1$ and $\mathbf{f}_2$ with a shared plane in between; (b) the triangular faces of the force diagram corresponding to the applied loads; (c) the resultant forces perpendicular and proportional to the areas of the closing faces of the force diagram in (d).

shows the completed force diagram and the angles of the forces $\mathbf{r}_1$ and $\mathbf{r}_2$ in the equilibrated form. Note that the length of the vectors in the form does not correspond to the magnitude of the force.

Three or More Concurrent Forces in 3D

The next example includes three concurrent forces in 3D with undefined magnitudes, as illustrated in Figure 2.7a. The three concurrent forces share face planes that are used to construct the geometry of their corresponding faces in the force diagram. The two edges of each force face can be found using the right-hand rule, as shown in Figure 2.7b. As noted, the force faces do not have a third edge defining the area for each face.

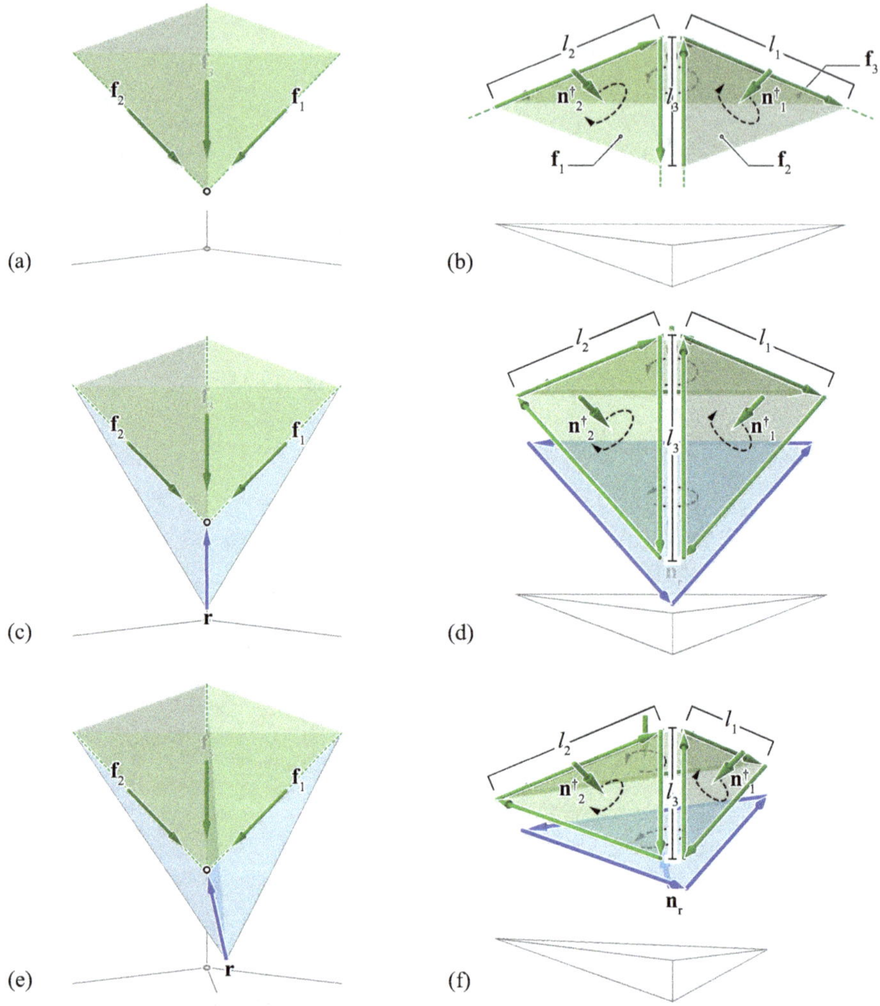

Figure 2.7 (a) Three concurrent forces and their shared planes; (b) the open triangular faces with edge vectors perpendicular to the shared planes in the form; (c) the resultant force perpendicular to the closing face of the force polyhedron in (d); (e) another resultant force perpendicular to the closing face of the force polyhedron in (f).

The designer can decide this edge by controlling the length of the shared edges, l_1, l_2, and l_3, in the force diagram.

These edges are perpendicular to the shared planes of the applied loads in Figure 2.7a. The lengths of the three edge vectors of Figure 2.7b define the magnitudes of the forces $\mathbf{f}_1$, $\mathbf{f}_2$, and $\mathbf{f}_3$. The direction of the faces in all examples is preserved, but as the length of l_i changes, the area of the faces and, thus, the magnitude of applied loads are altered. Once the edge lengths l_{1-3} are prescribed, the faces are completed, and the open polyhedron of forces is constructed, as shown in Figure 2.7d,f. The free edges of this open polyhedron establish an open polygon that should get closed by adding the fourth face to the force diagram.

Remark 2.1. Note that in polyhedral graphic statics, precisely controlling the applied loads by using the areas of faces is not straightforward because controlling the area of a polygon is a quadratic function. Using triangles helps constrain the geometry of a face, but as a result, the areas of the neighboring faces will also be altered. This limitation could be avoided by adding more free edges to the faces and will be addressed in Figure 2.16 in this chapter.*

The closing face of the force diagram provides the direction and magnitude of the resultant or reaction force **r** that keeps the applied loads in equilibrium (Figure 2.7c,e). In all examples of Sections 2.1 and 2.2, equilibrium could be achieved by simply closing the force diagram. This is because all the forces were concurrent, and the system had no rotation or moment. So what if the system has nonconcurrent forces?

2.3 Nonconcurrent Force Systems

In a nonconcurrent system of forces, the lines of actions of the forces do not intersect at a single point. Although the magnitudes of the forces in the force diagram can be changed such that they would form a closed polygon, the equilibrium of a nonconcurrent system of forces requires an additional condition to be satisfied, where the moments in the system should be zero:

$$\Sigma \mathbf{m}_x = 0$$

$$\Sigma \mathbf{m}_y = 0$$

A moment is a rotation caused by a force acting at a perpendicular distance from a point. Equations 2.1 and 2.2 mean that the sum of the moments around any arbitrary point must be zero. This equation can also be translated into geometry-based equilibrium principles. For a 2D system of nonconcurrent forces to be in equilibrium, any *funicular polygon* constructed on the line of action of the applied loads must be closed. This technique is called *funicular construction* and has been used in 2D graphic statics to find the equilibrium of a system of nonconcurrent forces.[21,41]

2.3.1 Closed Funicular Polygon

Consider a flat pallet of Figure 2.8 supporting two differing weights at its ends. Since the weights are pulled down by gravity, the lines of action are vertical, applied at the center of mass, and do not intersect, making the system nonconcurrent.

* This is equivalent to adding to the geometric degrees of freedom (GDoF) of the face polygon that was discussed in Chapter 1 and will be discussed extensively in Chapter 6.

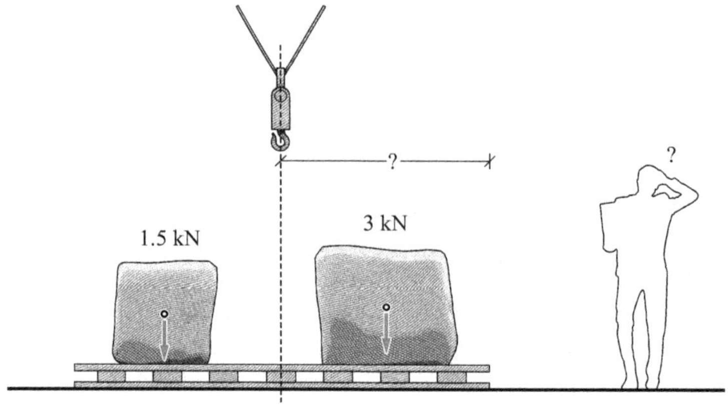

Figure 2.8 A flat pallet on which two weights are asymmetrically positioned. The exact location of the crane is needed to lift the pallet without causing any rotation that may overturn the parts.

Closing the force polygon in the concurrent system could immediately keep the system in equilibrium since the resultant is applied to the same point of concurrency (Figure 2.4b). For a nonconcurrent system, however, the location of the application of the resultant is unknown and has to be found. Simply put, the resultant force should be applied at a position such that the sum of the moments becomes zero.

Let us describe the pallet and its weights as a system of parallel forces $\mathbf{f}_1$ and $\mathbf{f}_2$ in Figure 2.9a. These two forces divide the space into three subspaces, *A*, *B*, and *C*. We can make a closed force diagram by drawing force vectors *AB*, *BC*, and the closing vector *CA* as shown on the right side of the form diagram of Figure 2.9a.

The closeness of the force diagram ensures that the sum of the forces in x- and y- is in equilibrium. So far, we know that the resultant force $\mathbf{r}$ has the magnitude proportional to the length of *CA*, which is 4.5 kN. But its line of action is unknown. *A closed funicular form* on the lines of action of the applied loads can give the location of the resultant and guarantee moment equilibrium. This is an ancient technique in the methods of graphic statics that is used to find the location of resultant.[21,41]

Constructing the Funicular Polygon

Before we start the funicular construction, we need to work with the closed polygon of forces. Let us extrude each vector of the force polygon into an arbitrary point in 2D. This point is shown by *O* in the force diagram of Figure 2.9b.

Extruding force vectors to a point makes triangular force polygons. Thus, the new force diagram will contain three closed triangles, each with an edge of either the applied load or the resultant. These triangles are shown as *ABO*, *BCO*, and *CAO* in Figure 2.9. We learned in Chapter 1 that each closed force polygon should represent the equilibrium of a node in the form diagram. Therefore, we should be able to find three nodes in the form diagram that are in equilibrium and reciprocal to the three

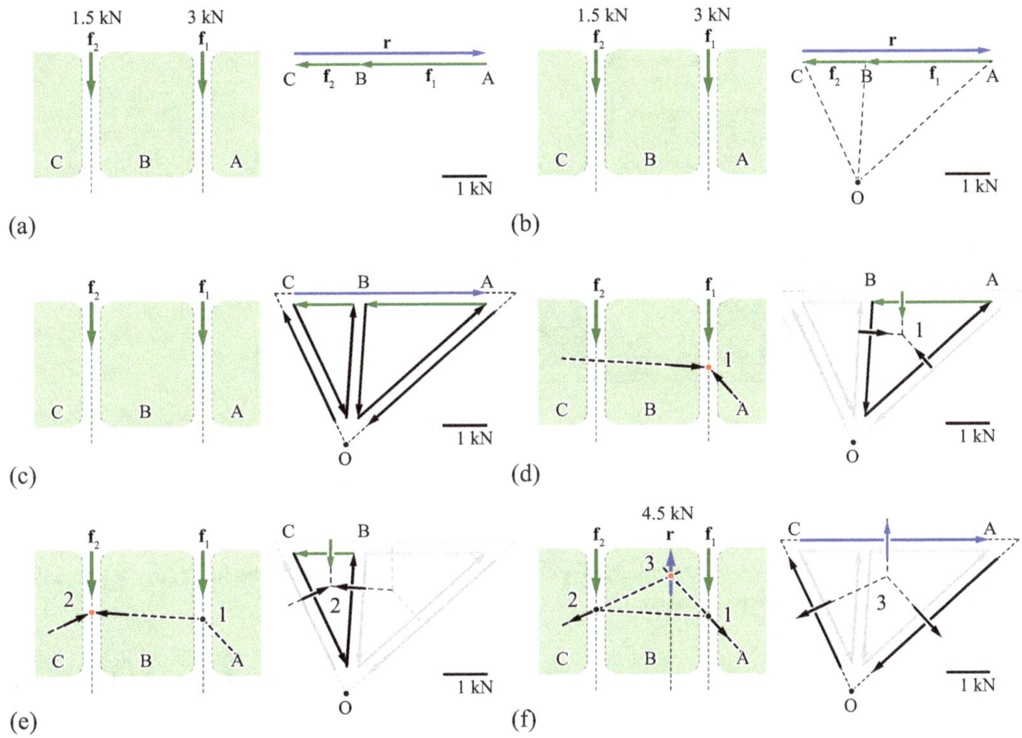

Figure 2.9 (a) A system of two parallel loads $\mathbf{f}_1$ and $\mathbf{f}_2$ and its closed force diagram; (b) extruding the force diagram into an arbitrary point O; (c) identifying the force triangles made by the edges of the initial force diagram; (d) to (f) finding the reciprocal nodes in the form diagram and closing the funicular polygon.

triangles of the force diagram. Moreover, each triangle shares an edge with its neighboring triangle, and their shared edge vectors are equal and opposite, as illustrated in Figure 2.9c.

Next, we will find the reciprocal nodes for each triangle in the form diagram. Let us start from the triangle *ABO*: the node reciprocal to this triangle should be on the line of action of $\mathbf{f}_1$ because triangle *ABO* has $\mathbf{f}_1$ as an edge. In the form diagram, we can choose an arbitrary point on the action line of $\mathbf{f}_1$ and draw lines normal to the other edges of the force triangle normal to *BO* and *OA* from the chosen point (Figure 2.9d). The next point is reciprocal to the triangle *BCO* and is on the action line of $\mathbf{f}_2$. Note that the triangles *ABO* and *BCO* share an edge *BO*; thus, their reciprocal nodes can also share an edge in the form diagram. Therefore, in the form diagram, we can extend the edge normal to *BO* from point 1 to intersect the action line of $\mathbf{f}_2$ and locate the new point reciprocal to the triangle *BCO* (Figure 2.9e).

The third triangle is *CAO* in the force diagram, for which the reciprocal node in the form diagram should be found. Note that the line of action of the resultant force $\mathbf{r}$ is not given, and in fact, locating the reciprocal node for this triangle will hint at the missing line of action. Since both triangles *ABO* and *BCO* share an edge with *CAO*, their reciprocal nodes should also share an edge. Thus, the node 3 location reciprocal

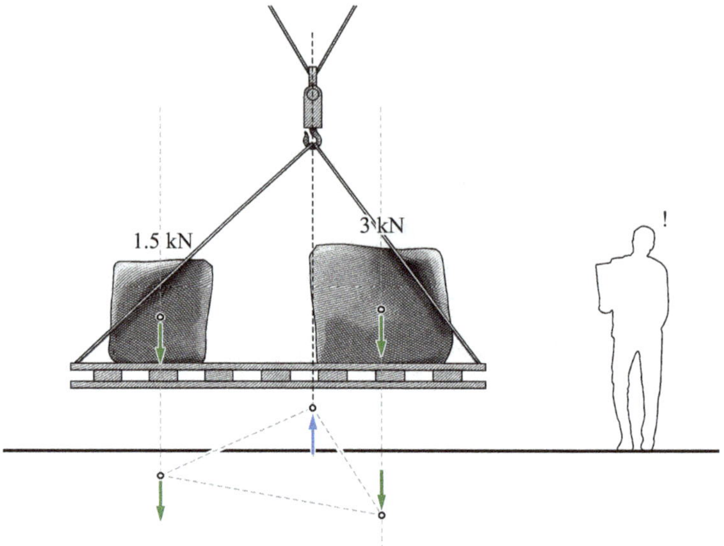

Figure 2.10 The closed funicular polygon gives the right location for the lift, **r**, to prevent any overturning moment in the system.

to triangle *CAO* can be found by intersecting lines drawn from nodes 1 and 2 normal to edges *AO* and *CO* of the force diagram. These two lines intersect at 3, which also identifies the line of action of **r** to keep the system in moment equilibrium.

Remark 2.2. Constructing a reciprocal geometry for the extruded force polygon also results in a closed polygon in the form diagram. This polygon is referred to as the *funicular polygon*. Its *closedness* is necessary to ensure moment equilibrium in the system, as the last node in the construction gives the line of action of the resultant force **r** (Figure 2.10).

The solution to having more applied loads on the system is similar. With more applied load vectors, there will be more subspaces. Whether the forces are parallel or not, a force polygon should get closed by adding the resultant vector, and the location of the resultant is found using closed funicular construction.

The Location of Extrusion

Changing the location of the point *O* does change the geometry of the funicular polygon, but it does not change the location of the resultant. Figure 2.11a–c shows two funicular polygons and the corresponding force diagrams. The number of nodes in the funicular polygon corresponds to the number of triangles/closed cells in the force diagram. The change in the angles of the edges of the force, as a result of choosing a different *O*, does change the geometry of the funicular polygon, as shown in Figure 2.11a. Nevertheless, the line of action of the resultant **r** always passes through the node reciprocal to the triangle *ACO*.

The earlier discussion can be used as a design technique to create Calder-like sculptures. The funicular polygons and the applied loads can be materialized artistically into

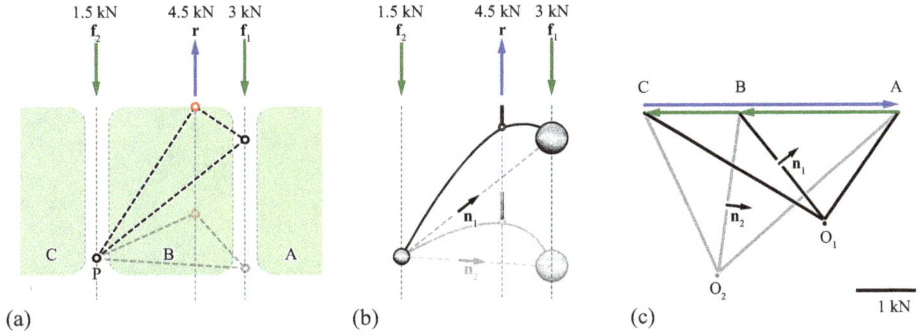

Figure 2.11 (a) A system of two parallel forces $\mathbf{f}_1$ and $\mathbf{f}_2$, the equilibrating forces $\mathbf{r}$, and two closed funicular polygons; (b) a materialization of the forces and the location of the equilibrating force; and (c) the arbitrary points o_1 and o_2 decomposing the force diagram into triangular faces.

two circular masses connected by a curved member, where the center of the masses precisely matches the location of the vertices of the funicular polygon (Figure 2.11b). Note that the masses can only go up and down along the line of action of the center of the mass and not sideways.

If each system is picked up from the line of action of **r**, then no rotation will happen to weights, and they will remain in their exact initial configuration. This result can be exploited creatively in the design of hanging sculptures similar to those by Calder,[95] to be explained in Section 2.5.

2.3.2 Closed Funicular Polyhedron

A similar scenario can be developed for the equilibrium of a set of forces in 3D. The equilibrium of a system of nonconcurrent forces in 3D necessitates that (1) the sum of the forces in x-, y-, and z- directions be zero and (2) the sum of the moments in x-, y-, and z- directions also be zero. The closedness of the force polyhedron in 3D guarantees that the sum of the forces in 3D space is zero, and it gives the magnitude and direction of the resultant forces. Similar to 2D graphic statics, a *closed funicular polyhedron* in the form diagram assures moment equilibrium and gives the locations of resultant forces.

Figure 2.12a shows two parallel forces in 3D and their faces. The force polyhedron corresponding to these forces consists of two triangular faces whose areas and the angles of the edge vectors are predefined by the faces connected to the applied loads in the form diagram. The triangular faces are co-planar and together have four open edges. We can use triangular faces to close the open polyhedron. Thus, two configurations of triangular faces may be used to close this force polyhedron.

As illustrated in Figure 2.12a,b, closing the force polyhedron with two faces implies the need for two resultant forces for the system's equilibrium. The resultant faces in the first configuration are identical to the faces of the forces $\mathbf{f}_1$ and $\mathbf{f}_2$. Thus, the

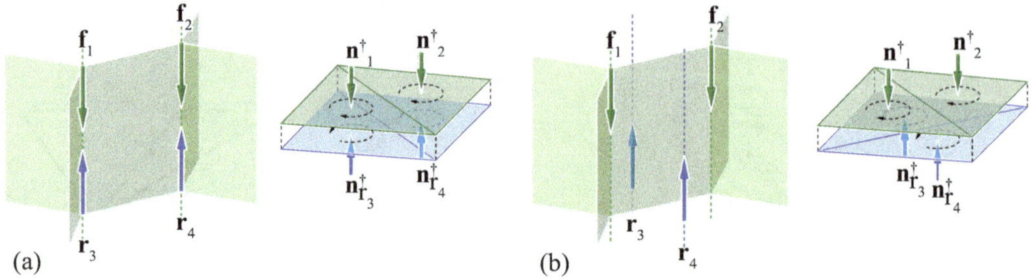

Figure 2.12 A parallel system of forces $\mathbf{f}_1$ and $\mathbf{f}_2$ in 3D and the closing faces $\mathbf{r}_1$ and $\mathbf{r}_2$ in two different configurations (a) and (b).

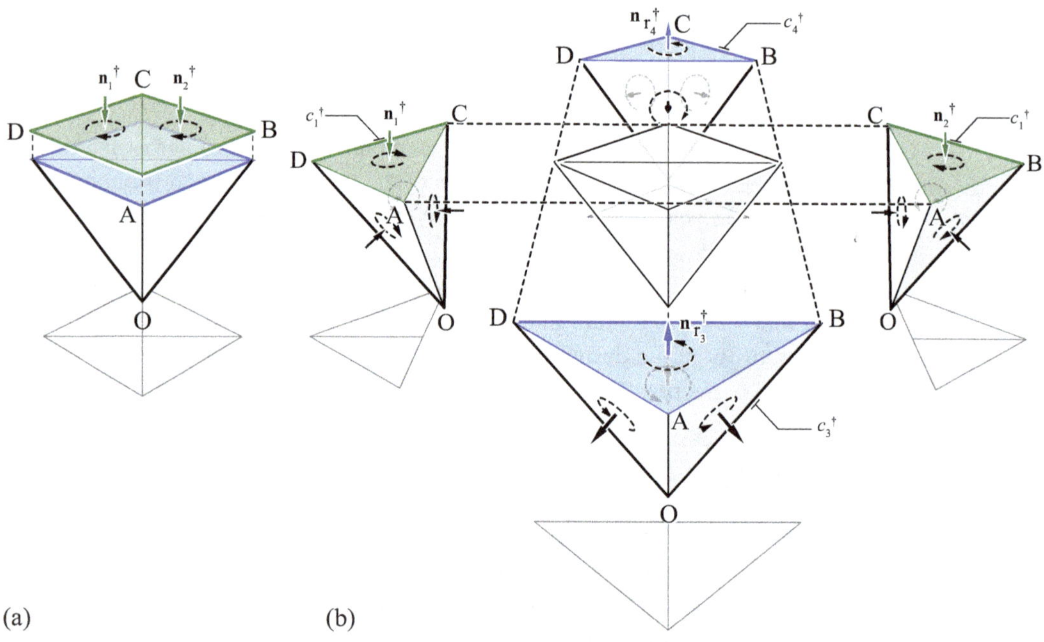

Figure 2.13 (a) The flat, closed force polyhedron *ABCD* is extruded to an arbitrary point *O*; and (b) the cells are constructed by the extruded faces and their normal directions.

funicular polyhedron is unnecessary since the system will not rotate in 3D. For the second configuration, the location of the resultants is yet to be found.

To find the locations of the equilibrating forces $\mathbf{r}_1$ and $\mathbf{r}_2$, we need to construct a closed funicular polyhedron on the lines of action of the forces.

Like the 2D case, the closed force polyhedron is extruded to an arbitrary point *O*, as illustrated in Figure 2.13a. Each face of the force polyhedron constructs a closed tetrahedral cell with the new faces made by extruded edges. For instance, face *ADC* makes a cell with faces *ABO*, *DAO* and *CDO*.

Each tetrahedron will have a consistent face normal direction defined by the face normal of the applied loads or the resultants. For instance, the directions of the tetrahedral cells *ABCO* and *ACDO* are toward the inside, and the directions of the cells *ABDO* and *BCDO* are towards the outside of the cell. Moreover, each cell shares faces with its adjacent cells. In addition, all the extruded faces are shared between the cells. For instance, the face *DAO* of the cell *ADCO* is identical to the face *ADO* of the cell *ADBO* with opposite normal directions (Figure 2.13b). In total, there are four tetrahedral cells that should be used for the funicular construction. Example 2.1 walks you through all the steps necessary to complete the closed funicular polyhedron.

Example 2.1 Construct a funicular polyhedron on the line of action of forces $\mathbf{f}_1$ and $\mathbf{f}_2$ of Figure 2.14 and find the location of the resultant forces equilibrating the system.

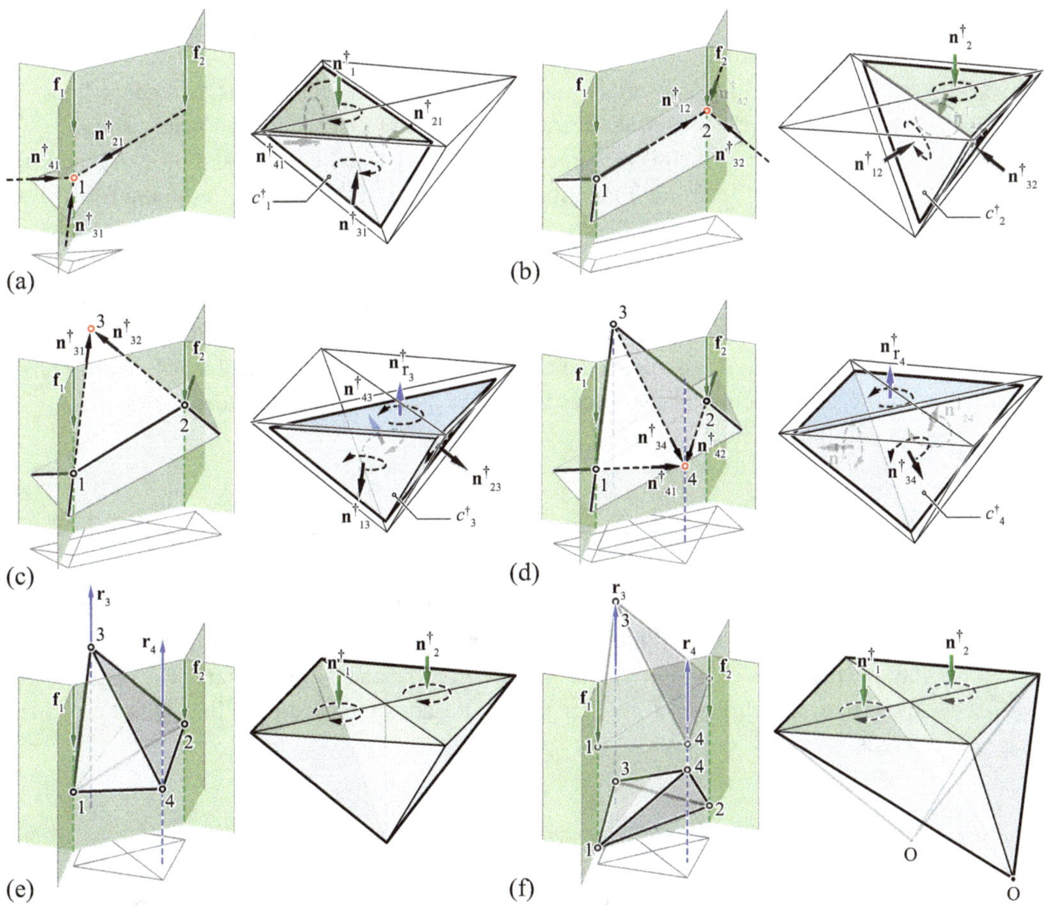

Figure 2.14 The step-by-step process of the construction of the funicular polyhedron to find the location of the equilibrating forces (a)–(f).

Construction steps:

1 Each tetrahedral cell in the force diagram has a reciprocal node in the form diagram that should be found. Let us start from cell *ADCO*, labeled as $c_1^\dagger$ with the force face reciprocal to the force $\mathbf{f}_1$ (see Figure 2.14a).
2 In the form diagram, pick an arbitrary vertex 1 on the line of action of the force $\mathbf{f}_1$.
3 Draw lines normal to the faces of $c_1^\dagger$ at vertex 1 (Figure 2.14a). This node is reciprocal to the cell $c_1^\dagger$ in the force diagram.
4 In the form diagram, from vertex 1, extend the line shown by a unit normal $\mathbf{n}_{21}$ to intersect the line of action of the force $\mathbf{f}_2$ (Figure 2.14b). Cell $c_1^\dagger$ shares the face *ACO* with cell $\mathbf{c}_2^\dagger$. Therefore, their reciprocal nodes in the force diagram should share an edge. We know this from the topological relationship explained in Chapter 1.
5 In the form diagram, from point 1, extend the line shown by a unit normal $\mathbf{n}_{21}$ to intersect the line of action of the force $\mathbf{f}_2$. This line is normal to the shared faces of cells $c_1^\dagger$ and $c_2^\dagger$, and its intersection with the line of action of the force $\mathbf{f}_2$ gives the location of vertex 2. Vertex 2 is reciprocal to the cell $c_2^\dagger$ of the force diagram (Figure 2.14b).
6 In the force diagram, cells $c_1^\dagger$ and $c_2^\dagger$ share faces with $c_3^\dagger$. Thus, extending lines normal to these shared faces from points 1 and 2 in the form diagram should intersect at 3, which is reciprocal to cell $c_3^\dagger$. These lines are labeled as $\mathbf{n}_{31}$ and $\mathbf{n}_{32}$ in Figure 2.14c. Point 3 identifies the location of the resultant $\mathbf{r}_3$ in the form diagram.
7 Continue the process to cell $c_4^\dagger$ in the force diagram and intersect the newly drawn edges from vertices 2, 3, and 1 in the form. Their intersection reveals the location of vertex 4, which gives the location of the resultant $\mathbf{r}_4$, as illustrated in Figure 2.14d. Vertex 4 completes the construction of a closed funicular polyhedron on the line of action of the forces $\mathbf{f}_{1-2}$ and $\mathbf{r}_{3-4}$ and thus guarantees the moment equilibrium in the system.

□

Similar to the 2D case, any arbitrary point *O* for decomposing the force diagram results in a closed but different funicular polyhedron. All these funicular polyhedrons lay on the lines of action of the given forces $\mathbf{f}_1$ and $\mathbf{f}_2$ and $\mathbf{r}_3$ and $\mathbf{r}_4$. Figure 2.14f illustrates two different funicular polyhedra constructed on the line of action of the forces.

Adjacent Parallel Loads

Figure 2.15a shows an example of three parallel loads $\mathbf{f}_{1-3}$ and their attached faces in the form diagram. A designer can define the magnitude of these three forces by choosing a closing edge in the reciprocal face in the force diagram. For instance, the face reciprocal to the force f_1 is the polygon *ABC* whose edge lengths *AB* and *CA*,

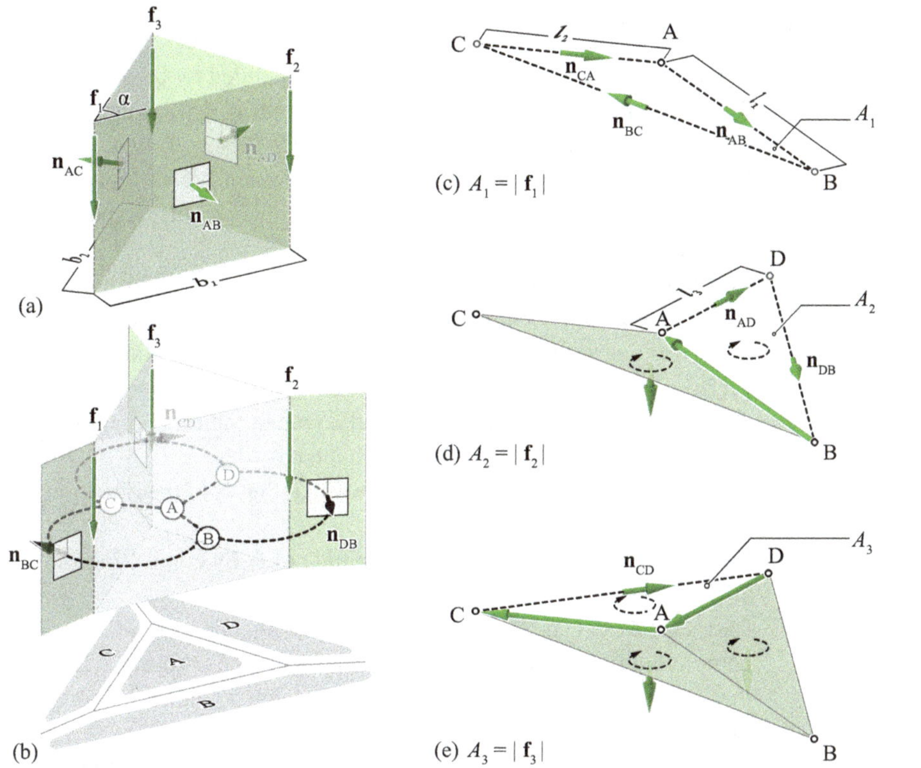

Figure 2.15 (a) A configuration of three parallel applied loads with their shared planes; (b) the geometric construction of the face corresponding to the load $\mathbf{f}_1$; (c) defining the area and the geometry of a face corresponding to the force $\mathbf{f}_2$; (d) the formation of the planes intersecting at the line of action of the forces by the external edges of the force diagram; and (e) the completion of the face.

l_1, and l_2 are chosen as the design input (Figure 2.15c). The face *ADB* is reciprocal to the force $\mathbf{f}_2$, and the length *AD*, or l_3, is chosen to close the polygon and to define the magnitude of the force (Figure 2.15d). Note that the third face *ACD* has all edges defined by the adjacent faces, and thus, the magnitude of the force is predetermined. This might be true at a cursory glance, but the number of edges to close the force polygon does not have to be three. Adding one more edge to the polygon *ACD* will liberate it from being constrained to its adjacent faces (Figure 2.15e). Example 2.2 will provide step-by-step procedures to construct reciprocal force faces for three parallel adjacent loads.

Example 2.2 Construct the reciprocal faces for the system of parallel forces of Figure 2.15 such that the magnitude of all faces can be separately controlled. In the form diagram, forces $\mathbf{f}_2$ and $\mathbf{f}_3$ are placed at distances b_1 and b_2 from force $\mathbf{f}_1$. Also, b_1 and b_2 make an angle α with each other.

Construction steps:

1 In the form diagram, demarcate the space bounded by the shared faces of the force vectors by A. Also, for each face separating A from the rest of the space, we can consider a label as B, C, and D (Figure 2.15b).
2 In the force diagram, start from an arbitrary point A and draw lines in the directions of normal $\mathbf{n}_{AB}$ and $\mathbf{n}_{AC}$ with lengths l_1 and l_2, respectively (Figure 2.15c).
3 The endpoints of lines l_1 and l_2 locate points B and C. Closing this triangle by connecting B to C completes the geometry of the face. The normal of the edge $\mathbf{n}_{BC}$ gives a new face with the same normal attached to the applied load $\mathbf{f}_1$ (Figure 2.15c).
4 The direction of the edge vectors of the face in the force diagram is found using the right-hand rule as discussed in Chapter 1.
5 We repeat the following steps by drawing a line l_3 in the direction of $\mathbf{n}_{AD}$. The endpoint of this line locates point D.
6 In the next step, we choose the magnitude of $\mathbf{f}_2$ by connecting B to D (Figure 2.15d).
7 Last, face CDA can also be completed by connecting DC in the force diagram and completing the construction of the reciprocal faces (Figure 2.15e). However, in that case, the geometry of ACD is predetermined, and its magnitude depends on the area of the adjacent faces.
8 From point C, draw a line with length l_4 and angle β from edge AC (Figure 2.16a–c).
9 The face $ACED$ controls the area of the force $\mathbf{f}_3$ independent of the neighboring triangles.

□

Note that in Figure 2.16c, the β angle is zero, and the edges CE and AC are co-linear. Nevertheless, all the adjacent faces share an edge, and a designer can control the areas by changing the parameters l_{1-4} and β in the force diagram.

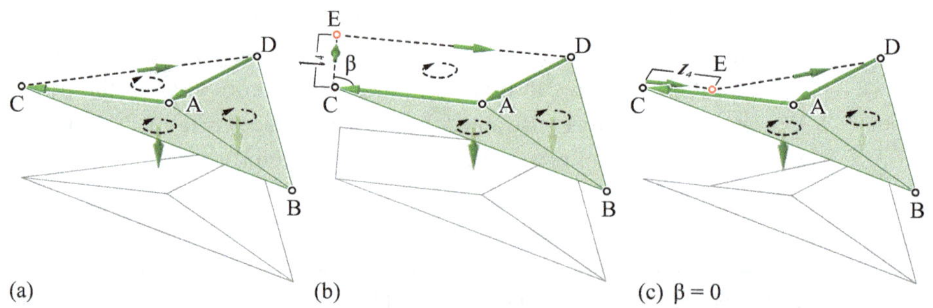

Figure 2.16 (a) Two adjacent triangles control the area of the face ACD; (b) an additional edge CE with length l_4 with an angle β has been added to the face $ACED$; (c) β zero is chosen for this case resulting in a nonconvex face.

2.4 Triangulating Open Force Polyhedrons for General Cases

Regardless of the configuration of the forces in 2D and 3D, the process of finding equilibrating forces involves the following steps: (1) closing the force polygon in 2D and the force polyhedron in 3D to ensure that the sum of force is zero, then (2) closing the funicular polygon in 2D and the funicular polyhedron in 3D to show the sum of the moments is zero. Figure 2.17a–h delineates the construction of a closed force polyhedron and a funicular polyhedron for various configurations, including concurrent and nonconcurrent systems in 3D.

The edges of the force faces in the force polyhedron shown by green color in the force diagram of Figure 2.17 are used to construct faces reciprocal to the resultant. These edge vectors are shown by $\mathbf{e}_{\bar{\mathbf{r}}}$ and make a closed polygon that is called the resultant polygon. The resultant polygon may not be planar (Figure 2.17b). Since all faces of the force polyhedron should be closed, we can use triangulation to close all planar or nonplanar resultant polygons. If there are multiple triangles to close a resultant polygon, then there will be multiple resultant components in the form diagram to keep equilibrium in 3D.

Figure 2.17b illustrates a concurrent system of forces and its open polyhedron of forces with four free edges. The number of triangles needed to close this polyhedron is two. Visiting other examples, including systems of nonconcurrent forces of Figure 2.17c–f, corroborates that the number of triangular faces to close the force polyhedron equals the number of free edges minus two, that is, $\mathbf{e}_{\bar{\mathbf{r}}} - 2$ (Figure 2.17c–f). The number of triangles closing the open polyhedron of force will determine the number of resultant components, keeping the system in equilibrium. These resultants in the form are located by constructing a closed funicular polyhedron, as shown in Figure 2.17c–f.

2.5 Designing for Equilibrium

The artist, Alexander Calder (1898–1976), used equilibrium principles to invent various hanging sculptures.[95] This section decodes his design strategy using graphic statics and the construction of closed funicular polygons and extends the method to 3D by constructing closed funicular polyhedrons.

Consider the sculpture of Figure 2.18 consisting of eight forces $\mathbf{f}_{0-7}$ and its formal configuration. The first step in designing such sculptures is to define the relationship between the parts in the system that stay invariant during and at the end of the design process. This relationship can be referred to as the *topology of the system* in which a simplified graph shows the connectivity between the weights and the hanging location.

2.5.1 Topology of a System of Forces

The topology of the sculpture can be defined in the following steps: (1) identifying the number of forces corresponding to the hanging weights in the system; for instance, in Figure 2.19a, there are eight weights, to begin with; (2) choosing the magnitudes of the

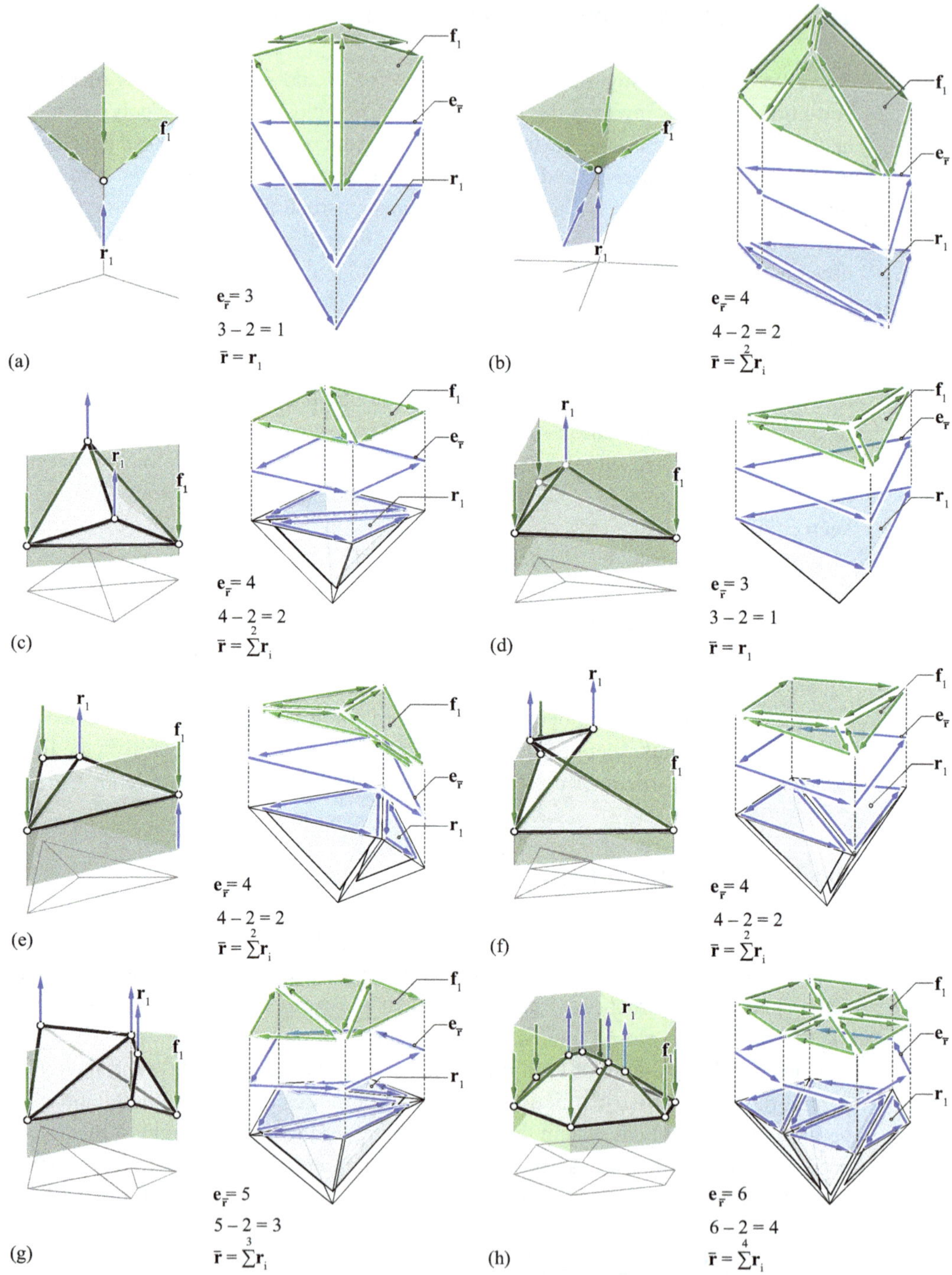

Figure 2.17 The funicular polyhedron and a closed polyhedron of force for (a) a concurrent system of forces; (b) a concurrent system with two resultants; (c) a system of two parallel forces; (d) to (f) a system of three parallel forces including nontriangular force faces; and (g) and (h) three and six parallel forces and their corresponding force diagram.

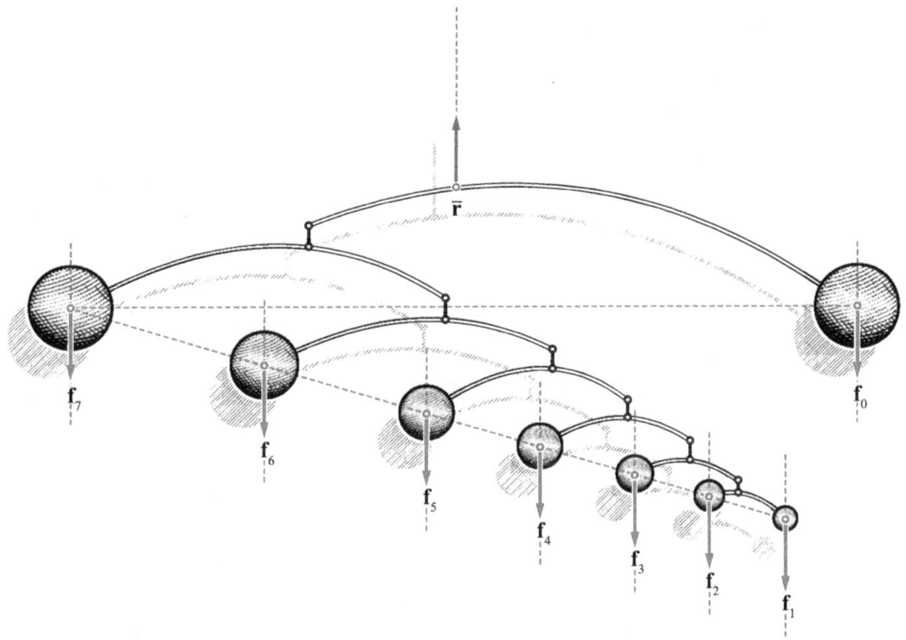

Figure 2.18 An example of a hanging sculpture designed by Calder and the equilibrating force $\bar{\mathbf{r}}$ keeping all weights at their exact height.

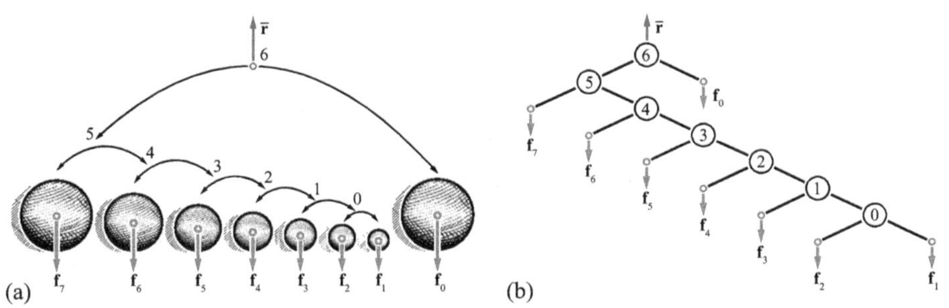

Figure 2.19 (a) The relationship between the weights in the sculpture; and (b) a topological graph representing the connectivity relationship between the forces in the system.

forces or weights; (3) deciding on the connectivity of the weights in the final sculpture; for instance, the weights of Figure 2.19a are divided into pairs. For instance, $\mathbf{f}_1$ and $\mathbf{f}_2$ are paired, and their resultant is paired with $\mathbf{f}_3$, and the resultant of this combination is paired with $\mathbf{f}_4$ and so on. The *topological* graph of Figure 2.19b describes the entire sculpture using simple edges and vertices.

Note that this graph contains no geometric information about the structure. Despite knowing how forces are related, the geometric location of the vertices of this graph should be found. Otherwise, simply connecting the weights and hanging the system from any arbitrary location in the system will not result in the design of Figure 2.18. The location where each pair is connected to the next should be found so the system does not experience any *rotation*.

2.5.2 Segmented Funicular Polygons

The next step in design is to decide on the location of the weights in the system. These locations are key because the rest of the design follows the initial location of the weights or the applied loads.

Each vertex in the graph represents the location of the resultant for its below forces. Thus, a force diagram can be constructed for the graph, and the equilibrium of each segment can be investigated separately, as shown in Figure 2.20a,b. The force diagram can be broken down into multiple equilibrated parts linked by a resultant force $\mathbf{r}_j$. The magnitude of the resultant for each segment can be found by constructing a closing force polygon.

The location of the resultant force $\mathbf{r}_j$ for each segment is found by funicular construction. The force diagram of the entire sculpture is extruded to an arbitrary point O, and the location of the resultant for each segment of the force can be found by constructing the corresponding funicular polygon on the line of action of the forces (Figure 2.21a,b). The number of nodes in the graph of sculpture is equal to the number of required funicular constructions. The location of each node is then found by the closed funicular polygon. Each graph segment is linked to another segment on the line of action of its connecting resultant. As long as the position of the weights and the resultant forces remain on their lines of action, the sculpture will maintain its equilibrium and not rotate.

2.5.3 Various Design Translation

Figure 2.22a–c shows two designs with the same connectivity of the weights and the exact location of the action lines of the loads. A weight can be located anywhere on the line of action passing through its center of mass. For instance, the mass ES

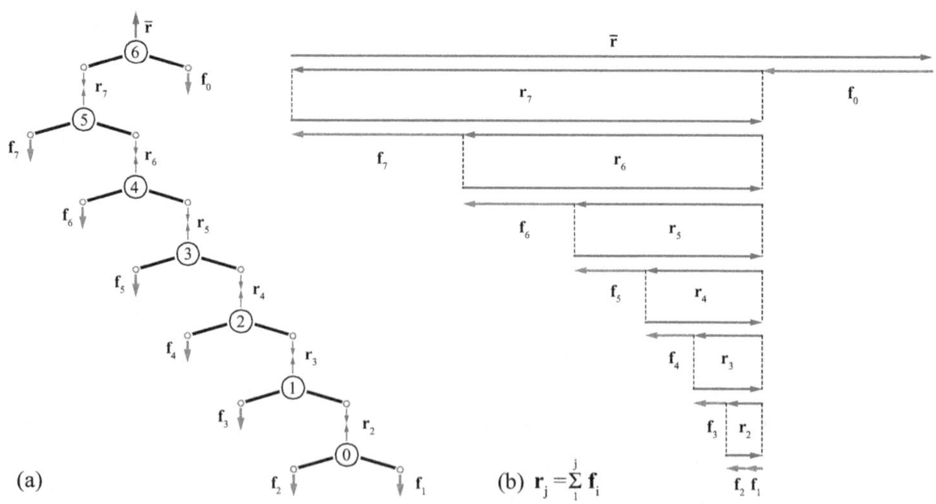

Figure 2.20 (a) The segmented graph and the equilibrium of each node and its connected members by drawing its force diagram and (b) the assembly of the force diagrams pertaining to each segment in the entire system.

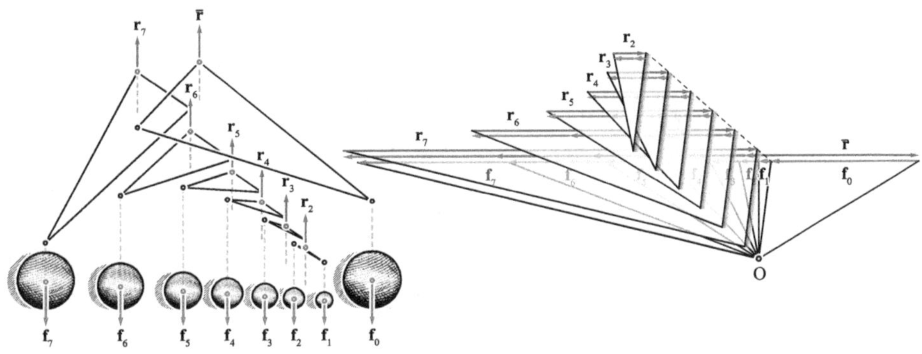

Figure 2.21 (a) The funicular construction for each segment of the graph and (b) overlaying them on the force diagram of the entire system to find the location of the

with the forces f_1 and f_2 in the two designs are at different heights, but they are only moving along the line of action of the forces. As shown in Figure 2.22a,b, since there is no rotation in the system, any location for the weights on the line of action of the forces can be chosen. This holds for the geometry of the connecting rods, linking the weights with the assumption that the self-weight of the bars is negligible and there is no bending in the rods. Thus, an infinite family of equilibrated forms can be designed for the same topological graph and the same location of the applied loads, as shown in Figure 2.22c,d. As is typical of mobile sculptures, the system can have out-of-plane rotations around the linkage points since there are no constraints in the third dimension. Figure 2.22e,f illustrates a completely different design for the same forces, while the location of the loads and their connectivity have been changed.

2.5.4 Subdividing and Segmenting Force Polyhedron

In the example of the 2D equilibrium sculpture, the important role of topology in design was discussed. This was a bottom-up approach where a designer starts from the last branches or leaves and moves toward the root in the connectivity graph. Another method can be a top-down approach, where the forces in a system are subdivided recursively to generate more force components. Consider the force diagram of Figure 2.23. Let us start from the force diagram that includes only the applied loads $\mathbf{f}_1$, $\mathbf{f}_2$, and $\mathbf{f}_3$. Take face $\mathbf{f}_3$ and isolate it with its connecting faces on the bottom of the force diagram. Subdividing the face $\mathbf{f}_3$ into three smaller faces does not change the global equilibrium of the system. But it would add three more forces to the system whose resultant equates to $\mathbf{f}_3$. We can repeat this step three more times to generate face $\mathbf{f}_6$.

Repeating the subdividing process for the new faces can add more force components to the system without disturbing its global equilibrium, as Figure 2.23a,b illustrates. In each design step, one of the new faces is considered the resultant for the next subdivision step. A new funicular polyhedron is constructed for each subdivision step to show that each segment is in equilibrium. The accumulation of all these funicular polyhedrons makes a larger closed funicular polyhedron on the top, showing

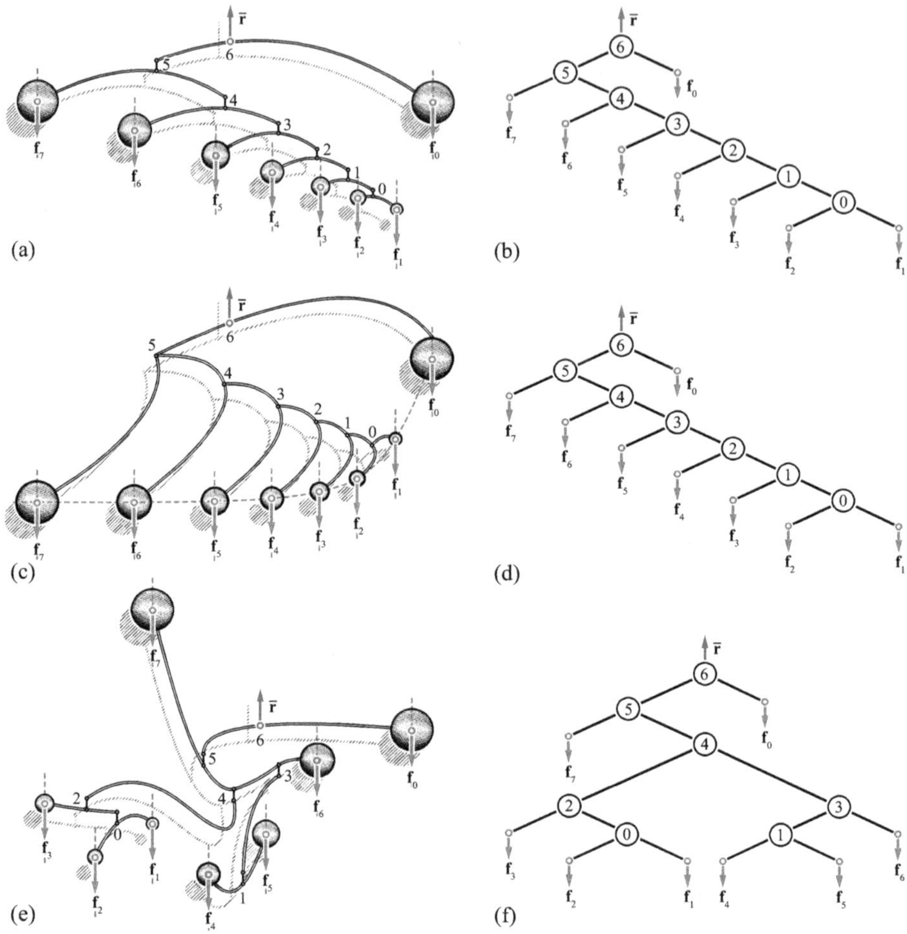

Figure 2.22 (a) The original sculpture by Calder; (b) its topological diagram; (c) and (d) another geometric interpretation of the same topology and the location of the loads; (e) a new design using the same force magnitudes placed differently in space; and (f) its associated topological graph.

that the entire force system is in moment equilibrium as long as the actual applied loads and resultants are located at their respective lines of action.

Locating the Subdivided Forces

In the earlier explanation, we suggested the subdivision of a face in the force diagram, and considering it as a resultant gives multiple force components in equilibrium. The question is how to find the location of these new force components and how we know that these forces are in equilibrium.

Take the example of a force in Figure 2.24a and its force diagram in Figure 2.24b. The resultant face is perpendicular to the line of action of the resultant in the form diagram. This resultant face can be subdivided into three applied loads, as shown in green in the figure.

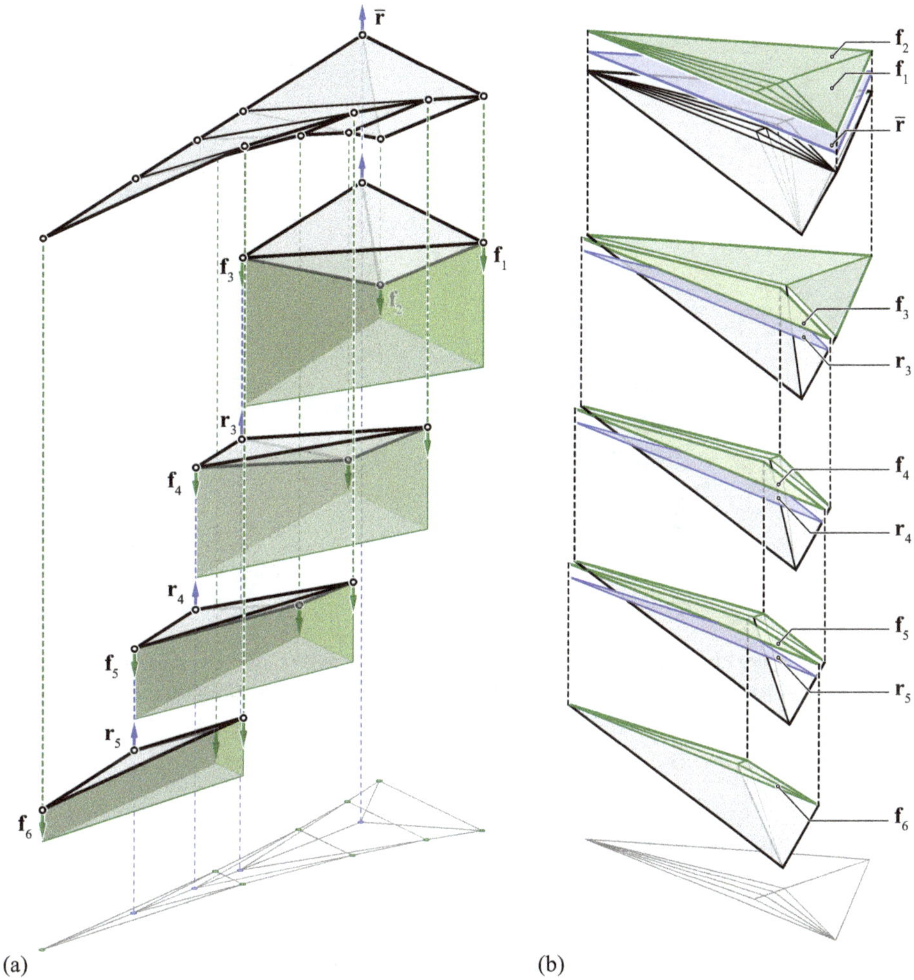

Figure 2.23 (a) Segmenting the funicular form into parts and considering a load as a resultant force and subdividing the reciprocal force face to get the force diagram in recursive steps (b).

Take the face with normal $\mathbf{n}_{AB}$ in the form diagram. This face is reciprocal to the edge *AB* of the face with normal $\mathbf{n}_1^\dagger$ and the edge *BA* of the face with normal $\mathbf{n}_r^\dagger$ in the force diagram. We can locate force f_1 in the form by picking a point from a distance l of $\mathbf{r}$. The location of the rest of the forces follows the geometric relationship between the form and force diagram as shown in Figure 2.24c. In fact, the entire system can be scaled uniformly up and down from the line of action of r in the form diagram by changing different lengths of l without changing the equilibrium of forces (Figure 2.24d). As shown, infinite sets of applied loads with the same magnitude but applied at different distances from the resultant are in equilibrium using the same resultant. This technique can be used in the design of the sculpture of Figure 2.25 by subdividing the faces of the force diagram step-by-step and locating the newly created forces in the form diagram.

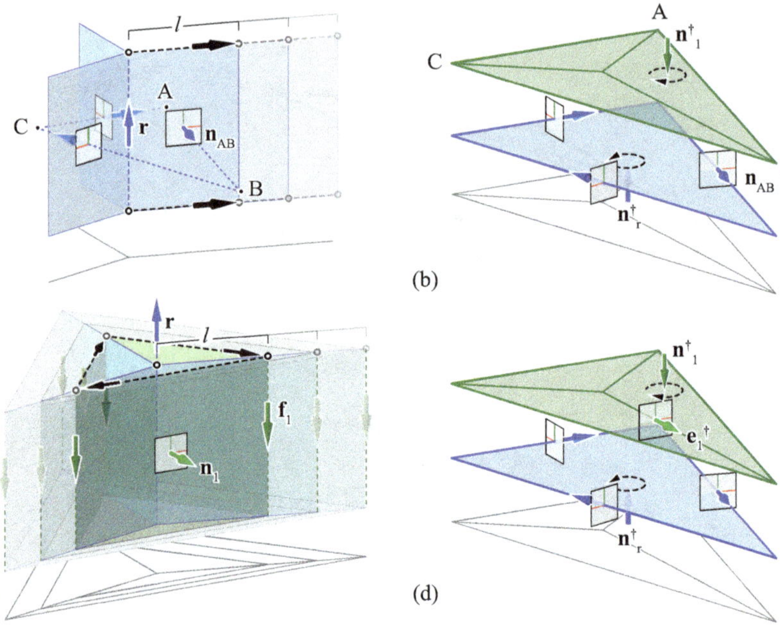

Figure 2.24 (a) The resultant vector and its attached planes; (b) the reciprocal force diagram; (c) the resultant planes are uniformly scaled to define the location of the applied loads while the force diagram stays intact (d).

2.6 Chapter Summary

This chapter introduced the geometric steps of construction to find the equilibrium for concurrent and nonconcurrent systems of forces in both 2D and 3D. It showed that constructing the force polygon in 2D and the closed force polygon in 3D as a force diagram ensures that the sum of the forces in the system is zero. In addition, we learned that constructing a closed funicular polygon in 2D and a closed funicular polyhedron in 3D in the form diagram guarantees moment equilibrium in the system. We need both constructions to have a complete equilibrium. Various configurations of forces in 3D were introduced, and the process of finding the resultant by constructing the closed force polyhedron and the funicular polyhedron was discussed. Ultimately, we showed how it is possible to design various hanging sculptures similar to those of Alexander Calder using geometric construction of forces and finding resultants in 2D. The design of such sculptures in 3D was pursued by subdividing the force diagram and locating equilibrating forces in the form diagram. With this introduction to equilibrium and funicular construction, we are ready to move on to Chapter 3, where the funicular construction for specific boundary conditions is discussed, and the principles of compression-only or tension-only structural form finding for architectural structures will be explained.

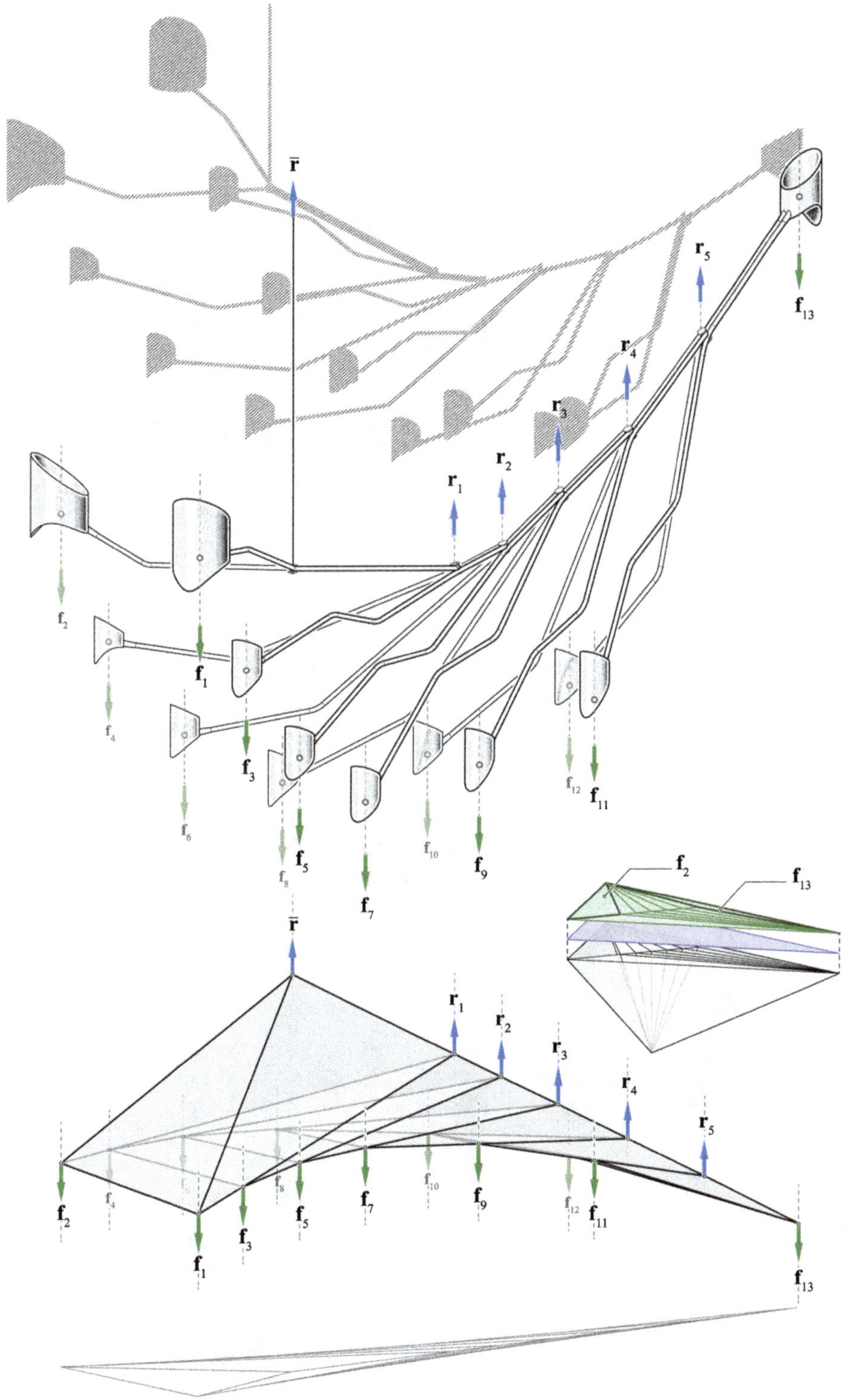

Figure 2.25 A conceptual design of a chandelier inspired by the works of Calder was designed using polyhedral graphic statics.

3 Compression-Only Form Finding

Masoud Akbarzadeh and Marton Hablicsek

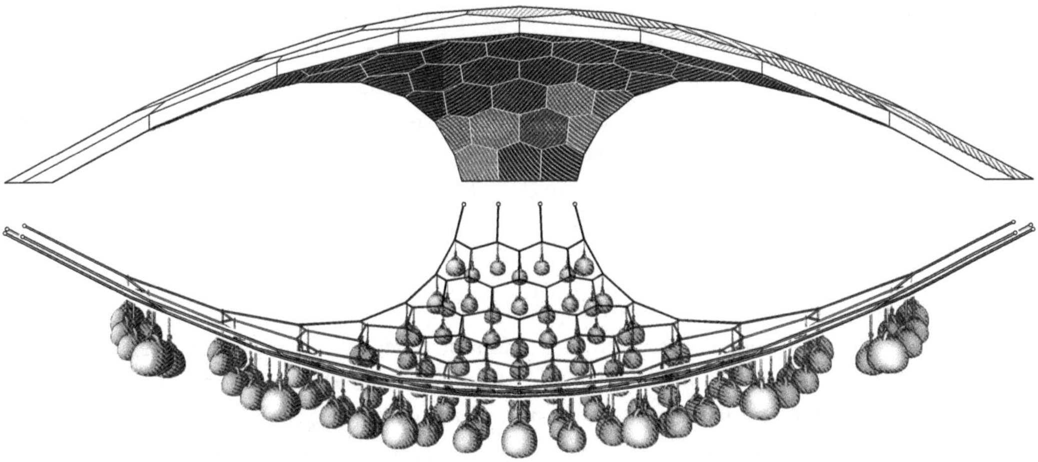

Figure 3.1 A compression-only shell and its reflected geometry as a tension-only network with hanging weights.

3.1 Funicular Systems: When Geometry Matches the Force Flow

As mentioned in Introduction and Chapter 1, funicular structural forms have a unique geometry that can carry external loads by only axial forces in their members. Figure 3.1 illustrates two seemingly different systems: one with compression-only and one with tension-only internal forces. In such systems, all internal members are either in pure tension or in pure compression. Moreover, the internal forces are perpendicular to the cross-section of the members. Thus, the form of the structure matches the geometry of the force flow, transferring the applied loads to the supports. Such structural forms are considered among the most efficient structural systems. Finding funicular geometries for a given set of loads and support locations has always interested researchers, architects, structural designers, and master builders.

Chapter Objectives

This chapter provides methods based on two-dimensional (2D) and polyhedral graphic statics to construct funicular geometries for a given set of loads and boundary conditions. The main objective of this chapter is to provide step-by-step procedures to find functional solutions in 2D and three-dimensional (3D). We show how, in 3D, we can extend funicular form finding to more than three support locations and what are the implications of choosing support locations with various heights. We will also visit some techniques to translate simple equilibrium solutions to more intricate designs by designing the load network. Ultimately, we visit a famous bridge design by Robert Maillart and explain funicular form finding with height constraints.

3.1.1 A Tension-Only Structural Form

In the context of structural design, a tension-only system is a system whose all members are in pure tension. Many structural systems around us carry tensile forces in their members. For example, a spider web is a natural structure where all the net members are in tension. In such systems, it is quite easy to observe that the geometry of the force flow is identical to the structure's geometry because we feel the tensile force in the ropes of the networks (Figure 3.2). So, if there is a sag in a rope, then it is not functioning well.

The Efficiency of Tension-Only Systems

In structural mechanics, there are different types of loadings that might occur in a member of a structure: axial force, shear force, bending moment, and torsion. The force carried by the member per unit area is called stress. Thus, four types of stress might occur in a member: axial stress caused by tension or compressive force, transverse shear stress caused by a shear force, bending stress caused by a bending moment, and torsional stress caused by torsion in a member. In a structural system, depending on the type of force it carries, a member experiences a single or a combination of all these stresses and thus should be sized to withstand these forces accordingly.

In a tension-only structure, the only force on the members is tensile, perpendicular to each member's cross-section. Consequently, the only type of stress is axial, and it

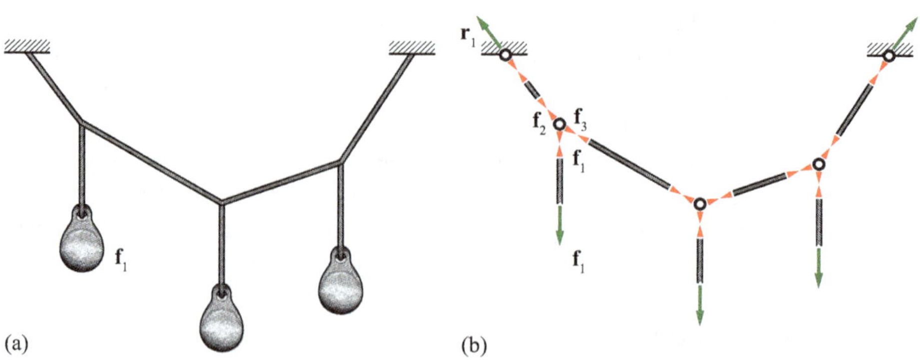

Figure 3.2 (a) A tension-only network of ropes keeping the hung masses in equilibrium; (b) the tensile forces around each node of the network and on both sides of the ropes.

is calculated by the internal force divided by the cross-section of the member. As a result, sizing the structure members with axial-only forces is much easier than sizing members with a combination of different types of stresses. In addition, in axial stress caused by tensile or compressive force, the entire cross-section of a member contributes uniformly to carrying out the load. Therefore, a smaller cross-section area is required to transfer tension or compression force than a cross-section needed for transferring shear, bending moment, or torsion. Thus, building a system with axial-only stress in its members might require less material. Such systems would be more efficient than those involving bending, shear, and torsional stress and transfer applied loads with the minimum use of materials.

3.1.2 A Compression-Only Structural Form

Compression-only systems are also very similar to tension-only systems. The *type* of internal forces in all members is the same, with the main difference: all members are in pure compression. Finding an example of a compression-only form is not as trivial as a tension-only one because the flow of a compressive force in a member is not as visible as tension in a cable. However, there is an exciting relationship between the form of a tension-only cable net and a compression-only system as proposed by Robert Hooke in 1675.[24,97] According to Hooke, if one hangs a chain or an elastic rope from two ends, the chain forms a tension-only geometry where all the chains are in tension under the self-weight of the system. If one inverts this geometry and rigidifies the chains, the form will be in pure compression under its self-weight. Hooke calls this geometry the True Mathematical and Mechanical Form of arches for buildings. This concept has been illustrated in Figure 3.3a,b. The method has been adopted by many architects and builders, such as Gaudi, to design compression-only structures.[29] There is another method to find a compression-only form; if all external forces, including the applied loads of a tension-only system, are inverted, the exact geometry becomes a compression-only form. For instance, for the tensile network in Figure 3.4a, if the direction of the applied loads flips, the form will become a compression-only form in Figure 3.4b. Section 3.1.3 will expand on the components of tension-only and compression-only systems to make this approach more tangible.

The Components of a Tension-Only or Compression-Only System

A tension/compression-only system consists of the following components: (1) external forces, including the applied loads and reaction forces at supports; (2) the members carrying either tensile or compressive forces; and (3) the nodes where members or external forces meet. The external forces are transferred to the nodes via the members or directly applied to the nodes in a system. Thus, in a compression/tension-only system, a node is a location where multiple forces meet and are in equilibrium. Flipping the direction of all external forces in such systems converts the direction of all internal forces from compression to tension or vice versa without changing the structure's geometry or the network.

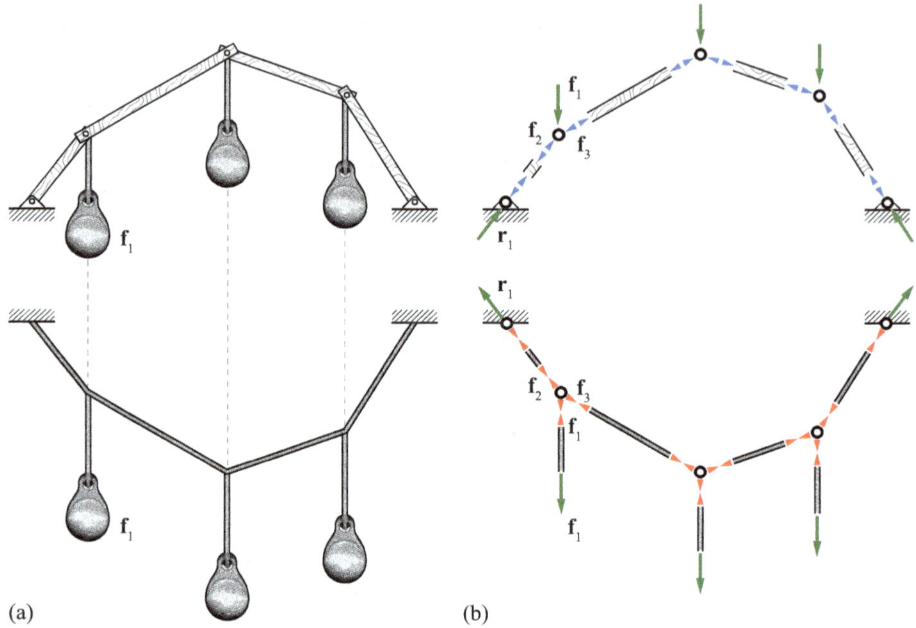

Figure 3.3 (a) A compression-only arch and its reflected geometry as a tension-only network with hanging weights; (b) the internal forces in the nodes and members of a compression-only and a tension-only network.

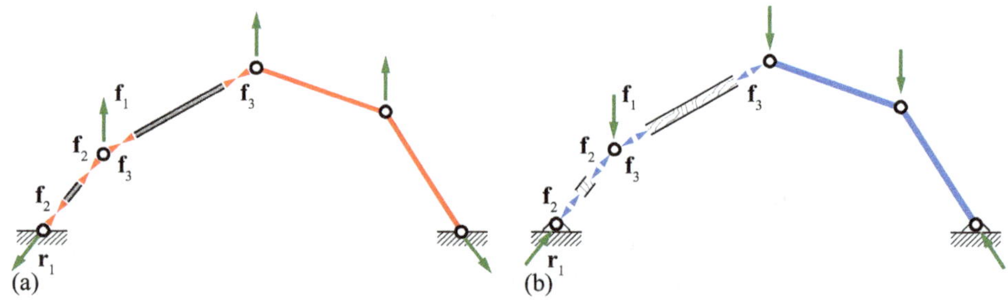

Figure 3.4 Flipping the direction of the external forces of (a) a tension-only network gives (b) a compression-only form with compressive forces in the members and nodes.

3.1.3 Equilibrium of a Compression/Tension-Only Network

In a compression/tension-only system, all internal and external forces that meet at a node should be in equilibrium. This equilibrium can be tested by making a closed, force polygon around a node in 2D and a closed, force polyhedron around a node in a funicular polyhedral form, as explained in Chapter 1. For example, consider the 2D tension-only and compression-only forms of Figure 3.5. Three force polygons can be constructed around three nodes of the form on both configurations. The magnitude of the force in both ends of a member connecting two adjacent nodes does not change. Thus, the length of the edges for every two adjacent force polygons is the

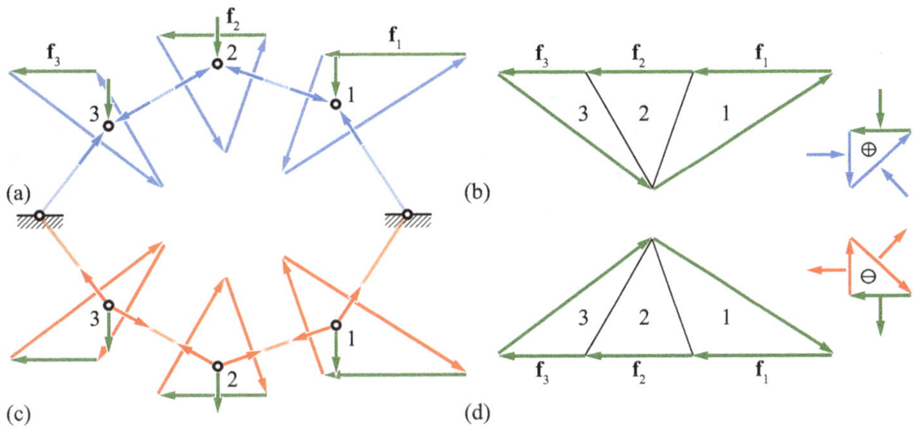

Figure 3.5 (a) Constructing closed force polygons around each node of a compression-only form (b) the aggregation of the force polygons making a complete force diagram with positive directions; (c) constructing closed force diagram around a tension-only form; (d) the reciprocal force diagram with negative direction of the cells.

same. Aggregating these polygons by their adjacent edges results in the force diagram of Figure 3.5b,d. The external edges of the combined polygons make a larger polygon, shown in green, representing the equilibrium of the external forces in the system.

The completed force diagram represents the equilibrium of both internal nodes and external forces: the closed convex polygons represent the equilibrium of forces around each node and are called *nodal* equilibrium. And the exterior faces of the force diagram make a larger cell representing the equilibrium of the external loads and reaction forces and is called a *global* force polygon.

A similar technique can describe the equilibrium of a compression- or tension-only polyhedral form in 3D. Consider the example of Figure 3.6. Around each node of a compression/tension-only funicular polyhedron, a closed polyhedral cell can be constructed such that each face of the cell is normal to a member or an applied load of the node, as shown in Figure 3.6. Furthermore, since the magnitude of the force at both ends of a member connecting two adjacent nodes does not change, the area of the two adjacent faces of the two neighboring cells is also identical. Therefore, all the cells around the nodes of the form can be aggregated to construct a force diagram for the compression/tension-only funicular form (Figure 3.6b,d). Similarly, each cell in this force diagram represents the equilibrium of a node, and the external polyhedron shows the global equilibrium of external loads and reaction forces in the system.

Remark 3.1. The force diagram of a compression- or tension-only form consists of convex-only polygons/polyhedrons. Each nodal force polygon in 2D and force polyhedron in 3D has a consistent normal direction. In this book, we use the following convention: for a compression-only form in both 2D and 3D, all nodal force polygons and polyhedrons have normal directions pointing inside the cell, and for tension-only forms, the normal directions of the nodal cells point toward the outside. In the context of this book, we call the cells with normals pointing *inside positive cells* representing

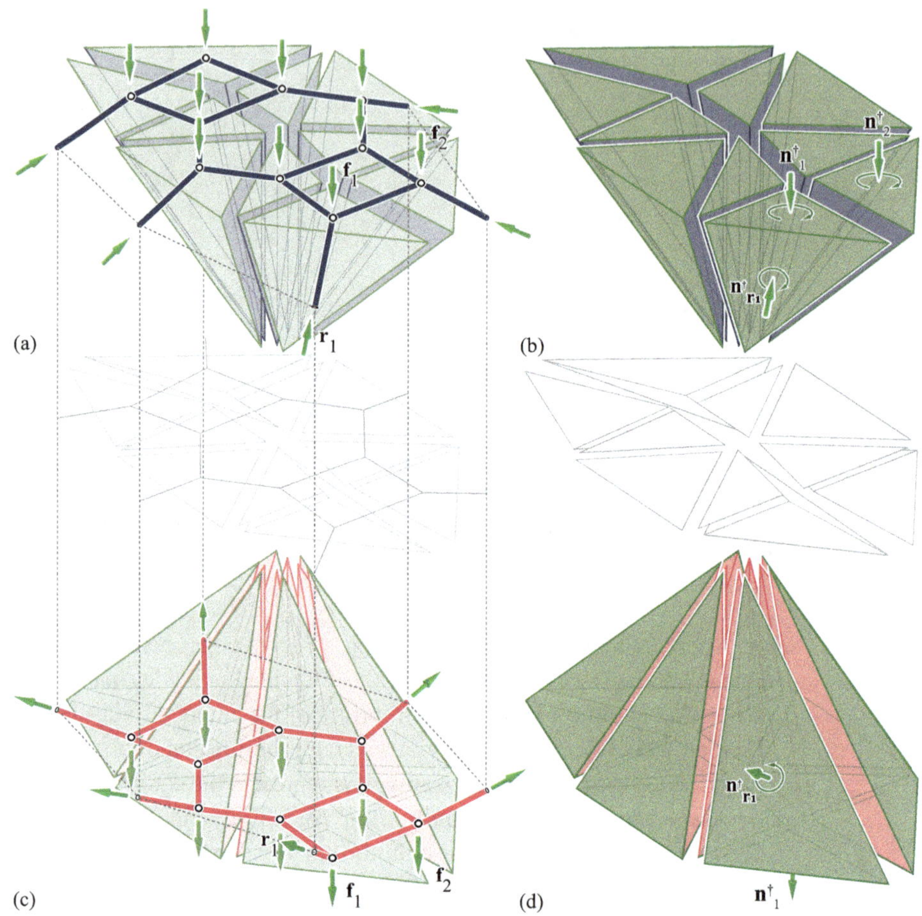

Figure 3.6 (a) Constructing closed force polyhedron around each node of a compression-only form (b) the aggregation of positive polyhedral cells as the force diagram; (c) closed force polyhedrons around the nodes of a tension-only form; and (d) the reciprocal force diagram with negative direction of the cells.

the equilibrium of a compression-only node. Similarly, the cells with *outside* normal are called *negative cells*, representing the equilibrium of a tension-only node.

3.1.4 Constrained Funicular Forms

In the examples of Section 3.1.1 to 3.1.3, the geometry of the funicular network was given and the equilibrium of the nodes was tested by making closed polygons in 2D and polyhedrons in 3D. In Chapter 2, the funicular construction was used to find the location of the resultant for a system of forces and to confirm the moment equilibrium in the system. It was also shown that infinite, *closed* funicular solutions could be constructed on the line of action system of forces in equilibrium. Nevertheless, the construction of funicular solutions for given applied loads and constrained to predefined boundary conditions was not addressed.

Often, a funicular solution does not exist in advance in the design process and must be found for a given system of applied loads and support locations. For instance, a designer may need to find a compression-only form of a bridge that spans two support locations with different elevations. This process is often referred to as *structural form-finding* in the literature and is the main objective of this chapter. Before we provide steps to find a constrained funicular geometry for given boundary conditions, let us visit examples of simple funiculars constrained to support locations and learn more about their properties.

3.2 A Simple Constrained Funicular in 2D

In the context of this book, constrained funicular geometry is a structural form confined to the lines of action of a set of applied loads and the exact location of supports. One can find infinite funicular solutions with various heights that satisfy the earlier criteria for a given boundary condition. Figure 3.7 can explain this concept further.

3.2.1 The Height of the Funicular

In Figure 3.7a, let us choose two points, 1 and 2, with the same height on a 2D plane. Force **f** is applied at point 3 in the space between 1 and 2 with the distances a and b from them, respectively. A simple tension-only structure can be made with three pieces of rope connected to points 1 and 2, and a mass hung from the strings at point 3 on the line of action of **f**. We can annotate the height of point 3 by h, which is the vertical distance of point 3 measured from a dashed line connecting points 1 and 2. In

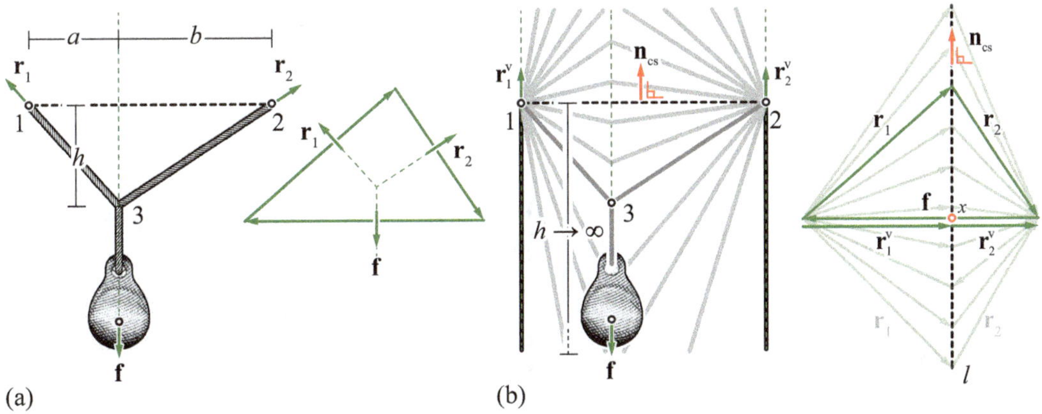

Figure 3.7 (a) The simplest funicular solution in the form of two ropes pinned at points 1 and 2 keeps the load **f** of the mass in equilibrium; and (b) multiple funicular solutions with various heights and their corresponding force diagrams.

this structure, node 3 is in equilibrium, and the system is a tension-only structure. To find the magnitude of the internal forces in the ropes, we can construct a closed force triangle around node 3, as shown in Figure 3.7a.

Changing the height h of point 3 gives another tension-only geometry with different slopes of the ropes that are connected to points 1 and 2 and still keeps $\mathbf{f}$ in equilibrium. Figure 3.7b illustrates multiple such geometries superimposed on the same drawing. A force triangle can be constructed for each funicular solution, as shown in Figure 3.7b.

Remark 3.2. The height of the force triangles changes as h changes in the form diagram: a taller h results in a shallower force triangle and vice versa. The following special occasions have particular importance: (1) if the height of h in the form diagram becomes zero, the force triangle cannot get closed, and thus the system would not be in equilibrium; and (2) if the height goes to infinity, the funicular form becomes very tall, and the force triangle collapses into a line.

3.2.2 The Change Rate of the Force Polygon

The change rate of the force triangles for the change of h in the form diagram can be shown by a line l in the force diagram passing through the apex of all force triangles (see Figure 3.7b). We may need another example to reveal the relationship between the line l and the form diagram.

Let us change the height of point 2, as shown in Figure 3.8a, without changing its horizontal distance b from the applied load $\mathbf{f}$. Similarly, height h is the vertical length of point 3 from the angled dashed line connecting the two supports. Changing the height h gives different funicular solutions. The superimposing of multiple constrained funiculars and their corresponding force triangles is illustrated in

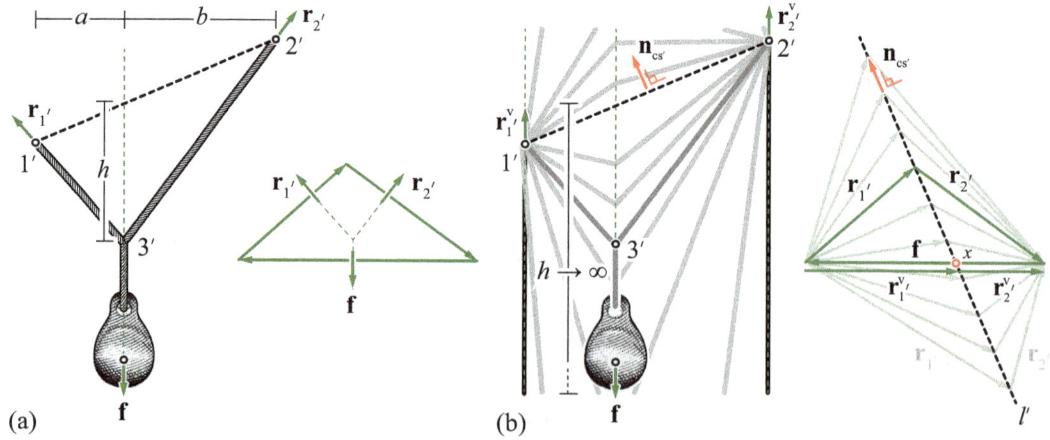

Figure 3.8 (a) A hanging mass and a funicular solution constructed from points with different elevations; and (b) all possible funicular solutions from the supports and the corresponding force diagram.

Figure 3.8b. Similarly, the rate of change of the geometry of the force triangles of the new configurations is shown by drawing a line l' in the force diagram.

3.2.3 The Closing String

Comparing the two examples, it is evident that the lines l and l' are perpendicular to the dashed line connecting the supports in both examples. The dashed line is referred to as the *closing string* since it closes the polygon of the funicular form.* This line is shown by normal $\mathbf{n}_{cs}$ in the form diagram, which is the unit vector of the line l in the force diagram.

Remark 3.3. A 2D funicular solution constrained to a given boundary condition is reciprocal to a force diagram that is also constrained to a line l in the force construction. Line l in the force diagram is perpendicular to the *closing string* connecting the support locations in the form diagram.

In both figures, lines l and l' intersect vector $\mathbf{f}$ in the force diagram. Let us take the funicular solutions of Figures 3.7 and 3.8 and superimpose them into a single drawing of Figure 3.9a. Overlaying their corresponding force triangles reveals another exciting property of constrained funiculars: line l from the first diagram intersects vector $\mathbf{f}$ at the same point as the line l' of the second force diagram does. This point is labeled as x in the drawings. One might ask whether this is a coincidence or if this relationship has more meaning.

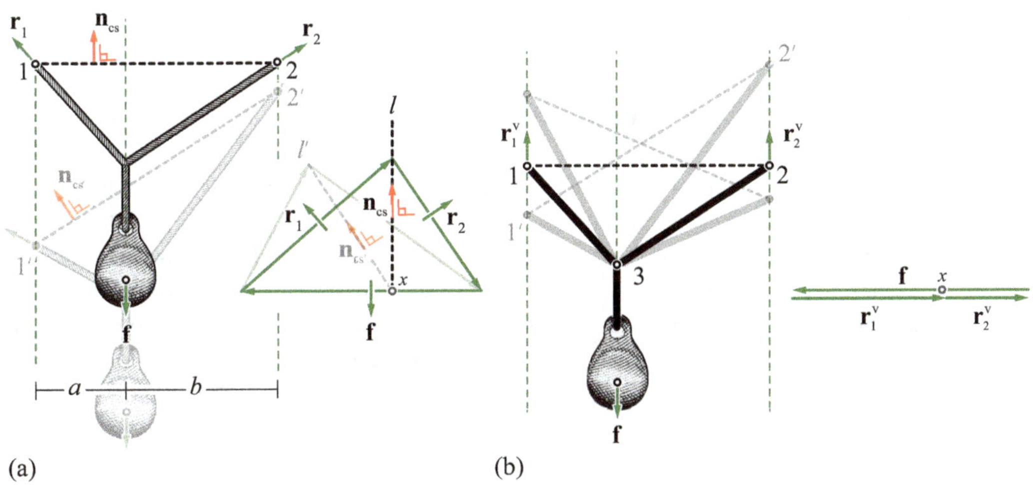

Figure 3.9 (a) Combining the funicular solutions and the force diagrams of Figures 3.8 and 3.9 into one diagram with the matching point x for both cases; and (b) point x determining the magnitude of the force components at the supports parallel to $\mathbf{f}$.

* The term closing string is borrowed from the book *Graphical Analysis A Text Book on Graphic Statics* by William Sidney Wolfe.[21]

3.2.4 Force Components Parallel to Resultant

Let us visit the boundary condition and the funicular form of Figure 3.7a. If the height of point 3 goes to infinity, the funicular geometry becomes very tall. Thus, the internal force in the ropes will become parallel to the applied load. Let us denote these forces by $\mathbf{r}_1^{\mathrm{v}}$ and $\mathbf{r}_2^{\mathrm{v}}$. The force polygon reciprocal to the very tall funicular collapses into a line, and the location of point x gives the lengths and magnitudes of the force vectors $\mathbf{r}_1^{\mathrm{v}}$ and $\mathbf{r}_2^{\mathrm{v}}$. The same procedure provides the lengths and magnitudes of the force vectors $\mathbf{r}_{1'}^{\mathrm{v}}$ and $\mathbf{r}_{2'}^{\mathrm{v}}$ of the example of Figure 3.8. The most important outcome of this experiment is summarized in the following remark.

Remark 3.4. For every boundary condition in 2D, including two supports and given applied loads, there is a force component, shown by $\mathbf{r}_i^{\mathrm{v}}$ at each support, parallel to the resultant of the applied loads. These parallel forces at the supports can equilibrate the applied loads regardless of the geometry of the funicular form (see Figure 3.9a,b). Point x in the drawings defines the length of $\mathbf{r}_i^{\mathrm{v}}$ vectors in the force diagram.

Remark 3.5. Every funicular solution, constrained to the applied loads and the lines of action of $\mathbf{r}_i^{\mathrm{v}}$ at the supports, is reciprocal to a force diagram constrained to a line l perpendicular to the closing string. For such funicular solutions, l intersects the force $\mathbf{f}$ at one and only one point x that gives the length and magnitude of the force vectors $\mathbf{r}_i^{\mathrm{v}}$ at the supports.

The earlier properties of the form and force diagrams are, in fact, a prelude to the concept of the *trial funicular construction* that was discussed for the equilibrium of systems of loads in Chapter 2 and will be used in this chapter to find constrained funicular solutions.

3.3 A Simple Constrained Funicular in 3D

The previous examples in Section 3.2 in 2D provided a foundation for the equivalent example in 3D. The main difference is that a funicular in 3D needs at least three support locations. Consider a boundary condition including three points with the same elevation and a force $\mathbf{f}$ applied at given distances from the supports. A simple funicular can be constructed using four pieces of rope connected to the support, and a mass is hung on the line of action of $\mathbf{f}$. As discussed both in Chapters 1 and 2, a line of action of a force in polyhedral graphic statics is the line of intersection of at least three planes, as shown in Figure 3.10a.

3.3.1 The Height of the Funicular

The point where all the ropes meet is labeled by 0 and is on the line of action of $\mathbf{f}$. The height of point 0 from the plane connecting the three supports can be represented by h. The resulting system of ropes is the simplest funicular form constrained to support locations keeping $\mathbf{f}$ in equilibrium. In this system, node 0 is tension-only, and its equilibrium can be shown by constructing a closed force polyhedron as explained in

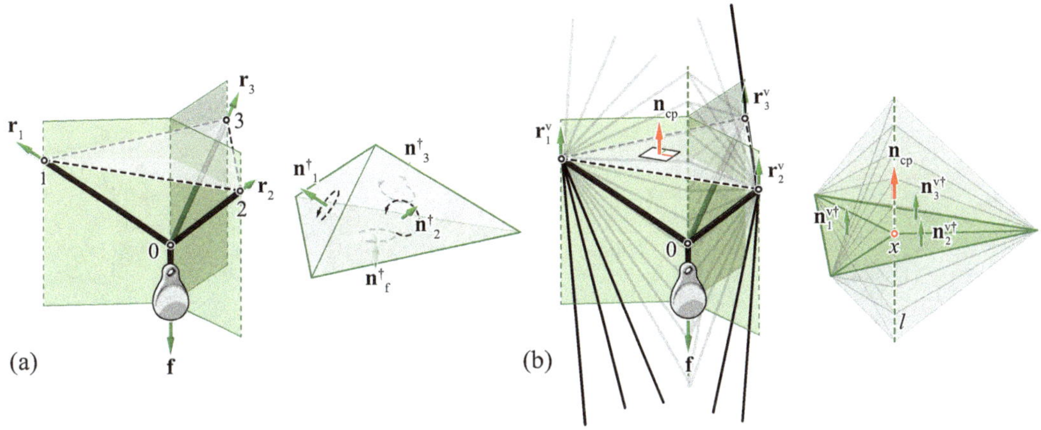

Figure 3.10 (a) A mass hung from three points represented by force **f** and its closed force polyhedron representing the equilibrium of the system; and (b) changing the angles of the ropes on the line of action of **f** changes the geometry of the force polyhedron revealing a line to which the change is constrained.

Chapter 1, as shown in Figure 3.10a. The area of each face of the force polyhedron represents the magnitude of the force in its related rope in the form diagram.

Similar to the 2D case, the change in the height of h in the form diagram can make a funicular solution with different heights constrained to the given support locations, as shown in Figure 3.10b. The reciprocal force polyhedron for all these examples can be overlayed in a single drawing, as shown in Figure 3.10b.

The following extreme scenarios might happen by changing h: if h goes to infinity, the funicular becomes very tall, and the edges become parallel to the applied load. Thus, the force polyhedron collapses to a single plane, and all the faces of the polyhedron will become coplanar. On the other hand, if h becomes zero, then the side faces of the polyhedron will become parallel and cannot close the polyhedron and thus cannot establish equilibrium in the node.

3.3.2 The Change Rate of the Force Polyhedron

Similar to 2D, the rate of change of the force polyhedrons can be shown by a line in the force diagram. This line is labeled by l in the drawing. Line l shows the rate of change of the force polyhedron for various heights and intersects the face of the applied load shown by normal $\mathbf{n}_\mathbf{f}^\dagger$ at the point x. Point x divides the face into three smaller triangles. Each triangle gives the magnitude of a force parallel to **f** at the supports. These forces are labeled by $\mathbf{r}_i^\mathrm{v}$ in the form diagram, and their corresponding faces are shown by normal $\mathbf{n}_i^{\mathrm{v}\dagger}$ in the force diagram.

To better understand the relationship between the form and force diagrams in a constrained configuration, let us change the height of support point 2. Figure 3.11a shows another configuration of a tension-only system where the height of point 2 has been altered for the given applied load **f**. Choosing different h results in funicular

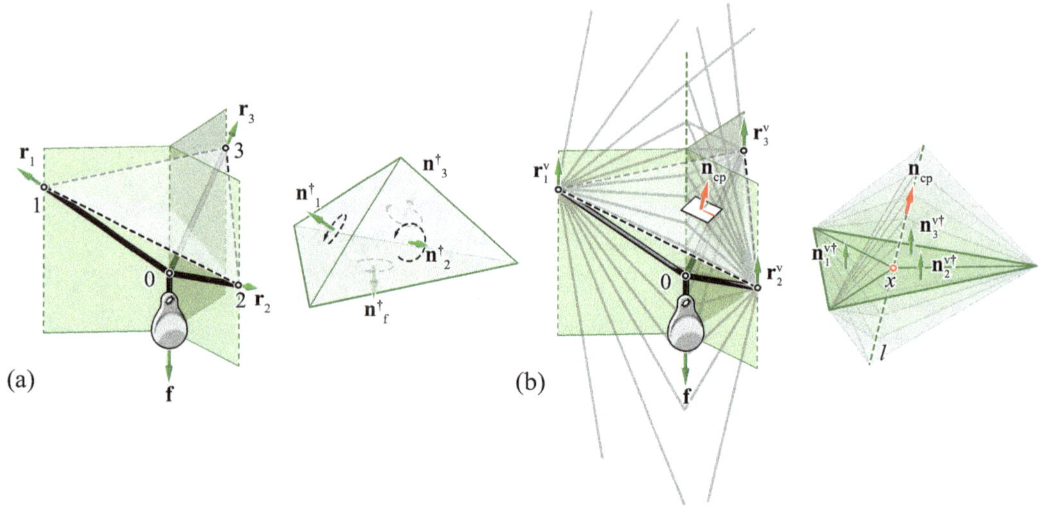

Figure 3.11 (a) Choosing another location for the supports on the planes passing through the line of action of the force **f** and the closed force diagram; and (b) changing the height of the funicular form and superimposing the force polyhedrons for each funicular solution.

solutions with different heights and their corresponding force polyhedron illustrated in Figure 3.11b. Similar to the previous example in Figure 3.10, line l' shows the rate of change of the force polyhedron for various heights and intersects the applied load face at the same point x.

3.3.3 The Closing Plane

By comparing Figures 3.10b and 3.11b, it is evident that the lines l and l' are normal to the plane connecting the support locations. For a constrained funicular polyhedron with three support locations, the force polyhedron is also constrained to a line l that is normal to a plane connecting three supports. This line is called the *closing plane* since it closes the bottom of the funicular polyhedron by connecting the supports. This plane is shown by the normal vector $\mathbf{n}_{cs}$ in the form diagram, and its normal vector is used as the unit vector for the line l in the force diagram.

3.3.4 Force Components Parallel to the Resultant

Let us overlay the funicular solutions of Figures 3.10a and 3.11 in the drawing of Figure 3.12. In the overlayed drawing of Figure 3.12a, we have two funicular solutions with different heights and closing planes constrained to the line of action of the applied load **f** and components of force $\mathbf{r}_i^{v\dagger}$ parallel to the applied load at the supports. The overlayed illustration shows that for every funicular form constrained to the same action lines of loads and reactions, there is a force diagram that is constrained to a line l_i. All l_is intersect at one and only one point in the force diagrams that is shown by x in the drawings (Figure 3.12a). The location of point x gives the faces reciprocal to the

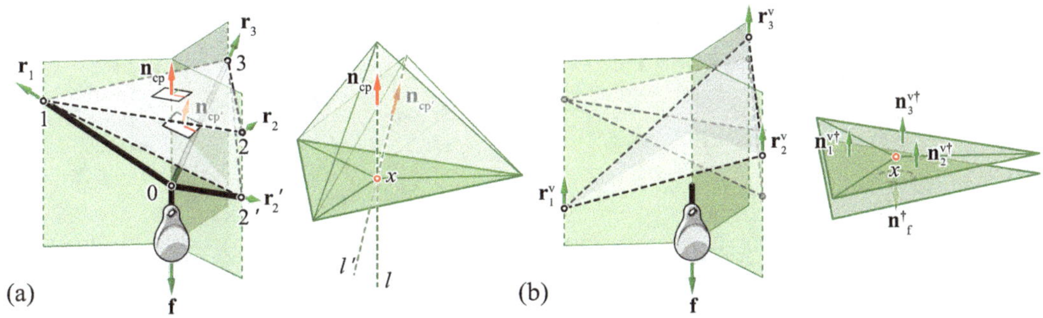

Figure 3.12 (a) Superimposing the boundary conditions from Figures 3.10 and 3.11 and their force polyhedron; and (b) the magnitude of the force components $\mathbf{r}^{v}_{1-3}$ parallel to the applied load **f** at the supports does not change.

magnitude of forces $\mathbf{r}^{v}_{i}$ at the supports that are parallel to the resultant of applied loads in the form diagram. This property is illustrated in Figure 3.12b and brings us to the following remark.

Remark 3.6. For the same applied loads and location of supports, the location of x in the force diagram never changes because the magnitude of the reaction forces parallel to the resultant stays unchanged.

3.4 Constrained Form Finding for a Single Applied Load in 2D

In the context of this book, the term *form-finding* is used when the funicular form of a structure and its force diagram should be found for a given boundary condition. In such cases, the boundary condition often includes the magnitude of the applied loads and the location of supports. Respectively, for this boundary condition, the funicular form of the structure, its force diagram, and the magnitude of reaction forces at the supports are all unknown, and a robust methodology is needed to find them.

Sections 3.2 and 3.2 emphasized the relationship between the components of the form and force diagrams constrained to a given boundary condition. Based on that knowledge, this section will provide a procedural method for finding a funicular form and its force diagram for a given boundary condition of a simple case in 2D. This procedure can be generalized and used for form-finding in 3D and nonconcurrent loads in 2D and 3D.

Example 3.1 Figure 3.13a shows a load **f**, 2 kN, applied at the space between two supports 1 and 2 with a distance a and b from each, respectively. (1) Find a compression-only funicular solution that is constrained to support 1 and 2 and keeps **f** in equilibrium; (2) show what parameter in the force diagram should be adjusted to change the height of the funicular form.

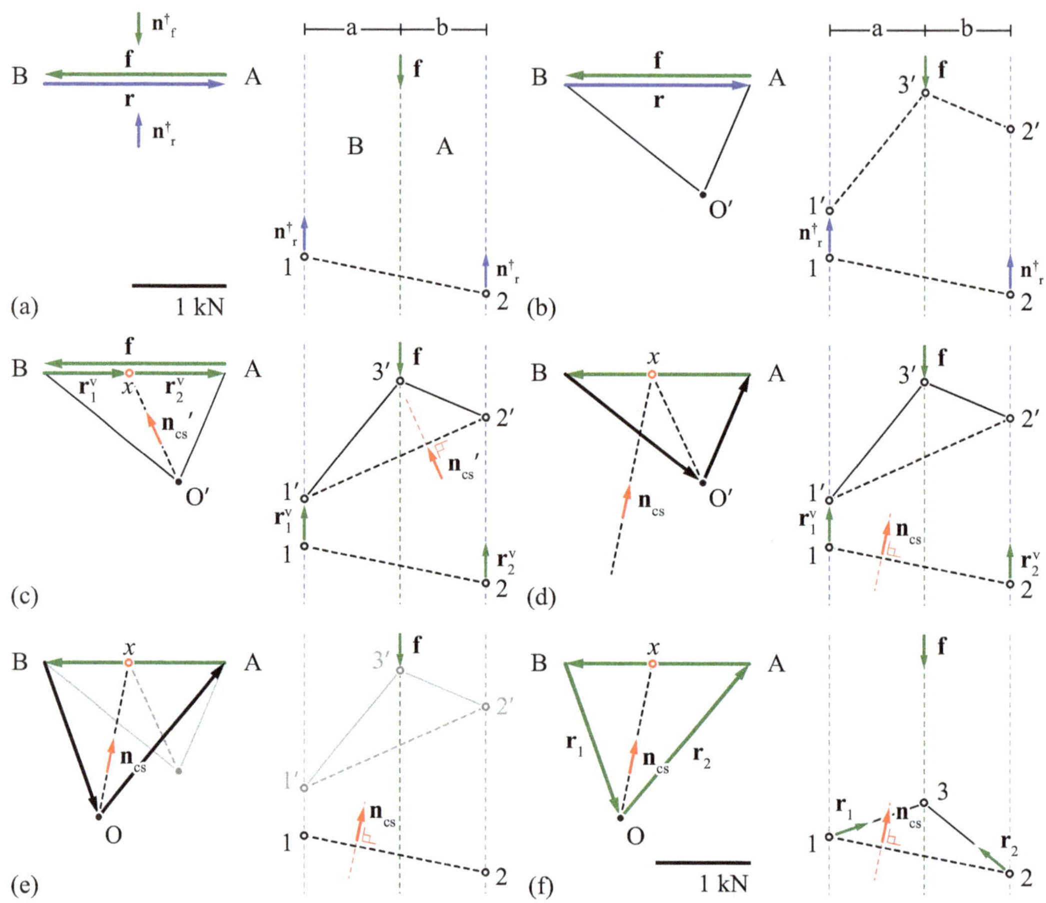

Figure 3.13 The process of constructing a constrained funicular and its force diagram for given boundary conditions: (a) closing the force polygon; (b) decomposing the force polygon and constructing a trial funicular polygon; (c) drawing the closing string of the trial funicular and drawing line *l′* in the force to find *x*; (d) drawing line *l* normal to the closing string of supports; (e) decomposing the force polygon to point *O*; and (f) completing the final constrained funicular solution.

Construction steps:

1 Divide the construction space into *two* parts: the left side will be used to construct the force diagram, and the right side will be allocated to the funicular form.

2 In the form diagram of Figure 3.13a, label spaces bounded by the supports and the applied load as *A* and *B*.

3 Put your thumb in the direction of the applied load; the extension of the four fingers gives the orientation of the force vector to be drawn in the force diagram of Figure 3.13a.

4 In the force diagram of Figure 3.13a, start from an arbitrary point *A*, follow the direction of the four figures from the previous step, and draw a vector with a

length of 2 cm, each cm equivalent to 1 kN, to locate point B. Vector $\vec{AB}$ represents the force edge reciprocal to the form diagram's applied load **f**.

5 In the force diagram of Figure 3.13b, close the force diagram by drawing the resultant vector $\bar{\mathbf{r}}$ connecting point B to A.

6 In the form diagram of Figure 3.13b, at the supports, draw lines normal to the resultant vector of the force: these lines are parallel to the resultant force of the applied loads in the form diagram and are shown by $\mathbf{n}_{\bar{\mathrm{r}}}^{\dagger}$.

7 In the force diagram of Figure 3.13b, extrude the force polygon to an arbitrary point O'.

8 In the form diagram of Figure 3.13b, pick an arbitrary point $1'$ on the line of action passing through point 1 and draw a line normal to BO' to intersect the line of action of **f**. Name this point $3'$.

9 In the form diagram of Figure 3.13b, from point $3'$, draw a line normal to $O'A$ to intersect the line, passing through point 2. Name this point $2'$. Thus, a funicular solution is constructed based on the lines of action of forces in the form diagram. Let us name this funicular the *trial funicular*.

10 In the form diagram of Figure 3.13c, on the trial funicular, draw a line connecting points $2'$ to $1'$. This line is the *closing string* as discussed in Section 3.2.3 and is shown by its normal vector $\mathbf{n}_{cs'}$.

11 In the force diagram of Figure 3.13c, from point O', draw line l' normal to the closing string to intersect vector **f** at point x. Point x gives the lengths of the vectors $\mathbf{r}_1^{\mathrm{v}}$ and $\mathbf{r}_2^{\mathrm{v}}$ at the supports parallel to the resultant. These parallel forces keep the applied load in equilibrium. As discussed in Section 3.2.4, the location of x does not change for any funicular solution constrained to a given set of loads.

12 In the force diagram of Figure 3.13d, draw line l from point x normal to the closing string that connects points 1 and 2 in the form diagram.

13 In the force diagram of Figure 3.13e, choose an arbitrary point O on line l and extrude/decompose vector **f** to the chosen point.

14 In the form diagram of Figure 3.13f, construct a funicular solution: draw a line from point 1 normal to BO to intersect the line of action of **f** at point 3; draw another line from point 3 normal to OA. This line should pass through point 2, and it completes the funicular construction.

15 Changing the location of point O on line l changes the height of the funicular solution constrained to the supports.

□

Following the above procedures, the form-finding process can be divided into a couple of essential steps. In the first step, regardless of the form of the structure, we need to ensure that the applied loads and reaction forces at the supports are in equilibrium. It would be sufficient to show that the sum of the applied loads and reaction forces is zero. Closing the force polygon is a geometric method that ensures the sum of the forces is zero. As discussed in Chapter 2, adding the resultant force to the force polygon will close it. Note that the form of the structure at this stage has not been developed. Thus, the reaction forces are unknown and cannot be used to close the force polygon. Nevertheless, we can assume that there are two components of reaction forces parallel to the resultant at the supports that keep the applied loads in equilibrium. This property was explained in Section 3.2.4.

In the second step, the moment equilibrium should be established. To guarantee the moment equilibrium of the system, a funicular solution should be constructed on the lines of action of all the external forces, including the parallel reactions of Figures 3.7 to 3.9. As discussed extensively in Chapter 2, a closed funicular construction on the lines of action of forces ensures the moment equilibrium in the system. As noted, an infinite number of funicular polygons can be constructed on the lines of action of a system of forces in equilibrium. But we are interested in only a particular group of funicular forms constrained to the support locations. To find the constrained funicular, we first find a family of force diagrams that are constrained to a line that is normal to the closing string of the form diagram. This line was shown by l in all previous examples and is drawn from the point x in the force diagram. To identify the location of the point x and the line l, we used the *trial funicular construction* process. The force diagram is then constructed by extruding the applied load vectors to a point on the line l. Every point gives a force diagram whose reciprocal funicular form is constrained to the support locations.

3.5 Constrained Form Finding for a Single Applied Load in 3D

Now that we have described the form-finding process for a simple case in 2D, the procedures can be extended for funicular form-finding for a simple case in 3D. In 3D form-finding, the boundary condition is usually given, and the funicular form of a structure and its force diagram should be found. In the form-finding process using polyhedral graphic statics, there are some critical remarks regarding boundary conditions.

Remark 3.7. Each applied load in the form diagram is defined by the intersection of at least three planes. This topic was explained in Chapter 2. Consequently, the intersection of a series of planes in the form diagram defines the line of action of the loads. For example, Figure 3.14 shows a single applied load **f**, which is defined by three planes intersecting at the line of action of **f** in the form diagram.

Remark 3.8. In the form diagram, the length of the vectors does not represent the magnitude of the force. Indeed, the magnitude of the force should be found by measuring the area of the corresponding face in the force diagram.

Remark 3.9. The supports in the form diagram are located on the planes of the applied loads. Therefore, the minimum number of supports is three because the minimum number of planes defining a force in 3D is three, as explained above.

Thus, similar to the 2D example, the form-finding process in 3D can be summarized into the following main steps:

1. Establishing the global equilibrium.
2. Finding reaction forces parallel to the resultant.
3. Constructing a constrained force diagram
4. Completing the constrained funicular solution.

Example 3.2 will elaborate on the earlier steps using a simple example in 3D.

Example 3.2 Figure 3.14 illustrates an applied load **f** and three support locations 1–3 located on the (green) planes intersecting at the line of action of **f**. For the applied load **f**, find a funicular form connected to the supports and its force diagram and show how

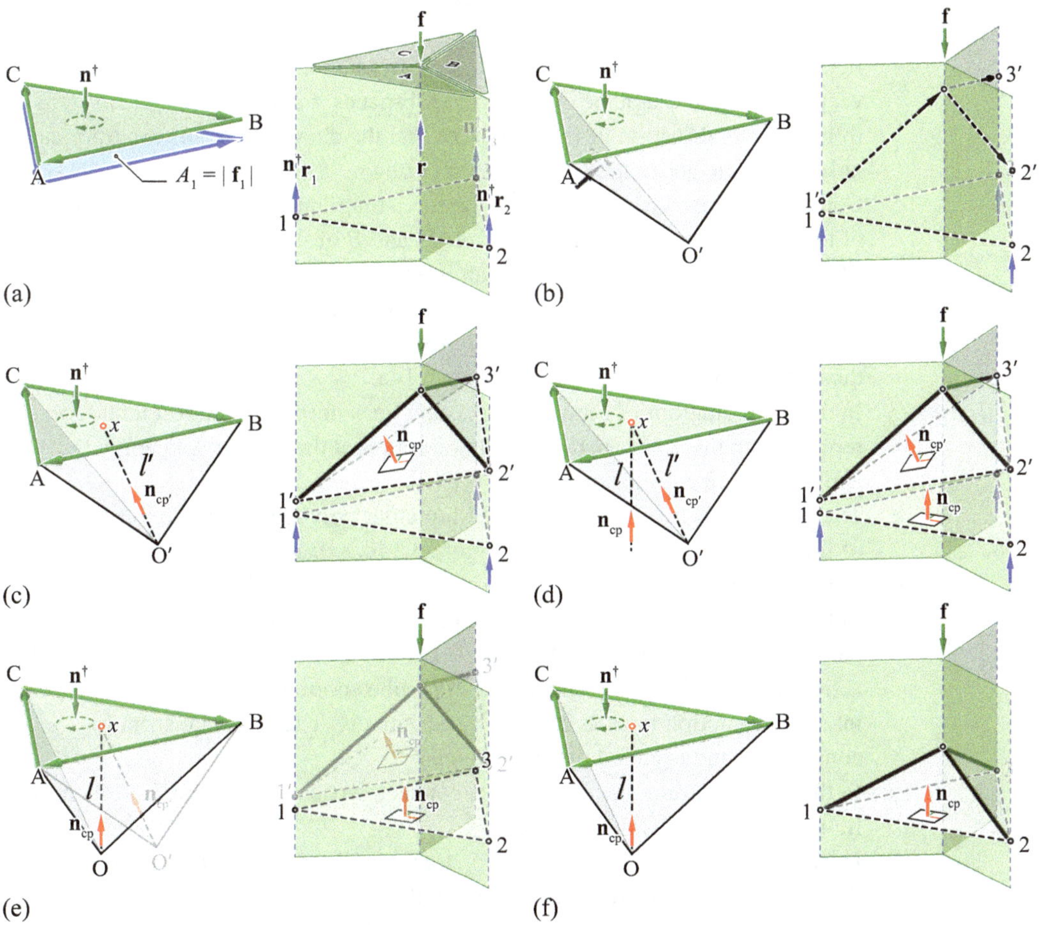

Figure 3.14 (a) Constructing and closing the force diagram for the given applied load and support locations; (b) decomposing the force polyhedron into an arbitrary point O' for trial funicular construction; (c) adding the closing plane to the trial funicular and locating x on the corresponding force diagram; (d) drawing a line normal to the final closing plane from x; (e) constructing the constrained force diagram; and (f) completing the constrained funicular form for the given boundary condition.

the height of the funicular form can be controlled by adjusting the height of the force diagram.

Construction steps:

1 Divide the construction space into two parts: the left part will be used to construct the force diagram, and the right part will be allocated to the funicular form.
2 In the form diagram of Figure 3.14a, label the subspaces created by the *action planes* of the force **f** as A, B, and C.
3 In the form diagram of Figure 3.14a, put your thumb in the direction of **f**, the curl of your fingers shows the order of passing the subspaces, and thus the direction of the edge vectors of the reciprocal face in the force diagram; this order is $A \rightarrow C \rightarrow B$ (Figure 3.14a).
4 In the force diagram of Figure 3.14a, start from an arbitrary point A and draw a vector normal to the plane dividing the subspaces A and C with a chosen length following the direction of the curl. Note that the chosen length is arbitrary and only scales the entire force diagram up or down.
5 Complete a closed triangle ACB in the force diagram using the normal direction of the planes dividing the subspaces A, C, and B of the form diagram. Triangle ACB is reciprocal to the form diagram's applied load **f**.
6 To ensure that the sum of the forces in the system is zero, *the force polyhedron should be closed*. Close the force polyhedron by adding face ABC (blue) with the opposite normal direction in Figure 3.14a.
7 In the form diagram of Figure 3.14b, draw lines normal to the resultant face representing the components of reaction forces at the supports, keeping the applied load in equilibrium.
8 The next step is to construct a closed funicular polyhedron on the lines of action of forces; in the force diagram of Figure 3.14b, extrude the face ACB to an arbitrary point O' in the 3D space.
9 In the form diagram of Figure 3.14b, pick an arbitrary point $1'$ on the line passing through support 1. Draw a line normal to the face CAO' to intersect the action line of **f**. From the new point, draw lines normal to faces ABO' and BCO' intersecting action lines at $2'$ and $3'$, respectively. The funicular form is constructed and is called the *trial funicular*.
10 In the form diagram of Figure 3.14c, connect vertices $1'$, $2'$, and $3'$ of the trial funicular by the closing plane $\mathbf{n}_{cp'}$.
11 In the force diagram of Figure 3.14c, draw line l' from point O' to intersect face ABC at x. As explained in Section 3.3.4, the location of point x in the force does not change for any funicular construction on the line of action of forces.
12 In the form diagram of Figure 3.14d, draw a closing plane 123 connecting the given support locations.
13 In the force diagram of Figure 3.14d, from point x, draw a line l normal to the closing plane 123.

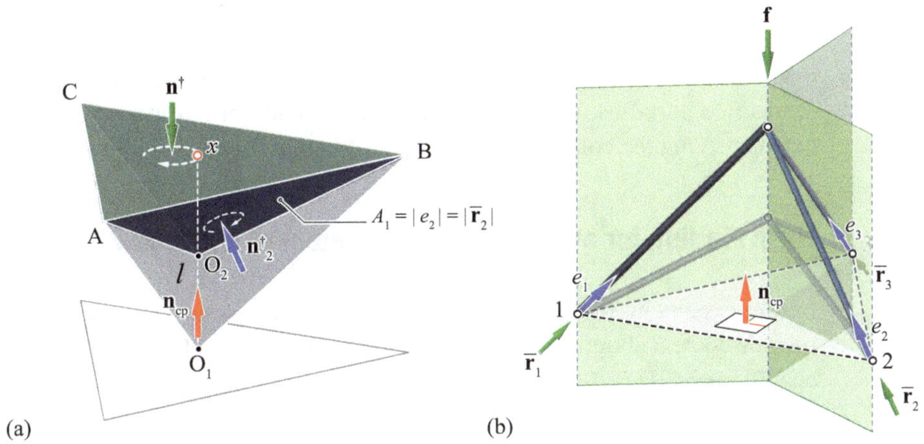

Figure 3.15 (a) The force diagram defined by the location of O_i on the line l; and (b) its reciprocal funicular form with different heights constrained to the supports and the line of action of **f**.

14 In the force diagram of Figure 3.14e, extrude face ACB to an arbitrary point O on line l. This step gives a new force diagram constrained to line l.

15 In the form diagram of Figure 3.14f, from point 1, draw a line normal to CAO to intersect **f**. From the intersection point, draw lines normal to ABO and BCO. These lines should pass through support locations 2 and 3. The funicular construction is complete.

16 Figure 3.15a,b shows changing the location of point O_i on line l in the force diagram that gives funicular forms with different heights.

□

As shown in Figure 3.15, the magnitude of the internal forces in the structure's members changes by changing the height of the funicular form. This is because the areas of the faces of the force polyhedron change by adjusting the location of point O. Therefore, the internal forces carried by the members decrease in a taller funicular form and vice versa. Moreover, the direction of the internal forces in the members attached to the supports matches the direction of the reaction forces at those supports for compression-only or tension-only structures.

The term *trial funicular* used during the construction process of the above example and in the 2D example was initially introduced in Chapter 2. The reason behind this naming is that a closed/completed funicular construction on the lines of action of given forces ensures moment equilibrium in the system and can be shown by any funicular construction constrained to those action lines. This property was utilized by choosing an arbitrary initial point O' to construct a force diagram and completing its reciprocal funicular for the given boundary condition.

Note that in the constrained form-finding process, from all possible funicular solutions constrained to the lines of action of forces, we are interested only in a particular solution that is constrained to the support points. As noted, the closing plane of the trial funicular polyhedron locates the point x that is used for constructing a particular force diagram whose reciprocal form is constrained to the supports. Thus, the trial funicular serves two purposes: (1) ensuring the moment equilibrium in the system is zero and (2) identifying the constraints of the force diagram for form-finding purposes.

3.6 Form Finding for a Nonconcurrent System of Loads in 2D

Before expanding on the form-finding process for a system of loads, it is helpful to provide context about nonconcurrent loads as a recap from Chapter 2 and the boundary conditions in 2D.

3.6.1 Nonconcurrent 2D Applied Loads

Often, in architectural or engineering applications, a funicular form of a structure like a bridge or a shell should be found for a group of applied loads. These loads either represent the self-weight of the structure or are the maximum applied load per square meter/foot of the system. Both distributed and point loads are shown by vectors with their corresponding lines of action and magnitude in graphic statics. Such groups of loads are often nonconcurrent. That is, their lines of action do not intersect like parallel loads; if they intersect, they meet at more than one point.

3.6.2 The Assumption about Self-Weight

In graphical methods, the self-weight of the structure is not considered directly in the form-finding process. If the self-weight of the structure is intended for the design, then the form-finding process should be pursued in multiple steps; in the first step, we need to make an assumption about the self-weight, pursue the form-finding process, and identify the geometry of the structure. In the next step, measure the new self-weight and check the design loads. If there is a discrepancy between the design loads and the self-weight, the design loads should be updated with the new self-weight. The iterative process should be repeated until the difference between the self-weight and the applied loads becomes insignificant. In this chapter, we avoid this complication by only finding the form of the structure for a given set of loads and assuming that the self-weight of the structure is negligible or the applied loads are significantly larger than the self-weight of the structure.

3.6.3 2D Support Locations and External Determinacy

In 2D, the structure's form is often constrained to two support locations. If there is only one reaction force with a magnitude and direction at each support for a load case and its resulting funicular form, then the system is *statically determinate*. 2D compression/tension-only forms are often constrained to two supports and thus

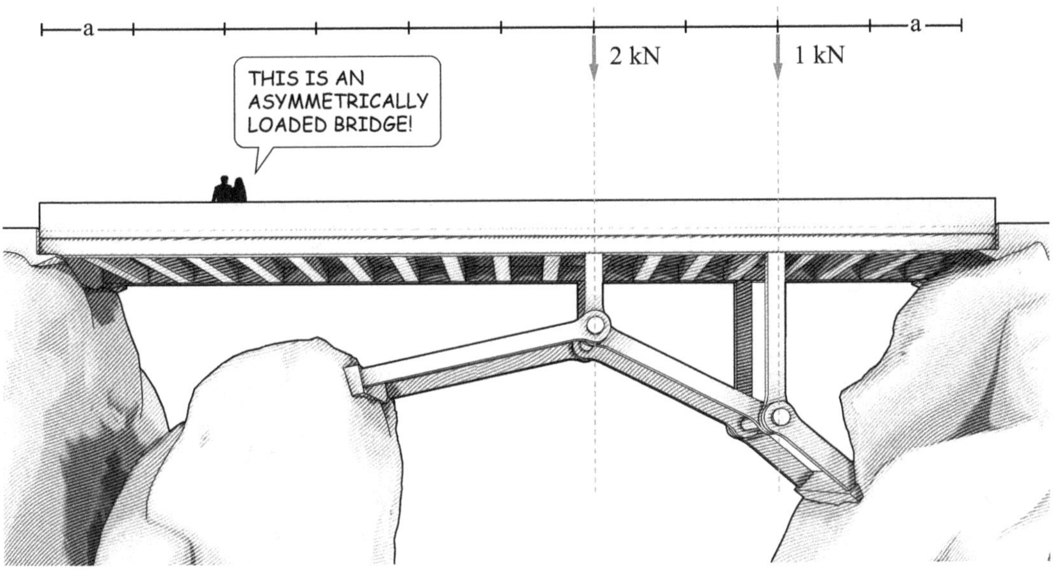

Figure 3.16 A concept of a prefabricated timber bridge supported asymmetrically by a compression-only timber arch.

statically determinate. Adding to the number of supports will result in the structure being externally indeterminate. In this chapter, we work with externally determinate, compression-only forms and leave the discussion about multiple support locations and the indeterminacy of the structures to Chapter 6. Now let us start the form-finding process for a nonconcurrent system of loads with Example 3.3.

3.6.4 An Example of Two Parallel Loads

Consider the form-finding process for the bridge illustrated in Figure 3.16. This is a simple example of a nonconcurrent load system and two supports. The objective is to design a compression-only form for a supporting arch that carries loads of a prefabricated timber bridge deck. Note that the applied loads are asymmetrical since the bridge's deck touches the arch only in two locations, and the load in one location is twice as big as the other. Example 3.3 will elaborate on the form-finding process of this bridge.

Example 3.3 Find a funicular solution for the parallel loads $\mathbf{f}_1 = 2$ kN and $\mathbf{f}_2 = 1$ kN of Figure 3.17 applied at distances a and c from supports 1 and 2, respectively. In addition, show multiple funicular solutions with different heights for the same boundary condition.

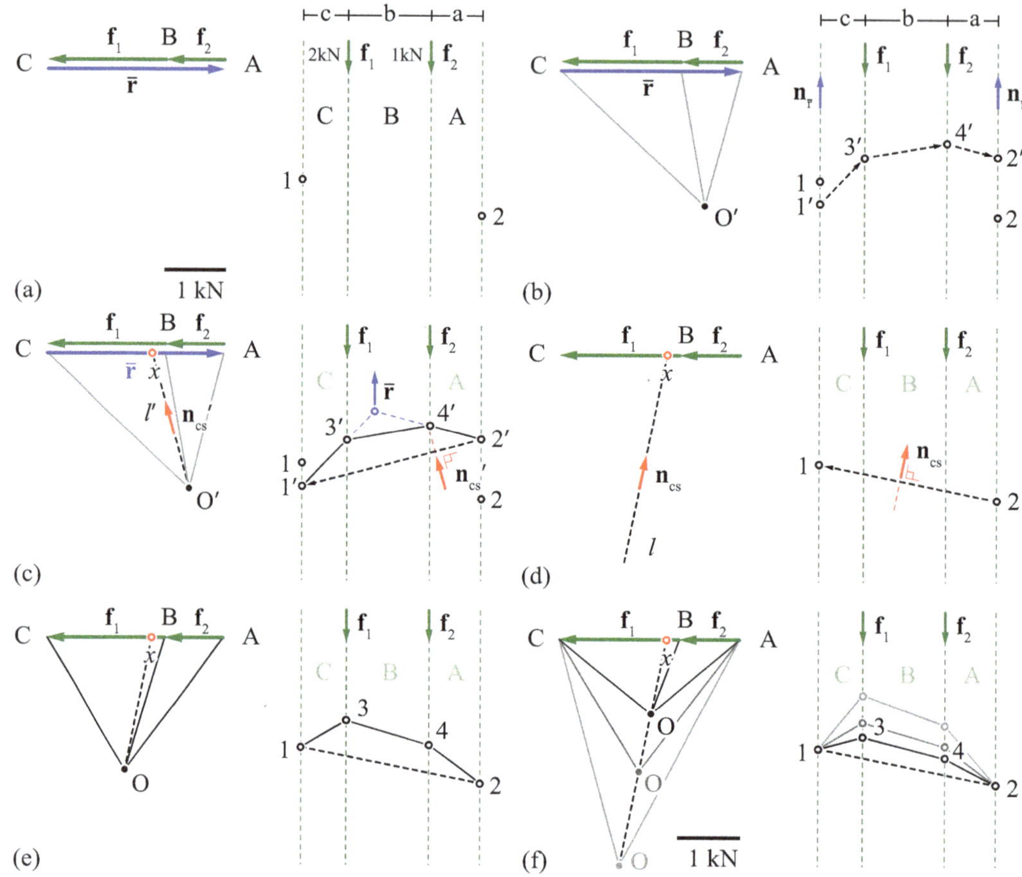

Figure 3.17 (a) Closing the force polygon by adding the resultant for the given system; (b) constructing a trial funicular; (c) closing the trial funicular by the closing string and finding the location of *x*; (d) connecting the support locations and drawing its normal from *x*; (e) decomposing the force vectors to a point on line *l* and finding a constrained compression-only form; and (f) changing the height of the funicular by adjusting the location of *O* of the force diagram.

Construction steps:

1. Let us allocate the right side of the construction space of Figure 3.17a to the form of the structure and its left side to the force diagram.
2. In the form diagram of Figure 3.17a, label the spaces between the lines of action of the applied loads and the support locations as *A*, *B*, and *C*.
3. In the form diagram of Figure 3.17a, from right to left, put your thumb in the direction of the applied load; the four fingers' direction shows the force vector's alignment to be drawn in the force diagram.

4 In the force diagram of Figure 3.17a, start from an arbitrary point A and draw a vector in the direction of the four fingers with 1 cm length representing 1 kN force and the magnitude of force $\mathbf{f}_1$. Then, name the endpoint B.
5 Repeat the same procedure for the load $\mathbf{f}_2$ and locate point C.
6 Based on equilibrium principles, the force polygon should be closed. Close the force polygon by drawing a vector connecting points C and A. This is the resultant vector named $\bar{\mathbf{r}}$. A closed force polygon guarantees that the sum of the forces in the system is zero.
7 Next, we need to ensure that the sum of the moments in the system is also zero. We use trial funicular construction for this purpose: in the force diagram of Figure 3.17b, pick an arbitrary point O' and extrude the force polygon to it.
8 In the form diagram of Figure 3.17b, draw a line of action normal to $\bar{\mathbf{r}}$ of the force at each support. The main objective is to substitute the resultant force with two parallel forces at the supports.
9 On the line of action of $\mathbf{n}_{\bar{\mathbf{r}}}$ of the form diagram of Figure 3.17b, start from an arbitrary point $1'$ and draw a line normal to CO' of the force to intersect the line of action of $\mathbf{f}_1$. Continue this drawing for BO' and AO' to find point $2'$ on the line of action of support 2. This process completes the trial of funicular construction.
10 In the form diagram of Figure 3.17c, the edge connecting $1'$ and $3'$ and the edge linking $4'$ and $2'$ intersect at a point, which is the location of the line of action of $\bar{\mathbf{r}}$.
11 Now, in the form diagram of Figure 3.17c, close the trial funicular by connecting points $2'$ to $4'$ and call it the closing string as shown by its normal vector $\mathbf{n}_{cs'}$.
12 In the force diagram of Figure 3.17c, draw a line l' normal to the closing string to intersect $\bar{\mathbf{r}}$. Name this point x. Point x gives the magnitude of the force components parallel to the resultant at each support.
13 Any funicular solution constructed on the lines of action of forces has a closing string that is normal to a line passing from point x in the force diagram. We are looking for a constrained solution attached to points 1 and 2. Thus, the closing string connecting 1 and 2 should also be normal to a line passing through x: In the form diagram of Figure 3.17d, draw another closing string connecting the two supports 1 and 2.
14 In the force diagram of Figure 3.17d, draw a line l normal to the new closing string.
15 Pick an arbitrary point O on the line l in the force diagram of Figure 3.17e and extrude the force polygon to it.
16 Construct a funicular solution in the form diagram of Figure 3.17e on the line of action of forces. As noted, the final construction is constrained to supports 1 and 2.
17 The height of this funicular form can be adjusted by moving the point O' up and down on the line l, as illustrated in Figure 3.17f.

□

3.6.5 Combining the Form and Force Diagrams into a Single Diagram

A simple technique can be used to link the two diagrams visually. This technique can combine the form and force diagrams and ensure the construction's correctness by immediately putting the force polygons close to the nodes in equilibrium in the form diagram. Every *internal* vertex of the form diagram corresponds to one and only one closed polygon in the force diagram. Also, every shared edge between two polygons in the force diagram is reciprocal to an edge of the form diagram. Thus, we can move every polygon of the force diagram close to its corresponding vertex in the form and extrude their faces to connect each polygon. Figure 3.18 shows an example of this process where the form and force diagrams are combined into a single diagram. In the following, we provide a rigorous procedure for the construction of the diagram of Figure 3.18:

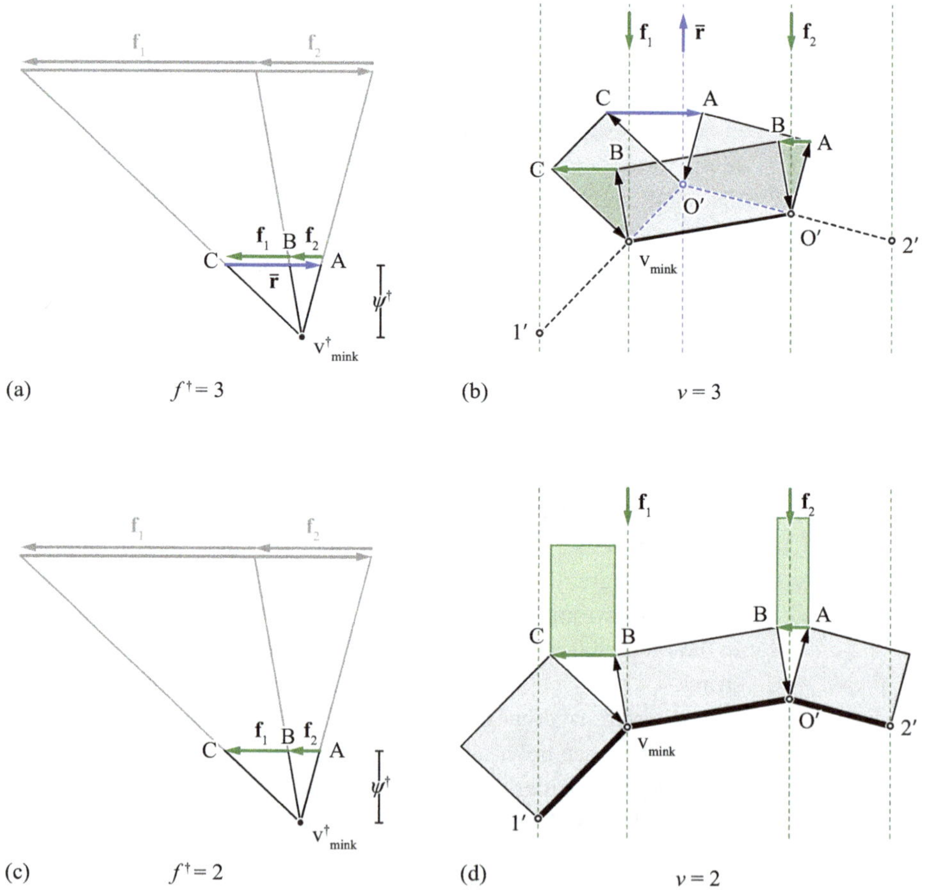

Figure 3.18 (a) Scaled force diagram including three force polygons; (b) relocating every polygon of the force diagram to vertex v_{mink} on the form and then moving each polygon $f^{\dagger}$ to its corresponding vertex in the form; (c) the force diagram excluding the cell of the resultant; and (d) relocating force polygons to their corresponding vertices in the form diagram.

1. Choose a scale factor $\psi^\dagger$ for the force diagram. This allows scaling the force diagram up or down for clarity during construction.
2. In the force diagram, pick a reference vertex as $v^\dagger_{mink}$ and scale the force diagram with a scale factor $\psi^\dagger < 1$.
3. In the form diagram, choose another reference vertex as v_{mink}.
4. Move the entire scaled force diagram from $v^\dagger_{mink}$ to v_{mink}.
5. In the relocated form diagram, move every force polygon $f^\dagger_i$ from vertex v_{mink} to its reciprocal vertex v_i.
6. For each of two adjacent force polygons, connect the start and end points of their shared edge using a quadrilateral.
7. Check whether the quadrilateral is a *rectangle*, that is, it has perpendicular edges. If not, the construction is not correct.

Figure 3.18a shows a scaled version of the force diagram of the trial funicular construction of Figure 3.17b. This force diagram is constructed by closing the polygon of applied loads using the resultant vector and extruding the closed polygon into an arbitrary point O'. This force diagram has three closed polygons reciprocal to three vertices in the form diagram, including the location of the application of the resultant force $\bar{\mathbf{r}}$. Vertex O' is chosen as $v^\dagger_{mink}$ in the force diagram, and vertex $3'$ is chosen as v_{mink} of the form diagram (Figure 3.18a,b).

By combining the form and the force diagrams, we can observe that for every closed polygon in the force diagram, there is a vertex in equilibrium in the form diagram. Moreover, each force polygon in the combined diagram is connected to its adjacent force polygon by a rectangle along the edge of the form. Two parallel edges of this rectangle belong to the force diagram, while the other two parallel edges belong to the form diagram. These two parallel sets of edges must always be perpendicular to each other for the entire construction. Thus, the edges' angle in the combined diagram can ensure the funicular construction's correctness.

Figure 3.18c shows the force diagram without the resultant vector that has only two closed force polygons. Figure 3.18d combines the form and force diagrams. Moreover, the external edges of the polygons are extended along the edges of the funicular form and the applied loads.

Remark 3.10. Comparing the diagrams of Figure 3.18b,d shows that the edges of the funicular construction that extend to the supports also intersect at the line of action of the resultant. In fact, it shows that the reaction forces or the internal forces in the members attached to the support will result in the resultant vector that keeps the applied loads in equilibrium. This observation is important because we can check the correctness of the funicular construction. We will use this technique also in complex 3D cases in Section 3.7.

Remark 3.11. This method is referred to as the *Minkowski sum* operation because it combines the form and force diagrams into a single diagram.[53] We will explain and use this technique in Chapters 4 to 8 to facilitate the visual link between the form and force diagrams for complex funicular forms and forms with tension and compression forces.

3.7 Form Finding for a Nonconcurrent System of Loads in 3D

The compression/tension-only funicular polyhedral solutions in 3D may take various configurations based on the type, the number of applied loads, and the number of supports. Figure 3.19 shows various configurations of a tension-only funicular network for different loading conditions and support locations. Although the form-finding process for all these examples may follow similar procedures, the important relationship between the supports and the applied loads should be explained in advance.

3.7.1 The Nonconcurrent Load Networks in 3D

The line of action of an external load or a reaction force in the form diagram of polyhedral graphic statics is defined by the intersection of planes (see Chapter 2). Each applied load or reaction force is reciprocal and normal to a face in the force diagram. The number of intersecting planes of a load vector in the form diagram equals the number of edge vectors of the reciprocal face in the force diagram. Therefore, at least three planes are needed to intersect and define an applied load or a reaction force in the form diagram so that its reciprocal face will be a closed polygon.

Figure 3.19a–c shows examples where three planes meet to define an applied load, and the reciprocal face is a triangle. In this case, the geometry and the area of a triangular face can be controlled by the length of an edge vector in the force diagram. If more than three planes intersect at the line of action of a load, the corresponding face in the force diagram will have more than three edge vectors. For a polygon with more than three edges, more than one edge length is needed to control its area. Figure 3.19d shows an example where the force face has four edges; thus, different geometries can be constructed with the same area.

A similar example was discussed in Chapter 1 as an indeterminate system. Such force diagrams have *geometric degrees of freedom* (GDoF), which can be used creatively in design, as explained in Chapter 6. Yet, it is important to know that a system of loads in 3D may consist of more than three planes intersecting at the line of action of forces, and their corresponding faces might be polygons with any number of edges.

Remark 3.12. The intersecting planes of the applied loads divide the 3D space into subspaces that are open polyhedral cells. These subspaces/cells are named as *A*, *B*, *C*, and so on, as shown in Figure 3.19. A system of loads in polyhedral graphic statics constructs a set of subspaces that might be convex or non-convex. However, the system of loads of a compression/tension-only polyhedral frame often consists of *convex* subspaces.

3.7.2 The Number of Supports in 3D

In polyhedral graphic statics, the supports are sitting on the planes, defining the loads in the form diagram. For a system of loads consisting of a series of intersecting planes,

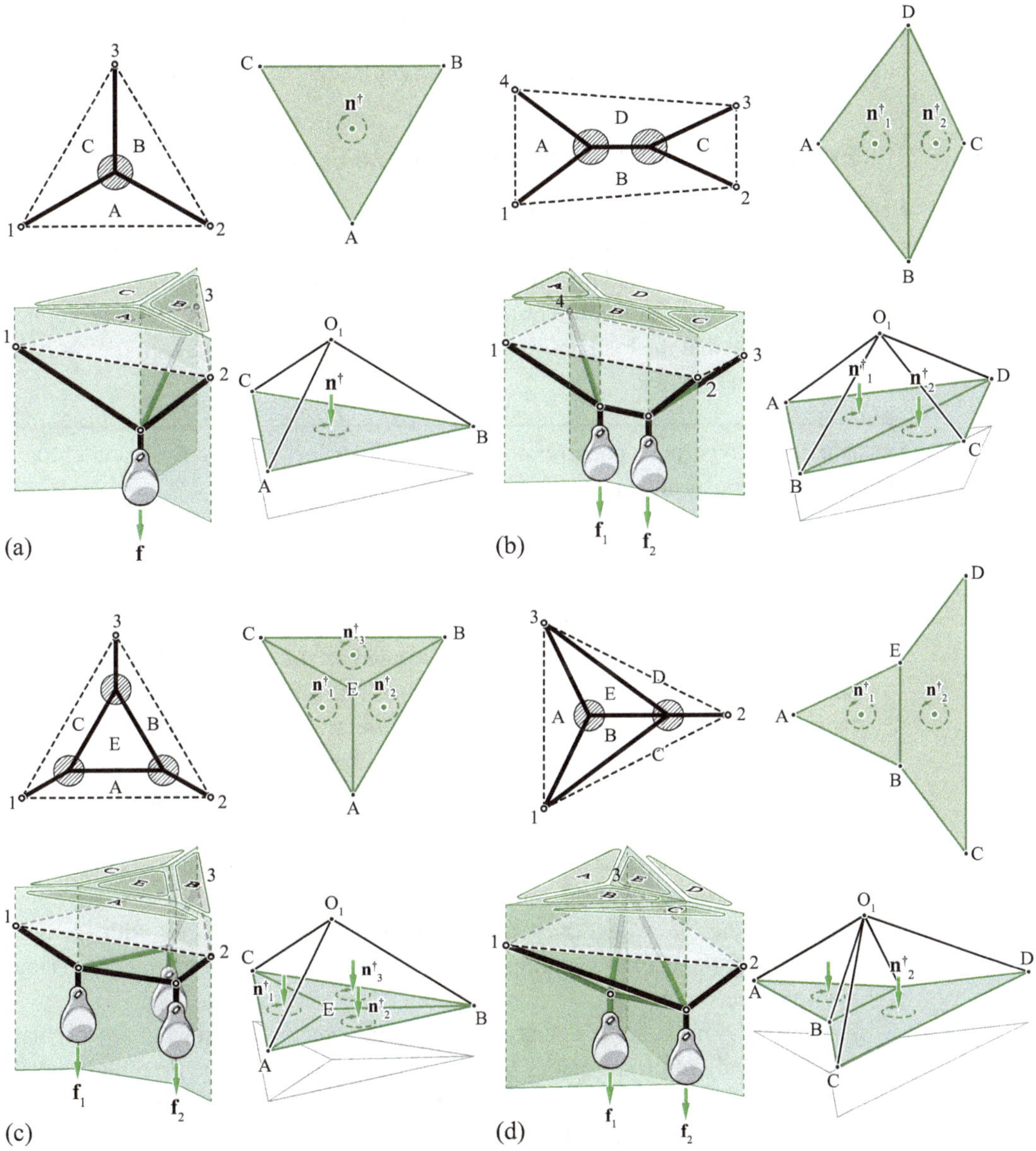

Figure 3.19 The applied loads, the number of support locations, and the configuration of the constrained funicular forms: (a) a single applied load and three supports; (b) two parallel loads and four supports; (c) three parallel loads and three support locations; and (d) two parallel loads and three support locations.

every plane should either pass through a support location or intersect another plane at the location of an applied load. A closer look at the configurations of Figure 3.19 corroborates this statement.

Remark 3.13. A funicular polyhedral solution in 3D follows the arrangement of the planes of the applied loads and the location of the supports. A funicular form always has edges lying on the planes shared by the supports and the applied loads.

3.7.3 The Example of Two Parallel Loads and Three Supports in 3D

Let us consider the example of Figure 3.19d and provide geometric procedures to find a funicular form for its boundary conditions. This example includes two parallel loads applied in the space bounded by three supports connected by a closing plane, as shown in Figure 3.21a. A procedure similar to those explained in Example 3.5 can be applied to this case. Note that the load network includes more than three planes intersecting at the line of action of a load.

Example 3.4 Construct a compression-only funicular form for the two parallel loads of Figure 3.20 constrained to support points 1, 2, and 3.

Construction steps:

1. Allocate the *right* side of the construction space to the *form diagram* and the *left* side to its reciprocal *force diagram.*
2. In the form diagram, identify subspaces formed by the planes of the load network. Label those as A, B, ... , E (Figure 3.20a, left).
3. In the form diagram, put your thumb in the direction of the applied load $\mathbf{f}_1$, the curl of the fingers gives the order to pass the subspaces around it as $A \rightarrow E \rightarrow D \rightarrow B \rightarrow A$. This is the order of the vertices of the reciprocal face in the force diagram (Figure 3.20b, left).
4. Let us construct the reciprocal face for the load $\mathbf{f}_1$. Since four planes intersect at the line of action of $\mathbf{f}_1$ in the form, the reciprocal face will have four edge vectors: in the force diagram, start from point A and draw a vector with a length d_1 normal to the plane separating subspaces A and E of the form diagram. Name the tip of the vector E.
5. From point E in the force diagram, draw another vector with a length d_2 normal to the plane dividing subspaces E and D of the form. Name the tip of the vector D. These two edge vectors are sufficient to find the location of B by intersecting lines normal to the planes $\mathbf{n}_{DB}$ and $\mathbf{n}_{BA}$. Note that this face is a trapezoid, thus requiring two edge lengths to be chosen to complete the face. The area of this face represents the magnitude of the force $\mathbf{f}_1$, and one can change d_1 and d_2 to adjust the area. The face reciprocal to force $\mathbf{f}_2$ can be completed since the edge vector DB that is shared by the two faces is known.
6. In the force diagram of Figure 3.20b, let us close the force diagram by adding triangular faces representing the resultant components, keeping the applied loads in equilibrium.
7. In the form diagram of Figure 3.20b, draw lines normal to the resultant faces of the force at each support as shown by unit vectors $\mathbf{n}^{\dagger}_{\mathbf{r}_1}$.

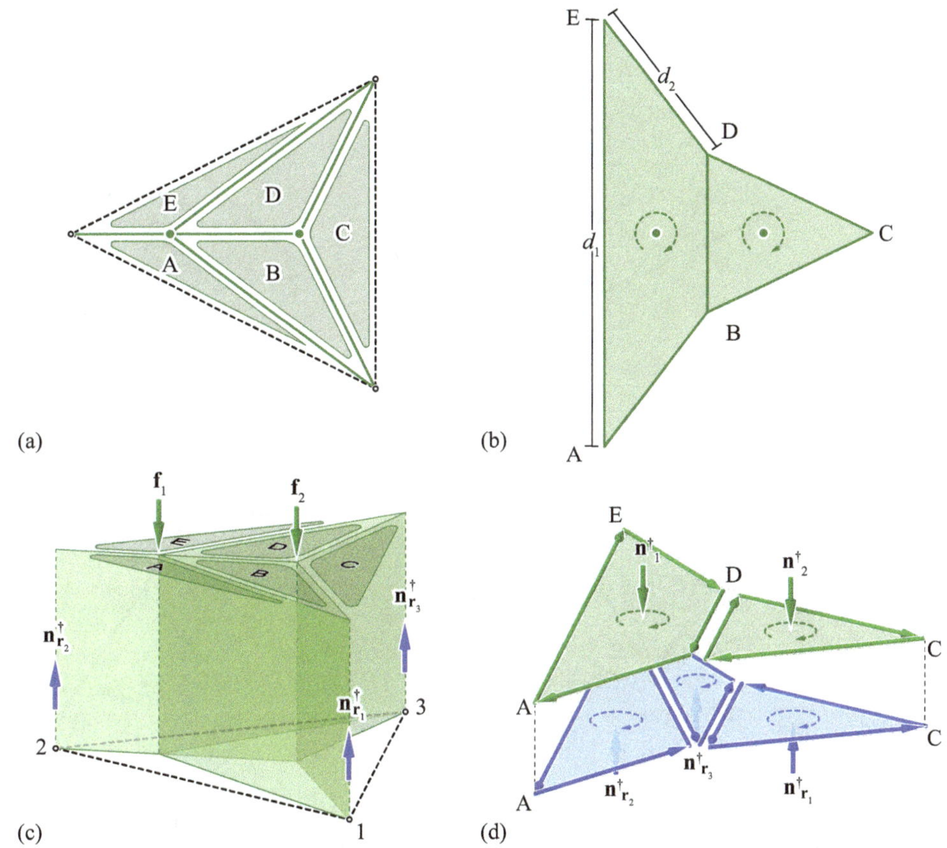

Figure 3.20 (a) The form diagram of two parallel loads and three supports on the planes intersecting at the applied loads; (b) the force faces and the resultant faces to close the force polyhedron; (c) the plan view of the form and the convex subspaces made by the applied loads and the supports; and (d) the corresponding force faces (plan view).

8 Let us move to Figure 3.21. In the force diagram, extrude the force faces to an arbitrary O'.
9 In the form diagram of Figure 3.21b, construct a trial funicular polyhedron starting from an arbitrary point $1'$ on the line passing through support 1.
10 As soon as all three supports of the trial funicular are found, close the funicular by adding a plane passing through points $1'$, $2'$, and $3'$ and denote this plane with normal $\mathbf{n}_{cp'}$.
11 Return to the force diagram of Figure 3.21a, and draw a line l' from point O' in the direction of $\mathbf{n}_{cp'}$ to intersect applied loads at x.
12 In the form diagram of Figure 3.21d, draw a closing plane denoted by $\mathbf{n}_{cp}$ connecting supports 1, 2, and 3.
13 In the force diagram of Figure 3.21c, from point x, draw a line in the direction of $\mathbf{n}_{cp}$ and name it l.

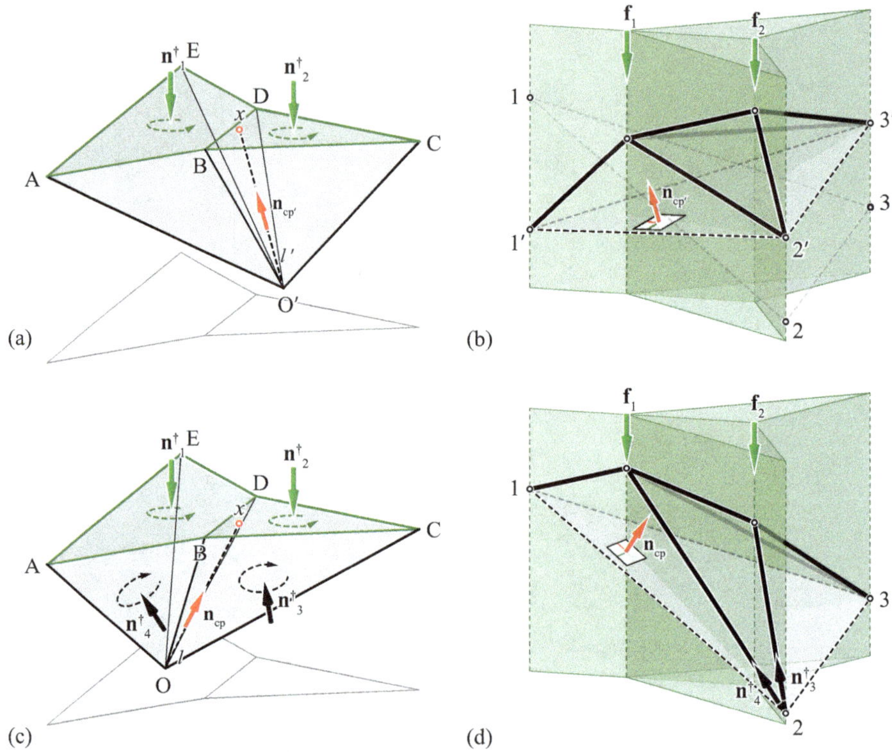

Figure 3.21 (a) A trial funicular construction and (b) its corresponding force diagram; (c) a constrained funicular with multiple members connected to the supports; and (d) its corresponding force diagram.

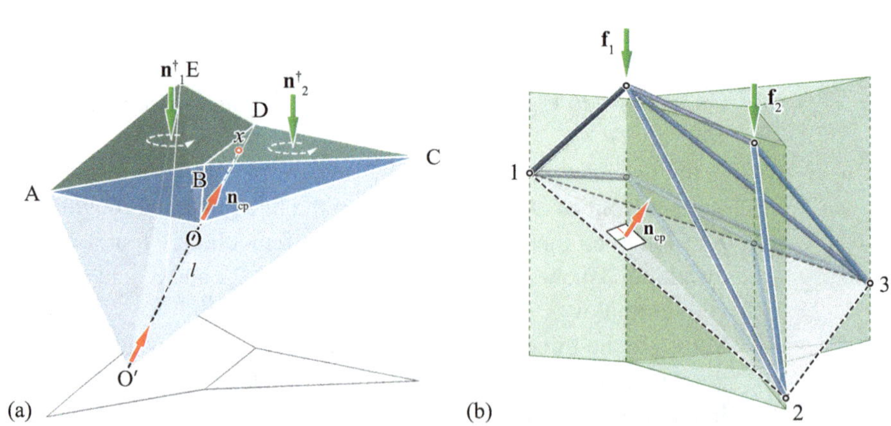

Figure 3.22 (a) Changing the height of the force diagram by moving the point O on the line l results in (b) a change in the height of the funicular solutions constrained to the support locations.

14 In the force diagram of Figure 3.21c, pick an arbitrary point *O* on *l* and extrude the applied loads to *O* to construct a force polyhedron of Figure 3.21c.
15 Complete the constrained funicular form on the line of action of the loads attached to the supports in Figure 3.21d.
16 Moving point *O* up and down along *l* changes the height of the funicular polyhedron, as shown in Figure 3.22a,b.

□

This example can be for the design of compression-only supporting structure of a branching bridge illustrated in Figure 3.23.

3.7.4 Eliminating Trial Funicular Construction in 3D

Let us reveal another interesting relationship between a closing plane and a constrained force diagram that may also shorten the geometric steps of form-finding. The force diagram reciprocal to a constrained funicular form is also constrained to a line named *l*. Multiple examples showed this concept in 2D and 3D. Line *l* of the force diagram is perpendicular to the closing edge in 2D and is normal to the closing plane in 3D of the form diagram. The location of line *l* in 2D and 3D was found using the trial funicular construction. This line is essential in the form-finding process because the force diagram for a particular location of supports is constrained to this line. And the height of the funicular form changes if the force diagram is adjusted along this line.

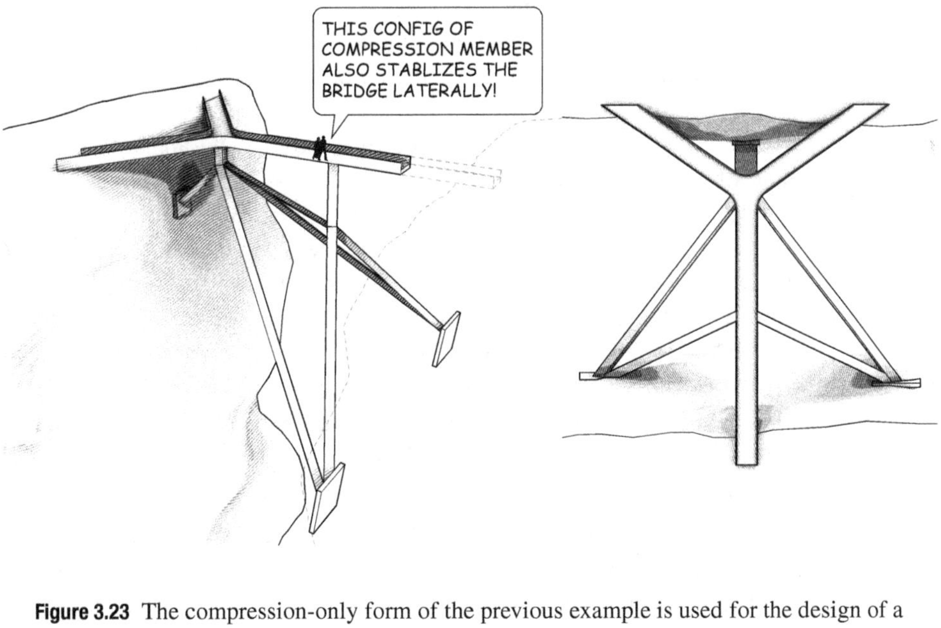

Figure 3.23 The compression-only form of the previous example is used for the design of a bridge with a branching deck.

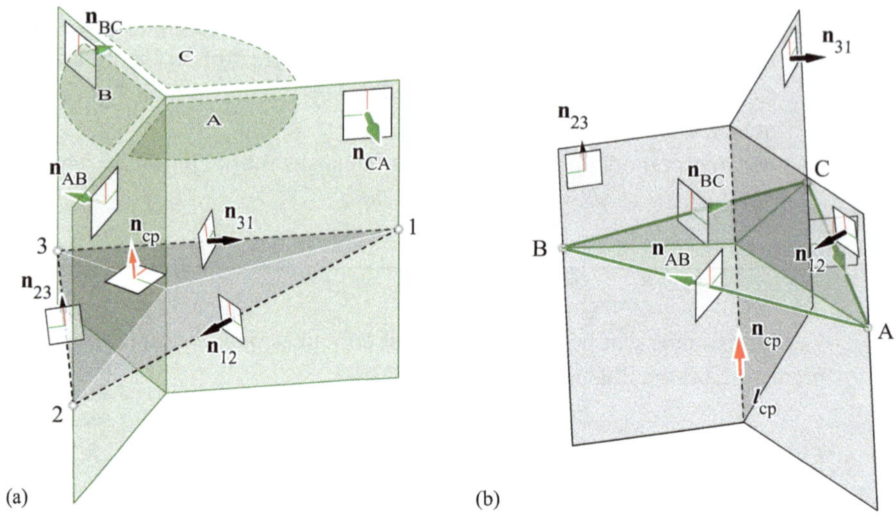

Figure 3.24 The relationship between (a) an applied load, a single closing plane, and planes normal to its edges with (b) the force face and intersecting planes of the closing edges at line l.

Moreover, the vertical reaction forces are found by x, which also lies on the line l in 2D and 3D.

While trial funicular construction is the best method to locate line l in 2D, it is not necessary for form finding in 3D.

Consider the example of Figure 3.24 where load $\mathbf{f}$ is applied in the space bounded by a closing plane $\mathbf{n}_{cp}$ connecting three supports with various heights in the form diagram. The subspaces surrounding the applied load $\mathbf{f}$ are labeled A, B, and C. Consequently, the planes dividing these subspaces are shown by their normal vectors $\mathbf{n}_{AB}$, $\mathbf{n}_{BC}$, and $\mathbf{n}_{CD}$.

First, let us construct the force face reciprocal to the applied load by using the normal of the subspaces, as shown in Figure 3.24. Next, consider planes normal to each edge of the closing plane in the form diagram and name them by the edge's vertices, for example, $\mathbf{n}_{12}$, $\mathbf{n}_{23}$, and $\mathbf{n}_{31}$. In the next step, take each of these planes from its subspace of the form diagram and place it on the vertex of the force face reciprocal to the same subspace in the force diagram. For instance, the plane with normal $\mathbf{n}_{12}$ is in the subspace A of the form and will be placed on the vertex A in the force diagram. If you put all planes on the vertices of the force face, they intersect at the line l, which constrains the force diagram for the given closing plane. Thus, the location of l can be found using the planes normal to the edges of the closing plane, and this method can eliminate trial funicular construction in 3D.

Remark 3.14. Each closing plane in the form diagram is reciprocal to a constraining line in the force diagram. Therefore, the force diagram for a boundary condition with n number of closing planes is constrained to n number of lines, each normal to its reciprocal closing plane.

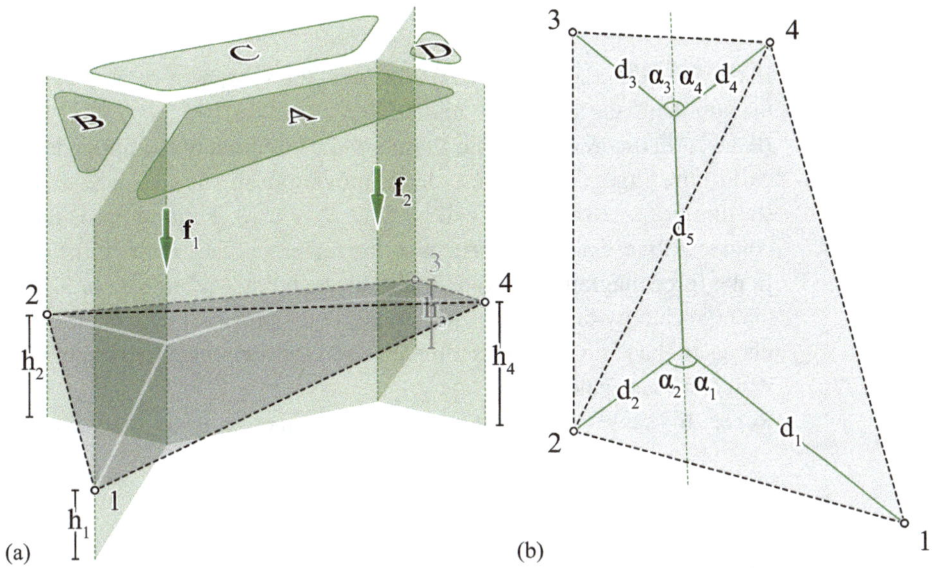

Figure 3.25 (a) A general boundary condition including two applied loads and four supports with different heights and two closing planes; (b) the plan view of the supports and applied loads with the parameters controlling the asymmetric geometry of the supports.

3.8 Form Finding for More than Three Supports

Unlike 2D graphic statics, in 3D, a boundary condition might include more than one closing plane. The number of closing planes has implications on the geometry and topology of the funicular form and necessitates a more general approach to from finding concerning the closing planes in 3D. Thus, we may need a generalized procedure to find funicular forms for nonconcurrent loads with more than one closing plane.

3.8.1 Parallel Loads and Multiple Closing Planes

Let us consider the four support locations of Figure 3.25 and two parallel applied loads, each attached to three planes/faces. As shown in Figure 3.25a, one plane/face is shared by two loads, and the other two planes pass through the supports. The angles of the planes are also shown by α_i, and their projected length is d_i. Each support has a different height shown by h_i, and two closing planes connect the four supports. Figure 3.25a,b shows this asymmetric boundary condition's perspective and plane view. We divide the form-finding problem for this boundary condition into multiple examples as shown in the Example 3.5.

Example 3.5 Construct the force diagram and find the location of the resultant forces using trial funicular construction for the boundary condition of Figure 3.25.

Construction steps:

1. In the form diagram of Figure 3.26a, label the subspaces created by the planes of the applied loads as *A*, *B*, *C*, and *D*.
2. In the form diagram, put your thumb in the direction of the applied load $\mathbf{f}_1$; the curl of the fingers shows the order of moving from one subspace to the next, starting from *A*. This order will be $A \rightarrow B \rightarrow C \rightarrow A$ and imposes the connectivity order of the vertices of the reciprocal face in the force diagram.
3. In the force diagram of Figure 3.26b, locate point *A* arbitrarily in 3D space. Using the connectivity order of vertices found in the previous step and the normal of the planes in the form diagram, complete the face reciprocal to the applied load $\mathbf{f}_1$ with a chosen scale factor.
4. Repeat the same procedures to complete the face reciprocal to the force $\mathbf{f}_2$.

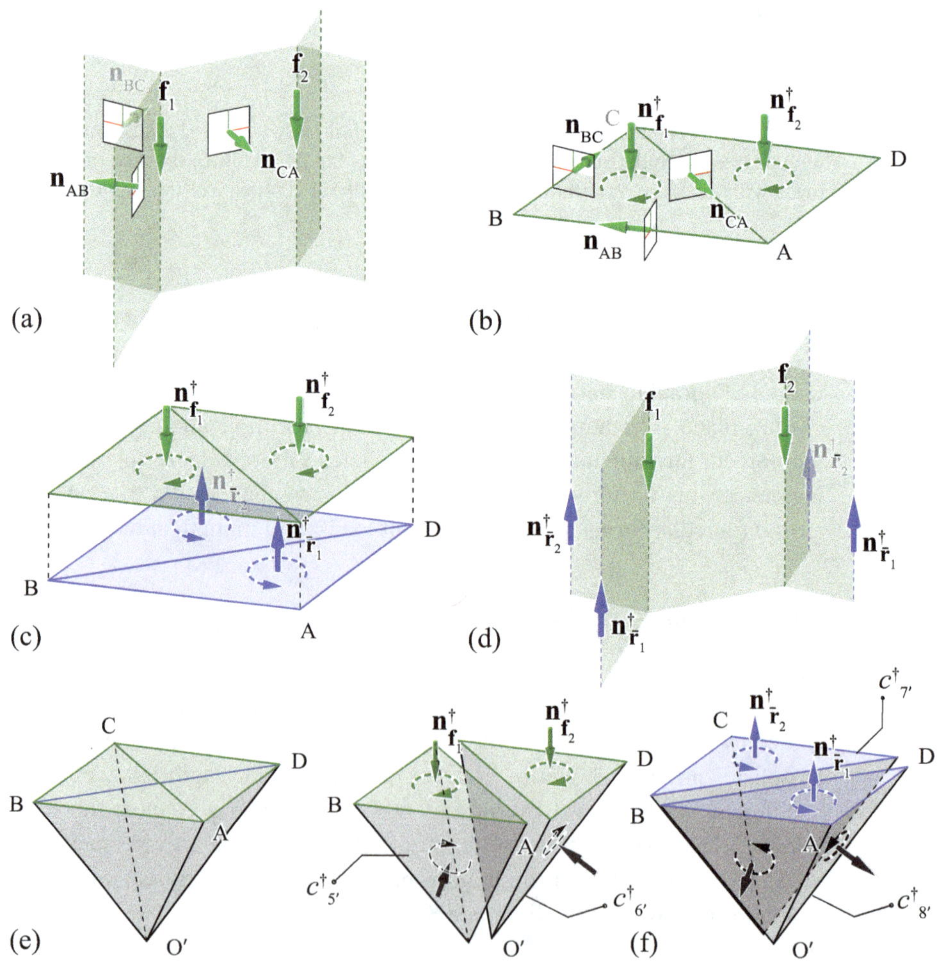

Figure 3.26 (a) The applied loads and the normal of their shared faces in the form diagram; (b) the reciprocal faces of the applied loads in the force diagram; (c) closing the force polyhedron by adding two resultant faces; (d) putting lines of action of resultant forces at the supports; (e) extruding the closed force polyhedron into an arbitrary point O'; and (f) the resulting closing cells with their related normals.

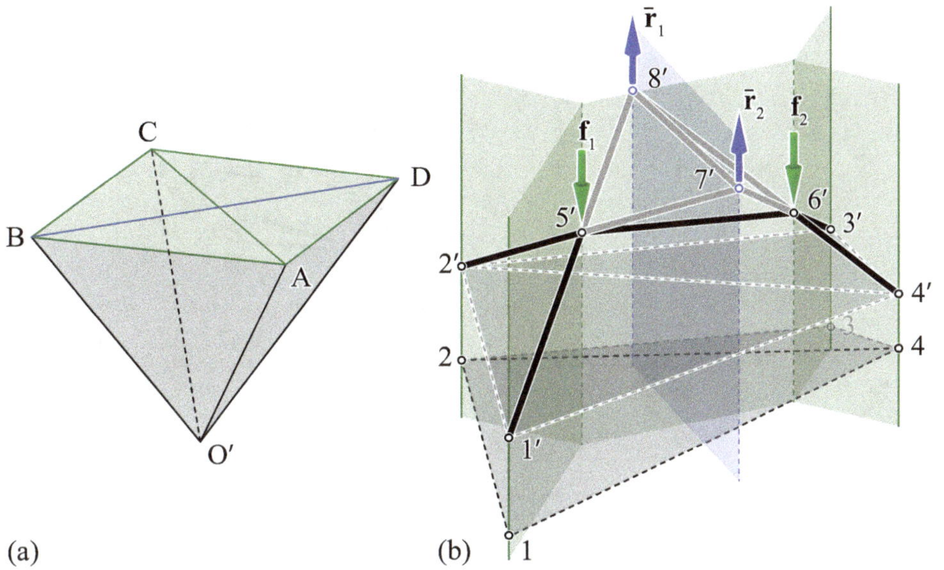

Figure 3.27 (a) The force diagram and (b) the construction of the trial funicular polyhedron on the lines of action of forces and supports locating the lines of action of resultants.

5 In the force diagram of Figure 3.26c, close the force polyhedron by adding two resultant faces, *ADB* with a normal $\mathbf{n}^{\dagger}_{\bar{r}_1}$ and *BDC* with a normal $\mathbf{n}^{\dagger}_{\bar{r}_2}$.

6 In Figure 3.26d, take the normal of the resultant faces from the force diagram and put them at the supports in the form diagram. These lines represent the component of reactions parallel to the resultant at the supports and are shown by a unit normal vector $\mathbf{n}^{\dagger}_{\bar{\mathbf{r}}_i}$.

7 In the force diagram of Figure 3.26e, extrude the closed force polyhedron to an arbitrary point O' in 3D space.

8 In the force diagram of Figure 3.26f, extruding the force polyhedron gives two pairs of closed cells: cells with face normals toward the inside and cells with face normals toward the outside.

9 Let us move to Figure 3.27a,b. The form diagram is the trial funicular polyhedron on the action lines of the applied loads, identifying the resultants' location.

□

3.8.2 Combining the Form and Force Polyhedrons into a Single Diagram

Let us draw a more explicit relationship between the closed trial funicular polyhedron and its force diagram. For this purpose, we scale the force diagram uniformly by a scale factor $\psi^{\dagger}$, and then move each force cell to its corresponding node on the form diagram of Figure 3.28a,b. In this case, we move the entire scaled force diagram from point O' to point $5'$, which is an arbitrary choice. Then, we can move every cell from point $5'$ to its corresponding node in the form diagram. As illustrated in Figure 3.28b,

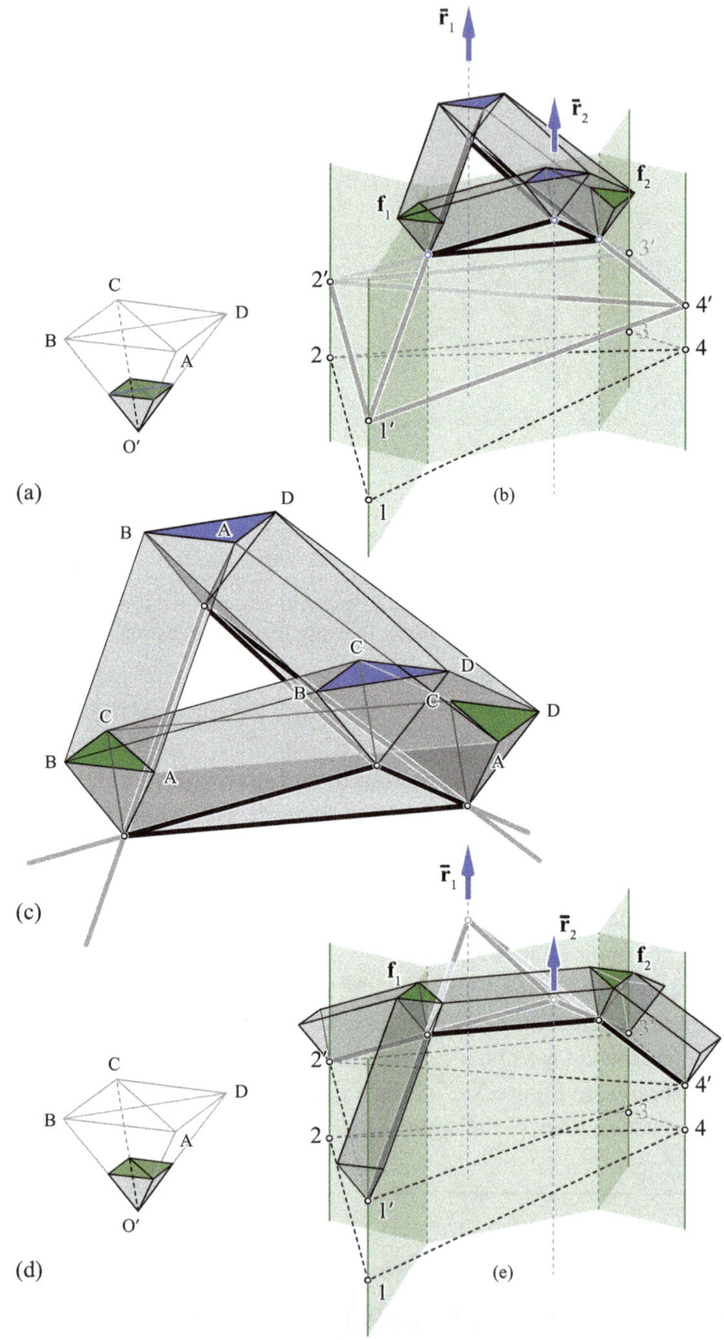

Figure 3.28 (a) The closed funicular construction and the force cells moved to their reciprocal nodes on the form diagram; (b) uniformly scaling the force diagram; (c) a close-up view of the closed funicular polyhedron thickened by extruding the faces of the reciprocal force cells; (d) the scaled force diagram; and (e) the thickened trial funicular as a compression-only form on the lines of action of the loads and the reactions.

the cells with resultant faces are reciprocal to the nodes on the line of action of the resultant force, and the cells with the applied loads are attached to the lines of action of loads. After moving the force cells, if we extrude each face of every cell in the direction of its normal, it should precisely meet the face of its adjacent cell at a normal angle. We exclude the faces reciprocal to the applied loads and the resultant forces. These extrusions make a closed funicular polyhedron thickened by the cells' extrusion along its edges, as illustrated in Figure 3.28c. This closed funicular polyhedron shows the moment equilibrium in the system, and the closed cells show the equilibrium of the forces at each node.

Suppose we exclude the cells connected to the resultant faces in the force diagram and assume that the force diagram is reciprocal to a compression-only funicular form. In that case, we will end up with a different funicular solution. Figure 3.28d,e shows the thickened funicular form where the faces of the cells are extruded along the edges of the funicular that intersect the lines of action passing through the supports. This funicular is a trial construction, a compression-only form constrained to the lines of action of loads and reactions but not precisely attached to the given support locations 1, 2, 3, and 4. Example 3.6 shows how the information of the trial funicular will be used to complete the form-finding process.

Example 3.6 Using the trial funicular construction of Figure 3.28, construct a force diagram and its constrained funicular solution for the given support locations.

Construction steps:

1 In the form diagram of Figure 3.29, close the trial funicular construction by connecting vertices $1'$, $4'$,$2'$ and $4'$,$2'$ and $3'$ with two triangular faces as the closing planes. These planes are shown by their normal vectors $\mathbf{n}_{cp'_1}$ and $\mathbf{n}_{cp'_2}$.
2 In the force diagram of Figure 3.29b, draw lines l'_1 and l'_2 using the normal directions $\mathbf{n}_{cp'_1}$ and $\mathbf{n}_{cp'_2}$ to intersect the applied load faces at points x_1 and x_2.
3 Let us move to the form diagram of Figure 3.29c to continue the construction. Identify the normal of the closing planes $\mathbf{n}_{cp_1}$ and $\mathbf{n}_{cp_2}$ connecting the support locations 1 to 4.
4 In the force diagram of Figure 3.29d, draw lines using normals $\mathbf{n}_{cp_1}$ and $\mathbf{n}_{cp_2}$ from points x_1 and x_2, respectively.
5 Let us move to Figure 3.30a. In the next step, we aim to construct a force diagram consisting of closed convex cells. In the force diagram of Figure 3.30b, extrude face *ABC* to a chosen point O_1 on the line l_1 passing through x_1 to make a closed cell.
6 Extrude face *CDA* to an arbitrary point O_2 on l_2 to make another closed cell for the force diagram.

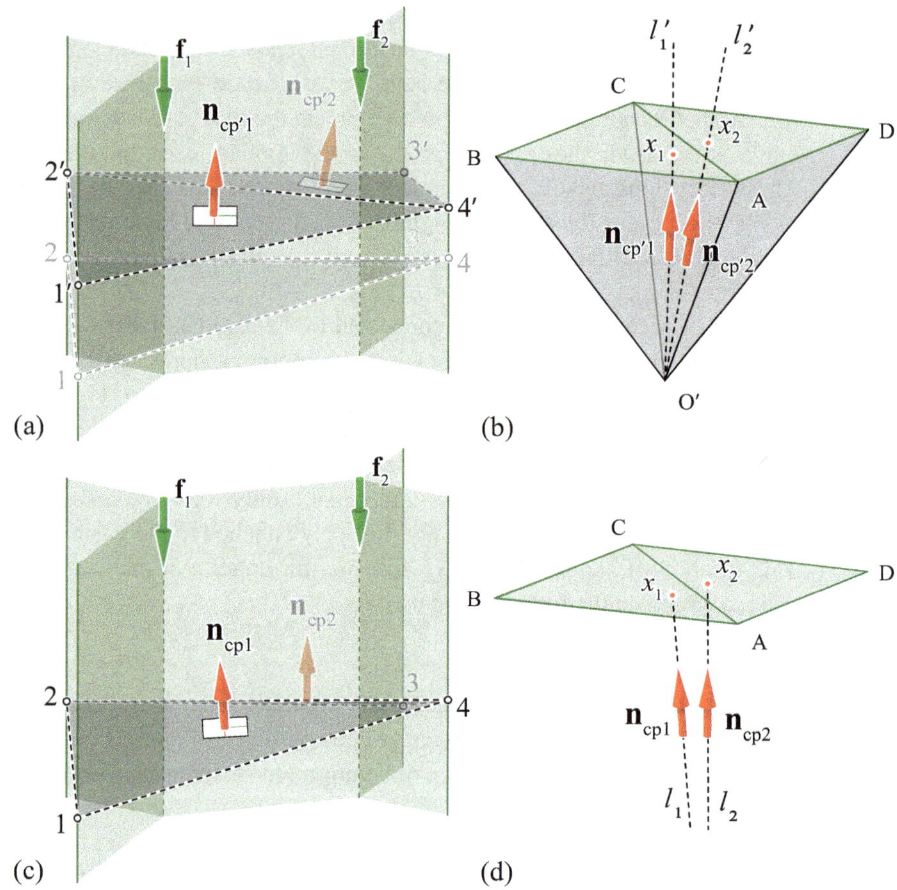

Figure 3.29 (a) Identifying the closing planes of the trial funicular with normals $\mathbf{n}_{cp'_1}$ and $\mathbf{n}_{cp'_2}$; (b) drawing lines using these normals from point O' to intersect the applied load faces at x_1 and x_2; (c) identifying the normals of the closing planes in the form diagram as $\mathbf{n}_{cp_1}$ and $\mathbf{n}_{cp_2}$; and (d) drawing lines l_1 and l_2 from points x_1 and x_2 in the force diagram.

7 The third cell can be constructed by connecting faces CAO_1 and ACO_2. Note that the location of O_1 and O_2 is arbitrarily chosen on the lines l_1 and l_2, as shown in Figure 3.30b,c.

8 Once the force diagram is constructed, we can complete the funicular form, as shown in Figure 3.30d. Note that you may begin from any support location and draw lines normal to the faces of the force diagram to intersect the lines of action of the applied loads. Once completed, there will be three vertices in this constrained funicular solution since there were three closed cells in the force diagram, as described in the previous step.

9 The funicular construction is complete, and changing the location of O_1 and O_2 on lines l_1 and l_2 will change the height of the funicular construction.

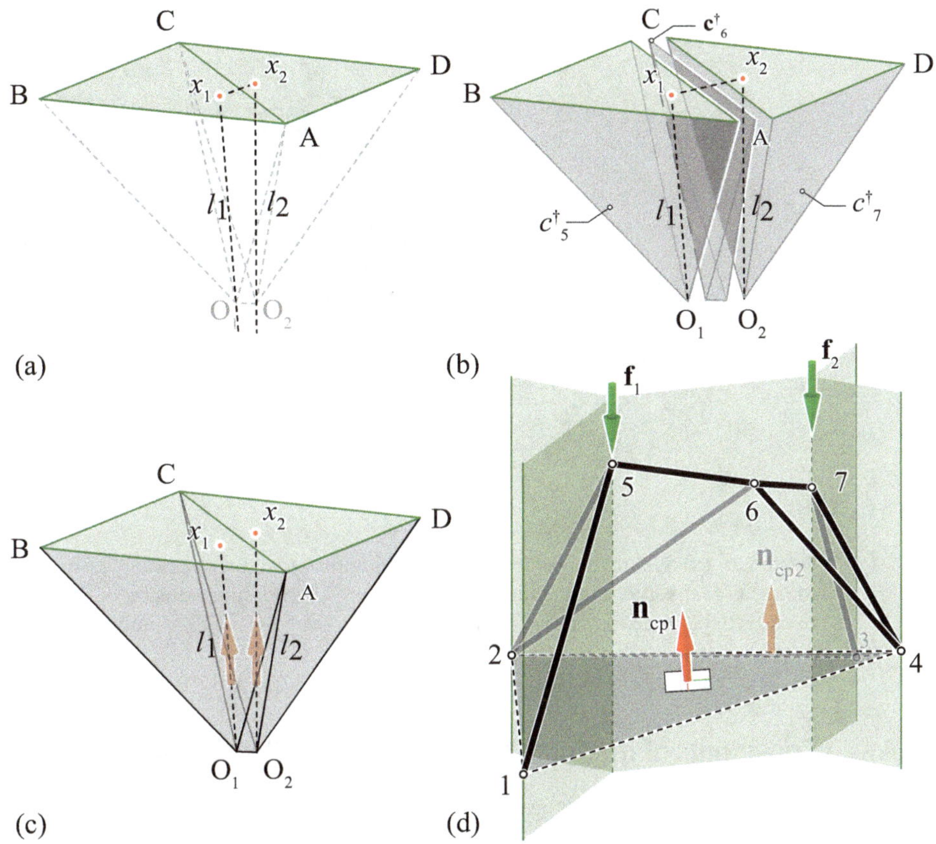

Figure 3.30 (a) Choosing arbitrary points O_1 and O_2 on the lines l_1 and l_2 to make (b) a force diagram with closed convex cells; (c) the completed force diagram; and (d) the reciprocal constrained funicular solution.

□

3.8.3 Checking the Validity of the Solution

To prove the correctness of our construction, let us start from the force diagram of Figure 3.31a and choose a scale factor $\psi^\dagger$. Next, we scale the force diagram from the reference point O_1 and move the entire force diagram from point O_1 to vertex 5 on the form. Note that any vertex of the form could be chosen for this transformation. From vertex 5, move every cell to its reciprocal vertex of the form. For instance, cell O_1O_2AB is moved to vertex 4 along the vector starting at vertex 5 and pointing toward vertex 4. Once all three cells are transferred, extrude the faces of each cell along the edges of the compression-only structural form, as shown in Figure 3.31b.

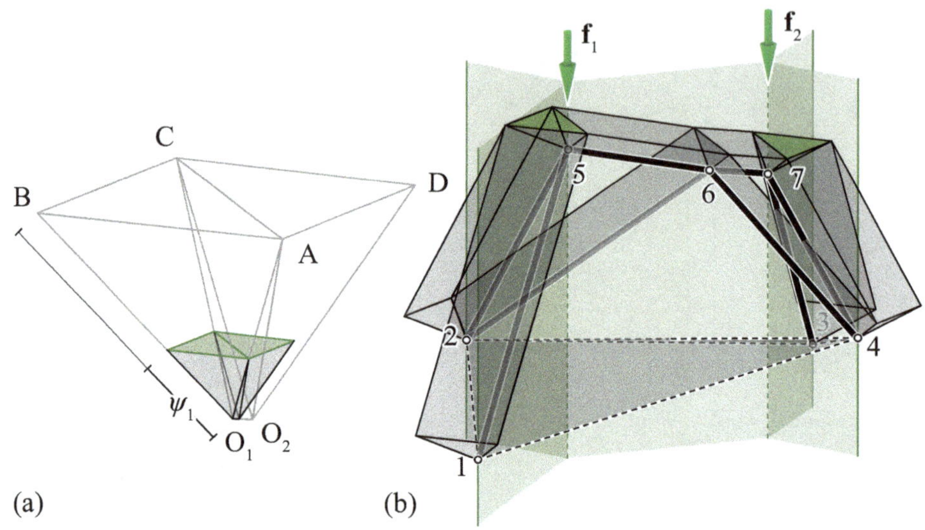

Figure 3.31 (a) Scaling the force diagram of compression-only forces and (b) moving the cells to the reciprocal vertices of the closed funicular polyhedron: the extrusion of the faces of the cells normal to the direction of the face is parallel with the edges of the compression-only form.

3.8.4 Constraints of the Force Diagram and the Effect of Boundary Condition

The reciprocal force diagram of a constrained funicular form is also constrained to a line or lines. Any extrusion of the applied load face to a point on lines l_i results in a force diagram reciprocal to a funicular form with different heights, as explained in all previous examples. The number of constraining lines equals the number of closing planes in the form diagram. The asymmetric boundary conditions, including the magnitude of the applied loads, the location of supports, and the dihedral angles of the closing planes, change the configuration of the constraining lines in the force diagram. These constraining lines may take various configurations, as shown in Figure 3.32, for both symmetric (a, c, and e) and asymmetric (b, d, and f) locations of the support. The following paragraphs expand on these configurations.

Symmetric Support Locations with Various Heights

Figure 3.32a shows a configuration where the boundary conditions are symmetrical and all four support locations are co-planar. In this case, there is only one constraining line and the load faces can be extruded to a point on the line l_1 to construct a force diagram for a constrained funicular solution. Figure 3.32c,e shows cases where the heights of the support change but the projection of the support maintains a symmetric geometry. These cases result in two closing planes and two lines l_1 and l_2 intersecting at a single point x in the force diagram. Note that the lines l_1 and l_2 share a plane with a normal $\mathbf{n}_c^\dagger$ that is parallel to the edge shared by the closing planes in the form diagram.

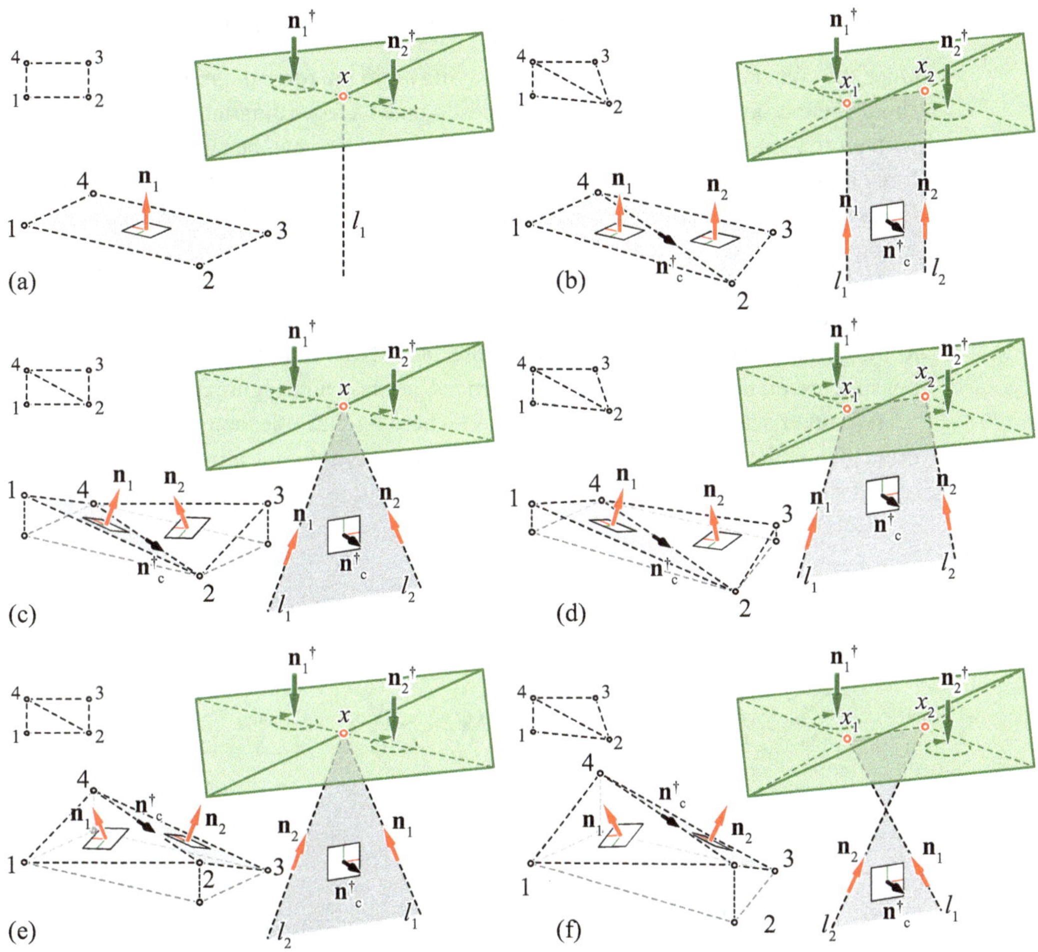

Figure 3.32 (a) A single constraining line as a result of a symmetric boundary and a single closing plane; (b) an asymmetric boundary with two closing planes and two constraining lines in the force diagram; (c) a symmetric boundary with closing planes intersecting at a ridge; (d) an asymmetric boundary with convex closing planes; (e) and (f) a symmetric and asymmetric boundary with concave closing planes.

Asymmetric Support Locations with Various Heights

Figure 3.32b,d,f shows the constraining lines for asymmetric configurations of supports where the lines intersect the force faces at two points x_1 and x_2 in the force diagram. The constraining lines l_1 and l_2 follow the angles of the closing planes. Similar to the symmetric boundary conditions, the edge shared by the closing planes in the form is perpendicular to the plane shared by the constraining lines l_1 and l_2 in the force. This plane is shown by the normal $\mathbf{n}_c^\dagger$ in the figure.

3.8.5 Decomposing the Force Diagram to More than One Point

Once the constraining lines are located and drawn in the force diagram, the force faces should be extruded to a point on those lines to make a force diagram and cells. In cases with more than one constraining line, the following steps can be taken to construct the force diagram for form-finding purposes. Figure 3.33 shows three different scenarios for decomposing the force faces for the same applied loads and support locations. For example, in Figure 3.33a, the force faces *ABC* and *ACD* are extruded independently to two points O_1 and O_2, respectively, on lines l_1 and l_2. This process makes two disjoint closed polyhedral cells connected and an additional cell in between. This is similar to the general example presented in Figure 3.30. Note that the funicular polyhedron constructed on the supports has three vertices, as shown in Figure 3.33a.

Two other scenarios may be considered where one face decomposition follows the other. Thus, the location of the second extrusion point is not arbitrary. Figure 3.33b

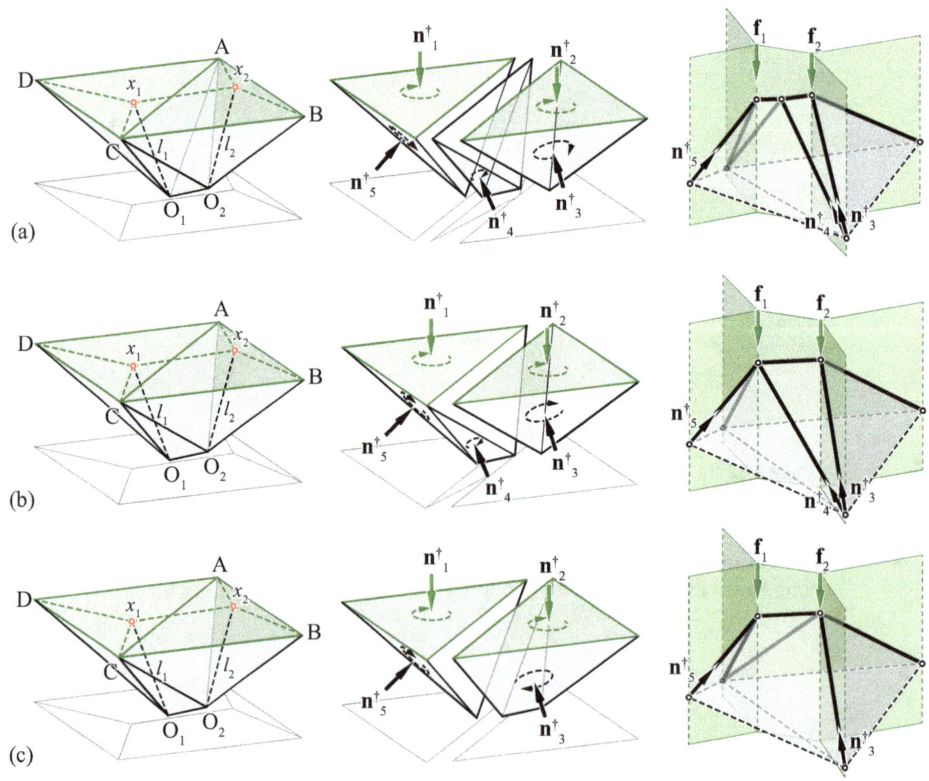

Figure 3.33 (a) Extruding the applied load faces into two arbitrary points O_1 and O_2 on lines l_1 and l_2 resulting in three cells for the force diagram and three vertices in the compression-only form; (b) extruding face *ACD* to point O_2 and extending face ADO_1 along *DA* to intersect l_1 in O_1 to make two cells with a compression-only form with two vertices; (c) extruding face *ABC* to point O_1 and finding the location of O_2 by intersecting the face CBO_1 and l_1 and its corresponding cells and funicular form.

illustrates a case where the load face ABC is extruded to a point O_1, and the location of O_2 is found by extending the edge AO_1 in the direction of AD to intersect line l_1 in O_1. This extension makes a face ADO_2O_1. If so, an additional triangular face CO_1O_2 should be added to close the newly created force cell (Figure 3.33b). Note that the same decomposition process can be started from the face ACD, which results in a different force diagram and funicular polyhedron. As a result, the configuration of the force diagram and the topology of its funicular polyhedron are not unique for boundary conditions with multiple closing planes. Thus, various design options exist for designers to explore alternative funicular solutions for the same boundary conditions.

3.9 Subdividing the Applied Loads

One of the limitations of polyhedral graphic statics is that not any given loading condition in 3D can be used in compression-only funicular form-finding. A system of loads in the form diagram must have the following characteristics in the form diagram:

- Each load in a system is attached to at least three noncoplanar faces – often shown by green faces/planes in the form diagram.
- Each load shares a planar face with one or more than one adjacent load in a system, which establishes a network of loads.
- For compression-only form finding, the subspaces created by the faces of the loads are convex.
- The supports are located on the faces of the applied loads.

These characteristics make it quite difficult for an inexperienced designer to come up with the right type of applied loads for a compression-only form finding. However, there is a simple design technique that can be quite helpful.

A single applied load in 2D may represent the sum of a system of loads. In the examples of Figure 3.34b,c, the sum of the applied loads is equal to the magnitude of a single load of Figure 3.34a. Comparing the force diagrams for all three cases shows that the magnitudes of the reaction forces at the supports do not change, and only the applied load is divided into smaller vectors and, thus, the topology of the funicular form. In fact, we can subdivide a single load to design a more articulated funicular form for the same boundary.

A similar technique can be used in 3D to replace a single load with a network of loads. If the face of a single applied load in the force diagram is *subdivided* into smaller polygons such that the sum of the areas of the smaller polygons is equal to the area of the initial polygon, then the equilibrium stays intact. Consequently, a single applied load in the form diagram can be replaced by a system of loads. Figure 3.35 shows multiple examples where a single resultant load $\bar{\mathbf{f}}$ is replaced by a network of applied loads in a space bounded by supports 1, 2, and 3.

In each case, the triangular face ABC is divided into smaller polygonal faces, as shown in Figure 3.35b–d. Note that each newly created polygon corresponds to a new load $\mathbf{f}_i$ in the form diagram, and the area of the face represents the magnitude of the

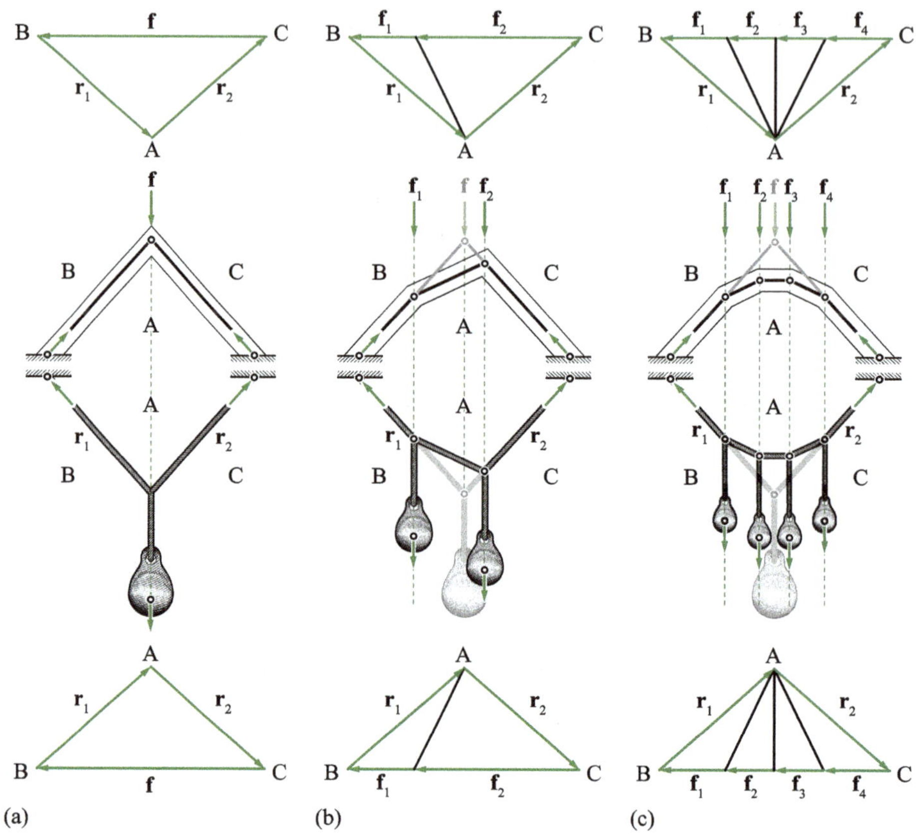

Figure 3.34 (a) A single load and its tension-only (bottom) and compression-only funicular forms (top) and their corresponding force diagrams; (b) and (c) multiple funicular forms made by subdividing a single applied load and the same reaction forces.

load in the system. The corresponding load $\mathbf{f}_i$ in the form diagram is at the intersection of planes normal to the edge vectors of the reciprocal face in the force diagram. Moreover, the planes attached to loads in the form diagram make convex subspaces named by the capital letters A–F. These subspaces correspond to the vertices of the faces of the force diagram. Every new vertex in the subdivided force diagram makes a subspace in the form diagram. For instance, in Figure 3.35b, the new vertex D in the force diagram corresponds to a subspace bounded by A, B, and C in the form diagram.

Subdividing Rules

Subdividing the load face in the force diagram to design a system of loads for a compression-only form finding requires the following geometric rules:

- Every newly created polygon should be convex, that is, the sum of the internal angles must not exceed 360°, and none of the internal angles are bigger than 180°.
- Each polygon should share one and only one edge with its adjacent polygons.

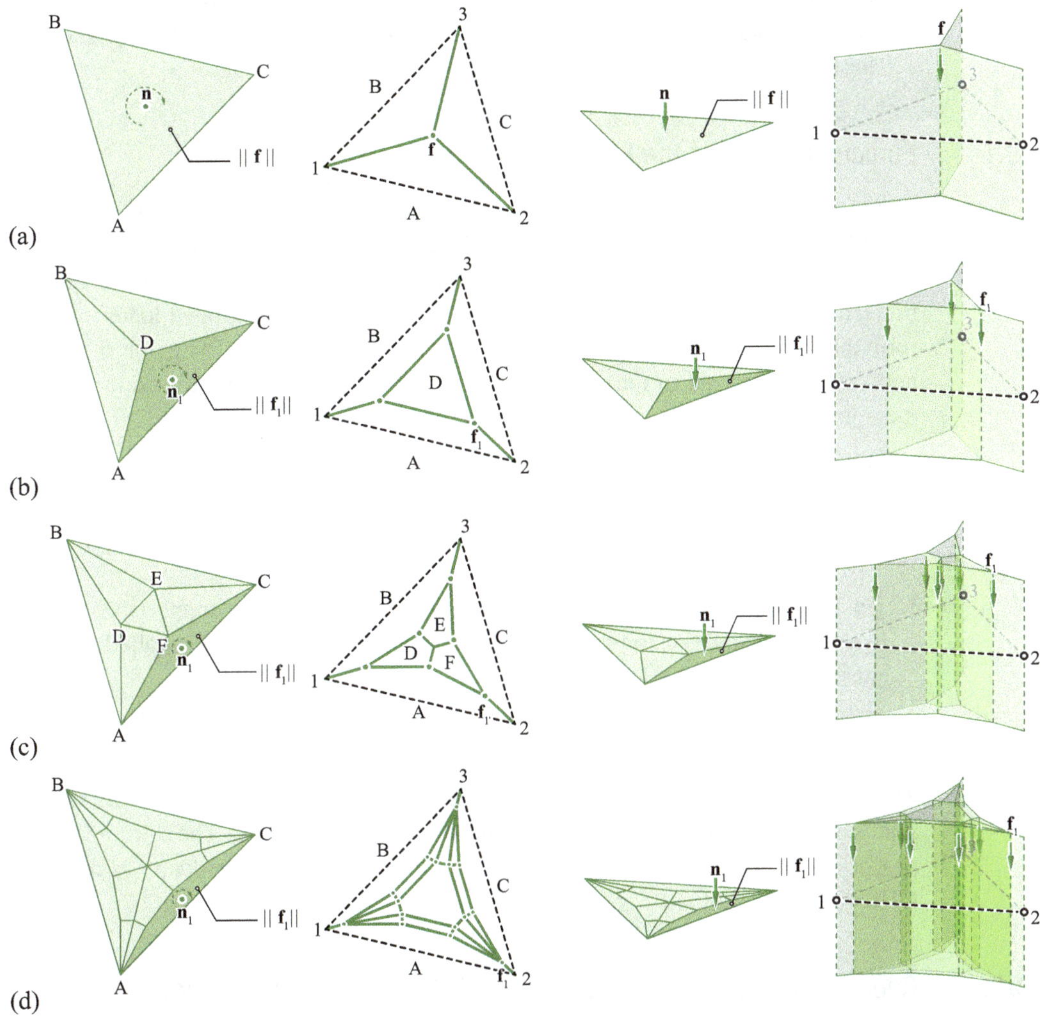

Figure 3.35 (a) A single applied load **f**, its force face, the top views of the connected planes and the boundary conditions, and its corresponding face; (b) to (d) the force systems as a result of subdividing the face corresponding to a single/resultant force and the network of connected planes intersecting at the line of action of the forces.

The load network and the subdivided faces have the following topological relationships:

- The number of loads in the network of the form is equal to the number of polygons or faces in the subdivided face of the force diagram.
- The number of faces in the load network of the form diagram and the number of edges in the subdivided face of the force diagram are equal.

- The number of subspaces/open cells created by the load network in the form diagram equals the number of vertices of the subdivided face in the force diagram. These subspaces are named by the letters A–E in Figure 3.35.

3.9.1 Funicular Graph of the Load Network

The load network, as a result of subdividing a single face, can also be abstracted into a 2D funicular graph if projected on a plane that is parallel to the subdivided face. The projected graph may represent a 2D network where the edges are, in fact, the projection of the faces, and the vertices are the location of the applied loads in the 3D form diagram as illustrated in the top views of Figure 3.35. The 2D graph of the load network is reciprocal to the force polygons made by subdividing a face in the force diagram. Thus, the following topological properties hold:

1. The load network of the form diagram is the topological dual of the graph of the subdivided force face of the force diagram.
2. Each edge of the 2D network of loads in the form diagram is perpendicular to its corresponding edge in the force diagram.
3. The load network of the form is constrained to the location of supports;
4. In a compression/tension-only funicular form, the network of the applied loads consists of convex polygons only.
5. The load network is not unique, and the length of its edges can change without changing the reciprocity.

The first two properties are the direct outcome of the reciprocity between the two diagrams, which imposes topological and geometric dependencies. The third and fourth properties are related to the boundary conditions, and the last property is related to GDoF of a 2D graph.

3.9.2 Geometric Degrees of Freedom of a Polyhedral Load Network

As shown in Figure 3.36a, the topology of the load network can be found from the graph of the edges of the subdivided face. Each closed polygon of this 2D force diagram corresponds to a vertex in the load network, and the edges of the force are reciprocal to the edges connecting the vertices in the load network. The vertices of the load network are indexed by numbers. Once the topology, that is, the connectivity of the form diagram, is found, the projected network of the loads can be constructed from the support locations named 1, 2, and 3 in Figure 3.36. Let us start from point 1 and locate vertex 4 by moving in the direction perpendicular to its reciprocal edge in the force diagram. The edge length connecting vertex 1 to 4 can be arbitrarily chosen as shown by l_1 in Figure 3.36b. From vertex 4, the location of the vertex 5 may be found by choosing another arbitrary length for the connecting edge as l_2 shown in Figure 3.36c. Continuing the process requires the third edge length l_3 to complete the entire network by intersecting the rest of the edges, as shown in Figure 3.36d–f.

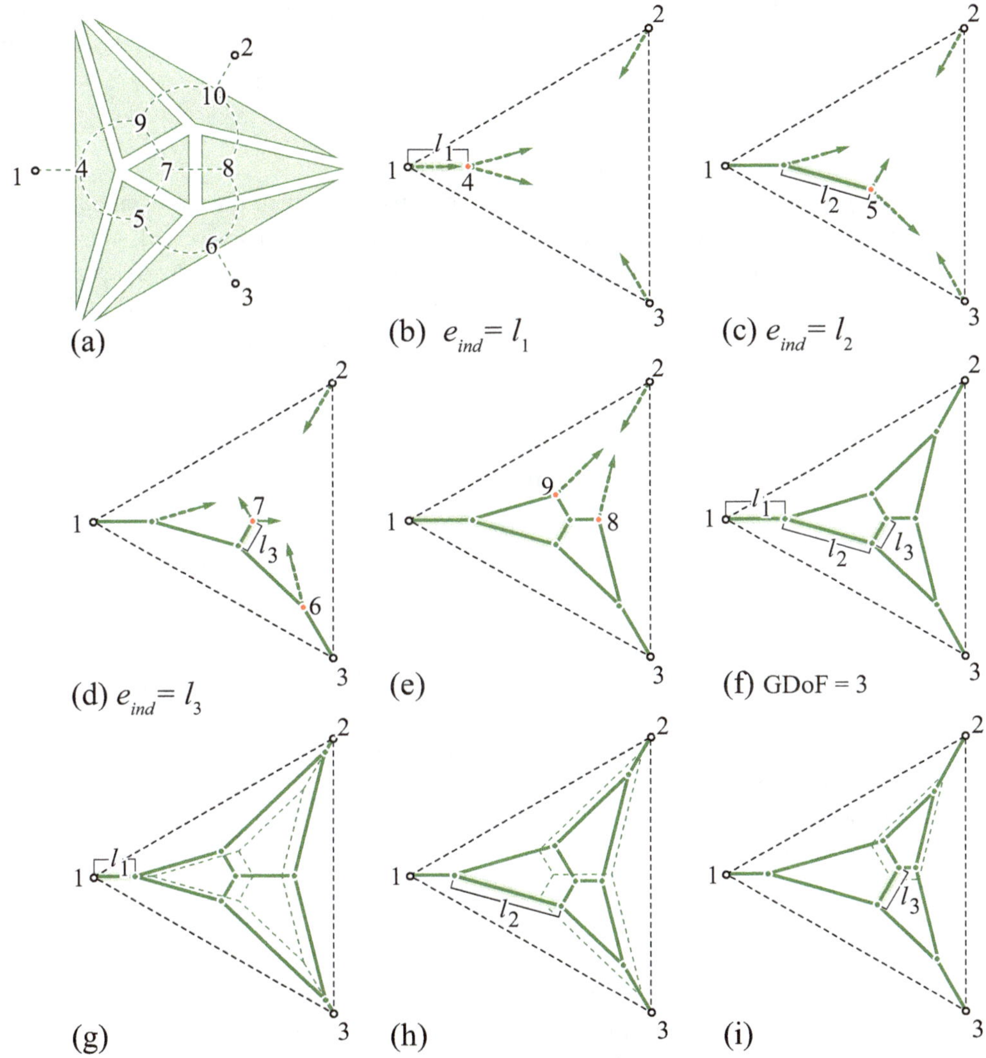

Figure 3.36 Constructing a load network for a subdivided force face: (a) extracting the topology of the network as a dual graph of the force polygons; (b) to (e) starting from a support location and choosing an edge length for an independent edge e_{ind} to find the internal vertices of the graph such that each edge is perpendicular to the corresponding edge in the force; (f) the number of independent edges and the geometric degrees of freedom of the graph; and (g) to (i) other families of the graph with different edge lengths for the independent edges.

As noticed, *three* edge lengths had to be chosen *independently* to complete the geometry of the entire network. Thus, the geometry of the network is not unique and could change if different lengths were used for l_{1-3}. Figure 3.36g–i shows different networks with the same topology and the same direction of the edges with different edge lengths.

These edges whose lengths were independently chosen are called the *independent* edges or the GDoF of the network and are shown by e_{ind}. The length of the rest of the edges is found by choosing the length of the independent edges.

The Number of GDoF of a Load Network

The geometric properties of the load network, including their precise location of application and their relative location concerning the adjacent loads and supports, can be revealed by completing the geometry of its 2D projected graph. As explained in the example of Figure 3.36, completing the 2D load network requires identifying the number of independent edges e_{ind} in the graph, which is equal to the GDoF of the graph.

The GDoF of a single polygon f_i is equal to $e - 2$, where e is the number of edges of the polygon. The GDoF for a 2D graph consisting of polygonal faces constrained to given support locations can be found using the following formula:

$$GDoF = e_{ind} = \sum_{i}^{n}(e_{f_i} - 2) - e_{sh} - (e_{ext} - 3) \tag{3.1}$$

where e_{f_i} is the number of edges of the ith polygon/face, e_{sh} is the number of shared edges between the faces or polygons of the graph, and e_{ext} is the number of external/peripheral edges connecting the support locations. For instance, Figure 3.37a shows a load network consisting of six faces, three of which have five edges, and the other three have four edges. Note that the edges connecting the supports are also included in this calculation. The number of shared edges is 12, which results in the GDoF being 3 (Figure 3.37b).

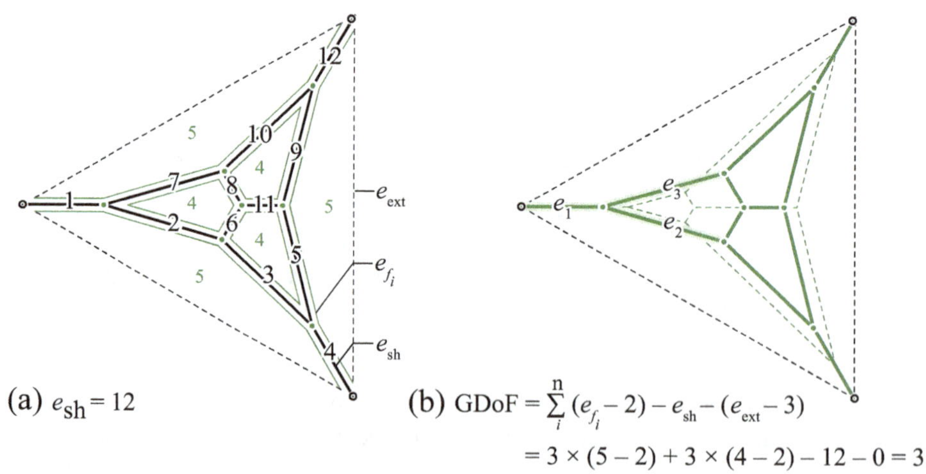

Figure 3.37 (a) A load network with six faces, three of which have five edges and another three have four edges; (b) the GDoF of the network is 3 and can be found using Equation 3.1.

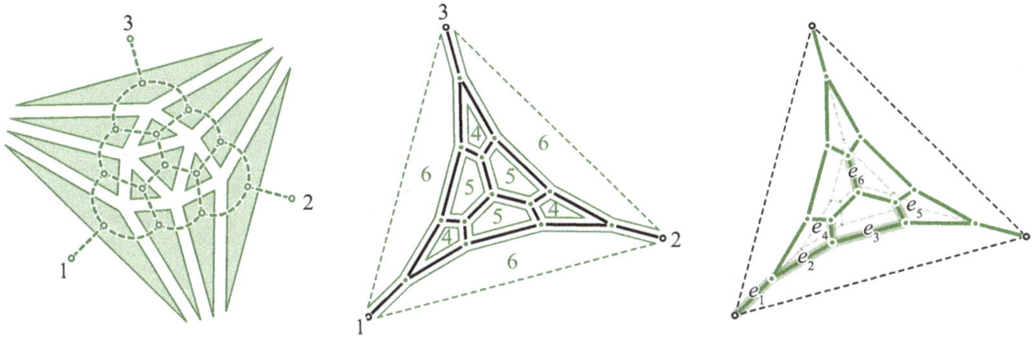

(a) (b) $e_{sh} = 21$ (c) GDoF $= 6$

Figure 3.38 (a) Another subdivision of the force face and its topological dual graph as the load network; (b) the number of polygonal faces and their edges together with the shared edges of the network; and (c) the six independent edges of the network as the GDoF of the system.

The independent edges can also be chosen arbitrarily in a network. Any edge of the network can be regarded as an independent edge as long as the number of independent edges does not exceed the GDoF of the graph. The rest of the network can be constructed following the assignment of the lengths of the independent edges. Figure 3.38 shows another subdivision of the applied load face and its dual graph as the load network and its independent edges.

3.10 Polyhedral Funicular for a Subdivided System of Loads

As explained, for compression-only form finding in 3D, a 3D system of loads can be designed by subdividing a single load face, and the location of the applied loads can be controlled using the GDoF of the load network. Once the load network is completed, a funicular solution can be constructed on the lines of action of the loads and the boundary conditions. For example, Figure 3.39 shows the geometric procedures of finding a compression-only form constrained to three support locations and the load network designed by subdividing the load face of the example of Figure 3.14.

The process of form finding is the same as the process for a single applied load, which includes the following steps: (1) constructing and closing the force polyhedron; (2) completion of a trial funicular construction; (3) locating the constraining line l for the force diagram; (4) extruding the force polyhedron to a point on the line l to construct a force diagram; and (5) constructing the constrained funicular form attached to the supports for the given lines of action of the applied loads. In the following paragraphs, we briefly expand on these steps.

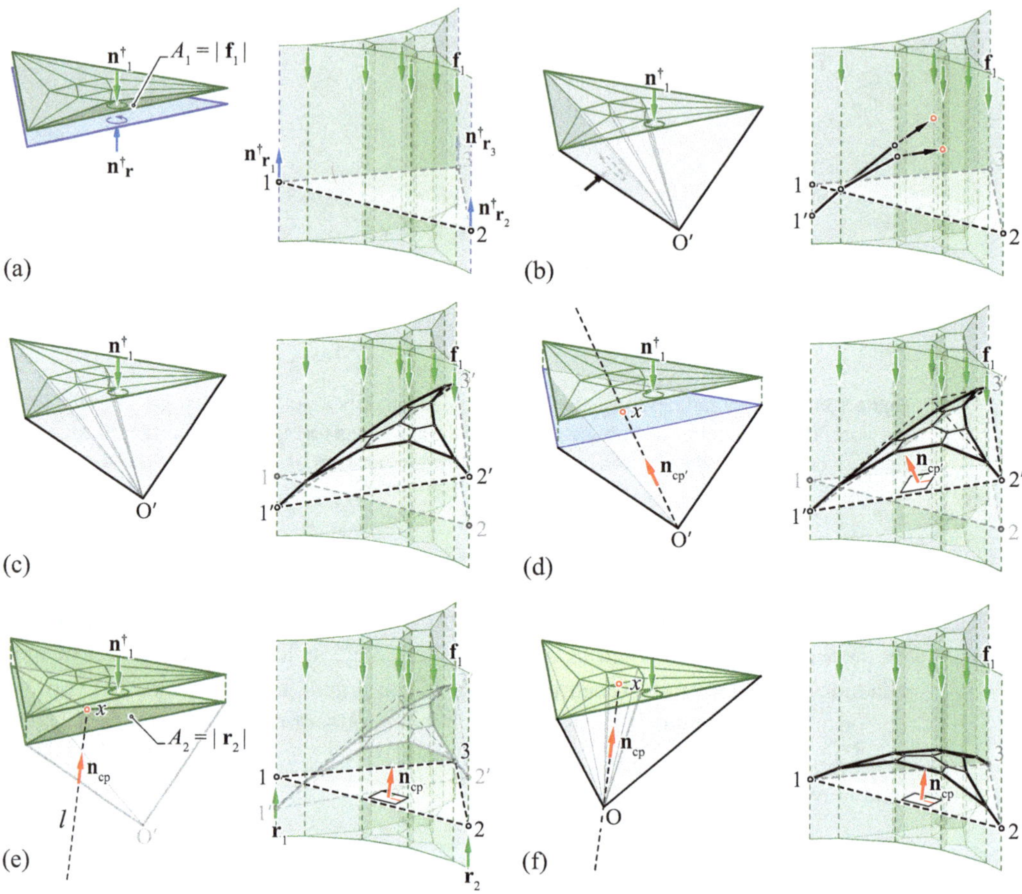

Figure 3.39 (a) The force face, its subdivision, and the corresponding load network in the form; (b) decomposing the force diagram into an arbitrary point O' and starting the trial funicular construction from point $1'$ on the line of action passing through the supports; (c) completion of the trial funicular; (d) the closing plane, its normal $\mathbf{n}_{cp'}$, and the location of x; (e) the line parallel to the normal $\mathbf{n}_{c}p$ drawn from x; and (f) decomposing the force to point O and completing the funicular solution.

Closing the Force Polyhedron

The first step in finding a funicular form constrained to a given boundary condition is to find the global equilibrium for the applied loads and the reaction forces at the supports. Let us look at Figure 3.39 for this purpose. The global equilibrium can be established by adding a resultant face to the faces of the applied loads. In the case of Figure 3.39, a blue triangular face with the normal $\mathbf{n}_{\mathbf{r}}^{\dagger}$ closes the force polyhedron. In this case, the closed force polyhedron will be flat, and the applied loads and the resultant force face are co-planar (Figure 3.39a). While the direction and magnitude of the reaction forces at the supports are yet to be found, it is a valid assumption to consider components parallel to $\mathbf{n}_{\mathbf{r}}^{\dagger}$ at the supports that can keep the applied loads in

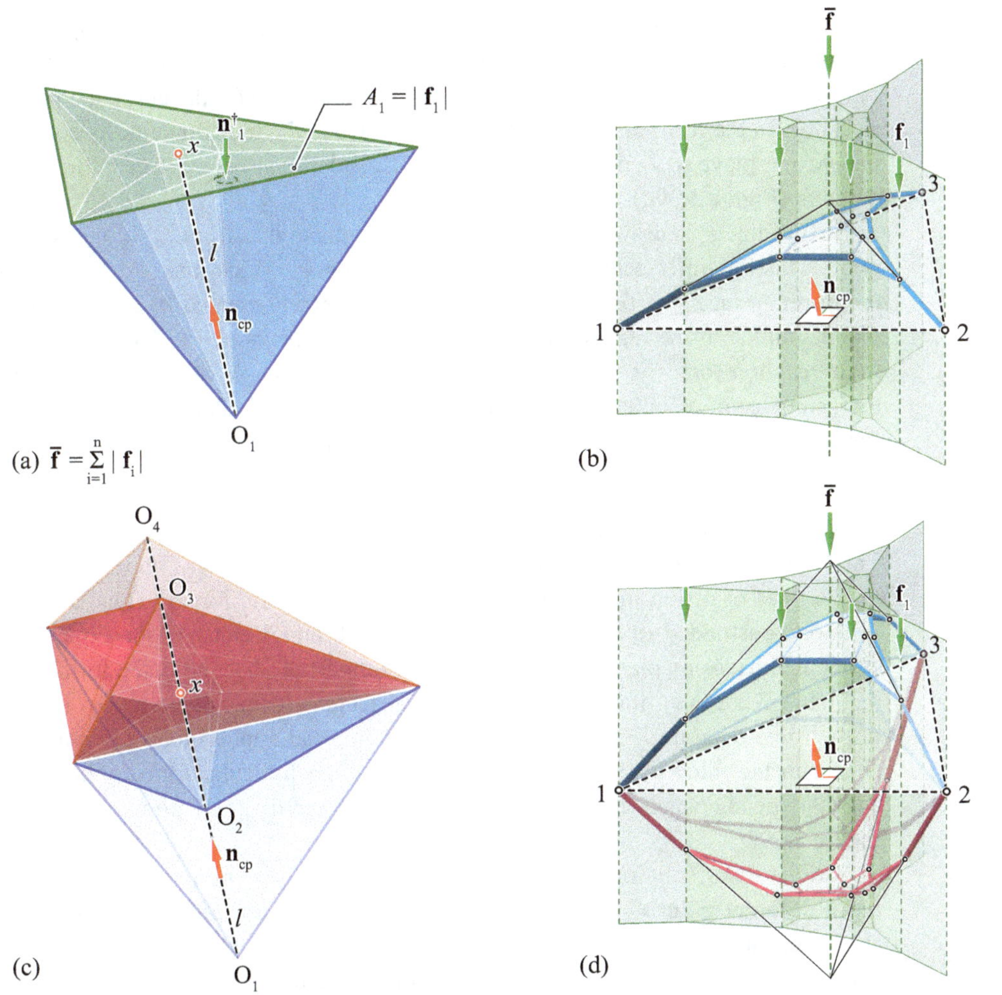

Figure 3.40 (a) The force polyhedron decomposed to point O_1 on the line perpendicular to the closing plane passing through the point *x*; and (b) multiple other funicular solutions can be explored by moving point O_1 on the line *l*. (c) moving point o1 on the line l above the point x results in the force diagram of a tension only system; and (d) the reciprocal tension-only shell constrained to the same boundary locations.

equilibrium. The line of action of these force components is used to construct the trial funicular in Figure 3.39b.

Trial Funicular Construction

The construction of the trial funicular is the next step in finding a constrained solution. Infinite funicular solutions can be constructed on the lines of action of the applied loads and the reaction forces parallel to $\mathbf{n}_\mathbf{r}^\dagger$ at the supports. Next, the force polyhedron is extruded to a point O' in 3D space. Starting from an arbitrary point $1'$ on the line of action passing through support 1, a funicular polyhedron can be completed on the line of action of the applied loads. It intersects the lines passing through support locations at 2′ and 3′ (Figure 3.39b). Extending edges connected to vertices 1′, 2′, and 3′ intersect at the location of application of the resultant force for the entire load network as shown in the form diagram of Figure 3.39d. Once the trial funicular is completed, the closing plane $\mathbf{n}_{cp'}$ connecting points 1′, 2′, and 3′ can be constructed in the form diagram (Figure 3.39d).

Next, a line l' is drawn normal to the closing plane $\mathbf{n}_{cp'}$ from O' in the force diagram. This line intersects the resultant face at point x, which makes closed faces by the resultant face's vertices (Figure 3.39d). These newly created faces correspond to the magnitude of the reaction forces parallel to $\mathbf{n}_r^\dagger$ at the supports. The magnitude of these reaction forces is invariant to any change in the angle of the closing plane or the geometry of a funicular form; that is, the location of x does not change for any funicular on the given lines of action (see the example of Figure 3.12).

Constrained Funicular Form

The location of x in the force diagram does not change for any funicular solutions on the lines of action of the applied loads and the reaction forces. Thus, a funicular solution constrained to the support locations 1, 2, and 3 can be found by: (1) constructing the closing plane $\mathbf{n}_{cp}$ connecting the supports 1, 2, and 3 in the form diagram of Figure 3.39e; (2) drawing a line l normal to the closing plane $\mathbf{n}_{cp}$ from point x in the force diagram (Figure 3.39e); (3) extruding the applied load faces to a point O on l in the force diagram; and (4) completing the funicular form by drawing the edges perpendicular to the faces of the extruded force diagram from the supports in Figure 3.39f.

Various other solutions with different heights can be explored by simply choosing a different point O on the line l of the force diagram of Figure 3.40a–d. The shallower the force diagram, the higher a funicular form will become and vice versa (Figure 3.40b,d). Besides, the compression-only funicular form can flip into a tension-only system if the force diagram is extruded to a point on the extended line l on the other side of the applied load faces, as shown in Figure 3.40b.

3.11 Subdividing Two Applied Loads and Four Supports

The same approach in designing a load network can be applied to the example of Figure 3.30 with two applied loads and four supports. The simple funicular polyhedron and four supports of that example can be expanded to a shell by subdividing the load faces into convex co-planar polygons, as shown in Figure 3.41a. The topology of the graph of the load network can be derived by connecting the centroid of the subdivided faces (Figure 3.41a). The funicular 2D graph of the load network can then be constructed on a 2D plane perpendicular to the applied loads, as shown in Figure 3.41b. The GDoF of the network can be calculated using Equation 3.1.

In constructing the force diagram, note that the newly created faces should be extruded to lines l_1 and l_2 normal to the closing planes. In the force diagram, take

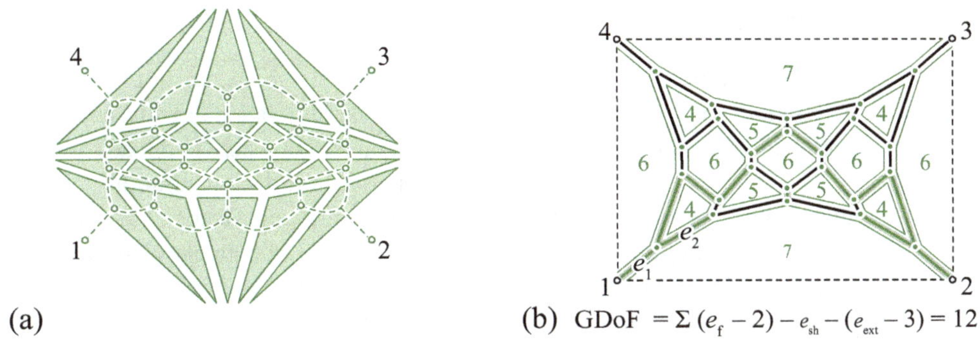

Figure 3.41 (a) The applied load faces of two loads are subdivided into a group of triangular faces on which the topology of the load network is superimposed; and (b) the projected graph of the load network and the number of independent edges (GDoF).

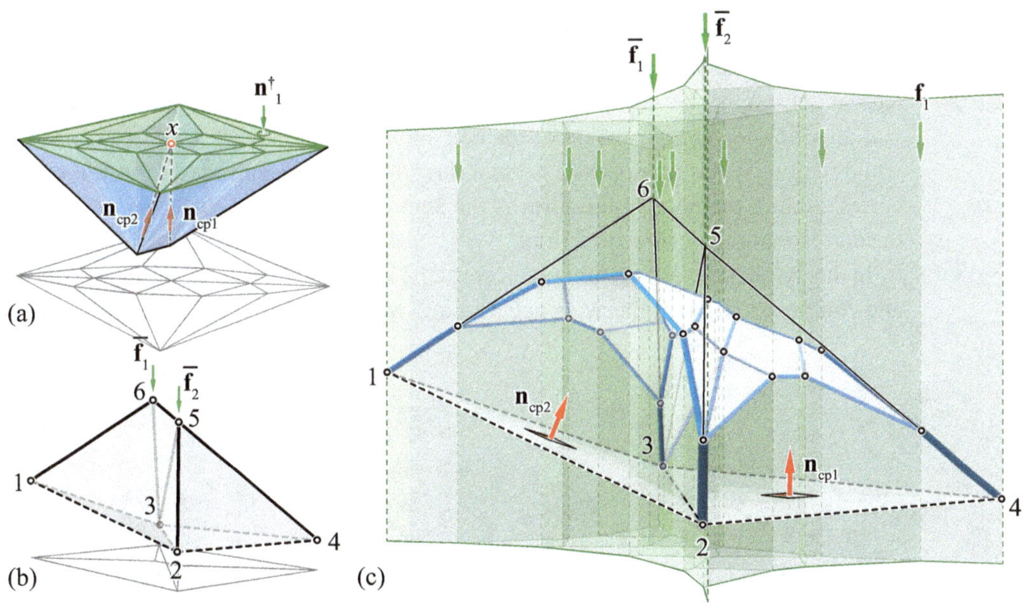

Figure 3.42 (a) Subdivided force diagram; (b) an unsubdivided funicular form with two main applied loads as $\bar{\mathbf{f}}_1$ and $\bar{\mathbf{f}}_2$; and (c) the final funicular solution with the subdivided load network and four support locations in different heights.

the faces whose reciprocal loads in the form diagram are applied to the space of the closing plane $\mathbf{n}_{cp1}$. Extrude those faces to a point on l_1. Also, extrude the rest of the faces to a point on the line l_2. There will also be cells connecting the faces of the two parts, resulting in a crease in the funicular form, similar to the case explained in Figure 3.33. Figure 3.42a–c illustrates the subdivided force diagram and its corresponding funicular form attached to four support locations with various heights.

3.11.1 Overlapping and Self-Intersecting Force Cells

Let us revisit the simple funicular compression-only geometry for two loads and four support locations in Figure 3.43. The supports' configuration directly affects the funicular form's geometry. Figure 3.43a shows a condition where two loads are applied in the space bounded by four co-planar supports. In this configuration, the force diagram is constrained to a single dashed line with a unit vector of $\mathbf{n}_{cp}$ normal to the closing plane. Figure 3.43b shows the same system with an elevated location for the support 4. Consequently, two closing planes appear in the form diagram that correspond to two constraining lines with unit vectors $\mathbf{n}_{cp1}$ and $\mathbf{n}_{cp2}$. This configuration makes a new edge in the form diagram that connects vertices 6 and 2. Changing the height of the support might also cause unexpected results in the force diagram. Figure 3.43c shows a different height for the support 2 that results in a force diagram with a self-intersecting face.

Let us take the force diagram of Figure 3.43c and analyze its geometry, as illustrated in Figure 3.44a. This force diagram includes a self-intersecting face CBO_1O_2. The direction of the edge vectors of the self-intersecting face is shown in Figure 3.44b. As illustrated, the self-intersecting face has a consistent edge direction, but it has two regions with opposite normal directions labeled as A_1 and A_2. Region A_1 has a normal direction toward the inside of the cell similar to the rest of the faces of the cell, while region A_2 has an opposite normal direction. The magnitude of the force in the corresponding member of the form is the sum of the areas of each region singed by the corresponding normal direction.

In this book, the normal direction toward the inside of the cell is considered positive and compression. It is otherwise negative and tension. In Figure 3.44b, region A_1 has

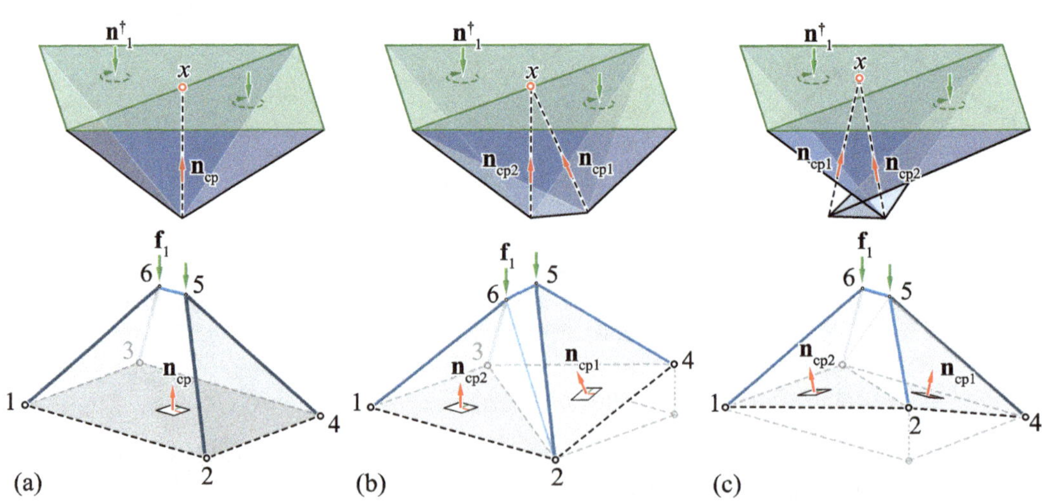

Figure 3.43 (a) Two applied loads on symmetric support locations and co-planar closing planes; (b) two closing planes with different surface normals result in the formation of a compression-only crease edge, the edge connecting 6 to 2; and (c) another configuration of the closing planes and a self-intersecting geometry of the force diagram.

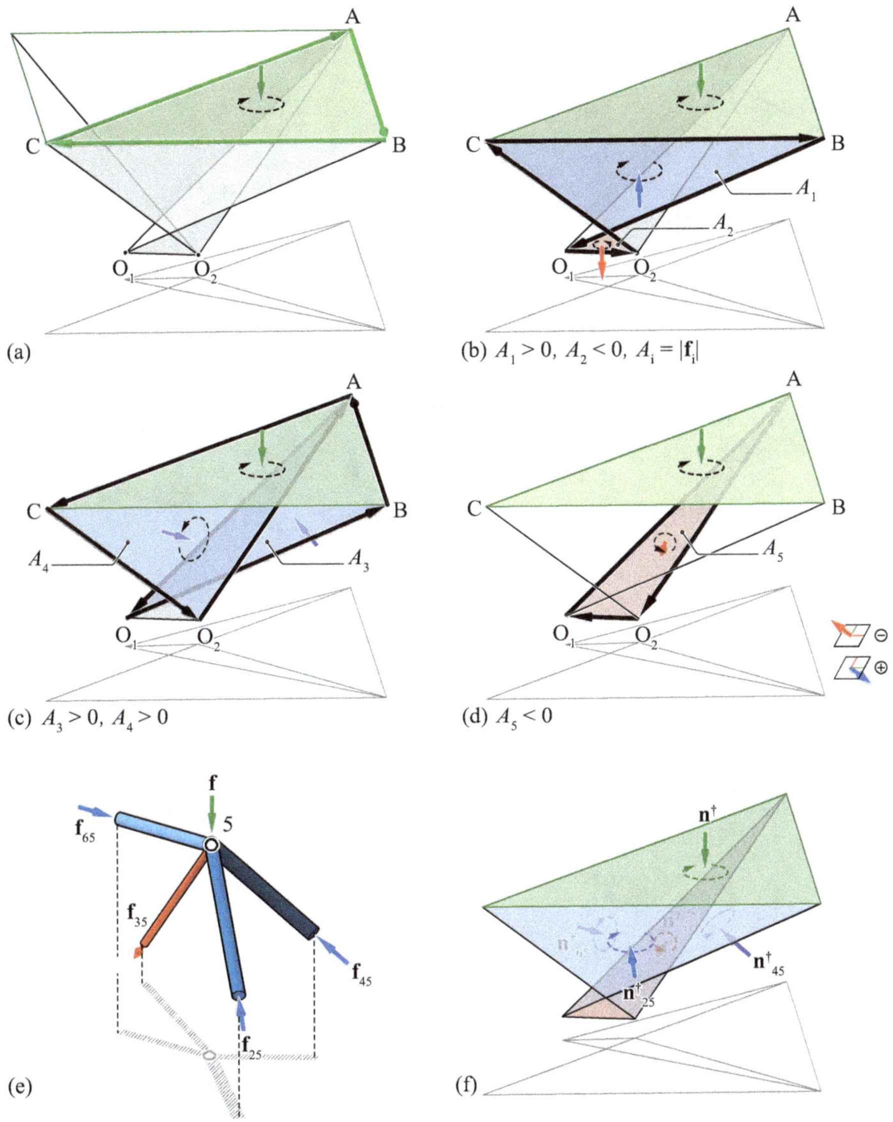

Figure 3.44 (a) The force diagram with a self-intersecting face; (b) the self-intersecting face, its areas, and the normal directions; (c) other faces of the cell; (d) the flipped face of the cell with a flipped normal direction; (e) a node of a structure reciprocal to (f) a force cell with a self-intersecting face.

a positive normal direction, while region A_2 has a negative normal direction. The sum of these signed areas results in a positive value since the area of A_1 is larger than A_2; thus, the corresponding member in the form diagram carries compression force. Figure 3.44e shows a node from the structure of Figure 3.43c that is reciprocal to the force diagram of Figure 3.44. In this figure, face CBO_1O_2 is reciprocal to the edge e_{25}, and the force $\mathbf{f}_{25}$ is compressive.

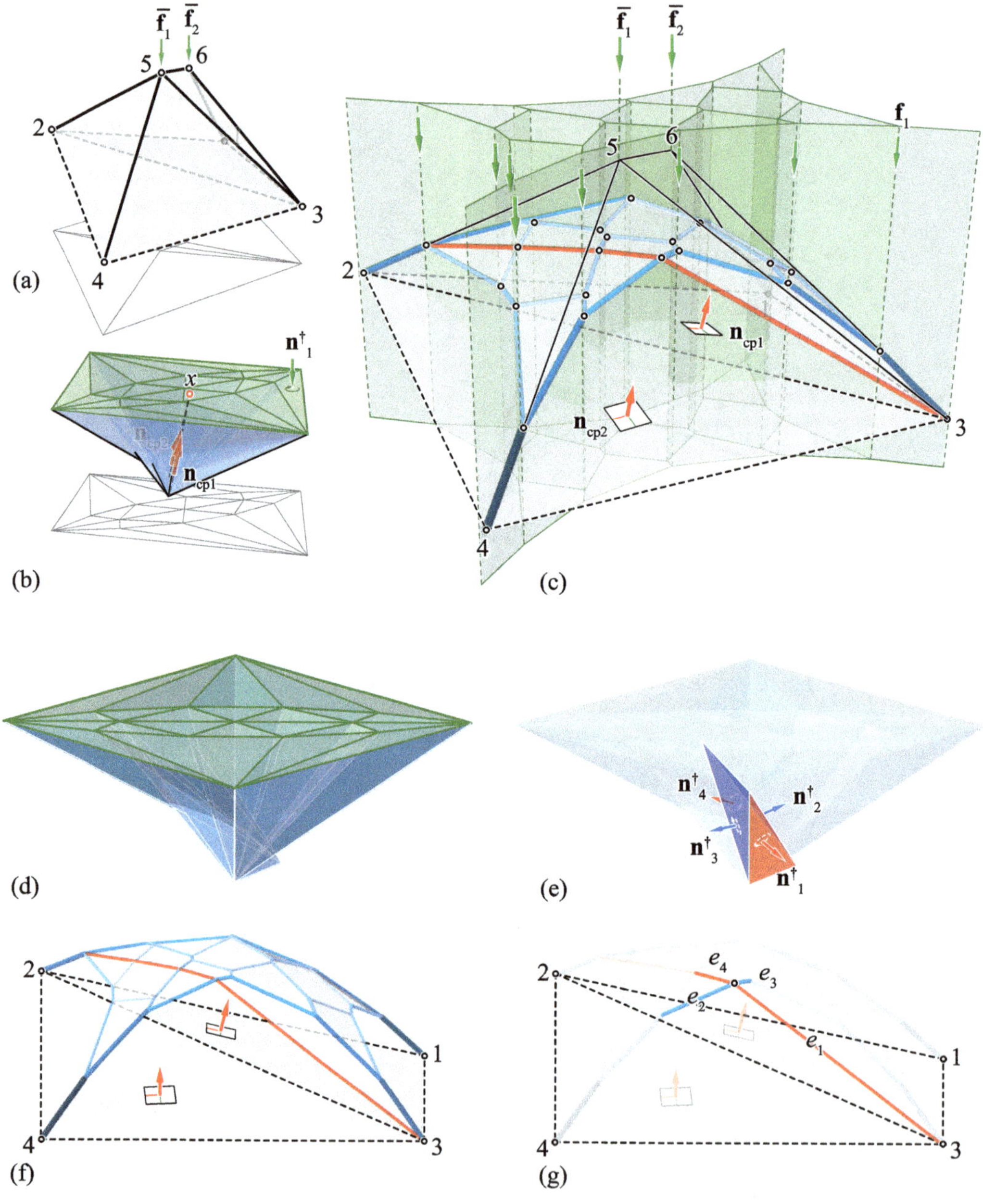

Figure 3.45 (a) A simplified funicular form with two resultant applied loads of Figure 3.43c; (b) a self-intersecting force diagram with subdivided applied loads; (c) the reciprocal compression-dominant form with a tension crease; (d) and (e) the force diagram and a highlighted internal cell with flipped normal directions; (f) and (g) another view of the reciprocal form and a highlighted node with mixed tension and compression members.

Faces with Flipped Normal

The self-intersecting polygon of a force cell affects the configuration of other faces. For example, a close look at the other faces of the force cell in Figure 3.44c,d shows that the remaining convex faces ACO_2 and BAO_1 maintain their normal directions, but the face O_2O_1A has a flipped normal. This change in the direction of the face results from the flip in the direction of the edge vector O_2O_1, which is caused by the self-intersecting face CBO_1O_2. This face is reciprocal to the edge e_{35}, and thus the internal force in this member is tensile, as illustrated in Figure 3.44e. This result is entirely unexpected since the form-finding process started with a compression-only form, and the change in support heights resulted in a self-intersecting face in the force diagram. Consequently, a member with a tensile force appeared in the structural form. This phenomenon can be pronounced in the subdivided force diagram of Figure 3.45 and its reciprocal form diagram.

Consider the simple funicular form of Figure 3.45a. The applied load faces of the reciprocal force diagram are subdivided into smaller convex polygons, as shown in Figure 3.45b. This force diagram is constrained to two lines with normals $\mathbf{n}_{cp_1}$ and

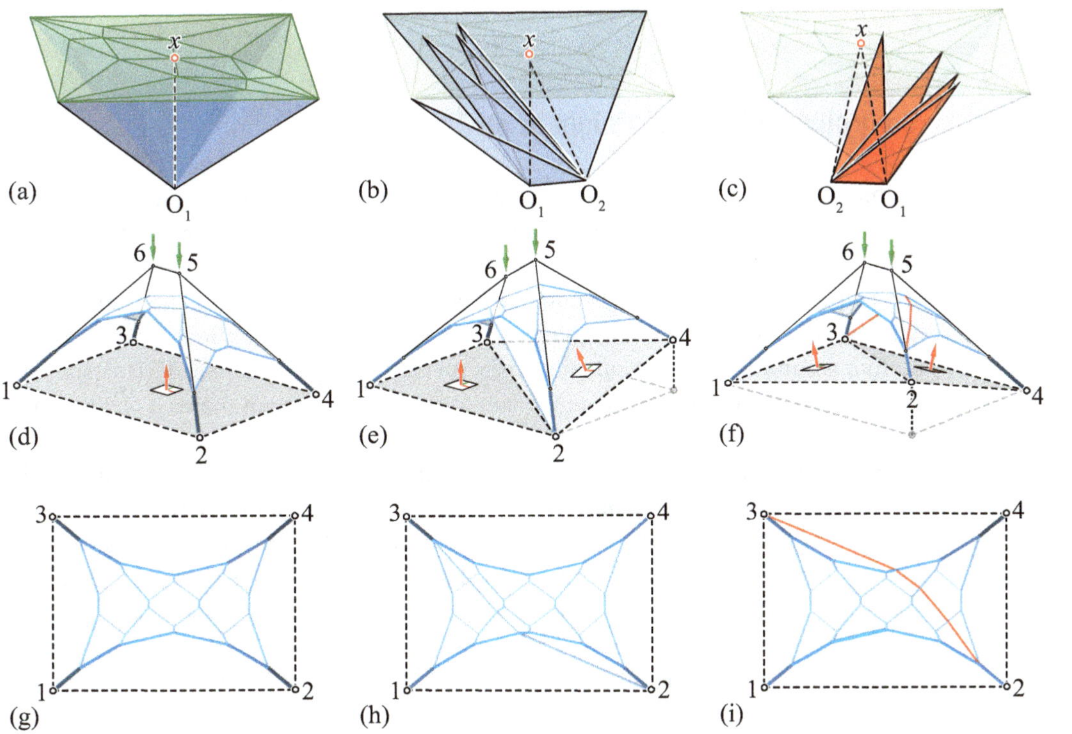

Figure 3.46 (a) A force diagram constrained to a single line and (d) its reciprocal compression-only form with flat supports and a single closing plane (g) and its top view; (b) a force diagram with two constraining lines and (e) its reciprocal form and (h) a crease passing through supports 2 and 3; (c) a force diagram with overlapping cells and faces with flipped normal (red) and (f) its reciprocal form diagram with (i) a tension crease connecting 2 and 3.

$\mathbf{n}_{cp_2}$. The change in the height of the supports results in overlapping force cells, and the reciprocal form diagram has tension members with all other compression-only members.

Note that the force diagram includes multiple internal cells whose directions have been flipped in this example. A closer look at the cells of the force diagram of Figure 3.45d,e shows that these cells have normal directions toward the outside of the cell. Consequently, relating their face directions to the corresponding members in the form diagram highlights nodes with mixed tension and compression forces, as illustrated in Figure 3.45f,g.

Figure 3.46 summarizes the previous discussions in a single illustration. The form and force diagrams shown in Figure 3.46a,d,g represent a compression-only funicular form constrained to four co-planar supports. As soon as the height of the supports changes, other internal cells may appear in the geometry of the force. In this example, these cells mainly share four faces that are reciprocal to the members establishing a crease or a valley in the compression-only form. Changing the support height may cause the force diagram to form overlapping cells, as shown in Figure 3.46c,f,i. In that case, the direction of the faces flips, and their reciprocal members carry a tensile force. The crease made by these structures will also be convex or a ridge.

3.12 Controlling Width and Height of an Arch in 3D

Although 2D graphic statics provides a simple and powerful technique to find funicular solutions for a given system of loads and support locations, in translating such solutions into a real structure, the third dimension often becomes a simple extrusion in the perpendicular direction. The main advantage of polyhedral graphic statics is that it allows us to design and tune the properties of a funicular form in three dimensions. Before we move to a 3D example, let us revisit a simple constrained 2D design and observe the relationship between the height or the *curvature* of the arch and the force diagram.

Figure 3.47 shows the entire procedural steps for constructing a constrained funicular form spanning from point 1 to 2. This process can be summarized as follows: (1) Figure 3.47a shows the load network and the location of the supports; (2) Figure 3.47b closes the force polygon and extrudes it into an arbitrary point O' to triangulate the force vectors; (3) Figure 3.47c shows the trial funicular construction to find the location of the resultant and the angle of the closing string; (4) in Figure 3.47d, the location of the point x is found and the magnitude of the reaction force components in the supports, $\mathbf{r}_1^v$ and $\mathbf{r}_2^v$, is identified that are parallel to the resultant; and (5) Figure 3.47e,f shows the construction of the form diagram for the given location of the supports and controlling the height/curvature of the funicular form by changing the height of the force diagram.

Remark 3.15. A deeper funicular form has high curvature and small radii, whereas a shallower funicular has low curvature and large radii. Respectively, a high-curvature

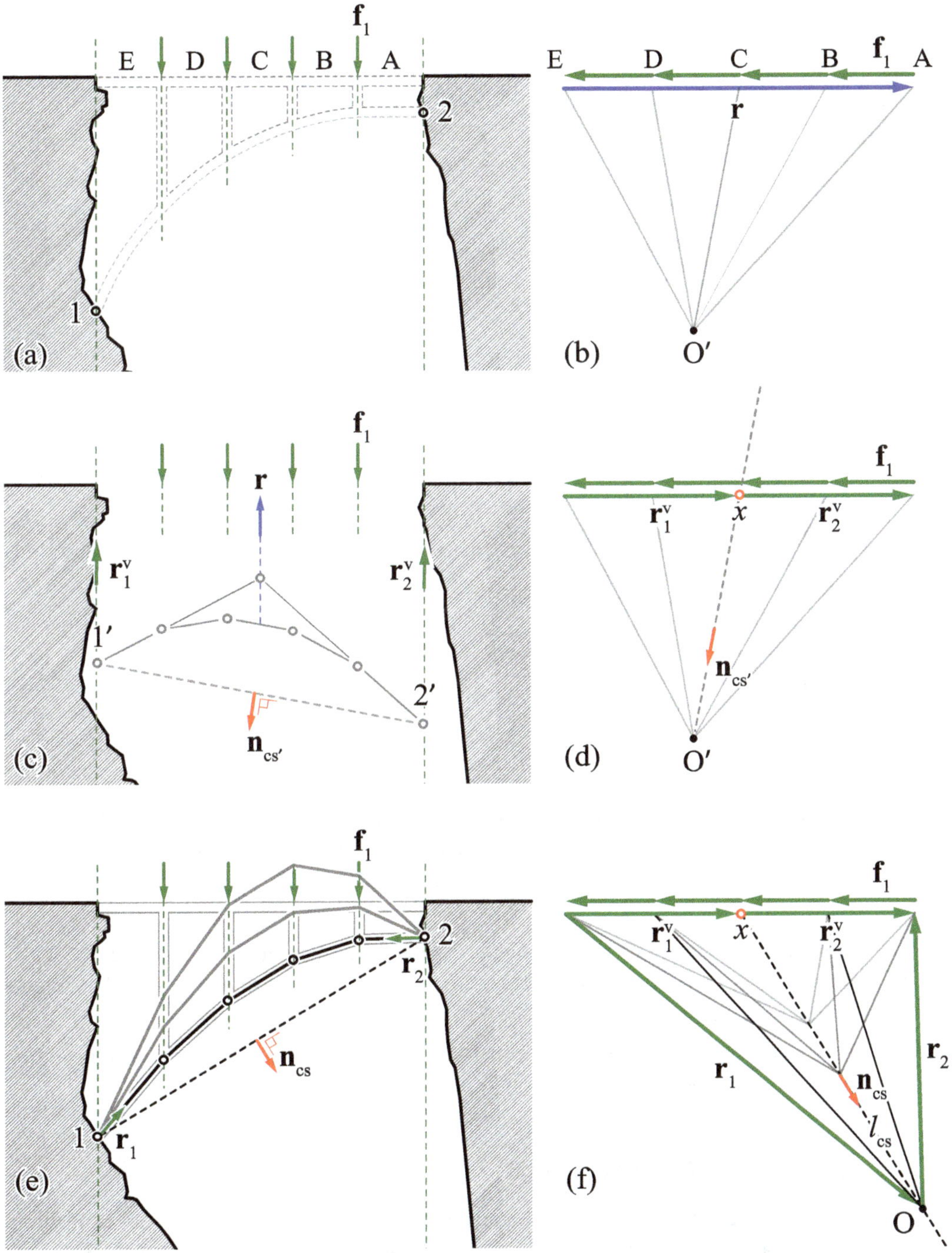

Figure 3.47 (a) Defining the load network and the location of the supports; (b) closing the force polygon and extruding it to an arbitrary point; (c) constructing the trial funicular; (d) identifying the location of x; and (e) and (f) changing the height of the funicular by controlling the depth of the force diagram.

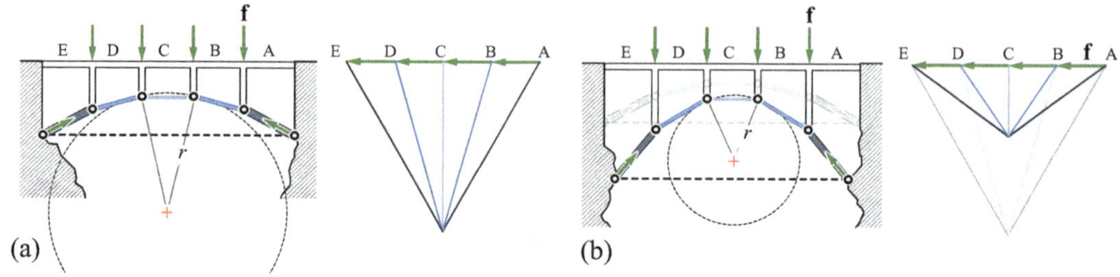

Figure 3.48 (a) A shallow compression-only funicular form with low curvature; and (b) a tall funicular solution for the same system of loads with high curvature.

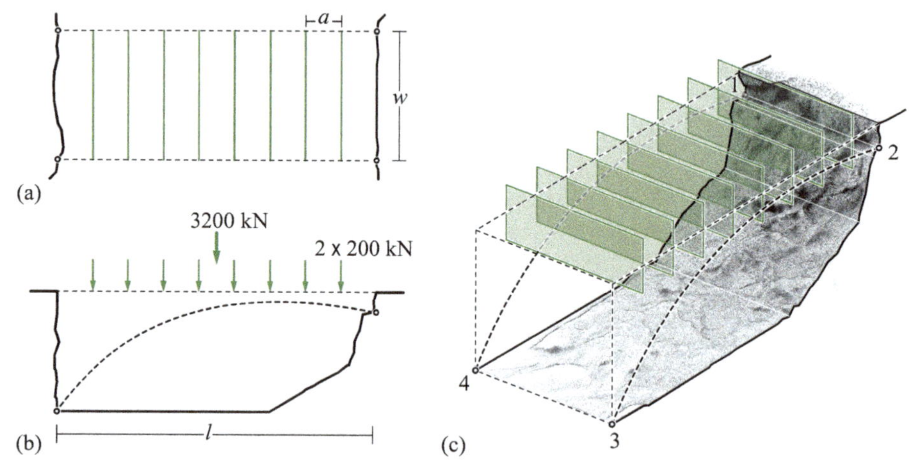

Figure 3.49 (a) to (c) The plan, elevation, and isometric view of the supports and applied loads of the given site. The loads in elevation are pairs, 2 × 200 kN, and the total is 3,200 kN.

funicular, which is taller in height, carries less internal forces, whereas a shallower/shorter funicular form carries more internal forces, as shown by the edge lengths of the force diagram.

Figure 3.48 shows two arches with high and low curvatures and their corresponding internal forces as edge lengths of the force diagram.

3.12.1 Designing a Pedestrian Bridge in 3D

With this recap in 2D, let us consider a design task to explain the process of controlling the curvature of an arch in 3D for a simple compression-only, funicular arch. This arch will act as a bridge connecting four support locations 90 m apart. All four supports are co-planar, but two have different elevations, as illustrated in Figure 3.49.

Designing the Load Network

Let us assume that the deck of the bridge consists of nine segments, and the live load and the dead load of the deck, approximately 3,200 kN, will be transferred to the arch of the bridge via eight pairs of vertical columns. These columns will lay on eight parallel planes that are equally spaced between the supports, as shown in Figure 3.49a–c. In fact, we intend to constrain the location of the columns/applied loads to these planes, as shown by the green color in the figure. Assuming each column carries 200 kN as an applied load to the arch, that makes a symmetric application of the loads in the bridge's elevation. But where exactly is the location of the applied loads or columns on those planes?

As discussed earlier, the first step in finding a polyhedral funicular form is to design the applied load network for the given boundary conditions. In Sections 3.9 and 3.10 of this chapter, we showed how an intricate network of applied loads could be constructed by subdividing a simple funicular form with one or two applied loads. Similarly, in this case, we can start the design process using a single or two applied loads that are in equilibrium by the reaction forces at the supports. Therefore, multiple scenarios can be developed, as shown in Figure 3.50. Figure 3.50a shows the boundary geometry of the bridge where a single applied load $\bar{\mathbf{f}}$ is at the intersection of the projected reactions. As a result, the face of the force diagram reciprocal to the applied load can be constructed using the angles of the projected reactions, as shown in Figure 3.50b. The applied load $\bar{\mathbf{f}}$ is in equilibrium with the resultant force $\bar{\mathbf{r}}$, which is equal and opposite to its direction. Figure 3.50c–f shows other scenarios where two applied loads are in equilibrium with pairs of resultants. In all scenarios, the applied loads and resultants intersect with the projected reaction forces at the supports.

Let us take the scenario of Figure 3.50e,f and develop a load network for our compression-only bridge. We can make a load network by subdividing the face of the applied loads. In this case, we divide the applied load face into smaller triangles with the same base length, as shown in Figure 3.51a,b. The load network of a 3D funicular form can be abstracted into a 2D funicular network on the *xy*- plane. Thus, the width of the bridge, *w*, is determined by the curvature of the 2D network of applied loads in the *xy*- plane.

Thus, the construction of the load network can be described as a 2D funicular form-finding on *xy*- plane, and the width of the bridge and the network can be controlled by the height of the triangular faces of the applied loads, as shown in Figure 3.51. Figure 3.51a,b shows the boundary conditions, the completed 2D funicular load network, and their reciprocal faces in the force diagram in both top and isometric views.

Closing the Force Polyhedron

The force polyhedron of the applied loads should be closed by adding the resultant faces. The free edges of the applied load faces in the force diagram make a parallelogram that can be closed by adding two triangular faces as resultant forces. The resultant faces are named $\bar{\mathbf{r}}_1$ and $\bar{\mathbf{r}}_2$ and are shown by blue color in Figure 3.52.

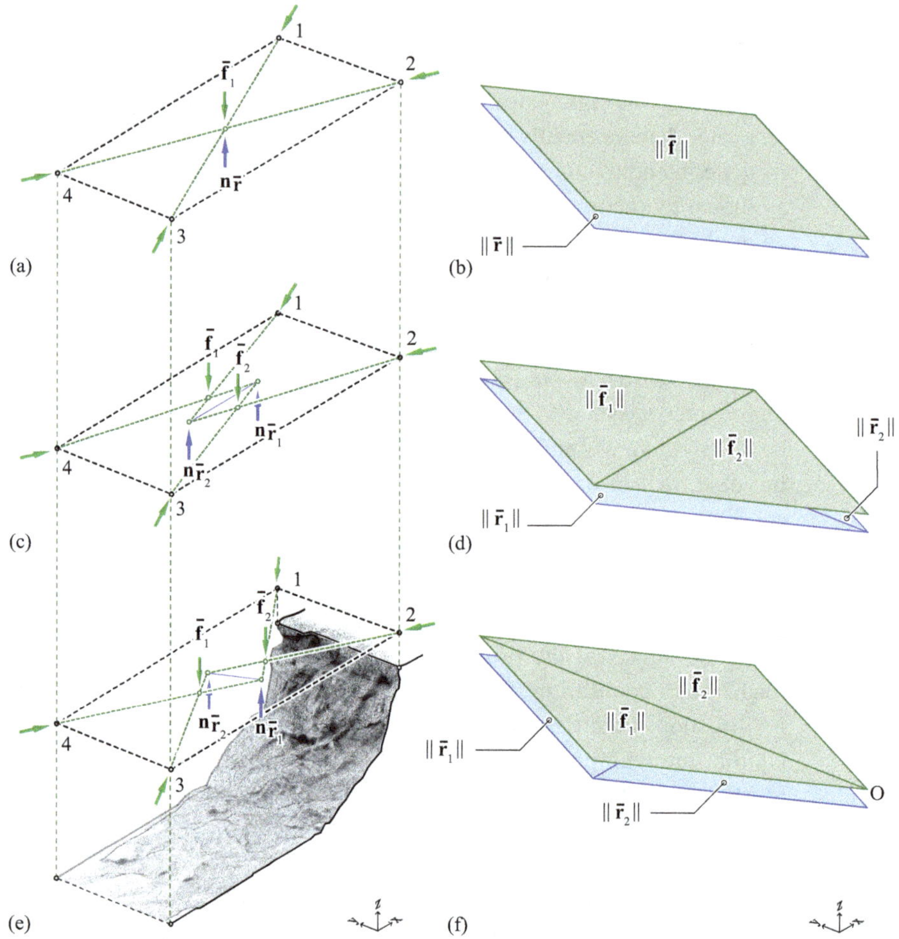

Figure 3.50 The boundary condition for the bridge and three scenarios for the applied loads: (a) and (b) a single applied load intersecting with all projected support reactions and its reciprocal force face; (c) to (f) two applied loads intersecting with reactions and resultants.

The Trial Funicular Construction in 3D

Next, the lines normal to the resultant faces are drawn at the supports on which the trial funicular form will be constructed. These lines are shown by vectors $\mathbf{n}_{\bar{\mathbf{r}}_1}$ and $\mathbf{n}_{\bar{\mathbf{r}}_2}$ in the form diagram of Figure 3.52b. In the next step, the closed force polyhedron is extruded to an arbitrary point O'', and the trial funicular form is then constructed on the lines of action of the load network in the form diagram, as shown in Figure 3.52d. The intersections of the starting and ending edges of the funicular give the locations of the resultant forces $\bar{\mathbf{r}}_1$ and $\bar{\mathbf{r}}_2$ in the form diagram. More importantly, the closing plane of the trial funicular gives the normal $\mathbf{n}_{cp'}$. Then, by drawing line l' from point O'' and intersecting the resultant faces, we can locate point x in the force diagram, as shown in Figure 3.52. Interestingly, the location of the x found by the 2D funicular construction matches the location of x found using 3D trial construction, as shown in Figure 3.52d.

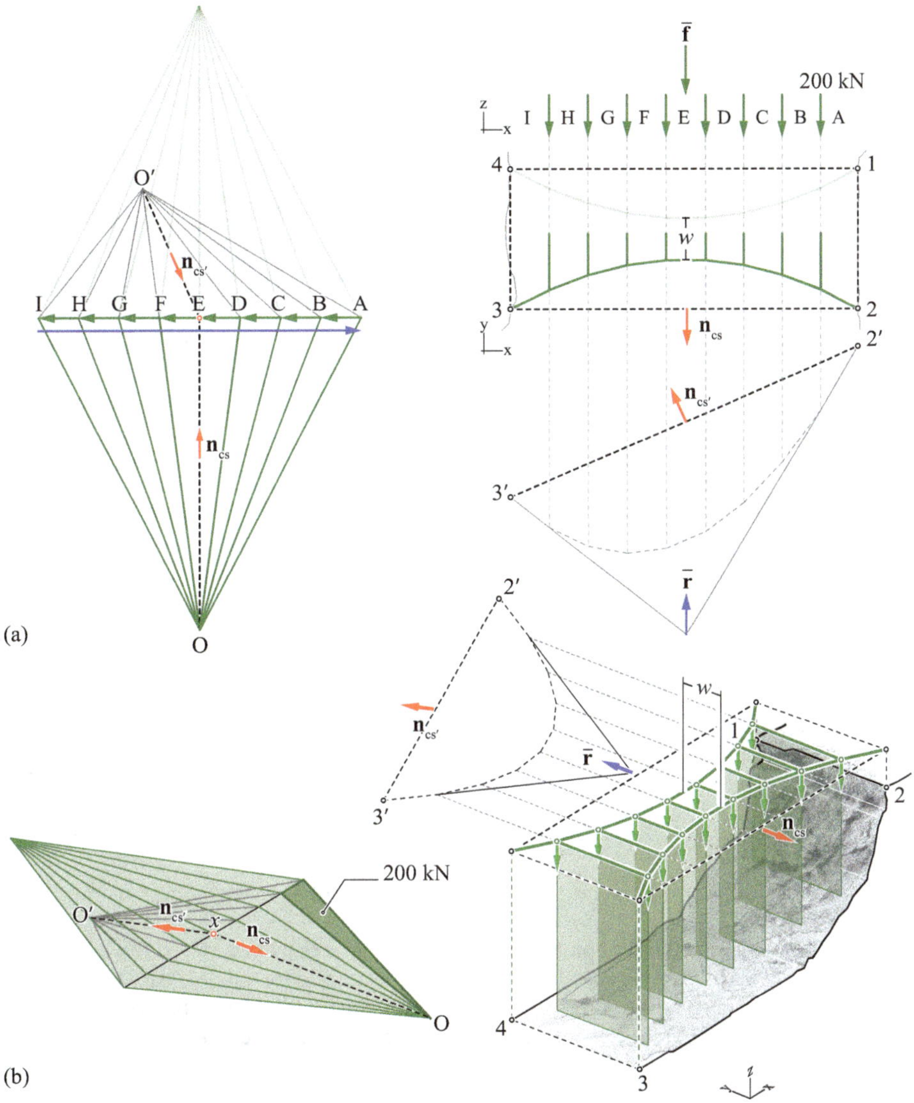

Figure 3.51 (a) The subdivided load network and its constrained 2D reciprocal load network with the width w including its 2D trial construction; (b) the isometric view of the subdivided faces of the applied loads and its 3D reciprocal load network with the width w and their shared faces.

Funicular Form of the Bridge

To construct a funicular form of the bridge for the given supports, its force diagram should be completed first. We start the process by constructing the closing plane connecting the structure's supports. The closing plane with its normal $\mathbf{n}_{cp}$ is the plane connecting the support locations 1–4, as shown in Figure 3.53b. Note that the elevation of the support locations has been deliberately chosen such that all four supports are co-planar. Otherwise, more than one closing plane must have been used to connect

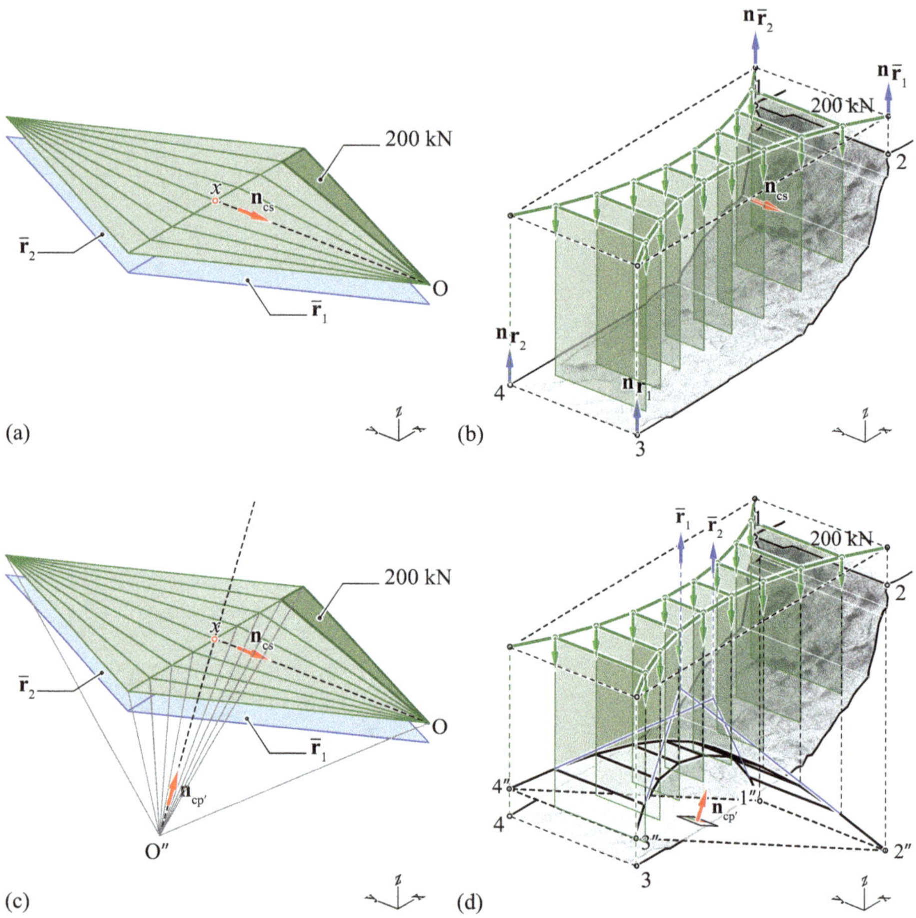

Figure 3.52 (a) The applied load and resultant faces make a closed force polyhedron; (b) the load network as a 2D funicular form is constructed for the boundary condition; (c) the trial force polyhedron constructed by extruding the faces into an arbitrary point O''; and (d) the trial funicular polyhedron and its closing plane connecting vertices $1'' - 4''$.

the support locations, and depending on the symmetry of the boundary conditions, the cases of Figure 3.32 could have happened.

Next, we draw line l parallel to the normal $\mathbf{n}_{cp}$ of the closing plane from point x in the force diagram (Figure 3.53a). Then, the applied load faces are extruded to an arbitrary point on the line l that results in new polyhedrons for the force diagram. This force diagram is reciprocal to a constrained funicular form of a bridge, and its height gives various funicular forms with different heights for the given boundary conditions, as shown in Figure 3.54. Note that the width w of the form stays invariant in all these funicular forms. In fact, as discussed earlier, the width of the bridge can be constrained by the 2D funicular network of the applied loads. At the same time, the height of the polyhedral force diagram controls the height. A shallower force diagram with smaller face areas results in a taller funicular form with smaller internal forces in the members

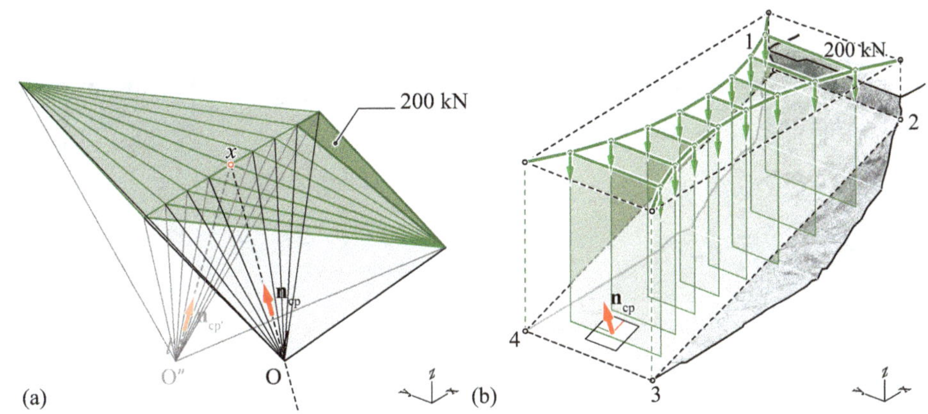

Figure 3.53 (a) In the force diagram, from point x, line l is drawn that is normal to (b) the closing plane $\mathbf{n}_{cp}$ of the form diagram. The force faces are extruded to a point O on line l.

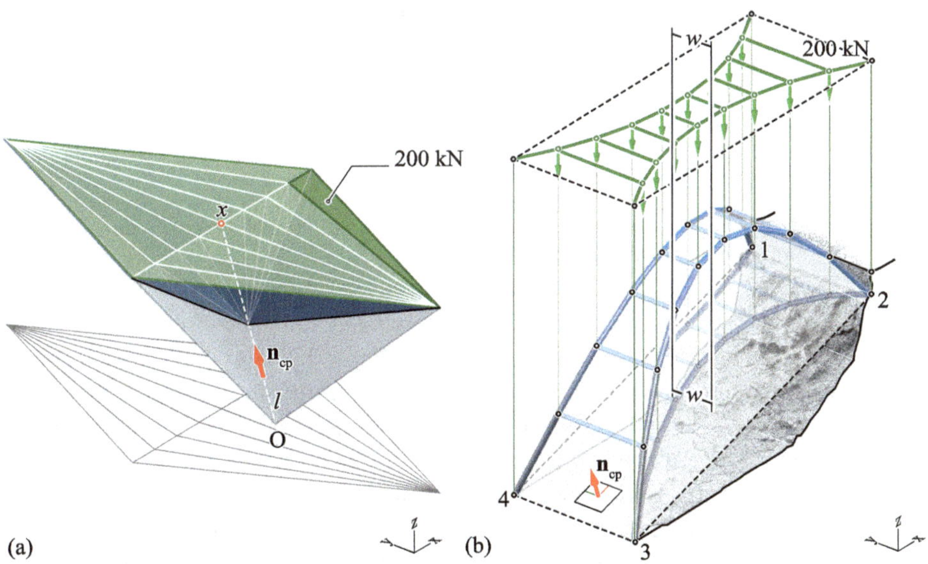

Figure 3.54 (a) The height of the extrusion on line l in the force diagram changes the height of the funicular form of (b).

of the form. In contrast, a deeper force diagram is reciprocal to a shallower arch with higher internal forces in the members (Figure 3.54).

3.13 Constraining the Height of a Funicular Form

In some design cases, it is needed to constrain a funicular form's height. For instance, one might wonder how to design a funicular form that passes through a target point in elevation while spanning between the supports.

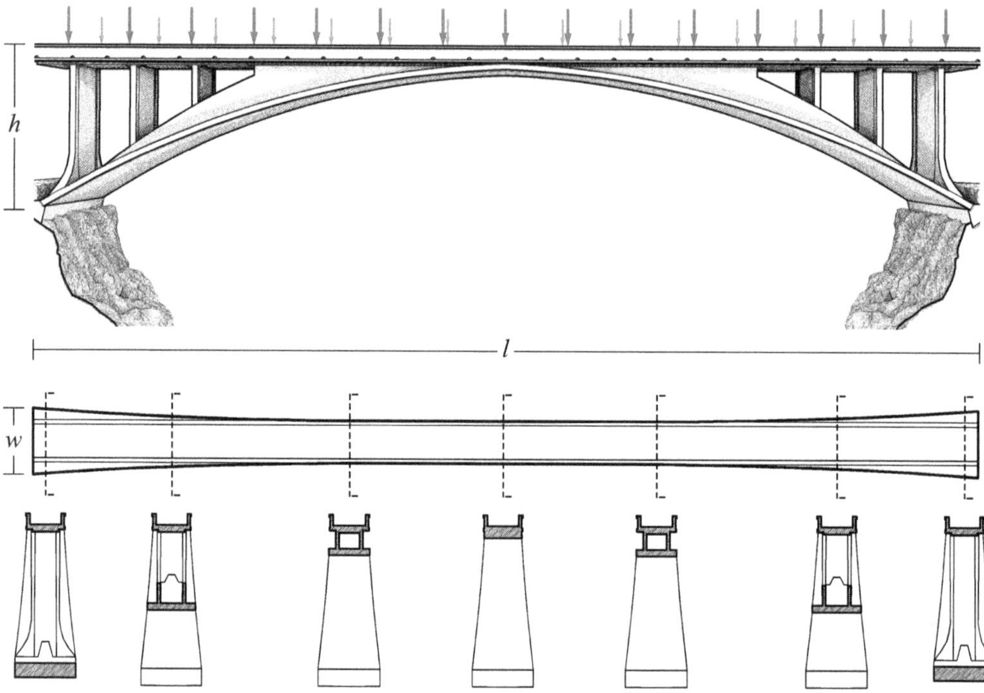

Figure 3.55 The Salginatobel Bridge by Robert Millart and its cross sections. The length l of the arch is 90 m and the height h is 10.4 m.

In 2D cases, we can achieve this task by dividing the form into two smaller funicular segments hinged at a target point. For simple 2D cases, with two supports and a target point, these systems are often referred to as *three-hinged arches*.[98] In 2D graphic statics, a single funicular form that is constrained to two support locations can be turned into a three-hinged funicular by dividing the form into two segments in space between the supports. Dividing the funicular into two segments allows us to precisely prescribe the location of the middle vertex to control the funicular's height.

Salginatobel Bridge in Switzerland, as illustrated in Figure 3.55, designed by Robert Millart in 1929, is an excellent example of a three-hinged funicular arch with a span of 90 m and a height of 10.2 m.[99–101] Millart used 2D methods of graphic statics in deriving the form of the bridge and used the concept of a three-hinged arch in controlling the height of the funicular in the crown of the bridge. In Section 3.13.1, the Salginatobel example will be used to explain the funicular form finding for a three-hinged arch.

3.13.1 A Three-Hinged Arch in 2D

Finding the funicular form of a three-hinged arch is quite similar to the form-finding techniques explained in earlier examples. The three-hinged arch consists of two funicular segments constrained to a shared hinge on one side and a support location on the other. For each segment, there is a force diagram representing the equilibrium of the segment such that these two force diagrams together construct a larger diagram for the equilibrium of the entire three-hinged arch.

Remark 3.16. There is only one and only one funicular solution that passes through a hinge and two supports (three hinges) on a 2D plane.

In Example 3.7, we will redraw the funicular form of the Salginatobel Bridge using the three-hinged arch technique.

Example 3.7 Derive the funicular form of the Salginatobel Bridge such that the height of the primary arch is precisely 10.4 m, and the arch is constrained to the points 1 and 3 that are 90 m apart from each other.

Construction steps:

1 In Figure 3.56b, divide the span of the bridge into two halves from point 1 to point 2 and then from point 2 to point 3, as shown. Each half contains eight applied loads aligned with the axes of the columns placed 6 m apart from each other. The magnitudes of the applied loads are equal and are shown by $\mathbf{f}_i$. The objective is to find a funicular form that connects point 1 to point 3 and passes from point 2.

   Dividing the bridge also divides its applied load in the mid-span into two halves: that is, 0.5 $\mathbf{f}_i$ is applied to one half, and the same load is applied to the other half in point 2.

2 Figure 3.56a,b shows the trial funicular construction for each half of the bridge to locate x_1 and x_2 in the force diagram.

3 In the form diagram of Figure 3.56b, we connect vertices 1 to 2 and then 2 to 3 using closing strings with normals of $\mathbf{n}_{cs1}$ and $\mathbf{n}_{cs2}$.

4 In the force diagram of Figure 3.56a, from points x_1 and x_2, draw lines l_1 and l_2 with unit normals $\mathbf{n}_{cs1}$ and $\mathbf{n}_{cs2}$. These lines intersect at the point O. In all previous examples of this chapter, the location of O was used to extrude the applied load in the force diagram to make a constrained solution. In this example, O is unique because the funicular solution should pass through all three designated points.

5 Figure 3.56c shows the force diagram constructed by extruding the applied loads to point O. As illustrated in Figure 3.56b, the form diagram is constrained to points 1 and 2. The same step can be repeated for the other half, giving the complete solution of Figure 3.57a,b.

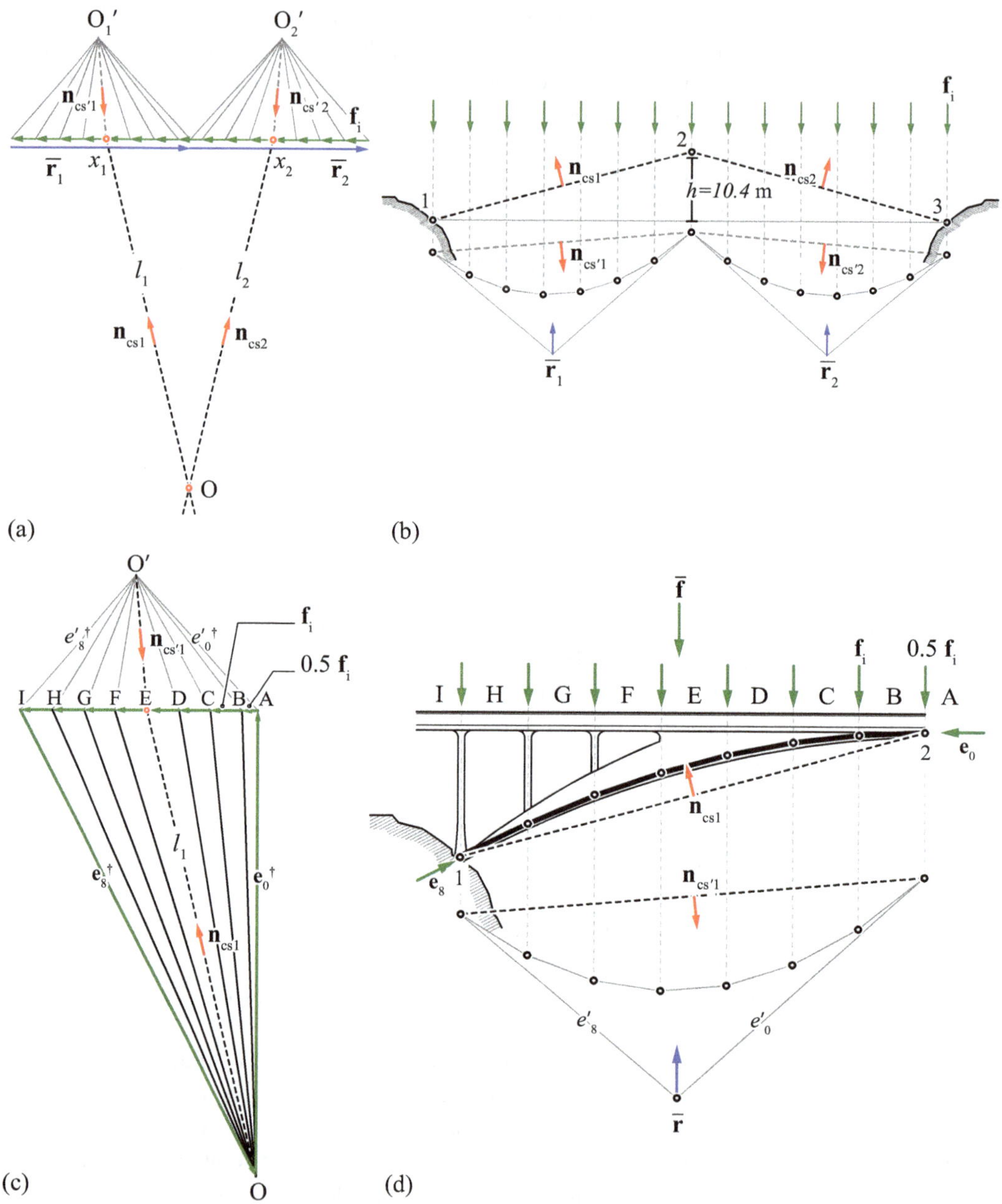

Figure 3.56 The process of completing the three-hinged arch for the Salginatobel Bridge; (a) the applied loads polygon and the trial force diagram; (b) the boundary conditions and the location of supports and trial funicular construction for each segment; (c) the first segment of the solution and the force diagram including the trial construction; and (d) the form and the trial funicular polygon.

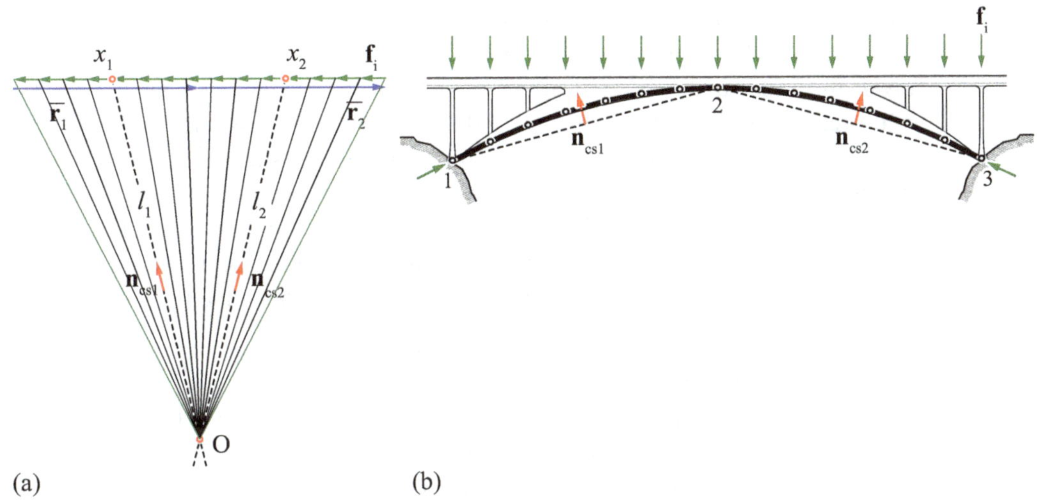

Figure 3.57 (a) The completed force diagram for both segments; and (b) the funicular form of the arch of the bridge passing through points 1, 2, and 3.

□

Remark 3.17. The location of the middle hinge for a three-hinged funicular can be chosen arbitrarily by the designer. Moreover, the direction of forces on both sides of the middle hinge is equal and opposite. This keeps the hinge node in equilibrium.

In constructing a three-hinged arch, if the trial funicular construction is pursued simultaneously and independently for each segment of the arch, then the lines l_i for each segment will intersect at one and only one point O (Figure 3.56c). Therefore, only one force distribution can be made by extruding the applied loads to the point O in the force diagram. Consequently, only one funicular solution passes through three points in 2D (Figure 3.56d).

3.13.2 Three-Hinged Arch in 3D

A similar method can be used in 3D to find a funicular polyhedral shell constrained to points in the mid-span. This section will explain the process of constructing a three-hinged funicular arch using 3D graphic statics and show if Millart could have used 3D graphic statics for his design and what could have possibly changed the geometry of the bridge.

Designing Half of the Bridge

Like the 2D example, the funicular arch in 3D can also be divided into two segments. The trial funicular construction can be made for each segment independently. The first step in designing a compression-only funicular arch constrained to the support locations is to develop the load network, as explained in the example of Section 3.12. Since we intend to constrain the bridge's height in the mid-span, we use symmetry and divide the bridge into two identical segments. Starting from one segment, the direction of the reaction forces at the supports 1 and 2 of Figure 3.58 is assumed to be horizontal both in plan and elevation. This assumption is necessary since the directions of the internal forces at the middle hinge in the mid-span are identical and opposite. They are also horizontal since the hinge is located at the arch's highest point. The distance between the support locations 3,4, and 1,2 is divided into eight segments by the planes of the applied loads, on which the load pairs with equal magnitude $\mathbf{f}_i$ lie.

Designing the Load Network for One Half

With the assumption that all the applied loads are vertical, the load network will be constrained to the projection of the support locations 1–4 to the *xy*- plane. The following steps are necessary to complete the geometry of the load network for half of the arch (Figure 3.58).

Example 3.8 Construct a load network for half of a 3D bridge that is constrained to points in its mid-span.

Construction steps:

1. In the form diagram of Figure 3.58b, on the *xy*- plane, identify the projected location of the supports and the points of the mid-span. We draw a bounding trapezoid connecting vertices 1 to 4 shown by the dashed line.
2. Divide the bounding trapezoid, which is the projected span of the half-arch, into seven segments by drawing lines in the *y*- direction. These lines are the projection of vertical *yz*- planes in 3D where the applied loads will lie. Label the subspaces of the form by the letters *A* to *I*.
3. We draw half of the load network and get the other half by mirroring the force diagram and its related form. Let us include the applied loads as loads $\mathbf{f}_i$ on *xy*-plane on the lines dividing the space between the supports.
4. In the force diagram of Figure 3.58a, draw vectors $\vec{AB}$, $\vec{BC}$, and so on, to complete the force vectors as shown in the force diagram.
5. Extrude forces to point O' to get the trial force diagram.
6. In the form diagram of Figure 3.58b, construct the trial funicular polygon and connect points $2'$ and $3'$ using a closing string with normal $\mathbf{n}_{cs'1}$.
7. Return to the force diagram of Figure 3.58a and draw line l'_1 with unit vector $\mathbf{n}_{cs'1}$ to locate point x_1.
8. From point x_1, draw line l_1 in the direction of $\mathbf{n}_{cs1}$ normal to the closing string connecting 3 and 2.
9. In the form diagram of Figure 3.58b, control the width w of the load network and the bridge by choosing point O on line l_1.

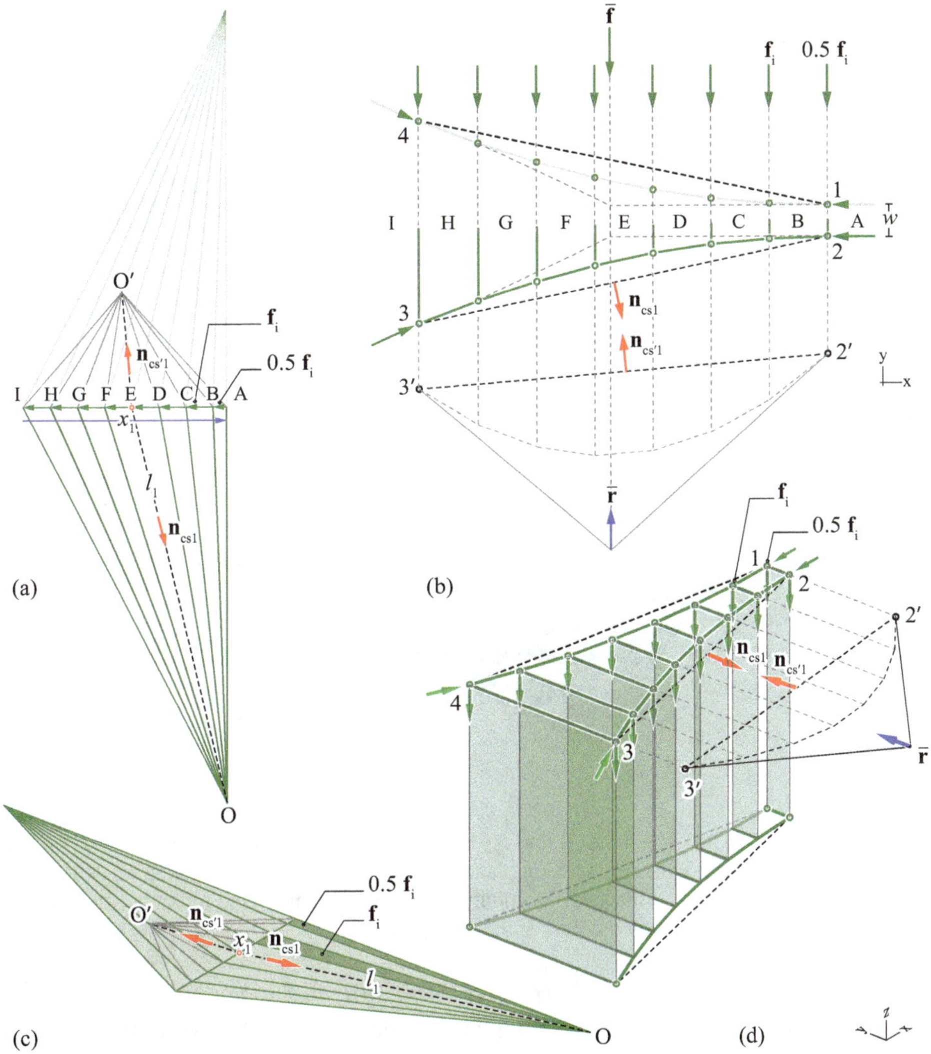

Figure 3.58 (a) Force diagram of the 2D load network; (b) the load network constrained to the projection of the supports on the *xy*- plane; (c) the faces of the force polygon as the force diagram for the 3D funicular form; and (d) the load network and the boundary conditions in 3D.

10 Extrude the applied load vectors of the force diagram into point *O* and complete the funicular network of the applied load for half of the bridge.

11 Figure 3.58c,d shows the isometric view of the applied load faces and the reciprocal network of loads in the form diagram.

□

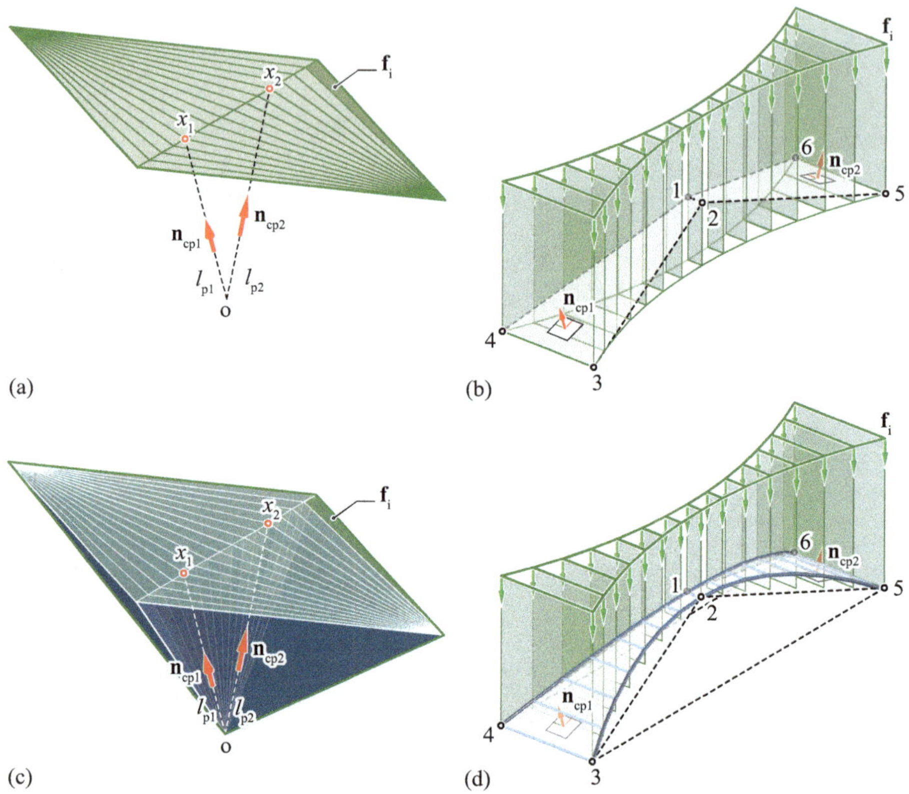

Figure 3.59 (a) The lines l_{p1} and l_{p2} for both segments intersect at point O; (b) two segments of the funicular with their closing planes and normals; (c) the force diagram cells made by extruding applied load faces to point O; and (d) the completed funicular form constrained to four supports and two points in the mid-span.

Completing the Three-Hinged Funicular

Once the force faces are completed, another line l_{p1} normal to the closing plane $\mathbf{n}_{cp1}$ can be drawn from point x_1. A similar process can be pursued for the second half of the bridge, and their force faces can be combined, as shown in Figure 3.59a. The intersection of the lines l_{p1} and l_{p2} from two halves of the force faces locates point O to which the force faces are extruded, and the force diagram for the entire constrained arch is completed (Figure 3.59b). The lines l_{p1} and l_{p2} intersect at one and only one point (Figure 3.59c). Thus, the force diagram is unique for the given applied loads and the boundary conditions. The funicular form can then be constructed starting from the supports by drawing edges perpendicular to the faces of the force diagram and intersecting them with the lines of action of the applied loads (Figure 3.59d). Unlike the famous Salginatobel Bridge, the bridge designed using 3D graphic statics has a funicular configuration in plan and elevation.*

* The height of the funicular forms in 3D can also be constrained to a plane. See the example of the three-hinged arch in *3D Graphical Statics Using Reciprocal Polyhedral Diagrams*, pages 139 to 143[9]

3.14 Overlaying Funicular Solutions for Multiple Loading Scenarios

Although funicular structural forms are designed for specific design loads, a structure might experience different loading conditions during its lifetime if it is built. For instance, a really large cargo may pass from a compression-only bridge. The change in the magnitude of the applied loads affects the geometry of the funicular form. For the new loads, the funicular geometry takes a different thrust to establish the equilibrium of compression/tension-only members under the newly applied loads. In such scenarios, it would be more practical if the final form of the structure is made by overlaying various funicular geometries formed by various load cases.

This technique was used in the design of the Salginatobel Bridge, as illustrated in Figure 3.60. Note that the primary structural form of the bridge was derived from a uniformly distributed load on the bridge's deck, as explained in Section 3.13. Maillart derived multiple funicular solutions for a loading scenario with a moving point load significantly larger than the uniform load placed at the location of each column pair to address the thrust change due to asymmetric loading. Consequently, a series of new funicular geometries were derived by considering asymmetric loading in each pair of columns. He then overlaid the resulting funicular solutions to get the final form of the bridge. Figure 3.60a,b illustrates an asymmetric load that results in a deeper funicular solution under the location of the applied load compared to the solution of the uniformly applied load. The overlying of the funicular solutions gives a thickened cross-section that can be used as a stiffener on the back of the shell. As a result, if the bridge experiences a point load significantly larger than the distributed design load, the thrust line will not fall off from the cross-section of the bridge (Figure 3.60c,d).

The same approach of overlapping funicular solutions can be used in 3D. Figure 3.61 shows a new force diagram for half of the constrained bridge of Section 3.13. In the new force diagram, the applied loads shared by the subspaces D and E in the form diagram have been enlarged. The area of the triangular face has been increased by a factor of seven by elongating the base vector DE to seven times its original length. As a result, the updated force magnitude is seven times bigger than the initial value for $\mathbf{f}_i$. This change in the load magnitude affects the geometry of the load network in the form diagram and, consequently, the width and height of the funicular solution in 3D. The load network's width decreases as the funicular networks' curvature on both sides increases. This change is shown in the top and axon views of Figure 3.61a,b. In the end, the geometry of a 3D bridge can be obtained by overlapping all funicular solutions with different widths and heights, as shown in Figure 3.62.

3.15 Chapter Summary

This chapter introduced compression-only form finding in both 2D and 3D. The properties of primitive constrained funicular forms in 2D and 3D were explained. The step-by-step procedures for constrained form finding for a single applied load in 2D and 3D were provided. The form-finding process was then extended from a single

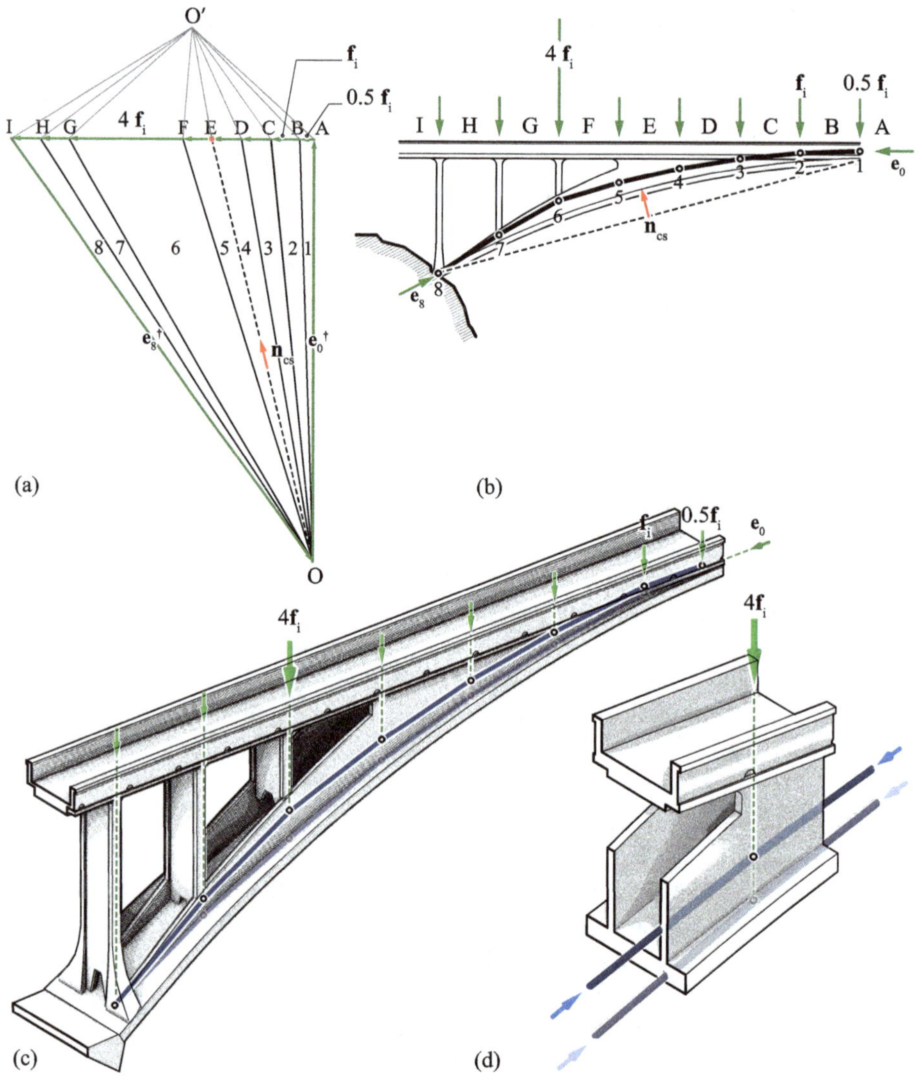

Figure 3.60 (a) Force diagram for the asymmetric loading including a point load as well as the distributed load of the bridge; (b) its funicular form overlaid on the section of the bridge; (c) all the funicular forms as a result of moving the point load along the deck of the bridge; and (d) the section of the bridge arch and the stiffener showing multiple funicular solutions and the change in the thrust depth in the section of the stiffener.

applied load to a nonconcurrent system of loads, for instance, a series of parallel applied loads. We showed how the form and force diagrams can be combined using the Minkowski sum to check the correctness of funicular construction. The relationship between the applied loads and the number of nonconcurrent loads in polyhedral systems is explained. A design technique by subdividing the applied loads was introduced that yields a variety of load networks and, thus, shells with various topologies. Besides, the effect of support locations in 3D was illustrated using multiple examples.

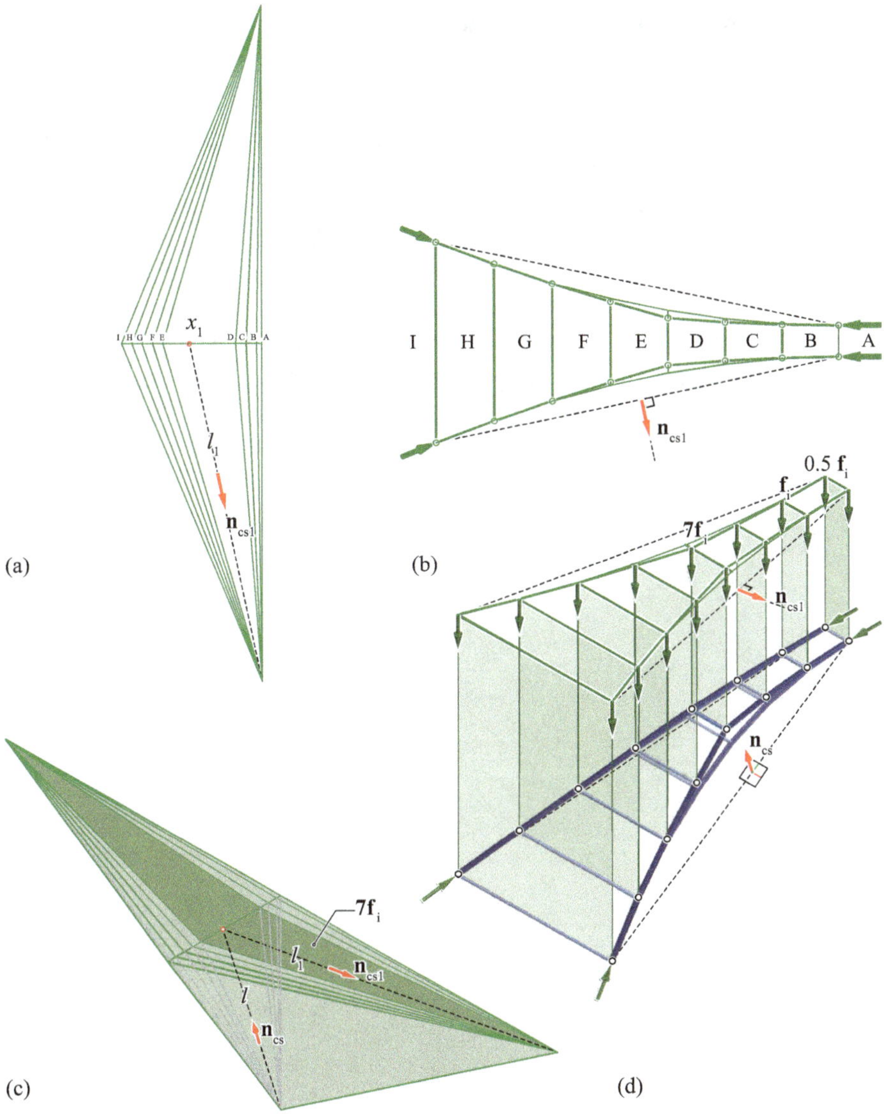

Figure 3.61 (a) The 2D force diagram for the construction of the load network with a point load seven times larger than the distributed load; (b) the new load network as a result of the change in the force magnitude; (c) the 3D force diagram and the location of the constraints l and l_1; and (d) the new form diagram with a spike under the point load overlaid on the form diagram of a distributed load.

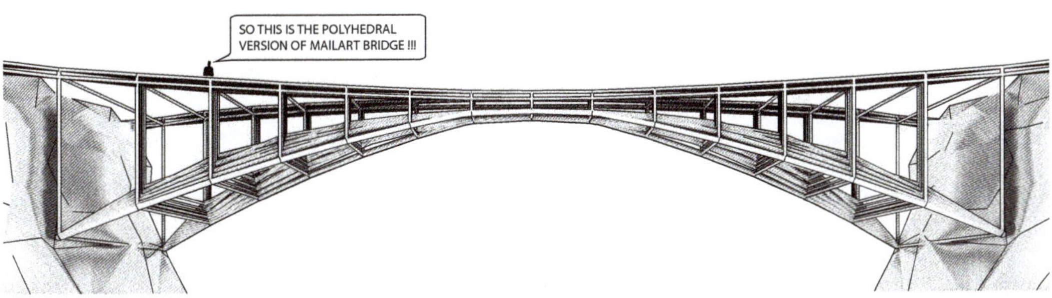

Figure 3.62 The overlaid funicular forms and overlapping geometries that establish the design of the bridge.

We provided examples to show an approach to controlling the width of funicular forms in 3D and constraining a funicular form's height. The topic of the three-hinged arch in 2D and 3D was introduced, and the construction of overlaying funicular geometries for asymmetric loading scenarios was explained.

Chapter 4 explains the computational methods for compression-only form finding, enabling us to develop more complex systems. In addition, some design techniques will be introduced to create particular geometric features, such as synclastic or anticlastic curvature, and creases to emphasize the design effects of polyhedral funicular systems.

4 Articulated Compression-Only Forms

Masoud Akbarzadeh and Mostafa Akbari

Figure 4.1 An intricate design for a lightweight compression-dominant floor system developed by subdividing the force diagram.

4.1 Introduction

The relationship between the funicular form and its force diagram in graphic statics provides more than just an explicit geometric definition of equilibrium. The fact that the force diagram consists of closed polygonal faces in two-dimensional (2D) and polyhedral cells in three-dimensional (3D) provides a unique design opportunity.

The closeness of a force polygon or polyhedron in the force diagram ensures the equilibrium of a node in the form diagram. This information is sufficient for a design enthusiast because assembling closed force polygons or polyhedrons can establish a larger force distribution that results in articulated funicular structural forms. These forms can further be fine-tuned and articulated by meticulously designing the force diagram (Figure 4.1). Graphic statics grants us a nonconventional design domain, which is the possibility of designing and manipulating *the geometry of the force diagram* and extracting surprising funicular solutions. This approach of design is completely nonconventional because a designer does not start from the geometry of the form but rather starts by designing the geometry of the force equilibrium. Designing the geometry of the force allows us to:

- assure equilibrium in the output funicular solution;
- control the magnitude of internal forces in the funicular geometry by manipulating the geometry of the force diagram; and,
- design and amend the topology of the funicular solution, which results in various design alternatives.

Chapter Objectives

This chapter will provide a deeper understanding of the relationship between the form and force diagrams and introduce the force domain as the primary design domain. This chapter will particularly introduce the following:

- The topological and geometrical relationship between the form and force diagrams in 2D and 3D.
- A brief computational implementation for the construction of reciprocal diagrams.
- Subdivision of the force diagram as a design technique for generating articulated structures in both 2D and 3D.
- Examples of complex polyhedral geometry, including synclastic, anticlastic curvatures, and shellular funicular systems.

4.2 Nodal and Global Equilibrium in 2D Funicular Forms

A closer look at the funicular form and force diagrams reveals important information about the form and force diagrams in 2D graphic statics. The geometry of the force diagram can further be analyzed by separating the external edges from its internal edges. If so, the external edges form a closed polygon of forces correlated with the form diagram's external loads and reaction forces (Figure 4.2a,b). The closeness of the external force polygon indicates that the sum of the external forces is zero, and we refer to it as the *global* equilibrium of external forces in the system. Similarly, the internal polygons formed by the internal edges correspond to *nodal equilibrium* in the system.[102] That is, each close polygon inside the external force polygon corresponds to the equilibrium of a single node in the system (Figure 4.2c,d). The fact that the internal and external equilibrium can be disintegrated geometrically is a unique property that can be creatively used to design various funicular forms for a given boundary condition.

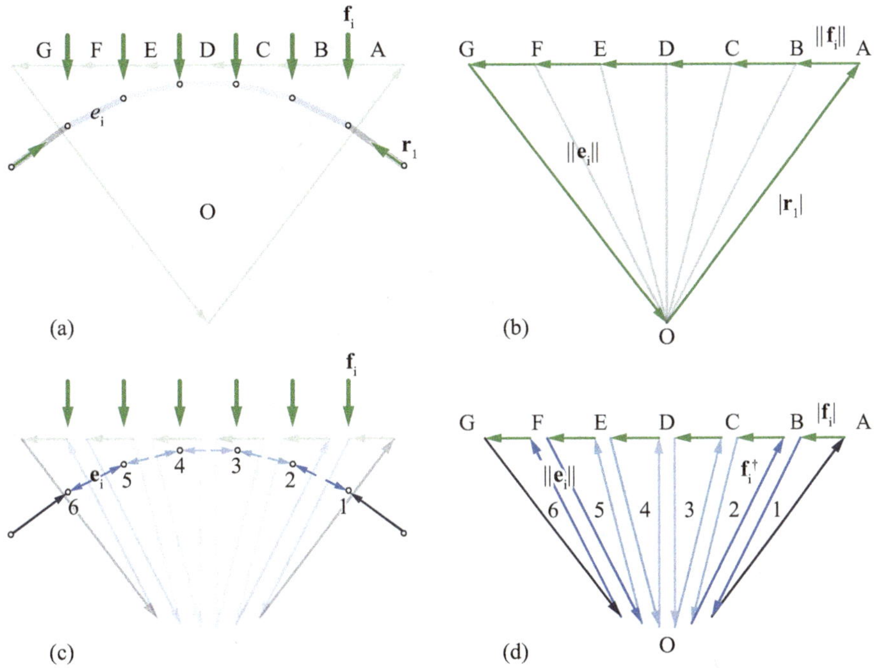

Figure 4.2 (a) The external loads and reaction forces in a funicular form; (b) the external force polygon indicating the global equilibrium; (c) the internal nodes of a funicular form; and (d) the internal polygons of the force diagram.

Remark 4.1. Note that the direction of the cells in the force diagram follows the right-hand rule and in a compression-only form, the force cells are all positive, that is, the normals point toward the inside of the cell.

Before we take advantage of this property, we should visit the topological and geometric relationship between the form and force diagram that allows us to build one from the other.

4.3 Topology of the Form and Force Diagrams in 2D

Consider the force diagram of Figure 4.3a as an example. The geometry of the force diagram may change due to various geometric deformations, as shown in Figure 4.3b,c. However, the connections between the vertices and edges of the graph of the force diagram stay intact. Moreover, the number of faces/spaces created by the vertices and the diagram's edges does not change. These properties can be referred to as the *structure* or the *topology* of the graph and are invariant in all geometric deformations. However, these deformations should not include breaking, disconnecting, or connecting parts.

The topology of a force diagram consists of information related to the number of vertices, edges, and faces, as well as the relation of these elements with respect to other

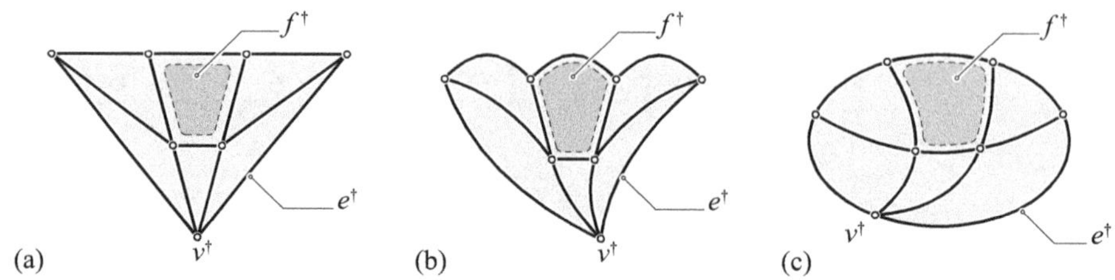

Figure 4.3 (a) The topology of the force diagram with its geometric properties; (b) and (c) multiple topological drawings of the graph of the force diagram detached from its geometric properties; the topological relation of the constructive components including vertices $v^\dagger$, edges, $e^\dagger$ and faces, $f^\dagger$ does not change in any configuration.

elements in the graph. This relationship can be described by connectivity information in a graph. For instance, every two vertices are connected by an edge, and thus, the vertices and edges in a graph are related to each other. The same relation can be used for edges and faces: every two faces share an edge. A similar relationship can be extended to faces, vertices, and so on. All these relationships establish the topology of the force diagram.

Similar to the force diagram, the force diagram can also be described topologically by its elements and their relationships with each other.

4.4 The Duality of the Form and Force Diagrams

There is a unique topological relationship between the form and force diagrams of graphic statics. Let us consider the force diagram of Figure 4.4a as a graph and take a closer look into the relationship between its faces and edges. Each closed polygon or face in this graph shares an edge with its neighboring faces and the graph's exterior space. The relationship between the faces and edges can be highlighted by (1) marking each face with a point and (2) connecting it to the points in the neighboring faces, as shown in Figure 4.4b. The new graph drawn by dashed lines shows the connectivity of the faces and their shared edges in the force diagram. More importantly, this new graph is also a topological description of the connectivity of the vertices and edges in the form diagram.

In fact, in graphic statics, the form and force diagrams are the *topological duals* of each other. That is, each element of one diagram can be represented by its dual element in the other diagram.

Let us use the v, e, and f letters to name the vertices, edges, and faces of two diagrams and use the † sign to specify force elements from the form elements. Thus, the force diagram is shown by $\Gamma^\dagger$ and the form diagram by Γ. Also, in the examples of this book, we name the vertices of the force diagram with capital letters and use numbers for the vertices of the form diagram. Also, capital letters are used to name the faces/polygons of the form diagram and numbers for the faces/polygons of the force diagram. The example of Figure 4.4 describes the relationship between the components of the form and force and shows the dual graph of each diagram superimposed on

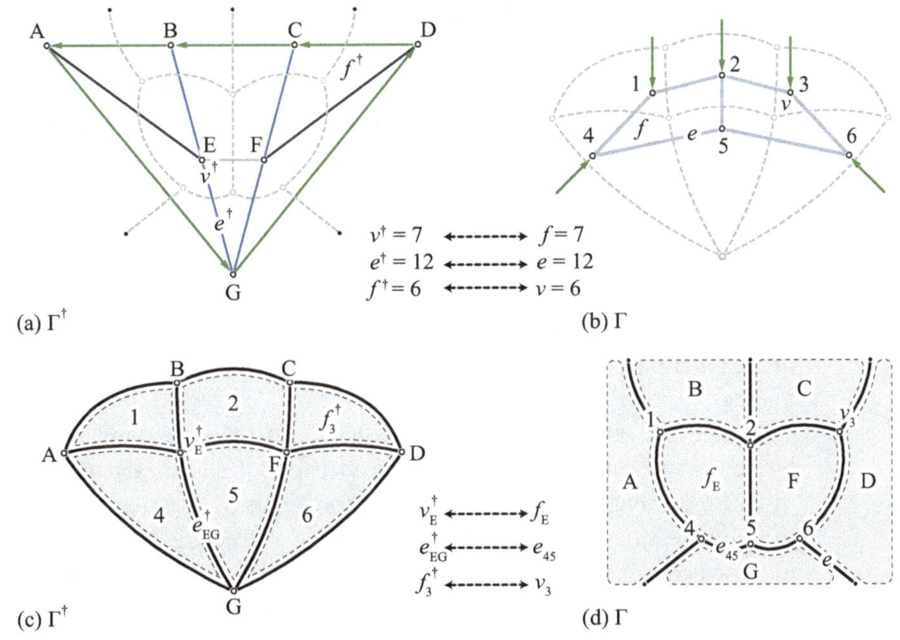

Figure 4.4 The topological relationships between the force and form diagrams are (a) the force diagram and the topology of the form as a dual graph; (b) the form diagram and the topology of the force as the dual graph; (c) the topological graph of the force diagram; and (d) the topological graph of the form diagram.

the other. The topological relationship between the two diagrams can be summarized as follows:

- Each face/polygon $f^\dagger$ in the force diagram is the topological dual of a vertex v in the form diagram. For instance, face $f_3^\dagger$ in the force diagram is the topological dual of vertex v_3 in the form diagram. As a result, the number of faces $f^\dagger$ in the force diagram is equal to the number of vertices v in the form diagram.
- This property can also be remembered as for every node/vertex v in the structure, there is a closed force polygon $f^\dagger$ in the force diagram, so their numbers match.
- Also, the vertices $v^\dagger$ of the force diagram correspond to the faces f of the form diagram. For instance, vertex $v_E^\dagger$ of the force diagram is reciprocal to the face f_E in the form diagram. Note that the faces of the form diagram might be open or closed. For instance, face f_A is an open face because it is not bounded entirely by the edges of the form diagram, while face f_E is a closed face.
- Every two faces in the force diagram share an edge $e^\dagger$ that is the dual of an edge e connecting two vertices in the form diagram. For instance, edge $e_{EG}^\dagger$ is the dual of the edge e_{45}.

These topological relationships are necessary to derive one diagram from the other but are not enough to complete the funicular construction. We need geometric information to complete the construction of reciprocal diagrams.

4.5 The Geometric Dependencies of the Dual Diagrams in 2D

In addition to the dual relationship between the components of form and force diagrams in 2D, there is a geometric dependency between the two diagrams' elements. Each edge of one diagram is perpendicular to an edge in the other diagram. This geometric dependency, together with the topological duality, makes these diagrams *reciprocal*. That is, any change in one diagram prompts a change in the other diagram. This reciprocity is at the core of graphic statics methods. It makes graphic statics a powerful design tool by relating the geometry of a funicular form to the geometry of its internal and external forces. Moreover, having one diagram allows the construction of the other one, which is a very important feature in geometry-based structural form finding.

Having the geometry of the force allows us to find the funicular structural form by knowing the reciprocal relationship between the form and force diagrams. For instance, by now, we know every closed force polygon can represent the equilibrium in a compression-only or tension-only node of a structure. Therefore, an aggregation of closed convex force polygons can represent a structural system in equilibrium, and the form of the structure can be extracted and constructed simply by knowing the reciprocal relationship between the form and force diagrams. In Section 4.6, we will show how the force diagram can be designed to derive articulated compression-only or tension-only structural forms.

4.6 Subdividing the 2D Global Equilibrium for Design Purposes

As mentioned, we can start by designing the force diagram as an approach in graphic statics. Because starting from the force diagram always guarantees that the structural form will be in equilibrium, whereas starting from the form diagram might not necessarily result in a compression-only or tension-only system. An interesting design approach is to design the nodal force equilibrium.

The global and nodal equilibriums in 2D funicular form-finding can be separated from each other for design purposes. In other words, the geometry or topology of a funicular form for a given set of loading and support locations is not unique. The global force polygon's interior space can be divided into smaller convex polygons. As a result, a nodal equilibrium and various structural forms with different topologies can be extracted. Note that all internal force polygons must be convex to ensure the compression-only behavior of the funicular form. There are infinite possibilities to subdivide the interior space of the global force polygon. Consequently, infinite families of compression-only funicular solutions for the same boundary condition will exist. Figure 4.5 shows various design options for the compression-only funicular form by subdividing the global force polygon into smaller convex nodal equilibrium.

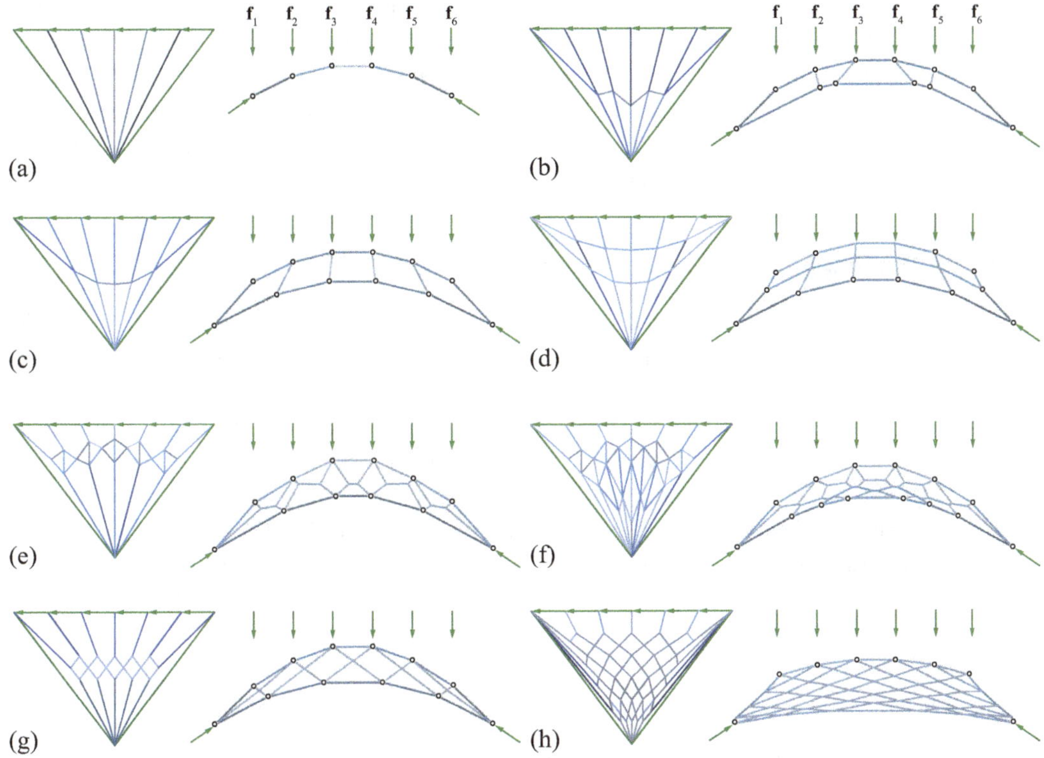

Figure 4.5 (a–h) Various funicular solutions for the same boundary condition by subdividing the global force polygon into convex polygonal faces.

4.6.1 Extracting the Graph of the Funicular Form

After the force diagram subdivision, constructing the funicular form requires revisiting the topological relationship between the form and force diagrams in 2D graphic statics. Let us name the force diagram $G^{\dagger}$ and its reciprocal funicular form G. As mentioned in Sections 4.4 and 4.5, these two diagrams are topologically and geometrically related. This relationship is referred to as the *reciprocal* relationship because (1) having one diagram allows us to find the other, and (2) any change in one diagram results in corresponding changes in the other.

4.6.2 The Families of Funicular Geometries for a Single Force Distribution

In some cases, there might be more than one funicular solution for a force diagram. Let us consider the force diagram of Figure 4.6 as a subdivided force diagram of a simple shell. The topology of the form can be found by connecting the centers of the faces of the force diagram shown by a dashed graph (Figure 4.6a). Let us assume that the location of the applied loads and reaction forces cannot change in the form-finding process. We can use the procedural form finding explained in Chapter 3 to find a funicular form. Skipping the trial construction process and jumping to the last step, a funicular solution can be found by choosing a length d_1 for the edge connecting vertices 1 and 6 in the form diagram. However, this is not the only funicular solution for this force diagram and the given boundary conditions. Any chosen length for d_1 leads to a new funicular form. In fact, for the same force diagram, there is an infinite number of solutions that have parallel edges, as shown in Figure 4.6c,d.

4.6.3 The Independent Edges of a Graph

In Figure 4.6, choosing the length of d_1 suffices for the completion of the form diagram construction. In other words, the lengths of other edges depend on the length of one edge in this example. This edge is often referred to as the *independent* edge and is shown by e_{ind} in the context of this book. The rest of the edges in the graph are then called *dependent* edges. An independent edge can be chosen arbitrarily in a graph, but finding the number of independent edges in a graph is not trivial. It needs further explanation of the form and the force diagrams' algebraic relationship. The number of independent edges is also called the *geometric degrees of freedom* (GDoF) in the context of funicular form-finding, and it has also been discussed in almost every

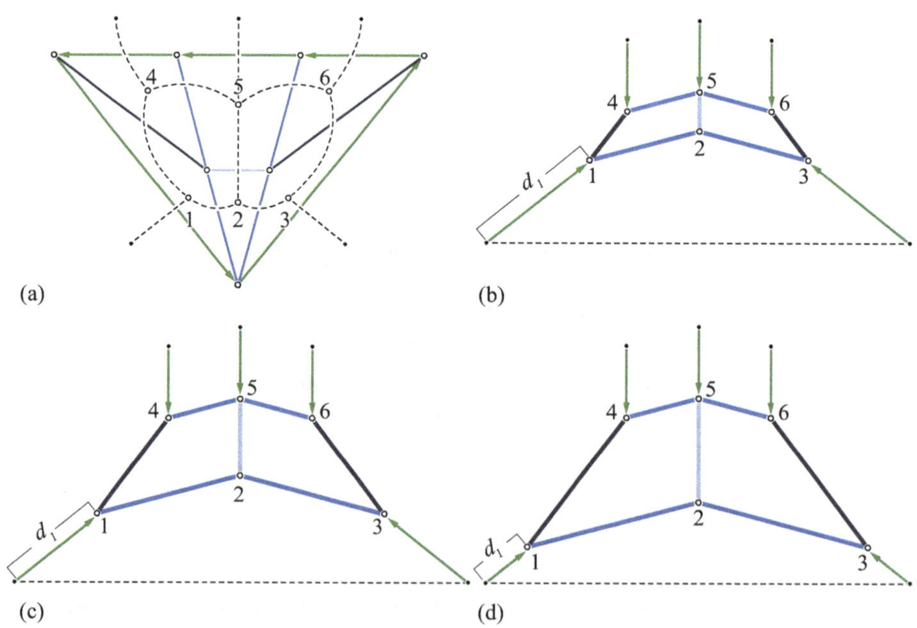

Figure 4.6 (a) A simple force diagram and the topology of the form; (b) the funicular form constrained to the supports and the line of action of the applied loads; (c) and (d) changing the length of the line of action of the form without changing the location of the supports. The location of the supports are at the end of the action line and the length of the dashed line does not change.

chapter of this book. Chapter 7 of this book will provide a comprehensive algebraic approach for finding the GDoF of a funicular graph. Section 4.6.4 will provide a *rule-of-thumb* to identify the GDoF of a 2D funicular form. However, this method *is not always reliable*, and an algebraic approach is preferred for a precise solution.

4.6.4 Finding the Geometric Degrees of Freedom of a Funicular Form

The GDoF or the number of independent edges of a single polygon with e number of edges is equal to $e-2$. For instance, the number of independent edges in the polygon of Figure 4.7 with five edges is equal to three, that is, three edge lengths of this polygon can be chosen independently from the rest of the edges to construct a polygon with parallel edges to the original one. We can use the formula provided in Chapter 3, Section 3.9.2, to calculate the GDoF of a simple funicular network in 2D. In addition, the following formula would also work.

Remark 4.2. Suppose the global equilibrium and the geometry of the external polygon in the form diagram, that is, the applied loads' location and the reaction forces, do not change. We choose an arbitrary length for the applied loads in the graph of the form diagram and connect the endpoints of the applied loads and the reaction forces at the supports to make closed polygons, as shown in Figure 4.8a.

Equation 4.1 can then be used to find the number of independent edges of a funicular form:

$$GDoF = e_{ind} = e - 2f - (\bar{e}_{ext} - 3) \tag{4.1}$$

where e is the total number of edges of the funicular form of Figure 4.8a, including the exterior edges, f is the number of faces in the diagram, and $\bar{e}_{ext}$ is the total number of external edges. Figure 4.8 shows an example of using this method in finding the GDoF of a graph of a form diagram.

Remark 4.3. Note that even though it might be possible to predict the number of independent edges of a funicular form using the earlier formula, the funicular construction might not be completed. In such cases, the algebraic approach is preferable (see Chapter 7).

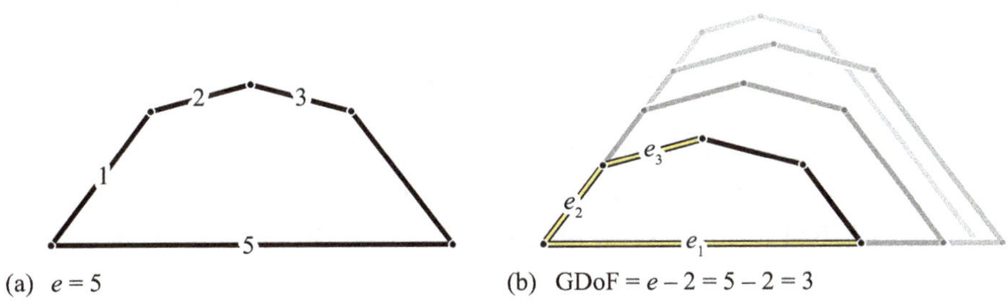

Figure 4.7 (a) A polygon with edges has a GDoF of three; and (b) three edge lengths can be chosen to construct another polygon with parallel edges to the original one.

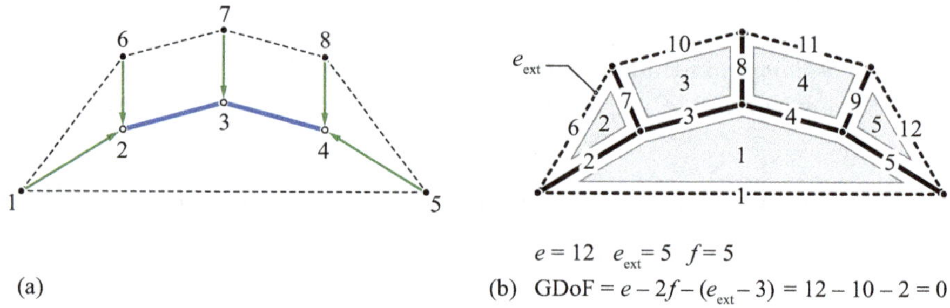

Figure 4.8 (a) Drawing a closed polygon around the funicular form; and (b) constraining the independent edges of the external polygon and finding the GDoF for the remaining polygons.

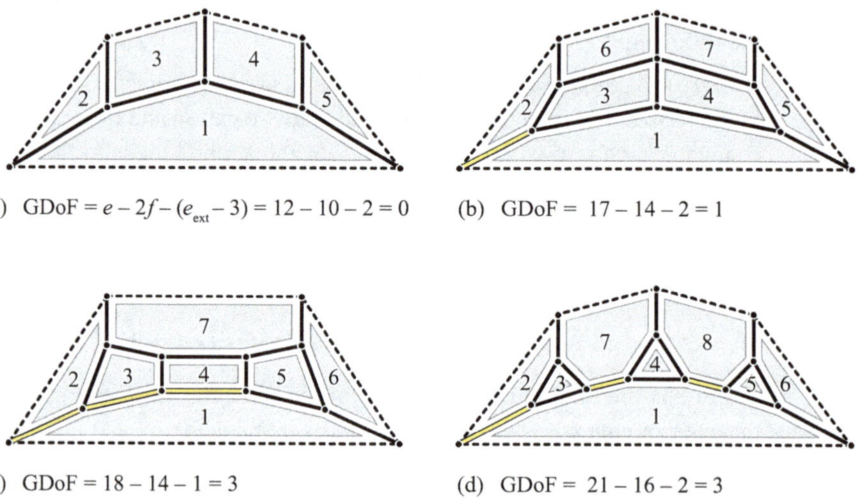

Figure 4.9 Calculating the geometric degrees of freedom (GDoF) or the number of independent edges in various funicular forms with (a) no independent edge, (b) one, (c), and (d) three independent edges.

Figure 4.9 shows multiple examples of funicular forms, the construction of which requires choosing the edge lengths of the independent edges. For instance, Figure 4.9a shows a simple funicular form constrained to the support locations. The GDoF of this form is zero. Given the supports and applied loads' geometry, there is no independent edge in the form diagram, and only one funicular solution exists. Similarly, in Figure 4.9b–d, there are one, three, and three independent edges, respectively. Figure 4.10 illustrates a compression-only funicular form designed by subdividing the force diagram. Can you count the number of independent edges in the form?

4.6.5 Computational Construction of Form Diagram in 2D

Although form finding using procedural graphic statics provides an unprecedented design approach, constructing the form diagram using the topological and geometric

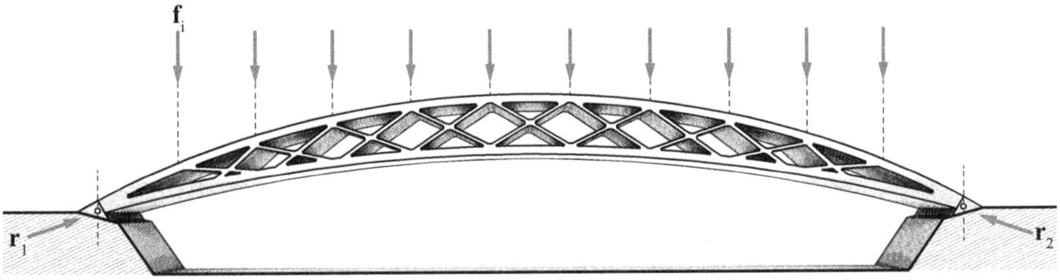

Figure 4.10 A double-layer, compression-only funicular form whose GDoF is zero.

properties of the force can become tedious and time-consuming quickly as the subdivision or the number of nodal force cells increases. A computational approach to constructing the form diagram can make this form-finding process quite efficient. There are two approaches in the computational construction of reciprocal diagrams: *iterative* methods and *algebraic* methods.

The *iterative approach* works as follows: in the first step, it constructs a dual graph from an input diagram and establishes the duality between the components of the two diagrams. In the following steps, every edge in the new graph is disconnected from its vertex and rotated to become perpendicular to its reciprocal edge in the other diagram. Next, the location of the vertices is calculated by the average of the endpoints of the rotated edges. Afterward, the rotated edges will be connected to the average vertex in the new location to preserve the graph's connectivity. This process is repeated in multiple steps until the edge becomes perpendicular to its reciprocal edge in the input diagram. For further information, we would invite readers to visit the works on the same topic by Akbarzadeh, Van Mele, and Block.[103]

The *algebraic approach* works based on the mathematical relationship between the geometry of the form and force diagrams and calculates a reciprocal geometry for an input geometry in a single step. This approach opens the door to various interesting possibilities in design by controlling the direction of internal force in the members of the form and many other design features, which will be discussed in detail in Chapter 7.

4.7 Nodal and Global Equilibrium in Polyhedral Funicular Forms

The great advantage of using polyhedral graphic statics over 2D graphic statics is the possibility of expanding the design space to three dimensions and exploring a range of different, novel spatial funicular geometries.

To use this unprecedented design potential, we need to learn more about the properties of the force diagram, the relationship between the geometry of the force diagram, and the equilibrium of internal and external forces in the form diagram. Consider the form and force diagram of Figure 4.11 for the following definitions and remarks.

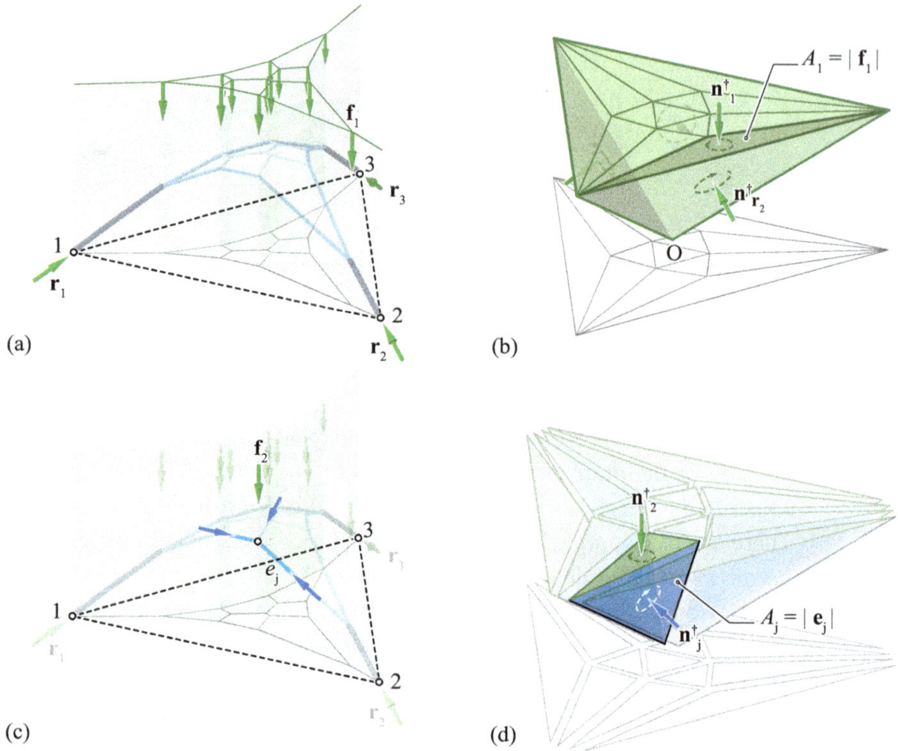

Figure 4.11 (a) The external faces of the force diagram make the global force polyhedron, which is reciprocal to (b) the external loads and reaction forces at the supports; (c) the closed polyhedral cells within the global force polyhedron are called the nodal force polyhedrons representing the equilibrium of (d) a node in the funicular form.

Definition 4.1. The external faces of the force diagram constitute a larger polyhedral cell, which is shown by the green color in Figure 4.11a. This larger polyhedron that might have co-planar faces is reciprocal to all the system's applied loads and reactions. This cell is called the *global force polyhedron*, and its closedness assures that the sum of the external forces in the system is zero (Figure 4.11b). Each face of the global force polyhedron will make a closed cell with the force diagram's internal faces. Each closed cell within the global force polyhedron is called a *nodal force polyhedron* and is reciprocal to a node in the form diagram, as shown in Figure 4.11c,d. The closedness of internal cells guarantees the equilibrium of their reciprocal nodes in the form diagram.

4.8 The Duality of the Form and Force Polyhedrons

The form and the force diagrams in the polyhedral graphic statics are the topological duals of each other (Figure 4.12a). Both form and force diagrams consist of vertices,

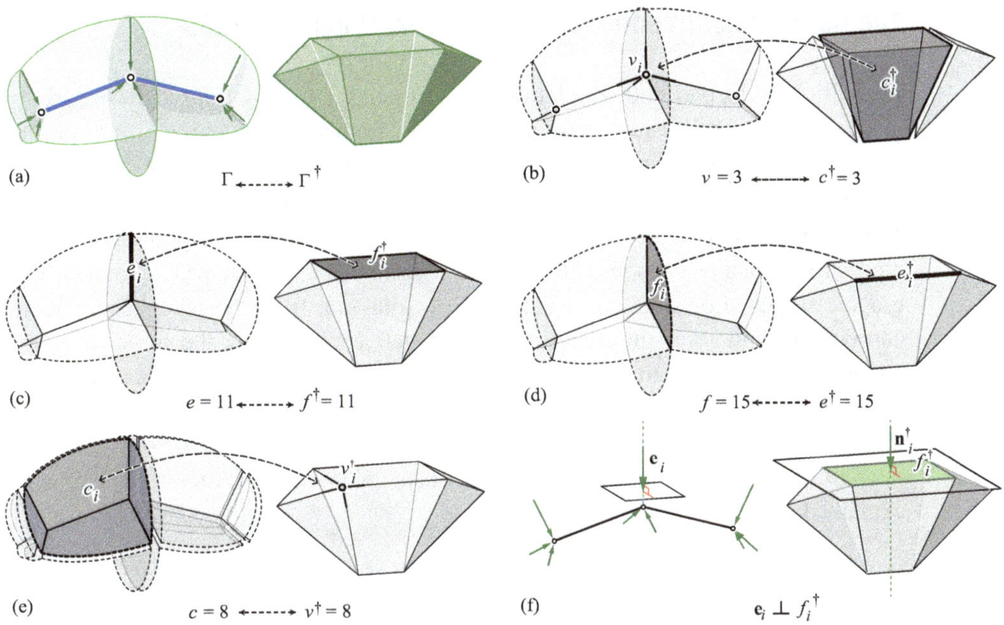

Figure 4.12 (a) The form diagram Γ and its reciprocal force diagram $\Gamma^{\dagger}$; (b) a vertex v_i of the form diagram and its corresponding cell $c_i{}^{\dagger}$ of the force diagram; (c) an edge e_i and its corresponding face $f_i^{\dagger}$; (d) a face f_i and its corresponding edge $e_i{}^{\dagger}$; and (e) an open cell c_i of the form diagram and its corresponding vertex $v_i{}^{\dagger}$ of the force diagram; (f) each edge in one diagram is normal to a face in the other diagram.

edges, faces, and cells. Let us denote the vertices, edges, faces, and cells of the diagrams by v, e, f, and c, and use superscript † for the elements of the force diagram. The following relationship holds for the components of the two diagrams:

- Each vertex $v^{\dagger}$ in the force diagram corresponds to a cell in the form diagram. The number of vertices in the force diagram equals the number of cells in the form diagram (Figure 4.12b).
- The number of edges $e^{\dagger}$ in the force diagram is equal to the number of faces f in the form diagram (Figure 4.12c).
- Each face, $f^{\dagger}$, in the force diagram corresponds to an edge e in the form diagram, and their numbers are equal (Figure 4.12d).
- Each cell $c^{\dagger}$ in the force diagram has a dual vertex v in the form diagram, and their numbers match (Figure 4.12e).

Remark 4.4. Only the internal vertices of the form diagram are included in the reciprocal relationship, and the vertices at the free end of the applied load vectors are not considered in the reciprocal relationship. Moreover, the edges connecting the endpoints of the open faces of the form diagram are also not included in the reciprocal relationship.

4.9 The Geometric Dependencies of the Dual Diagrams in 3D

In addition to the earlier dual relationship between the form and force diagrams, there is a geometric relationship between the two diagrams: every edge e_i of the form diagram is normal to its corresponding face $f_i^\dagger$ in the force diagram. The edge vector is shown by $\mathbf{e}_i$, which is parallel to the normal of the face $\mathbf{n}_i^\dagger$ (Figure 4.12f). The form and force diagrams in polyhedral graphic statics are *reciprocal* if both dual and geometric relationships between the diagrams are satisfied. This reciprocity in polyhedral graphic statics allows the construction of one diagram from the other. Therefore, we can take advantage of this relationship and start by designing the force diagram as an aggregation of closed polyhedral cells from which we can extract a spatial funicular form in equilibrium. One more important piece of information is missing from this reciprocity, which is related to the direction of the face normal of the cells and their correlation with the internal force in the members of the structure, which will be explained next.

4.10 Identifying Polyhedral Cells from a Group of Faces

The earlier discussions about reciprocity and duality of the form and force are valid if we can identify the cells in the force diagram and relate them to the equilibrium of their reciprocal nodes in a structural form. Although it is possible to construct each cell in the force diagram as a closed polyhedron and add them to the collection of cells, the task might be very tedious and time-consuming. It is always better to have a rigorous approach or algorithm to identify closed cells from a collection of planar polygonal faces that share edges.

Let us consider a group of polygonal faces in 3D that can contain multiple closed polyhedral cells. These cells are not identified, and our task is to find them from this given group of faces. We assume that this collection represents a force diagram with no open cells. Thus, each face in this collection shares an edge with other faces.

Let us take an edge from this force diagram and find its attached faces. Figure 4.13a shows an edge $e_i^\dagger$ connected to three faces $f_i^\dagger, f_j^\dagger$, and $f_k^\dagger$ in an imaginary group of polyhedrons.

As shown in Figure 4.13a,b, the faces have initially random normal directions. In the next step, we give a particular direction to the edge, for instance, going from vertex A to B. Let us use the following naming convention for the face normals: if the direction of the edge vector of a face matches the direction of $e_{AB}^\dagger$, then we use a positive sign for its normal and otherwise negative. Accordingly, the edge $e_{AB}^\dagger$ is initially connected to three faces with $+\mathbf{n}_j^\dagger$, $-\mathbf{n}_k^\dagger$, and $+\mathbf{n}_l^\dagger$. Next, we amend the direction of the face normals to match the direction of the edge $e_{AB}^\dagger$. Consequently, the direction of the faces will be changed using the right-hand rule, as shown in Figure 4.13c. If we look through the edge from vertex B to A, the faces normals are sorted in a counter-clockwise manner around the edge (Figure 4.13c,d). Note that we could have chosen a

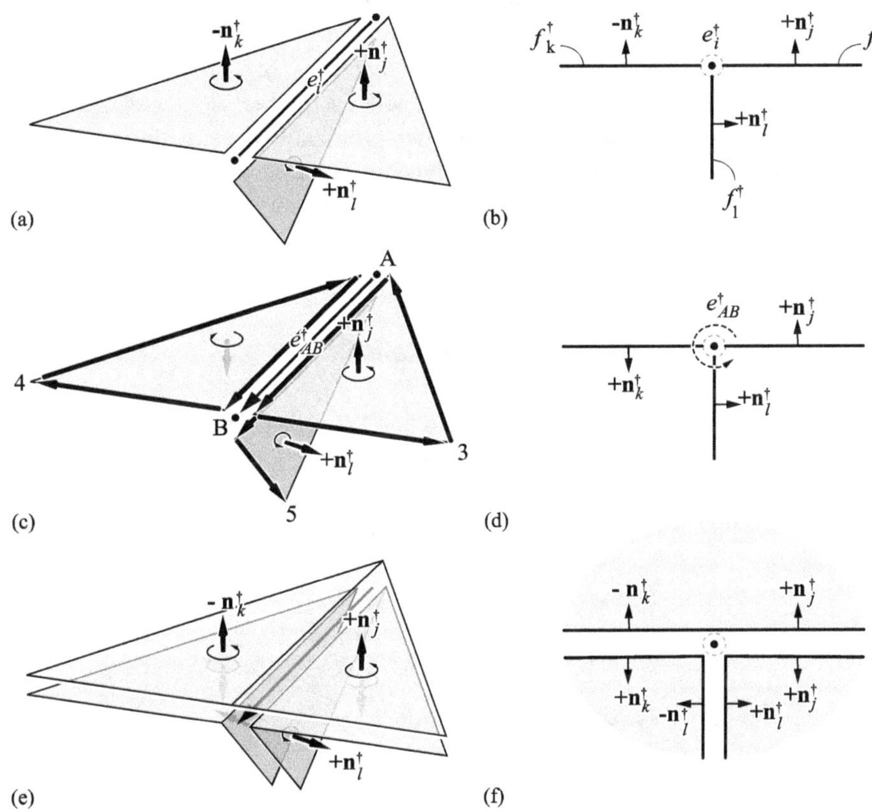

Figure 4.13 (a) and (b) An edge connected to three faces with random face normals; (c) and (d) giving a direction to the edge going from vertex 1 to vertex 2 and aligning the direction of the edge vectors with the directed edge; and (e) and (f) Adding half-faces to the initial faces.

reversed direction for the edge, which could result in opposite face directions. To find the internal spaces of polyhedral cells, we will replace each face with two *half-faces* that are identical but have opposite face normals (Figure 4.13e).

In a single closed polyhedral cell, each face shares an equal and opposite edge vector with its adjacent face. This property was explained in Chapter 1. Thus, in this group of half-faces around the chosen edge, we can identify face pairs with equal and opposite edge vectors, as shown in Figure 4.13e,f. These face pairs show the interior space of a polyhedral cell. By finding similar face pairs of other edges and matching their identical faces, the identification of the polyhedral cells will be completed, and we can find all possible cells in a group of polygonal faces connected by their edge. This method of storing data to identify polyhedral cells is called *half-face* or *winged-edge* data structure and was proposed by Baumgart[104] in computer graphics. This method was used later in the computational construction of reciprocal polyhedral diagrams.*

* See *On the Equilibrium of Funicular Polyhedral Frames and Convex Polyhedral Force Diagrams* by Akbarzadeh, Van Mele, and Block.[75]

Remark 4.5. This method is quite helpful in finding convex cells from a group of faces. Note that this method might not work well with configurations where the polyhedral cells are not convex or have self-intersecting faces. In such cases, the interior spaces of the polyhedral cells, if found, overlap with each other, and the earlier method cannot identify the closed cells. This is an open problem in constructing reciprocal diagrams from an input of self-intersecting polyhedral cells. Nevertheless, since in compression-only form finding, the cells of the force diagram are convex, the earlier method works very well.

4.11 The Direction of the Polyhedral Cells

Let us use the mentioned approach to find the cells of the force diagram for the group of faces that are illustrated in Figure 4.14a. This group includes seven faces with random normal directions, as shown in the exploded drawing of Figure 4.14b. Let us assume that we have a computational implementation of the half-face methodology

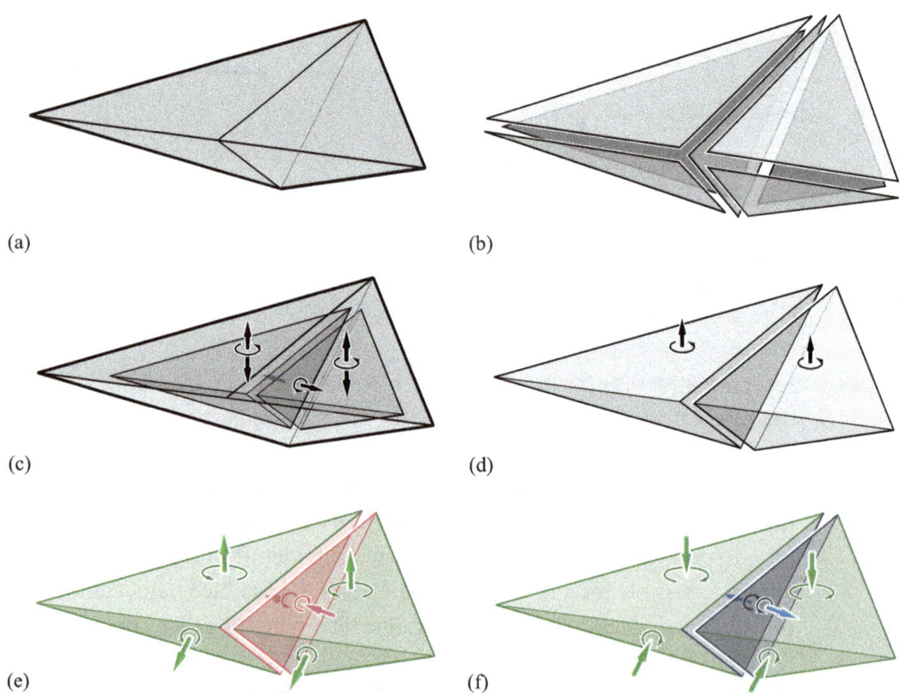

Figure 4.14 (a) An input group of faces as a force diagram; (b) an exploded view of all single faces; (c) adding half-faces to the system and finding all the closed polyhedral cells with opposite normal directions; (d) removing the faces directly connected to the external polyhedron; merging the faces of the global polyhedron with immediate nodal cells; (e) aligning the direction of the external polyhedron with the tension-only applied loads; and (f) the direction for compression-only force diagram.

that was explained in Section 4.10 to find the cells of this group of faces. Applying that method to the input of this example results in finding three polyhedral cells, as shown in Figure 4.14c. There is a large polyhedron that includes all the external faces and two smaller cells that are inside the larger cell. As shown in the illustration, the external polyhedron has a normal direction toward the outside, and the interior cells have normals toward the inside of the cell. Although we have found all the cells from this group of faces, we still need to make some adjustments before using this new configuration as a force diagram.

External Cell versus Global Force Polyhedron

In the force diagram of polyhedral graphic statics, the *external cell* is often referred to as the *global force polyhedron*, and the direction of the faces should match the direction of applied loads and reaction forces. Thus, the external cell in this configuration needs to be preserved, and the normal direction of faces should be altered based on the direction of applied loads.

Internal Cells versus Nodal Force Polyhedrons

In the context of polyhedral graphic statics, the faces of the global force polyhedron do not have half-faces. Therefore, the internal cells that are adjacent to the global force polyhedron have faces of the global cell. Thus, in the example of Figure 4.14, for those cells that are directly adjacent to the external cell, we remove their shared half-faces with the external cell and connect them directly to the faces of the external cell, as shown in Figure 4.14d,e. In addition, the normal direction of the cell follows the direction of the face of the global force polyhedron. In a force diagram of a compression-only form, the cells connected to the global force polyhedron have normal directions toward the inside of the cell, and the normals face outside if the force diagram belongs to a tension-only system, as shown in Figure 4.14e,f.

Visiting the global and nodal polyhedrons and their reciprocal components in the form diagram highlights the following properties:

- The faces of the global force polyhedron do not have half-faces.
- In the force diagram, each face that does not belong to the global force polyhedron can be called an *internal face*.
- Each face of the global force polyhedron makes a nodal cell with internal faces.
- Each face of the global force polyhedron belongs to one and only one nodal cell.
- Each internal face of the force diagram consists of two *half-faces* with opposite normal directions. Each half-face belongs to a nodal cell, and the normal direction follows the direction of the cell.
- Each of the two adjacent nodal cells in the force diagram shares a half-face.
- The direction of the normals of a nodal cell is consistent and is defined by the faces of the global force polyhedron.
- If the direction of one face is known, the direction of the cell is defined since the direction of the normals for each cell is consistent (Figure 4.11c).

- For a compression-only system, the direction of the normals of the faces of the *global* force polyhedron is toward inside the polyhedron. This direction is toward the outside if the system is tension-only.
- For a compression-only funicular form, the direction of the normals of *nodal* cells in the force diagram is always toward the inside of the cell.
- The nodal cells are always convex for a compression/tension-only system of forces.
- Each internal face is shared by two and only two adjacent nodal cells in the force diagram (Figure 4.11c).
- In a force diagram with multi-layered cells, only the cells in the out-most layer have a face of the global force polyhedron.

4.12 Computational Construction of the Form Diagram in 3D

Once the force diagram is constructed and the nodal and global force polyhedron and the directions of the cells are identified, the reciprocal form diagram can be constructed, which completes the form-finding process. For complex force diagrams, including thousands of faces and cells, the form-finding process needs a computational implementation to save time and effort. Similar to 2D, the computational construction of the reciprocal polyhedrons can be executed using either *an iterative* or *algebraic* approach.

Iterative Construction of Reciprocal Diagrams

In a previous study, the author showed an iterative construction of a reciprocal diagram by first finding the force diagram cells and connecting the adjacent cells' centroids to establish the topology of the form.[75] In the next step, the edges of the new graph are rotated to become normal to their reciprocal faces while maintaining the connectivity of the original graph. The algorithm's efficiency was later improved by Nejur and Akbarzadeh.[105] We do not cover the computational implementation of iterative methods in this book and refer the readers to the mentioned articles.

Algebraic Construction of Reciprocal Diagrams

The reciprocal diagrams of polyhedral graphic statics can be constructed in a single step using the mathematical formulation that connects the geometry of the form and force diagrams. We will expand on the algebraic methods of construction in Chapter 7 and talk about the advantages and disadvantages of these methods over the iterative approach.

4.13 Funicular Typologies and Subdivision Techniques

The key element in designing spatial funicular geometries is the arrangement and configuration of the nodal cells within the global force polyhedron. The simplest possible approach in the design of spatial funicular forms is to translate the 2D diagrams into 3D by extrusion.

4.13.1 Spatial Branching Geometries

We discussed subdividing global equilibrium into smaller nodal polygons in 2D as a design technique. Let us use the same approach in designing 3D structures by designing the force diagram as a 2D drawing and then extruding the drawing to create polyhedral cells. Figure 4.15a–d shows multiple form and force diagrams in 2D constructed by subdividing the global force polygon. If the 2D drawing is extruded to two

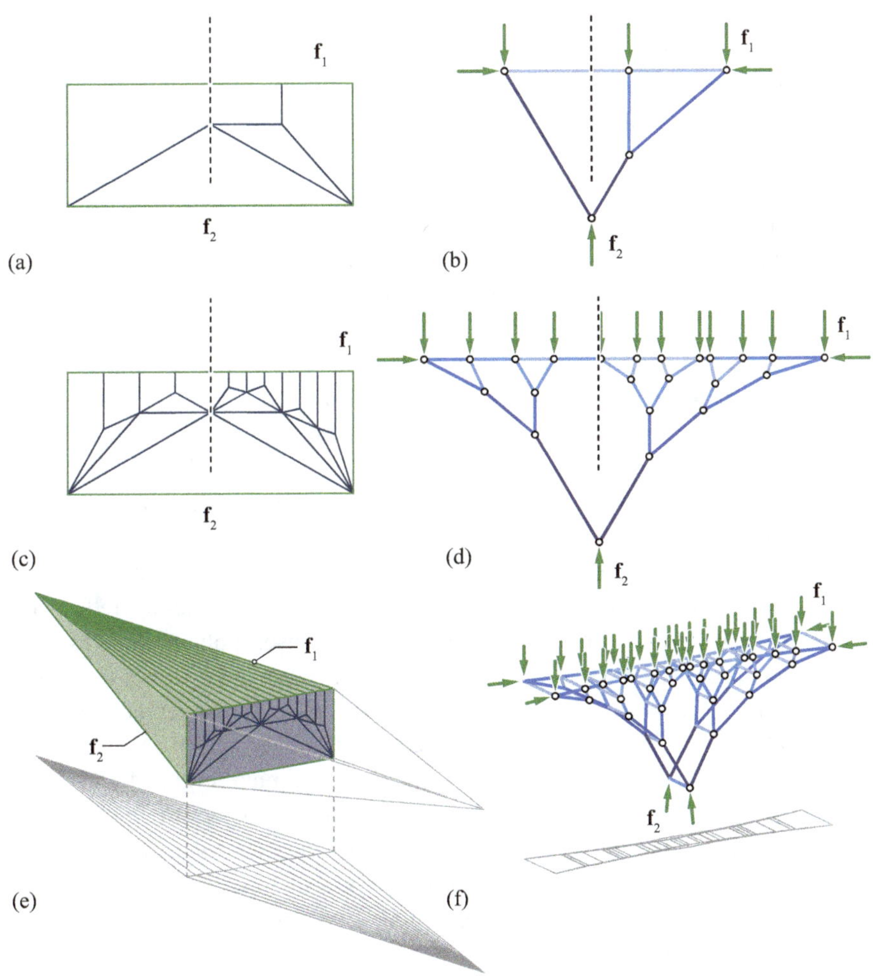

Figure 4.15 (a) A simple force diagram that is subdivided on one side showing the branching formation in (b); (c) and (d) show the force and form diagrams subdivided with different levels of articulation; (e) extruding the force diagram to two points equidistant from the 2D diagram makes a polyhedral force diagram with its reciprocal branching form in (f).

points equidistant from the original drawing in the out-of-plane direction, the result will be polyhedral cells reciprocal to a spatial funicular geometry. A closer look at the generated form and force shows that the extrusion to a point in 3D creates a curvature in the plan of the funicular form. Sections 4.13.2 to 4.13.6 will show how this curvature can be controlled in developing various features in the structural form.

The branching funicular form can be translated into volumetric geometry for construction purposes (Figure 4.15e,f). Figure 4.16 shows a schematic design of a compression-only bridge pier that branches on two sides and can make multiple arches if connected to similar modules in length.

4.13.2 Single-Layered Shells

Various polyhedral funicular forms can be explored by designing the arrangement and configuration of the nodal and global force polyhedrons in the force diagram. As mentioned in the definition of the global force polyhedron, the number of faces represents the number of applied loads and the reaction forces at the supports. Figure 4.17a–f shows a global force polyhedron in 3D and a global force polygon in 2D to provide an equivalent comparison between the 2D and 3D graphic statics. The force and form diagrams are shown by $\Gamma^\dagger$ and Γ, respectively. In 2D examples, the number of edges $e^\dagger$ in the global force polygon corresponds to the number of external loads e in the form diagram. In 3D examples, the number of faces $f^\dagger$ is equal to the number of external loads e in the form diagram. Figure 4.17a–c shows a process in which the corners of a triangular force polygon are truncated to increase the number of sides and thus the number of external loads or reaction forces in the system. The equivalent of this system in 3D is a tetrahedron, as shown in Figure 4.17d–f, where the corners of the polyhedron are carved in each step, resulting in more reaction forces in the form diagram.

The most intuitive typologies of funicular polyhedral systems are single-layered, compression-only shells. The force diagram of a single-layered shell can be constructed by extruding a 2D aggregation of convex polygonal faces to a point below the paper, as shown in Figure 4.18a–f. The left column shows the global force polygons of Figure 4.17 with internal subdivisions and their reciprocal compression-only networks.* This 2D pattern can be considered as a starting geometry to make the load network for 3D funicular forms in the right column of Figure 4.18d–f. Comparing this figure with the Figure of 4.17 shows that the top face of the global force polyhedron can be divided into a group of co-planar, convex faces as applied loads. Extruding these faces to the lowest vertex of the global force polyhedron makes nodal cells that together make a force diagram ($\Gamma^\dagger$) reciprocal to a compression-only shell (Γ) of Figure 4.18d–f.

* See also the works by Whiteley, Ash, Bolker, and Crapo.[106]

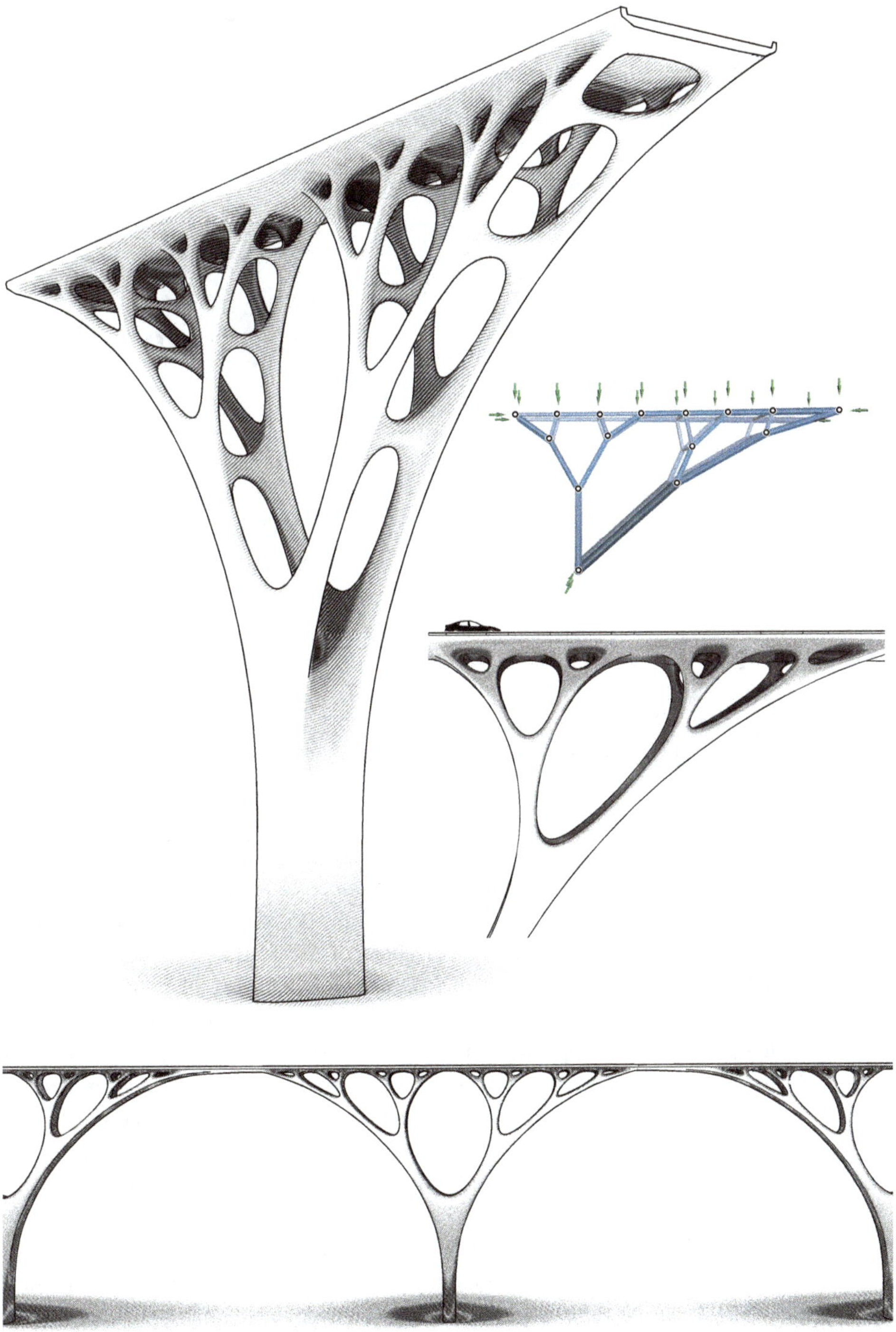

Figure 4.16 The branching columns of a compression-only bridge and its multi-span application in design.

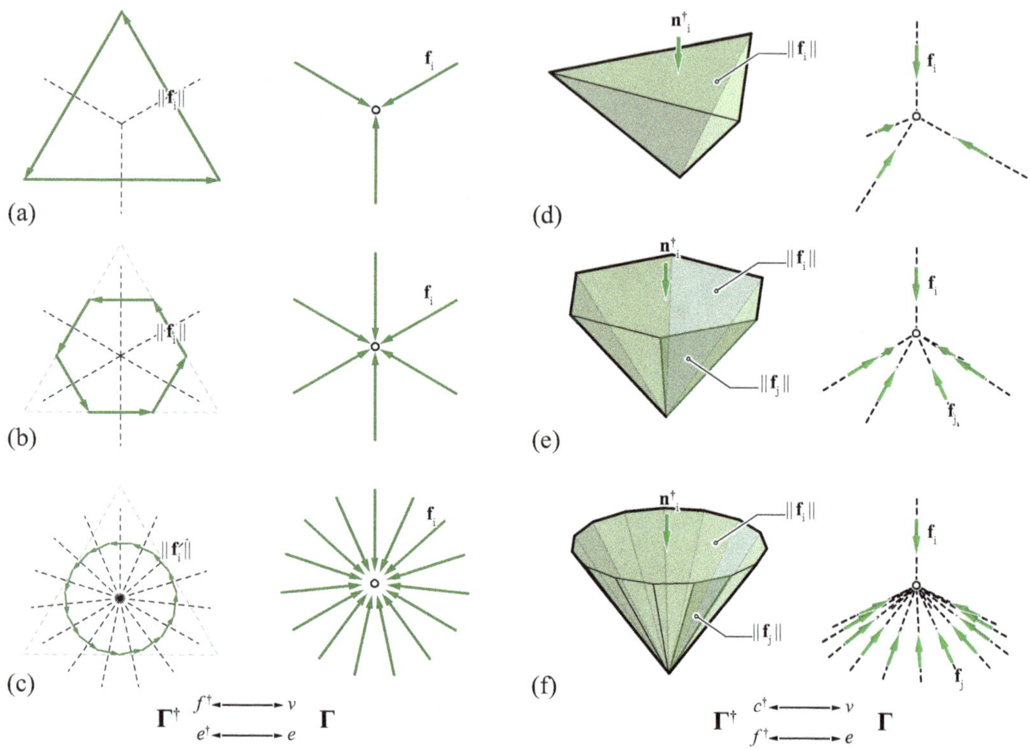

Figure 4.17 Truncating the corners to add to the number of reaction forces: (a), (c), and (e) show the global force polygon in 2D; and (b), (d), and (f) show the force polyhedron in 3D.

4.13.3 Creating Compression-Only Shells with Side Openings

Shell structures have tremendous applications in architecture and the design of functional spaces.* One of the key features in the design of compression-only shells is the free edges and openings at the periphery of the shell as entrances to the shell's internal space. Let us begin with the 2D and 3D examples of Figure 4.19. Consider the edge that is reciprocal to the force $\mathbf{f}_i$ in the 2D force diagram of Figure 4.19a. The length of the edge represents the magnitude of the force $||\mathbf{f}_i||$ in the form diagram. Let us remove the edge from its global force polygon and extrude all remaining edges to a chosen point on the *xy*- plane, as shown in Figure 4.19c. The newly created force diagram is reciprocal to a newly formed arch in the form. The same procedure can be used for force diagrams with more edges, as shown in both Figures 4.19 and 4.20. If these 2D networks are extruded to a vertex below the paper, it makes a 3D global force polyhedron representing the force equilibrium in a compression-only shell with an opening on its side (Figures 4.19e–h and 4.20e–h). The edge-removing and point

* See the "Shell Structures for Architecture: Form Finding and Optimization" by Adriaenssens, Block, Veenendaal, and Williams.[59]

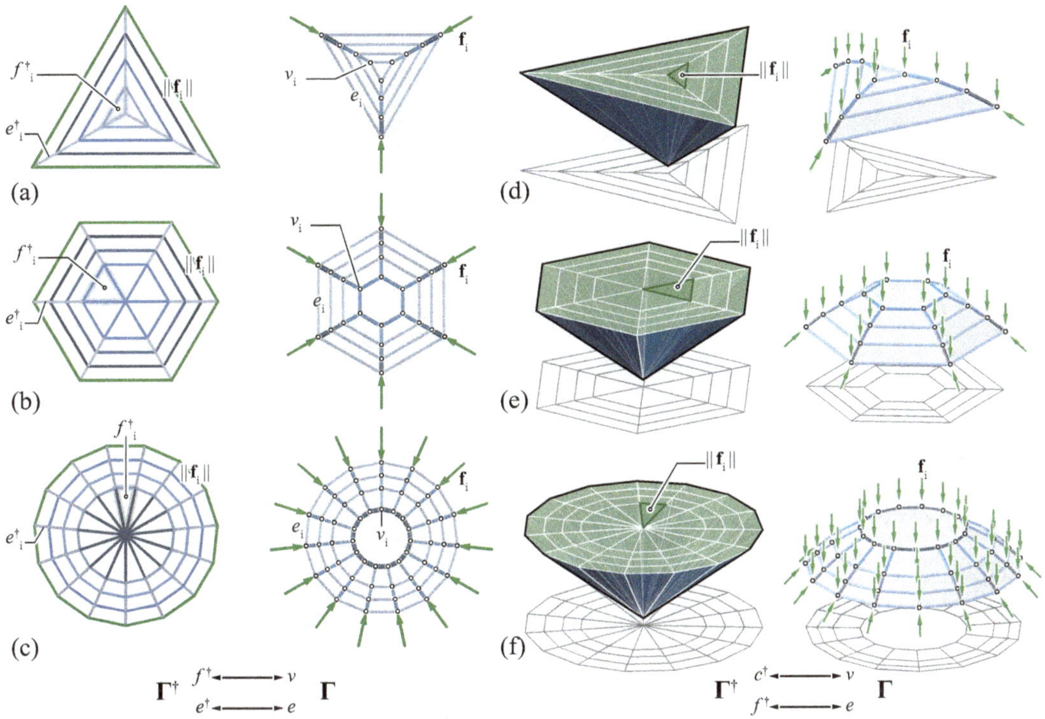

Figure 4.18 (a), (c), and (e) Subdividing the global force polygon with convex faces corresponds to compression/tension-only networks in 2D; (b), (d), and (f) extruding the 2D networks to a point results in 3D force diagrams reciprocal to compression-only polyhedral shells.

extrusion processes can also be applied to multiple adjacent edges and polygons within a force diagram, as shown in Figure 4.20.

4.13.4 Designing Unidirectional Compression-Only Systems

Developing certain features in the compression-only funicular forms requires particular techniques in designing the force's geometry. Figure 4.21 shows a simple approach to developing a 2D funicular network and a directional arch in 3D. An aggregation of closed quadrilateral polygons in 2D constructs a force distribution that is reciprocal to a linear network in equilibrium with the lateral and vertical external forces, as shown by green vectors in Figure 4.21a.

Let us assume that the same force polygon of Figure 4.21a consists of faces in 2D; extruding the faces to a single point on a 2D plane makes a force diagram comprising oblique triangles (see Figure 4.21b). This force diagram is reciprocal to a funicular, compression-only network in 2D. Each node of this funicular form is held in equilibrium by the vertical, lateral, and reaction forces in the form diagram, with their magnitudes as edge lengths in the force diagram. We can direct the lateral forces to the supports to stabilize the funicular form by extruding the side edges of the force diagram into a point (see Figure 4.21c). This technique in both 2D and 3D directs

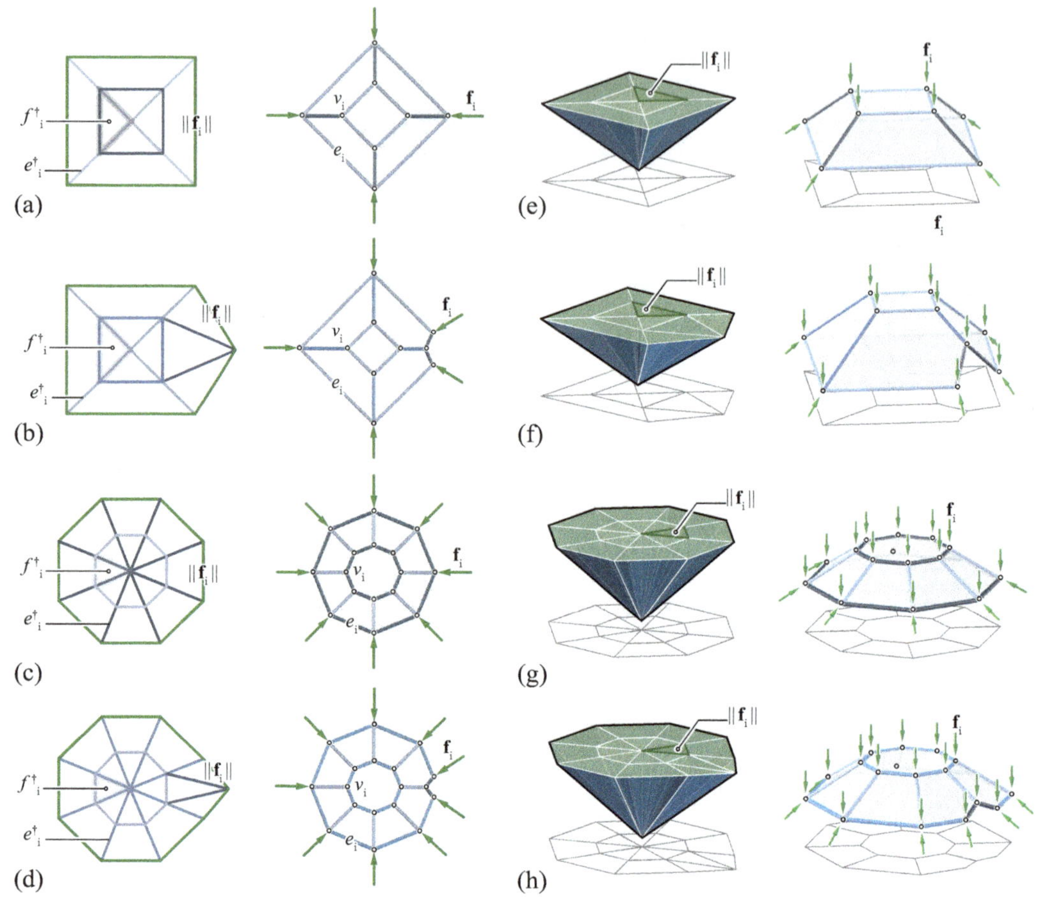

Figure 4.19 (a), (c), (e), and (f): A subdivided global force polygon and its reciprocal compression-only form with an opening on the side; (b), (d), (g), and (h) the 3D equivalent of the same network extruded to a point below the network and their reciprocal forms with openings on the side.

the forces toward the supports and eliminates the lateral forces in the system. As a result, a compression-only network is developed that is supported at the six corners (see Figure 4.21c). The same technique can be extended in 3D by extruding the faces of the applied load to a point underneath the plane of the faces. Note that the term *unidirectional* in this context is merely used to describe the spanning direction of the force flow.

Triangulating the quadrilateral polygons of the force diagram of Figure 4.21a results in an inclined network in 2D, as shown in Figure 4.22a. We can use the same extruding technique for the edges and the faces of both 2D and 3D force diagrams, resulting in a network with nodes of three valency. In addition, the triangulation of the force faces results in the staggered polygonal faces in the form diagram, as shown in Figure 4.22a–f.

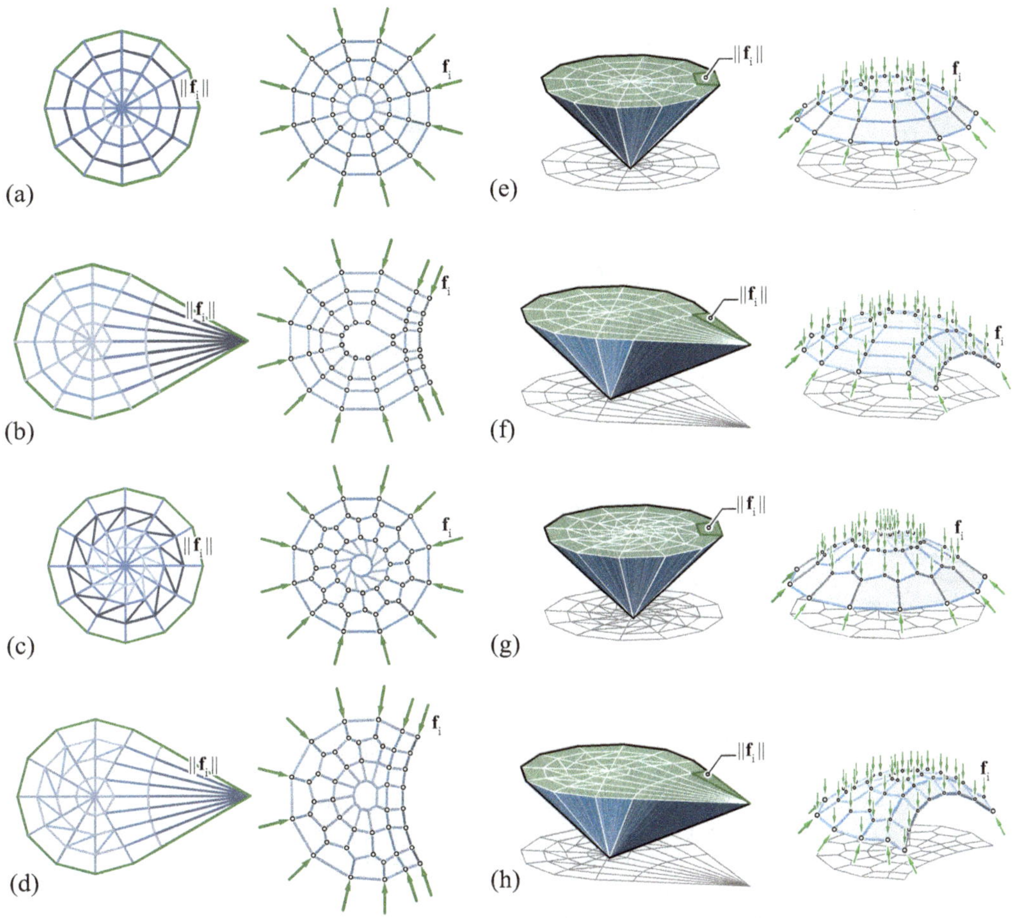

Figure 4.20 (a), (c), (e), and (f): A subdivided global force polygon and its reciprocal compression-only form with an opening on the side; (b), (d), (g), and (h) the 3D equivalent of the same network extruded to a point below the network and their reciprocal forms with openings on the side.

4.13.5 Creases in Compression-Only Polyhedral Forms

Extruding the edges to a single point in 2D makes closed force triangles reciprocal to funicular arches in 2D, as shown in Figure 4.23a. This method has been used extensively in various examples of 2D funicular form finding in this book and other related literature covering the methods of 2D graphic statics (see, for instance, Cremona,[41] Bow,[42] and Wolfe[21]). Let us consider the same force triangles in 3D, as shown in Figure 4.23b. Extruding the triangular set to two points underneath in the z- direction results in two sets of force cells sharing an edge, as shown in Figure 4.23d. Each force set is reciprocal to a spatial funicular curve, and these funicular networks are disconnected.

These two separate funicular arches of Figure 4.23d can be connected by adding faces linking two groups of polyhedral cells in the force diagram, as shown in

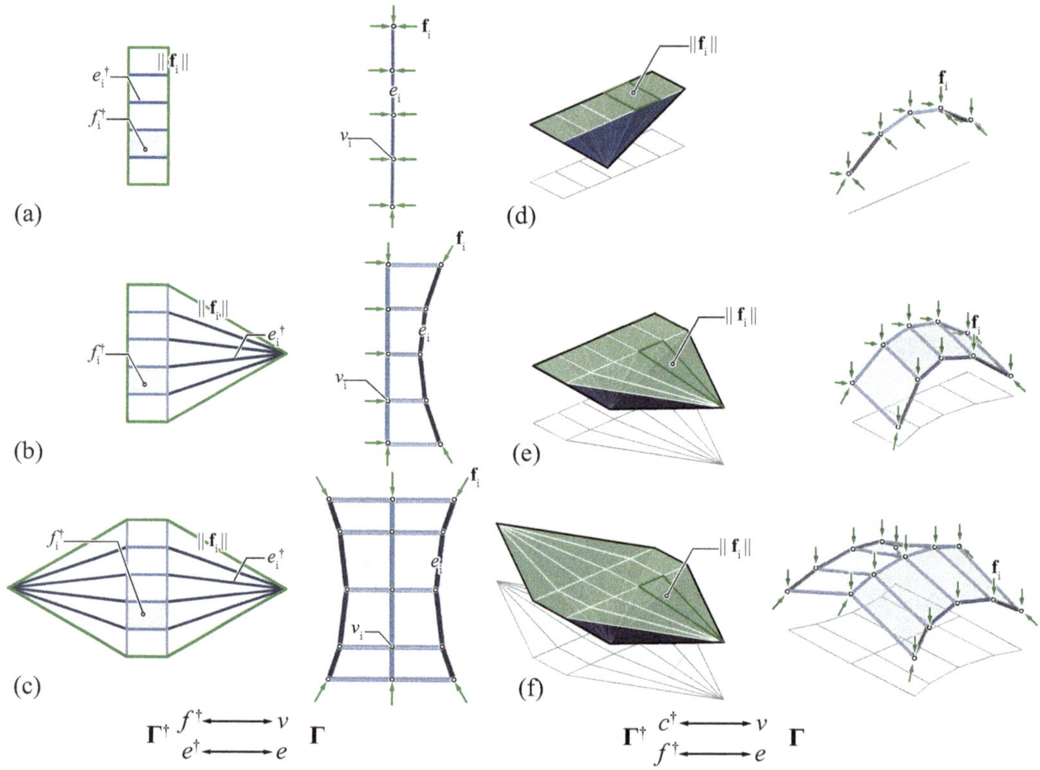

Figure 4.21 (a) An aggregation of quadrilateral force polygons in 2D and its reciprocal form diagram; (b) the force diagram resulted from the extrusion of the same quadrilaterals to a single point and its reciprocal form with lateral forces; (c) the extrusion to two points on two sides; (d), (e), and (f) extruding the 2D force polygon to a single point below the paper to make a force polyhedron and its reciprocal, unidirectional compression-only arch.

Figure 4.23e. These new faces are highlighted in Figure 4.23e. Consequently, a reciprocal form is constructed where the two previously disconnected funicular arches are connected via a *crease* in the reciprocal form. The magnitude of the internal force in the members of the crease is identical to the area of the connecting triangular faces in the force diagram.

The force triangles of Figure 4.23a could also get connected by inserting a strip of quadrilateral polygons in between them (Figure 4.23c). This additional column of force polygons corresponds to an additional column of edges between the funicular arches in 2D. These quad faces in 2D result in trapezoidal connecting faces in the force diagram in 3D (Figure 4.23f).

4.13.6 Combining the Design Techniques

The previously mentioned design techniques can be creatively used to articulate the geometry of the force diagram from which a compression-only funicular polyhedral shell with specific design features can be developed. Although the same procedural

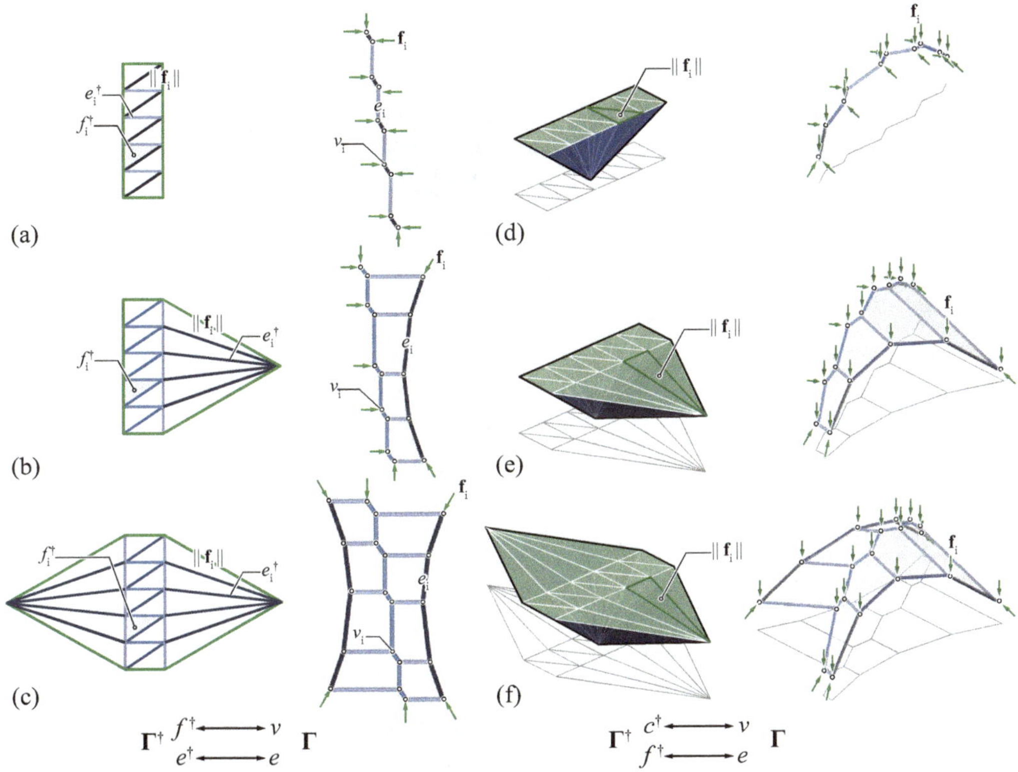

Figure 4.22 (a) Triangulated polygons in the force diagram and its reciprocal form network with three-valency nodes; (b) the equivalent force polyhedron and its reciprocal form; (c) to (f) 2D and 3D force diagrams constructed by extruding the edges to points on the sides and their reciprocal forms.

form-finding methods explained in Chapter 3 are sufficient to derive the funicular form, a computational approach is preferable for its time efficacy.*

Figure 4.24a–l shows multiple design steps for transforming a simple compression-only polyhedral form into a shell with four openings on the sides and creases around the openings. The left column of Figure 4.24 shows a 2D force and its corresponding form diagram on the *xy*- plane, and the right column illustrates the 3D version of the same force diagram constructed by extruding the 2D diagram to a single point in the *z*- direction. In each step of the design, multiple elements are added to the force diagram, and the resulting form is illustrated to show the effect of each element in the development process. In Figure 4.24c,d, the additional edges in the 2D force diagram and their similar faces in the 3D force diagram result in curved corners in the 2D funicular form and side openings in 3D.

* Please see Akbarzadeh, Van Mele, and Block,[75] Akbarzadeh and Nejur,[107] and Lee[11] for computational implementation of the methods of 3D/polyhedral graphic statics.

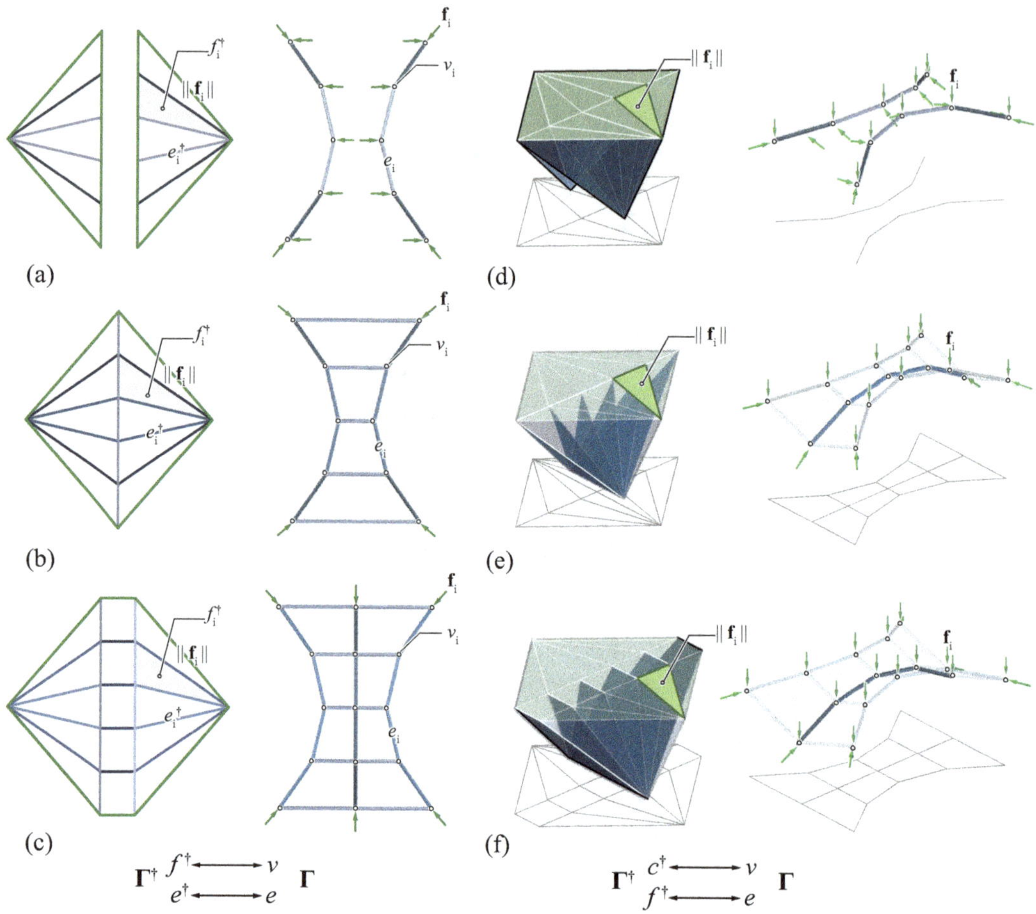

Figure 4.23 (a) Two force systems consisting of triangles reciprocal to two funicular forms; (b) the 2D networks interpreted as faces extruded to two different points, constructing force polyhedrons and their disconnected form diagrams; (c) combining the 2D force diagrams to make a single diagram and its reciprocal form; (d) connecting the edges of the force cells to create connecting triangular faces in the force diagram and its reciprocal form; (e) adding a strip of quadrilateral polygons in 2D and (f) trapezoidal faces in 3D to make a single force diagram.

Figure 4.24e,f shows how additional support locations around the funicular form can be created for both 2D and 3D. A similar technique is used in Figure 4.24g,h not only to divide the support locations but also to expand the form by creating parallel edges. Figure 4.24i,j divides the newly created edges and adds a group of edges in the opposite direction. In Figure 4.24k, the edges connected to the vertices of the 2D global force polygon are divided, resulting in an additional edge parallel to the curved edges on the sides of the network. Similarly, dividing the triangular faces, connected to the apex and the four corner vertices of the global force polyhedron in 3D results in triangular faces as shown in Figure 4.24l. These triangular faces create similar creases in the funicular form, as discussed in Section 4.13.6. Figure 4.25 shows a high-resolution

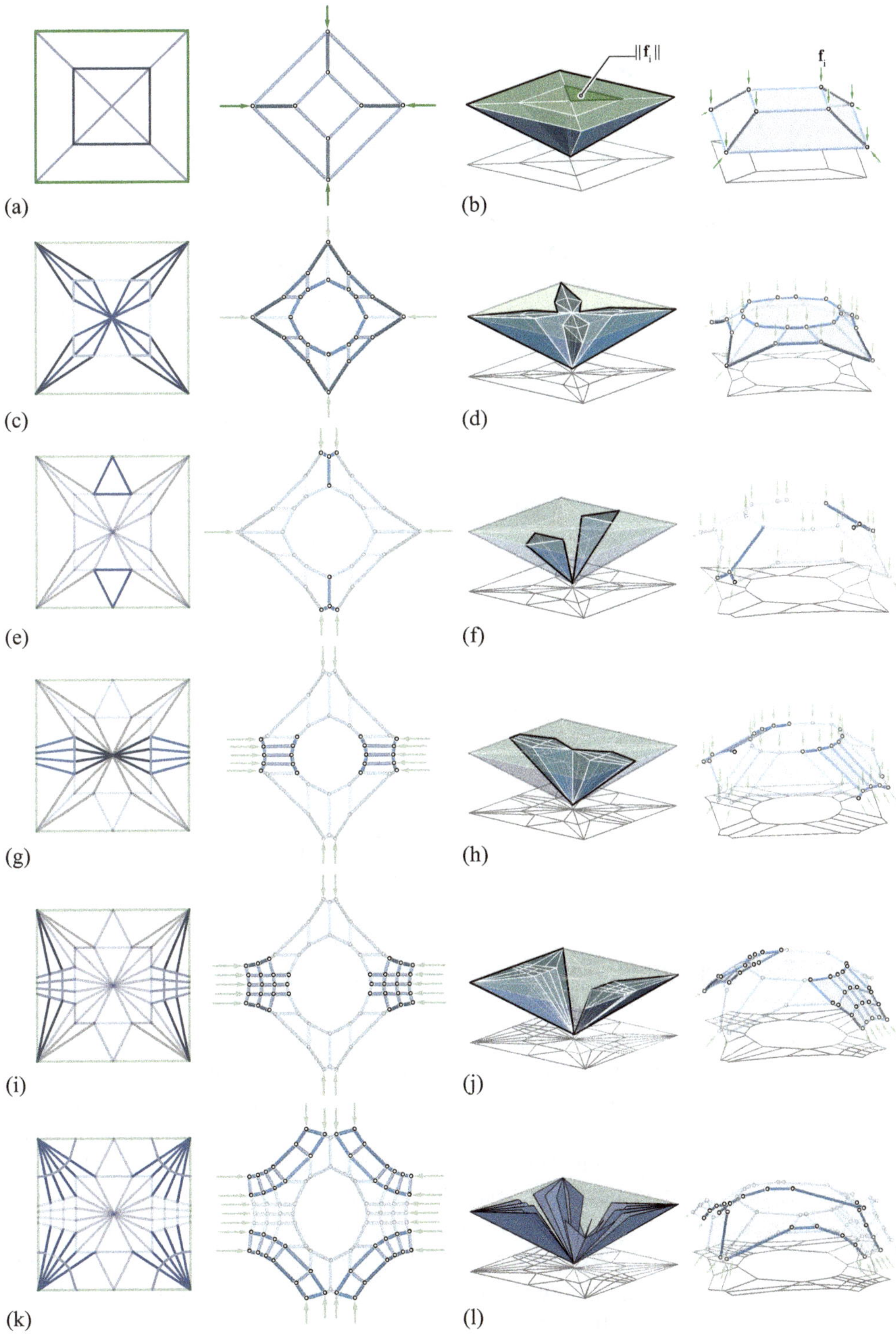

Figure 4.24 (a) to (l) A step-by-step design process for a force diagram and its corresponding form to develop a complex compression-only shell.

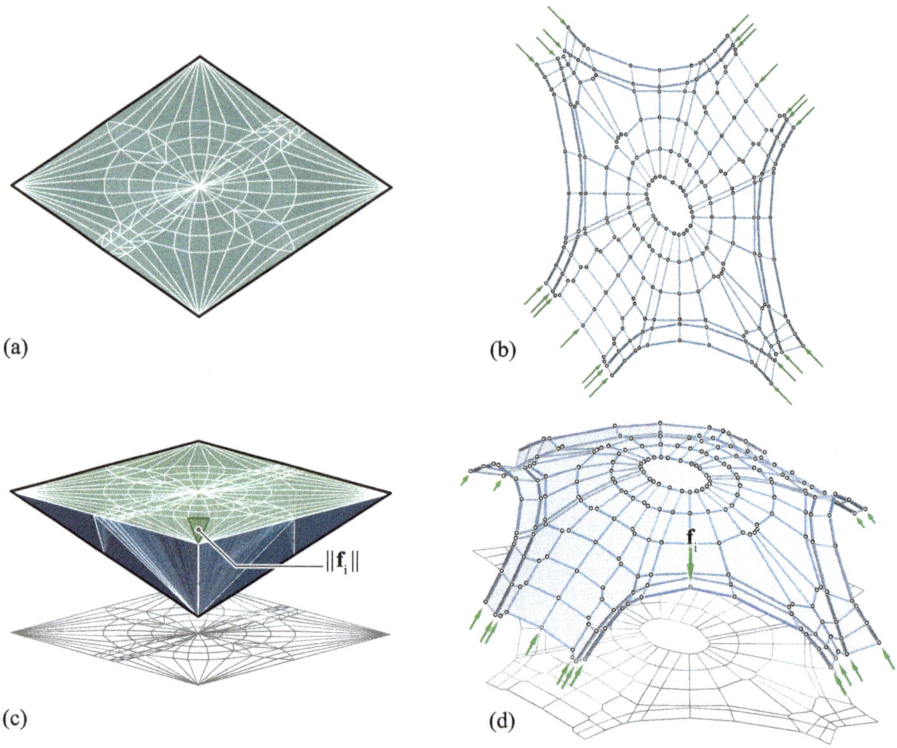

Figure 4.25 (a) The top view of the force; (b) its corresponding compression-only network from above; (c) the axonometric view of the force diagram; and (d) the reciprocal compression-only polyhedral shell.

force diagram designed using the described steps and its corresponding funicular form with creases around the openings.

4.14 Gaussian Curvature Approximation for Polyhedral Surfaces

In differential geometry, the *principal curvatures* measure the maximum and minimum bending of a continuous surface at a given point. The normal vector at the given point and a chosen tangent vector on the surface can make a normal plane. The normal plane intersects the surface, creating a curve on the surface. The curvature at the chosen point is called the normal curvature κ_n, which is the reciprocal radius of the osculating circle. The normal curvature may change since the tangent vector on the surface can take different directions. The direction of a tangent for which the κ_n has a minimum or maximum value is called the *principal* curvature direction. A tangent direction with the maximum κ_{max} and minimum κ_{min} curvature are perpendicular to each other on each point on a continuous surface.[108]

The principle curvature direction is positive if the surface bends away from the normal and negative if it bends toward the normal. The dot product of a surface's maximum and minimum curvature is called *Gaussian* curvature K. The sign of the Gaussian curvature defines whether a surface at that point is *synclastic*, *anticlastic*, or *monoclastic*. The surface is synclastic if the Gaussian curvature on every point on the surface is positive, $K > 0$; it is anticlastic if negative, $K < 0$; and monoclastic (developable) if it is zero, $K = 0$.

Similar surface properties can describe the geometry of funicular polyhedrons with some assumptions and approximations. Note that the polyhedral geometries of the methods of this book consist of discrete planar elements and thus can only approximate a continuous surface. If these discrete elements become infinitesimally small, they can describe curvature on polyhedral geometries more precisely than larger elements.

4.14.1 A Two-Manifold Polyhedral Subsystem

The principal curvature directions and the Gaussian curvature are often defined for a *two-manifold* geometry. Simply put, infinitesimally small polygonal faces with linear edges can describe a two-manifold surface on which each edge is connected to two and only two faces. In contrast, polyhedral systems have *nonmanifold* geometries, that is, each edge is often connected to more than two faces. Nevertheless, the curvature analysis and surface properties can also be described where there is a two-manifold geometry subsystem.

Figures 4.26, 4.27, and 4.28 show multiple examples of such a subsystem, which is a continuous single-layer polyhedral surface with two-manifold geometry. The common process of finding principal curvature directions on a continuous surface can be applied to the polyhedral surface of these figures. In all cases, the polyhedral surface should be approximated by a continuous surface and analyzed to find its principal curvature directions, but that is not the main purpose of this section. The main intention is to realize the approximate surface curvature of a polyhedral geometry by analyzing the corresponding elements in its reciprocal force diagram. Subsequently, the outcome will be applied in designing a force diagram to control surface curvature in the reciprocal funicular forms.

4.14.2 Synclastic Polyhedral Surfaces

Consider the polyhedral surface of Figure 4.26a. The curvature at the node i of the form diagram can be approximately described by the direction and the alignment of the edges connected to the node. The direction of edges is shown by the unit vectors $\mathbf{n}_{\mathrm{v}_i}$ and $\mathbf{n}_{\mathrm{u}_i}$. The edges connected to the node of Figure 4.26 are reciprocal to the faces of the reciprocal cell in the force diagram. Thus, the normal of the faces $\mathbf{n}^{\dagger}_{\mathrm{v}_i}$ and $\mathbf{n}^{\dagger}_{\mathrm{u}_i}$ are identical to the unit vectors of the edges in the form diagram (Figure 4.26a). Therefore, the curvature of the polyhedral surface in the form can be measured by fitting oscillating circles tangent to the unit vectors of the edges of the form or the normals of the force diagram's faces, as shown in Figure 4.26b. The curvature in two directions, u and v, on the polyhedral surface can be represented by κ_1 and κ_2,

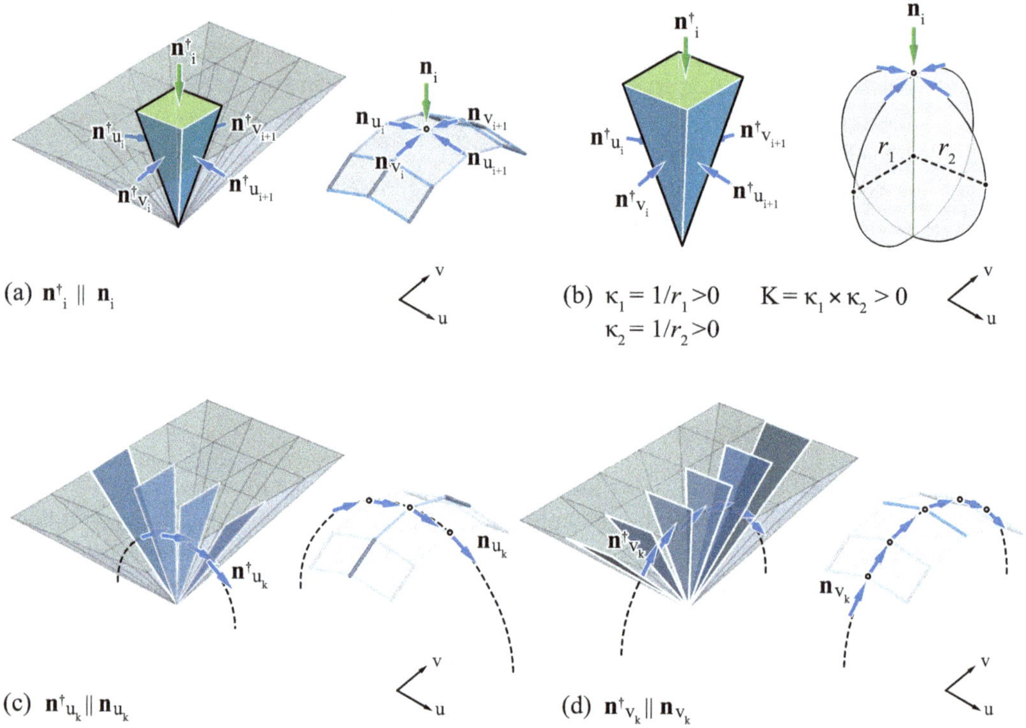

Figure 4.26 (a) A closed polyhedral cell in the force diagram and its reciprocal node in the form diagram; (b) curvature in two directions of the polyhedral surface and the oscillating circles; (c) a series of faces in the force diagram and their corresponding row of edges in the form diagram; and (d) another series of faces responsible for the curvature in the opposite direction.

measured by the radius of their oscillating circles. The curvature κ_1 is positive if the edge moves away from normal $\mathbf{n}_i$ and is negative otherwise. The curvature K close to the Gaussian curvature measured on continuous surfaces can be found by the multiplication of κ_1 and κ_2. For a *synclastic* polyhedral surface, $K > 0$.

Remark 4.6. The aggregation of convex cells in the force diagram can control the overall curvature of the corresponding polyhedral surface of the form diagram. An assembly of cells consisting of five faces (pyramid) in multiple rows such that all the cells converge to a single point is reciprocal to compression-only, *synclastic* polyhedral form diagram. The curvature in each row (e.g., v) of edges in the form diagram is controlled by a series of normal vectors $\mathbf{n}^{\dagger}_{\mathrm{v}_k}$ of faces converging to a point in the force diagram (Figure 4.26c). Similarly, the curvature in the perpendicular direction (u) is controlled by the normals of another series of faces converging to the same point in the force diagram (Figure 4.26d). Thus, the force diagrams made by extruding a 2D network to a single point are often reciprocal to a synclastic compression-only form.*

* The same force diagram can be reciprocal to an anticlastic form with combined tension-and-compression forces; see Chapter 5 for related examples.

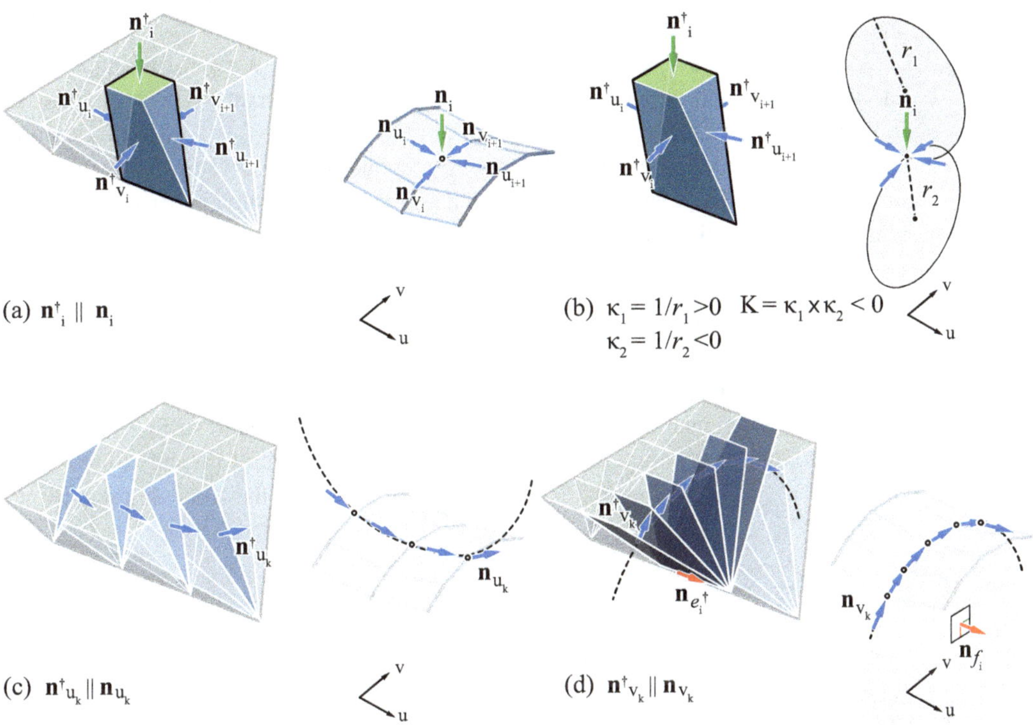

Figure 4.27 (a) A force diagram with cells converging to a line and its reciprocal form diagram; (b) a single force cell and its corresponding vertex and the associated curvatures κ_a and κ_2; (c) curvature in a nominal v direction and the associated faces in the force; and (d) the curvature in the v direction and the associated faces in the force.

Remark 4.7. Finding the curvature for polyhedral subsurfaces with high-valence vertices might be less straightforward for the following reasons: (1) the maximum and minimum curvature in all directions should be found and compared, and (2) using edges connected to a vertex to define the two nominal directions u and v on a polyhedral surface may have the following drawbacks: the edges in two directions might not be necessarily normal to each other. Thus, the curvature found using the edges on the surface cannot be called the principal stress directions as it is called on a continuous surface.

4.14.3 Anticlastic Polyhedral Surfaces

Let us consider another case where the force diagram consists of cells where each of the two side faces of a single cell shares an edge at the bottom of the force diagram. Figure 4.27a shows a single cell in the force diagram and its reciprocal vertex in the form. The curvature on the polyhedral surface can be measured in two directions of the surface (e.g., u and v), as shown in Figure 4.27b. In this example, κ_1 is positive in the u direction, while κ_2 is negative in the v direction. Thus, the curvature of the surface as a result of the multiplication of these two values will be negative ($K < 0$), and the polyhedral surface is *anticlastic*.

The geometric analysis of the force diagram shows that a row of triangular faces and their associated normals $\mathbf{n}^{\dagger}_{\mathrm{v}_k}$ are responsible for the curvature in a row of edges in the v direction on the polyhedral surface (Figure 4.27c). Besides, the faces in the opposite direction of the force diagram correspond to the curvature in the form's u direction. These faces share an edge with a unit vector $\mathbf{e}^{\dagger}_i$, which is perpendicular to a plane in the form diagram with a normal $\mathbf{n}_{f_i}$ on which all the edges with a unit normal $\mathbf{n}_{\mathrm{v}_k}$ lie (Figure 4.27d).

Definition 4.2. If, in a force diagram, multiple faces share a common edge $e^{\dagger}_i$, then the reciprocal face f_i in the form diagram has edges whose directions are controlled by the normals of the faces connected to the edge $e^{\dagger}_i$ in the force diagram. The edges of the face f_i are coplanar and control the curvature of the polyhedral surface on the plane of the face, which is called the *curvature plane*. The normal of the curvature plane is shown by $\mathbf{n}_{f_i}$ in the figures and is parallel to the edge $e^{\dagger}_i$ in the force diagram (Figure 4.27d).

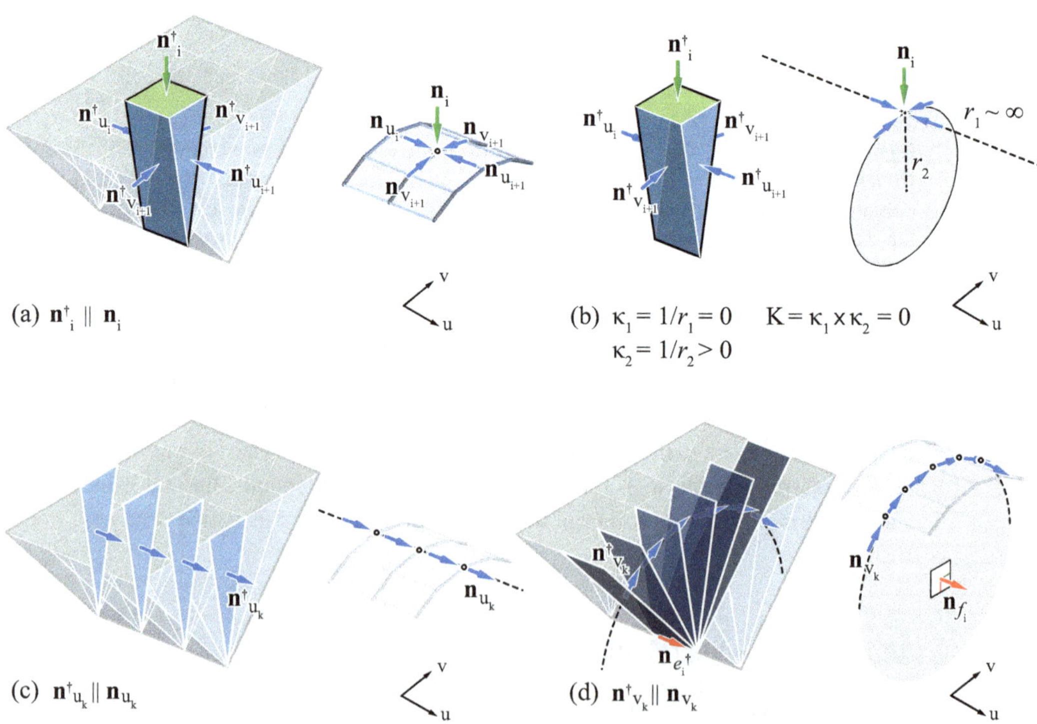

Figure 4.28 (a) The form and force diagrams in a monoclastic polyhedral surface; (b) a single force cell and its face normals in calculating the curvature in two directions in the reciprocal node; (c) the alignment of parallel faces in a nominal v direction of the surface resulting in a zero curvature; and (d) a series of faces in the u direction of the surface responsible for the curvature in the dual polyhedral form.

4.14.4 Monoclastic Polyhedral Surfaces

The third case is related to monoclastic surfaces where the curvature in one direction is zero. Figure 4.28a–d shows a force and form diagram in which the curvature in each node of the form diagram in the v direction is zero. These systems may represent a developable surface with numerous applications in fabricating complex geometries using flat-sheet materials. If the edge lengths of the form in the *u* direction will become very small, then the polyhedral geometry can be replaced by strips bent in the u directions, as shown in Figure 4.28c. Similar to the anticlastic case, the edges in the u direction also lie on a plane with a normal parallel to the edge shared by the faces in the force diagram (Figure 4.28d).

4.15 Understanding the Structural Geometry through Design Instances

In this section, some examples of famous structural systems built by renowned master builders are chosen, together with some new designs using the methods of polyhedral graphic statics. The historical examples are used to show how the funicular polyhedral form findings explained in this chapter can be used to (1) understand the structural geometry and force equilibrium in such systems and (2) reproduce particular design features, for instance, creases, anticlastic geometry, and double-layered configurations in compression-only forms. In the last part of this section, some novel concepts are introduced to highlight the potential of using polyhedral-based structural form-finding in the design of lightweight structures.

4.15.1 Guastavino Tile Vaults

The *Guastavino Vaulting* (after Ochsendorf and Freeman[109]) is considered one of the most efficient methods of masonry construction, in which thin ceramic tiles, sized about 15 × 30 × 2.5 cm, are laid in multiple layers to create arches, shells, and domes.[109,110] Guastavino Jr. often used 2D graphic statics to find the vaults' geometry and to calculate the forces in his vaulted structures. His design methodology is remarkable because of his ability to design efficient 3D structural forms using 2D methods of graphic statics. Indeed, his extensive construction experience allowed him to find such 3D structural solutions, although he used only 2D theoretical methods.

A synclastic funicular polyhedral shell and its force diagram can describe the force distribution in a 3D structural geometry of Guastavino shells, as shown in Figure 4.29. The design of the funicular form for Guastavino's shell can start with the 2D force and form diagrams of Figure 4.29a. Extruding this 2D force polygon into a point below its 2D plane will make a force polyhedron reciprocal to a synclastic funicular form (Figure 4.29b).

It is evident from the vault's geometry that the curvature in the short direction is larger than the one in the long direction. The force diagram of such a funicular system can be designed as a series of square polygonal faces arranged in a rectangle matrix.

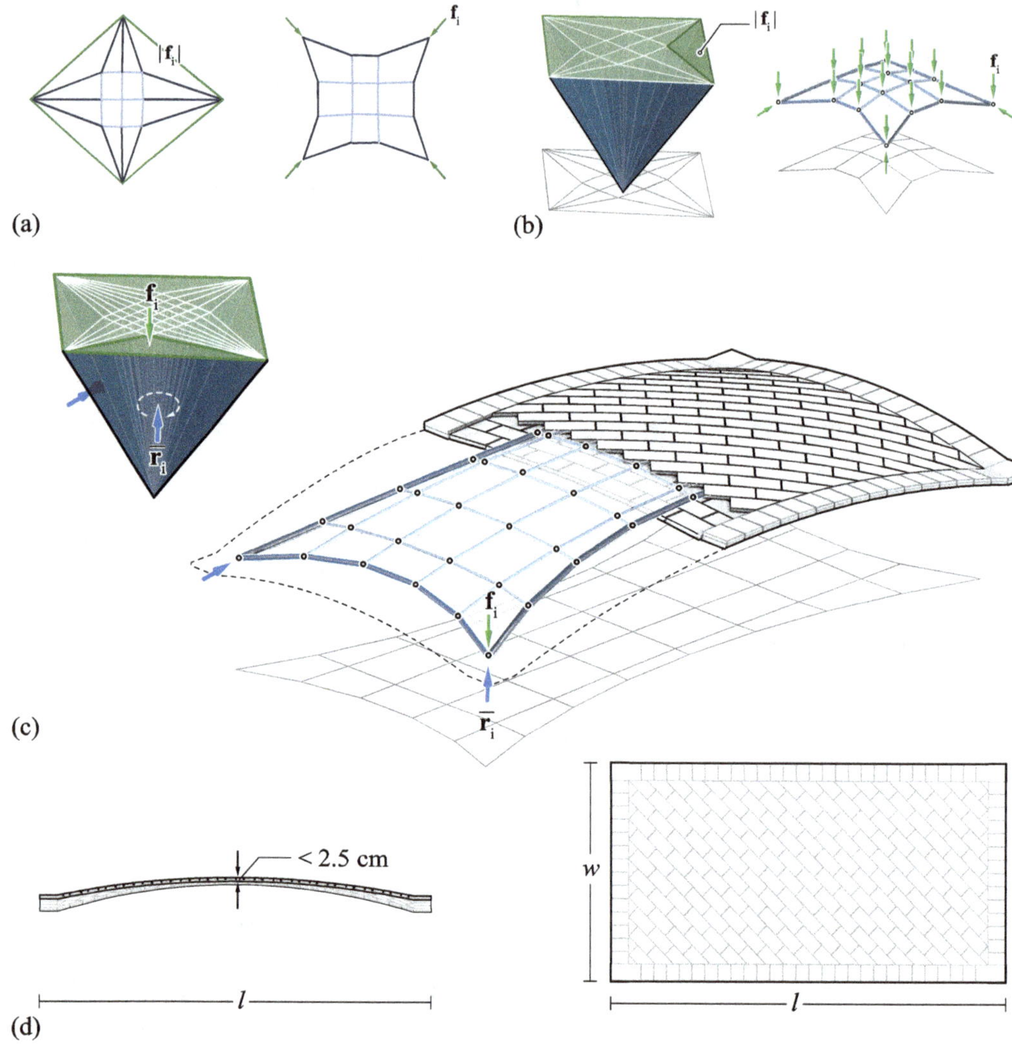

Figure 4.29 (a) A 2D force and force diagram with free edges on four sides; (b) an extruded 2D pattern to generate force polyhedrons with a synclastic compression-only form; (c) an unequal distribution of the square polygons in the force diagram that creates higher curvature in the form in the shorter direction and lower curvature in the longer direction; (d) the plan and section of the arch materialized by tile vaults.

Fewer polygons are placed, for example, in the x- direction, compared to the number of polygons in the y- direction, as shown in Figure 4.29c. Extruding the external edges of this aggregation of polygons into points on the sides eliminates horizontal side forces, as explained in Section 4.13.3. The force diagram's depth controls the funicular form's height, as discussed extensively in Chapter 3. The fewer triangular faces in the short direction result in a funicular form with a higher curvature in the width of the vault. This property is also shown in Figure 4.26.

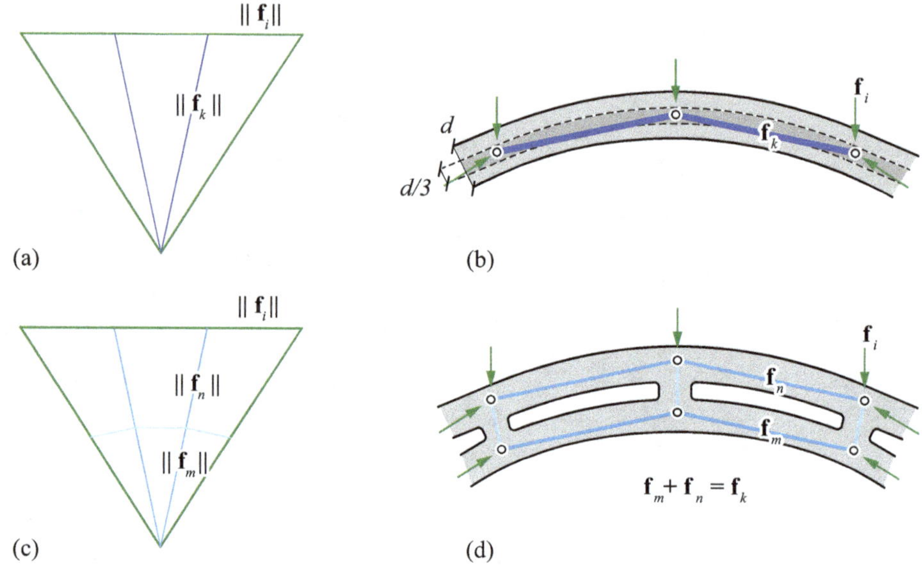

Figure 4.30 (a) An initial funicular form and (b) the cross-section with the thrust lines at the middle-third of the section; (c) a subdivided force diagram; and (d) a multi-layer funicular network with the same volume of material but different internal forces distributed in two layers of the funicular network.

4.15.2 Multi-layered Polyhedral Shells

In materializing a funicular form, the section of material is proportioned such that the thrust network of the form-finding stays within the cross-section for the design loads. An asymmetric loading or a settlement in the abutments might cause the thrust line to fall out of the section. In such cases, the structure will experience tension in its members instead of all compression. As a rule of thumb, if the thrust line stays within the *middle-third* of the geometry of a section, then the structure will be a compression-dominant structure.[54] The middle-third rule is an approximation for a region in the section where, for any compressive force, the average stress will be compression. This region is called the *kern* of the section, and it can be computed for various geometries of a section.[111,112]

There is a simple technique in the design of the force diagram that results in a multi-layered funicular form. Cutting the cells of the force polyhedron in 3D or the force polygon in 2D results in a multi-layered funicular system, as shown in Figure 4.30. The advantage of this approach is that the same applied loads are redistributed among more internal members, and thus, the magnitude of the force in the members and the stress in the cross-sections will be decreased. The resulting funicular form will have multiple layers, as shown in Figure 4.30. In multi-layered systems, for the same volume of material, the depth of the cross-section increases, which may increase

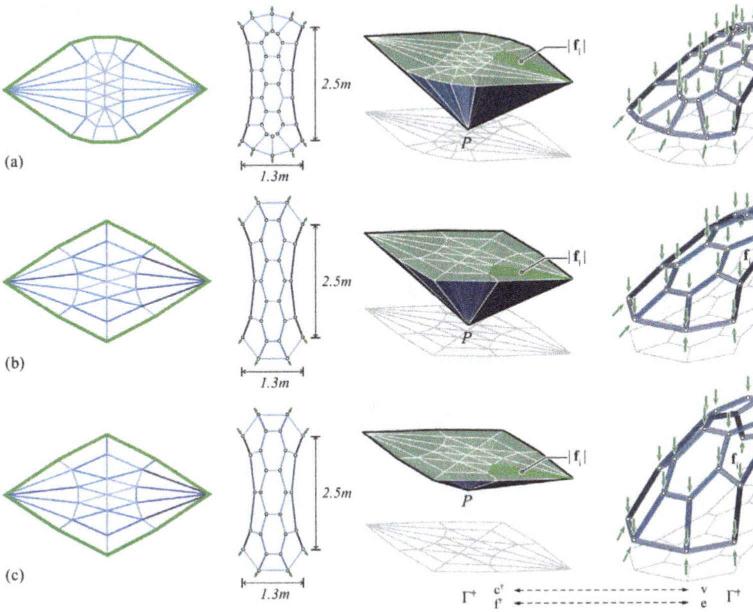

Figure 4.31 The form-finding of a single-layer bridge form. (a) and (b) have different topologies but the same rise; (b) and (c) have the same topology but different rises.

the load-bearing capacity and improve the resiliency of the structure for asymmetric loading scenarios.

Toruca, which is a compression-only hollow-glass unit structure, is a prototype designed and fabricated to show how the inherent planarity of the polyhedral geometries and multi-layered funicular network can be harnessed for efficient construction processes.[81]

The form-finding starts with constructing a force diagram of a single-layer funicular bridge. The force diagram of the glass bridge is made by extruding a 2D aggregation of polygonal faces to a point (Figure 4.31), where the 2D polygonal faces determine the bridge topology as well as applied load distribution, and the extrusion height determines the bridge rise and internal force magnitudes. The total area of the faces maps to the total applied load of 1 kN, and the area of each face maps to the magnitude of the tributary load on the corresponding node. By changing the polygonal faces' areas, the applied load's distribution can be manipulated. The derived form diagram with 13 polygonal faces is constrained on four anchor points that control the bridge's span (2.5 m) and width (1.3 m).

Adjacent edge angles and edge lengths are two key geometrical parameters that need to be tuned in this stage as these are related to the fabrication constraints. The angles between any pair of connecting edges are greater than 60°, and the lengths of the internal edges (the edges shared by two polygonal faces) are set to be smaller than 500 mm. The edge intersection angles are controlled by adjusting the edge angles of

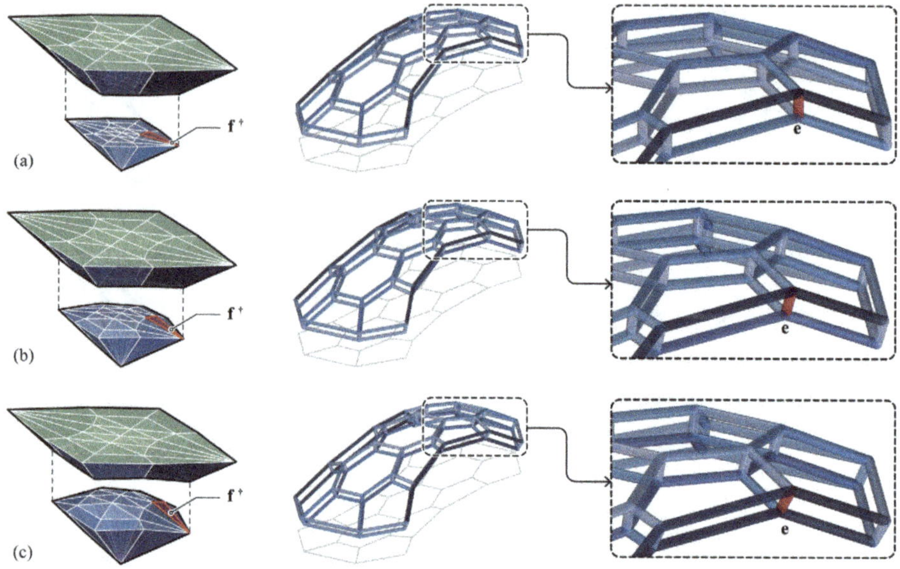

Figure 4.32 Transforming the single-layer form to double-layer by subdividing the force polyhedrons: (a) to (c) shows that different angles of the subdividing face change the edge orientation in the form.

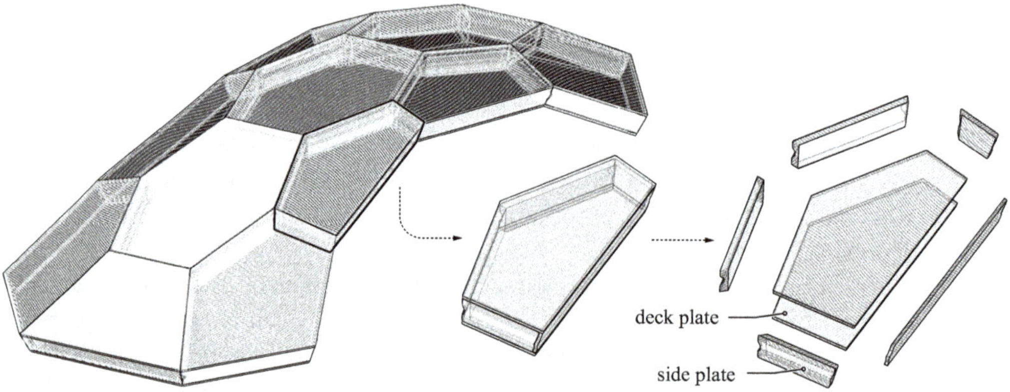

Figure 4.33 The glass bridge prototype and a highlighted hollow glass unit and its constructive components.

the initial 2D polygons in the force diagram, and the edge lengths can be constrained in the form diagram using the PolyFrame software.[107]

After having the single-layer funicular form and force diagrams, the double-layer form can be derived by subdividing the force polyhedrons of the single-layer form. We can take a horizontal surface and cut all the cells, as shown in Figure 4.32. Changing the curvature of the cutting surface will change the angle of the connecting members, *e*, as shown in the figure. The construction block of this bridge of this double-layer funicular form is a hollow-glass unit that is shown in Figure 4.33.

In the form-finding process, the total applied load (the area of all the top faces of the force diagram shown in green) is considered to be 1 kN. As a result, the members' internal force and the reaction forces at the supports can be found as a fraction of the applied load. The ultimate design load, including the dead load and live load, can then be multiplied by the total applied load and the members' internal forces. The members are then sized based on the allowable stress in the glass. The structure needs further analysis for asymmetric loading scenarios to finalize the sizing of the members (Figure 4.32).

4.15.3 Undulating Polyhedral Shells

The combination of synclastic and anticlastic geometry in the design of corrugated shells has been explored in the extraordinary works of Eladio Dieste,[113,114] González Zuleta, and Sánchez del Río.[115,116] These structures are made of reinforced-brick shell structures. Although there is no evidence of using graphic statics in the form finding of such structures by Dieste, Zuleta, or del Río, their configurations can be used as an example to show how such undulating funicular geometries can be constructed by designing the geometry of the force cells.

Figure 4.34a–d shows multiple examples of corrugated funicular shells and their corresponding force diagrams. Figure 4.34a is quite similar to Figure 4.23, where two sets of faces extruded to a single point are added together by connecting triangular faces in the middle. The triangular faces are, in fact, the cause of the formation of creases in the geometry of the shell. This shell includes two sets of arches: the arches at the peaks of the shells and the ones in the valleys of the shell (Figure 4.34b,c). The curvature at any point on the peak arch is synclastic, while it is anticlastic at any point on the valley arch. If the number of faces reciprocal to the applied loads increases in the width of the arch, the synclastic part of the shell becomes smoother, as shown in Figure 4.34d–f.

Suppose the connecting triangular faces corresponding to the crease of the shell are subdivided into smaller triangles. In that case, the crease will be turned into smooth anticlastic regions, as shown in Figure 4.34g–i. Once a single-layer funicular shell is constructed, its force diagram can be cut by a horizontal plane into an aggregated set of cells of Figure 4.34j. The resulting funicular form is a corrugated, multi-layered shell of Figure 4.34k,l.

4.15.4 Angles of the Curvature Planes

A designer can control the angle of multiple curvature planes in the anticlastic section of the shell. As shown in Figure 4.28, the form diagram's curvature plane is perpendicular to an edge of the force diagram where multiple faces are connected. A change in the angle of the edge in the force diagram will change the curvature plane's normal, resulting in a sectional variation in the funicular form. Consider the example in Figure 4.35 where the two edges in the force diagram are shown by their unit normals $\mathbf{n}_1^\dagger$ and

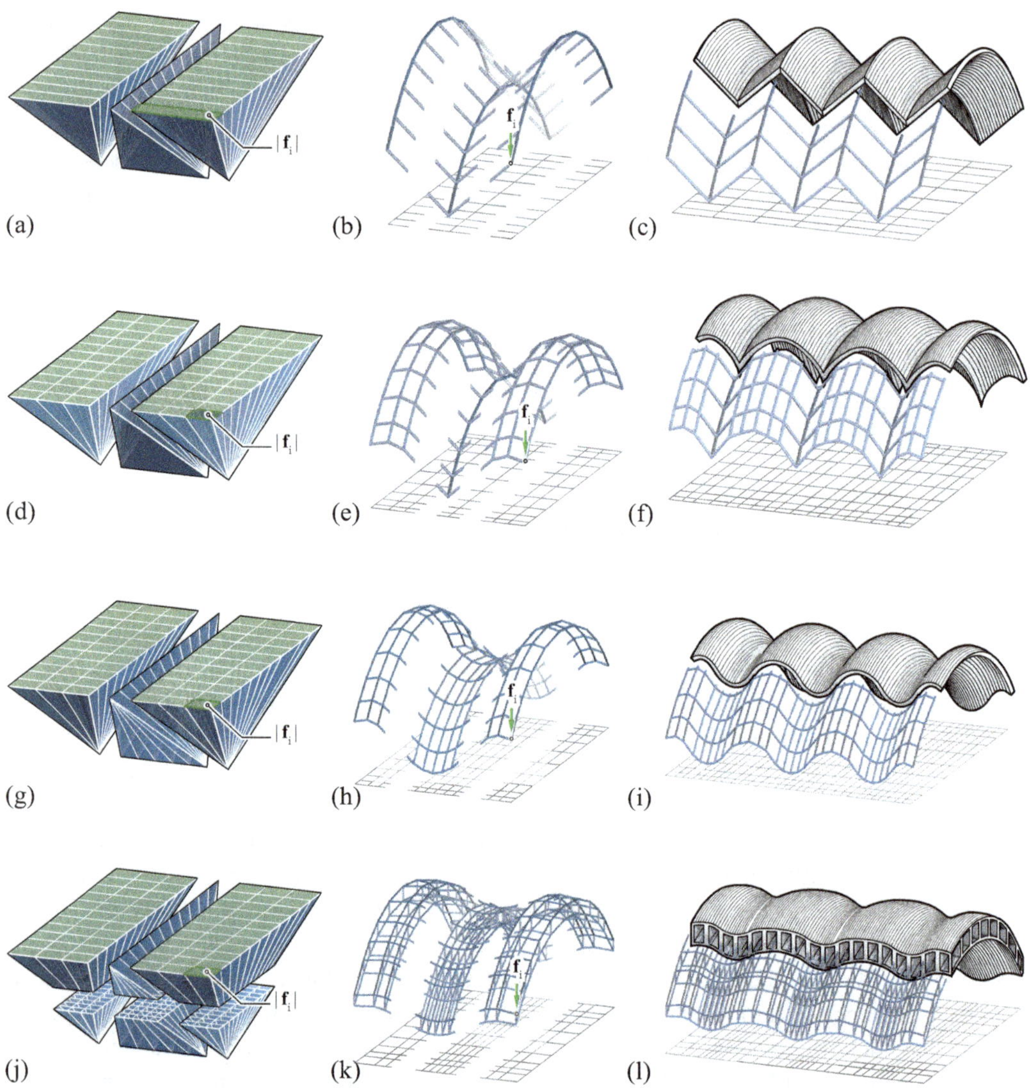

Figure 4.34 Various corrugated shells and their corresponding force diagrams: (a) the force diagram consisting of a three groups of cells (b) the reciprocal form diagram with single funicular arches at the peak and valleys; (c) a materialized version of a corrugated shell; (d) a subdivided force diagram reciprocal to (e), (f) arches with smoother synclastic regions; (g) further subdivided force diagram reciprocal to (h), (i) arches with smooth synclastic and anticlastic regions; and (j) multi-layered force diagram reciprocal to (k), (l) multi-layered corrugated shells.

$\mathbf{n}_2^\dagger$. Their reciprocal faces in the form diagram have parallel normals $\mathbf{n}_1$ and $\mathbf{n}_2$. As shown in the figure, the geometry of the funicular form and its overall curvature can be controlled by the angles of the edges of the force diagram and the faces that are connected to them.

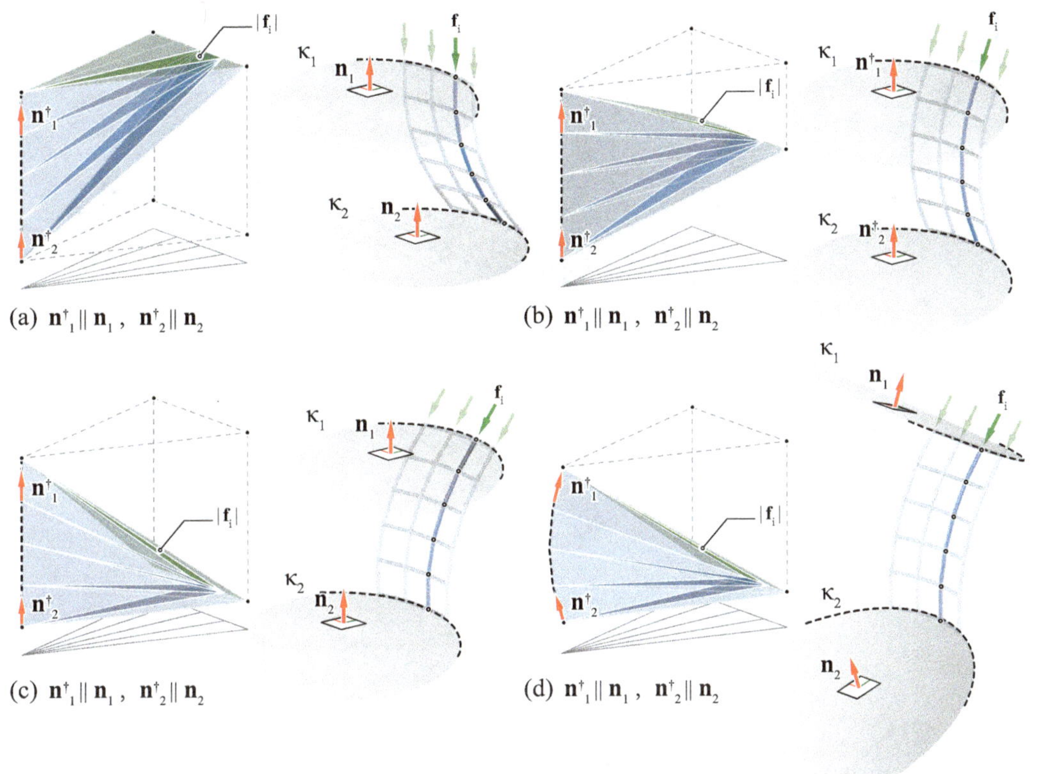

Figure 4.35 Aggregation of tetrahedral cells converging to (a) a vertical line lower than the skew bases of the tetrahedrons (b) higher than the bases; (c) a line with equal distance of its both ends to the skew bases of the tetrahedron; and (d) to a segmented curved polyline.

Change in the curvature plane of the anticlastic polyhedral geometry has application in controlling the section geometry of a compression-only shell. Figure 4.36 shows a compression-only shell with an anticlastic region similar to the roof structure designed by renowned engineer Eladio Dieste for the Gimnasio Don Bosco in Montevideo, Uruguay.[113] The force diagram of the design is divided into three segments to show each segment's properties and their effect on the geometry of the shell. The middle part of the force diagram makes the anticlastic part of the shell. The changes in the angle of the bottom edges in the middle segment of the force diagram result in an "S" shape section in the middle of the shell. This S-shaped shell was used to create a series of long-span roofs with openings for natural light at the Gimnasio Don Bosco in Montevideo (Figure 4.36).

4.15.5 Fan Vaults of King's College Chapel

King's College Chapel at the University of Cambridge is a fascinating example of Gothic architecture, including one of the world's largest fan vaults (Figure 4.37). The structure consists of fan vaults with a thickness of about 12–15 cm spanning 12.66 m,

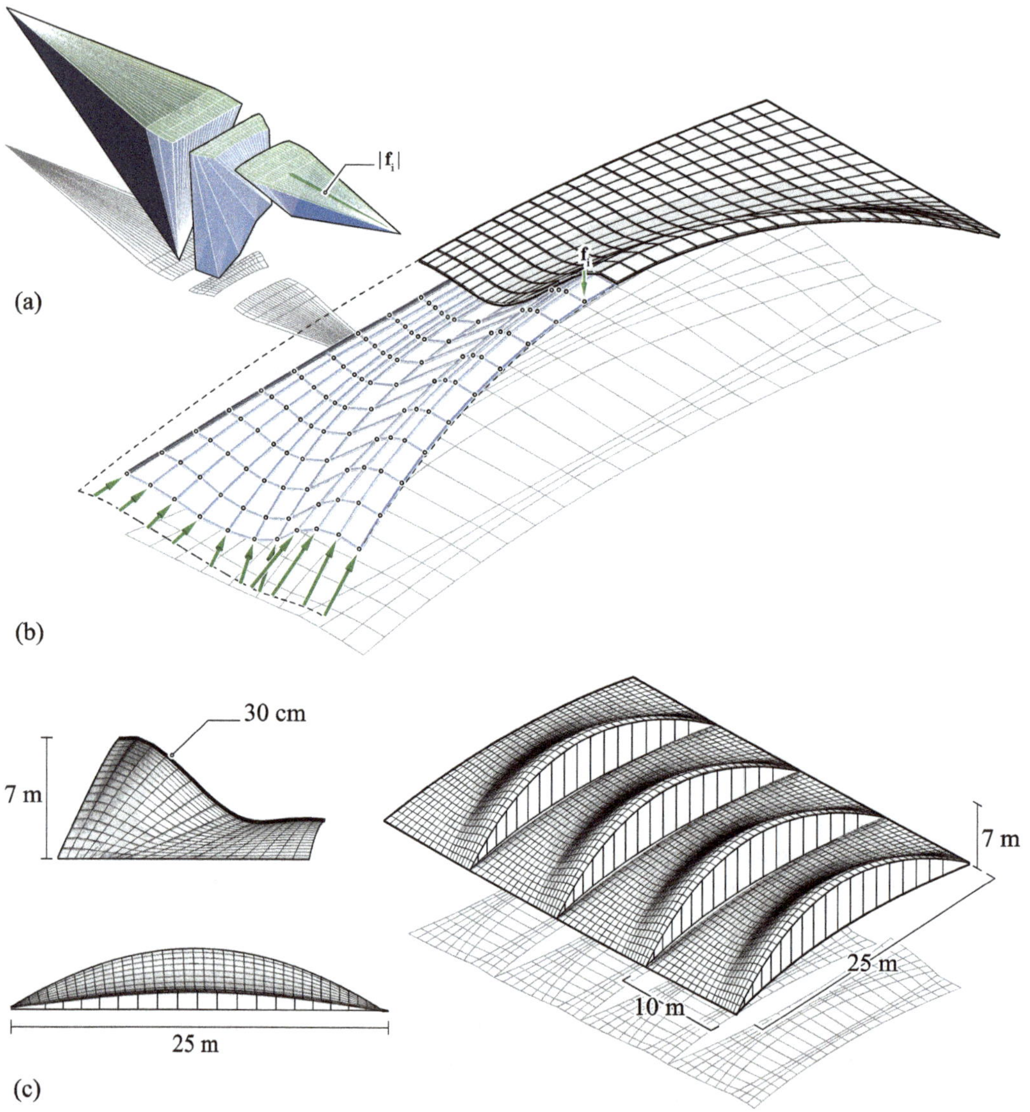

Figure 4.36 (a) controlling the section undulation by sculpting the force diagram and changing the direction of curvature planes in the anticlastic part (b) of the force diagram; (c) the repetition of the shell modules results in a roof system similar to the arches of the Gimnasio Don Bosco in Montevideo, Uruguay, designed by Eladio Dieste.[113]

which make a roughly 1% thickness-to-span ratio.[117] The geometry of the fan vaults and almost the entire spanning structure can be constructed using polyhedral graphic statics. To understand the techniques that can be used to construct the structural form of the vaults, let us divide the structure into multiple segments and explain the design of the force diagram for each segment. Then, these segments will be added together to complete the entire structure.

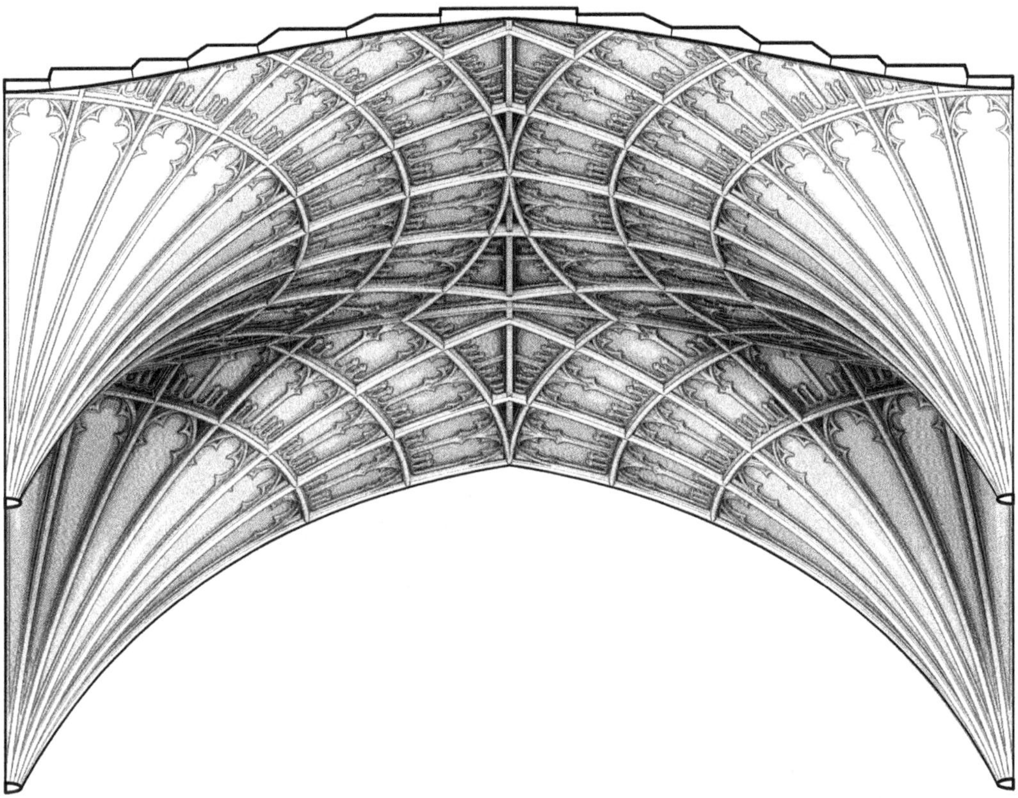

Figure 4.37 The drawings of the fan vaults of the King's College at the Cambridge University.[117]

Figure 4.38 shows the construction of a quarter of the fan vault. It divides the quarter geometry into three segments. The figure shows these segments by numbers 1, 2, and 3. Segment 1 is the fanning geometry, and segments 2 and 3 link geometry to two other quarters of the vault with different elevations.

Segment 1 of the fanning geometry can be constructed by aggregating a series of tetrahedral cells radially repeated around a vertical edge (Figure 4.38a). Aggregating these tetrahedral cells in the vertical direction gives the force diagram for the fanning geometry in Figure 4.38a. Note that this quarter fanning vault should be connected to two other quarter vaults on two sides to complete a module for the structure. These connections can be made by developing two types of connecting force cells, as shown in Figures 4.38b and 4.38c. The cells of group 2 consist of triangles.

Figure 4.38b shows the force group made by extruding the triangle faces from *ABF* to point *D* and then extruding face *DAF* to point *E*. The third segment is made by extruding the edges of the triangles *AHF* to a plane *AGE* and then connecting the rectangular bases from *GH* to *EF*, as shown in Figure 4.38. The rest is the aggregation of the force segments to complete one-quarter of the fan.

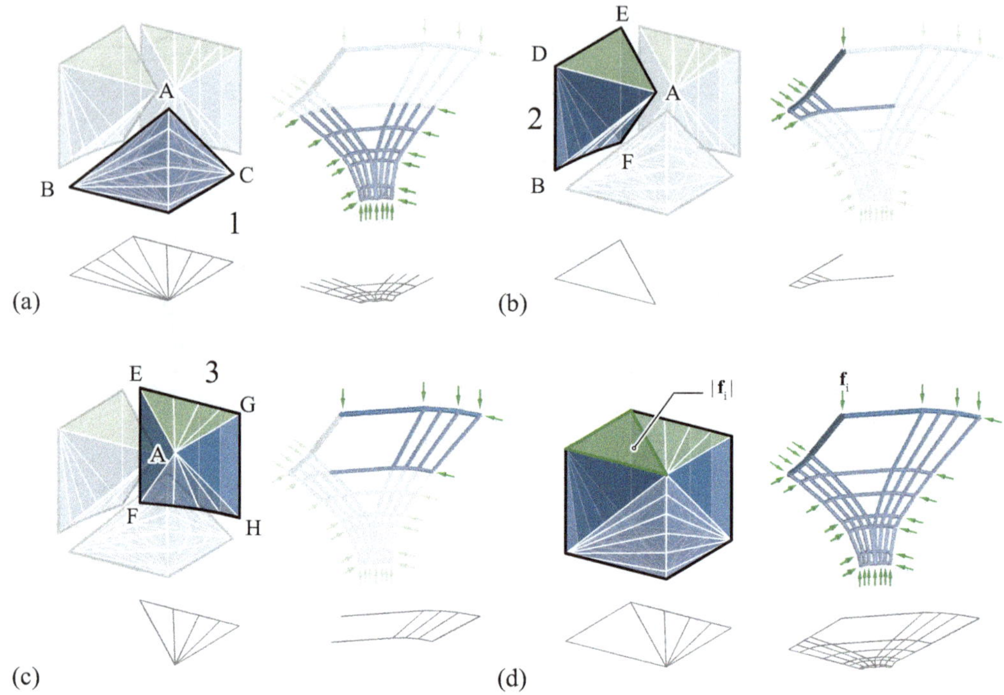

Figure 4.38 The three segments of the force diagram to complete a quarter of the fan vault: (a) segment 1 reciprocal to the fanning geometry; (b) segment 2 reciprocal to the connection with a lower height; (c) segment 3 reciprocal to the segment with a higher reach; and (d) the assembled force diagram and its reciprocal form.

Once the quarter force diagram is constructed, we can perform symmetrical mirroring to complete the geometry of the force of the fanning structure for King's College Chapel. Aggregation of the force cell groups can make multi-span funnel shells, as shown in Figure 4.39. The funnel shells span in two directions with different lengths. The funnels meet at the main span at a higher elevation, while they meet at a lower height on the sides. This change in the intersection geometry is very well represented in the form diagram of Figure 4.39b. The intersection of the funnels on the side makes an anticlastic ridge, while they make a synclastic geometry on the front.

Figure 4.40 shows the materialized version of the shell using masonry construction as it is in real life adjacent to its polyhedral approximation.

Looking at the polyhedral model and comparing it with the existing structure, we can observe two interesting characteristics: (1) the form of the structure needs lateral support to maintain compression-only force equilibrium, and (2) the magnitude of the vertical force in the supports is significantly larger than the applied loads on other nodes of the shell. In the actual structure, the architect or builder also filled the funnel's inside with crushed stones, as shown in Figure 4.40.

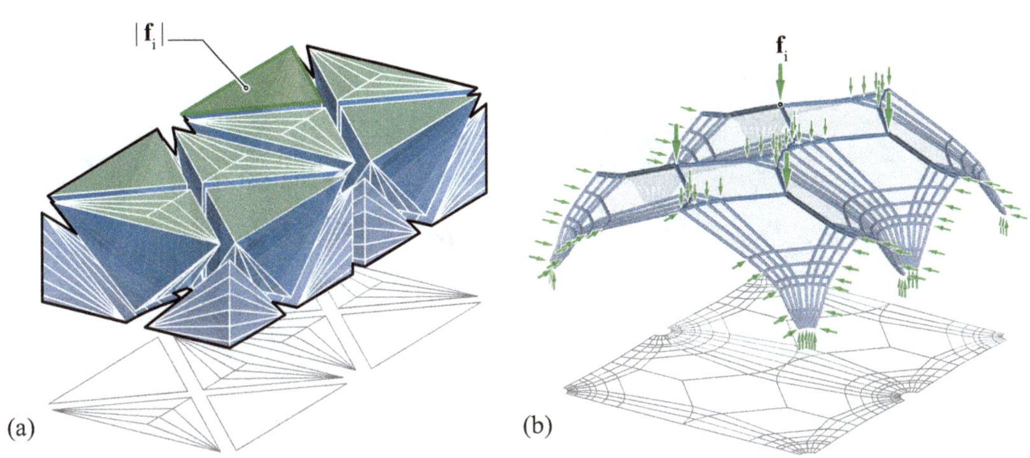

Figure 4.39 (a) The force cells and their aggregations to make multi-span funnel shells of (b).

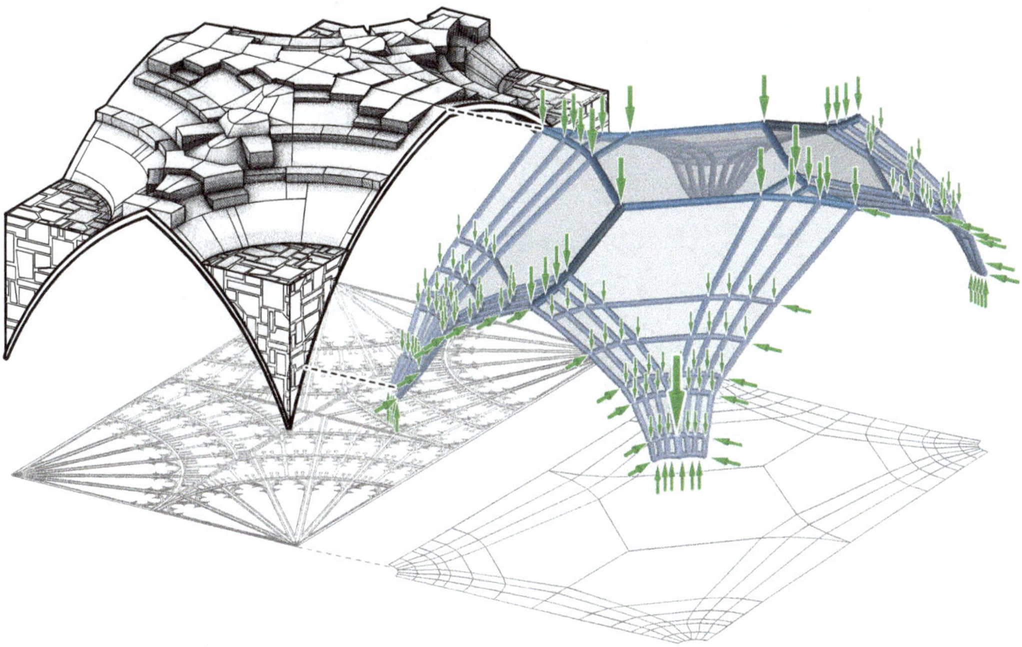

Figure 4.40 The masonry structure of the King's College Chapel adjacent to its approximate polyhedral form.

Different Force Diagrams for the Same Funicular Form

An example of this funnel geometry is the best representative of describing a statically indeterminate load path. The indeterminacy has been discussed on multiple occasions in Chapters 1 and 3 and will be discussed in detail in Chapter 6. Nevertheless, this example can make an excellent introduction to this topic. Consider the force and form

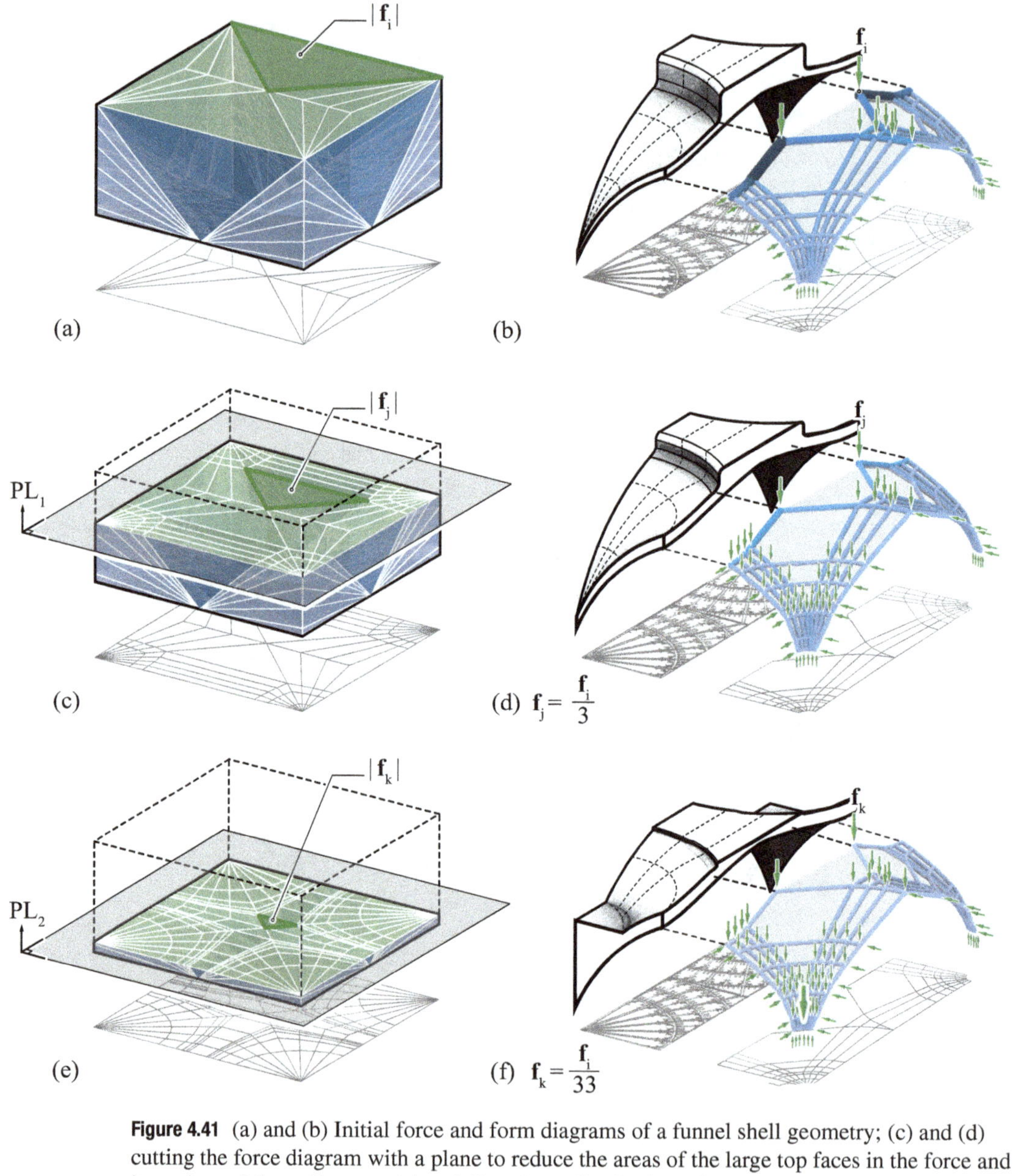

Figure 4.41 (a) and (b) Initial force and form diagrams of a funnel shell geometry; (c) and (d) cutting the force diagram with a plane to reduce the areas of the large top faces in the force and its reciprocal form diagram; (e) and (f) a closer distribution of the area for the applied loads resembling the self-weight of the shell.

diagrams in Figure 4.41a. The diameter of the pipes represents the magnitude of the forces in the funicular network of the form. This force equilibrium is based on applied loads on the top and side vertices of the form and does not include any applied load on other vertices. If we materialize this design using masonry construction, the shell will have a thicker cross-section on the top ridge than the rest (Figure 4.41b).

Now, without changing the force or form diagram, take a plane PL_1 parallel to the ground and cut the force diagram, as shown in Figure 4.41c. At the location of the cut, cap all the polyhedral cells so that the new force diagram will have new applied loads with different areas, as shown in the figure. Note that in this operation, the number of cells in the force diagram has not been changed, and only the face area has been adjusted. Rendering the new applied loads and internal force magnitude as the diameter of pipes in the form diagram shows a new loading condition and equilibrium for the form, as shown in Figure 4.41d. We have adjusted the cutting plane to give a third of the area of the applied load $\mathbf{f}_i$ of the first step. As noted, the geometry of the funicular did not change, but the magnitude of the applied loads and the point of application of loads have changed. That, in turn, changes the distribution of the internal forces in the funicular network.

This phenomenon shows that this particular funicular form can be in equilibrium under various magnitudes of applied loads and represents a statically indeterminate system because the equilibrium state is not unique. We are continuing the cutting process by PL_2, resulting in a new form with a more even distribution of applied loads, which may represent the self-weight of the shell and, therefore, be closer to the built structure, as shown in Figure 4.41e,f. Thus, a funicular geometry can stay intact while external and internal forces can change. Note that the cross-section of the shell in the last materialized version is thicker than the ones in the previous versions.

4.16 Shellular Funicular Forms

As described in Sections 4.13 to 4.15, increasing the number of subdivisions in the force diagram results in a form diagram with shorter edges and smaller internal forces. This technique can reduce the effective length of the members in the funicular form and thus reduce the probability of buckling in the members. There is a particular subdivision approach in polyhedral graphic statics that makes the edges of the form diagram small enough that they approximate a discrete spatial shell. This subdivision results in shell-like funicular systems referred to as *shellular* funiculars. Shellular systems are single-layer, continuous, smooth-curved shells with anticlastic curvature.[118]

In shellular funicular systems, the system's behavior changes from discrete elements to a continuum shell-like system.[84] Such spatial shell systems exhibit enhanced shear capacity compared to cellular systems and, thus, stiffer structures with higher load-bearing capacity[83,84] (Figure 4.42).

4.16.1 From Node to Anticlastic Shell

Translating a cellular form to a shellular version requires an in-depth understanding of anticlastic subdivision in polyhedral graphic statics. Figure 4.43 shows a subdivision process in which a single node with four applied loads in equilibrium is transformed into a discrete anticlastic surface. In such a discrete anticlastic surface, at each node,

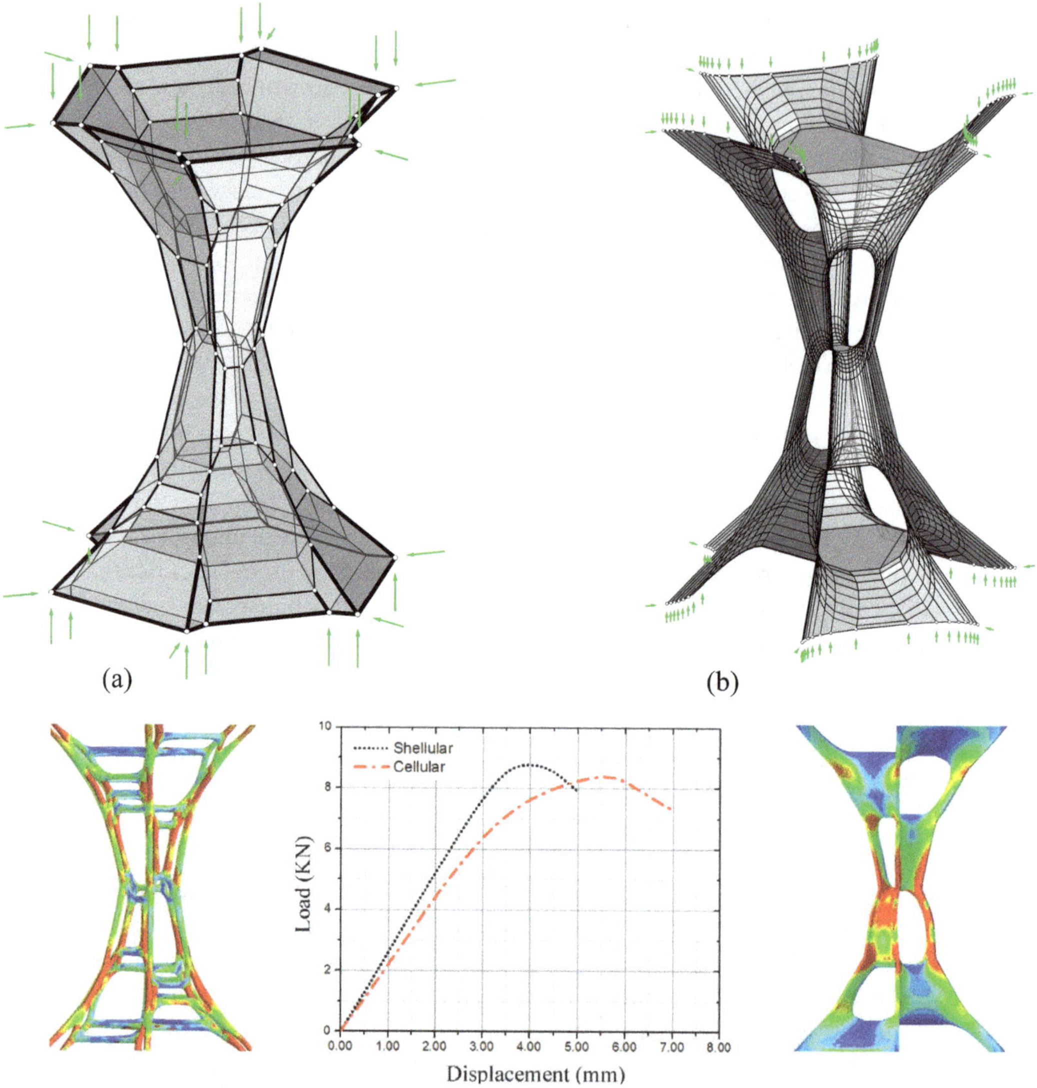

Figure 4.42 (a) A cellular funicular form versus (b) its shellular version with the same total magnitude of applied loads on the boundaries and the finite element analysis results showing stiffer performance of the shellular version.[83]

two edges are *hanging* that are curved upwards, and two are *standing* and curved downward.

Note that the starting force polyhedron is a tetrahedron with four faces corresponding to four applied loads in the form diagram, two upwards and two downwards (Figure 4.43a). Subdividing this tetrahedron step-by-step gives a discrete anticlastic surface with a force diagram that contains only tetrahedrons, as shown in Figure 4.43a–d.

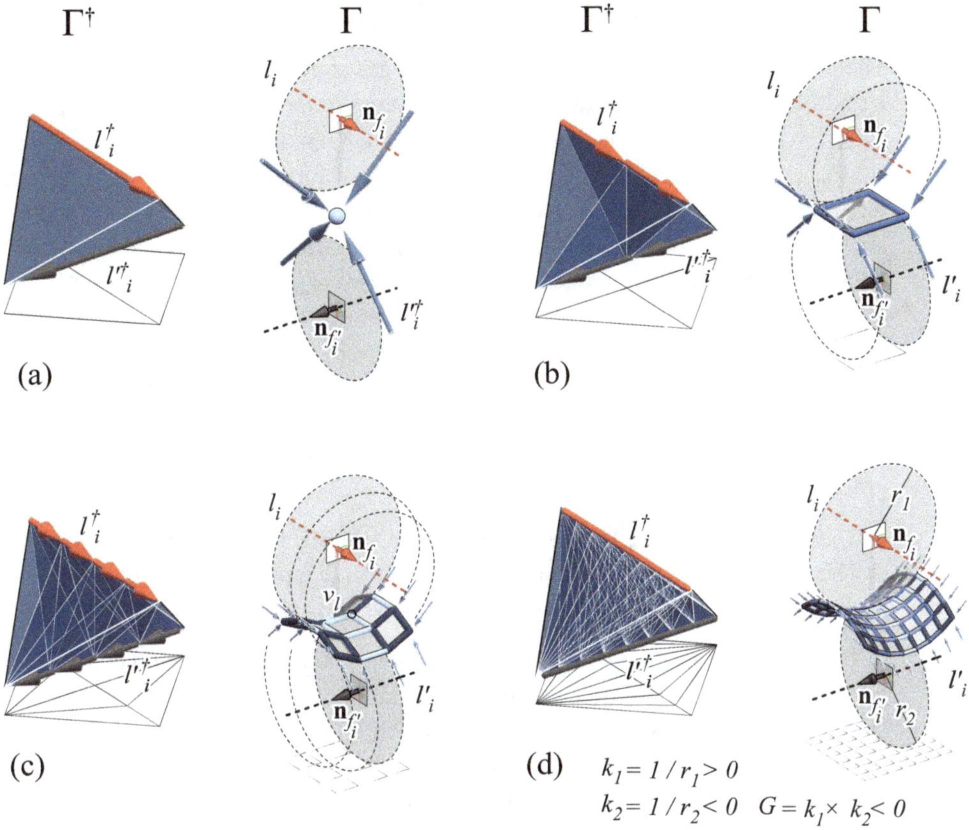

Figure 4.43 (a) to (d) The process of transforming a node into a discrete anticlastic surface by subdividing tetrahedrons.

In this example, the force tetrahedron has two skew edges, $l_i^\dagger$, red, and $l_i'^\dagger$, black, that are in skew positions to each other. These edges are subdivided in each step to make smaller tetrahedrons. In this process, the skew edges $l_i^\dagger$ and $l_i'^\dagger$ are divided into an equal number of segments such that each segment of the former will make a tetrahedron with its associated segment in the latter (Figure 4.43b). Further subdivision of those edges results in a smoother surface as a form diagram.

This particular type of subdivision is referred to as *anticlastic subdivision* in the context of this book. In this type of subdivision, the nodal cells of the force diagram are only tetrahedrons, which result in a reciprocal, valency-four vertex in the form diagram (Figure 4.43d). For instance, vertex v_1 is connected to two *hanging* and two *standing* edges, establishing a negative Gaussian curvature. The resulting anticlastic surface in Figure 4.43d is only subjected to the applied loads in its boundaries.

4.16.2 Curvature Axes and Labyrinths in Shellular Geometries

The curvature axes on an anticlastic surface are shown by dashed lines and named l_i and l'_i. These axes are parallel subdivision axes in the reciprocal force diagrams $l_i^\dagger$ and $l_i'^\dagger$. An anticlastic surface geometry in Figure 4.43d divides the 3D space into two subspaces, above and beneath. These subspaces can also be identified by the curvature axes l_i and l'_i. These axes are referred to as *labyrinths*[119] in shellular systems and are shown by red and black colors. The labyrinths in anticlastic geometries are *two* intertwined graphs whose edges are always in a *skew* position with each other. These axes may represent the connectivity graph of the sub-spaces in complex shellular geometries.

The two labyrinths interpenetrate each other, while the anticlastic surface is in between. The geometry of the surface changes when the angle of these labyrinths changes. Interestingly, the labyrinths in the form and force diagrams of polyhedral graphic statics are related as follows:

- The labyrinth l_i of the form is parallel to the $l_i^\dagger$ of the force diagram.
- The labyrinth l_i is the curvature axis in the form diagram, while its parallel labyrinth $l_i^\dagger$ in the force diagram is the subdivision axis.

Note that the force diagram of shellular systems contains tetrahedral cells, and each tetrahedron contains one red and one black labyrinth segment $l_i^\dagger$ and $l_i'^\dagger$ that are in skew positions, as shown in Figure 4.44. Each tetrahedron with six edges can have three pairs of skew edges. That is, in each tetrahedron, there are three different possibilities for choosing a pair of labyrinths with skew edges. Figure 4.45a–c shows these options of labyrinth graphs subdivided, resulting in a different discrete anticlastic surface (Figure 4.45a–f). As illustrated, these surfaces had different curvature directions as the directions of the labyrinths and the curvature axes differed in each anticlastic example.

Aggregating Tetrahedrons to Extend Labyrinths

We can aggregate multiple tetrahedrons to extend the network of labyrinths and thus generate a more complex shellular form diagram. Figure 4.46 displays three different design possibilities for the same group of tetrahedrons as a force diagram. Choosing a different set of skew edges in the first tetrahedron propagates to the adjacent tetrahedrons, making different labyrinth graphs and, thus, three different form diagrams. However, the main question is how to identify these labyrinths in a complex aggregation of cells and how to translate almost any cellular to its shellular possibilities.

4.16.3 Translating Any Cellular to Its Shellular Funicular Version

Before discussing a general method to translate a cellular system to a cellular geometry, we need to better understand the properties of labyrinths.

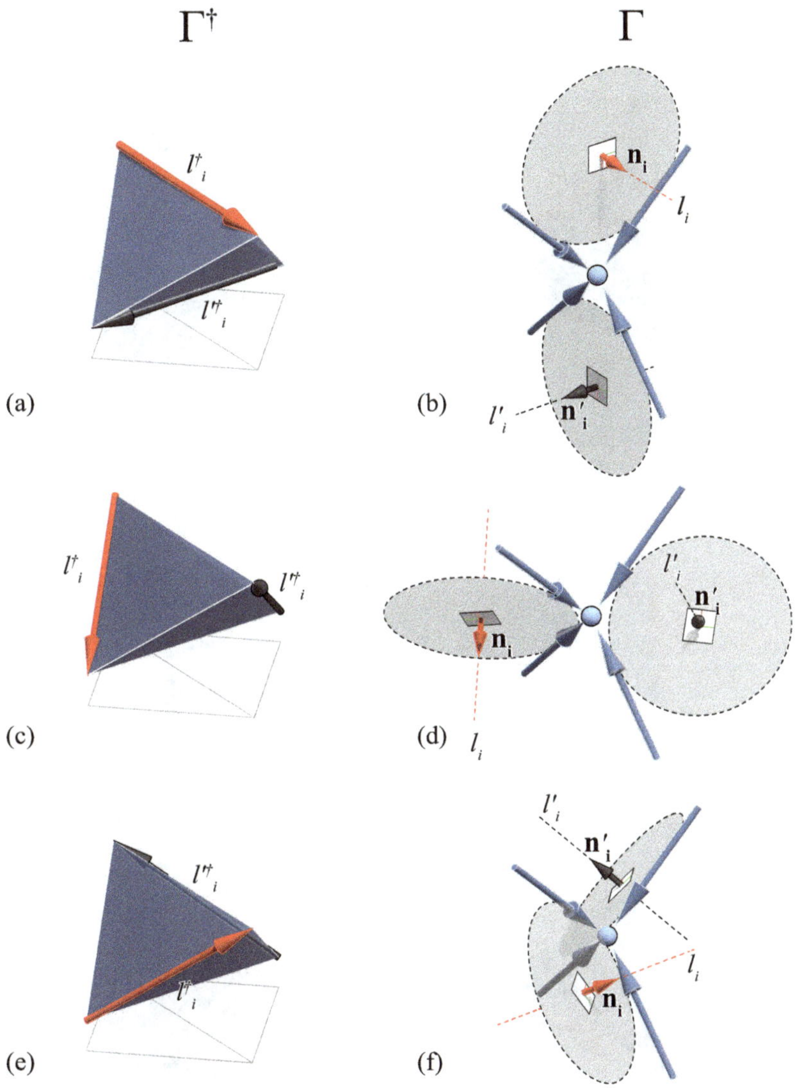

Figure 4.44 (a) to (f) Three different possibilities for designing a shell as a form diagram corresponding to a tetrahedron as a force diagram: three pairs of skew edges in a force diagram.

Labyrinths Properties

The properties of the labyrinths in the force diagram for design purposes are as follows:

- Each pair of labyrinth segments in the force diagram belongs to one and only one tetrahedron, corresponding to a node with an anticlastic curvature in the form diagram.
- There are only two labyrinths in the force and form diagrams, shown by red and black graphs. Each segment or edge of these graphs is always skew to the other

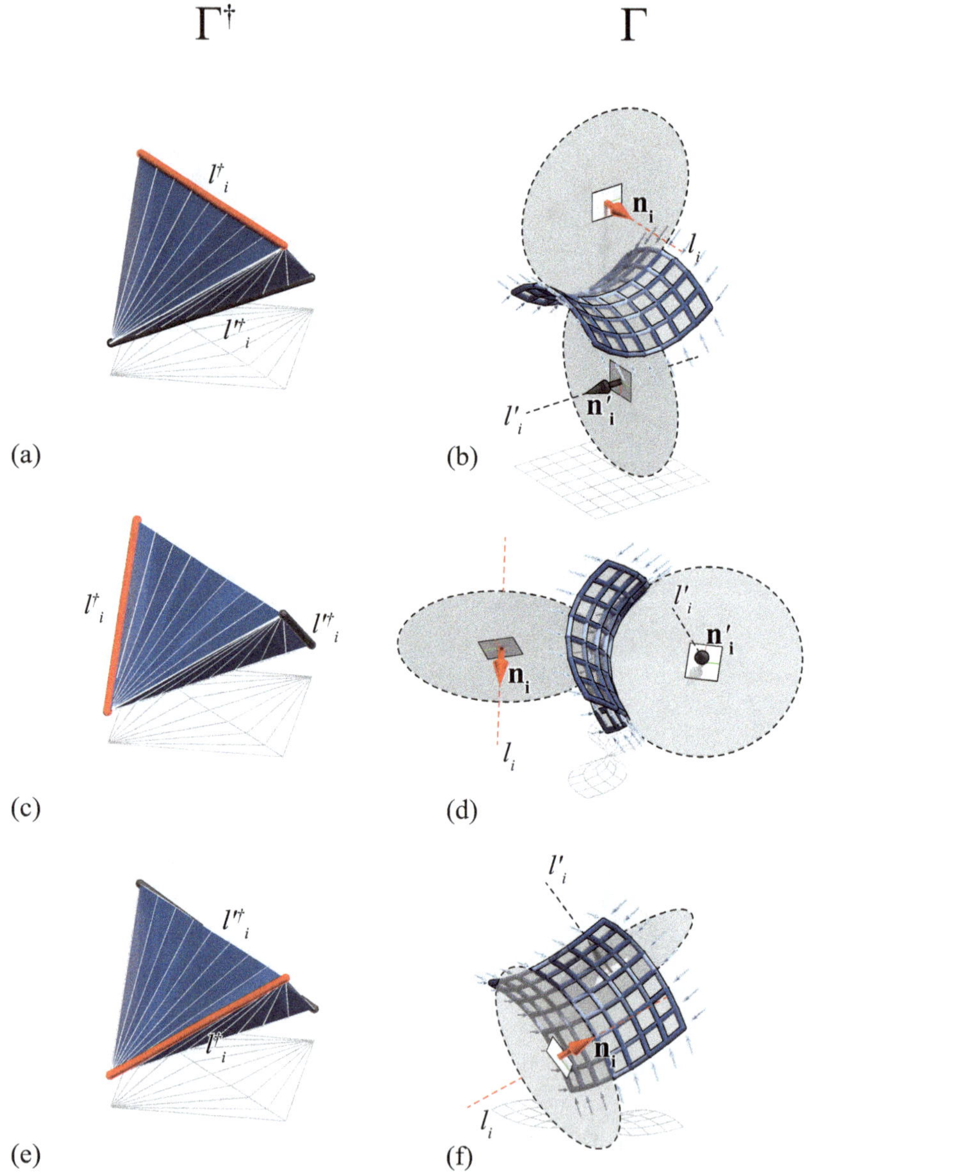

Figure 4.45 (a) to (f) Dividing the force diagram based on the labyrinth edges results in three different shells with curvatures in different directions.

graph, and these two graphs cannot be connected. This assures that the surface between the labyrinths' sets in the form diagram is continuous and divides the space into two continuous sub-spaces.

In general, to design a shellular geometry, the following steps are necessary:

1. Having a tetrahedral force diagram;

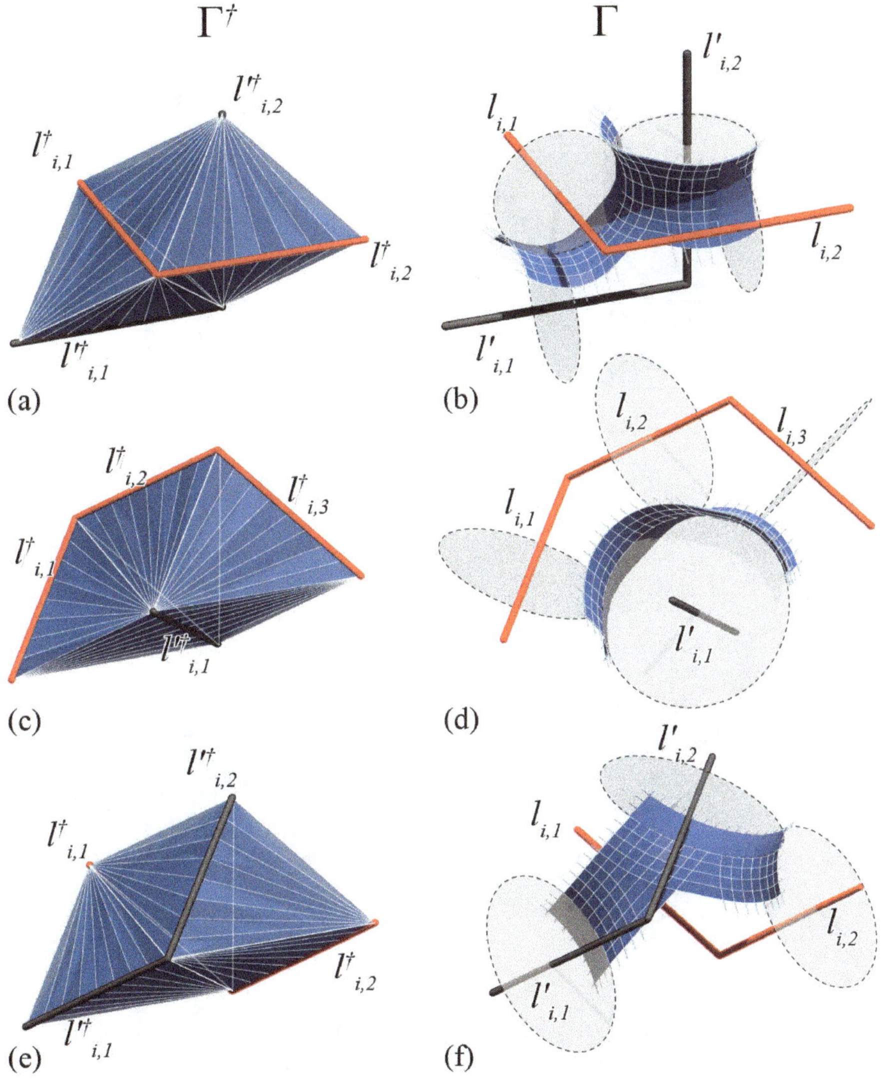

Figure 4.46 (a) to (f) Three different possibilities for designing a shellular surface from the same aggregation of tetrahedrons but choosing different labyrinths for subdivision resulting in three different shellular surfaces.

2. Identifying the labyrinth pairs in each tetrahedron and connecting them to the adjacent labyrinths of the neighboring tetrahedrons; and
3. Subdividing the labyrinths to generate various surface geometries with different resolutions.

Remark 4.8. The force diagram for shellular systems should consist of tetrahedral cells. However, any force diagram with tetrahedral cells may not necessarily be translated into two continuous labyrinths with skew edges. Random tetrahedral cells

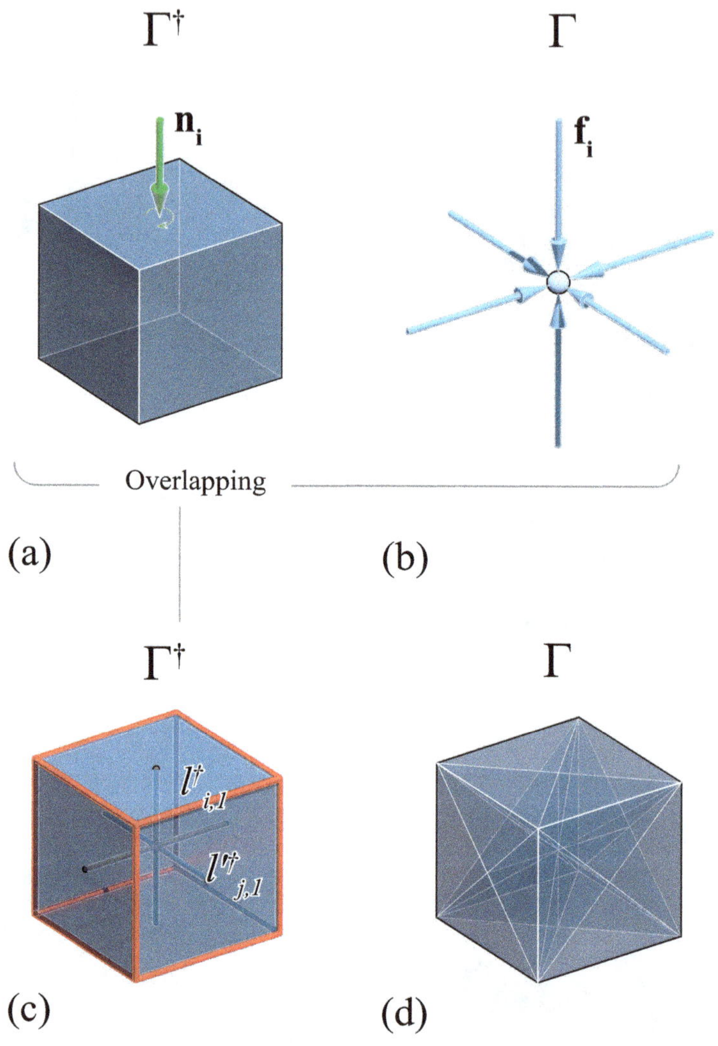

Figure 4.47 (a) A simple force diagram; (b) the reciprocal form; (c) the form diagram is superimposed on the force diagram; (d) the tetrahedral cells constructed by the pairs of edges of the form and force diagram.

may result in singularities in which one cannot find two graphs of continuous labyrinths.

Superimposing Form and Force Diagrams

To avoid singularities in tetrahedral force diagrams in finding labyrinths, we have developed an approach that assures the existence of continuous labyrinths in the force diagram. In this approach, the topology of the form diagram is superimposed onto its reciprocal force polyhedron with general polyhedral cells. The edges of the initial

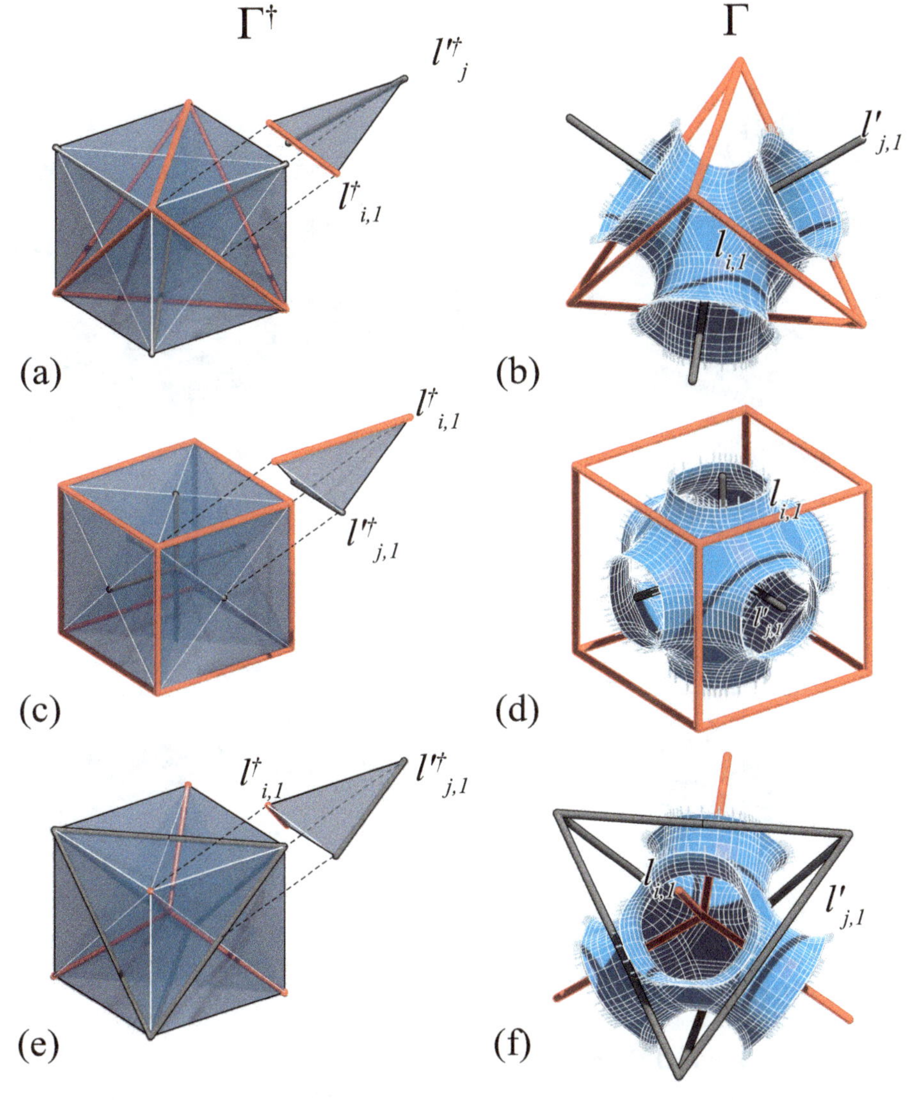

Figure 4.48 (a) to (f) Three different shellular force and form diagrams by choosing three different possibilities of labyrinths in the force diagram of a simple cube.

force diagram give one set of labyrinths, and the edges of the superimposed dual give another set. Consider the example of Figure 4.47a,b, which shows a simple force diagram $\Gamma^{\dagger}$ that is a cube and its reciprocal form diagram as Γ. By superimposing the two, we simply move the reciprocal vertex of the form diagram into the space of the force diagram and then connect each edge of the form to its reciprocal face in the force diagram, as shown in Figure 4.47c.

Figure 4.49 The plan and elevation of a shellular bridge concept based on a cellular funicular form.

Tetrahedral Force

Once the form and the force diagrams are superimposed, the force tetrahedrons should be constructed. Each edge of the form graph makes a tetrahedron with every edge of its reciprocal face and gives a tetrahedral force polyhedron with two graphs of labyrinths for subdivision. Figure 4.47a illustrates a force diagram and its reciprocal node with six external forces. The superimposition of form and force is equivalent to finding the centroids of the cells in the force diagram and connecting them to the centroids of the faces of each cell, as shown in Figure 4.47c. Let us use the color red for the edges of the force diagram and black for the edges of the dual graph.

Next, each edge of the dual graph makes a tetrahedron, with the edges of its reciprocal face highlighted in red. This approach translates the force diagram into a group of nonoverlapping tetrahedral cells (Figure 4.47d). The main advantage of this approach is that since it starts with two sets of labyrinths, the existence of the labyrinths is guaranteed in the force diagram and results in a shellular geometry.

Other Possible Labyrinths

The advantage of superimposing the form and force diagrams is that it results in a force diagram with valid tetrahedral cells. Thus, for each tetrahedron, we can accommodate three different possibilities of labyrinths, resulting in three different cellular design options for the same cellular geometry (Figure 4.48a–f).

Subdividing the Labyrinths

After identifying the labyrinths in the tetrahedralized force diagram, one may apply the anticlastic subdivision between pair edges in each tetrahedron to get a smoother subdivision of the form diagram. For further discussion on this topic, we recommend visiting the following publications.[83,84,120,121] Figure 4.49 shows a shellular design of the cellular funicular form. Three other possibilities can be extracted from the same design.

4.17 Chapter Summary

This chapter covers various topics on the design of funicular forms with an emphasis on particular features in the structural forms. It starts by introducing the topological and geometrical relationship between the form and force diagrams in 2D and 3D and describes nodal and global equilibrium in the context of both 2D and polyhedral graphic statics. It explains how to find closed polyhedral cells from a group of faces and the computational approach that could be used for this purpose. It argues for the necessity of implementing computational tools to explore the realm of intricate funicular forms with pronounced design features. To give instances of how computational implementation can help in design, various design examples are introduced, and their related design topics are explained. Note that the design domain in this approach is the force diagram. Thus, all these design features are explained with respect to the techniques of design incorporated in the force diagram that give the desired feature in the funicular form. In the end, the translation of the cellular to shellular systems is explained and discussed as an interesting design approach that results in a more resilient structural form by creating anticlastic curvature in 3D space.

5 Systems with Combined Tension and Compression

Masoud Akbarzadeh, Hua Chai, and Mathias Bernhard

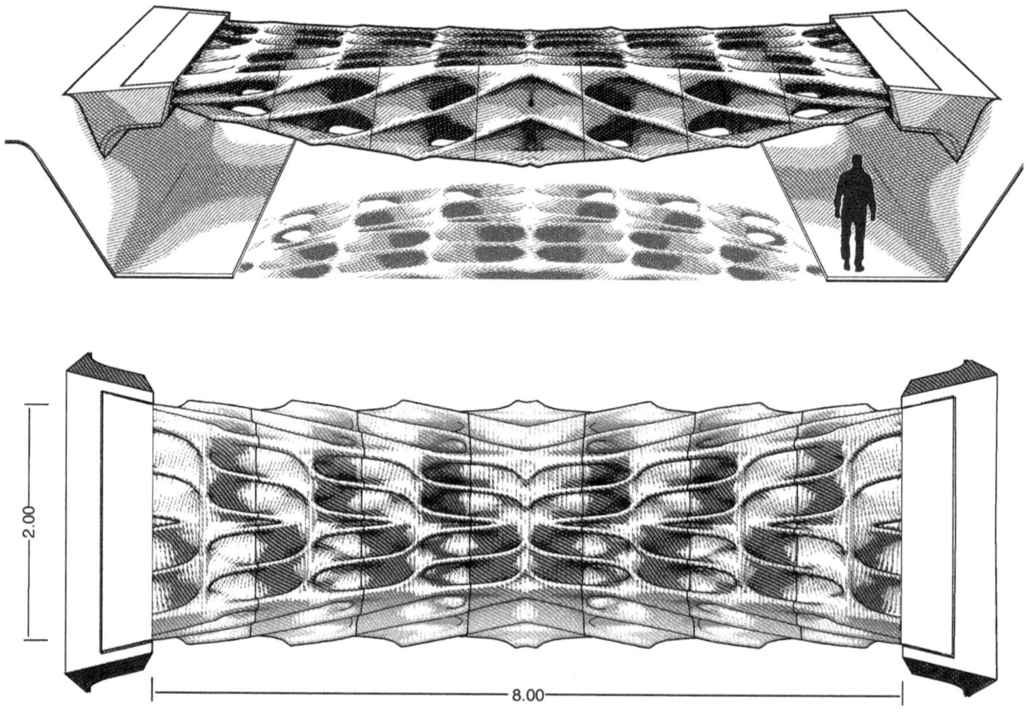

Figure 5.1 A concrete structure that is designed as a funicular solution with tension and compression internal forces embedded within anticlastic spatial geometry.

5.1 Introduction

Chapters 3 and 4 covered topics related to the global and nodal equilibrium for compression-only systems and showed how one could aggregate or subdivide convex force polygons in two-dimensional (2D) and polyhedrons in three-dimensional (3D) to design intricate compression-only funicular structural forms. But funicular systems can have both tensile and compressive members. This chapter provides insight into the design of members' systems with mixed compressive and tensile forces. In this book, we have always used the force diagram as the main design domain; thus, we need to learn the properties of the force diagram for systems with tension and compression force. To understand the properties of the force diagram in such a system, we introduce a simple analysis approach for a simple truss system in both 2D and 3D and then show how the force diagram of such systems can be manipulated to design funicular forms with mixed types of forces.

Chapter Objectives

This chapter intends to:

- Unravel the relationship between the form and force diagram in systems with both tension and compression.
- Explain the procedures to analyze simple truss systems in both 2D and 3D using geometric methods to better understand the properties of force diagrams with mixed tension and compression forces.
- Provide design methodology for the design of the force diagram reciprocal to funicular systems with tension-and-compression-combined internal forces.
- Introduce additional geometric constraints to optimize internal forces in systems with mixed types of forces.

Figure 5.1 shows a funicular form of a compression and tension combined structure with constant force in the tensile members. This structure is materialized using an anticlastic periodic shell, as shown in Figure 5.2. To understand the principles and steps behind the design of this structure, we will first visit the geometric method of analyzing a truss system and then explain how using some techniques, it is possible to translate a simple truss into a funicular form with mixed types of forces. We will

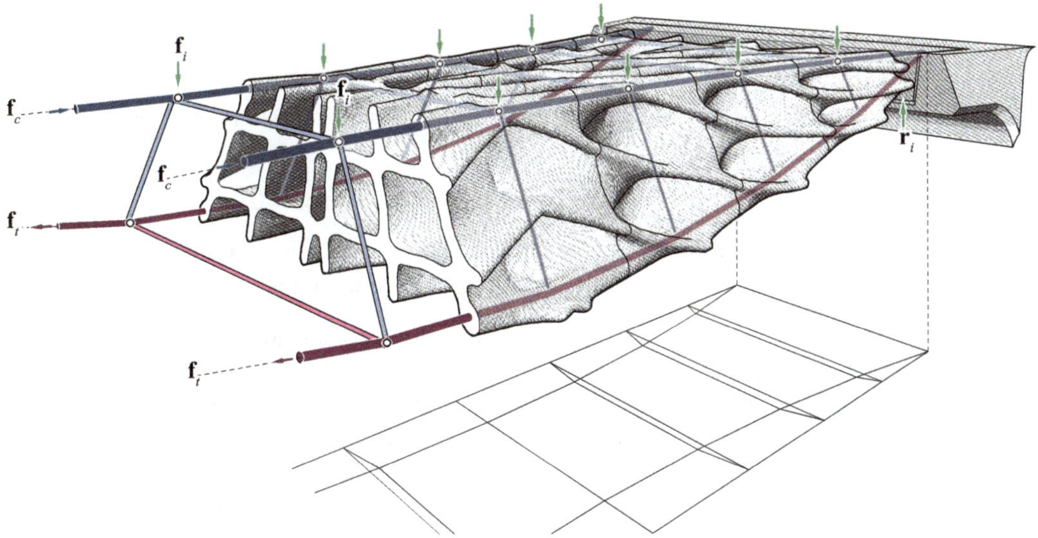

Figure 5.2 A section cut of a funicular polyhedral form with both tension, $\mathbf{f}_t$, and compression, $\mathbf{f}_C$, internal forces within a body of anticlastic periodic shell.

also show that the force diagram of such systems contains both convex and nonconvex polyhedral cells.

5.2 Analyzing a Simple 2D Truss System

Truss systems are structures of linear elements or bars connected at both ends to form triangular configurations. The node where the bars meet is often called a joint with no bending resistance. That is, the members can freely rotate around the joints. However, the members' triangular configuration prevents them from rotating around the joints, thus making the system rigid. Moreover, the bars' internal forces are axial, either tensile or compressive. This section explains the geometric steps necessary to construct a force diagram for a truss system and find the reaction forces, the magnitude, and the type of internal forces in the members connected to the truss nodes.

The truss analysis procedures using graphic statics can be summarized into the following major steps:

- *Finding global equilibrium* of the system: that is, identifying reaction forces at the supports of the truss that keep the applied loads in equilibrium. This step is independent of the truss's geometry and only needs the boundary conditions, including the applied loads, location, and direction of the reactions.
- *Constructing nodal equilibrium* for each node of the truss to identify the magnitude and type of forces in the members. This step relies on the previous step and needs the geometry of the global force polygon and the geometry of the members meeting at each node to find the internal forces.

These steps are subsequently divided into multiple substeps that will be explained in Section 5.2.1

5.2.1 Establishing Global Equilibrium in 2D

The numerical definition of global equilibrium in two dimensions involves three statics equations that must be satisfied:

$$\sum \mathbf{f}_x = 0, \quad \sum \mathbf{f}_y = 0, \quad \sum \mathbf{m}_{xy} = 0$$

where $\mathbf{f}_x$ and $\mathbf{f}_y$ are the sum of the forces in the x- and y- directions and $\mathbf{m}_{xy}$ is the sum of the moments on the xy- plane. The geometric definition is much more intuitive:

- The external forces of the truss, including the applied loads and reaction forces, should make a closed polygon of force in the force diagram. When the external force polygon is closed, the sum of forces in the x- and y- directions equals zero. This closedness satisfies the first two numerical equations.
- A funicular construction on the line of action of the applied loads and the reaction forces in the form diagram should be closed. A closed funicular construction satisfies the moment equilibrium in the system.

These two essential steps will be expanded next.

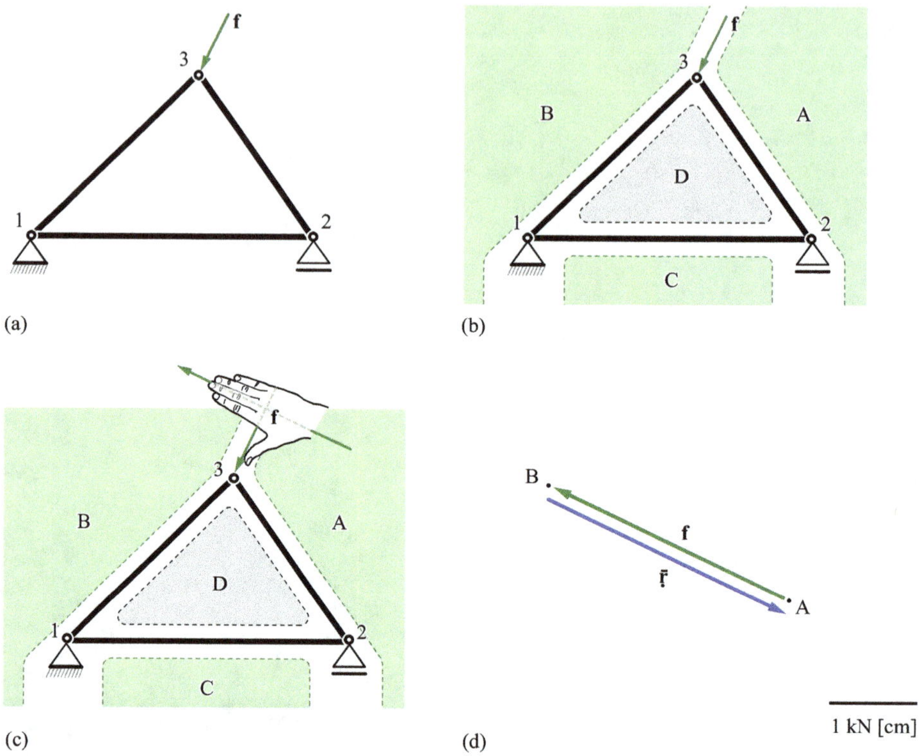

Figure 5.3 (a) A simple truss with three nodes and an applied load; (b) its subspaces defined by the members of the truss, external loads, and reaction forces that are yet to be found; (c) the thumb placed on the applied load; and (d) the force vector is drawn normal to the direction of $\mathbf{f}_i$ in the force diagram with a length proportional to 3 kN.

Closing External Force Polygon

Consider the truss of Figure 5.3a with an applied load **f**. Let us make a graph representation of the truss, including vertices, edges, and faces. This representation is key in establishing the topological and geometrical relationship between the form and the force diagram.

The first step in solving for global equilibrium is to demarcate the subspaces in the form diagram. The subspaces divide the entire 2D space around the truss with their applied loads and reaction forces. The external subspaces are open faces and colored green, and the edges of the truss and colored gray bind the internal subspaces.

Traversing the external subspaces in the form diagram gives us the connectivity of the vertices of the external force polygon in the force diagram. Note that the direction and magnitude of the reaction forces are yet to be found, and additional steps are needed to identify them. Nevertheless, a resultant force and its magnitude can be found to keep the applied loads in equilibrium. The resultant will subsequently be replaced with the reaction forces. Example 5.1 provides step-by-step procedures to construct a closed force polygon for the truss of Figure 5.3.

Example 5.1 Construct a closed force polygon for the truss of Figure 5.3.

Construction steps:

1 In the form diagram, label the subspaces defined by the edges of the truss, the applied loads, and the reaction forces at the supports. Note that the reaction forces are yet to be found, and an imaginary boundary is drawn for the reaction forces as if they exist in the form diagram (Figure 5.3b). These subspaces are labeled *A*, *B*, *C*, and *D*.
2 Place your thumb on the applied load; the palm should be between two subspaces. The alignment of the four fingers gives the direction to traverse subspaces around the truss (Figure 5.3c).
3 Start a new force diagram by drawing a force vector from a chosen point *A* in the direction of the outstretched four fingers with a length proportional to the magnitude of the applied load, for instance, 3 cm proportional to 3 kN (Figure 5.3d).
4 In the force diagram, label the start and end of the applied load vectors *A* and *B* to be the subspaces that the right-hand rule came in contact with in the form diagram.
5 Add a resultant vector $\bar{\mathbf{r}}$ that connects *B* back to *A* to close the force polygon (Figure 5.3d).

□

The resultant makes the force polygon closed and thus assures that the sum of the forces in *x*- and *y*- is zero. Nevertheless, the external equilibrium means that the applied loads should be in equilibrium with the reaction forces, and the reaction forces have not yet been found. It can be deduced that reaction forces in the supports should produce the same resultant to keep the applied loads in equilibrium. The trial funicular construction locates the resultant in the form diagram and identifies the magnitude and direction of the reaction forces.

Closing the Trial Funicular Polygon

To satisfy the moment equilibrium in the system, the resultant force in the form diagram should be located such that it prevents any rotation. A closed funicular construction on the line of action of the applied loads and reaction forces locates the resultant in the system and guarantees that the system is in moment equilibrium.

This concept was explained extensively in Chapters 2 and 3. For the case of the simple truss of Figure 5.4 with a single applied load, the line of action of the resultant force will be identical to the line of action of the load **f** and the funicular construction might seem redundant. However, to make a close connection between the analysis of a truss in 2D and 3D, we will provide a step-by-step explanation for this simple case. For any other system with multiple applied loads, the same procedure can be used to locate the line of action of the resultant. Once the resultant is located, the reaction forces will be found accordingly. Example 5.2 expands on this process.

Example 5.2 Construct a trial funicular for a 2D truss of Figure 5.4.

Construction steps:

1 On the form diagram of Figure 5.4a, draw two lines of action at the support that are parallel with the line of action of the resultant. Because the resultant force can be substituted with two parallel force components at the supports (Figure 5.4a). We spend the next steps finding the magnitude of these parallel forces.
2 On the force diagram, extrude the force polygon *AB* to an arbitrary point *O*′ to make two overlapping triangles *ABO*′ and *BAO*′ (Figure 5.4b).
3 In the form diagram, pick an arbitrary point, for example, 1′, on the line of action passing through support 1.
4 Draw a line normal to *O*′*B* to intersect the line of action of the applied load and name it 3′ (Figure 5.4a).

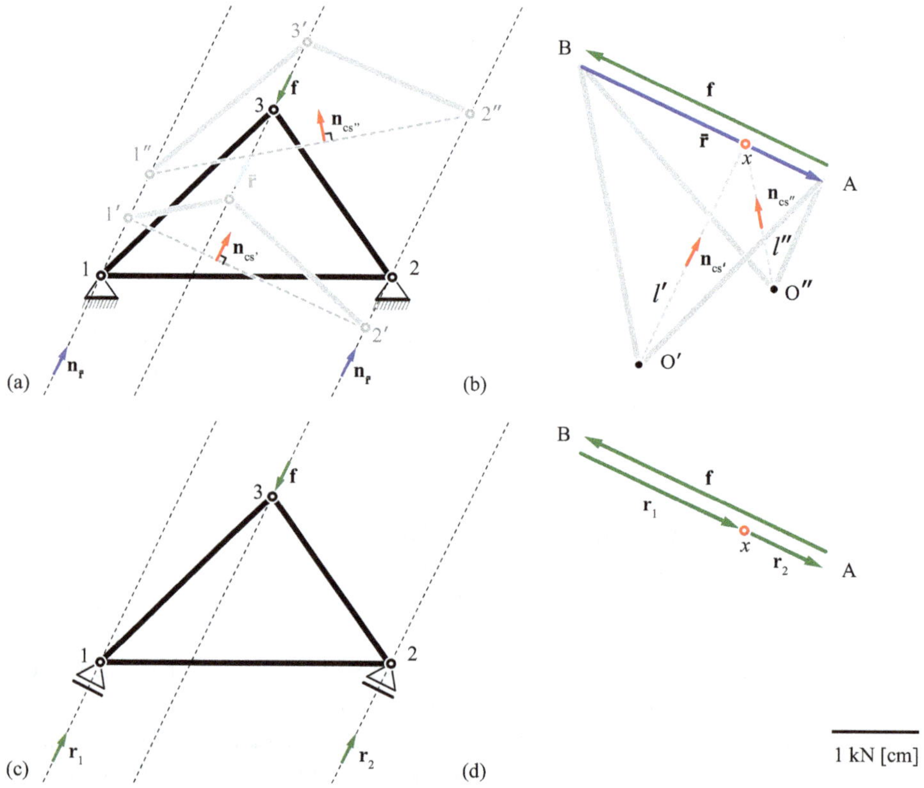

Figure 5.4 (a) Many trial funiculars that can be constructed from an extrusion of force polygon (b) to point *O*′ or *O*″, all producing the same location for *x* in the force diagram, which denotes the magnitude of reaction forces in the form diagram (c) and the force diagram (d).

5. In the form diagram, from point $3'$, draw a line normal to the edge AO' to intersect the line of action passing through support 2. Since there is only one applied load, the location of the resultant is known, and the rest of the construction identifies the magnitude of the force components parallel to the resultant at the supports.
6. Connect point $1'$ to point $2'$ by a line that is referred to as the *closing string* with its normal vector $\mathbf{n}_{cs'}$ (see Chapters 2 and 3).
7. In the force diagram, draw a line l' from point O' parallel to $\mathbf{n}_{cs'}$ of the form.
8. Intersect the line l' with the resultant vector and name it x (Figure 5.4b).
9. Divide the resultant vector $\bar{\mathbf{r}}$ into two force components $\mathbf{r}_1$ and $\mathbf{r}_2$ at the point x. Note that the force diagram could be extruded to any arbitrary point; for instance, O'' and the location of x on the resultant do not change. See Chapter 3 for more examples on this topic.
10. These two force components parallel to $\bar{\mathbf{r}}$ at the support keep the applied load in equilibrium.

□

Direction and Magnitude of Reaction Forces in 2D

The trial construction showed that the resultant force could be replaced by two parallel forces at the supports, keeping the applied load(s) in equilibrium. Although these two parallel forces can be considered a solution for reaction forces, their direction might not be ideal for the site.

The boundary conditions and site constraints often impose the direction of the reaction forces. It can also be defined as a design parameter affecting the system's force flow. This section explains how simple geometric operations can control the magnitude of the reaction forces at the supports.

For any funicular construction in the form diagram, the location of x on the resultant does not change in the force diagram. It was shown in Chapter 3 how this property is used to control the magnitude and the angles of reaction forces and the height of a funicular solution for given support locations. Similarly, the magnitude and the angles of the reaction forces of a truss can be controlled by the point x in the force diagram: picking any point on the line l that is normal to the closing string decomposes the resultant into two components. The two components close the external force polygon and thus keep the applied loads in equilibrium. These two vectors are reaction forces at the supports, and their angles can be controlled by the chosen point on the line l. The following of the Example 5.3 will provide the step-by-step procedures to find the direction and magnitude of the reaction forces for the truss of Figure 5.5.

Example 5.3 Find the direction and magnitude of the reaction forces of the truss of Figure 5.5 assuming that the support 2 is a roller support, that is, it can only have a vertical reaction force.

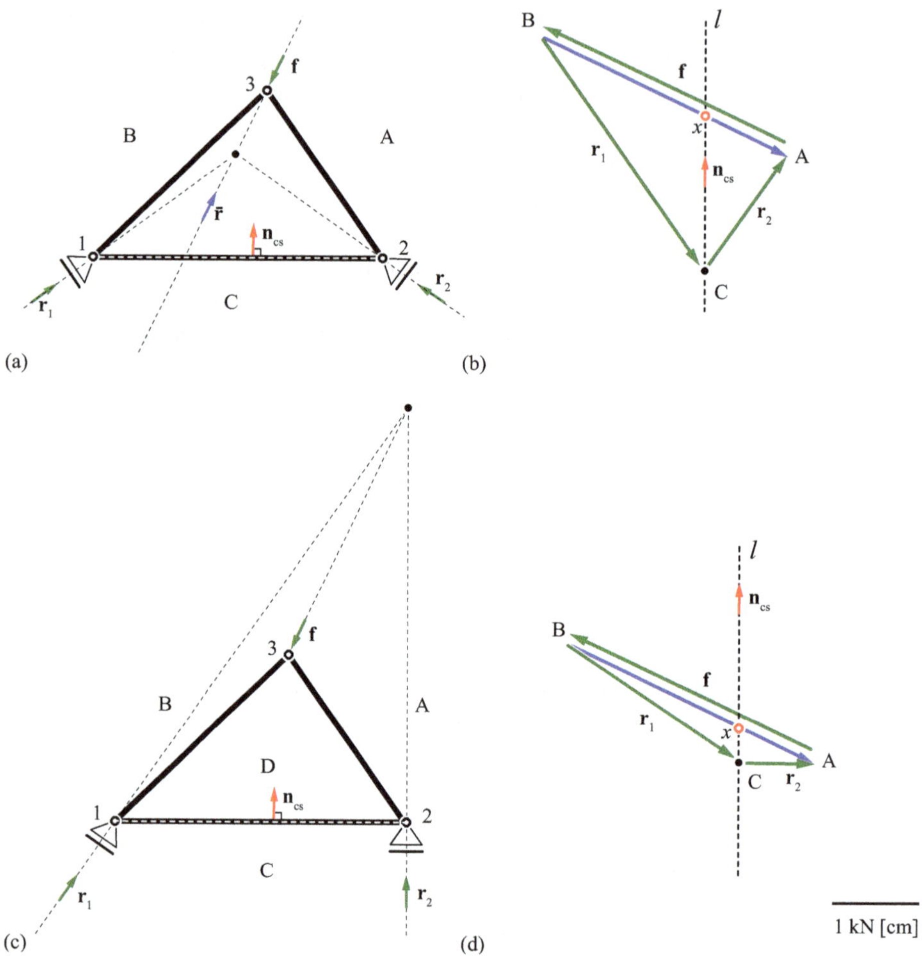

Figure 5.5 Regardless of the direction of the supports, their intersection always lies on the line of action of **f**: (a) and (b) reactions meet inside the truss; and (c) and (d) reactions meet outside of the truss.

Construction steps:

1 In the form diagram of Figure 5.5a, draw the closing string connecting vertices 1 and 2 with a normal $\mathbf{n}_{cs}$.
2 From point x in the force diagram, draw line l normal to the closing string with the direction of $\mathbf{n}_{cs}$ (Figure 5.5b).
3 Pick a point C on the line l and connect it to both ends of the resultant. These two force components are the reaction forces, $\mathbf{r}_1$ and $\mathbf{r}_2$, that keep **f** in equilibrium. Subspace C in the form diagram is reciprocal to the chosen point C on the line l, and it completes the external force polygon.
4 On the form diagram, draw reaction forces normal to the edges of the force polygon as $\mathbf{r}_1$ and $\mathbf{r}_2$. These reactions intersect at the line of action of the resultant.

5 To get the vertical-only reaction at the support 2, its reciprocal force vector must be horizontal. If point C has the same y- value as point A in the force diagram, the reaction force $\mathbf{r}_2$ will become horizontal (Figure 5.5d). The location of C can give the magnitude and direction of both reactions.

6 Drawing the reciprocal reaction forces on the form diagram shows their intersection on the line of action of the resultant (Figure 5.5c).

□

Measuring the lengths of the reaction forces in the force diagram gives their required magnitude to keep force **f** in equilibrium, as shown in Figure 5.6a,b. After the global force polygon is completed and the reactions are found, we can proceed to solve the internal forces in the truss members.

5.2.2 Nodal Force Polygon

The global equilibrium assures the equilibrium of external loads and the reaction forces applied to multiple truss nodes. Consequently, every node in the structure should also be in equilibrium to maintain the global equilibrium. That is, all internal forces carried by the members should also be in equilibrium. Simply put, each node in an equilibrated form diagram should be reciprocal to a closed force polygon in the force diagram, which is called the *nodal force polygon*. The closedness of the nodal force polygon guarantees equilibrium in the node, and the length of its edges gives the magnitude of the force. The process of constructing nodal force polygons and finding the internal forces of a truss starts with rounding every node of the structure with its known elements. For instance, the applied loads are given, and the reaction forces are found in the previous step; thus, their lengths are known.

The first node to consider is a node that has a reaction force. Note that the number of unknown internal forces at each node should not exceed two. For each node of the

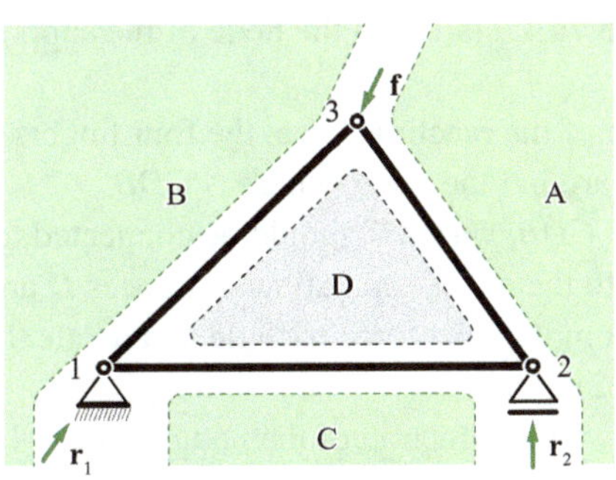

(a)

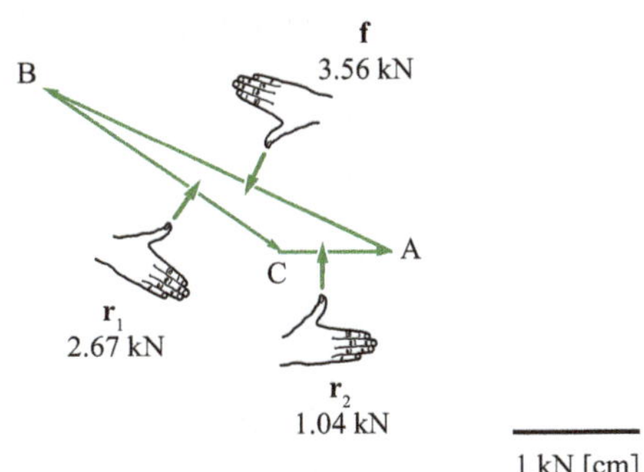

(b)

Figure 5.6 (a) The truss and its applied load and reaction forces; and (b) its constructed global force diagram. The applied load vector **f** drawn to scale in the force polygon and the magnitudes of the reactions $\mathbf{r}_1$ and $\mathbf{r}_2$ are measured from the force polygon's edges BC and CA, respectively.

structure, a closed force polygon should be constructed in the force diagram. Once the nodal force polygon is completed, the magnitude of the internal forces can be found by measuring the edge lengths of the closed polygon. The construction of the nodal force polygons finishes when the number of nodes in the structure equals the number of closed force polygons in the force diagram.

Directions of Nodal Force Polygons and Type of Internal Force

The magnitude of the internal force is found by measuring the edge length of the force polygon, but the type of internal force, for example, tension or compression, within a member is defined by the direction of the force polygon in conjunction with the orientation of the bar connected to the node.

Definition 5.1. A nodal force polygon can have a positive or negative direction. A force polygon is called *positive* if all the normals are toward the inside of the polygon, and it is *negative* if its normals are toward the outside. The direction of the nodal polygons is defined by the direction of applied loads and reaction forces and propagates through the entire force diagram. In each nodal force polygon, vectors are connected from tip to tail. One edge vector's direction can give the entire polygon's direction.

Matching the direction of the normal for each edge of the polygon and the orientation of the reciprocal member connected to the node defines the type of internal force in the member. If the normal pushes toward the node, the member is in compression. Otherwise, it is in tension. The following procedures explain the construction of the nodal force polygons for the truss in Example 5.6.

Example 5.4 Construct nodal force polygons for the truss of Figure 5.6.

Construction steps:

1 In the form diagram of Figure 5.7a.1, start with the node of the support 1 and label the subspaces around it.
2 Put your thumb in the direction of the reaction force; the four fingers show the direction of cycling the spaces around the node, which is *CDB*.
3 Following the cycling direction *CDB*, vertex *C* should be connected to *D* and then back to *B*. Use vectors normal to the edges separating subspaces *D* and *B* and *C* and *D* in the form diagram to complete the force triangle and locate the point *D* in the force diagram (Figure 5.7a.2).
4 The direction of the thumb shows the direction of the nodal force polygon, which is toward the outside of the polygon and thus negative (Figure 5.7a.3);
5 Place the right hand on the edge of the force triangle and move it to the corresponding edge in the form diagram (Figure 5.7a.4). The direction of the thumb on the member shows the direction of its internal force: if it is pulling the node, the force is tensile; otherwise, it is compressive.

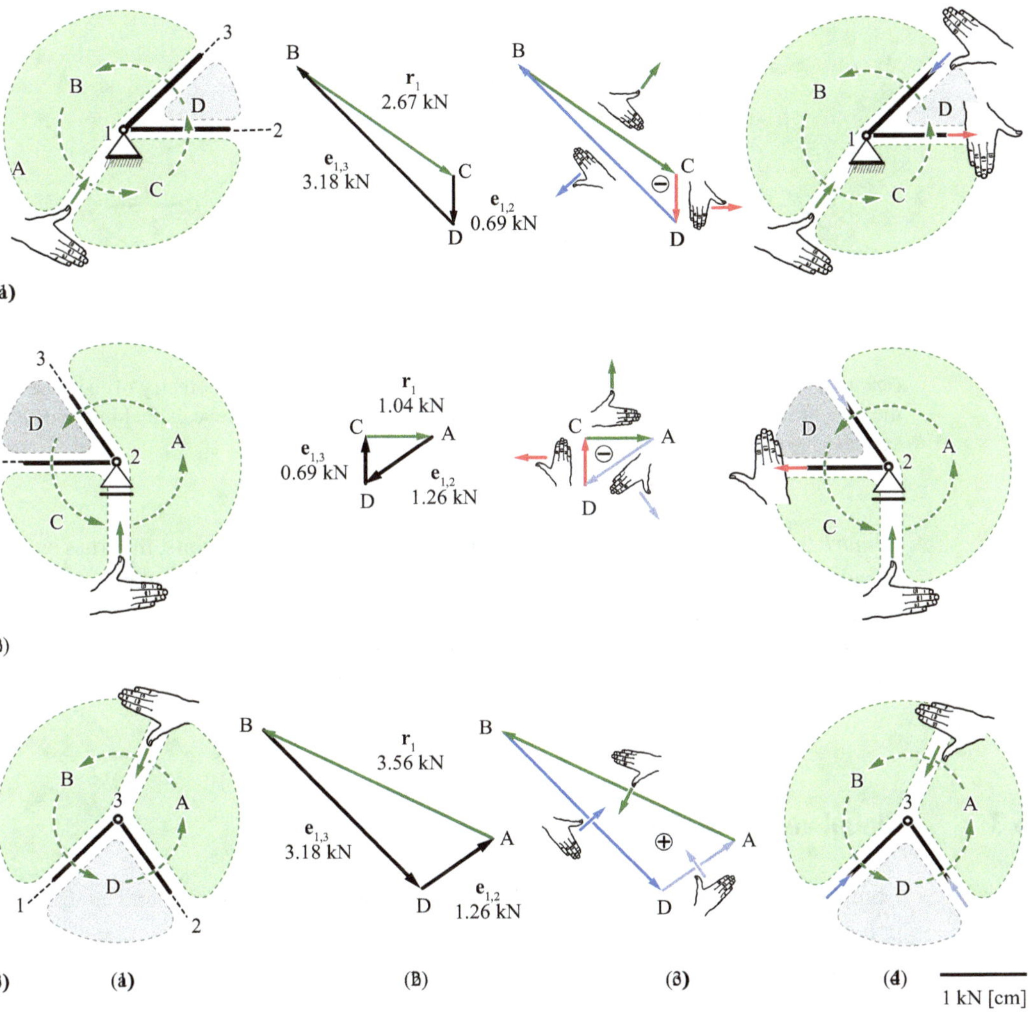

Figure 5.7 (1) to (3) The three nodes of the truss and the direction of rounding each; (a) to (d) constructing a closed polygon of the force around each node and finding the direction of the internal forces by matching the right hands in the force and form diagram.

6 Continue the construction of the nodal force polygons for nodes 2 and 3, as shown in Figure 5.7b,c until the number of nodal force polygons equals the number of vertices in the form of the truss.

□

All the rest of the nodal force polygons of the force diagram can be found using the earlier procedure. The force diagram for the truss is completed. The magnitude of the internal forces can be measured geometrically from the force diagram, as shown in Figure 5.8a,b.

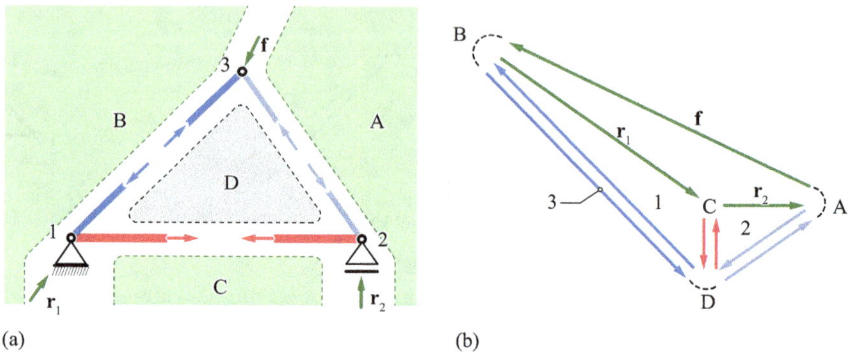

Figure 5.8 A simple truss analyzed with graphic statics reveals (a) the geometry of the truss and the type of internal forces in the members; and (b) the force diagram reveals that internal force vectors align with the global force diagram.

Remark 5.1. In the first step of constructing the nodal force polygons, the truss system is indeterminate if all nodes have more than two members with unknown internal forces. This chapter briefly covers discussions on indeterminate truss systems in 3D in Section 5.14. However, Chapter 6 will provide a more in-depth discussion related to the static degrees of indeterminacy and their application in design.

5.3 Global versus Nodal Force Polygon

Completing the force diagram, including the global force polygon and nodal force polygons, can unravel the relationship between the two:

- Each edge of the global force polygon is connected to one and only one nodal force polygon.
- Consequently, each edge of the nodal force polygon either *belongs* to the global force polygon or is *shared* by an adjacent nodal force polygon.
- The direction of the nodal force polygons, either positive or negative, is defined by the direction of the edge vectors of the global force polygon. Because an edge of the global force polygon constructs the first nodal force polygon.
- Each nodal force polygon shares an edge vector with its adjacent nodal force polygon of equal length but opposite direction. Each internal edge of the force polygon can be divided into two halves, each belonging to a nodal force polygon (Figure 5.8b). This definition is also called a *half-edge* of a network. It is used in developing a data structure intended for a computational implementation of graphic statics in 2D.[122]

Section 5.4 will clarify the topological relationship between the form and the force diagram in systems with compression and tension.

5.4 Topology of Form and Force Diagrams in 2D

Previously, Chapter 4 explained the topological relationship between the form and force diagrams for compression-only systems in 2D. Revisiting the topological properties of the form and force in this chapter enhances our understanding of the relationship between the different components in systems with tensile and compressive forces. The topological relationships between the two diagrams are illustrated in Figure 5.9 and are summarized as follows:

- The number of vertices, v_i, in the form diagram G is equal to the number of faces $f_i^\dagger$ in the force diagram $G^\dagger$; note that the global force polygon is not counted as a closed face in the force diagram.
- The number of edges e_i and $e_i^\dagger$ in both diagrams is equal.
- The number of faces and subspaces f_i in the form diagram G corresponds to the number of vertices $v_i^\dagger$ in the force diagram $G^\dagger$.

Note that the graph of the force diagram in Figure 5.9b has overlapping faces as the face ABD overlaps with faces CAD and BCD. The location of vertices C and D

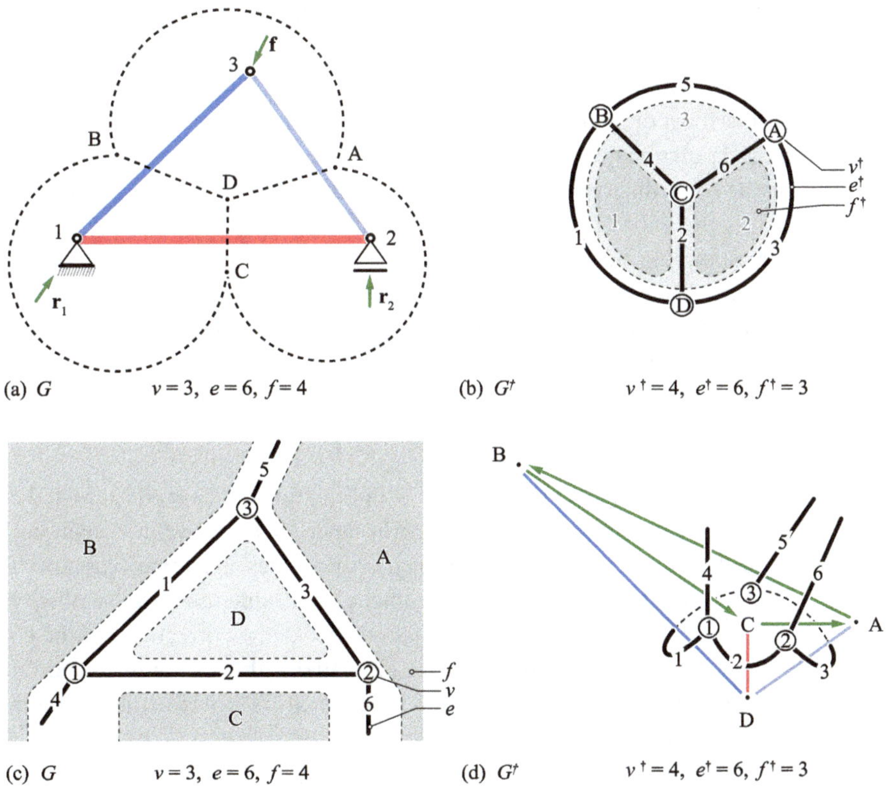

Figure 5.9 (a) The form of the truss and the number of its components with an overlaid graph of the force diagram; (b) the components of the dual graph; (c) the graph of the form diagram; and (d) the force diagram with an overlaid graph of the form.

is adjusted in the graph to resemble the actual geometry of the force diagram without changing the graph's topology, as shown in Figure 5.9d. Such overlapping faces do not exist in the force diagrams of compression or tension-only systems. Indeed, force diagrams with overlapping cells yield structural systems with both compression and tension internal forces.

5.5 Minkowski Operation in 2D

As explained in Chapters 1 to 3, the form and force diagrams can be combined into a single diagram and morph into each other using a method called the *Minkowski sum*, or *dilation*.[53] Combining form and force diagrams into a single diagram allows for a quick transformation of one diagram into another using only a single parameter. This property has many educational benefits as it can check the correctness of the diagrams and relate the force polygons to their reciprocal nodes in the form diagram.

The Minkowski sum operation can be intuitively explained as follows. As explained in Section 5.4, each nodal force polygon/face $f_i^\dagger$ in the force diagram is reciprocal to a vertex v_i in the form diagram, and this reciprocity is used for the operation. The operation starts with choosing a base point for the force diagram as $v_{mink}^\dagger$ and another base point for the form diagram as v_{mink}. Next, each polygon/face $f_i^\dagger$ of the force diagram is moved from the base point $v_{mink}^\dagger$ on the force diagram to the base point v_{mink} on the form diagram using a translation vector $\mathbf{t}^\dagger$ between the two points. Then, the moved polygon $f_i^\dagger$ is moved again from the base point v_{mink} to its corresponding node v_i on the form diagram using another translation vector $\mathbf{t}_i$. Note that the first translation moves the entire force diagram on the form, and the second moves every cell to its corresponding node. Then, every edge of the moved force polygon is linked to its adjacent force polygon via a rectangular connection. In addition, for each diagram, a scale factor is considered so that each can be scaled independently from its base points. These scale factors are ψ for the form diagram and $\psi^\dagger$ for the force diagram. These two scale factors can be linked to one another such that:

$$\psi = 1 - \psi^\dagger \tag{5.1}$$

If so, choosing a single value for ψ in Equation 5.1 between zero and one can transform one diagram into another. For instance, if ψ is one, the diagram becomes the form diagram. If it is zero, the diagram becomes the force diagram. This transformation includes all other combinations of both diagrams for numbers between zero and one. Also, this quick transformation checks the correctness of the construction of the diagrams and is an intuitive tool for visualization of the form and force diagrams. Example 5.5 provides step-by-step procedures for constructing a combined diagram using the Minkowski sum.

Example 5.5 Combine the form and force diagrams of a 2D truss of Figure 5.10a,b via Minkowski sum.

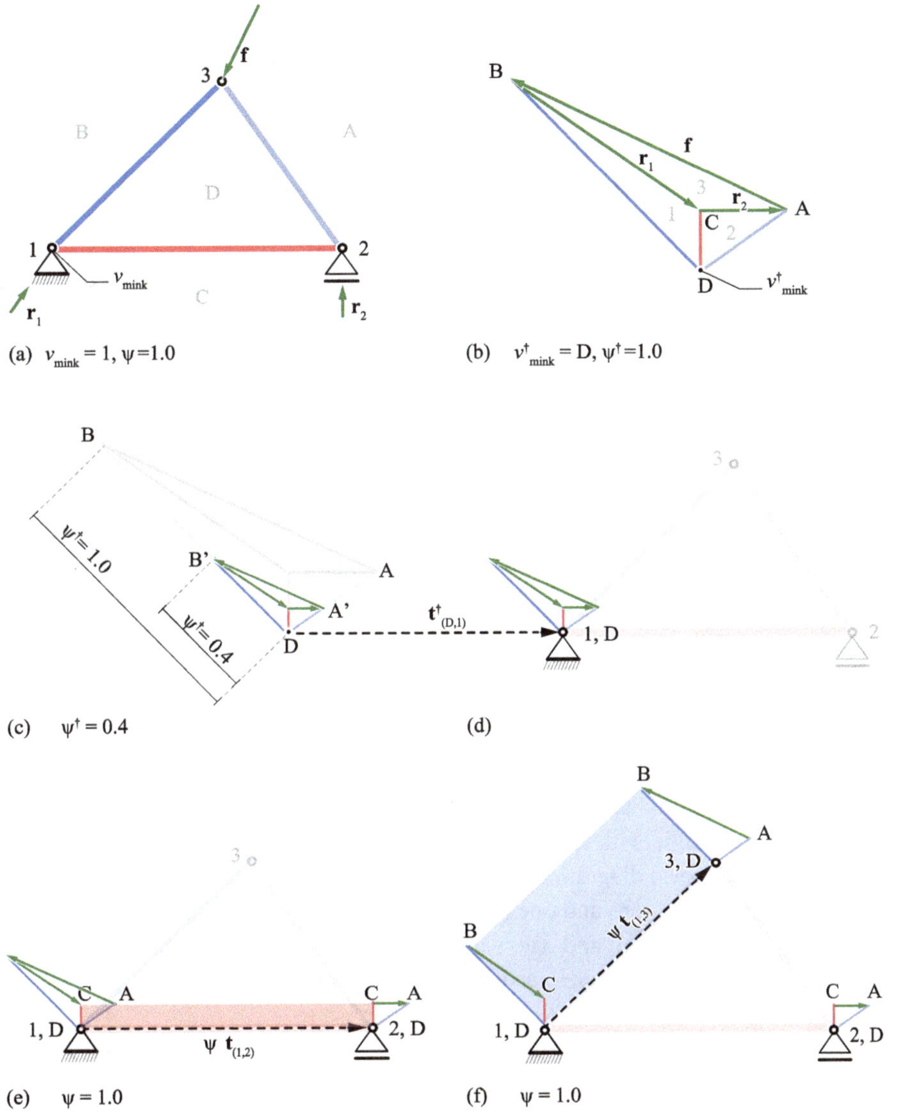

Figure 5.10 The process of combining force and form diagrams using the Minkowski addition: (a) the form diagram of the truss and the chosen base point v_{mink}; (b) the force diagram and its base point $v^{\dagger}_{mink}$; (c) scaling the force diagram using $\psi^{\dagger}$; (d) moving the scaled force diagram to the base point of the form; (e) and (f) moving each force polygon from v_{mink} to their reciprocal nodes in the form.

Construction steps:

1 Pick a base point for both form and force diagrams, for example, vertex D in the force diagram and vertex 1 in the form diagram.
2 The force diagram can be scaled uniformly in both $x-$ and $y-$ directions from the base point D using the scale factor $\psi^{\dagger}$, which is a real number between zero and one (Figure 5.10c).

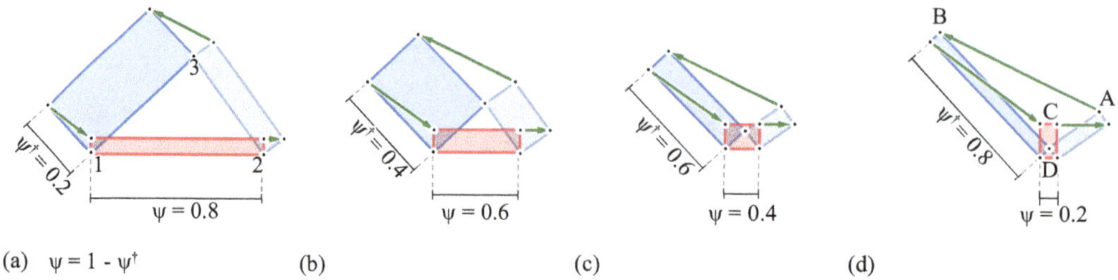

Figure 5.11 (a) to (d) The force diagram can turn into the form diagram using the Minkowski operation using the scale factor ψ.

3 Move the scaled force diagram from the base point D to the base point 1 of the form diagram (Figure 5.10d).
4 Relocate each nodal force polygon in the moved force diagram from base point 1 to its corresponding vertex in the form diagram (Figure 5.10e). For instance, the force triangle DCA is reciprocal to vertex 2 and is moved from vertex 1 to vertex 2 using the translation vector $\mathbf{t}_{1,2}$ (Figure 5.10e).
5 Continue the previous steps for all force polygons of the force diagram (Figure 5.10f).
6 Construct a rectangular face between the *half-edges* of two adjacent force polygons that are now placed at two adjacent vertices at both ends of an edge in the form diagram. For instance, edge BD from the ABD polygon makes a rectangle with edge DB from the BCD polygon.
7 The form diagram can also be scaled using a scale factor ψ, a real number between zero and one (Figure 5.10f). Using Equation 5.1, for any value of ψ between zero and one, various transformations of the form and force diagrams can be constructed, as illustrated in Figure 5.11a–d.

□

Remark 5.2. The translation vectors $\mathbf{t}_{mink,v_i}$ are not identical to the edges of the form diagram. In fact, they are a group of vectors with a start point as the form reference point v_{mink} and all other vertices v_i of the form diagram. All these translation vectors can be directly scaled by the scale factor ψ of the form diagram in the process of transformation (Figure 5.11).

5.6 Analysis of a Simple Truss in 3D

The techniques for analyzing a simple truss in 2D can be extended to the third dimension. This section introduces the procedures in 3D by which a simple truss can be analyzed using 3D graphic statics.

3D Truss Assumptions

A 3D truss system in the context of this book has the following properties:

- It is made of linear elements where all members are connected to joints at both ends.
- The members can freely rotate around the joints. Thus, the bending resistance in the truss nodes where all members meet is assumed to be negligible.
- Although the term truss in 3D can be used for any 3D system that consists of triangular faces, the term 3D truss in this book is used for systems constructed of tetrahedral elements.
- The tetrahedral configurations prevent members from freely rotating around the joints, thus making a rigid system, that is, the system is kinematically stable.
- Such a truss system consists of vertices, edges, planar faces, and closed polyhedral cells. This is similar to the polyhedral funicular structures in Chapter 3.
- Each edge of the truss is shared by at least three faces that are either internal or external.
- The inner cells of the truss do not overlap with each other, and each face is shared by two and only two cells.

Similar to 2D analysis, the truss analysis in 3D can be summarized in two major steps:

1 *Finding global equilibrium* using geometric methods. This process includes identifying the direction and magnitudes of the reaction forces that keep the applied loads in equilibrium.
2 *Finding the internal forces* in the members of the truss that are in equilibrium with the external loads and reaction forces. This procedure includes constructing all nodal force polyhedrons to complete the force diagram for the truss.

Section 5.6.1 will expand on these two major steps.

5.6.1 Establishing Global Equilibrium in 3D

The external forces exerted on the truss nodes and reaction forces at the support should be in equilibrium, regardless of the topology of the truss. The global equilibrium in 3D translates to the following sets of numerical equations:

$$\sum \mathbf{f}_x = 0, \quad \sum \mathbf{f}_y = 0, \quad \sum \mathbf{f}_z = 0, \qquad (5.2\text{–}5.4)$$

and,

$$\sum \mathbf{m}_x = 0, \quad \sum \mathbf{m}_y = 0, \quad \sum \mathbf{m}_z = 0. \qquad (5.5\text{–}5.7)$$

The geometric equivalents of the above relations are as follows:

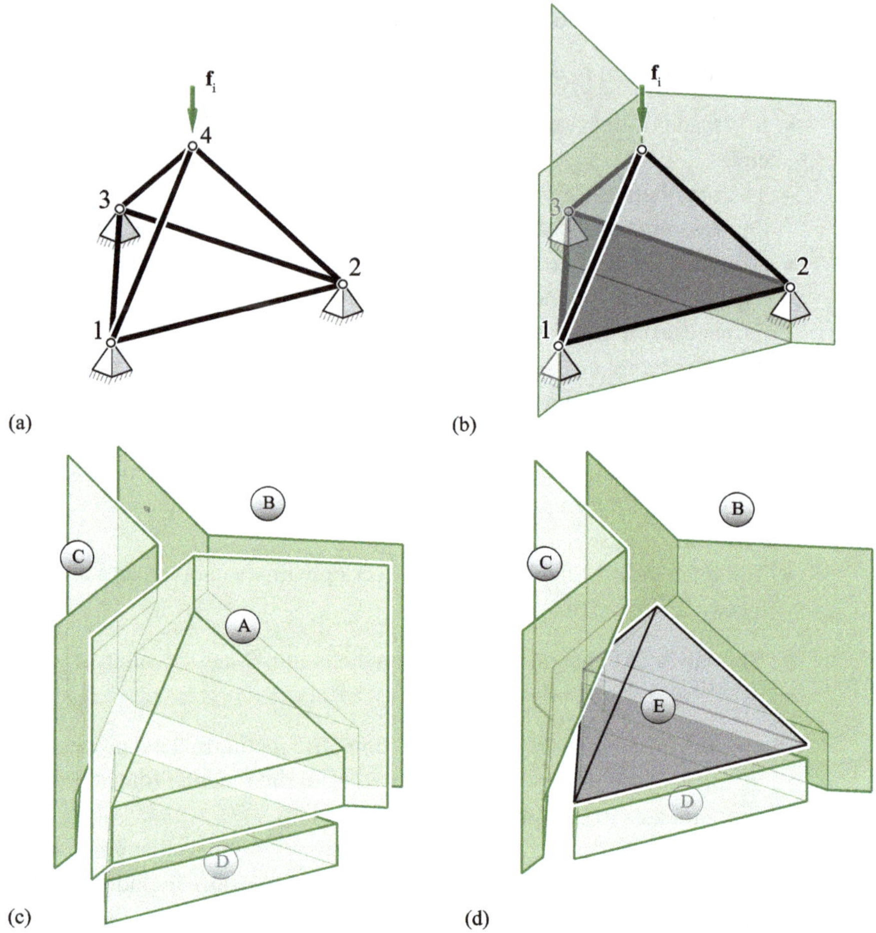

Figure 5.12 (a) A simple truss in 3D and its three supports with an applied load; (b) identifying the open and closed faces in the geometry of the truss; (c) constructing the open subspaces by combining the open faces shared by the applied load and an assumed reaction forces; and (d) the internal and external cells of the truss.

- The truss's external forces, including the applied loads and reaction forces, should make a closed polyhedron called the *global force polyhedron*. When the global force polyhedron is closed, the sum of the forces in x-, y-, and z- is zero. Thus, the closed force polyhedron satisfies the Equations 5.2–5.7.
- A funicular construction on the line of action of the applied loads and the reaction forces in the form diagram should be closed. A closed funicular construction satisfies the moment equilibrium in the system.

Closing the Global Force Polyhedron

The first step in analyzing a truss using polyhedral graphic statics is to develop a graph representation of the system identifying vertices, edges, faces, and cells in 3D. This graph representation is necessary to establish the topological relationship between the form and force diagrams. The form diagram in a 3D truss consists of open and closed polyhedral cells. The cells connected to the external load and reaction forces are often open. These open cells are labeled by *A*, *B*, *C*, and *D* in the example of Figure 5.12c. The polyhedrons inside the truss are closed and are tetrahedral cells (Figure 5.12).

The open subspaces/cells around the edges and applied loads of the form diagram are reciprocal to the vertices of the global force diagram. These open subspaces consist of open faces. Thus, identifying the subspaces in a 3D truss demands recognizing the open faces around the truss.

Remark 5.3. An open face in a truss is constructed by an externally applied load, an edge/member of the truss, and a reaction force at the support. A reaction force at a support vertex is co-planar with an applied load if connected to the other end of the same member. Since the reaction forces' direction is yet to be found, a co-planar direction can be assumed with its adjacent applied load. For instance, the reaction force at the support 1 is assumed to be co-planar with the applied load in Figure 5.12c.

Once the open cells/subspaces around the truss are identified, the topology of the force diagram is found, and the force faces can be constructed starting from the applied load. The subspaces around the applied load are labeled with capital letters.

Constructing the face reciprocal to the applied load starts by traversing the subspaces around the load using the right-hand rule. Similar to a single node example of Chapter 1, if the thumb is put in the direction of the load, the curl of the fingers gives the cycling direction of the subspaces around it, and the reciprocal face can be completed similarly to the examples of Chapter 1. The rest of the process is explained in Example 5.6.

Example 5.6 Construct the external force polyhedron for the truss of Figure 5.13a and find the resultant face that closes the polyhedron.

Construction steps:

1 In the form diagram, if the external faces are not constructed, extrude planes from the line of action of the applied load to intersect support vertices.
2 Assume an arbitrary direction for the reaction forces at the supports co-planar with the applied load. Note that each reaction force at the support makes an open face with another reaction force if they share a truss member (Figure 5.12b).
3 Label the internal and external spaces with capital letters (Figure 5.12c,d); pay attention to the subspace *D* made by the underside face of the truss and the faces shared by assumed reaction edges at the supports.
4 Move to Figure 5.13, put the right-hand thumb in the direction of the applied load of Figure 5.13a, and curl the four fingers to show the direction of cycling subspaces around the load, from *A* to *C* and then to *B*.
5 In the force diagram, start from point *A*. Draw a vector in the direction of the normal $\mathbf{n}_{AC}$ dividing subspace *A* from *C* of the form diagram. Choose a length, for example, *l*, for vector *AC* in the force diagram (Figure 5.13a,b).

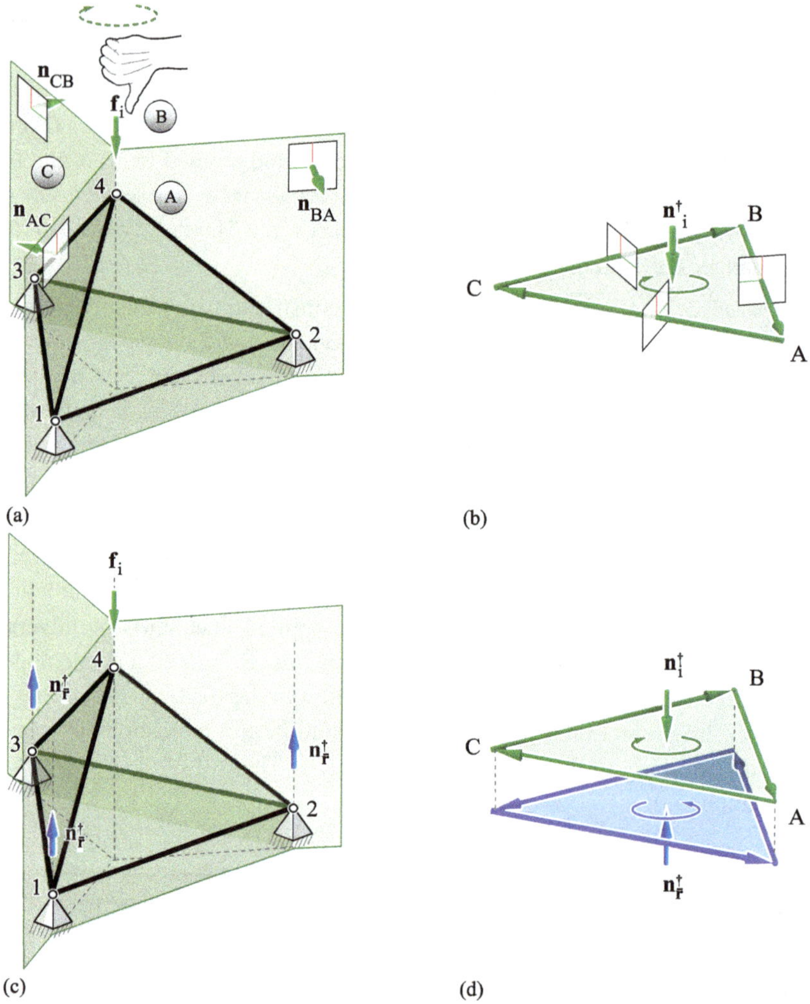

Figure 5.13 (a) In the form diagram, use the right-hand rule to find the order of connectivity of the subspaces to find the topology of the face reciprocal to the applied load; (b) the reciprocal face *ACB*; (c) placing the parallel reaction force component normal to the resultant face at the supports; (d) the closed force polyhedron.

6 Find the location of vertex *B* by intersecting a line drawn from *A* in the direction of $\mathbf{n}_{BA}$ of the face dividing subspaces *A* and *B* of the form and another line drawn from vertex *C* in the direction of the normal $\mathbf{n}_{CB}$ of the form diagram (Figure 5.13b). The face *ACB* is completed in the force diagram and is reciprocal to the applied load $\mathbf{f}_i$.

7 Face *ACB* establishes an open polyhedron, although this polyhedron has only one face. In the force diagram, draw a resultant face using the edge vectors opposite the face *ACB*. The new face with normal $\mathbf{n}_{\bar{\mathbf{r}}}^{\dagger}$ is equal to the applied load but has the opposite direction and closes the external force polyhedron.

□

After the completion of the resultant face, its magnitude and direction are determined. The trial funicular construction can identify the location of the application of the resultant in the form diagram. In the simple truss of Figure 5.13, the line of action of the resultant is identical to the one of the applied load. Thus, using funicular construction to locate the resultant force is redundant. Nevertheless, the trial funicular construction is needed to show the system has moment equilibrium and find the magnitude and direction of the support's reaction forces.

Trial Funicular Polyhedron

The process of trial funicular construction will be explained in Example 5.7, as illustrated in Figure 5.14.

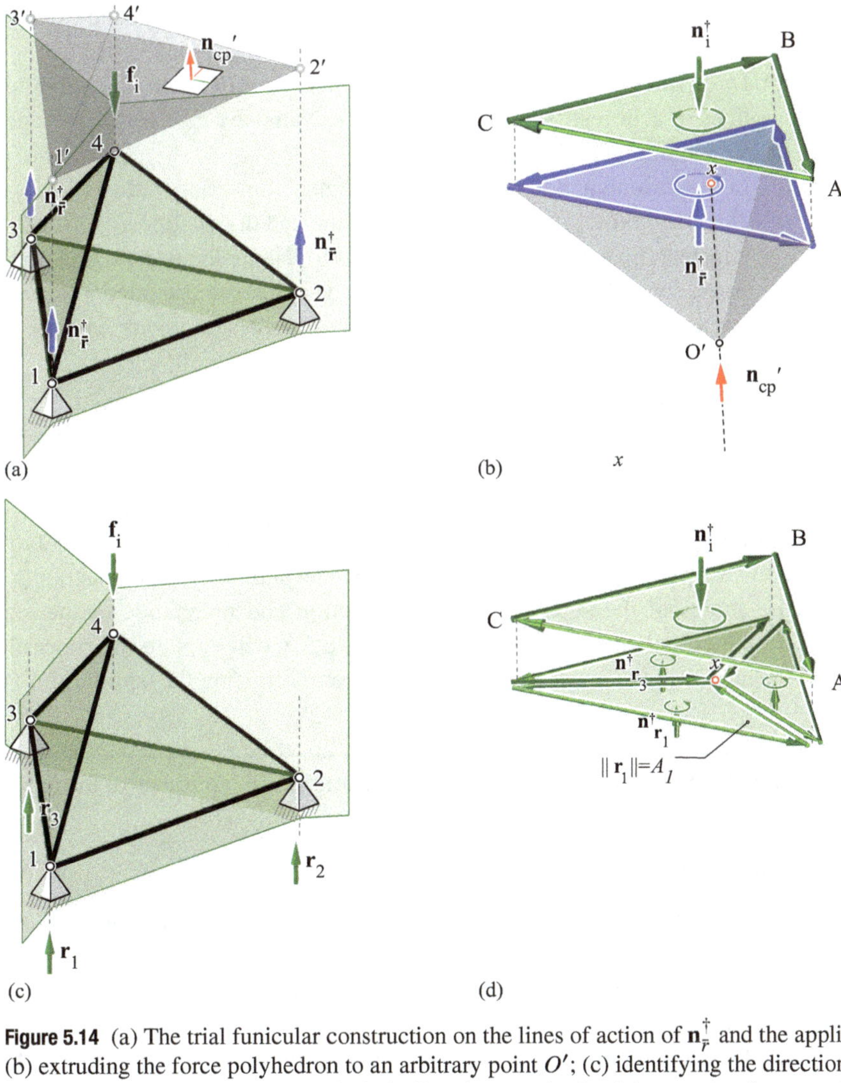

Figure 5.14 (a) The trial funicular construction on the lines of action of $\mathbf{n}^{\dagger}_{\bar{r}}$ and the applied load; (b) extruding the force polyhedron to an arbitrary point O'; (c) identifying the direction of the reaction forces at the supports; and (d) finding the magnitude of the reaction forces and their reciprocal faces.

Example 5.7 Construct a trial funicular for the truss of Figure 5.14 and make faces reciprocal to the reaction forces parallel to $\bar{\mathbf{r}}$.

Construction steps:

1 In the form diagram, at the location of each support, draw three lines of action normal to the resultant face of the force diagram.
2 In the force diagram, extrude the two faces of the closed force polyhedron, including the resultant face and the face of the applied load, to an arbitrary point O' in 3D space. Since the starting polyhedron had only two faces, the extrusion made two closed tetrahedrons.
3 In the form diagram, start from the line of action passing through support 1: from an arbitrary point $1'$, draw a line normal to the face ACO' and extend it until it intersects the line of action of the load $\mathbf{f}$. Name the new vertex $4'$.
4 Draw lines from vertex $4'$ normal to the faces ABO' and CBO' to intersect the lines of action passing through 2 and 3. Name the new vertices $2'$ and $3'$, as shown in Figure 5.14a.
5 In the form diagram, draw a closing plane connecting vertices $1'$, $2'$, and $3'$.
6 Take the normal $\mathbf{n}_{cp'}$ of the closing plane and draw a line, $l_{cp'}$, from point O' to intersect the resultant face (Figure 5.14b). Name the intersection point x.
7 In the force diagram, extrude the edge vectors of the applied load face ACB to point x to make three faces of the reaction forces co-planar with the applied load and the resultant, as shown in Figure 5.14.

□

Direction and Magnitude of the Reaction Forces in 3D

In Figure 5.15, we determined three reaction force components parallel to the resultant $\bar{\mathbf{r}}$ to keep the applied load in equilibrium. The site's physical constraints often dictate the angles of the supports, and the direction and magnitude of the reaction forces should be calculated accordingly. Example 5.8 will explain the process of changing the angles of the reaction forces without compromising the equilibrium of the system.

Example 5.8 Find and control the directions and magnitudes of the reaction forces of the truss of Figure 5.15.

Construction steps:

1 In the form diagram, draw a closing plane connecting the support locations 1, 2, and 3 and name its normal $\mathbf{n}_{cp}$.
2 In the force diagram, draw a line l_{cp} from point x normal to the closing plane of the form.
3 Pick a point on line l_{cp}, then extrude the edges of the load face, as illustrated in Figure 5.15d.

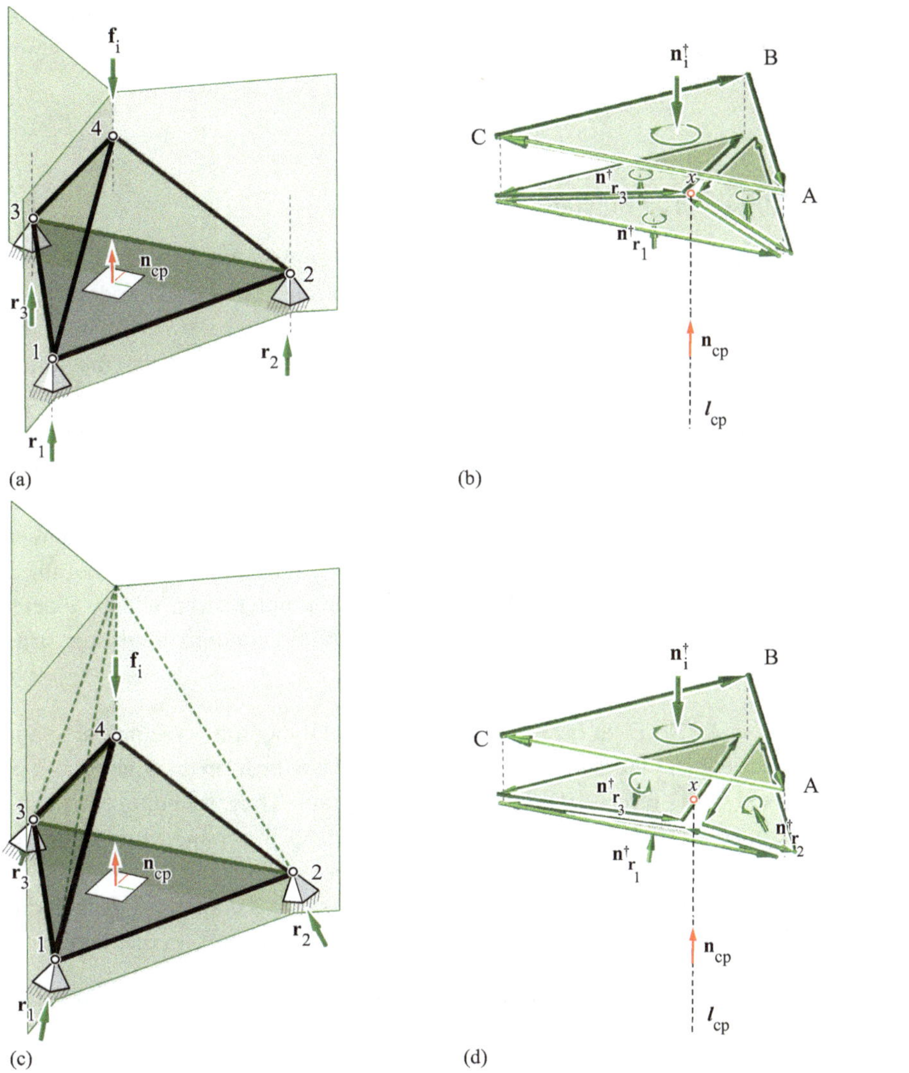

Figure 5.15 (a) A closing plane with normal $\mathbf{n}_{cp}$ for a simple truss with three supports; (b) line l_{cp} drawn from point x in the force diagram; (c) and (d) The direction and magnitude of the reaction forces change by extruding them up and down on the line l_{cp}.

4 The newly constructed faces in the force diagram represent the direction and magnitude of the reaction forces. Note that the location of the extruded point on the line l_{cp} controls the angle and geometry of the reaction forces in the form diagram.

5 In the force diagram, find the direction of each face from its edge vectors that are equal but opposite in direction to the edge vectors of the applied load.

6 In the form diagram, the new reaction forces intersect on the line of action of $\mathbf{f}_i$.

□

The global force polyhedron gets closed by the applied load and reaction faces. Each face of the global force polyhedron will make a closed nodal force polyhedron reciprocal to the nodes of the truss and will be explained next.

5.6.2 Nodal Force Polyhedrons

Each node with its internal members and applied loads or reaction forces should be in equilibrium for a truss to be in equilibrium. Thus, each node of the form should have a reciprocal, closed polyhedral cell in the force diagram. This force polyhedron is referred to as a *nodal force cell*, and its geometry reveals the magnitude and type of internal forces in the truss members.

We can construct the nodal force cells in the force diagram starting from a node with at least one known force. The global force polyhedron gives the faces reciprocal to the applied loads and reaction forces. Therefore, a support node or a node receiving an applied load should be chosen for the first step. Moreover, the number of unknown forces in the chosen node of a determinate truss should not exceed three – this number is 2 for 2D trusses. Otherwise, the force polyhedron will be indeterminate. The construction of nodal force polyhedrons should continue until every truss node has a reciprocal force cell in the force diagram.

Remark 5.4. Each face of the nodal force polyhedron is reciprocal to a truss member. Each edge of the truss is surrounded by cells in the form diagram. These cells are reciprocal to the face vertices in the force diagram. Thus, rounding each edge and cycling around its subspaces in the form diagram gives the connectivity of the reciprocal face vertices in the nodal force polyhedron. Moreover, the number of faces in a nodal force polyhedron equals the number of members/edges connected to its reciprocal vertex, including the applied loads and reaction forces in the form diagram.

Example 5.9 will clarify the step-by-step procedure for constructing a nodal force polyhedron.

Example 5.9 Construct a nodal force polyhedron for vertex 1 of truss of Figure 5.15a.

Construction steps:

1 Move to Figure 5.16.
2 In the form diagram, isolate node 1, which is one of the three supports of the truss with its connected edges and faces.
3 Identify all the subspaces/cells around the edges and the reaction force in Figure 5.16. As illustrated in Figure 5.16, three open subspaces *A*, *C*, and *D* and a part of a closed cell *E* encompass node 1.

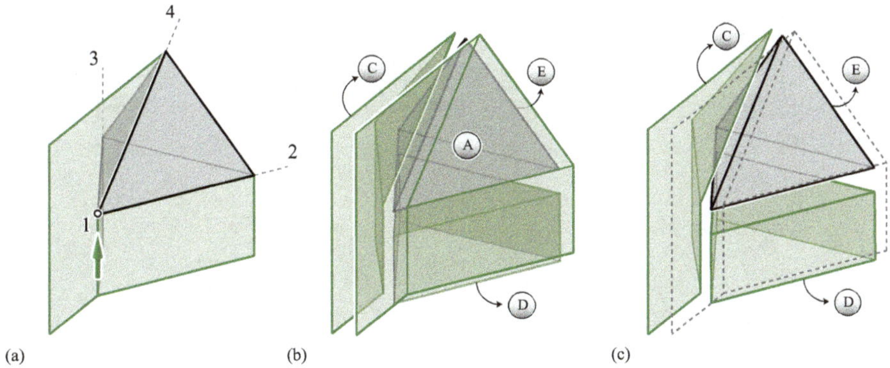

Figure 5.16 Isolating vertex 1 of the truss with all its components, including (a) edges and faces; and (b) and (c) exterior and interior cells.

4 Now move to Figure 5.17a. In vertex 1 of Figure 5.17a, the geometry of the face corresponding to the reaction force $\mathbf{r}_1$ has been given by the global force polyhedron in Figure 5.17b. Using the right-hand rule and putting the thumb in the direction of the reaction force, the curl of the fingers shows the order of cycling subspaces *A*, *D*, and *C* around the reaction edge $\mathbf{r}_1$ (Figure 5.17a). This order matches the order of the edge vectors around the face *ADC* whose normal $\mathbf{n}^{\dagger}_{\mathbf{r}_1}$ is parallel to the reaction vector $\mathbf{r}_1$ (Figure 5.17b).

5 In the force diagram, each edge vector of *ADC* is equal and opposite to the edge vector of an adjacent face. Take vector *AC* as an example. This vector belongs to an adjacent face whose third vertex is yet to be found (Figure 5.17c).

6 In the force diagram, find an edge shared by two subspaces, *A* and *C*. This edge is $e_{(1,4)}$.

7 Rounding the edge $e_{(1,4)}$ using the direction of *AC* gives the third subspace *E* and thus the topology of the face *ACE* (Figure 5.17d).

8 In the force diagram, draw lines from *A* and *C* normal to the faces separating subspaces *A* and *E*, and *C* and *E* in the form diagram (Figure 5.17d).

9 The intersection of these lines gives vertex *E* (Figure 5.17e,f).

10 The remaining faces of the nodal force cell can be completed by similar procedures (Figure 5.17g). Once the nodal force polyhedron is completed, the type of internal forces in the truss members can be found next.

11 Place the right hand on each face such that the thumb aligns with the direction of the face normal (Figure 5.17h).

12 Move the right hand from the face and place it on the reciprocal member in the form diagram.

13 The direction of the thumb shows the type of force in the member: if it pulls the node, the member carries tensile force; otherwise, it carries compressive force (Figure 5.17i).

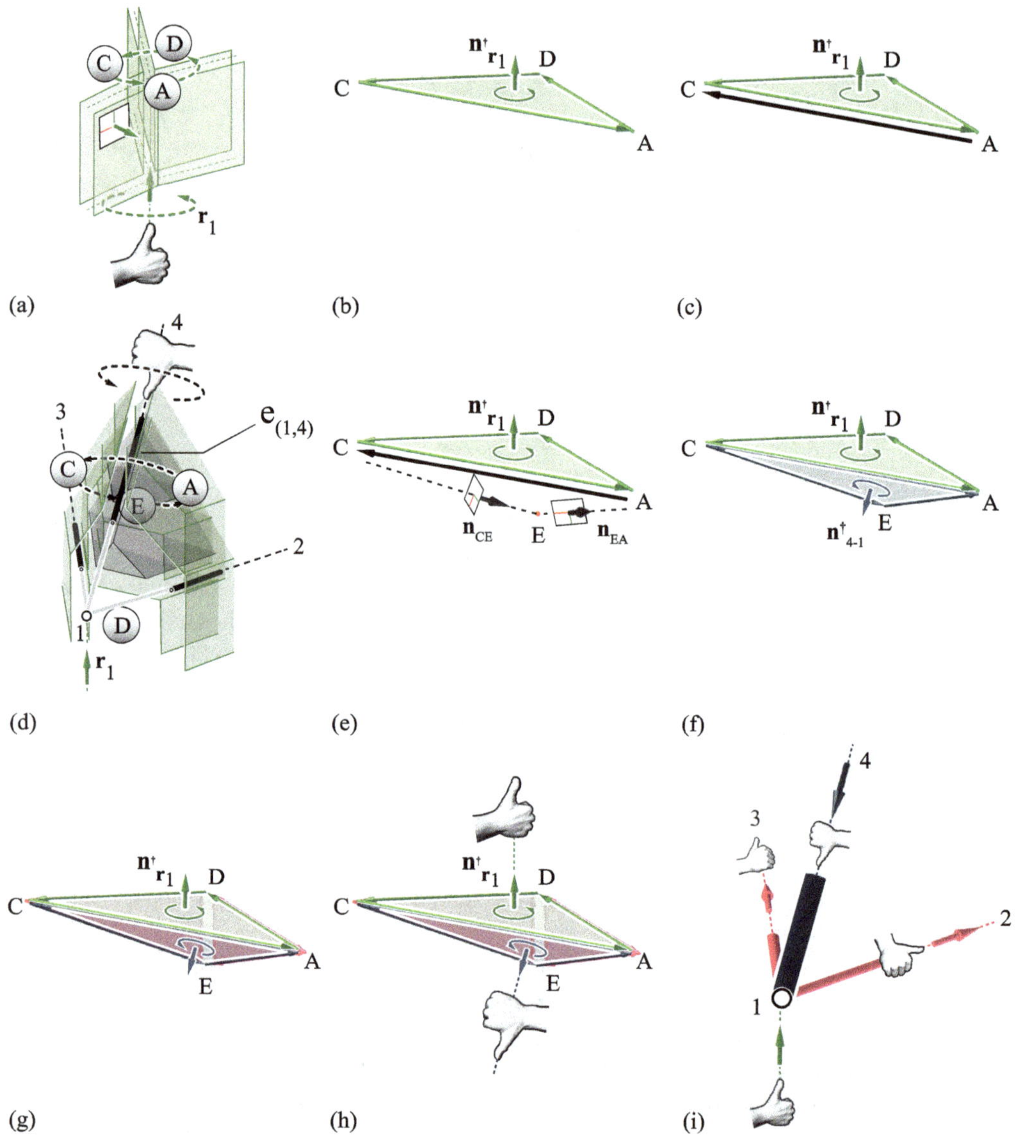

Figure 5.17 Constructing a nodal force polyhedron for vertex 1: (a) rounding the edge $\mathbf{r}_1$ and its subspaces A, D, and C; (b) the geometry of the face ADC is given by the global force polyhedron; (c) the edge AC is used to build the adjacent face ACE; (d) rounding the edge $e_{(1,4)}$ and its subspaces gives connectivity of A to C to E; (e) finding vertex E by intersecting lines from A and C normal to faces separating subspaces around the edge $e_{(1,4)}$; (f) the completed geometry of the face ACE; (g) the completed geometry of the nodal cell; (h) placing the right hand on the faces of the nodal cell; and (i) finding the direction of the internal force in the members of the form.

□

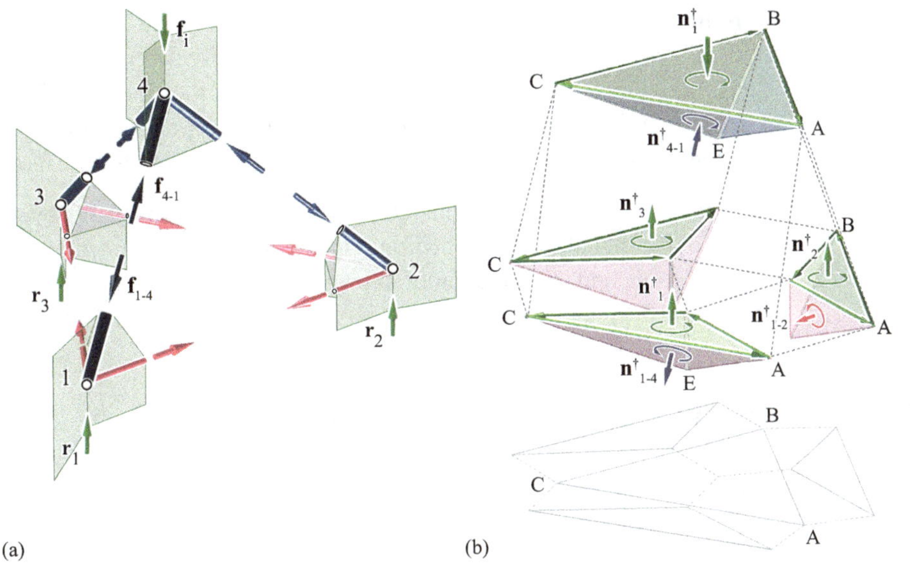

Figure 5.18 (a) and (b) The internal force directions in the members connected to each node of the truss and their reciprocal nodal cells in the force diagram and their face directions.

Remark 5.5. If, in a 3D truss, every node receiving an applied load or reaction force is connected to more than three internal members, then the system is statically indeterminate. This feature is a rule of thumb for identifying indeterminate trusses before the construction of nodal equilibrium.

We will visit the static indeterminacy of a simple truss and its related discussions in Section 5.14 of this chapter.

The same procedure is valid for the rest of the nodes of the truss to complete the construction of the nodal force cells and the entire force diagram. Figure 5.18 shows all internal forces in the members of the structure in an illustration with separated nodes and their reciprocal nodal force polyhedron. The closedness of each nodal cell assures the static equilibrium of each node. In contrast, the green faces of the force polyhedron together make a closed global cell representing the equilibrium of the external load and the reaction forces in the system.

Directions of Nodal Cells

The directions of nodal cells reveal the type of internal force in the truss members. This direction can be found once the cell's geometry is completed and closed. Each closed, convex nodal cell has a consistent normal direction toward inside or outside the cell.

In a single cell, each face has a shared edge vector with its adjacent face. This edge vector is equal in length and opposite in direction. The edge vectors around a face are connected tip-to-tail, and their cycle is consistent with the normal of the face. Therefore, a single face with a known direction is sufficient to find the direction of the rest of the faces and the entire cell.

Definition 5.2. Each of the two adjacent nodal cells in the force diagram shares an identical face. The face shared between two cells is divided into two *half-faces*; each belongs to a single cell and thus follows the direction of the cell. The directions of these two half-faces are opposite of each other. Hence, if one cell's direction is known, its adjacent cell's direction can be found by the direction of the half-face.

As mentioned in Section 5.6.2, the construction of nodal cells always starts from the face of the global force polyhedron. Thus, a global face's direction defines its nodal cell's direction, and the direction of known half-faces defines the direction of the rest of the cells.

5.7 Global versus Nodal Force Polyhedron

Similar to the relationships summarized for 2D form and force diagrams in Section 5.3, the following relations exist between the global and nodal force polyhedrons:

- Each face of a global force polyhedron makes one and only one nodal force polyhedron/cell.
- Consequently, each face of a nodal force polyhedron either *belongs* to the global force polyhedron or is *shared* by an adjacent nodal force cell.
- If a nodal force cell has a global face, then the direction of the cell is defined by the direction of the global face. For instance, if the direction of the faces of the global force polyhedron is toward inside the cell, then the nodal cell made by the face of the global force polyhedron will also have the direction toward inside.
- Each nodal force polyhedron shares a face with its adjacent cell. The shared face is divided into two *half-faces* equal in area and opposite normal direction.
- The direction of a nodal cell that does not have a global face is found by the direction of the half-faces. For example, suppose the direction of a nodal cell is inside. In that case, the normal of all its faces is pointing toward inside the cell, and all half-faces shared by the neighboring cells will have opposite directions that can give the direction of the neighboring cells.

Now that we know the relationship between the global and nodal force polyhedrons, we should look at the topological information embedded in the form and force diagrams of a 3D truss.

5.8 Topology of the Form and Force Diagrams

A truss system comprises vertices, edges, faces, and cells in polyhedral graphic statics. Each face in the polyhedral truss system should be planar, but the cells can be closed or open. The following topological relationships exist between the components of the form and force diagrams in 3D, where the force components are marked with a † sign, as illustrated in Figure 5.19.

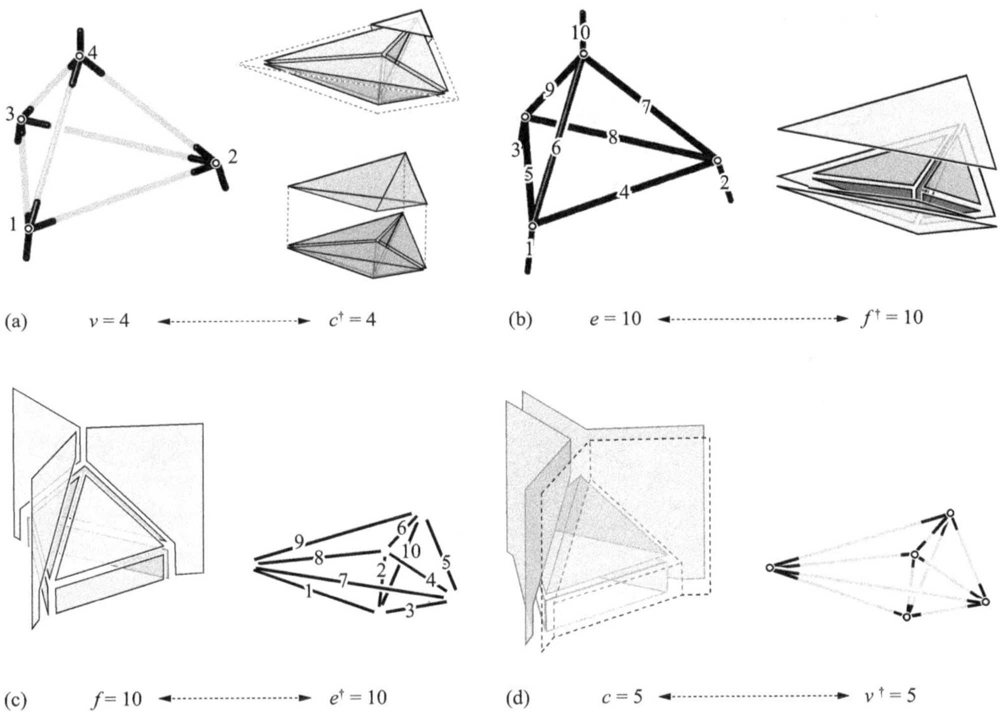

Figure 5.19 (a) to (d) The topological relationships between the form and force diagrams: the number of vertices v, edges e, faces f, and cells c in the form diagram equals the number of cells $c^\dagger$, faces $f^\dagger$, edges $e^\dagger$, and vertices $v^\dagger$ correspondingly.

- Each vertex v of the form diagram corresponds to a cell $c^\dagger$ of the force diagram. Thus, the number of vertices equals the number of cells in the force diagram (Figure 5.19a).
- Each edge e of the form diagram corresponds to a face $f^\dagger$ of the force diagram. Therefore, the number of edges of the form and the faces of the force are equal (Figure 5.19b).
- Each face f of the form diagram corresponds to an edge $e^\dagger$ of the force diagram.
- Each cell c, open or closed, of the form diagram corresponds to a vertex $v^\dagger$ of the force diagram (Figure 5.19d).

Remark 5.6. The global force polyhedron is not counted as the cells of the force diagram. Because there is no node in the form diagram that corresponds to the global force polyhedron. This topological relationship contrasts with the ones proposed by Maxwell[17,37] in his definition of the reciprocal diagrams in 3D. Maxwell's definition includes the global force polyhedron as a cell. Thus, there is a reciprocal node in the form diagram where all the externally applied forces intersect. This description is more suitable for self-stressed systems and might be a bit limiting in designing structures with external loads that might not intersect at a single point. Therefore, we do not consider a global force polyhedron as a cell in the topological relations of the reciprocal diagrams of graphic statics.

5.9 Minkowski Addition in 3D

The Minkowski operation in 3D is very similar to the one in the 2D method explained in Section 5.5. An arbitrary base point is chosen for each form and force diagram as v_{mink} and $v^{\dagger}_{mink}$, respectively. In addition, scale factors ψ and $\psi^{\dagger}$ are assigned to form and force diagrams that uniformly scale each diagram in x-, y-, and z-. Figure 5.20a,b

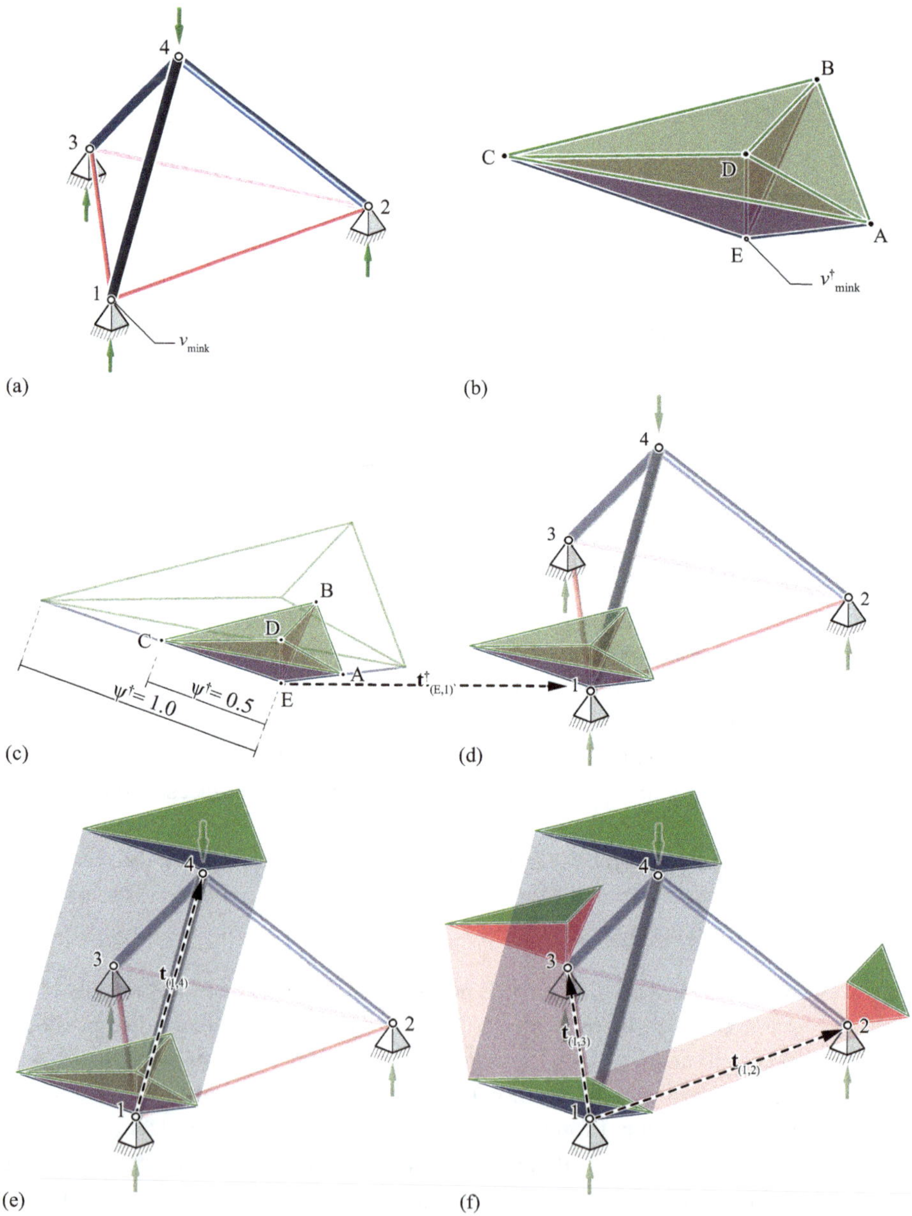

Figure 5.20 Minkowski operation in 3D: (a) and (b) identifying base points in both form and force diagrams; (c) and (d) scaling and moving the force diagram to the form; (e) and (f) moving each force cell to its reciprocal vertex in the form diagram.

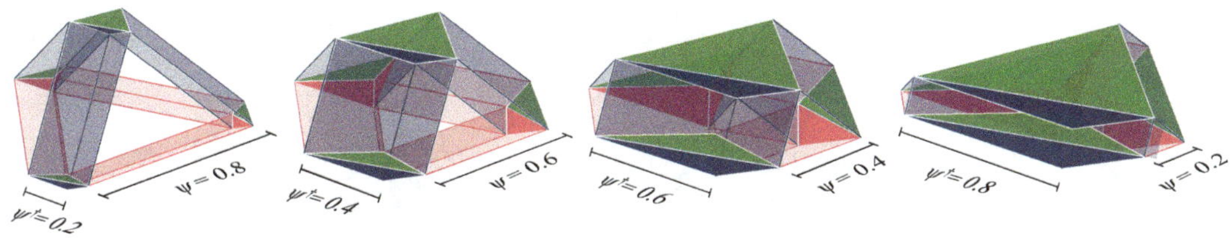

(a) $\psi = 1 - \psi^{\dagger}$, $\psi = 0.8$ (b) $\psi = 0.6$ (c) $\psi = 0.4$ (d) $\psi = 0.2$

Figure 5.21 (a) to (d) Linking the scale factor of the form and force and transforming the diagrams into one another.

shows a simple truss, its force diagram, and its chosen base points for the Minkowski operation.

The force diagram is scaled and moved from its base point $v^{\dagger}_{mink}$ to the base point of the form diagram v_{mink} (Figure 5.20c,d). Subsequently, each cell of the force diagram is moved from the base point to its reciprocal node in the form diagram (Figure 5.20e,f). The topological relationship between the cells of the force diagram and the vertices of the form diagram, as mentioned in Section 5.8, is used for this operation.

Then, the edge vectors around each half-face of a cell are extruded to connect to its opposite half-face in the adjacent cell. If the scale factors are linked based on Equation 5.1, then the form and force diagrams can transform into one another using values from zero to one, as illustrated in Figure 5.21.

5.10 The Effect of Boundary Conditions on Internal Forces

The site constraint and construction detailing of the project often impose boundary conditions on a structure. These conditions include the angle of the supports and the rotational or transnational constraints they might have in *x*-, *y*-, or *z*- directions. Any change in the angles by which the external loads are transferred to the support can change the internal distribution of forces within the structure members. This property can be explicitly controlled and observed by polyhedral graphic statics during the design and analysis stages of the project.

A designer can change the angle of the reaction forces by changing the angle of the reciprocal faces in the global force polyhedron, thus altering the internal distribution of the forces in the truss. For instance, consider the force and form diagram of the truss of Figure 5.22. The faces of the reaction forces in a global force polyhedron share a vertex D that is constrained to the line l_{cp}, which is normal to the closing plane $\mathbf{n}_{cp}$ connecting the supports. Vertex D can move up and down the line l_{cp}. Consequently, the angle of the reaction faces changes, which alters the angles of the reaction forces at the supports in the form diagram. Changes in the angles of the reaction faces also alter the areas of the internal faces of the force diagram. However, their angle stays intact because the geometry of the truss does not change.

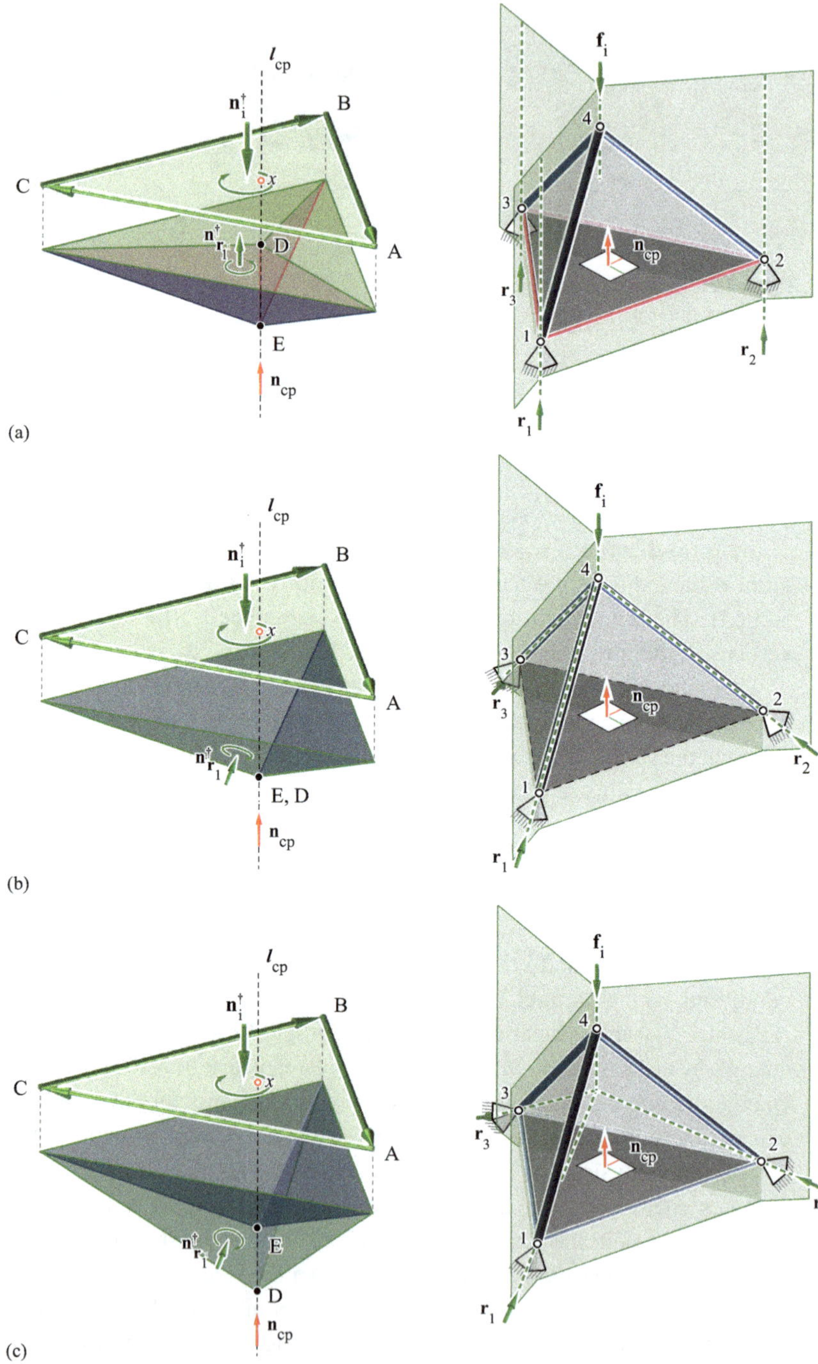

Figure 5.22 Modifying the geometry of the global force polyhedron and altering the type of internal forces in the members: (a) reaction forces parallel with the applied load or resultant (vertex D is on the face ACB); (b) reaction forces aligned with the compression members (vertex D is on vertex E); and (c) reaction forces intersecting inside the truss (vertex D is below E in the force diagram).

For instance, if in the force diagram, vertex D lies on the resultant plane BCA, which is the same as the plane of the applied load, the area (and thus the magnitude) of the reaction faces will decrease. The reaction directions in the form diagram become parallel to the resultant/applied load. This effect is visible in Figure 5.24a.

If vertex D moves toward E on the line l_{cp}, the area of the internal face ADE decreases, and so does the magnitude of the internal force in the member $e_{(1,2)}$ connecting vertices 1 and 2 in the form diagram. If vertex D falls on E in the force diagram, the area of the face ADE will become zero, and thus the edge $e_{(1,2)}$ and all other edges connecting the supports will carry no internal force (Figure 5.22b). In such a case, the reaction forces align with the truss's compression members, and no additional force will be transferred to the members connecting the supports.

If D moves below E on the line l_{cp}, the system becomes a compression-only system, and all members will carry compressive force (Figure 5.22c). This control over the geometry of the global equilibrium in polyhedral graphic statics provides an explicit design and educational tool to understand the force flow in 3D structures.

5.11 Applied Load Outside of the Closing String in 2D

It is not always the case where the external load is applied between the two supports. In some cases, the line of action of the applied load falls off the closing string that connects two supports in 2D. Consider the truss of Figure 5.23a as an example. The load is applied outside the space defined by the closing string. Note that the edges connected to nodes are free to rotate. The translation in x, y is restrained for node 1 and node two can only move in the direction of x.

One can follow the same procedures explained in the examples of Figures 5.6 and 5.7 to construct a force diagram. Figure 5.23 shows the completed force diagram and the direction of the internal forces in the members of the form. Using the Minkowski sum, the form and force diagrams can be combined, as shown in Figure 5.24. The

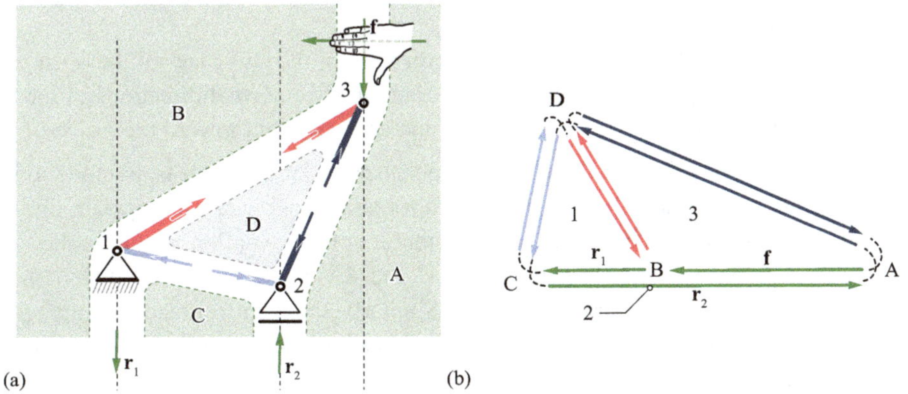

Figure 5.23 (a) A simple truss with an applied load outside of the closing string; and (b) the force diagram with its closed polygons.

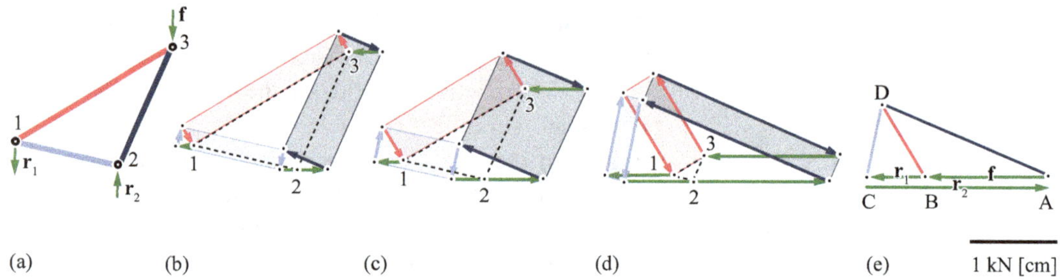

Figure 5.24 (a) to (e) The use of the Minkowski sum to combine the form and force diagrams into one.

process of constructing force for such cases in 3D includes one additional step, which will be explained in Section 5.12.

5.12 Applied Load Outside of the Closing Plane in 3D

Figure 5.25a illustrates a 3D truss where the applied load is outside of the closing plane of the support locations. The first step in constructing the force diagram is to specify the subspaces in the form diagram. Figure 5.25b shows the open cells around and beneath the given truss example. Note that these open cells are constructed under the assumption that the reaction forces at the supports will be parallel to the direction of the applied load. The direction of the reaction forces can then be adjusted, as explained earlier in this chapter.

The objective is to construct a reciprocal force diagram for this truss and find the magnitude and type of internal forces in the truss members. The construction of the force diagram starts by constructing the face reciprocal to the load face, followed by the completion of the global force polyhedron. Rounding the applied load using the right-hand rule shows the direction of cycling the subspaces *A*, *B*, and *C* around the load.

Figure 5.26a,b illustrates a closer view of the half-faces of the open cells surrounding the applied load in the form diagram. The normal directions of the half-faces are equal and opposite to each other and are consistent toward the inside of the cells.

Remark 5.7. The direction of the subspaces of the form is assumed to be toward the inside of the cell. This is just a convention; one can use opposite directions, but all cells' directions should be consistent. In our convention, the normal of the face of the subspace *C* and shared between *C* and *A* is named $\mathbf{n}_{AC}$. The direction of the normal is from space *A* toward space *C*. Similarly, the identical face with the opposite normal direction belongs to subspace *A* and is named $\mathbf{n}_{CA}$ pointing in the direction from *C* to *A* (Figure 5.26b).

Example 5.10 will expand on the procedures to construct the face reciprocal to the applied load of the truss with an overhang.

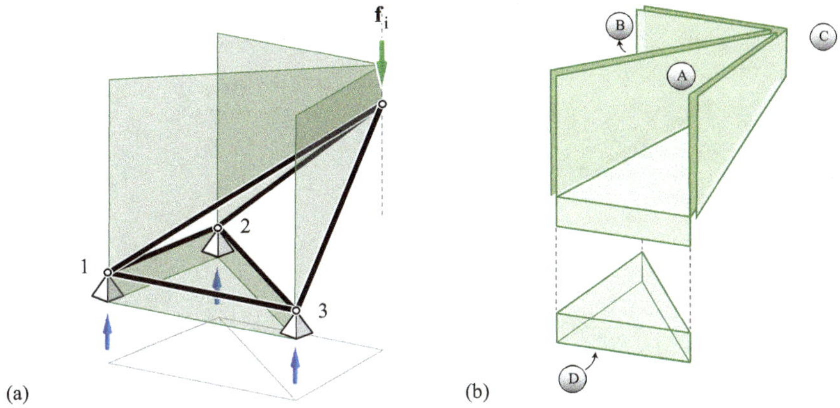

Figure 5.25 (a) A polyhedral truss with three supports with an overhang; and (b) the subspaces around and beneath the truss.

Example 5.10 Construct a face of the force diagram reciprocal to the applied load of the truss of Figure 5.25.

Construction steps:

1. In the form diagram of Figure 5.26, use the right-hand rule to find the cycling direction around the applied load. We use the normal of the half-faces matching the direction of cycling. For instance, space A is connected to B in the direct normal $\mathbf{n}_{AB}$, and so on (Figure 5.26a–d).
2. Let us move to Figure 5.27. In the force diagram, start from an arbitrary point A and move in the direction of the normal $\mathbf{n}_{AB}$ with a length l to find the location of point B. This process is illustrated in Figure 5.27a. The length l_{AB} is used as a scaling factor to control the size and area of the face.
3. From point B, draw a line in the direction $\mathbf{n}_{BC}$. Also, from point A, draw a line in the direction of $\mathbf{n}_{CA}$. The extensions of these two lines intersect at C.
4. Once point C is located, the direction of the edge vectors is found, which must be consistent from A to B and C (Figure 5.27a). Matching the direction of the edge vectors shows that vectors $\mathbf{n}_{BC}$ and $\mathbf{n}_{CA}$ should take the opposite direction to close the face. Thus, they are shown with a negative sign in Figure 5.27a.
5. The curl of the right-hand fingers around the face ABC of Figure 5.27a shows a normal direction toward the outside of the page, which does not match with the direction of the applied load in the truss.
6. To fix this issue, start from point A and move in the opposite direction of l_{AB} to locate point B, as shown in Figure 5.27b. The normal direction of the new face matches the direction of the applied load, and the construction is completed.

□

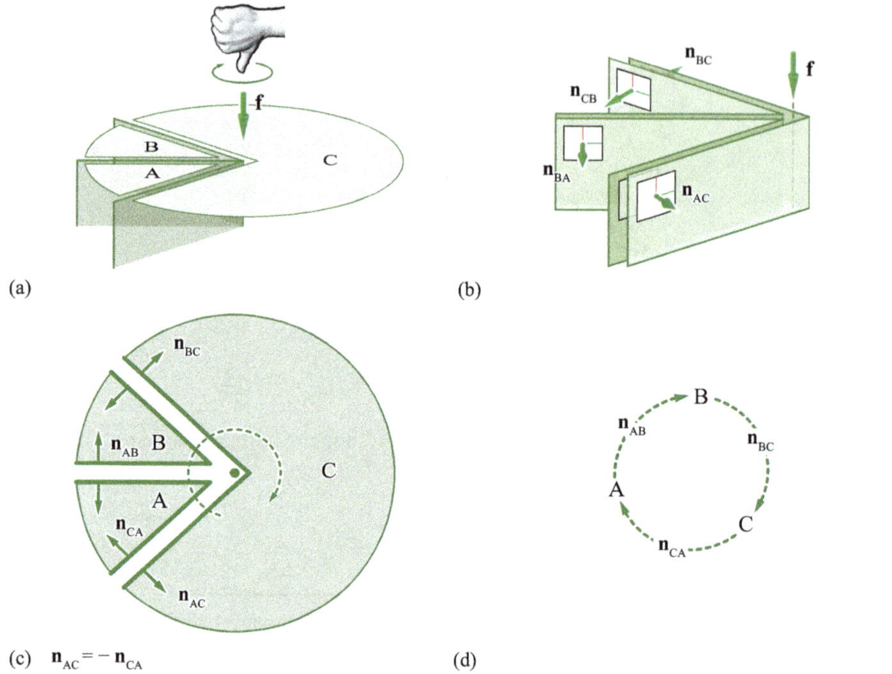

Figure 5.26 (a) The subspaces around the applied load and the right-hand rounding it; (b) the half-faces surrounding the applied load with consistent normal directions pointing toward the inside of the subspaces; (c) a top view of the subspaces around the load; and (d) the topological connection of the vertices of the reciprocal face.

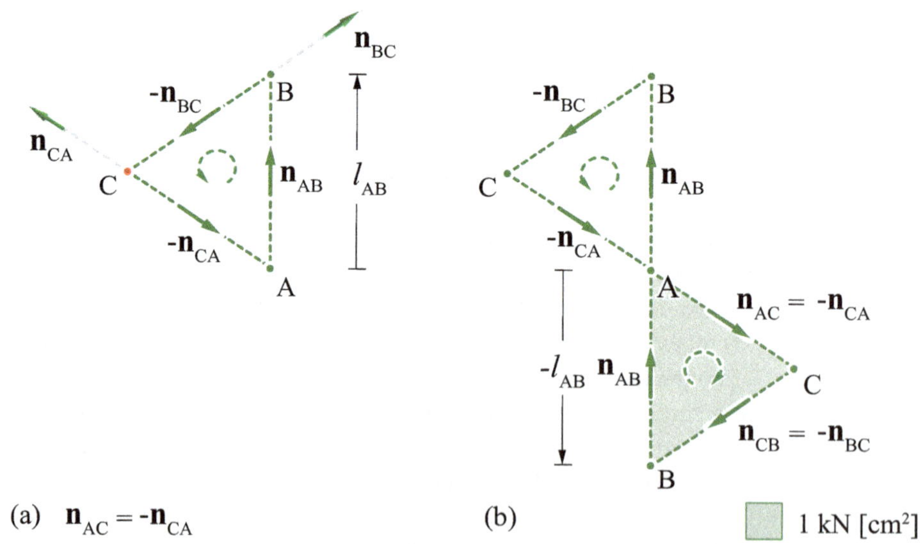

Figure 5.27 (a) Initial construction of the reciprocal face with the opposite normal direction; (b) choosing a negative length for the initial length without changing the direction of the unit vector gives a face normal, aligned with the applied load.

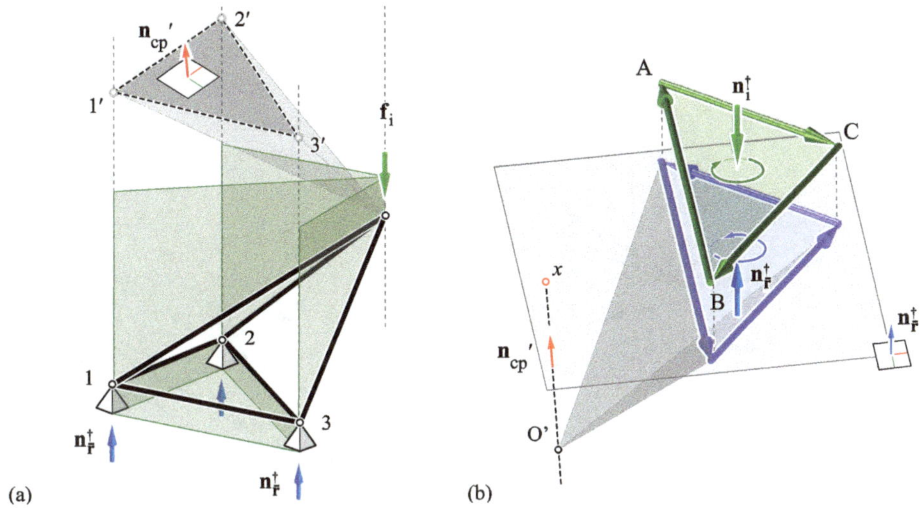

Figure 5.28 (a) A trial funicular construction on the line of action of the applied load and three reactions parallel to the resultant at the supports; and (b) the extruded force polyhedron including the applied load and the resultant face.

Remark 5.8. The cycling direction of the right hand in the form diagram gives the connectivity of the vertices of the reciprocal face of Figure 5.26d. We can refer to this connectivity to the topology of the vertices and edges of the reciprocal force face. However, this cycling direction might not always match the actual direction of the edge vectors of the reciprocal face, as shown in the truss example with the overhang. A similar case was discussed for a single node with compression and tension in Chapter 1. In constructing the reciprocal face, the direction of the face normal *must always match* the direction of the applied load.

Therefore, if the initial direction of the normal of the reciprocal face is *opposite* to the direction of the applied load, then a *negative edge length*, in the beginning, should be used to correct the geometry of the face. Chapter 7 will expand on the algebraic construction of the reciprocal diagrams and the definition of the positive and negative edge lengths in both form and force diagrams.

The Trial Funicular Construction

Similar to the previous examples in this chapter in both 2D and 3D, the global force diagram should be completed to find the magnitude and direction of the reaction forces of the truss. The force polyhedron gets closed by adding the resultant face, as shown in Figure 5.28b. Then, it is extruded to an arbitrary point to make tetrahedral cells for the construction of the trial funicular. Figure 5.28a shows the trial construction and its closing plane. A line normal to the closing plane drawn from the extrusion point O' intersects the plane of the resultant and locates point x in the force diagram that will be used to find the reaction forces in the next step.

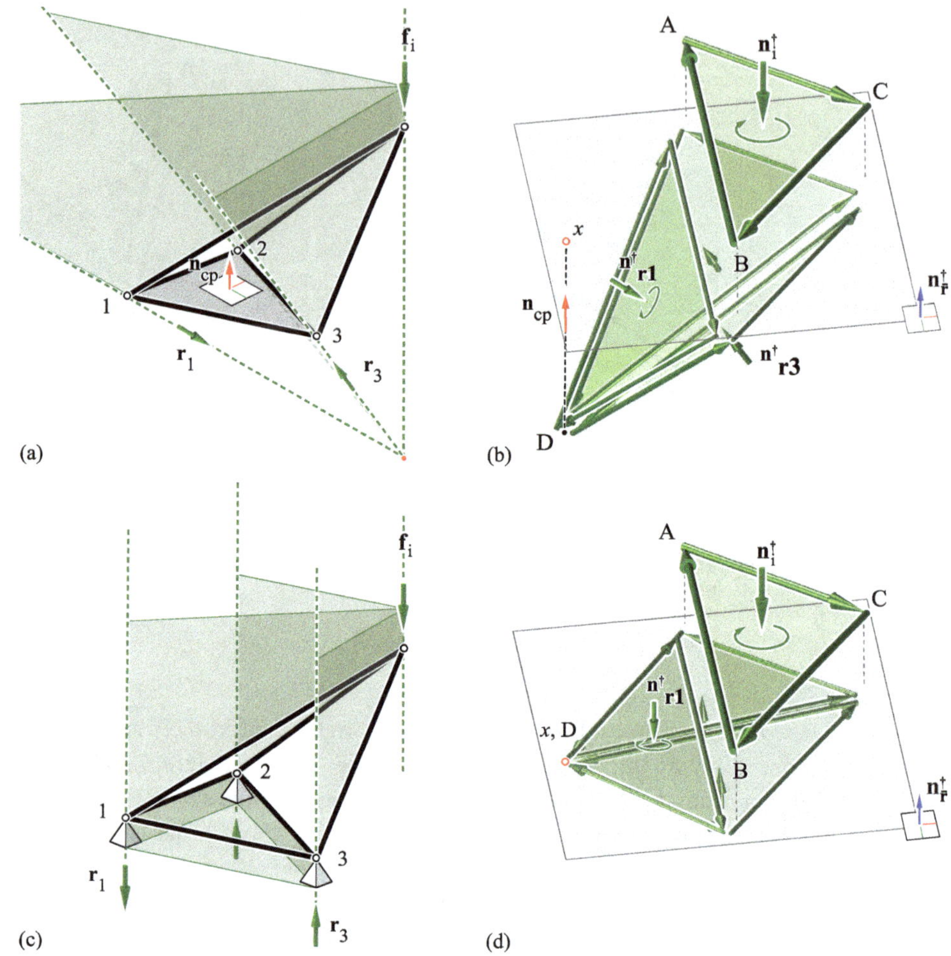

Figure 5.29 (a) and (b) A condition for the reaction forces that intersect the applied load and its reciprocal global force polyhedron; and (c) and (d) reaction forces parallel with the applied load and its flat global force polyhedron.

Finding the Reaction Forces

As explained on multiple occasions in this chapter and before, the global force diagram of a constrained form is also constrained to a line normal to the closing plane drawn from point x. Let us move to Figure 5.29. The closing plane of the truss connects supports 1, 2, and 3 with a normal $\mathbf{n}_{cp}$. We draw the line l_{cp} from x and extrude the edge vectors of the face ACB at a point on the line l_{cp}. This extrusion decomposes the applied load into three faces representing the magnitude and direction of the reaction forces at the three supports.

Figure 5.29a,b shows two configurations of the reaction forces. In Figure 5.29a, the reaction forces intersect with the line of action of the applied load in the form diagram. In Figure 5.29b, the reaction forces are parallel with the applied load, and the global force polyhedron is flat.

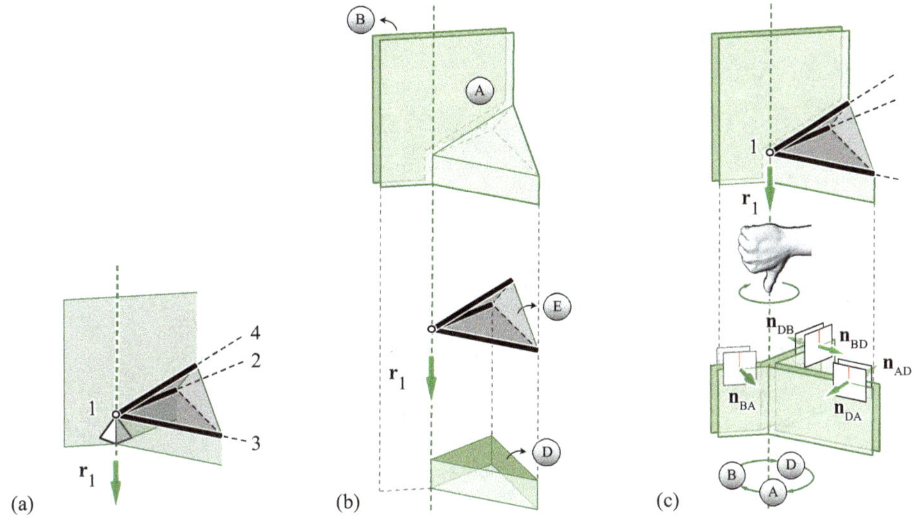

Figure 5.30 (a) Node 1, its connected edges, including the reaction $\mathbf{r}_1$, faces, and subspaces; (b) subspaces around vertex 1; and (c) the curl of the fingers and the normals of the half-faces of subspaces A, B, and D.

The Nodal Equilibrium

Once the global force polyhedron is constructed and the magnitude and direction of the reaction forces are found, the nodal force polyhedron can be completed step by step. Let us begin with node 1, which is one of the supports for the structure. Figure 5.30 shows the subspaces around node 1. The subspaces A, B, and D are open cells. Although the subspaces A and B are neither convex nor concave, the direction of the normal stays consistent on their faces.

The normal direction always points from the adjacent space toward the inside of the space. For instance, $\mathbf{n}_{BA}$ shows the normal on subspace A pointing from subspace B toward A. The following of Example 5.11 and steps are necessary to construct the form's nodal force cell for vertex 1.

Example 5.11 Construct nodal force cells for the truss of Figure 5.29.

Construction steps:

1 In the form diagram of Figure 5.30, use the right-hand rule to round the edges connected to vertex 1.

2 Start with reaction force $\mathbf{r}_1$. The global force polyhedron gives the corresponding face ABD; we can double-check the process by putting the thumb of the right hand on the reaction force and rounding the subspaces A, B, and D. Figure 5.31a–d shows a connected graph from A to B to D back to A using $\mathbf{n}_{AB}$, $\mathbf{n}_{BD}$, and $\mathbf{n}_{DA}$. To match the edge vector AB with its adjacent edge vector BA of the face CBA, we multiply all normals with negative edge length without changing the direction of the face ABD, as shown in Figure 5.31d.

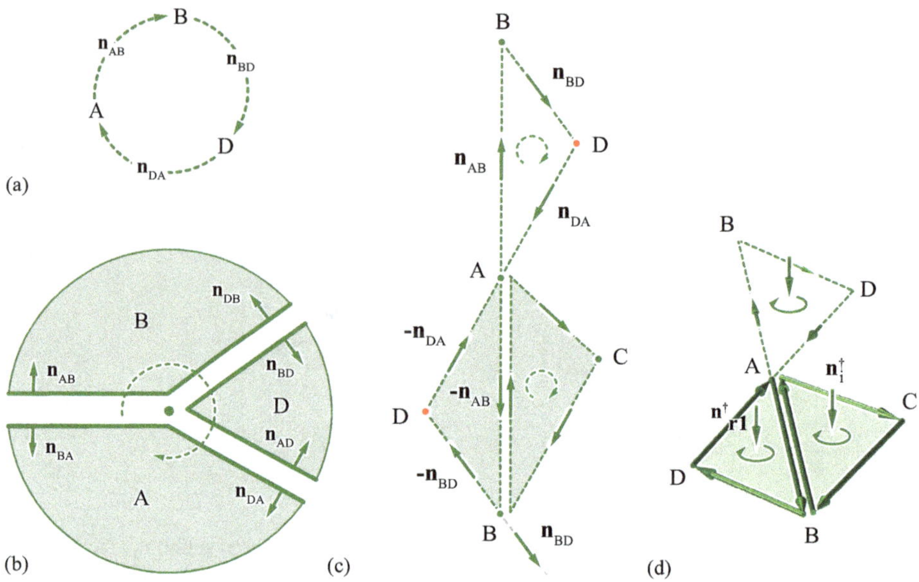

Figure 5.31 (a) and (b) Cycling around the reaction force $\mathbf{r}_1$ and the subspaces *A*, *B*, and *D*; (c) the construction of the face *ADB* using the edge vector *AB* given by the load face *BAC*; and (d) the axonometric view of the completed face.

3 Move to Figure 5.32a,b showing all the subspaces around the edge $e_{1,3}$. The edge vectors of the face *ABD* reveal the direction and magnitude of the edge vectors of all its adjacent faces.
4 Edge vector *AD* is equal to the *DA* of face *ABD* and belongs to an adjacent face *ADE* that is yet to be constructed.
5 Look at the edges connected to vertex 1 of the form diagram and their surrounding subspaces. In the force diagram, we started from the edge *AD* reciprocal to the subspaces *A* and *D* in the form diagram. Edge $e_{1,3}$ in the form diagram is shared by subspaces *A*, *D*, and *E*, and its rounding should complete the face *ADE* of the force cell (see Figure 5.32a,b).
6 In the force diagram, from vertex *D*, draw a line with the direction of $\mathbf{n}_{DE}$. Draw another line in the direction of $\mathbf{n}_{EA}$ from *A*. The intersection of these two lines locates vertex *E*.
7 Once vertex *E* is located, the face *ADE* can be constructed, which gives two other faces *BAE* and *BED* (Figure 5.32c,d).
8 Once the nodal force cell is completed, the face normals should be matched with the truss members to find the type of internal forces in the members.
9 In the force diagram, put the right-hand curl around a face. Then, move your right hand to the form diagram and align the thumb with the corresponding edge. If the thumb points toward the vertex, the force is compressive ($\mathbf{f}_{1,3}$, and $\mathbf{f}_{1,2}$), otherwise it is tensile ($\mathbf{f}_{1,4}$). Figure 5.33 illustrates this step.

□

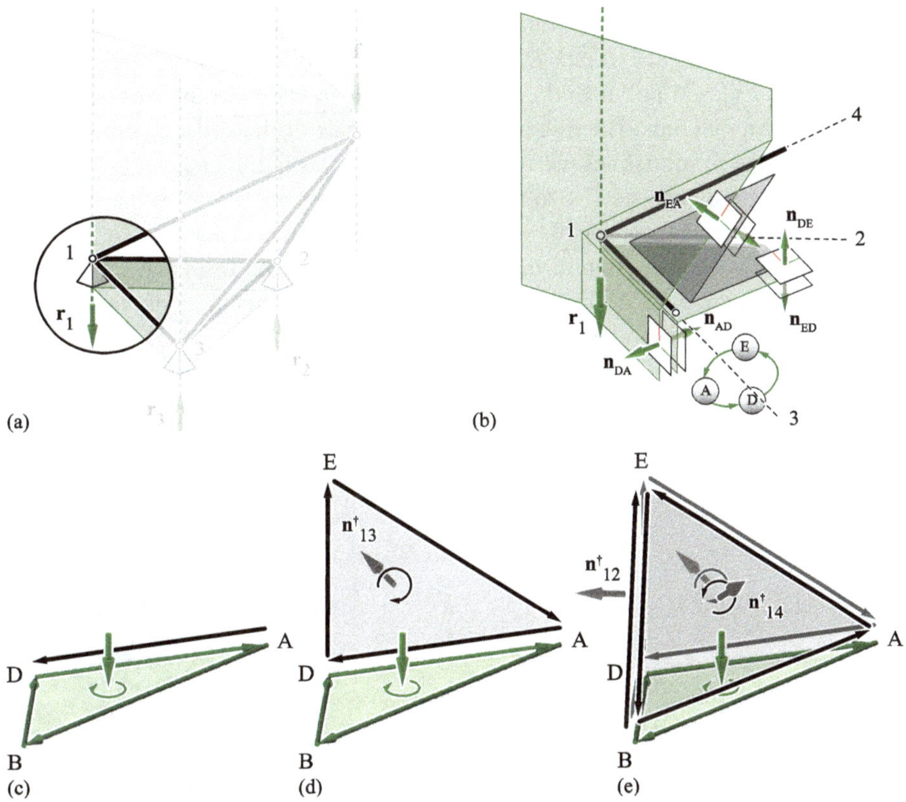

Figure 5.32 Constructing the nodal force polyhedron for vertex 1; (a) completing the reaction force face and finding the adjacent edge vector AD; (b) rounding edge $e_{1,3}$ and the subspaces A, D, and E, (c) drawing the edge pair AD in the force diagram; (d) completing the face ADE by intersecting the normals from the faces shared by the subspaces in the form diagram; and (e) the completed force cell.

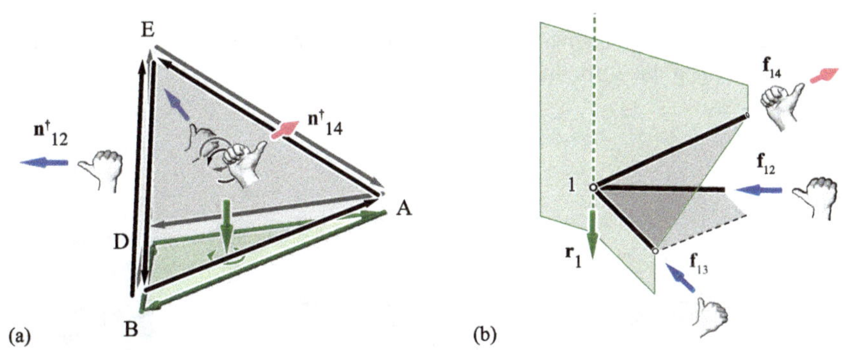

Figure 5.33 Matching the direction of the faces (a) with the geometry of the members (b) to find the direction of the internal forces; for instance, normal $\mathbf{n}_{1,3}$ pushes toward node 1, and thus edge $e_{1,3}$ carries compression force.

For the truss of Figure 5.34, the rest of the nodal cells can be easily constructed by locating the vertex E in the force diagram. The entire force diagram consists of four closed polyhedral cells representing the equilibrium of the four vertices of the form diagram. The direction of the normals of the cells, together with the geometry of the form, defines the type of internal forces in the members of the truss. This force diagram consists of two positive cells with normals toward the inside of the cell and two negative force cells with normals toward the outside.

The construction of the Minkowski sum for the form and force diagram is shown in Figure 5.35. This construction confirms the correctness of the force diagram and its reciprocal relations.

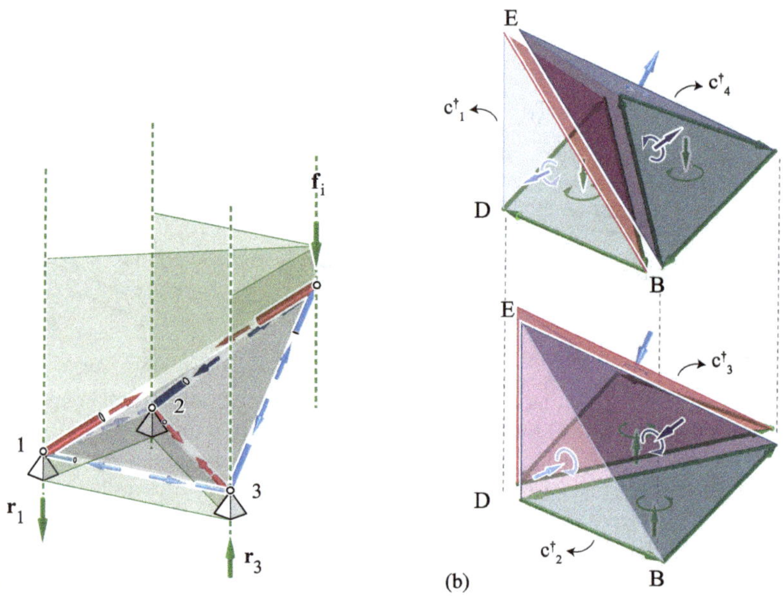

Figure 5.34 (a) The internal forces in the members of the truss and (b) the exploded axon of the nodal cells of the force diagram.

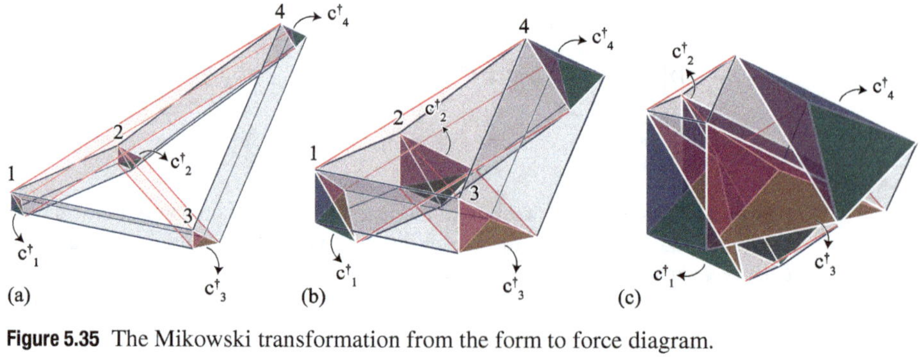

Figure 5.35 The Mikowski transformation from the form to force diagram.

5.13 Nonconvex Cells in the Force Diagram

In almost all examples of this chapter, the nodal force polyhedrons were convex. We briefly discussed the nonconvex force cells in Chapter 1. The force diagram of systems with mixed tension and compressive forces can include nonconvex and even self-intersecting cells. Sections 5.13.1 include 2D and 3D examples of nonconvex cells to explain the properties of the equilibrium geometries.

5.13.1 A Self-Intersecting Nodal Force Polygon

Figure 5.36 shows a 2D truss and its force diagram, which was constructed following the procedures explained in this chapter. The force diagram consists of a global force polygon shown in green color and four nodal force polygons. Note that the number of nodal force polygons equals the number of vertices of the form diagram. The polygon *ABDE* in the force diagram shows the equilibrium of node 4 in the form diagram. This polygon is self-intersecting with a consistent direction of normals around its edges. The self-intersecting polygon forms two regions: one with a positive normal direction toward the inside of the cell and another with a negative normal direction toward the outside. The direction of the internal force in the truss member is found by matching the right hand with the normal of the edge in the force diagram and the alignment of the edge connected to the node in the form diagram.

Remark 5.9. Each edge of the global force polygon makes a closed nodal force polygon in the force diagram. However, the number of nodal force polygons might be more than the number of edges of the global force polygon. As shown in Figure 5.37a, the force diagram consists of three edges, each of which constructs a nodal force polygon. There is an additional force polygon *CDE* that does not have any global edge and yet represents the equilibrium of an internal node in the truss.

Remark 5.10. The direction of the normals of the global force polygon is consistent and matches the direction of the applied loads in the form diagram. Moreover, the direction of the nodal force polygons is defined by the direction of the edges of the global. For those nodal force polygons with an edge belonging to the global cell, the direction of the polygon can be easily found using the right-hand rule. These force polygons define the direction of the rest of the cells that do not have a global edge by their half-faces.

Figure 5.37 shows an exploded illustration of all cells of the force diagram. For instance, the direction of the polygon *CED* can be found by identifying the direction of its edge *CE*. The direction of *CE* is opposite to that of *EC* in the adjacent polygon (Figure 5.37a,b). An interesting observation can be made regarding the force polygon *CED*; this polygon has a positive direction with normals facing toward the inside of the cell, but its reciprocal vertex in the forms diagram is a tension-only node. Can you justify the reasoning for this anomaly? Note that the geometry of the self-intersecting polygon *ABDE* has flipped the force polygon *DCE* in the force diagram.

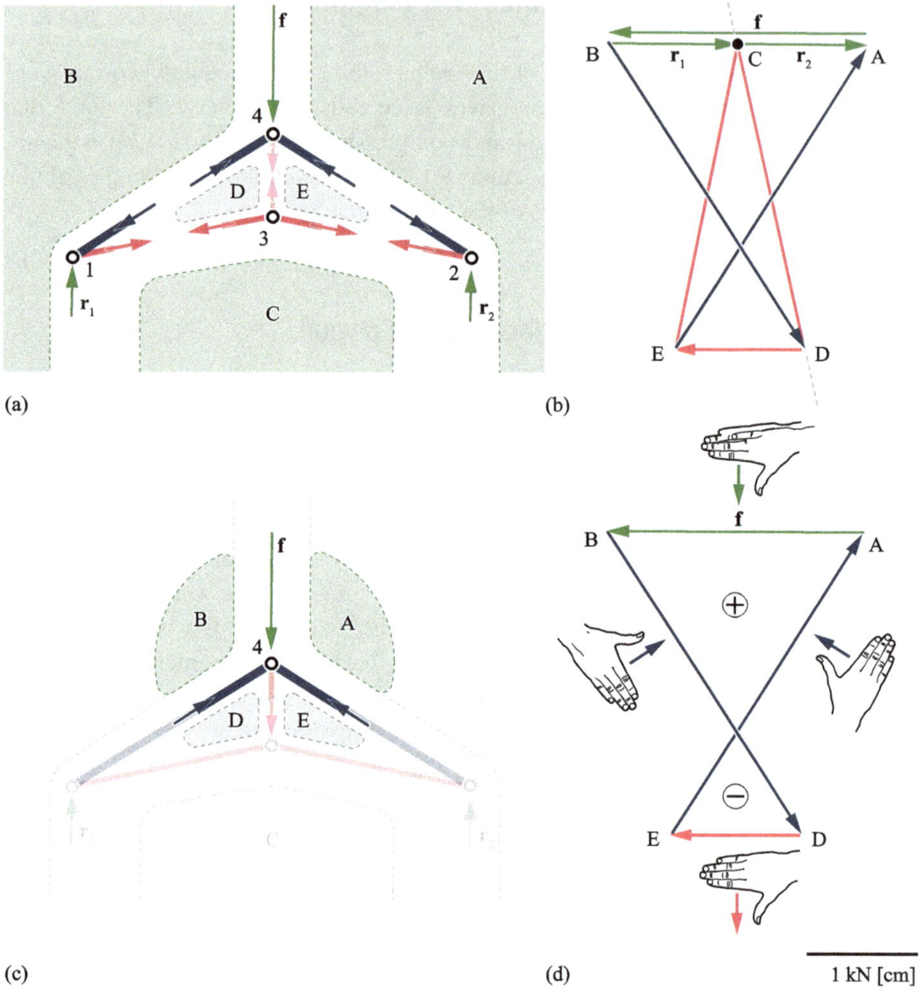

Figure 5.36 A 2D truss and nonconvex nodal polygons in its force diagram: (a) the geometry of the form; (b) the completed force diagram; (c) a single node with three members and an applied load; and (d) the corresponding closed polygon with self-intersecting geometry in the force diagram.

Maxwell Truss with Self-Intersecting Force Polygons

We briefly discussed Maxwell's significant contribution to the field of graphic statics in Introduction. Maxwell, in his groundbreaking paper called "*On Reciprocal Figures, Frames, and Diagrams of Forces*,"[37] for the first time formulated the relation between the reciprocal diagrams of graphic statics in both 2D and 3D. He concludes the article with an exciting truss design and its reciprocal force diagram. We have illustrated this truss in Figure 6.7 and will slightly modify its geometry in the next steps to highlight the effect of geometry on the magnitude, internal force flow, and efficiency of the structure.

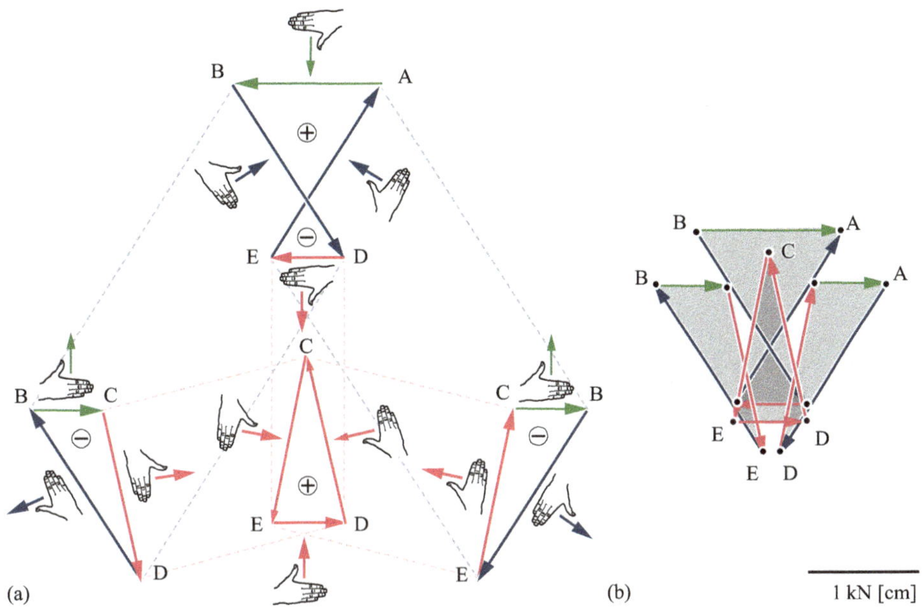

Figure 5.37 (a) The nodal force polygons of the truss of Figure 5.36 pulled a part from each other with the direction of the polygons and their normals; and (b) the overlapping force polygons with their equal and opposite shared edges.

Maxwell's truss consists of a flat chord on the top connected to an arch in the bottom chord of the truss by triangulated members. In Figure 6.7a, we put the applied loads on the top chord of the truss and chose the direction of the reaction forces to *be parallel with* the last member of the arch connected to the supports. We constructed the force diagram, and surprisingly, the internal forces of the truss represent a compression-dominant structure, as all nodal cells of the force diagram are convex and are placed inside the global force polygon. The Minkowski operation combines the form and force diagrams visualizing the proportional force magnitude in the members of Figure 6.7a.

In Figure 6.7b, an additional set of loads is added to the bottom arch of the truss, and the reciprocal force diagram is constructed accordingly. Still, the structure is compression-only, as all the force polygons are inside the global force polygon, and only the magnitude of the internal forces has been increased in the arch members.

We have discussed earlier in this chapter that the change in the angles of the reaction forces will change the type and magnitude of internal forces in the members. In the example of the Maxwell truss, if the reaction force is not aligned with the last member of the arch, the internal force flow changes (Figure 5.38). This misalignment is induced by changing the arch height from h_0 to h_1 in Figure 5.38a,b. The new force diagram is constructed based on the altered truss geometry and shows that it has combined tensile and compressive members. Figure 5.39e,f show the self-intersecting force polygons *BADE*, *CBEFG*, and *DAHI* corresponding to the mixed forces in the reciprocal node. These self-intersecting polygons have a flipped edge representing a tensile force in the reciprocal member. For instance, the edge *DE* in *BADE* is flipped, showing tensile force in the truss member that separates subspaces *D* and *E* of Figure 5.39d.

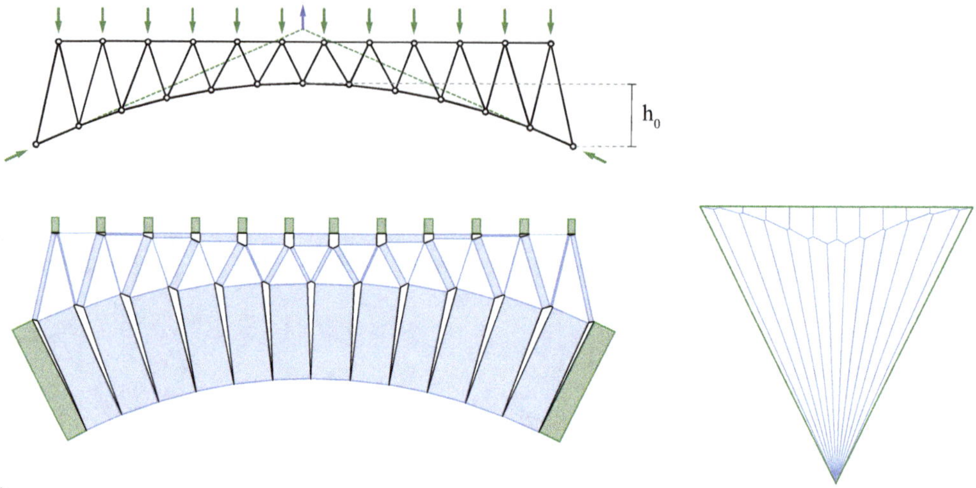

(a)

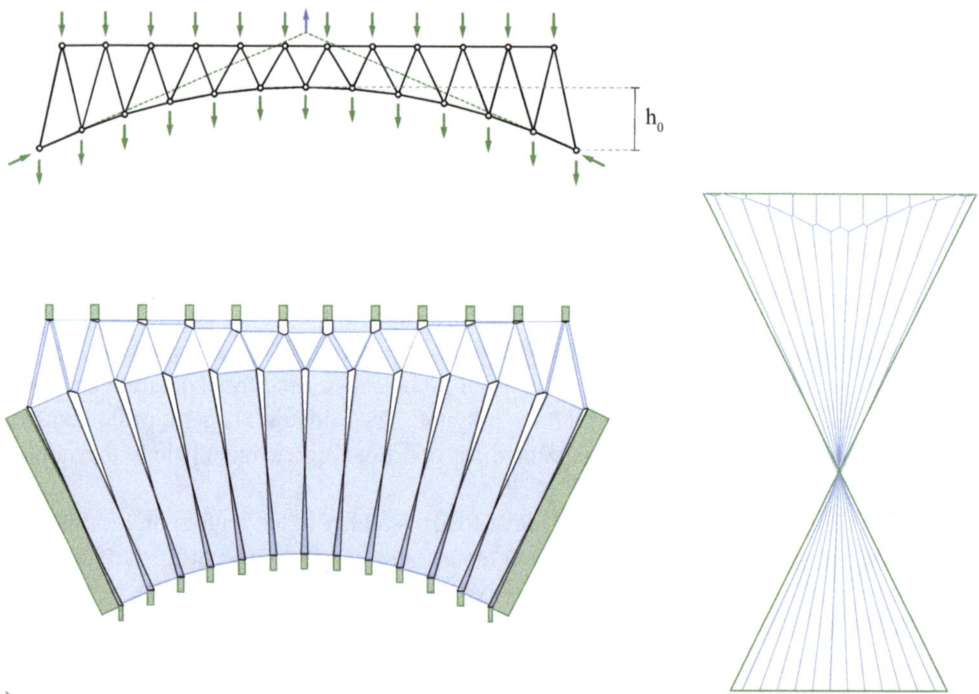

(b)

Figure 5.38 The form of a truss and its reciprocal force diagram (a), with supports aligned with the compression members; there are no tensile forces. All members of the truss experience compressive force if the global force polygon encompasses the nodal force polygons, even if a series of loads are added to the bottom chord (b). The force polygons self-intersect to accommodate the pulling forces of the additional applied loads.

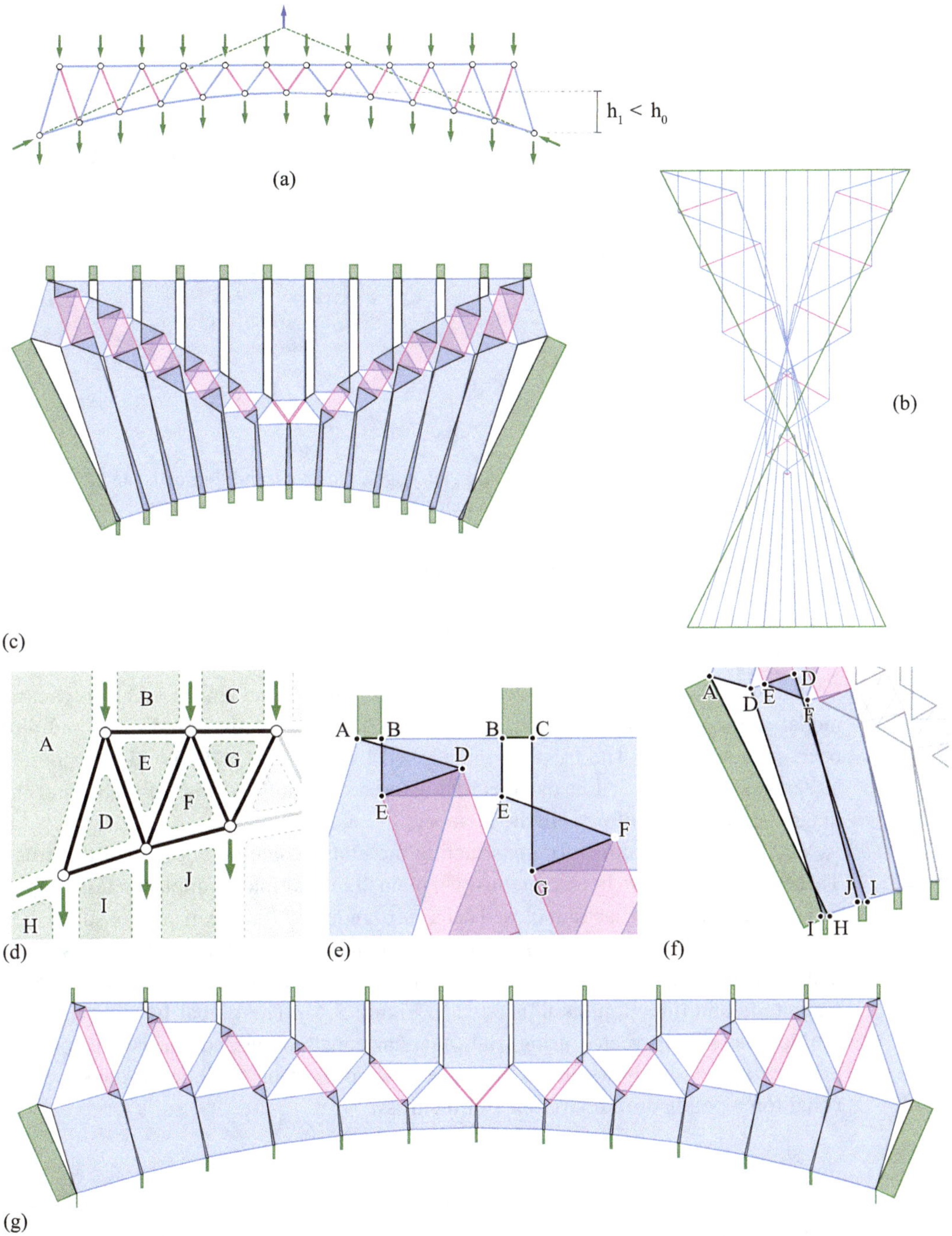

Figure 5.39 (a–c) The change in the height of the arch makes more members flip into tension; (d–g) the top and bottom force polygons are self-intersecting with a flipped edge that results in tension force in the reciprocal member.

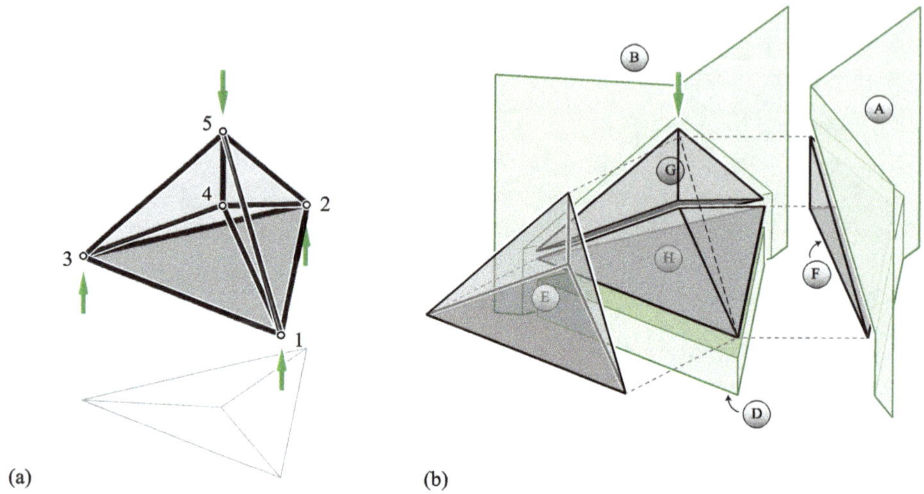

Figure 5.40 (a) A 3D truss with five vertices, ten edges, ten faces, and four cells; and (b) the applied load and reaction forces will add four open subspaces around the truss.

A Self-Intersecting Nodal Force Polyhedron

Figure 5.39 shows a truss geometry in 2D and its reciprocal force diagram involving self-intersecting force polygons. What would the self-intersecting force cell look like in 3D?

Figure 5.40a shows a 3D truss consisting of five nodes and ten bars. The polyhedral model of the truss has five vertices, ten edges, ten faces, four closed cells, and four open cells around it. The truss has an applied load $\mathbf{f}$ on its node 5 and is supported at vertices 1, 2, and 3. Let us construct a force diagram for this truss and find the magnitude of internal forces in the members.

We skip the procedures for constructing the global force polyhedron of this truss. These procedures have been previously explained in multiple examples in this chapter and before. This truss has four closed cells surrounded by four open cells or subspaces, as shown in Figure 5.40b. The applied load and the reaction forces are the edges of the open cells in the form diagram. The global force polyhedron gives the reaction forces' magnitude and direction, as illustrated in Figure 5.41. The global force polyhedron of this truss is constructed using trial funicular construction and consists of four co-planar triangles, as shown in the figure. Following Example 5.12, let us construct the nodal force polyhedron for vertex 1 of the truss.

Example 5.12 Construct the nodal force polyhedron for node 1 of the truss of Figure 5.41a.

Construction steps:

1 In the truss of Figure 5.41a at the location of vertex 1, locate reaction force $\mathbf{r}_1$ and its reciprocal face in the force diagram *ADC*, given by the global force

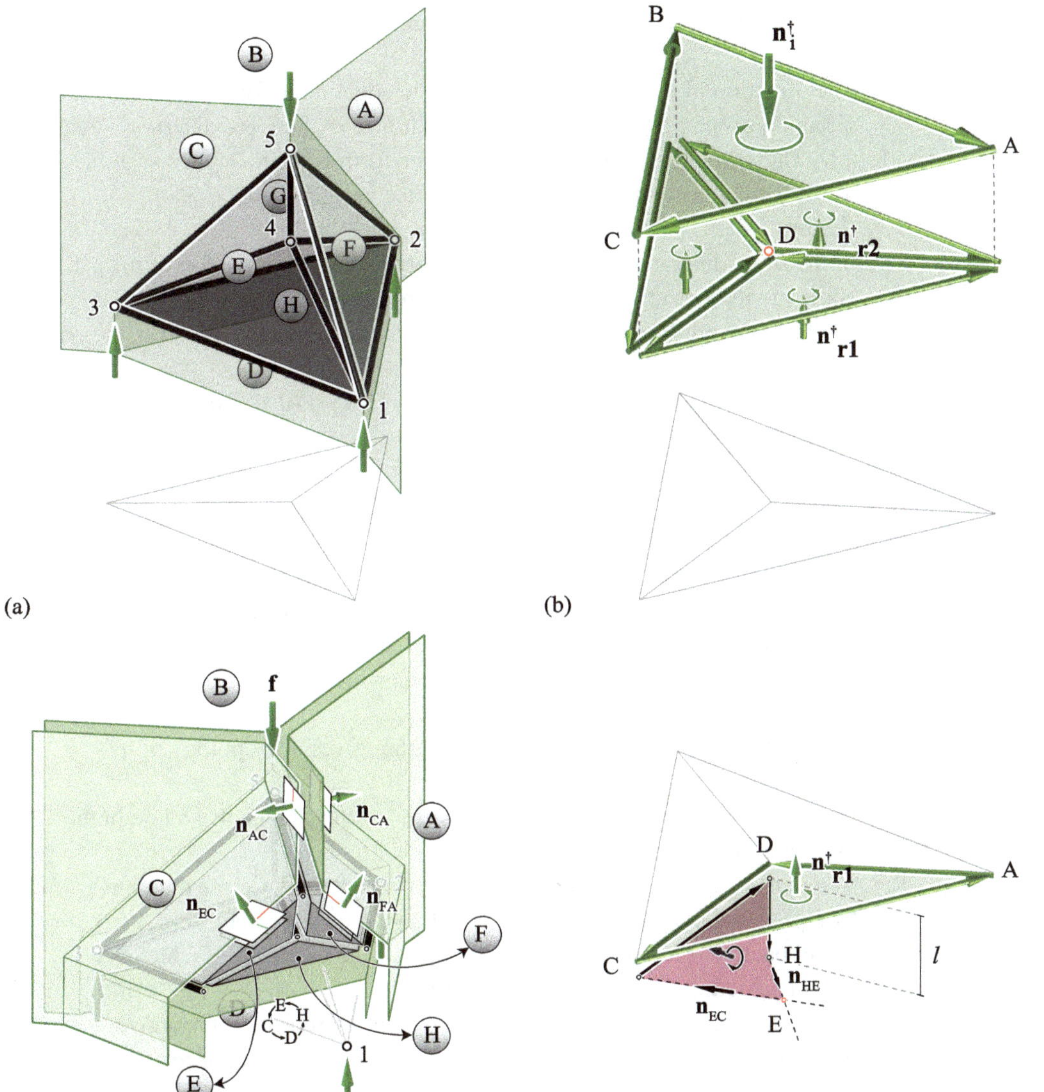

Figure 5.41 (a) The form of a truss with five vertices, nine edges, and four internal and four external cells; (b) the global force polyhedron; (c) cutting the spaces and edges around vertex 1; and (d) the construction of a nonconvex face *CDHE* reciprocal to the edge $e_{(1,2)}$ of the truss.

polyhedron (see Figure 5.41a,b). The goal is to complete the rest of the faces of the nodal force cell reciprocal to vertex 1.

2 Let us complete the face corresponding to the edge $e_{(1,3)}$. Cut the cells and subspaces around vertex 1 in the form diagram.

3 In the force diagram, find the edge *CD* that is opposite to the *DC* of *ADC*. This edge belongs to the forthcoming face adjacent to face *ADC*.

4 In the form diagram, identify cells around the edge $e_{(1,3)}$, which are C, E, H, and D.
5 The direction of cycling spaces around the edge in the form is given by the direction of the first edge of the reciprocal face in the force diagram, which is CD. Therefore, the cycling direction in the form is $C \rightarrow D \rightarrow H \rightarrow E$. Thus, the connectivity of the reciprocal face is found as *CDHE* (Figure 5.41c).
6 In the force diagram, draw a line from vertex D in the direction of $\mathbf{n}_{HD}$. To complete the face, the length of the DH should be chosen.
7 Choose a length l for the edge DH and locate vertex H at the end of the edge vector (see Figure 5.41d).
8 From vertex H, draw a line in the direction of $\mathbf{n}_{HE}$. Also, from vertex C, draw a line in the direction of $\mathbf{n}_{EC}$.
9 Label the intersection of the two lines as vertex E.
10 Complete the face *CDHE* by drawing the edge vectors DH, HE, and EC.
11 As shown in Figure 5.42, choosing different edge lengths for DH results in different face geometries.
12 Let us move to Figure 5.43. Continue constructing the rest of the faces of the nodal force by taking another member of vertex 1. For instance, let us look for a member surrounded by subspaces A and C, which is $e_{(1,5)}$, and it also has cells E and F around it.
13 Rounding the edge $e_{(1,5)}$ in the form diagram gives the topology of its reciprocal face *ACEF* in the force.
14 Draw a line in the direction of the $\mathbf{n}_{FE}$ from vertex E and also a line in the direction of $\mathbf{n}_{EC}$ from vertex C.
15 The intersection of these two lines is the vertex F in the force diagram.
16 Complete the geometry of the face by drawing edge vectors from A to C to E to F and back to A to complete the geometry of the face *ACEF*, which is a self-intersecting face (see Figure 5.43b).
17 The rest of the faces can be completed by simply repeating the above procedures or connecting the found vertices to complete their related faces.

□

You noted that you could choose the edge length DH independently of face *CDHE* in Example 5.12, which may result in *convex* or *nonconvex* face geometries. As shown in Figure 5.42, there are many possibilities for the geometry of the face *CDHE*, which varies as the length of the chosen edge length DH varies.

Also, the face *ACEF* in this polyhedron is a *self-intersecting* face. The direction of the edge vectors around a self-intersecting face is consistent (from tip to tail), similar to convex or nonconvex faces. A closer look at the face *ACEF* shows that the face consists of two regions with opposite normal directions.

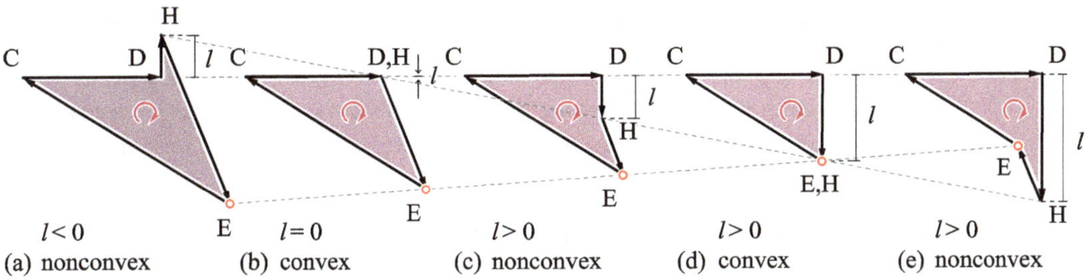

Figure 5.42 Mulitple nonconvex and convex geometry for the face *CDHE* as a result of choosing various edge lengths for *DH*: (a) negative edge length and nonconvex face; (b) zero edge length and a convex face; and positive edge length and a nonconvex (c), a convex (d), and a nonconvex face (e).

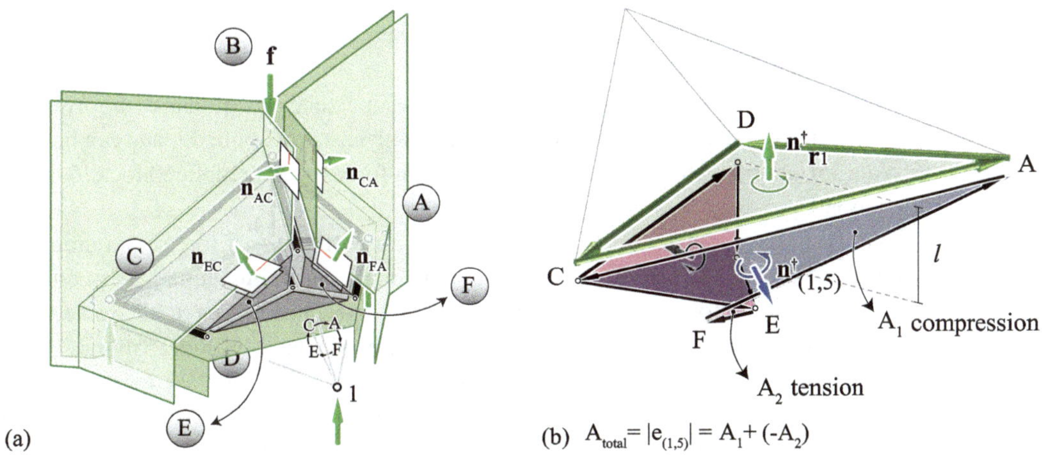

Figure 5.43 (a) Rounding edge $e_{(1,5)}$ in the form diagram and cycling its surrounding subspaces to find the connectivity of the vertices of the force; and (b) completing the geometry of the self-intersecting face *ACEF*.

Remark 5.11. The magnitude of the force in the self-intersecting faces is the sum of the *signed areas* of the regions formed by the edge vectors of the face. The signed area is the area times the direction of the normal of the region. There are two types of regions: the ones with normals toward the outside of the cell and those with normals toward the inside. We sum all the areas of these positive and negative regions to find the face's total area. The greater area and its sign will define the final direction of normal in the face.

For instance, in Figure 5.43b, the total area of the self-intersecting face is $A_1 - A_2$. Also, since A_1 with a normal toward the outside of the cell is greater than A_2 with a normal toward the inside, the sum of the areas represents a value with a normal toward the outside. This concept was also shown with a simple example in Chapter 1.

Let us return to the various face geometries of *CDHE* by choosing different lengths for an edge. One might ask the meaning of the force diagram when a face can take various geometries with different face areas. This is the definition of the *indeterminacy* of a force system.

Definition 5.3. If multiple force diagrams with different face areas can be constructed for a given set of external loads, then the structure is *statically indeterminate* or has various states of self-stress.

Let us discuss this definition in further detail in Section 5.14.

5.14 The States of Self-stress in a 3D Truss Systems

The term *indeterminacy* in structural systems refers to the condition where there is more than a single static equilibrium condition for a structure with given applied loads and boundary conditions. A statically indeterminate structure can be *internally* and/or *externally* indeterminate.

The *external indeterminacy* refers to the various states of global equilibrium of forces. The system is globally indeterminate if a given set of applied loads can be held in equilibrium by various magnitudes of reaction forces with predefined locations and directions.

In 3D, to keep a given set of loads in equilibrium, the sum of forces and moments in x-, y-, and z- should be zero, giving six numerical equilibrium equations to find six unknown reaction forces. If the number of unknown reaction forces in 3D exceeds the number of equilibrium equations, then the system is externally indeterminate. That is, multiple force distributions at the supports can keep the applied loads in equilibrium. Thus, the system is externally indeterminate. We will discuss the external indeterminacy and its application in design using polyhedral graphic statics in depth in Chapter 6.

A system might be externally determinate but internally indeterminate. Consider the truss of the example of Figure 5.41. In this example, the global equilibrium of the truss could be constructed, but the nodal equilibrium required an input edge length l to complete the nodal polyhedral cell.

As shown and discussed in Example 5.12, the input edge length l can change the geometry of a single face that, in turn, changes the geometry of its surrounding faces in the force diagram and, thus, the distribution of the internal forces in the members of a truss without changing the global equilibrium of the system. Figure 5.44a,b shows two nodal force polyhedrons with various face areas due to a change in the length l of the edge DH in the force diagram. This property is called the *internal indeterminacy* of a system and is also referred to as the state of self-stress.[123]

Without any applied load, a structure might experience states of self-stress caused by an imperfection in the construction of its members. For instance, one of the members might be made longer/shorter than what it is expected to be. As a result, after the assembly, this imperfection exerts internal stress on the system members aside from the externally applied loads.

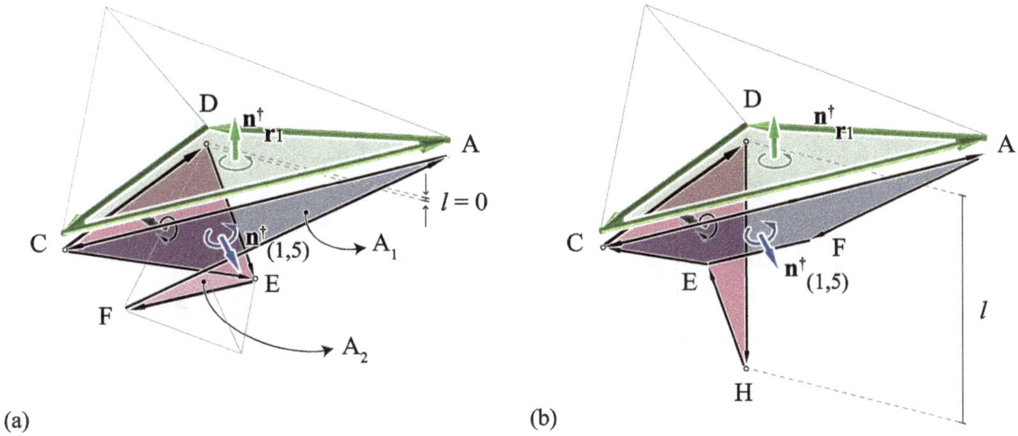

Figure 5.44 (a) The length l of DH is zero that results in a convex face $CDHE$ and self-intersecting face $ACEF$; and (b) the positive length for l for DH that results in nonconvex $ACEF$ and convex $ACEF$.

Remark 5.12. The internal degrees of indeterminacy of a truss can be found using the Maxwell-Calladine formula:[124]

$$3v - e - 6 = m - s \tag{5.2}$$

where v is the number of vertices including the support locations, e is the number of bars or edges of the truss, 6 is the number of equilibrium equations in 3D, m is the number of mechanisms (kinematic degrees of indeterminacy), and s is the states of self-stress or the internal degrees of static indeterminacy.

Since the structure consists of tetrahedral elements and all the faces of the form are triangles, there is no mechanism in the system, that is, $m = 0$, and the truss is kinematically determinate. Using the above formula for the truss of Figure 5.43 results in $m-s = -1$. Thus, the state of self-stress or internal static degrees of indeterminacy is 1.

This property can also be proven and shown using polyhedral graphic statics. The static degrees of indeterminacy in a structural form can be related to the geometric degrees of freedom (GDoF) of its reciprocal diagram, that is, the force diagram. For instance, in Figure 5.45, one additional length must be chosen to complete the force diagram. This additional edge length made it possible to close the geometry of a face, which was necessary to complete the force diagram. In other words, the force diagram has the GDoF of one (GDoF = 1).

Having various possibilities for the face $CDHE$ in the truss of Figure 5.41 means that the magnitude of an internal force within a truss member can be independently chosen, and the rest of the equilibrium can be found accordingly. Thus, the equilibrium state is *not unique*. This is the exact meaning of the static degrees of indeterminacy, and it was shown by various force diagrams for the truss system of Figure 5.45.

The relationship between the GDoF of the force diagram and the static degrees of indeterminacy of the form in polyhedral systems is rather complex. It will be visited again in Chapters 6 and 7. For now, the following statement can be quite useful.

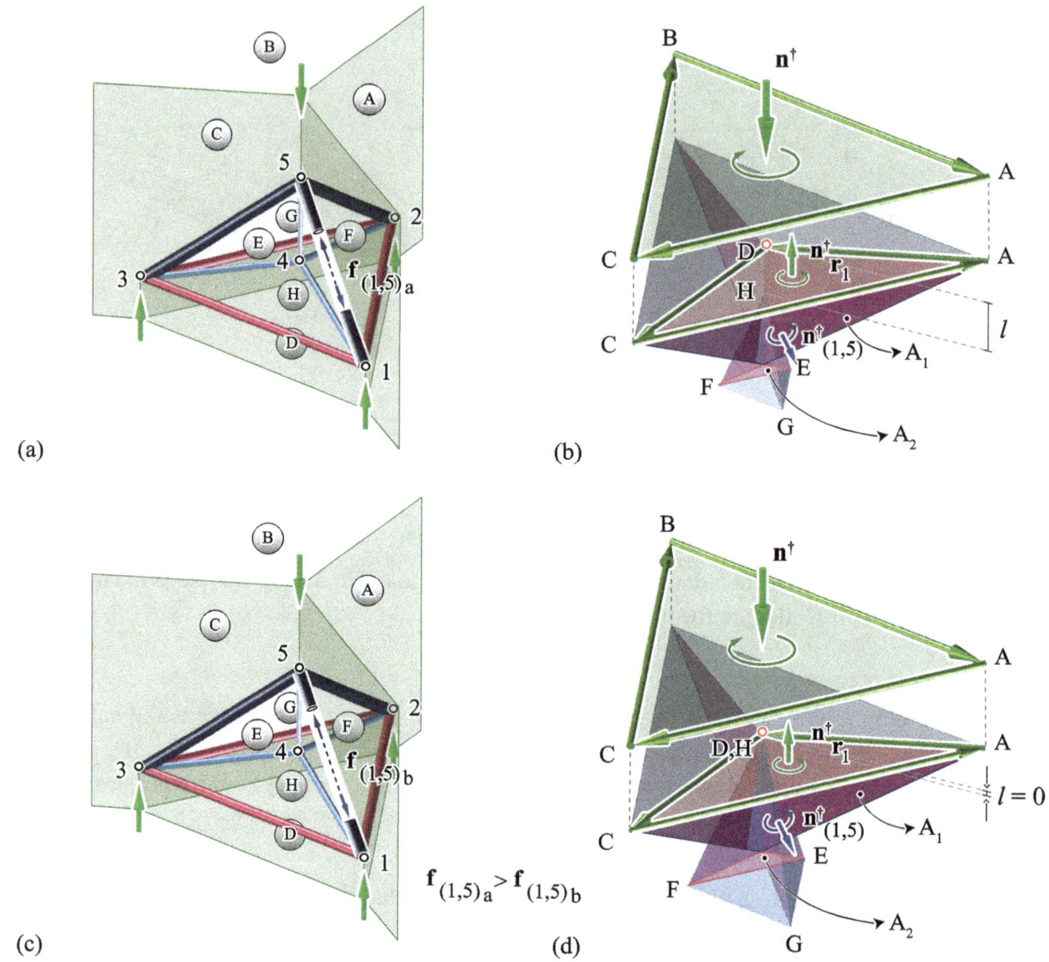

Figure 5.45 Two different force equilibrium for an indeterminate truss system with the same applied load: (a) and (b) a positive value is chosen for the edge length DH; and (c) and (d) zero edge length is chosen for the edge DH.

Remark 5.13. Polyhedral graphic statics represents a subset of all equilibrium possibilities in 3D space. Thus, the GDoF of the force diagram represent a minimum number of static degrees of indeterminacy in polyhedral systems.

5.15 Harnessing the Potentials of the Reciprocal Diagrams

Suppose the form and force diagrams are constructed in a parametric environment. In that case, any changes in the structure's geometry can immediately update the force as long as the topology of the structure does not change. This explicit relation between the form and force diagram can be harnessed for design and educational purposes. This section aims to design *funicular structures* with members carrying both tension

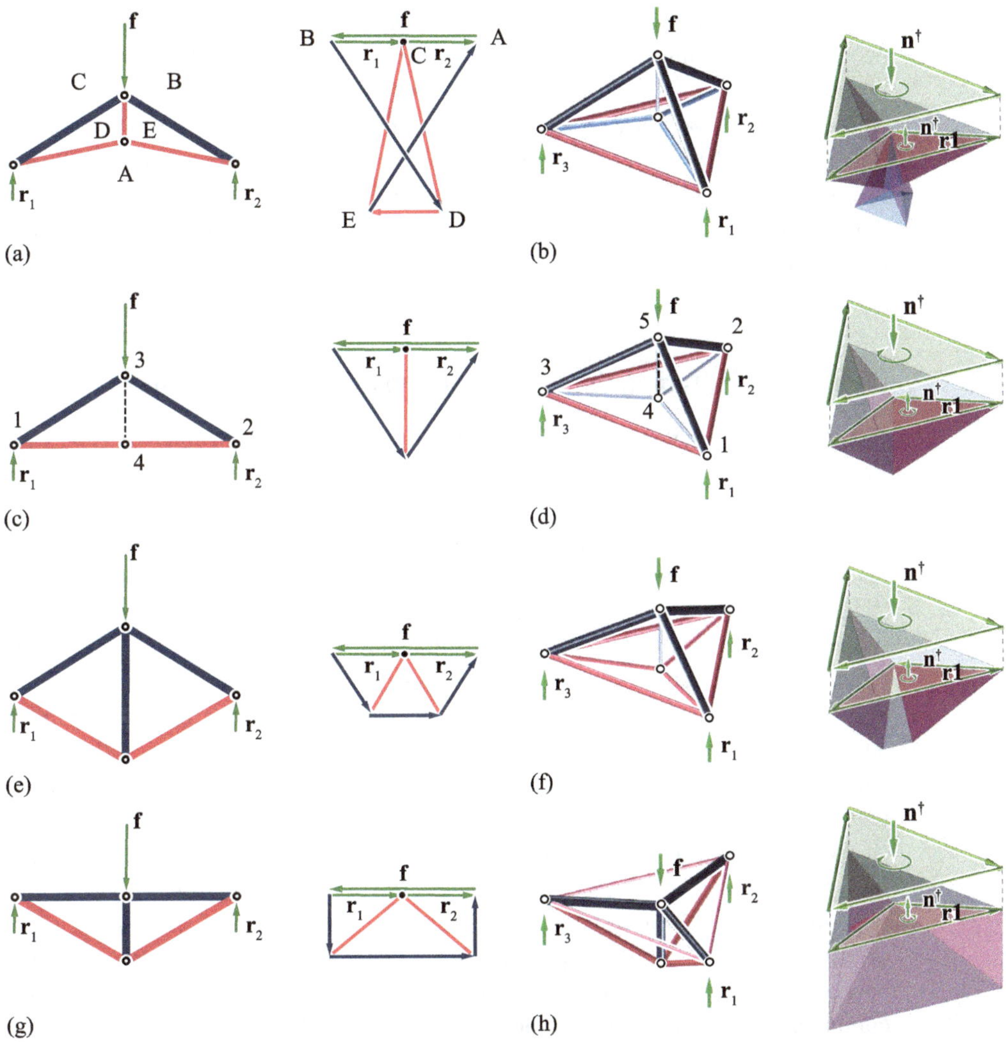

Figure 5.46 (a) A 2D truss with 5 edges, 5 faces, and 4 vertices and its corresponding force diagram; (b) a similar truss system in 3D with 5 vertices, 10 edges, 10 faces, and 4 cells; (c) a configuration where three vertices of the truss are co-linear in 2D and zero force members in the truss; (d) four vertices co-planar and zero force elements; (e) moving vertex 4 under 1 and 2 and its reciprocal force diagram; (f) moving vertex 4 below the plane of vertices 1, 2, and 3 and its force diagram; (g) making vertex 1,2, and 3 co-linear and (h) moving vertex 5 to the plane of 1,2, and 3 and its reciprocal force diagram.

and compression forces, building upon understanding the equilibrium of forces in the previous truss example.

Figure 5.46 shows the trusses of Figures 5.36 and 5.45 and their corresponding force diagrams side by side. In each row, the geometry of the truss is manipulated, and the force diagram is updated accordingly. Figure 5.46a,b shows the starting configuration of the truss as described in detail in Sections 5.13 and 5.14. In Figure 5.46c, node 4 in 2D becomes co-linear with nodes 1 and 2. As a result, the force in the member $e_{(3,4)}$ becomes zero. Similarly, in Figure 5.46d, node 4 becomes co-planar with vertices 1, 2, and 3. As a result, the force in the edges $e_{(1,4)}$, $e_{(2,4)}$, $e_{(3,4)}$, and $e_{(5,4)}$ will disappear since their corresponding faces in the force diagram either collapse to a point or to a line. Note that the initial force diagram used for this truss is the distribution illustrated in Figure 5.45b, where the length of the edge *DH* is zero. Figure 5.46e–h also show other configurations of the bars and nodes with the same location of the applied load and the supports.

5.15.1 Cutting and Extruding the Force Diagram as a Design Technique

Designing funicular structures with tension and compression internal forces is not trivial. It requires understanding the relationship between the tensile and compressive elements in the force system.

We use the *cut and extrude method* as an instance of a design technique in 2D and 3D to generate funicular systems with both tensile and compressive members. In this technique, we start from a simple truss, for instance, the one of Figure 5.47a, and separate the compression nodes and their nodal force polygon or polyhedron from the tension nodes and reactions, as shown in Figure 5.47b. Let us focus on the force polygon *ABD*, which is reciprocal to compression node 3. The force polygons *BCD* and *CAD* can be constructed by taking the edges *BD* and *DA* from *ABD*, reversing them, and extruding them to point *C*. This process will give the two force polygons reciprocal to nodes 1 and 2.

We use the same approach to generate structures with tension and compression members. The only additional step is cutting the initial compression-only node to produce more edges. This process is illustrated for polygon *ADB* and results in *ABEF* of Figure 5.47c, which is reciprocal to a new node of Figure 5.47d. By reversing the new edges and extruding back to point *C*, we get the cells reciprocal to the nodes with combined tension and compression forces (Figure 5.47e).

In the end, tension and compression nodal force polygons are combined into a single force diagram reciprocal to a structural form with tensile and compressive elements (Figure 5.47f,g). We refer to this method as *cutting and extruding* the force diagram. Note that this concept is based on an important premise in which the global equilibrium of the structure remains intact. The magnitude of the reaction forces in the force diagram and the location of the supports and applied loads in the initial structure do not change (Figure 5.47h).

This technique suggests how the force diagram can be designed to get systems with mixed types of forces and is not the only way that can be used to design structural systems with tension and compression elements by modifying and manipulating their force diagrams. The works by Lee, Fivet, and Mueller,[125] Lee, Meled, and Block[126] and Timo, Masoud, and Per[127] are suggested for further reading.

The *cut and extrude* method can also be used for a 3D truss of Figure 5.48a. The truss has five nodes and four supports. Figure 5.48b shows the exploded axon of the

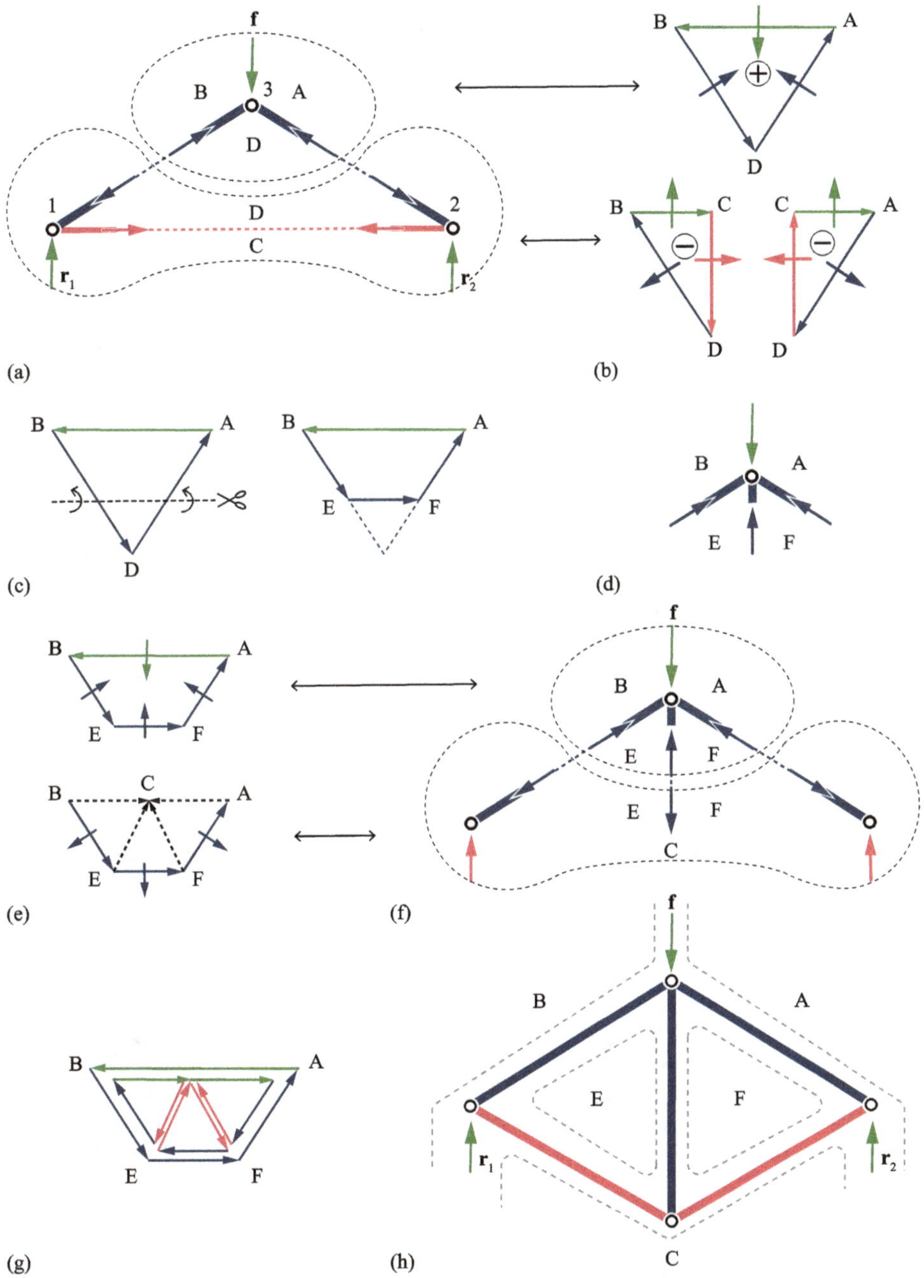

Figure 5.47 (a) A simple truss with three nodes and three bars; (b) the nodal force polygons; (c) cutting the positive force polygon *ABD* and making a new polygon *ABEF* with an additional edge; (d) the reciprocal node has an additional member; (e) reversing and extruding the internal edges of *ABEF* to point *C* to make a new polygon; (f) considering the edges of the new polygon as applied loads for the new nodes of the structure; (g) combining the new force polygons; and (h) the new structure with an additional edge and mixed tensile and compressive members.

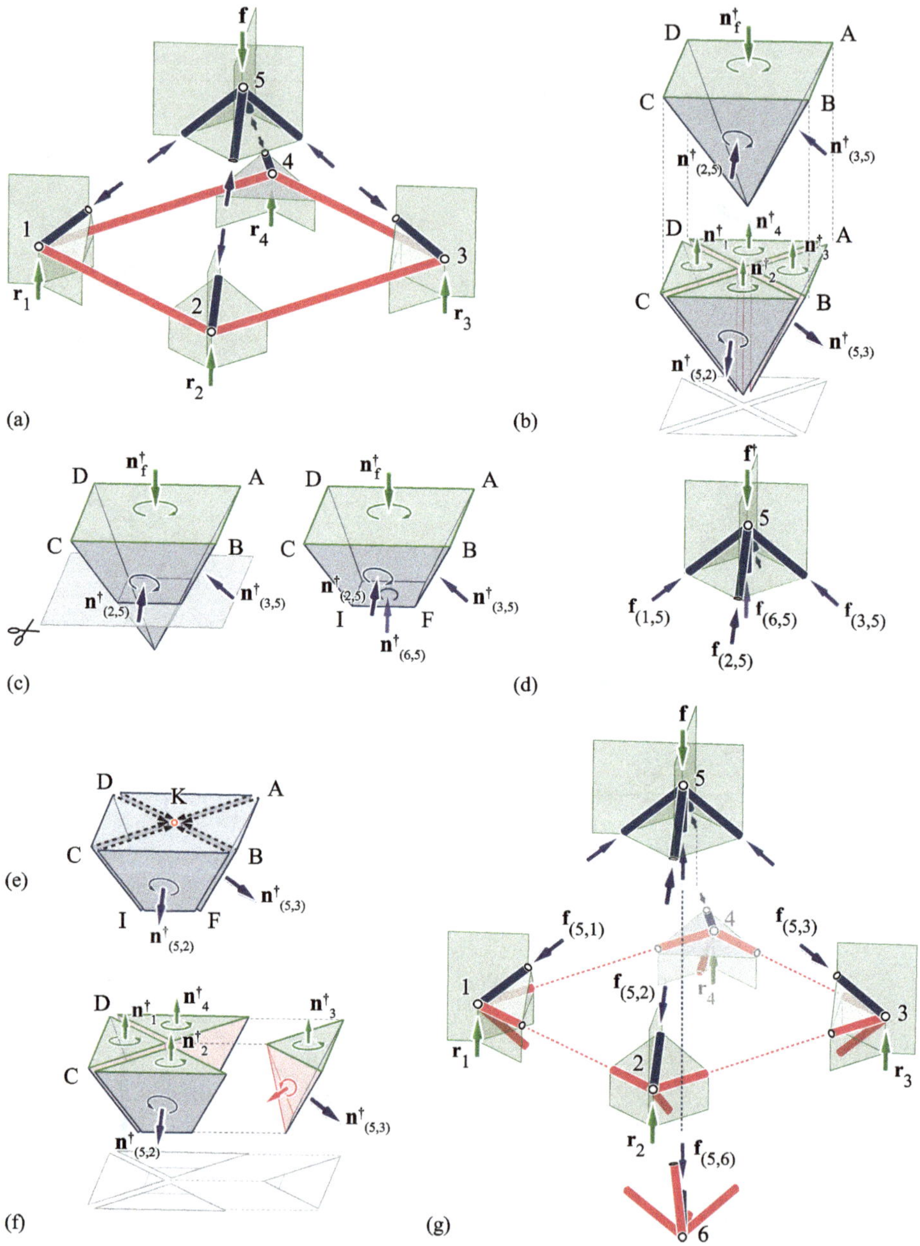

Figure 5.48 (a) A simple truss with five nodes, eight members, and four supports; (b) the exploded force diagram showing positive and negative nodal force cells; (c) cutting the positive cell with a plane and adding a new face to close the cell; (d) the reciprocal node; (e) and (f) extruding the internal faces of the cut cell to point K of the global force polyhedron to make new nodal force polyhedrons; and (g) the reciprocal nodes with mixed types of forces.

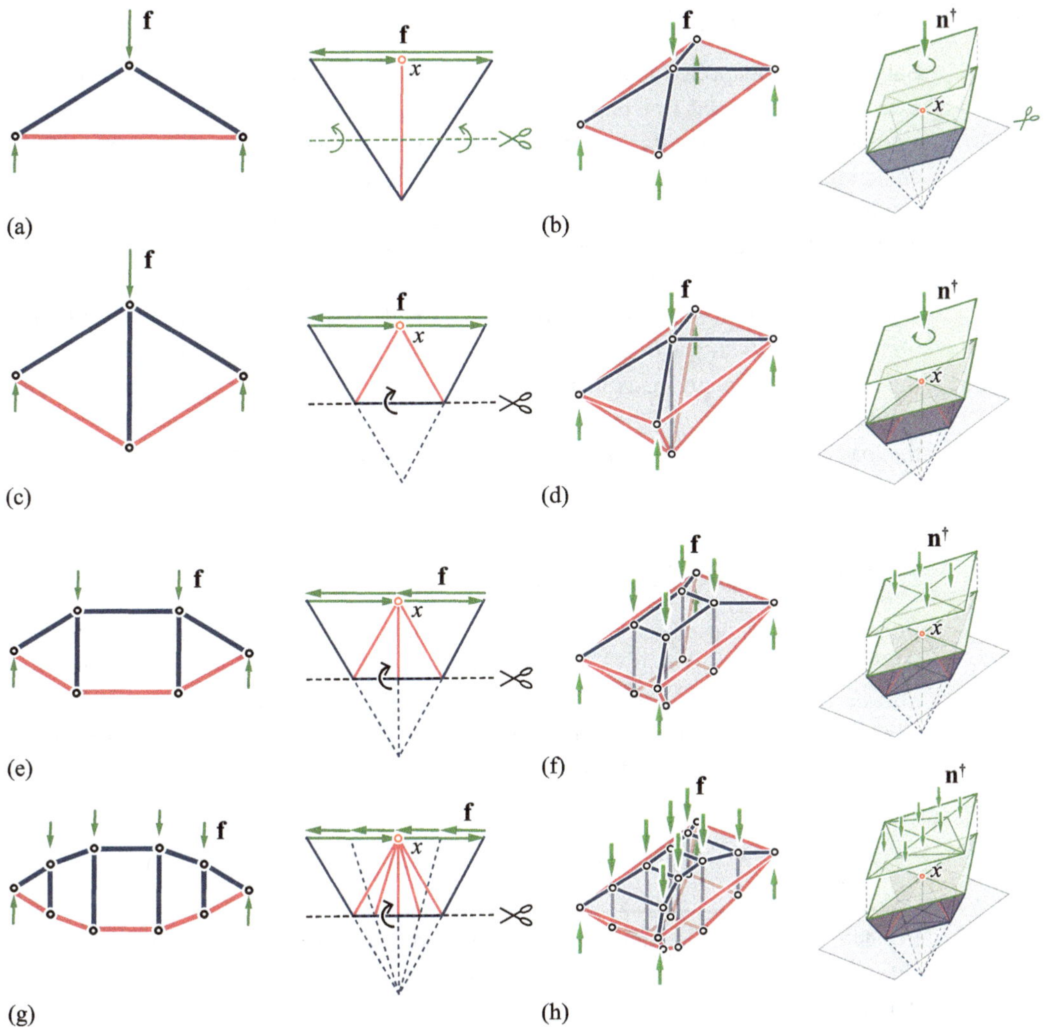

Figure 5.49 (a) A 2D truss configuration with a single applied load **f** and its force diagram; (b) a 3D truss with four supports and a single applied load and its reciprocal force diagram; (c) to (h) the applied load face is subdivided in each step, and the force diagram is manipulated by the *cut and extrude* method to drive the final structural form.

force polyhedron with positive and negative normal directions. Let us take the positive force cell connected to the applied load, cut it by a plane, and close the cut edges with a new face, as illustrated in Figure 5.48c. The new node reciprocal to this positive cell has a new edge. Now, take the faces of the cell, reverse their direction, and extrude them to point K of the global force polyhedron. This cutting and extruding results in new force cells shown in Figure 5.48e,f that are reciprocal to the nodes with mixed forces in the form diagram. This process gives us the truss of Figures 5.48g and 5.49d.

5.15.2 Subdividing the Applied Load Network

Subdividing the applied load network is another important design technique that was extensively discussed in Chapter 3. Subdividing the force diagram can increase the resolution of the structural geometry by adding more members and vertices without changing the global equilibrium in the system. The load network for a funicular system can be subdivided to design compression-only structural forms with various topological properties. Designing the force diagram by combining the cut and extrude method with subdividing the applied loads can yield interesting results.

Let us take a simple truss example in 2D and 3D as an initial design configuration. Figure 5.49 shows both 2D and 3D versions of the truss in multiple steps where the number of applied loads increases in each step, and so does the resolution of the compression-only part of the structural form. Note that, in each step, after subdividing the load network, the *cut and extrude* method is used to design the tension part of the system.

5.15.3 Further Carving the Force Diagram

In some 3D design cases, removing the tension elements of Figure 5.49 connecting the support locations is desired. In multiple steps, we can use the *cut and extrude* method to further articulate the structural form's geometry and remove or adjust the tension elements. To explain the process explicitly, let us start with the form and force diagrams of Example 5.12 and carve the force diagram in Example 5.13.

Example 5.13 For the funicular form and force diagrams of 5.49f, replace the two side tension members connecting the support locations by articulating the force diagram.

Construction steps:

1. Start from the nodal force polyhedrons reciprocal to the compression-only nodes of the structure with the applied loads (Figure 5.50a).
2. Take a cutting plane PL_1 and place it at the location of vertex B in the force diagram.
3. Change the dihedral angle of the plane such that it would truncate the corner F of the nodal cells.
4. Cut the corners of the cells by the plane and add faces to close the cells (Figure 5.50b).
5. Take the external faces of the new cells and reverse the direction of their face normals (Figure 5.50c).

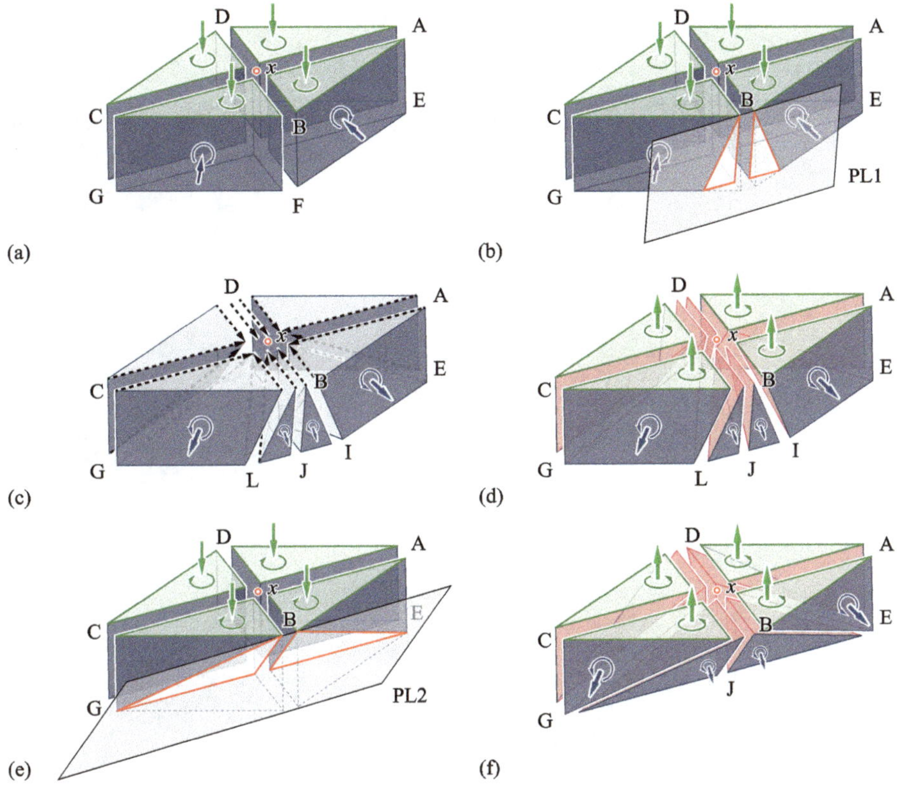

Figure 5.50 (a) The positive nodal cells of the force diagram reciprocal to the compression nodes of the truss of Figure 5.49f; (b) cutting the positive cells by a plane PL_1; (c) and (d) reversing the normal of external faces and extruding them to the point x to get new cells; (e) changing the angle of the cutting plane as PL_2; and (f) the new cells reciprocal to the tension part of the structure of Figure 5.51.

6 Extrude the new faces to the point x of the global force polyhedron and construct negative pyramidal cells (normals point outside of the cell) (Figure 5.50d).
7 The new force diagram corresponds to the form of Figure 5.51b.
8 The angle of the cutting plane PL_1 could be changed in step 2 such that it passes through vertices B, G, and E, as illustrated in Figure 5.50e.
9 The new force diagram is reciprocal to the structure of Figure 5.51c with no horizontal tension ties connecting the supports.

□

Note that the *cut and extrude* is not the only design technique that can be used to sculpt the force diagram. It is merely an example to provoke the development of many other ideas that can be used to design force and form diagrams, respectively.

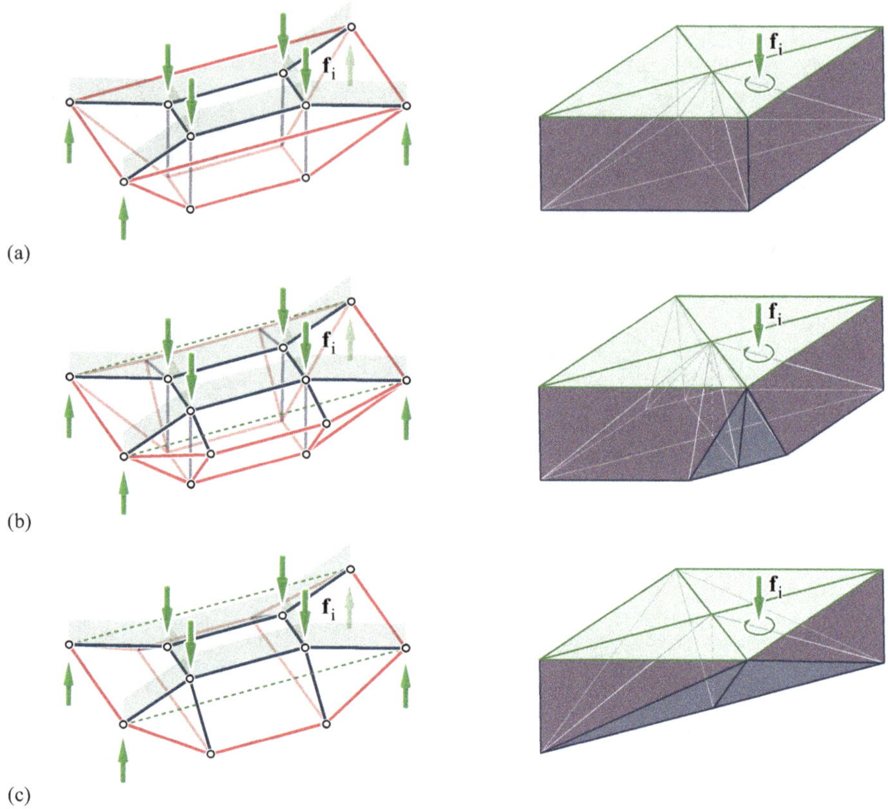

Figure 5.51 (a) The form and force diagrams with tension ties connecting the support locations; (b) connecting the tension ties to the compression-only nodes of the structure; (c) further reducing the tensile and compressive elements by sculpting the force diagram.

5.16 Force Optimization

Sections 5.15 of this chapter demonstrated techniques by which the force diagram for systems with mixed compression and tension forces can be designed and how particular features in the form diagram can be articulated. Nevertheless, less attention was given to how the magnitude of the internal forces can be controlled and even optimized to address the economy of construction. For instance, can the force diagram be designed such that certain members of this structure carry equal internal force? If members carry the same force, they can be made of the same cross-section and material, making the construction process more economical.

For compressive elements with the same internal forces, the length of the elements is an important parameter that might result in bucking the member. Thus, buckling criteria should also be considered in the design process. In contrast, the internal force can be used directly to design the cross-section of the tensile elements. Thus, working with tensile elements is easier.

In graphical methods, explicit geometric techniques can be developed to design the internal forces to obtain optimized structural forms. This section shows a simple geometric procedure in conjunction with the *cut and extrude* technique that can be used to optimize a funicular form further.

2D Force Optimization

In 2D graphical methods, if we can design a force diagram with equal edge lengths, then the corresponding elements in the structural form will carry equal internal forces. To achieve this goal, some constraining geometries are needed to control the edge lengths of the force diagram. These constraining geometries can be as simple as a circle imposed on the force diagram such that particular edges of the force diagram are radii of the same circle. Jürg Conzett used the same technique in the design of Traversina Bridge.[128] Robert Millart also used the same approach in the design of the Chiasso Shed structure[129] to get members with equal force in the system.

Example 5.14 provides step-by-step procedures to constrain a force diagram to a circle using the *cut and extrude* technique explained in Section 5.15.

Example 5.14 Construct a force diagram for the structure of Figure 5.49g such that its tensile elements carry equal internal force.

Construction steps:

Start with the compression-only nodes of the force diagram of Figure 5.52a:

1 Draw a circle from point F passing through the positive/compression force polygons.
2 Use the circle's circumference as a guide to cut the force polygons. This path is illustrated as path (1) in Figure 5.52a.
3 Draw the compression-only nodes for the newly constructed force polygons (Figure 5.52b).
4 Duplicate the external edges of the new force polygons, reverse their direction, and extrude them to point F in the force diagram to make new negative force polygons (Figure 5.52c); the extruded edges have equal lengths because they are the radii of the circle.
5 Combine the new negative cells with the previously made positive cells and complete the force diagram.
6 The reciprocal structural form will have constant internal forces in the tensile members (Figure 5.52d).

The force in the tensile members of the structure is equal since their reciprocal edge lengths l are equal. □

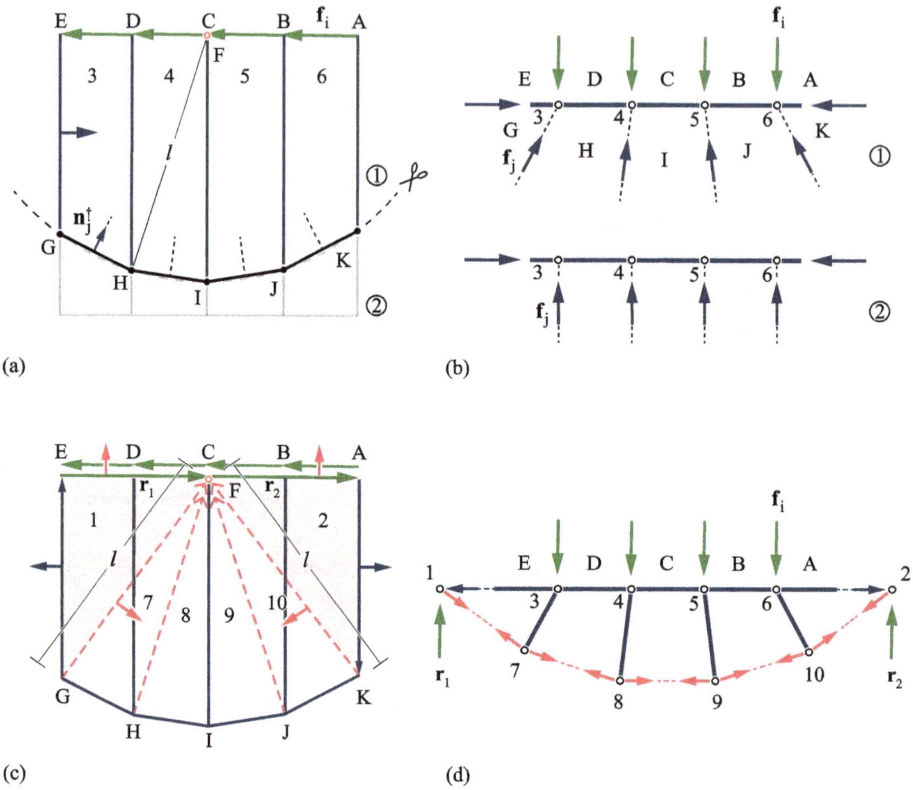

Figure 5.52 (a) Cutting positive nodal force polygons in (1) circular, (2) straight, and (3) zig-zag patterns; (b) the angles of the reciprocal compression-only nodes in the form; (c) extruding the circular cut edges to vertex *F* and constructing new force polygons; (d) and the resulting structural form with constant force in the tensile members.

3D Force Optimization

The same approach can be used in 3D to optimize the force magnitude in the structure's members. In polyhedral graphic statics, the magnitude of the force in a member of a structure is represented by the area of its corresponding face in the force diagram. Controlling and optimizing the areas of faces is not a trivial task compared to optimizing edge lengths in 2D graphical methods. In mathematical terms, the area is defined by a quadratic function, and its optimization requires sophisticated approaches, which will be discussed in detail in Chapter 7. Nevertheless, simple geometric techniques can be adopted from 2D to optimize internal forces in 3D.

The optimization might be relatively simpler for a force diagram in 3D that consists of triangular faces. Following a simple geometric rule, the area of all triangles with the same bases and heights is the same despite the difference in their angles in 3D. With a simple 2D circle in 3D, we can constrain the triangular faces to have the same height and base to make the internal forces in the members of the form equal. Figure 5.53a,b

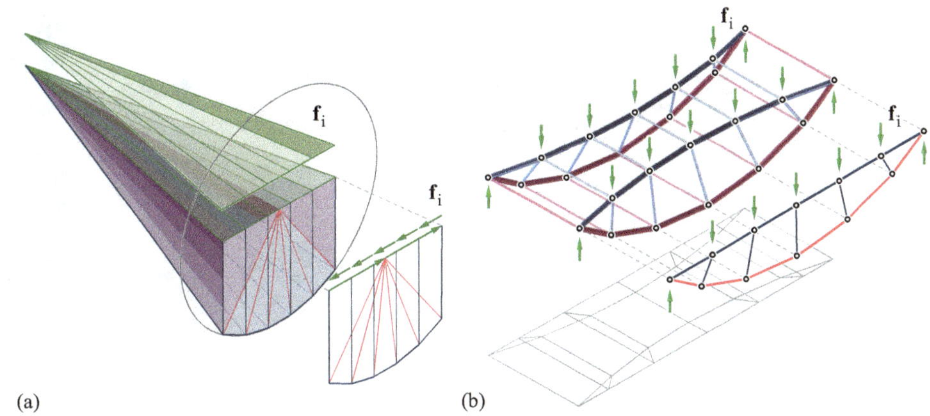

Figure 5.53 (a) A constrained 2D force diagram can be extruded in a direction normal to its construction plane to make a 3D force diagram with polyhedral cells; (b) the structural form with equal force magnitude in the tensile members reciprocal to triangles with equal lengths in the force diagram.

shows this approach where the optimization method in 2D can be directly used to generate optimized structural form in 3D.

Note that in all cases, the outer edges of the compression force cells are extruded to make the force polygons for the nodes with mixed forces.

Subdividing Compressive Elements

In the previous design examples, we subdivided the applied load network that defined the number of compression nodes, and we used those compression nodes to generate nodes with mixed types of forces using the *cut and extrude* method.

Subdividing the compression cells changes the topology of the compression members connected to the bottom tensile chord. Figure 5.54 shows four different configurations of forms generated by subdividing the compression part of the force polygons into smaller convex polyhedral cells. Subdividing the nodal force polygons of the compression members reduces the magnitude of internal forces in the members. It distributes the internal force into more members and nodes. It thus enhances the structural performance by reducing the effective lengths of the members and the chance of local buckling and global failure in the system.[102,130]

5.17 Increasing Geometric Stiffness

The bar-node models of the funicular structures shown in all the examples in this chapter still need to be translated into volumetric dimensions of materials to be realized as a real structure. These systems can be immediately translated into real structures by using tubular sections with various thicknesses. Although such structures are quite efficient for the loads similar to the design loads of graphic statics, their performance

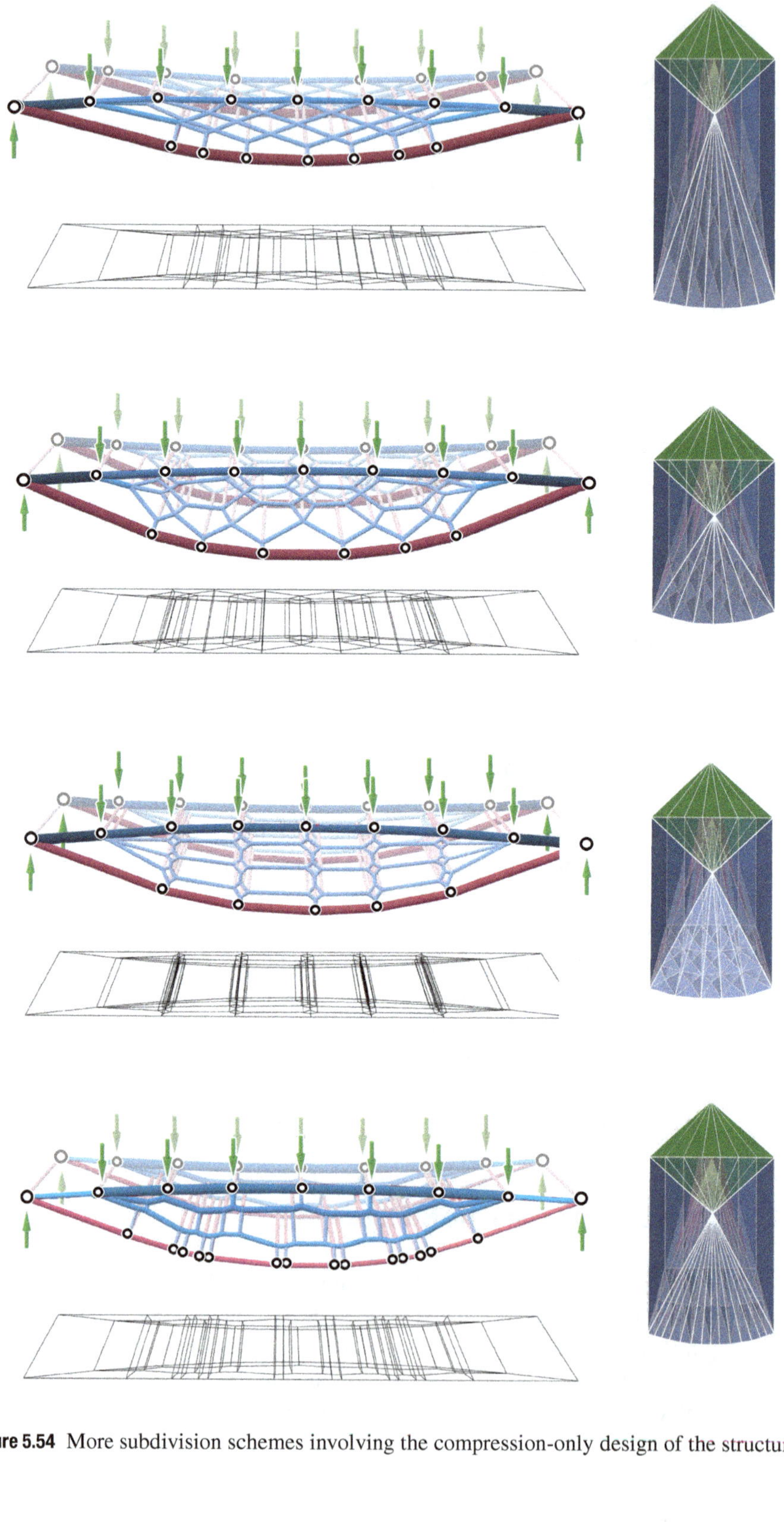

Figure 5.54 More subdivision schemes involving the compression-only design of the structure.

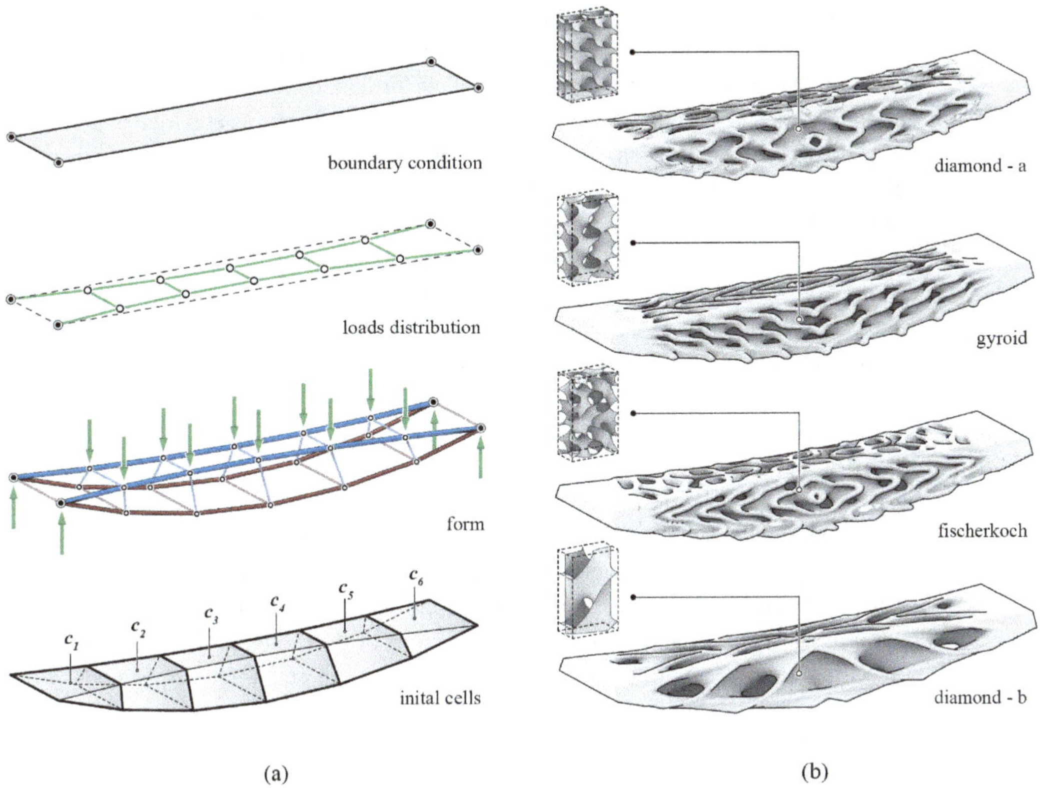

Figure 5.55 (a) The form-finding process through polyhedral graphic statics; and (b) bridge designs based on the different TPMS units.

might be compromised if the loading scenario changes in real life. We can introduce additional geometric stiffness to the system to mitigate this shortcoming by embedding the funicular form's thrust lines within a volumetric, modular geometry with anticlastic curvature.

Figure 5.55 shows an example of a funicular form embedded within various periodic anticlastic surfaces. It shows the process where each cell of the form diagram is populated with these surfaces such that the geometry of the surface is contorted to follow the outline of the cell. The superimposed anticlastic geometry gives additional stiffness to the system in the case of asymmetric loading to enhance the performance of the funicular structural form.

The anticlastic geometry of the periodic surfaces populates the 3D space with the maximum area and minimal mass, preserving the porosity and lightness of the design. Volumetric modeling[131] and Signed Distance Functions (SDF)[132] were used for molding the geometry. SDF describes space as a continuous scalar field, and geometry is defined as mathematical formulae instead of individual points, lines, and faces. Triply Periodic Minimal Surface (TPMS), a formula-based generated geometries, is tested for these designs. Adapting the chosen TPMS unit to fit the cells in the form diagram involves adjusting the unit's scale, orientation, and position. Figure 5.55b shows the four TPMS units morphed into the initial funicular form.

We determined the appropriate scale for the TPMS unit to fit within the cells, ensuring that it maintains its periodicity and anticlastic surface properties while providing sufficient coverage for the desired structural performance. Second, we adjusted the orientation of the TPMS unit to align with the geometric constraints and load paths within the cells. Proper orientation ensures efficient load distribution and optimal structural behavior. Positioning may require minor adjustments to the unit's location to ensure a seamless fit.

The whole process presents a computational methodology for innovative structural design. By uniquely combining polyhedral graphic statics and volumetric modeling, we can design and produce highly porous, structurally efficient systems for large-scale architecture, engineering, and construction applications.

5.18 Chapter Summary

In this chapter, we started with a graphical analysis of a simple truss system to explain the properties of force diagrams and the geometry of the equilibrium of systems with both tensile and compressive members. We showed the procedures to analyze polyhedral truss systems and explained the conditions of overhangs and their effect on the force diagram. We continued by visiting systems with self-intersecting nodal force polygons in both 2D and 3D.

We also showed examples of structures with static degrees of indeterminacy and their reciprocal force diagrams. The GDoF in the force diagram of a polyhedral truss show the minimum number of states of self-stress in a system. We then introduced design techniques such as subdividing the applied load network and the *cut and extrude* method that can manipulate and carve the force diagram of a simple truss and derive articulated funicular systems with combined tensile and compressive members in 2D and 3D.

We also introduced the topic of force optimization by containing the force diagram to particular geometries. Finally, we showed an approach to realizing a funicular network for real-world application by embedding the thrust in anticlastic periodic surfaces using volumetric modeling and SDF to enhance the performance of such systems under asymmetric loading.

6 Geometric Degrees of Freedom in Design of Form and Force Diagrams

Masoud Akbarzadeh, Marton Hablicsek, and Hua Chai

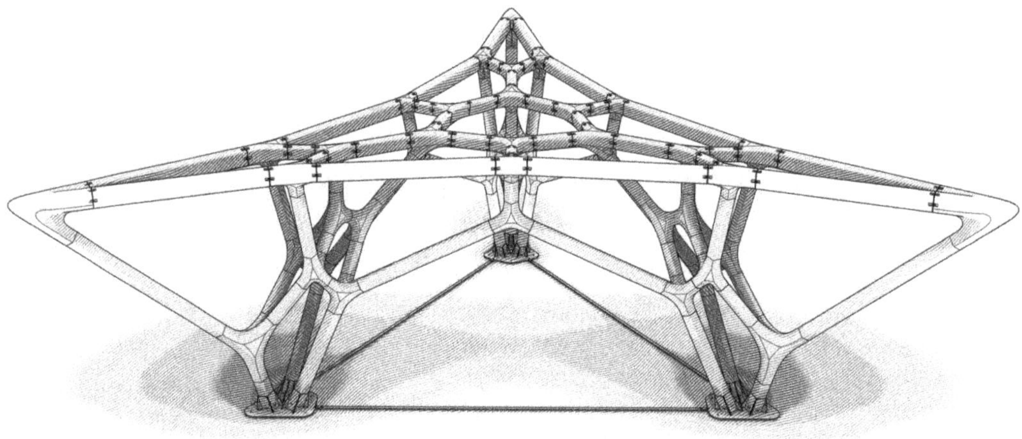

Figure 6.1 Hedracrete project was designed by controlling the geometric degrees of freedom of the force diagram; the spatial branching structure has tensile members on the top chord and compression branching members in the bottom.[76,77]

6.1 Introduction

Let us start this chapter with an example of a built structure. Hedracrete is a unique research project addressing important digital design and fabrication topics, including funicular form finding in three dimensions, fabrication of complex spatial systems, and the innovative detailing of conventional material. The form of the structure is a funicular polyhedral frame with both compressive and tensile members with a maximum height of 3.33 m, spanning from three supports located 5.4 m apart from each other (Figure 6.1). The structure consists of 45 prefabricated joints, 54 compressive, and 30 tensile members sitting on steel supports connected by steel rods.

Hedracrete was initially designed as a compression-only form using polyhedral graphic statics. What makes the form-finding process unique is that the magnitude of the forces in the boundary conditions was altered to change the internal forces in the structure from compression-only forces to a system with both compressive and tensile forces[77,78] (Figure 6.2).

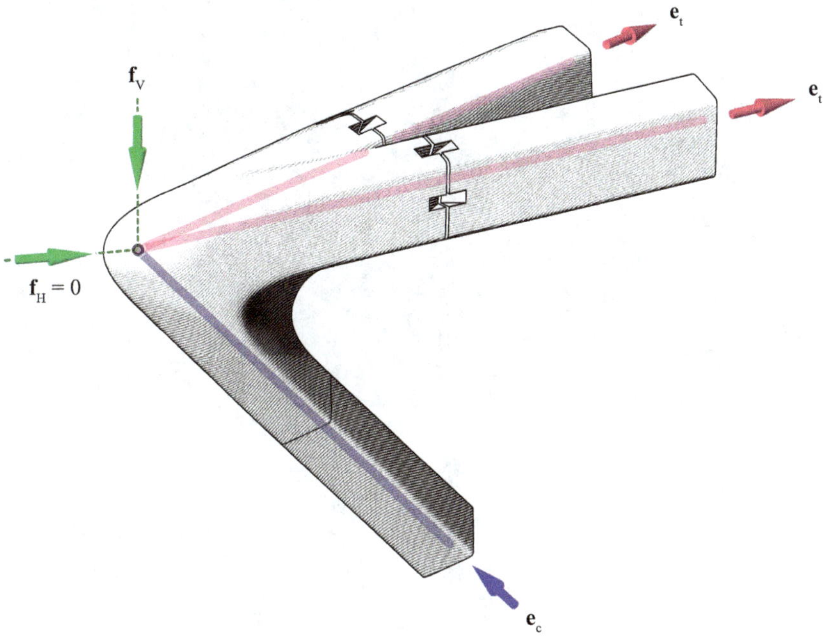

Figure 6.2 A single node of the Hedracrete project in which the magnitude of the horizontal external load decreased to zero in the design stage by manipulating the geometry of the force diagram without changing the geometry of the funicular form.

This approach was made possible by manipulating the edge lengths of the force diagram, removing several applied loads in the boundary of the structure, and redistributing the forces in the internal members. This approach in the design of funicular systems will be explained using simple examples in this chapter.

6.1.1 Chapter Objectives

This chapter introduces an important geometric property in form and force diagrams that can be used to design creative funicular systems. In particular, we will discuss the possibilities of changing edge lengths in reciprocal diagrams and understand the effects of this change on the equilibrium of forces and the typology of the structural forms. In this regard, we will discuss the geometric degrees of freedom (GDoF) or our freedom as a designer to adjust the reciprocal diagram's edge lengths without compromising the forces' equilibrium.

6.1.2 Geometric Degrees of Freedom: The Definition

The concept of *GDoF* in this book has been introduced briefly in previous Chapters 1, 3, and 4.

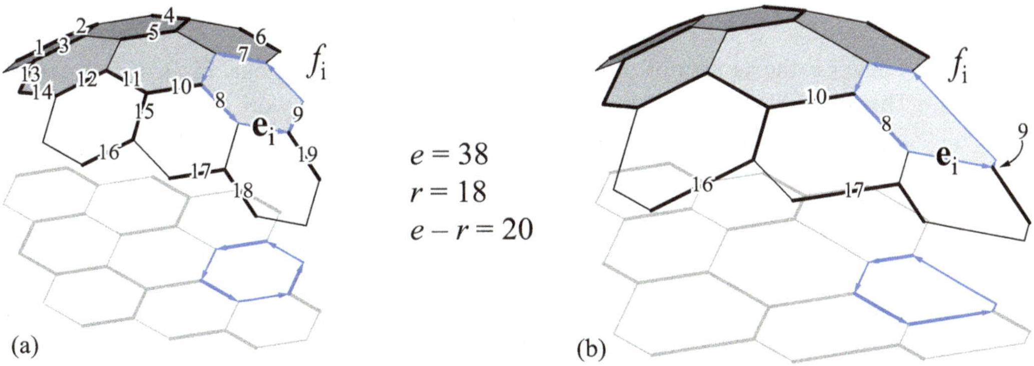

Figure 6.3 (a) Simple network in 3D with planar faces and its projection to 2D highlighting the geometric degrees of freedom; and (b) a transformed graph based on changing the lengths of the independent edges where the highlighted members represent the independent edges in the network.

Let us assume we want to make a closed polygon where the number of edges and interior angles are given. We cannot choose all the edge lengths and internal angles to complete the polygon since the problem will become over constrained. For instance, for a triangle, if all the internal angles are given, we can only choose the length of one edge, and the length of the rest of the edges will be found by intersecting lines drawn from both ends of the chosen edge with the given interior angles. Thus, for a single polygon, the GDoF is the number of edge lengths that can be chosen *independently* from the rest of the edges to complete the geometry of the polygon with given interior angles.

For instance, three edge lengths out of five must be chosen for a five-sided polygon to complete the geometry if we want all the angles to be equal. If the three edge lengths are equal, the result will be a pentagon. These three edge lengths can be independently chosen. Thus, they are called *independent edges*, and the rest of the edges can be referred to as the *dependent edges* since they follow the independent edges in completing the polygon.* The GDoF of a polygon is, in fact, the number of independent edges (e_{ind}) whose various lengths give a family of polygons with parallel edges to the initial polygon, that is,

$$GDoF = e_{ind} \tag{6.1}$$

The same concept can be extended to a network of two-dimensional (2D) or three-dimensional (3D) polygonal faces, as shown in Figure 6.3. Similar to the case of a single polygon, there are multiple edges on a network that can be chosen independently to complete a network with different edge lengths but parallel angles. The GDoF of a single polygon equals $e - 2$ where e is the number of edges. Finding the GDoF of a network in 2D and 3D is not a trivial task, and a proper mathematical definition of the network is needed. However, the following inequality holds

* Please visit Algebraic Graph Statics by Tom Van Mele and Philippe Block.

$$GDoF = e_{ind} \geq e - 2f \tag{6.2}$$

where e is the number of edges and f is the number of polygonal faces of the network. This formula applies to simple cases in 2D but does not work in complex scenarios. Equation 6.2 can be rewritten in terms of the edges that are shared between the faces of a network:

$$GDoF = e_{ind} \geq \sum_{i}^{n} (e_{f_i} - 2) - e_{sh} \tag{6.3}$$

where $e_{f_i} - 2$ is the GDoF of the i-th polygon/face and e_{sh} is the number of shared edges between the faces of the graph.

Remark 6.1. The GDoF of 2D/3D networks that consist of only triangular faces is exactly one because a triangle has a determinate geometry, and thus having a single edge length is sufficient to reconstruct the entire network.

Finding the GDoF of polyhedral systems is not a trivial task. In the following, we will provide a brief mathematical formulation to calculate the GDoF of a network. However, a more comprehensive discussion on this topic requires an algebraic definition of the problem, which will be provided in Chapter 7. This chapter aims to work with the GDoF of simple polyhedral systems and shows various examples in designing funicular spatial structures where the system does not have many GDoFs.

6.2 Mathematical Definition for GDoF

Mathematically, we can write equations for each polygon f_i of the network in 2D or 3D. Around each polygon f_i, we choose a consistent orientation of the edges. We will denote the direction vector corresponding to the edge e_j by the vector $\mathbf{e}_j$. Since each face provides a closed loop of edges, the sum of the edge vectors has to be the zero vector. Thus, we obtain a vector equation for the edge lengths $|e_j|$ as

$$\sum_{e_j} \mathbf{e}_j |e_j| = \mathbf{0}$$

where the sum runs over the edges e_j of the face f_i.

This vector equation is satisfied if and only if the x-, y- coordinates (and z- coordinates for 3D) of the left-hand side of the equation are equal to zero. Hence, each polygon f_i provides two (or three in 3D) equations for the edge lengths, and we obtain a $[2f \times e]$ matrix (and $[3f \times e]$ matrix in 3D), $\mathbf{A}$ describing the geometry of the network. Here f denotes the number of closed polygons and e denotes the number of edges in the network. In other words, we have a linear equation system

$$\mathbf{Aq} = \mathbf{0} \tag{6.4}$$

where $\mathbf{q}$ denotes the vector of edge lengths.

Each solution of the linear equation system represents a network whose edges are parallel to the associated edges of the original network with a different edge length.

The dimension of the solution space of Equation 6.4 is the geometric degrees of freedom of the network (GDoF). This dimension is $e - r$, where r is the rank of the matrix $\mathbf{A}$. In other words, there are $e - r$ edges in the network whose edge lengths can be independently changed, resulting in a unique network with parallel edges to the original network. If $e - r = 1$, a unique network is constructed up to scaling with the given edge directions. If $e - r > 1$, multiple, significantly different (different up to scaling) networks exist.

6.2.1 The Importance of GDoF in the Design of Funicular Structures

The GDoF allows parallel transformations of polyhedrons such that the dihedral angles between the faces do not change. This property will enable us to transform the geometry of both form and force diagrams for design purposes.

Working with the GDoF of the *force* diagram gives new polyhedral geometries with different face areas. That is, a designer can manipulate the force magnitude without altering the number of faces or the topological relationship with the structural form. This property can control the force magnitude in the boundary or the internal distribution of forces within a structural form to address particular design needs or site constraints. GDoF in the *form* diagram allows exploring a variety of geometric configurations of structural forms by changing the edge lengths of the system without disturbing the equilibrium of forces.

Remark 6.2. The parallel transformation in the *form* diagram allows finding various structural forms without changing the magnitude of forces or equilibrium. In contrast, parallel transformation in the *force* diagram allows better control over the boundary or internal forces without changing the form of the structure.

In Sections 6.3 to 6.6, we visit the use of GDoF in the force and form diagrams and show examples of structures designed by manipulating GDoF.

6.3 GDoF of the Force Diagram

To work with the GDoF of the force diagram, we need to know whether a force diagram has a geometric degree of freedom larger than one. As discussed extensively in Chapters 3, 4, and 5, the force diagram consists of nodal force polygons in 2D, polyhedrons in 3D, and the global force polygon and polyhedron. Thus, both nodal and global force polygons and polyhedrons contribute to the GDoF of the entire force diagram.

The GDoF of a 2D force diagram can be found in simple cases using Equations 6.2 and 6.3. However, algebraic formulations are needed for polyhedral systems to find the precise number of GDoF, which will be discussed in Chapter 7. Nevertheless, the GDoF in small systems can be found in the procedural construction of the force diagram. The number of edge lengths a designer must choose to complete the force diagram corresponds to the number of GDoFs of the system, which will be explained next.

6.3.1 GDoF of a Nodal Force Polygon

In Chapter 1, we explained procedures to construct indeterminate force diagrams for a given system of concurrent forces in 2D and 3D. The force magnitude could be manipulated by changing the edge lengths of the force diagram. The force diagram of such systems has independent edges whose number equals the GDoF. Thus, finding the number of GDoF of a force polygon or polyhedron is the same as finding the number of independent edge lengths in the geometry of the force.

The independent edges can be found procedurally in the construction process of the force diagram because a designer has to choose an edge length during the construction process to complete the process. The number of input edge lengths is the same as the number of GDoFs of the force diagram. Example 6.1 in both 2D and 3D will elaborate on this concept.

Example 6.1 Construct a force polygon for the concurrent system of loads of Figure 6.4. Assume that the magnitude of the force $\mathbf{f}_1$ is given as 1 kN, and the magnitude of the rest of the forces should be found geometrically to maintain equilibrium.

Construction steps:

1 Start by identifying the subspaces of the form diagram and labeling them as *A*, *B*, *C*, and so on.
2 In the form diagram, put the thumb of the right hand on the applied load; the stretch of the four fingers gives the cycling direction around the subspaces.

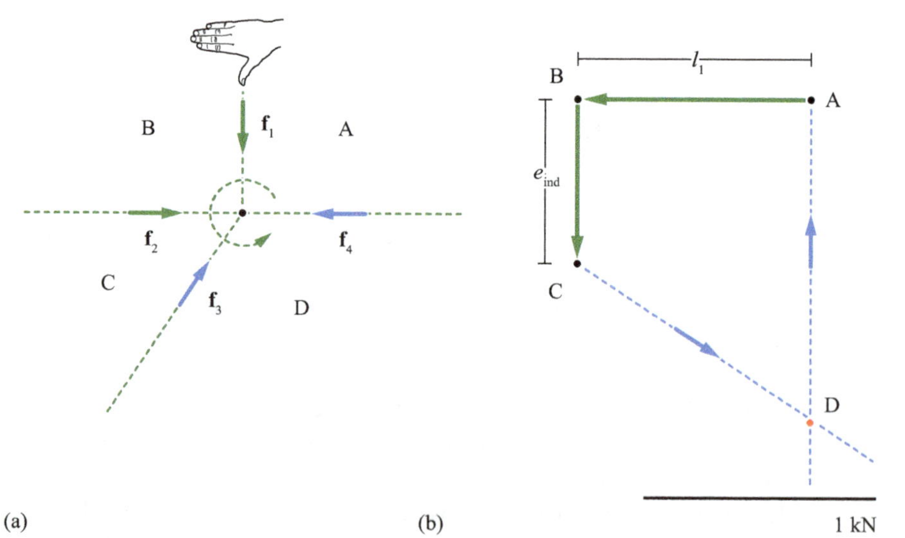

Figure 6.4 (a) A system of concurrent forces and (b) the reciprocal indeterminate force polygon with one independent edge.

3 In the force diagram, start from an arbitrary point A and move perpendicular to the force $\mathbf{f}_1$ by choosing the length l_1; for instance, every 1 cm is equivalent to 1 kN force.
4 In the force diagram, label the end of the force vector B.
5 From point B in the force diagram, move perpendicular to the direction of the force $\mathbf{f}_2$ of the form diagram. To continue the construction and find the location of point C, you should choose a length for the vector BC in the force diagram. Thus, vector BC is an independent edge in the system (Figure 6.4).
6 From point C in the force diagram, draw a line normal to $\mathbf{f}_3$ of the form.
7 Also from point A, draw another line normal to $\mathbf{f}_4$ of the form.
8 Find the intersection of the lines drawn from A and C and name it D in the force diagram.
9 As noted, the length of the rest of the edges of this force polygon could be found just by using their direction, and thus, the rest of the edges are dependent.
10 The construction of the force diagram is finished, and the GDoF is 2.

□

Two edge lengths have been chosen in constructing the force diagram of Figure 6.4; thus, the GDoF of the force polygon is two. The first edge length defines the scale of the applied load, and the second length is necessary to complete the force polyhedron. Usually, in constructing the force diagram, the first chosen edge length is related to the magnitude of the applied load and scales the diagram.

Also, note that in this system, with the given configuration of the forces, that is, the angles of the line of action, only the designer can choose the magnitude of two forces. To maintain equilibrium, the magnitude of the rest of the forces should be found by the construction of the force diagram (Figure 6.4). Thus, the rest of the edges are dependent edges.

6.3.2 Manipulated Force Polygon

Figure 6.5 shows various equilibrium states for the node of Example 6.1 by choosing different edge lengths for the force $\mathbf{f}_2$ in the force diagram. For instance, Figure 6.5a.1 shows a positive edge length for the force $\mathbf{f}_2$, whereas Figures 6.5a.2 and 6.5a.3 show zero and negative edge lengths, respectively. In all cases, the line of action of forces in the form does not change, but the magnitude of the forces is controlled using GDoF to maintain equilibrium.

6.3.3 GDoF of a Nodal Force Polyhedron

Similar procedures can be used for a system of forces in 3D. For example, Figure 6.6 shows a concurrent system of forces in 3D and its force polyhedron. Let us assume the intersection of these forces is a node belonging to a funicular structure where forces

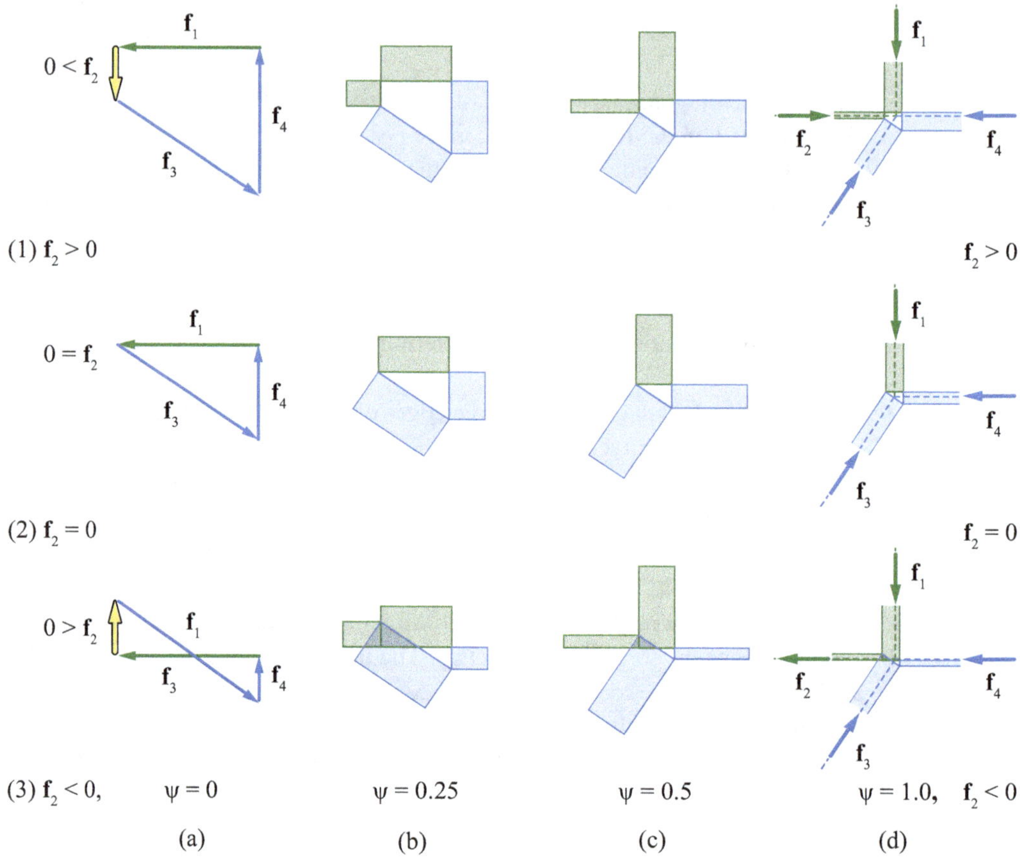

Figure 6.5 (1) to (3) The Minkowski sum of the form and force diagrams of Figure 6.4 with various edge lengths for $\mathbf{f}_2$ that keep the node in equilibrium; (a) to (c) various scale factors from force to form diagram.

$\mathbf{f}_1$ and $\mathbf{f}_2$ are the externally applied loads and the rest are the internal forces in the members attached to the node.

The force polyhedron of Figure 6.6 has GDoF. That is, multiple polyhedral geometries can be constructed with different face areas to represent the equilibrium of the forces in the form diagram. To identify the GDoF of the force polyhedron, it is sufficient to find the independent edges of the polyhedron. Example 6.2 will provide a step-by-step procedure to find the GDoF of the force polyhedron for the given direction of applied loads.

Example 6.2 Construct the force polyhedron for the concurrent loads of Figure 6.6 and find the independent edges in the system such that the force polyhedron can be modified to represent the equilibrium with various force magnitudes. Note that the forces $\mathbf{f}_2$, $\mathbf{f}_5$, and $\mathbf{f}_4$ are co-planar in this configuration.

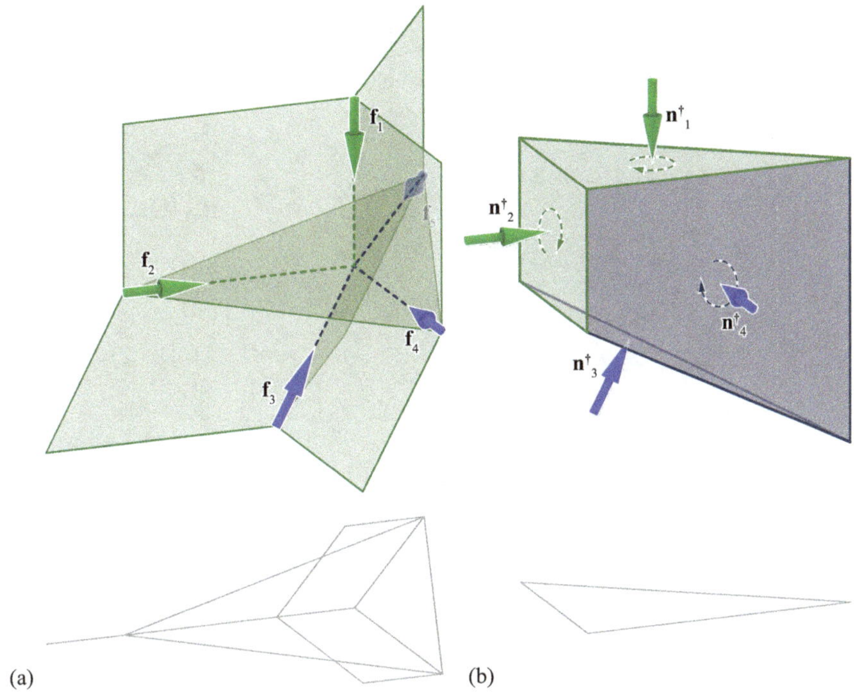

Figure 6.6 (a) A configuration of concurrent loads with their shared faces; and (b) the reciprocal force polyhedron with geometric degrees of freedom.

Construction steps:

1 In the form diagram, start by identifying the faces shared by the applied loads.
2 Using the shared faces, find the subspaces around each edge of the form and name them as *A*, *B*, *C*, and so on. (Figure 6.7).
3 Start from the applied load $\mathbf{f}_1$ on which you put the thumb of the right hand: the curl of the fingers shows the connectivity of the vertices of the force diagram by going around the subspaces in the form.
4 In the force diagram, start from an arbitrary point *A* and move perpendicular to the plane shared by the subspaces *A* and *B* (as discussed in Chapters 1 and 2).
5 Choose a length for the vector *AB* as l_1 and the first independent edge in the force diagram (Figure 6.7b).
6 Complete the rest of the face *ABC* reciprocal to the force $\mathbf{f}_1$ by finding the intersection of *BC* and *CA*.
7 In the force diagram, start from vertex *B* and move normal to the face shared by subspaces *B* and *D* to find the location of *D*.
8 Note that an additional edge length, l_2, is needed to locate vertex *D* in the force diagram. Thus, edge *BD* is another independent edge in the force diagram.
9 Round all the edges of the form diagram and complete their reciprocal faces in the force polyhedron.

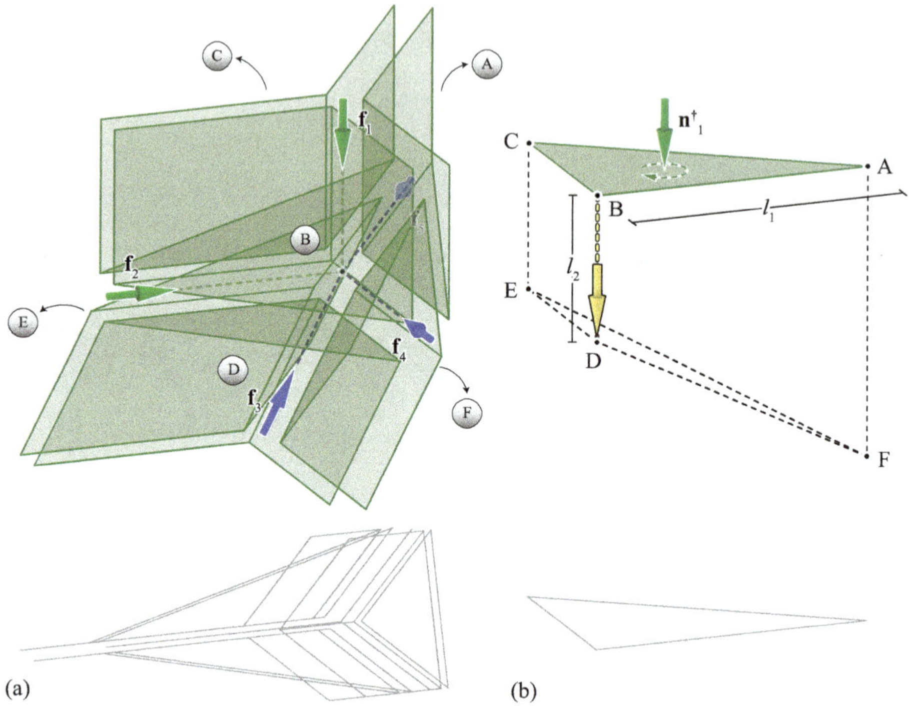

Figure 6.7 (a) Identifying the subspaces in the form diagram; and (b) constructing the force faces by finding the independent edges in the system: l_1 is the edge length of the starting face, and l_2 is the edge length necessary to complete the force construction.

10 Note that the GDoF of the force diagram is two, considering the chosen lengths l_1 and $_2$.

□

Changing the edge length of *BD* in the force diagram as one of the independent edges shown by vector $\mathbf{e}_{ind}$ in Figure 6.8 changes the geometry of the force. The change in the force diagram does not change the form configuration; it only gives various force magnitude combinations that are in equilibrium for the given form diagram. Figure 6.8a.1–d.1 shows the Minkowski sum of the form and force diagrams for three different lengths of $\mathbf{e}_{ind}$. The first value is positive in the negative direction of z- and describes the equilibrium of multiple compressive forces applied to the node in the form. Figure 6.8a.2–d.2 shows a force diagram if the length of $\mathbf{e}_{ind}$ is zero. In this case, the reciprocal face of the force $\mathbf{f}_2$ in the force diagram transforms into a single line with zero area. Thus, the force $\mathbf{f}_2$ disappears from the system of forces. In Figure 6.8a.3–d.3, the length of $\mathbf{e}_{ind}$ is negative (in the positive direction of z-) and results in $\mathbf{f}_2$, $\mathbf{f}_4$, and $\mathbf{f}_5$ becoming tensile forces at the node.

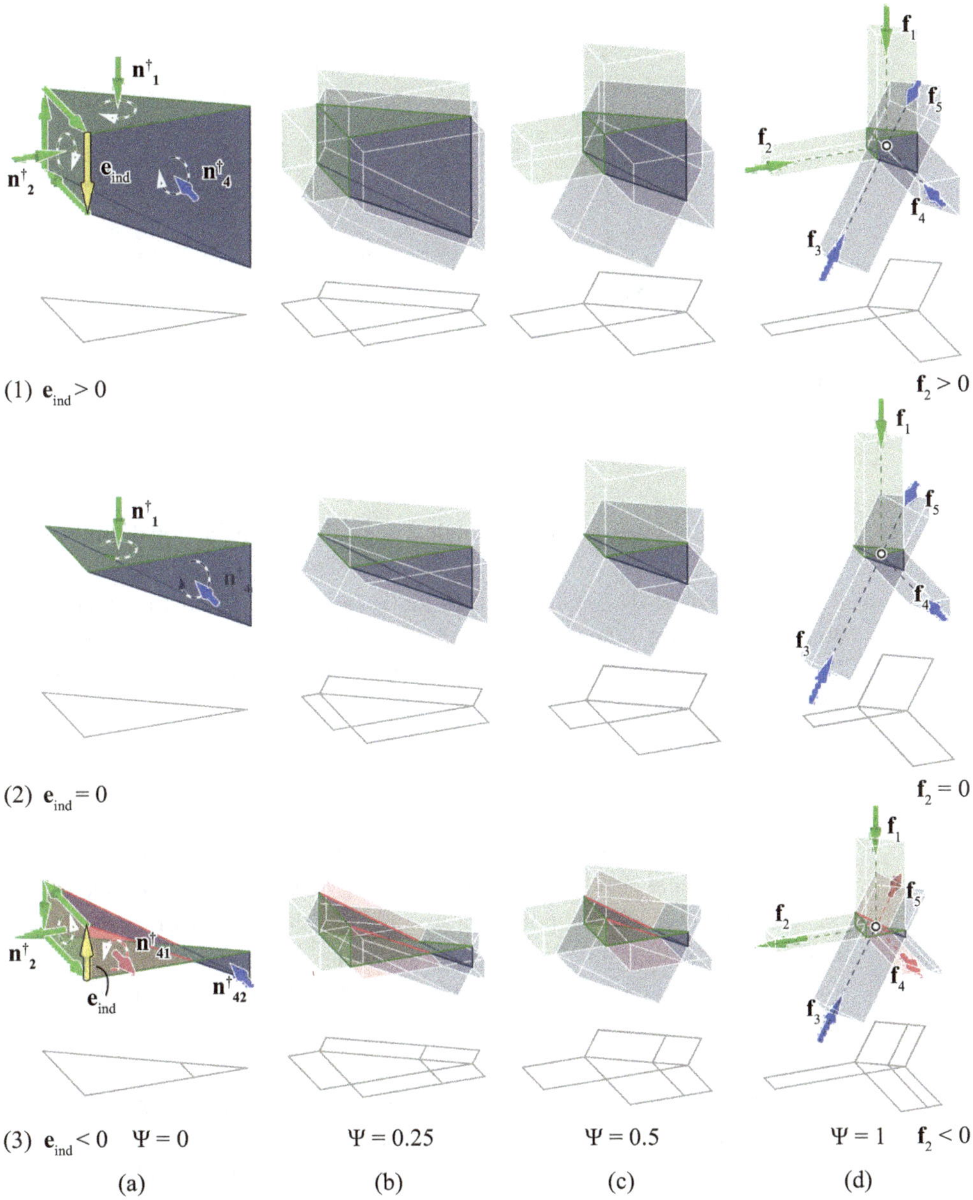

Figure 6.8 (a) to (d) Manipulating the edge length of the independent edge in the force diagram and generating different force equilibrium for the system of forces applied to a single node.

6.3.4 GDoF of a Global Force Polygon (2D)

The global force polygon that consists of multiple nodal force polygons can also have GDoF. We learned in Chapter 3 that a single convex force polygon can be subdivided into smaller convex polygons, resulting in various funicular solutions. In that case, the external force polygon is the global polygon representing the external force equilibrium, and the internal polygons represent the nodal equilibrium. Thus, the force

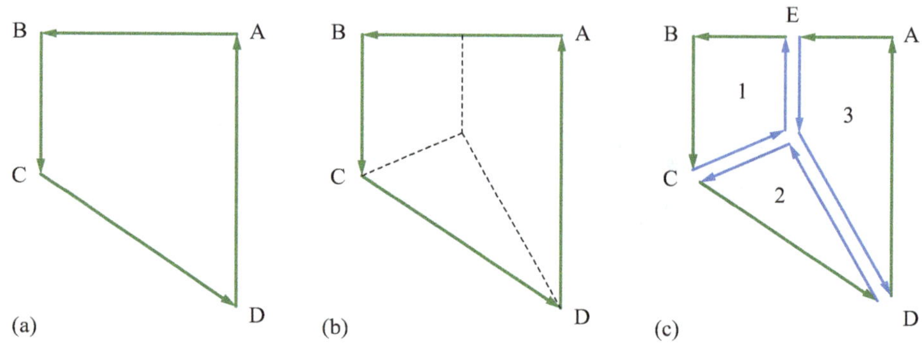

Figure 6.9 (a) A global force polygon; (b) subdividing the internal space of the polygon; and (c) the nodal polygons after subdivision.

subdivision technique can turn a single force polygon into a global force polygon with GDoF.

Consequently, the change in the global geometry alters the internal polygons' geometry without changing the topology or dihedral angle of the faces. In that case, the change in the edge lengths of the global polygon modifies the distribution of the loads in the internal members of the structure. For instance, the single force polygon of Figure 6.9a can be subdivided into multiple internal faces, as shown in Figure 6.9b,c. Note that every new polygon resulting from the subdivision must be closed and convex.

Remark 6.3. Although the GDoF of the global force polygon can be calculated separately, the internal polygons' geometry may restrict the global network's freedom. For instance, if all the internal polygons are triangles, then the GDoF of the entire network equals one, so the system is unique up to scaling. If the global polygon is subdivided, then the GDoF of the subdivided network is calculated by considering all the polygons in the system.

For instance, in the case of Figure 6.9, the internal polygons are two trapezoids and a triangle, and using Equation 6.2, the GDoF of the system is 2. A subdivided global force diagram has internal faces and is thus reciprocal to a form with edges and vertices. The new form diagram can be constructed on the initial lines of action of the concurrent forces reciprocal to the global force polygon. We explain this procedure with Example 6.3.

Example 6.3 Construct a reciprocal form diagram for the subdivided force diagram of Figure 6.9.

Construction steps:

1 Start from the node reciprocal to the global force polygon where all the external forces are applied (Figure 6.10a).

2 In the force diagram, start from a single force polygon, for instance, polygon 1, and identify the direction of the edges and their normals (Figure 6.10c). Note that the edge reciprocal to the force $\mathbf{f}_1$ has also been divided into $\mathbf{f}_{11}$ and $\mathbf{f}_{12}$. Thus, we will have a force system in the form diagram whose sum is equivalent to the magnitude of $\mathbf{f}_1$. The location of this force system will be defined next.

3 The goal is to find a node reciprocal to the polygon 1 of the force diagram. On the line of action of $\mathbf{f}_2$ in the form diagram, pick a point with distance l_1 from force $\mathbf{f}_1$. This point locates node 1 of the form, which is the location of the force

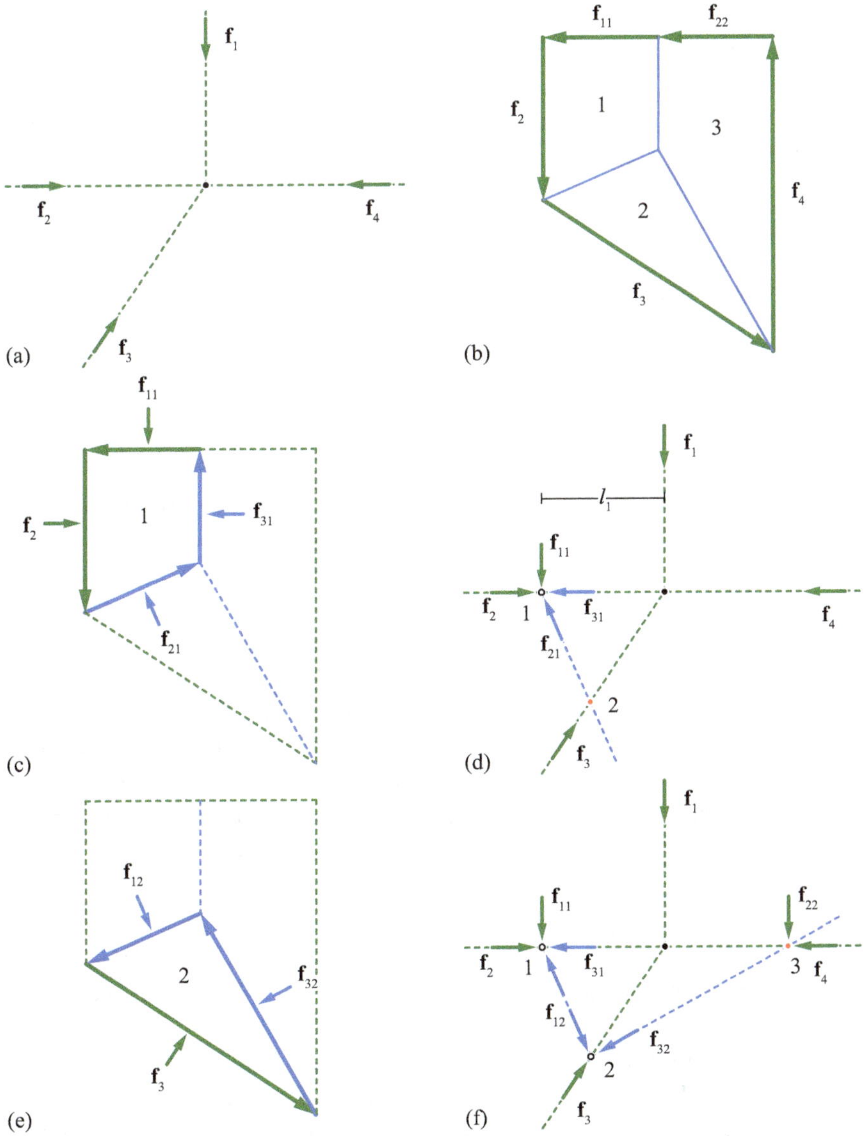

Figure 6.10 (a) to (f) The construction process of the reciprocal form on the lines of action of a force system reciprocal to the global force polygon.

$\mathbf{f}_{11}$ in the form diagram (see the examples of Chapter 2 for a system of parallel loads).

4 In the force diagram, choose an adjacent force polygon, for instance, polygon 2 (Figure 6.10c), and locate its reciprocal node 2 in the form by extending the edge $\mathbf{f}_{21}$ and intersecting it with the line of action of $\mathbf{f}_3$ (Figure 6.10f).

5 Similarly, in the force diagram, choose the last remaining force polygon 3 (Figure 6.10e) and locate its reciprocal node in the form using vectors $\mathbf{f}_{32}$ and $\mathbf{f}_{13}$ (Figure 6.10f).

6 Complete the form diagram by putting the load $\mathbf{f}_{22}$ at node 3.

□

Note that the force edge $\mathbf{f}_2$ was the independent edge in the force diagram. Thus, the length of the independent edge of the force $\mathbf{f}_2$ can be manipulated to redistribute the internal forces in the members of the form diagram without changing the structure's geometry. Figure 6.11 shows multiple conditions in which the edge length of the force $\mathbf{f}_2$ as an independent edge in the system is modified to change the boundary forces for the simple reciprocal structure.

Figure 6.11a.1 shows a positive value for the length of the force $\mathbf{f}_2$. This force diagram can represent a compression-only structural form. Choosing zero edge length for the force $\mathbf{f}_2$ in the force diagram results in a compression and tension combined system with no applied load on the outer left side of the system. The top member of the structure will carry tensile force to keep the equilibrium of the nodes with no change in the location of the supports or applied loads (Figure 6.11d.2). The edge length of $\mathbf{f}_2$ can also be chosen such that the length of another edge becomes zero. In the case of Figure 6.11a.3, a negative value is chosen for the edge length such that the length and magnitude of force $\mathbf{f}_4$ become zero. In this case, the structural form will also include both tensile and compressive forces internally, but the type of force $\mathbf{f}_2$ changes from compressive to tensile external load in Figure 6.11d.3.

Figure 6.12 illustrates multiple architectural applications of the simple structural forms introduced in Example 6.3. In Figure 6.12, the top chords of the structures are in tension, while the reaction force $\mathbf{f}_4$ in *a* is tensile and in *b* and *c* is compressive. In all cases, one of the side reaction forces is designed to be zero so that the structure can lean over the pathway and cover the architectural space.

6.3.5 Different Subdivisions

Different force diagram subdivisions result in various funicular forms in equilibrium with different topologies.

Figure 6.13 shows multiple examples of subdivided force diagrams and their corresponding forms. Figure 6.13a shows a force diagram with a GDoF of two. This force diagram internally consists of triangles and trapezoids. The triangular network is determinate (GDoF = 1), but the trapezoids of the system have a GDoF of one as the height of the force diagram.

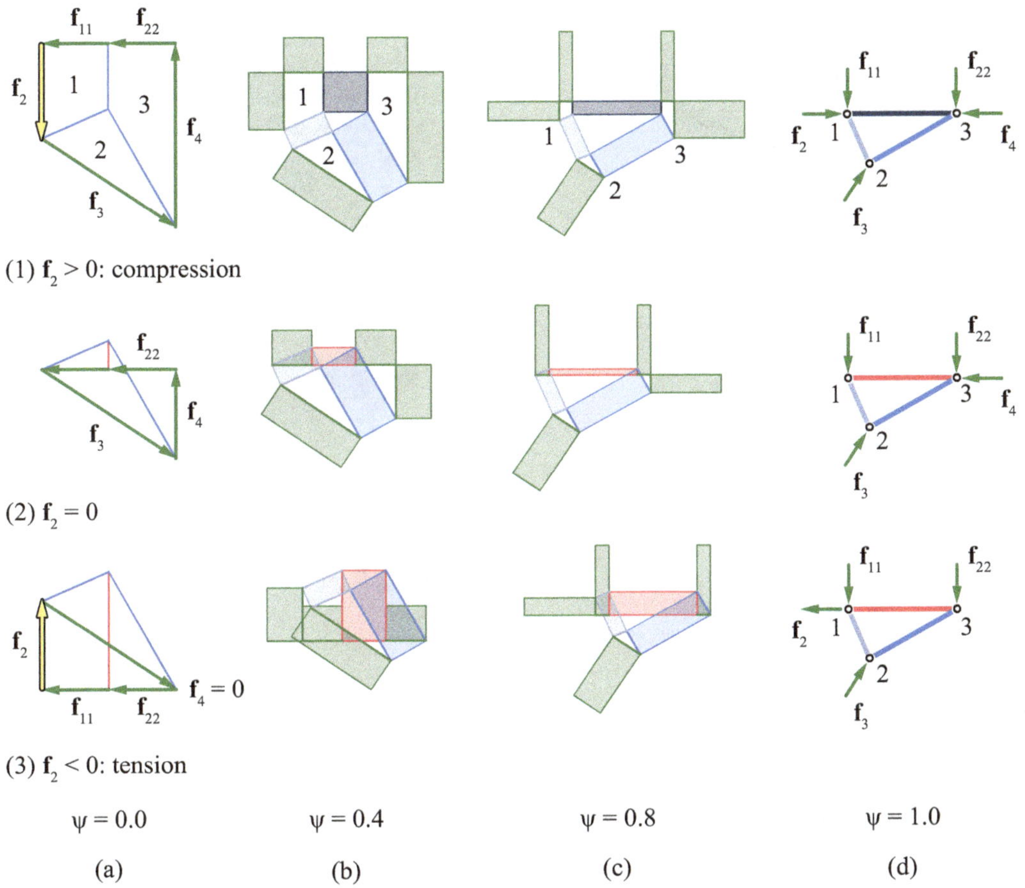

Figure 6.11 Changing the length of the independent edge in the force diagram $\mathbf{f}_2$ and observing the effects on the internal force distribution in the form in (1) the edge length is positive; (2) the edge length is zero; and (3) the edge length is negative; (a) to (d) shows the Minkowski transformation of the force to form diagram for each case.

In each step of Figure 6.13a–d, although the force diagram is subdivided and the system becomes more complex, the GDoF of the force diagram remains the same. Indeed, in each step, only the internal triangular faces are subdivided; thus, the GDoF of the entire network remains intact. This is an interesting design approach since various structural forms can be found for the same boundary conditions. Still, the boundary forces can be adjusted by changing the independent edge lengths in the global force polygon. Figure 6.14 illustrates a conceptual design of a pedestrian bridge based on the subdivision of Figure 6.13d.

6.3.6 GDoF of a Global Force Polyhedron

A single force polyhedron can be subdivided internally to make a group of polyhedral cells. Note that there are infinite ways that a single polyhedron can be subdivided into

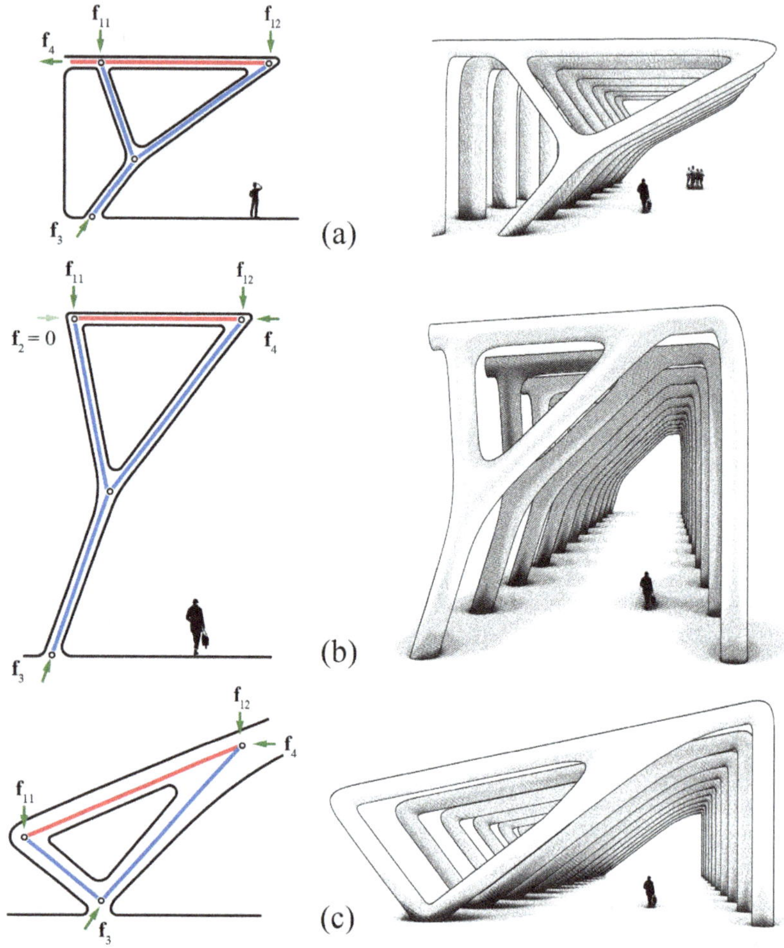

Figure 6.12 (a) A canopy with a tensile support on the top; (b) a canopy with a compression support on the top; and (c) a canopy with a compression support on the top and no vertical column.

smaller convex polyhedrons. However, in all these subdivisions, the faces of the newly created polyhedrons should remain planar.

The resulting polyhedral system represents a complete force diagram with an external cell encompassing the internal cells. Thus, the outer cell is the global force polyhedron, and the inner cells contribute to the nodal equilibrium of the system. Therefore, such systems' GDoF include all the independent edges in both nodal and global polyhedrons.

Finding the exact number of GDoF for a group of polyhedral cells is not a trivial task and requires algebraic formulation for the faces of the polyhedrons, which will be discussed in Chapter 7. As a rule of thumb, the entire system will become geometrically determinate if all the subdivided cells are tetrahedrons. The tetrahedral subdivisions may constrain the global force cell's GDoF and leave no free-edge length

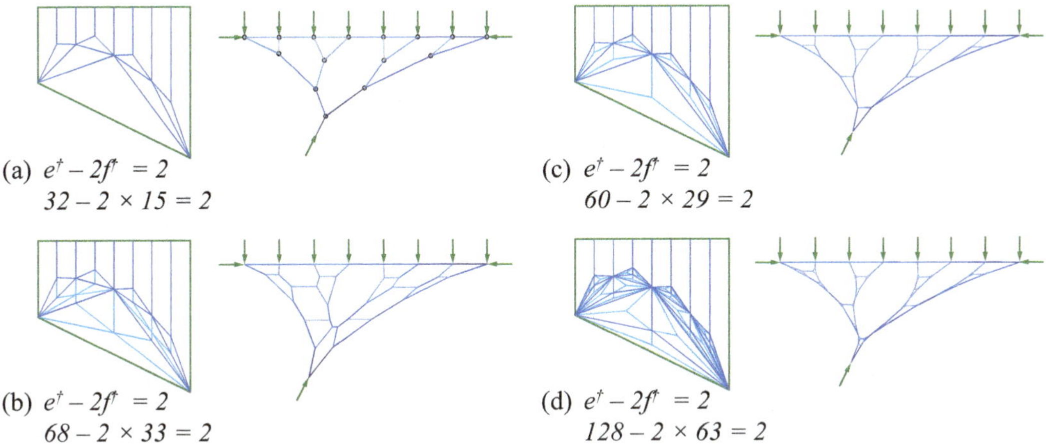

Figure 6.13 (a)–(d) The form and force diagrams of a branching structure with two additional pre-stressing forces (blue) at the top chord.

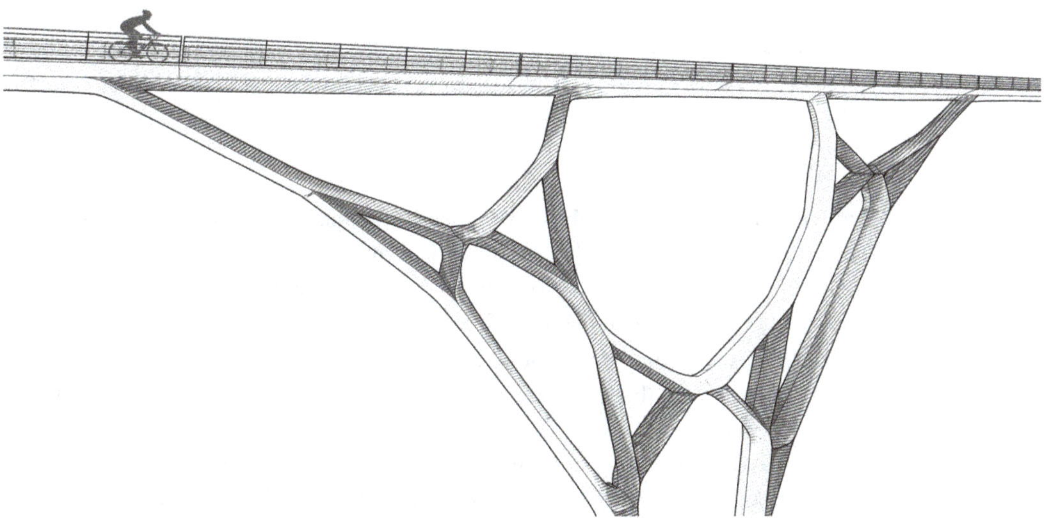

Figure 6.14 The subdivided branching structure used for the conceptual design of the pedestrian bridge.

to manipulate for design purposes. Thus, if the subdivision includes cells of more than four faces, it will ensure that the nodal cells have independent edges.

So why is the GDoF of a global force polyhedron so important? To answer this question, consider the single polyhedron of Figure 6.15. This force polyhedron has a GDoF of two, as explained in Figure 6.8. Next, let us subdivide the interior space of the polyhedron with polyhedral cells that are not all tetrahedrons, as shown in Figure 6.15b,c. Note that all the faces of the subdivided polyhedrons must be planar. Hence, the faces *AFGH*, *CEGH*, and *BDGH* all share the edge *GH*.

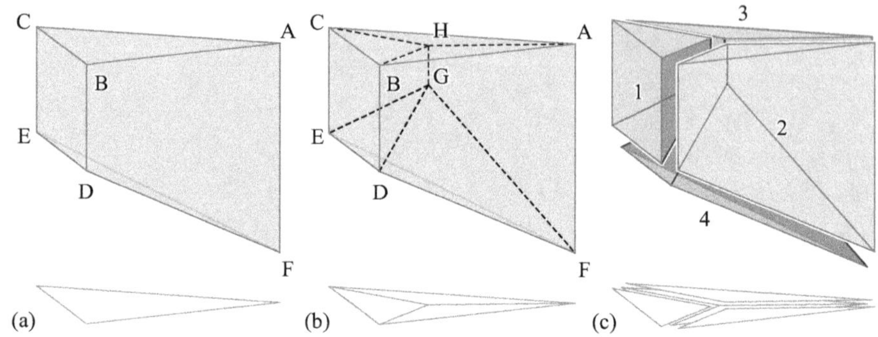

Figure 6.15 (a) to (c) Subdividing the force polyhedron into convex polyhedral cells with geometric degrees of freedom.

To explain the importance of the GDoF of a global force polyhedron in design, let us construct its reciprocal structural form by following in the Example 6.4.

Example 6.4 Construct a reciprocal structural form for the subdivided force diagram of Figure 6.15c.

Construction steps:

1 Start from the force polyhedron of Figure 6.15a that is yet to be subdivided. The reciprocal node to this polyhedron is a concurrent set of forces, as shown in Figure 6.16a. Force $\mathbf{f}_1$ is reciprocal to the face *ABC*, subdivided into three smaller triangles in the new force diagram. These faces are shown by normals $\mathbf{n}^{\dagger}_{11}$, $\mathbf{n}^{\dagger}_{22}$, and $\mathbf{n}^{\dagger}_{33}$. Thus, the number of applied loads in the form diagram has changed. The first step is to locate the load network and the lines of action of $\mathbf{f}_{11}$, $\mathbf{f}_{22}$, and $\mathbf{f}_{33}$ in the form diagram.
2 In the force diagram, start from the force face *HBC* reciprocal to $\mathbf{f}_{11}$ (Figure 6.16c). Also, in nodal cell 1, the faces corresponding to $\mathbf{f}_{11}$ and $\mathbf{f}_2$ have a shared edge (Figure 6.16b). Thus, the line of action of their reciprocal force in the form diagram should intersect.
3 In the form diagram, start from an arbitrary point on the line of action of $\mathbf{f}_2$ with a distance l from the force $\mathbf{f}_1$. This point represents the location of the line of action of $\mathbf{f}_{11}$. Label this point as node 1 in the form diagram.
4 From node 1, draw *planes* normal to the edges of the face *HBC* of the force diagram and extend until they intersect the lines of action of forces $\mathbf{f}_4$ and $\mathbf{f}_5$ in the form diagram, as shown in Figure 6.16d. The load network is constructed.
5 In the force diagram, start from cell 1 and its face normals, as shown in Figure 6.16e.
6 In the form diagram, start from node 1 at the intersection of forces $\mathbf{f}_2$ and $\mathbf{f}_{11}$ and draw lines in the direction of the normals of the faces of cell 1 to intersect the

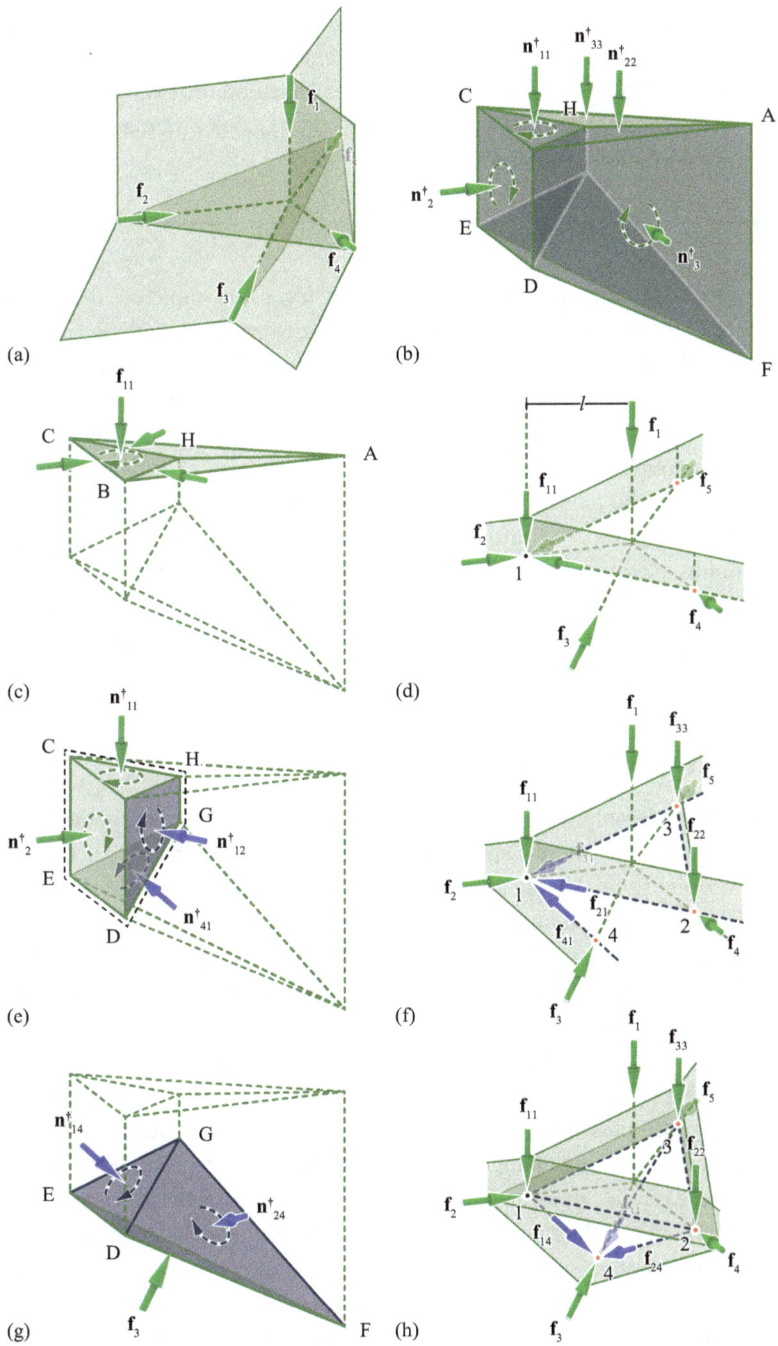

Figure 6.16 (a) The concurrent force system reciprocal to the single force polyhedron pre-subdivision; (b) the global force diagram of the subdivided force cell; (c) starting from a face belonging to the applied loads; (d) locating the load network reciprocal to the faces of the applied load in the form diagram; (e) to (h) constructing the structural form on the line of action of the initial concurrent forces using face normals of the nodal cells.

lines of action of the forces $\mathbf{f}_{1-4}$ (Figure 6.16f). These intersection points locate vertices 2, 3, and 4 in the form diagram.

7 In the form diagram, connect vertices based on the shared faces of the adjacent cells in the force diagram (Figure 6.16g,h). This process completes the construction of the form diagram.

□

Once the construction of the funicular form is completed, the GDoF of the global force diagram can be used to adjust the forces in the boundaries of the form and consequently alter the internal forces in the structural form. Figure 6.17 shows three steps in adjusting the length of $\mathbf{e}_{ind}$ as the GDoF of the global force polyhedron.

In Figure 6.17a.1, the edge length is positive, and the corresponding structural form is a compression-only system. Reducing the length of $\mathbf{e}_{ind}$ changes the geometry of the trapezoidal face attached to it and makes a self-intersecting face with two opposite surface normals. The sum of the signed areas of the faces is positive, which yields a compressive force in the members connected to the node for the case of Figure 6.17a.2. Reducing the edge length to zero will remove the external force $\mathbf{f}_2$ from the system and change the internal forces from compressive to tensile in the top members connected to node 1. Figure 6.17a.3 shows this transformation.

6.3.7 Subdividing the Global Force Polyhedron

Global force polyhedron can be used for a variety of design explorations. For example, subdividing the interior space of the global force cell into smaller convex cells results in various funicular forms with similar performance but different topologies. Figure 6.18 shows two examples of a branching structure with the same global force cell but different internal subdivisions. The global force polyhedron for both examples is the same as the global force polyhedron in the example of Figure 6.17. Both global force polyhedrons have the GDoF of two.

Interestingly, the GDoF of all of the internal subdivisions remains the same. Indeed, the internal tetrahedral network has a GDoF of one, and the height of the other cells contributes one more to the GDoF of the force diagram. Consequently, these subdivisions do not change the GDoF of the global force polyhedron.

The external loads and reaction forces in both of these examples can be manipulated to control the internal forces in the system. For instance, as it was shown for the structural system of Figure 6.17, the horizontal reactions can be decreased to zero as a design parameter to release the support of the structure on one side that results in tension-only members on the top of the branching structure.

6.4 Aggregating Global Force Polyhedrons

As briefly discussed, subdividing a global cell gives topologically different funicular forms. Multiple global force polyhedrons can also be aggregated to construct a more

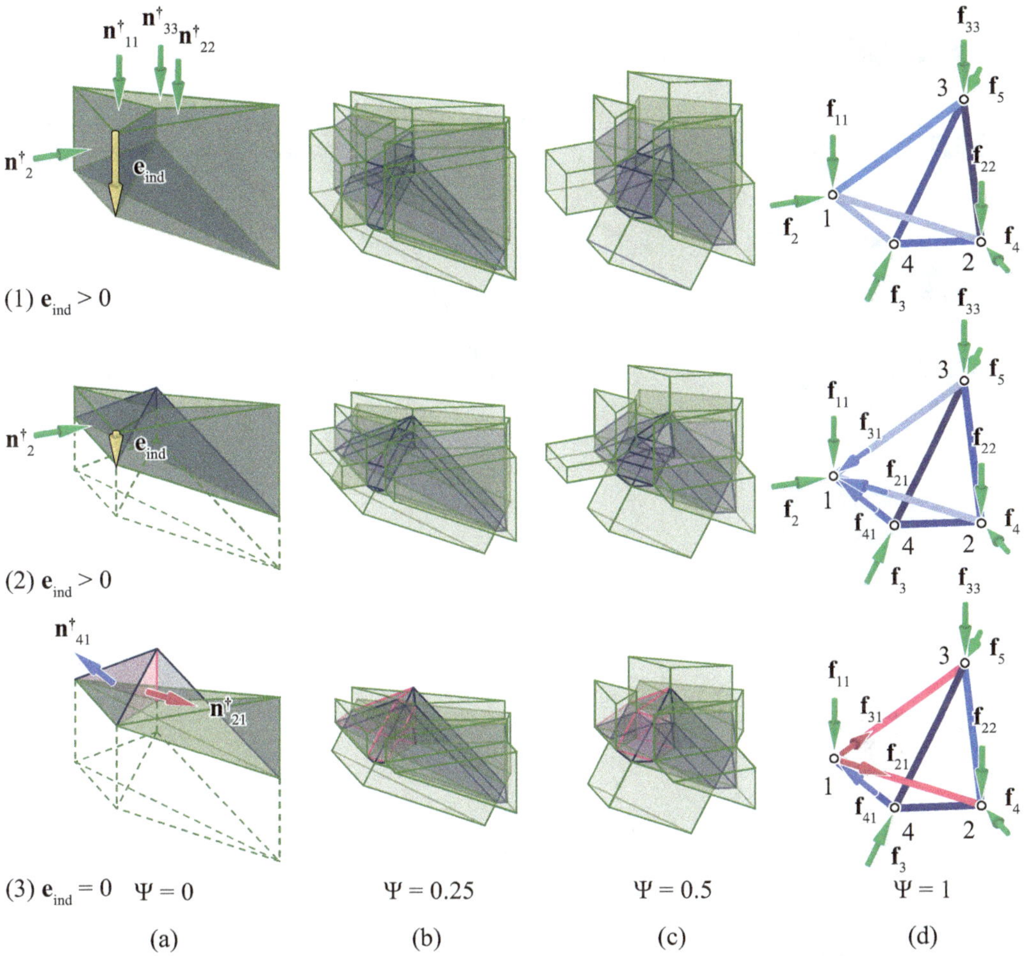

Figure 6.17 The transformation of a single cell within the force diagram as a result of changing the edge length of an independent edge and its effects on the areas of the external and internal faces: (1.a–d) positive edge length; (2.a–d) reducing the length of the edge to get a self-intersecting face; and (3.a–d) zero edge length making a face with a flipped normal.

extensive system with GDoF. To successfully combine global cells, they all should have identical faces from which they can be attached to build a larger global polyhedron. Hedracrete is a built structure designed using this approach, and its structural design process will be explained in Section 6.4.1.

6.4.1 Hedracrete; Prefab, Funicular, Spatial Concrete

Building a self-supporting branching structure is an excellent exercise since it should be laterally supported at its periphery. These supports can be removed if multiple branching structures are combined such that internal members transfer the lateral forces within the network and finally to the ground. Hedracrete, as we introduced at the beginning of this chapter, consists of multiple branching systems supporting

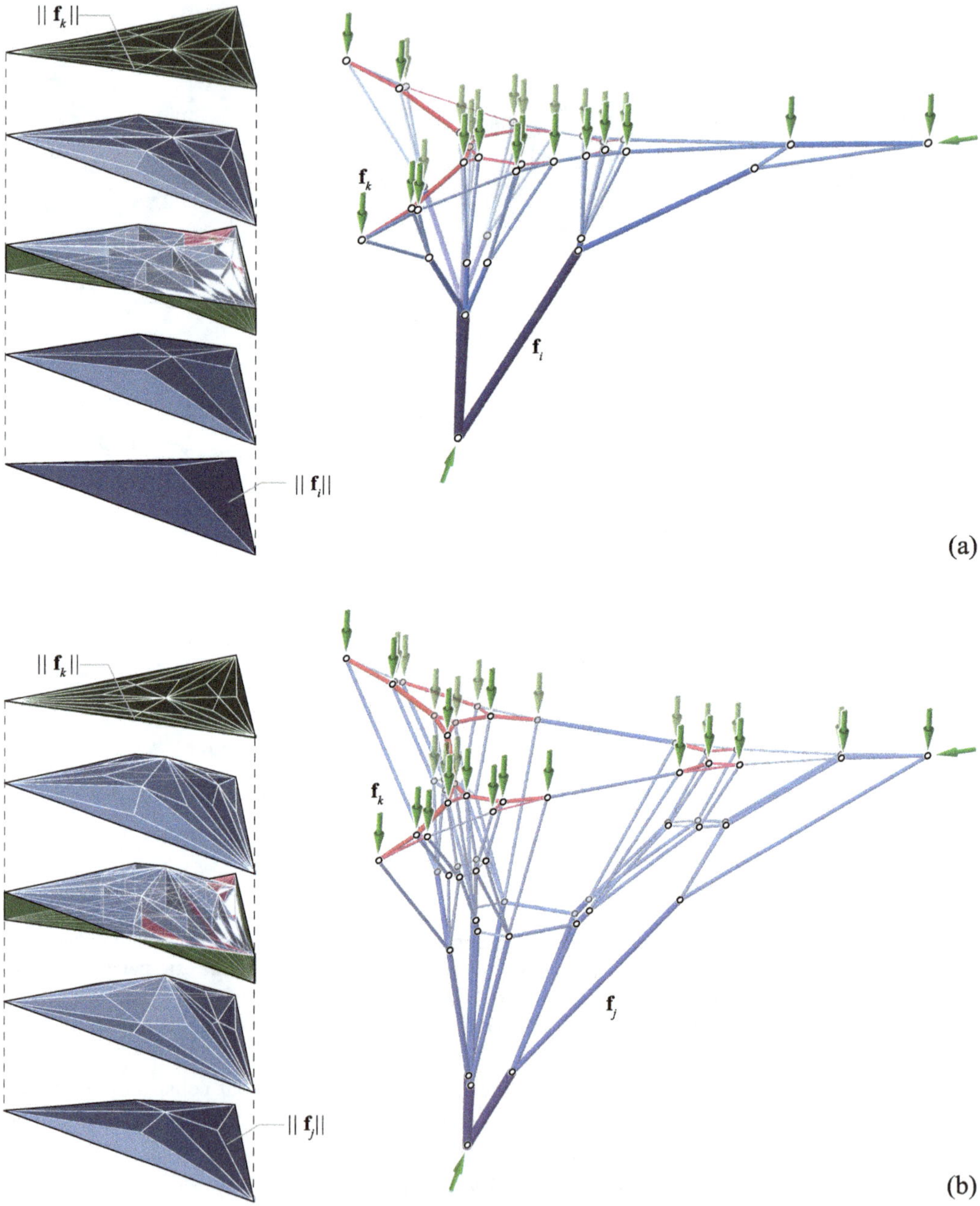

Figure 6.18 The exploded force diagram and corresponding branching form diagram: (a) before and (b) after subdividing the internal cell of the global force polyhedron. The areas of highlighted faces in the polyhedral force diagram represent the magnitude of forces in the labeled edges of the form diagram.

each other to build a compression-only thrust with tensile members on its top. What makes the form-finding process unique and intuitive is that the GDoF of the global force polyhedron is manipulated to remove the lateral supports in the structure that resulted in a system with both compressive and tensile members.[77,78]

Structural Form Finding

Figure 6.19a shows the design of Hedracrete by combining three global force polyhedrons similar to the example of Figure 6.17. Each force polyhedron has a GDoF of 2. Subsequently, the internal volume of the combined global force polyhedron was subdivided into smaller, convex polyhedral cells to derive the topology of the structural form.

In Example 6.5, we reconstruct the global force polyhedron for the entire structure in a procedural approach to show the GDoF of the force that can be manipulated in later design stages.

Example 6.5 Construct the global force polyhedron and its GDoF of the aggregated force polyhedrons of Figure 6.19 using procedural modeling.

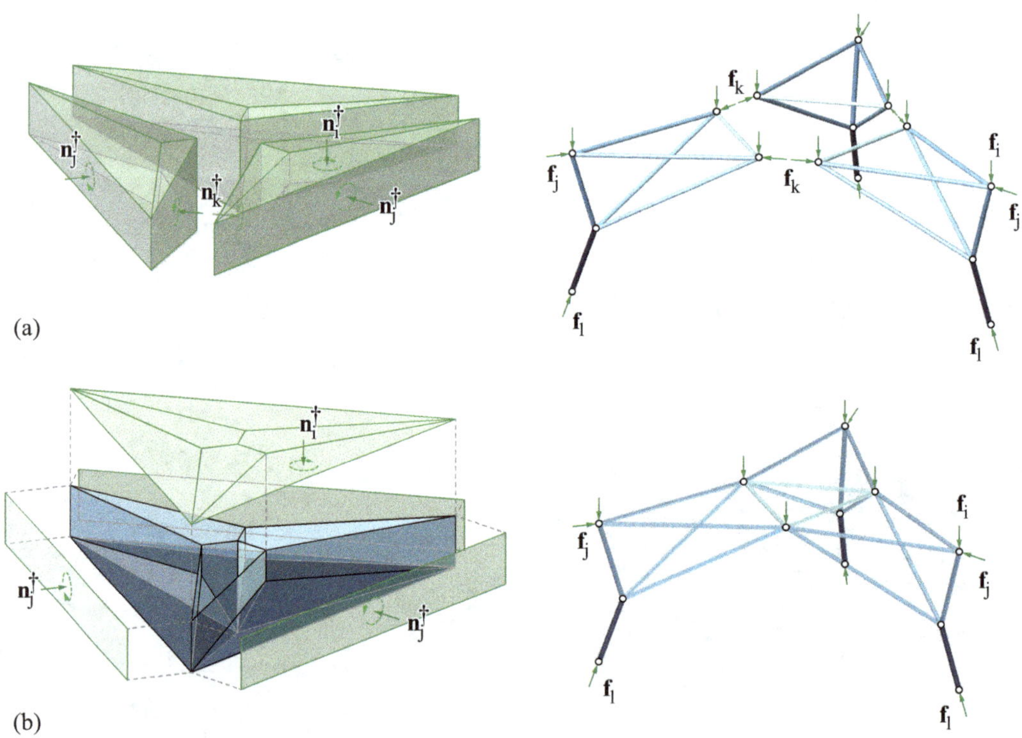

Figure 6.19 (a) The aggregation of the global force polyhedrons and their reciprocal form diagram; (b) the combined force diagram with its internal subdivision and its reciprocal form.

Construction steps:

1 In the form diagram of Figure 6.20a, identify the three supports, 1, 2, and 3, establishing a closing plane $\mathbf{n}_{cp}$.
2 In the form diagram, locate the primary applied load **f** in the space bounded by the supports 1 to 3.

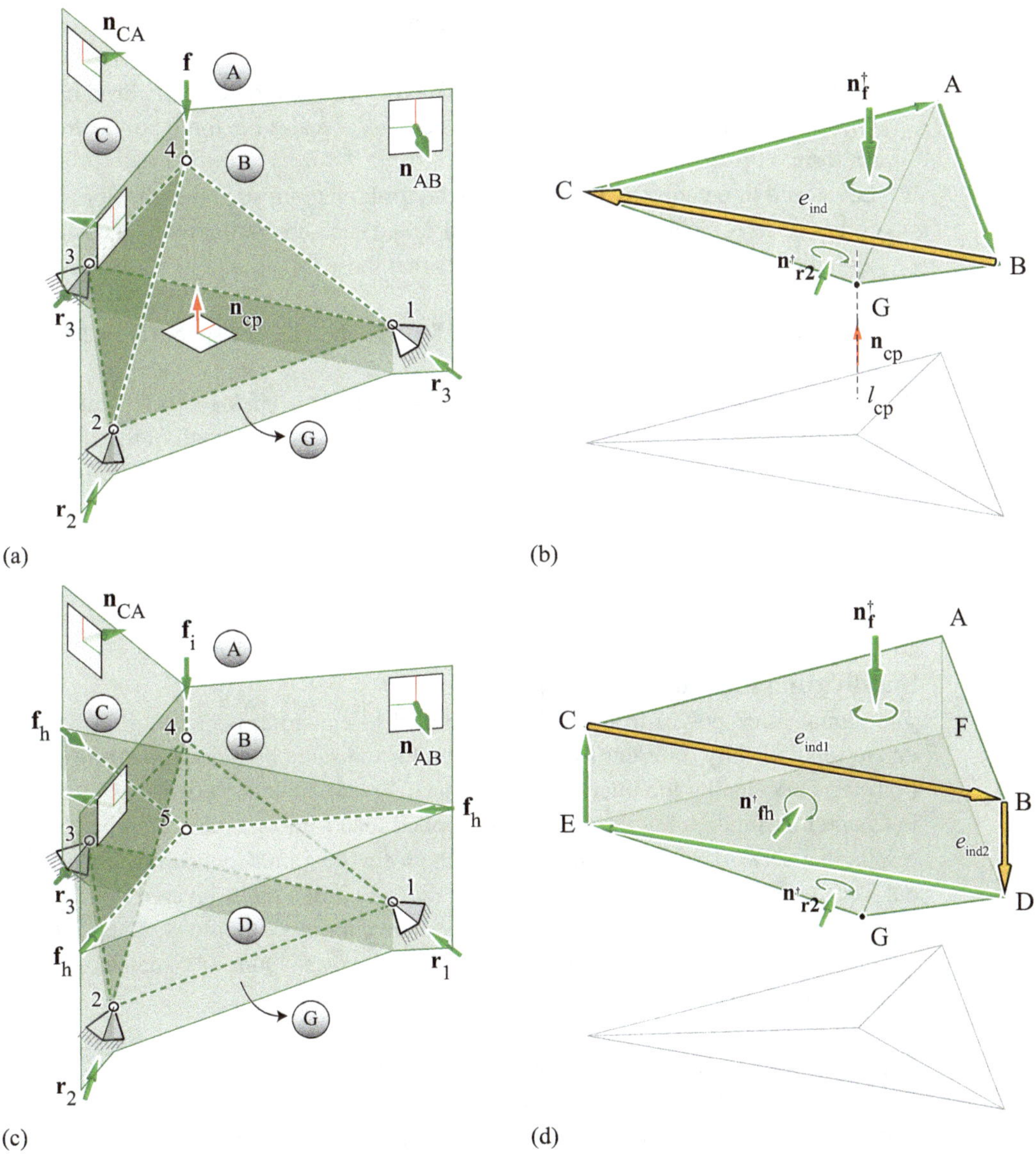

Figure 6.20 The construction of the global force polyhedron of Hedracrete: (a) three reaction forces of the intersecting on the line of action of **f**; (b) the reciprocal force polyhedron with a single GDoF $\mathbf{e}_{ind_1}$; (c) adding three horizontal forces $\mathbf{f}_h$ to the system intersecting the line of action of **f**; and (d) the new global force polyhedron with an additional GDoF shown by $\mathbf{e}_{ind_2}$.

3 We can choose a height where the supports meet on the line of action of **f** and construct the force accordingly. Let us name this point as 4.
4 As discussed in Chapters 2 and 3, the applied load and the reaction forces in the form diagram are co-planar, and a line of action of a load is shared by at least three faces dividing the 3D space into three subspaces.
5 In the form diagram, identify the subspaces and name them. For instance, the space bounded by the supports 1 and 2 and point 4 is called *A*. The other subspaces are shown by *B*, *C*, and *D* (Figure 6.20a).
6 Construct the force diagram of Figure 6.20b using the right-hand rule as explained in other examples in Chapter 3. Note that this force polyhedron has GDoF of one as the length of the edge vector $\mathbf{e}_{ind_1}$ scales the force polyhedron uniformly up and down.
7 Let us add three horizontal forces $\mathbf{f}_h$ to the form diagram such that the horizontal forces intersect with the line of action of **f**, and each horizontal force is co-planar with the load **f** and a reaction force. Note that these forces should not necessarily meet at point 4.
8 The open faces shared by the horizontal forces subdivide the space into additional subspaces *F*, *D*, and *E* (Figure 6.20c).
9 Construct the force polyhedron by considering the newly added horizontal forces. This force polyhedron has a GDoF of two as shown by the length of the edge vector e_{ind_2} that can change the areas of the faces *BAFD*, *CBDE*, and *ACEF*, as shown in Figure 6.20d.

□

Subdividing the Face of the Applied Load

The global force polyhedron can be subdivided to construct nodal equilibrium and derive topologically different funicular forms for the same boundary conditions. The first step in designing the internal topology is to create the load network as discussed in Chapter 3. Subdividing the face of the applied load **f** into smaller convex polygons results in a network of the applied loads in the form diagram. Connecting these external faces to internal cells can turn the funicular form into a multi-layered system. The same method has been used to design the Hedracrete project's topology. These internal cells are shown in Figure 6.21, which uses the Minkowski sum to visualize the cells and the transformation from form to force diagram.

Eliminating the Horizontal Force by Controlling the GDoF

The purpose of identifying the GDoF in the force diagram is to use it in design. Now that we have constructed the form and force diagrams for a compression-only branching structure, we can use the GDoF to manipulate the magnitude of the forces in the system without disturbing the form of the structure and the equilibrium.

The horizontal forces we added during the form-finding process certainly add to the spatial quality of the system. Nevertheless, the horizontal reaction forces at the top of

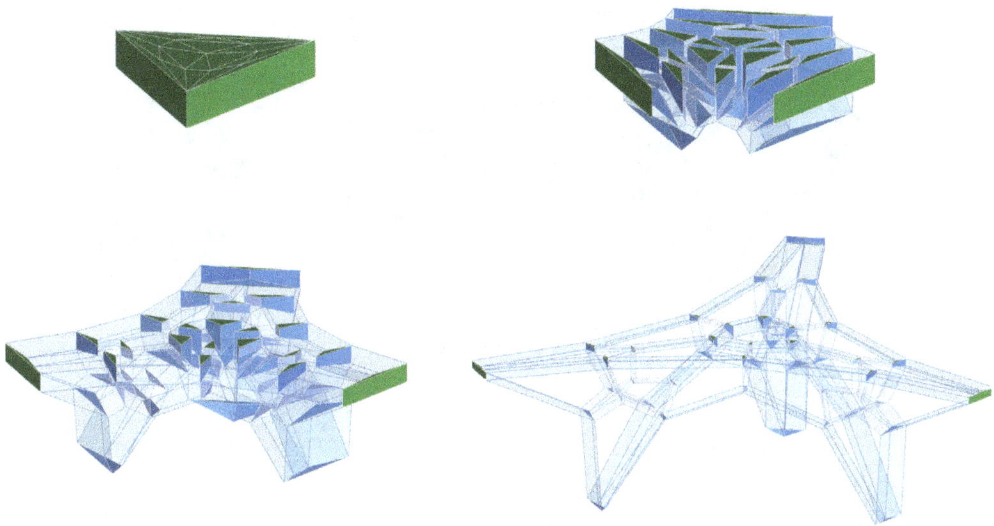

Figure 6.21 The Minkowski sum of the form and force diagrams shows the internal cells of the force diagram and their relation with the nodes of the form.

the structure might be a bit restrictive. That is, to have a compression-only system, the structure should be supported laterally on the top three corners. We can manipulate the force diagram using the GDoF of the global force polyhedron to liberate the structure from these support requirements. Let us revisit Figure 6.20d. Changing the length of the edge $\mathbf{e}_{ind_2}$ alters the area of the side faces of the global force polyhedron, which corresponds to the magnitude of the horizontal forces $\mathbf{f}_h$. Choosing zero length for the edge $\mathbf{e}_{ind_2}$ makes the rectangular side faces of the force polyhedron collapse into a line with zero area, consequently making the horizontal forces' magnitude zero. By eliminating the horizontal forces, the form of the structure stays intact, but the type of internal forces in the top compression members changes into tension. This is because the directions of the faces of the internal cells flip, and their normal will point toward the outside of the cell, as shown in Figure 6.22.

Figure 6.23 summarizes the three main steps in the structural form finding of the Hedracrete project. Figure 6.23a shows the process of defining the boundary conditions, including supports, horizontal forces, and a vertical applied load in the form diagram. The areas of the faces reciprocal to the horizontal forces are turned into zero by controlling the GDoF of the global force polyhedron. The resulting form and force diagrams in Figure 6.23c show a compression-only branching structure with tensile members on the top.

Materialization

This book did not cover much about translating the funicular forms into actual structural members for construction. But Hedracrete is a good example of how to go

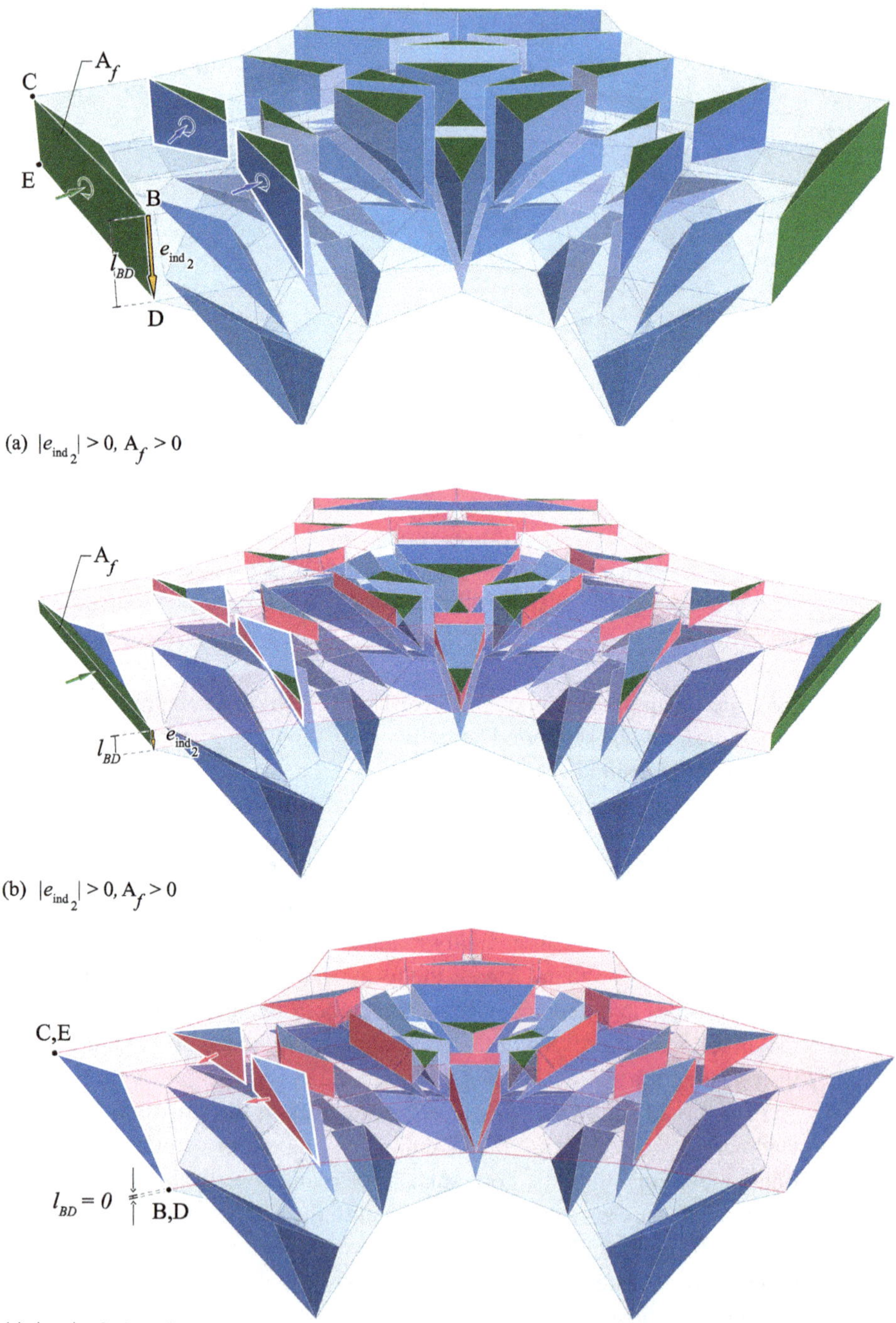

(a) $|e_{\text{ind}_2}| > 0$, $A_f > 0$

(b) $|e_{\text{ind}_2}| > 0$, $A_f > 0$

(c) $|e_{\text{ind}_2}| = 0$, $A_f = 0$

Figure 6.22 Changing the GDoF of the global force polyhedron and eliminating the side faces by turning them into a line: (a) compression-only initial configuration; (b) reducing the length of the e_{ind_2}; (c) zero length of the edge e_{ind_2} and collapsing the rectangle of *CBDE* into a line.

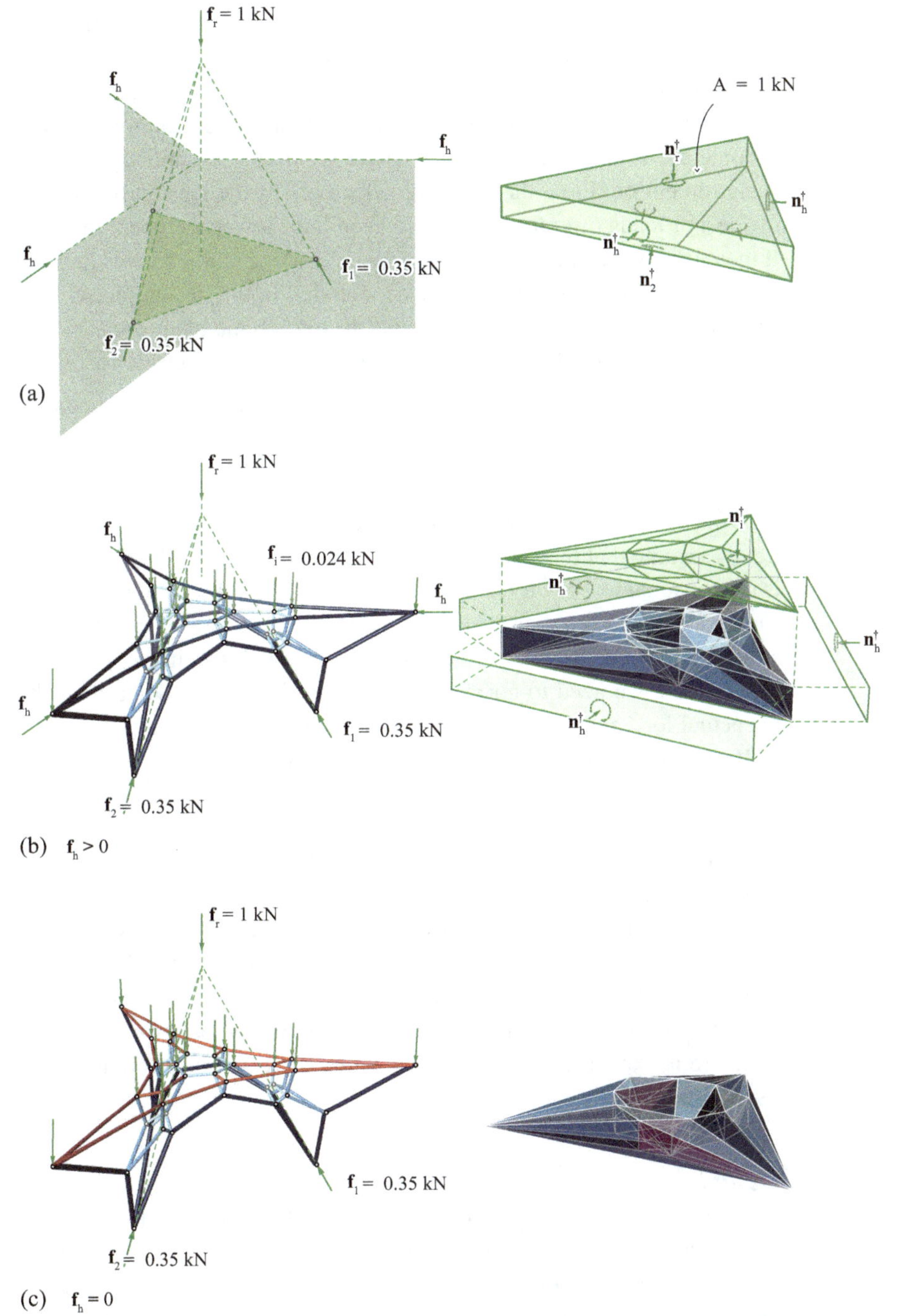

Figure 6.23 Three main steps of the structural form finding: (a) defining the boundary forces and constructing the global force polyhedron; (b) subdividing the global force polyhedron and deriving the compression-only network; and (c) changing the magnitude of the horizontal forces into zero to get a combined tension and compression system.

through this process. Once the form and force diagrams are completed, the next step is to design the cross-section of the members and choose a construction material to build the structure. Hedracrete structure includes tensile members on the top supported by compression-only members at the bottom. In the design phase, it was intended to make the tension members also from concrete so that the tension members' weight would become the system's applied load. There are multiple implications as a result of this design intention. First, the applied load/weight of the tension members should match the design loads of graphic statics to keep their lower members in pure compression. If there is a discrepancy between the design and the applied loads in the built structure, the funicular form will not behave as intended in the design phase. Second, the construction detailing should be different if the tension members are also made of concrete because unreinforced concrete cannot take a significant tensile force.

Remark 6.4. The actual applied load on a built structure should match the design load so the funicular structure behaves as intended in the design phase.

Finding the Actual Load of Each Node

Let us assume that the sum of all areas of the top faces of the global force polyhedron is proportional to 1 kN force. Therefore, force $\mathbf{f}_i$ is a fraction area of this total force on each node on the top. For instance, $\mathbf{f}_i = 0.029$ kN is a design load applied to vertex v_i, as shown in Figure 6.24a,b. To compare the applied load in the built structure with the design load, we need to normalize the values for each node based on the design and the actual load and find the difference Δ_i between the two values.

If the total applied load in the built structure is the weight of the tensile members, then the tributary area for each node on the top of the structure is equal to the sum of half of the weight of all its connected members, as shown in Figure 6.24b,c. The tributary load values per vertex of the top chord are calculated and normalized with respect to the total weight of the tensile members, which is $1,765$ kg.

The normalized value for node v_i based on the tributary area is 0.026 kN. Thus, the difference between the design load and the actual load is $\Delta_i = 0.003$ kN, which is insignificant.[133] Calculating the difference for all nodes shows a negligible difference between the design and applied loads for the Hedracrete project. Therefore, the built structure is expected to behave as a compression-dominant branching system tied by its tensile members on the top.

Designing the Connection Details

After checking the design loads and the applied loads in the system, it is time to design the connection details of the tension members. A tensile member has minimal reinforcement embedded in its cross-section to carry the tensile load. Moreover, the connection detailing of the node is designed such that all the tensile forces would form a tension-only network on the top of the structure to preserve the equilibrium of the structure under the applied loads (Figure 6.25). This detailing includes a steel plate at the end of each tension member connected to an internal rod. The plate is then connected to the plate of the node to create a tensile network on top of the structure.

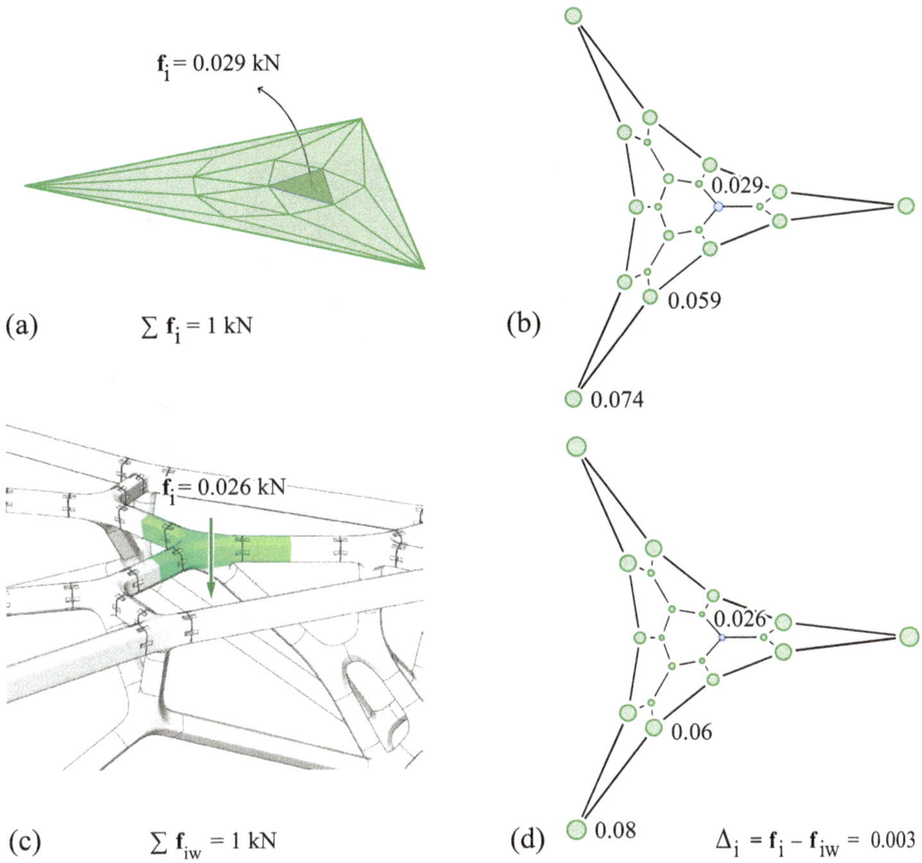

Figure 6.24 The negligible differences between the graphic statics' design load per node (a and b) and the weight of the tensile members for each tensile node (c and d) and the difference between the two.

Efficiency

To understand the structure's efficiency, let us assume that the structure was intended to function under dead and live loads. We can propose covering the top chords of the structure with a layer of structural glass. If so, the dead load for the glass will be 48 kg/m^2, and the live and snow loads will be 196 kg/m^2 and 78 kg/m^2, respectively – following the *ASCE/SEI7* – 10.[134] Therefore, the maximum loading combinations are 410 kg/m^2. The total area of the top part is 20.5 m^2. Therefore, the structure should be designed for 82 kN resultant force on top instead of the weight of the tensile members, 17.6 kN.

Analyzing the structure under 82 kN applied load shows that the 3.1 m^3 concrete distributed in 50 m^3 of space in a funicular configuration can transfer 1,000 kN to the supports. Simply put, the structure has a factor of safety of 12 for the design load 82 kN.

To highlight the structure's efficacy, let us design and analyze a conventional reinforced concrete frame with the length, width, and height of 5.4, 2.8, and 3.3 m,

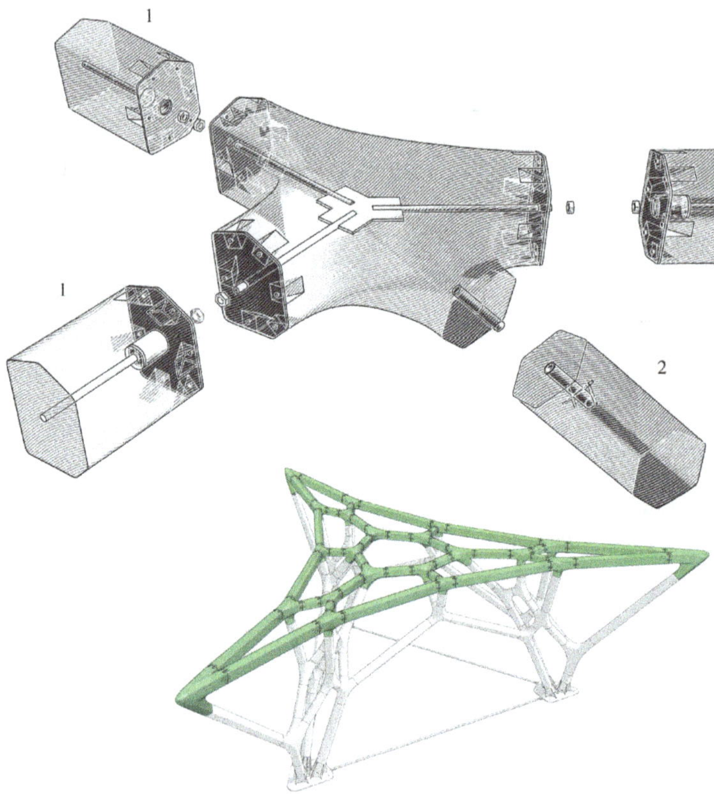

Figure 6.25 Top: exploded axon of a joint with three adjacent tensile members (1) and a compressive member (2) revealing the tensile rebars, connecting plates, and the custom cut plate to control the angles of the rebars in the middle of the joint; bottom: the weight of the tensile members is considered as the applied load for the built structure.

respectively, to cover the same space. The frame should be designed for 410 kg/m^2 (82 kN in total) based on a recommendation by the *American Concrete Institute's* (ACI-318) minimum requirements.

A conventional frame satisfying the requirements for covering the same space consists of four beams with a span of 2.8 m, placed 1.8 m apart from each other and subjected to 7.38 kN/m load. These beams sit on two girders with a span of 5.4 m supported by four columns carrying a 15 kN load to the ground. Figure 6.26a shows a typical cross-section designed to fulfill the requirements for the given design load. The conventional frame designed for 82 kN design load requires approximately 1.5 m^3 concrete and 163 m rebars in addition to stirrups. Although the concrete volume is less than Hedracrete's, the frame cannot take up to 1,000 kN of distributed loads. Therefore, the design of the frame should be updated to provide a better comparison. As shown in Figure 6.26b, the updated frame requires 3.5 m^3 concrete and 207 m rebar and stirrups, which is around 10% and 330% more concrete and steel compared

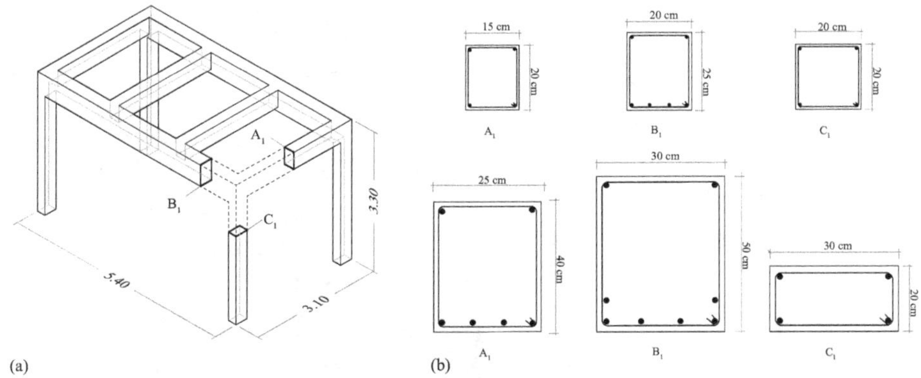

Figure 6.26 The design of conventional concrete structure; (a) details based on required loads and the same volume; and (b) details based on the ultimate load capacity of Hedracrete.

to Hedracrete. This increase in the use of the material is mainly due to the bending moment in the conventional frame that can be avoided by using spatial funicular geometries designed by polyhedral graphic statics.

6.5 Projecting Polyhedral Faces to Make GDoF

There are other design approaches to building a force diagram with large GDoF, which is useful for design purposes. Projecting a face on a particular plane and completing the side faces usually results in a polyhedron with a GDoF of at least two in the direction of the projection. Saltatur is a prefabricated structure that has been designed using this technique.

6.5.1 Saltatur: Node-Based Assembly of a Funicular Systems

The *Saltatur* demonstrates innovative research in the design and fabrication of a prefab, discrete, spatial composite structure consisting of a spatial, compression-only concrete body, tensioned steel rods, and an ultra-thin glass structure on its top (Figure 6.27). This project was exhibited at the American Institute of Architects in Philadelphia as a spatial furniture. The horizontal components of the reaction forces in the structural form were eliminated by controlling the GDoF of the global force polyhedron. The areas of the side faces of the global force diagram were turned into zero. Consequently, the internal members of the structure on the top and bottom turned into tensile elements, keeping the entire concrete body in compression.[135,136]

Design of the Force Diagram

The following paragraphs will explain the process of constructing a force diagram and controlling its GDoF to design the structural form of Saltatur. The main intentions of this design task were to (1) design a structural form that can be assembled

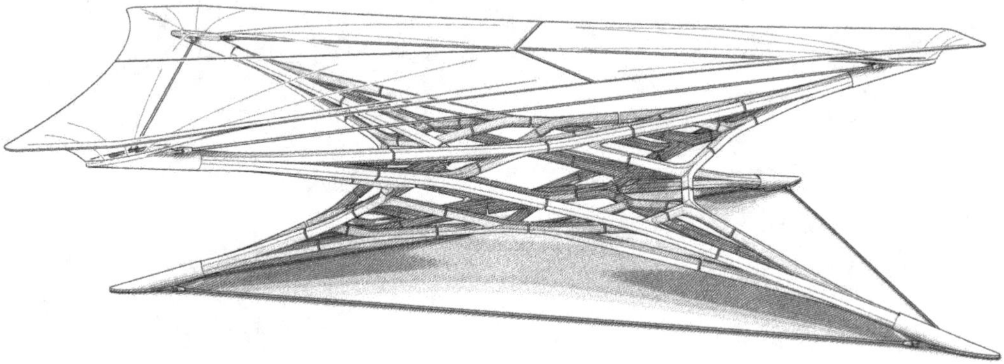

Figure 6.27 Saltatur is a polyhedral funicular structure with a compression-only body and tension ties on the top and bottom of the structure with vertical-only reaction forces, as a part of the *Spatial Efficiency* Exhibition at the Center for Architecture and Design in Philadelphia in January 2020 and the Venice Architecture Biennale 2021 virtual exhibition.

and disassembled with vertical-only reaction forces at the supports and (2) rotate the top branches of the structure to the bottom one and create a twist in the body of the structure.

Example 6.6 The aim of the following construction is (1) to build a force polyhedron for a concurrent system of loads and identify its GDoF and (2) to create additional polyhedral cells by projecting the faces of the constructed force polyhedron on two parallel planes that, in turn, decompose the applied loads in three planes in the form diagram.

Construction steps:

1. Let us begin with two concurrent systems of forces in the form diagram, as shown in Figure 6.28a. Each system has four forces, including a vertical force meeting at a point. The top system is a mirrored image of the bottom, except it has been rotated 180° on the xy- plane.
2. Combine these systems into a single node and remove the vertical loads. This combination gives us six forces applied on the node whose open faces construct two triangular bases on the node's top and bottom.
3. Connect the lines of action of the forces with shared faces as shown in Figure 6.28b. This should make a convex hull using the tip of the applied loads around the central node.

4 Extrude the convex hull to the central node and make the faces shared by the forces in the form diagram (Figure 6.28b). Note that the magnitude of the forces is unknown and will be identified by constructing a closed polyhedron of force for this system of loads.
5 Once the convex hull is constructed, the shared faces in the form diagram define the subspaces around each applied load.
6 Round each applied load and its connected face normals in the form diagram to make a face with a closed loop around its edges in the force diagram, as shown in Figure 6.28c.
7 The single force polyhedron for this configuration of forces has six faces and a GDoF of three that can transform the polyhedron into multiple configurations with various face areas (see Figure 6.29a–c). This variation shows that many combinations of force magnitudes and face areas can keep the node in equilibrium given the direction of the applied loads.

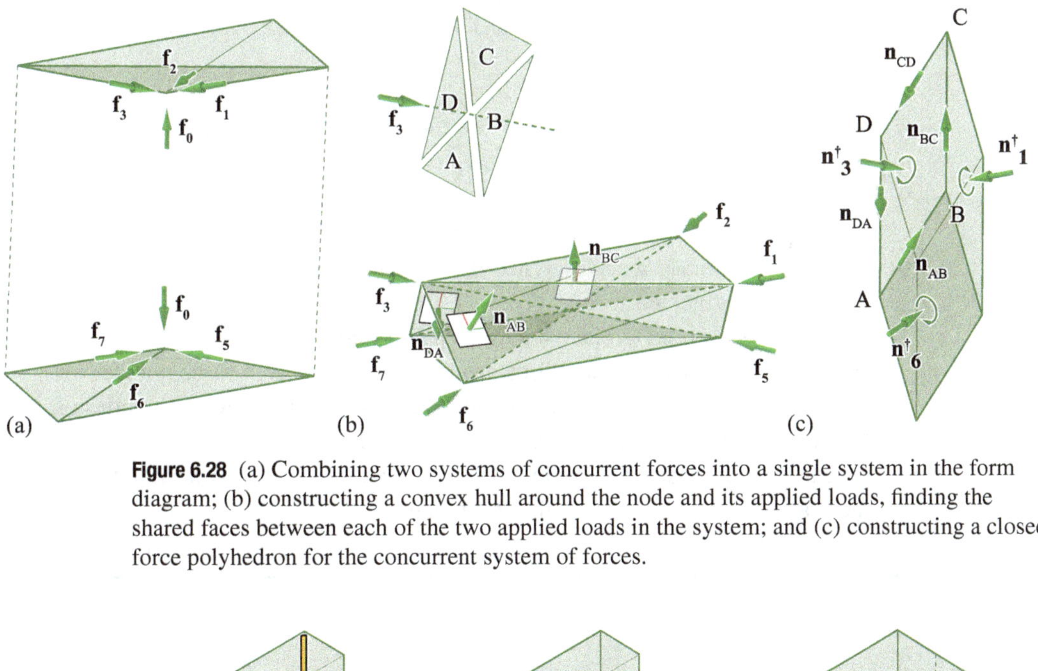

Figure 6.28 (a) Combining two systems of concurrent forces into a single system in the form diagram; (b) constructing a convex hull around the node and its applied loads, finding the shared faces between each of the two applied loads in the system; and (c) constructing a closed force polyhedron for the concurrent system of forces.

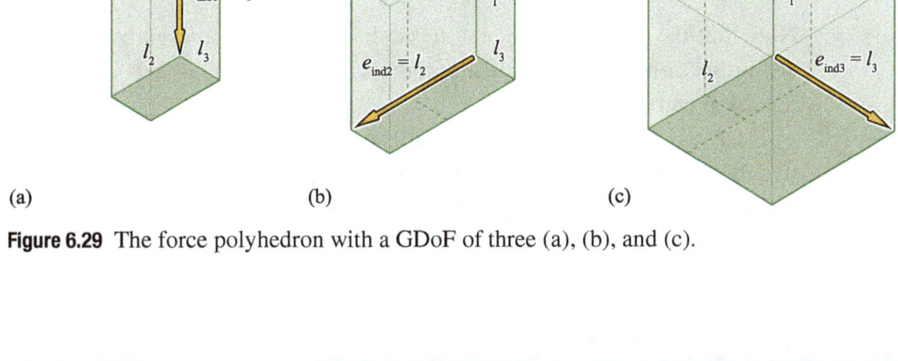

Figure 6.29 The force polyhedron with a GDoF of three (a), (b), and (c).

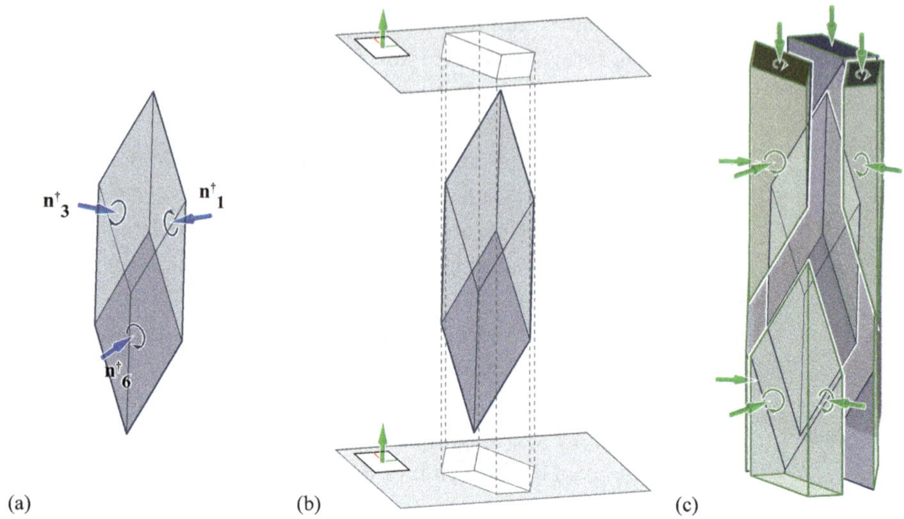

Figure 6.30 (a) A single force polyhedron for concurrent forces; (b) projecting the faces of the force polyhedron on two planes on the top and bottom; and (c) constructing new force polyhedrons using the projected and extruded faces.

8 Take two horizontal planes, one on the top and one on the bottom of the force polyhedron at an arbitrary distance from the force polyhedron (Figure 6.30a,b).
9 Project the top three faces of the force polyhedron to the top plane and the bottom three to the bottom plane.
10 For each face, make a new polyhedron by connecting the edges of the original face and the projected face.
11 Note that each projected face makes a new closed-force polyhedron in the system, and we will end up with a group of closed, convex force polyhedrons.
12 The external faces of this group of cells are a new global force polyhedron for our force diagram (Figure 6.30c). In addition, the newly constructed force diagram decomposes the applied loads into force components normal to the new faces and additional nodes in the form diagram.

□

6.5.2 Removing Horizontal Reactions Using GDoF

Note that the GDoF of the new force diagram has increased by two, one for each of the projections. Because all the heights of the newly added faces in the projection direction can be manipulated.

First, we explain how to manipulate a single force polyhedron, and then we will use this in the design of our structure. In Figure 6.31, the length of $\mathbf{e}_{ind}$ as an independent edge can be manipulated. Figure 6.31a shows a positive length for $\mathbf{e}_{ind}$ that is reciprocal to a compression-only node. If the edge length of $\mathbf{e}_{ind}$ is set to be zero

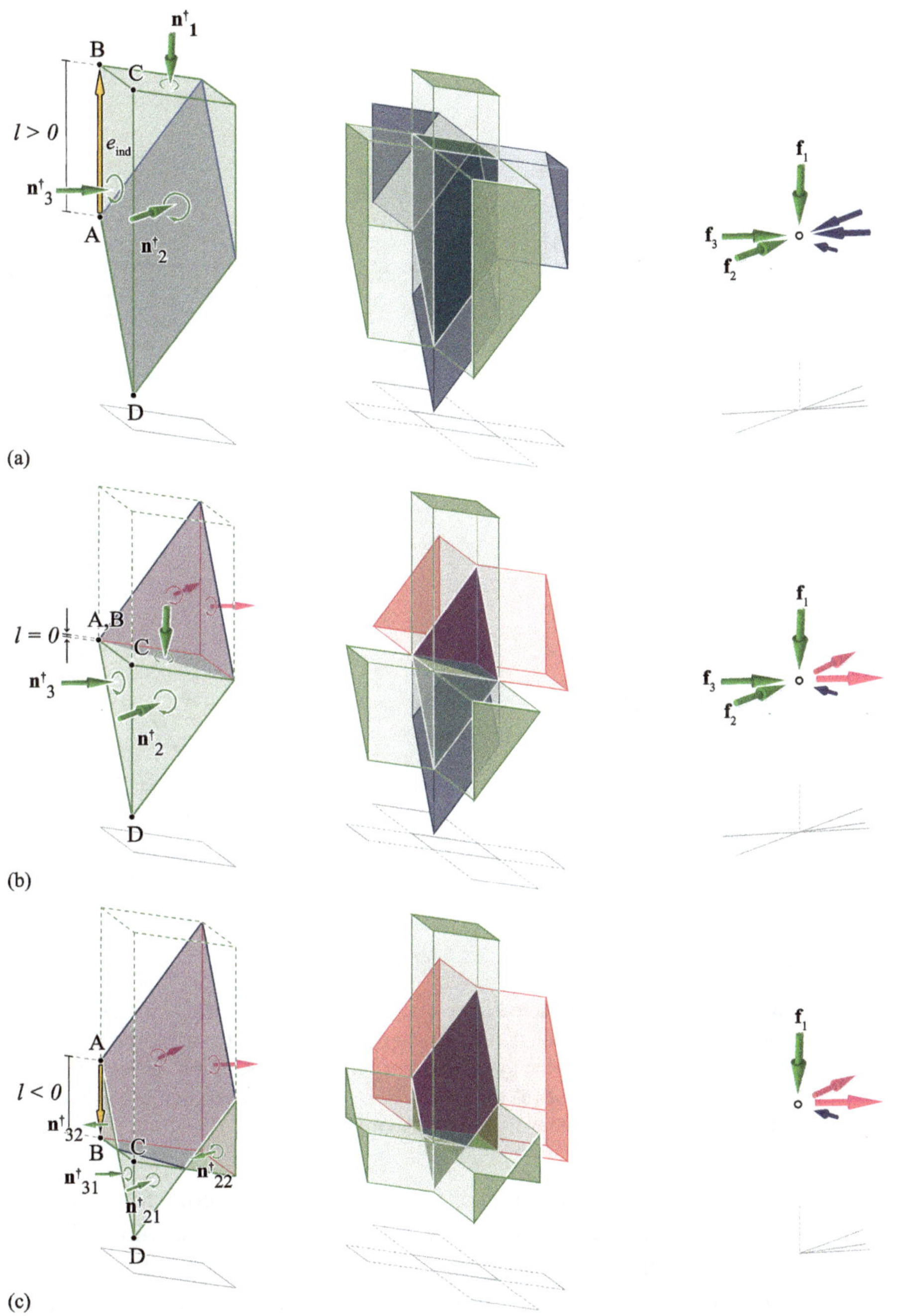

Figure 6.31 (a) A force polyhedron constructed by projecting a face, its Minkowski sum, and its reciprocal node in the form diagram; (b) changing the length of an independent edge to manipulate the areas of the side faces and the reciprocal loads; and (c) eliminating the horizontal forces by making self-intersecting faces with zero area.

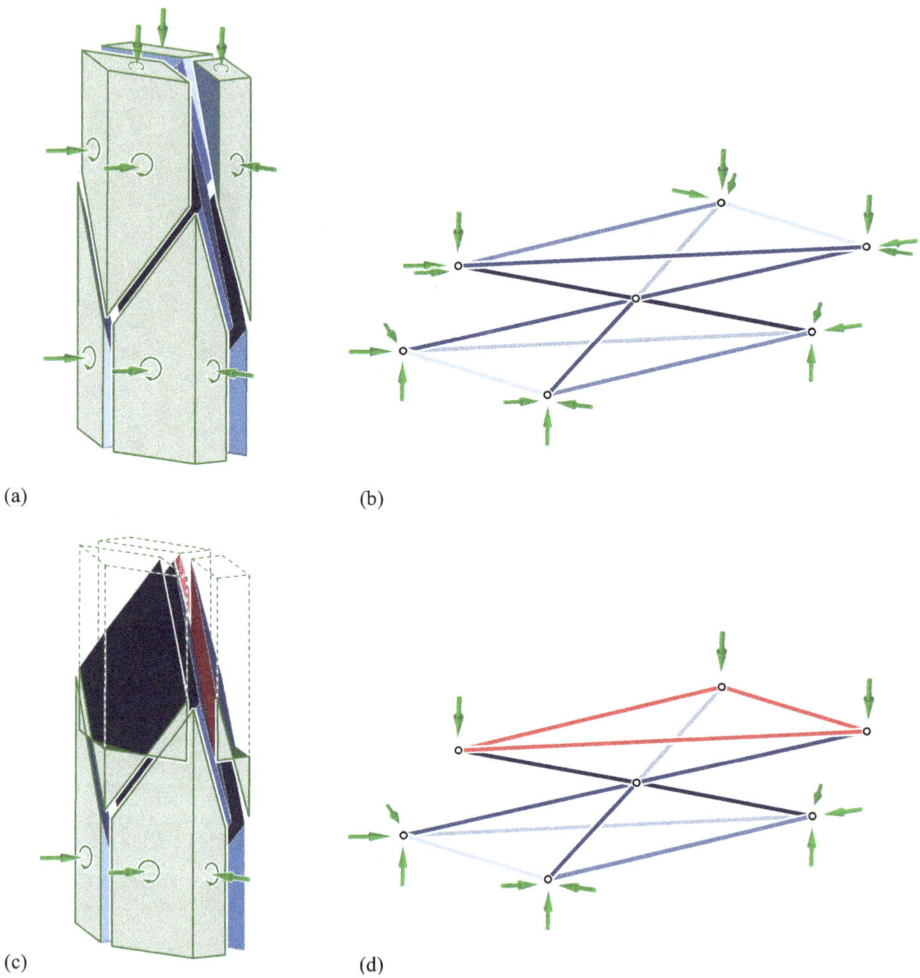

Figure 6.32 (a) The constructed force diagram; (b) its reciprocal compression-only form diagram with both horizontal and vertical reaction forces; (c) using the geometric degrees of freedom, we manipulate the system to make the face area of the vertical faces on the top three force polyhedra zero; and (d) the elimination of the top horizontal forces in the form diagram and the formation of the tension tie on the top.

in Figure 6.31b, the direction of the corresponding internal force in the form diagram will be flipped into a tensile force. In Figure 6.31c, we obtain self-intersecting faces by changing the edge length of $\mathbf{e}_{ind}$ into a negative value. If the sum of the areas of the self-intersecting regions is zero, then we have a polyhedron with zero-area faces that eliminates the forces in the form diagram.

We use the same method to manipulate the global force diagram of Figure 6.30c, as shown in Figure 6.32a–d. We begin with the projected polyhedral system, Figure 6.32a, and its dual diagram, Figure 6.32b. The force diagram is then manipulated using its GDoF to achieve zero-face area for the vertical faces of the above three cells. These faces become self-intersecting faces. In the corresponding form diagram, the top horizontal forces are eliminated, as shown in Figure 6.32d, and the force in the internal members turns into a tensile force. The same technique can be used to remove horizontal reactions at the bottom of the table.

6.5.3 Subdividing the Force Polyhedron

Section 6.5.2 showed how projecting the faces of a single polyhedron can make a force diagram with an additional geometric degree of freedom. This GDoF was manipulated to change the magnitude of the externally applied loads in structural form. We also learned in Chapter 3 that a single force polyhedron could be subdivided into smaller convex polyhedral cells that result in various funicular forms with different subdivisions for the same boundary conditions. It is certainly possible to combine these two techniques in the design process.

The initial force polyhedron we used in projecting the faces can further be subdivided to make a more intricate structural form with further spatial complexity. Figure 6.33a,b shows the base force diagram and a subdivided force diagram and their reciprocal funicular structural form in a Minkowski transformation. Figure 6.33 illustrates how the cells with faces reciprocal to the external loads and reaction forces do not change in the subdivision process, and only the main internal cell is subdivided into smaller convex polyhedrons to make a new structural form.

In the subdivision process of the force diagram, the cells that have the faces of the external loads stayed intact. Thus, their GDoF can manipulate the face geometry and eliminate the horizontal forces and reactions in the structural form without disturbing the equilibrium. Figure 6.34 shows how the GDoF of the global force polyhedron can

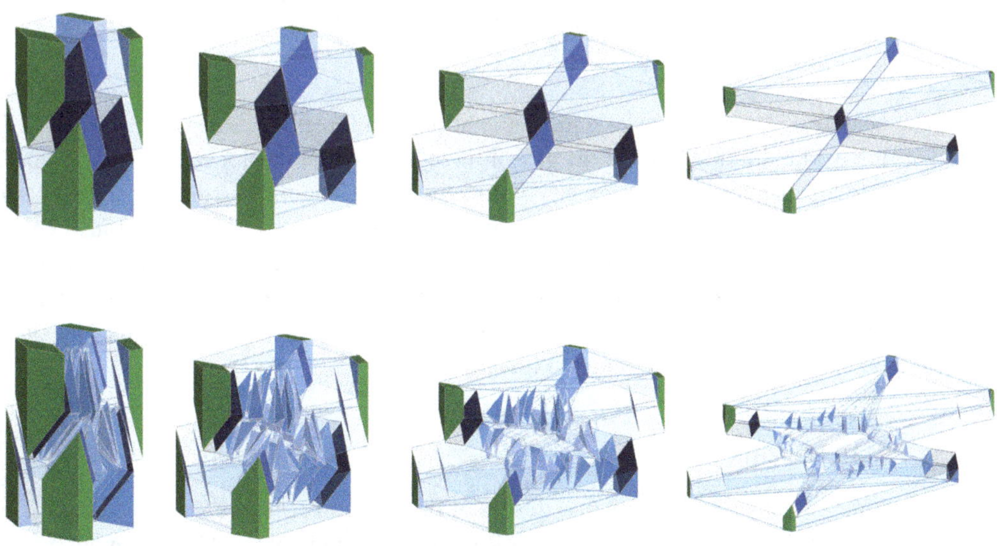

Figure 6.33 (a) Shows the initial force polyhedron and its Minkowski transformation to the form diagram; and (b) a subdivided force polyhedron and its transformation to the form diagram.

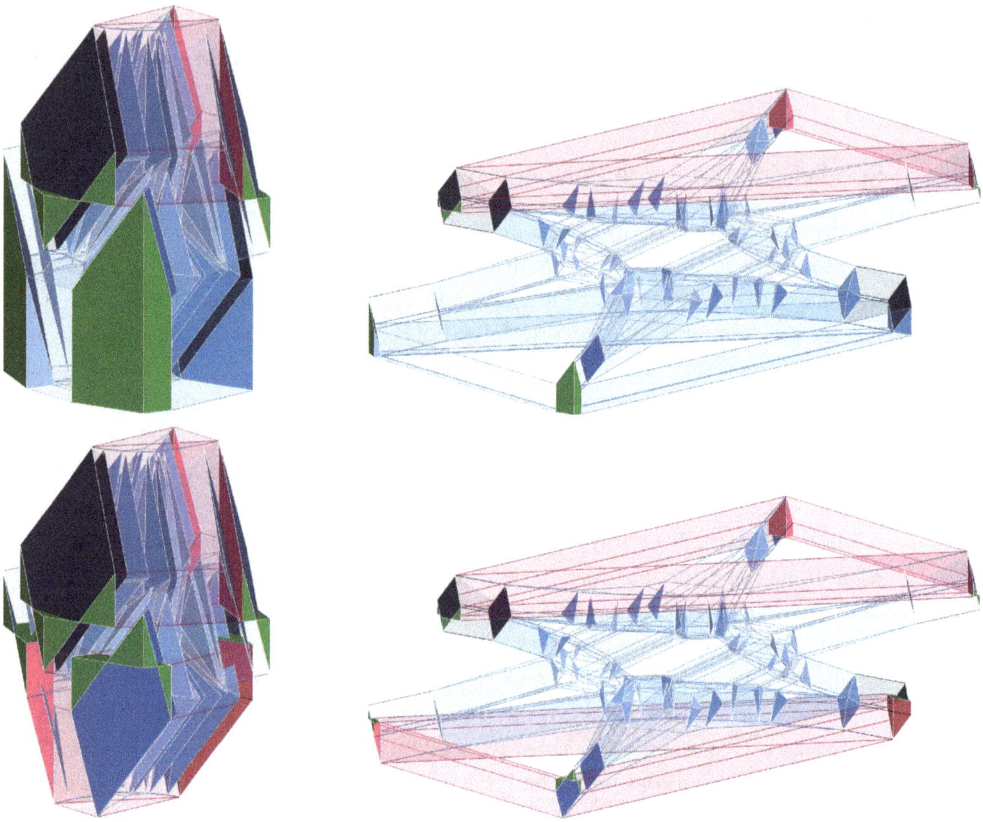

Figure 6.34 (a) Changing the areas of the upper vertical faces of the force diagram in the subdivided version and eliminating the horizontal forces on the top nodes that form a tension tie on the top of the structure; and (b) similar manipulation on the bottom chord of the structure.

be adjusted in two steps to remove the horizontal forces in the top nodes. First, in Figure 6.34a, the side faces of the global force diagram on the upper part are transformed into a self-intersecting geometry with zero areas. The transformation of faces eliminates the horizontal external loads on the top nodes of the structure. It also changes the direction of the internal force in the top member of the structure from compression into tension. In the second step, the same transformation is used for the lower faces of the force diagram to eliminate the horizontal reactions and make another tension tie on the bottom of the structure (Figure 6.34b).

6.5.4 Materialization

Figure 6.35 shows the realized design of the structure, including ruled-surface geometry for the architectural detailing of the nodes and members. High-strength concrete was used for the compression-only body of the structure. The entire volume of concrete used for this structure is 0.06 m^3 distributed in 4.44 m^3 (3.1 m × 0.8 m × 1.9 m)

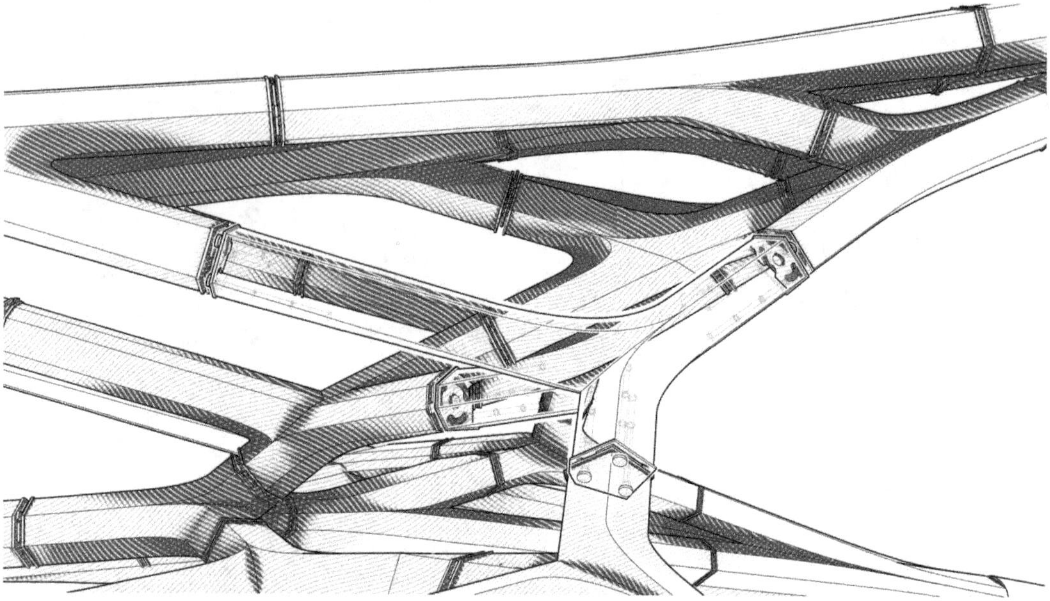

Figure 6.35 A node-based assembly was used to construct the structure: the highlighted node is connected to the four adjacent nodes using a rotatory lock mechanism that was devised to prevent deadlocking the members for the assembly.

of space. This volume makes the relative density of this structure as small as 1% (0.013).

A node-based assembly was considered as a method of construction. To facilitate the assembly and to avoid the possible spatial collision of the branches of the nodes during the assembly, innovative detailing was developed that allowed fixing the members in their exact locations in the structure by amending the conventional male-female connection details (Figure 6.35). Although the concrete structure has been designed to act in pure compression, some members experience tensile stress in the cases of asymmetric loads, especially during the assembly process. The proposed bespoke steel connection effectively transfers the tensile force between the concrete members. The spatial concrete nodes of the structure were fabricated by injecting high-strength concrete in 3D-printed polylactic acid (PLA) molds holding the bespoke steel connections in their precise location at the ends of each node.

6.6 GDoF of the Form Diagram

Since the form diagram also consists of closed polygonal faces, it can have GDoF. The GDoF in the form diagram allows us to find a family of funicular solutions for a single force diagram. In other words, if a form diagram has GDoF larger than one, its force diagram is reciprocal to a large family of funicular solutions, each with different edge

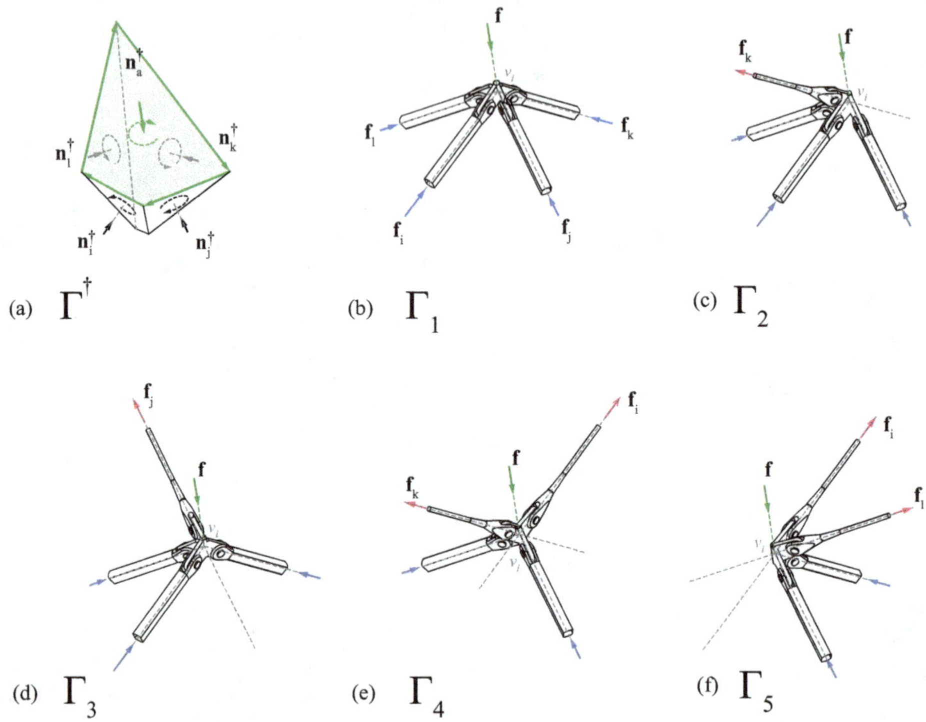

Figure 6.36 (a) to (f) A single indeterminate force polyhedron can also represent the equilibrium of multiple systems of forces, including compressive and tensile forces.

lengths. The independent edge lengths of the form diagram can take various lengths. In addition, as shown in Section 6.5, the edge lengths of a geometrically indeterminate diagram can also take negative values. Using the same technique for the form opens the door for various exciting solutions for a single force diagram.

6.6.1 A Form Node and Its GDoF

A single-force polyhedron can describe multiple equilibrium states of a single node in 3D, including compression and tension forces. This phenomenon was explained briefly in Chapter 1. That explanation was a prelude to the GDoF in the form diagram, as we discuss now.

Figure 6.36a shows a closed force polyhedron $\Gamma^\dagger$ with five sides where the normal directions $\mathbf{n}_{i-l}^\dagger$ are toward the inside of the cell. This force diagram is reciprocal to a compression-only node, v_i, as shown in Figure 6.36b. Each face is reciprocal to a force in the member connected to the node, and the area gives the magnitude of the internal force in the member of Γ_i.

If all the members have positive edge lengths, the form represents a compression-only node, for example, Γ_1. However, if any of the members of the Γ has a negative

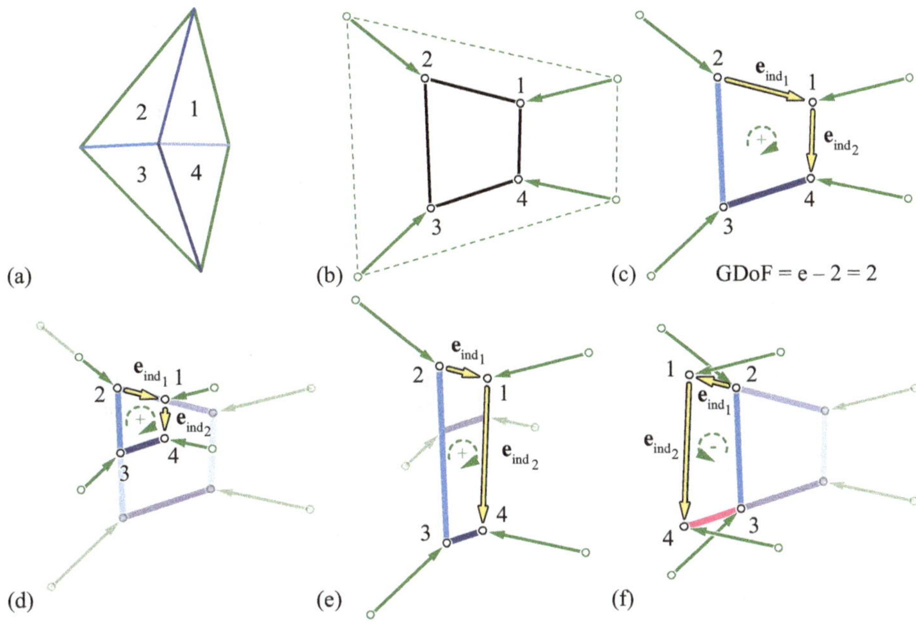

Figure 6.37 (a) A force diagram that corresponds to (b) a trapezoidal structural form with applied loads; (c) to (f) shows various structural manipulations of the trapezoidal form using the two independent edges $\mathbf{e}_{ind_1}$ and $\mathbf{e}_{ind_2}$.

length, the node will have tension and compression forces. Thus, $\Gamma^{\dagger}_{2\text{–}5}$ represents various configurations in the node v_i that can be in equilibrium with the same absolute magnitude of the forces but positive and negative edge lengths (Figure 6.36c–f). This property allows the design exploration of various structural solutions within the same equilibrium state.

6.6.2 A 2D Funicular Form and Its Closed Boundary Polygon

Consider the compression-only, trapezoidal structural form with its applied loads and reciprocal force diagram shown in Figure 6.37a,b. The GDoF of the trapezoid is two, and a pair of independent edges $\mathbf{e}_{ind_1}$ and $\mathbf{e}_{ind_2}$ are illustrated in Figure 6.37c. The lengths of these independent edges, l_1 and l_2, can be manipulated to generate various families of funicular solutions, as shown in Figure 6.37d–f. Notably, in Figure 6.37f, a new structural form with tensile and compressive forces is constructed by choosing a negative edge length for l_1. However, this transformation neglected the location of the external loads or reaction forces. To include the location of the boundary forces, one may add a polygon to close the endpoints of the external loads, as shown in Figure 6.38a.

The GDoF of the new network remains two using Formula 6.2:

$$GDoF = e - 2f = 12 - 10 = 2.$$

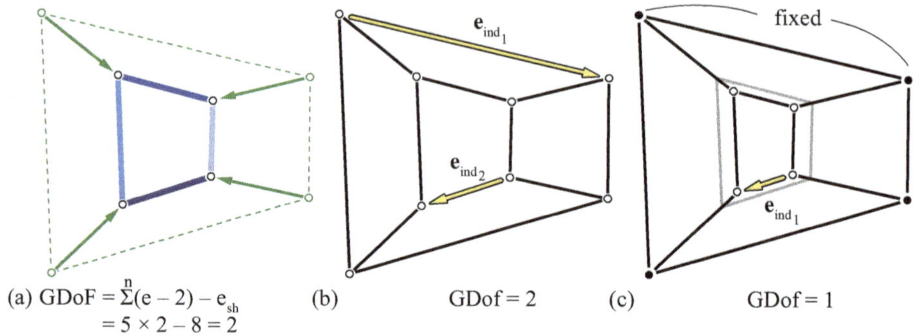

Figure 6.38 (a) Adding a closed polygon to the endpoints of the external loads; (b) calculating the GDoF of the new network; and (c) constraining the external polygon and the new GDoF of the constrained network.

Figure 6.38b provides a choice of independent edges. For instance, the top edge of the boundary is independent, and its length can be freely chosen. Therefore, given any prescribed distance, we can alter the initial boundary location of the structure without changing the force magnitudes and the equilibrium of forces in the system. If we assume that the boundary is fixed, then the GDoF will be creased to one, as shown in Figure 6.38c, and the funicular network can only be scaled up and down in the boundary.

Remark 6.5. Although with Equation 6.2, it would be possible to predict the GDoF for a 2D network for simple cases, the assignment of the independent edge to the network is not an arbitrary process, and it is not unique. The edge assignment should either follow procedural steps to complete the network or be derived using a mathematical definition of the problem that will be visited in depth in Chapter 7. Similarly, controlling multiple boundary conditions is a complex task, which will also be addressed in Chapter 7.

6.6.3 Unidirectional Transformation of a 2D Form

Figure 6.39a,b illustrates the compression-tension combined structure from Chapter 5 with four applied loads and two vertical reaction forces. We can get a different structural form by manipulating the GDoF of the form diagram without disturbing the force equilibrium. This form diagram has a GDoF of four:

$$GDoF = e - 2f = 14 - 10 = 4.$$

That is, the form diagram has four independent edges, e_i whose lengths l_i can be adjusted to make a new equilibrated form with different edge lengths. As discussed, these edge lengths can be positive or negative based on design decisions. Figure 6.39c–f shows independent edges' positive and negative edge lengths. As illustrated, the initial form diagram with compressive members on top and tensile members at the bottom is transformed into a new form with a compression arch on top and

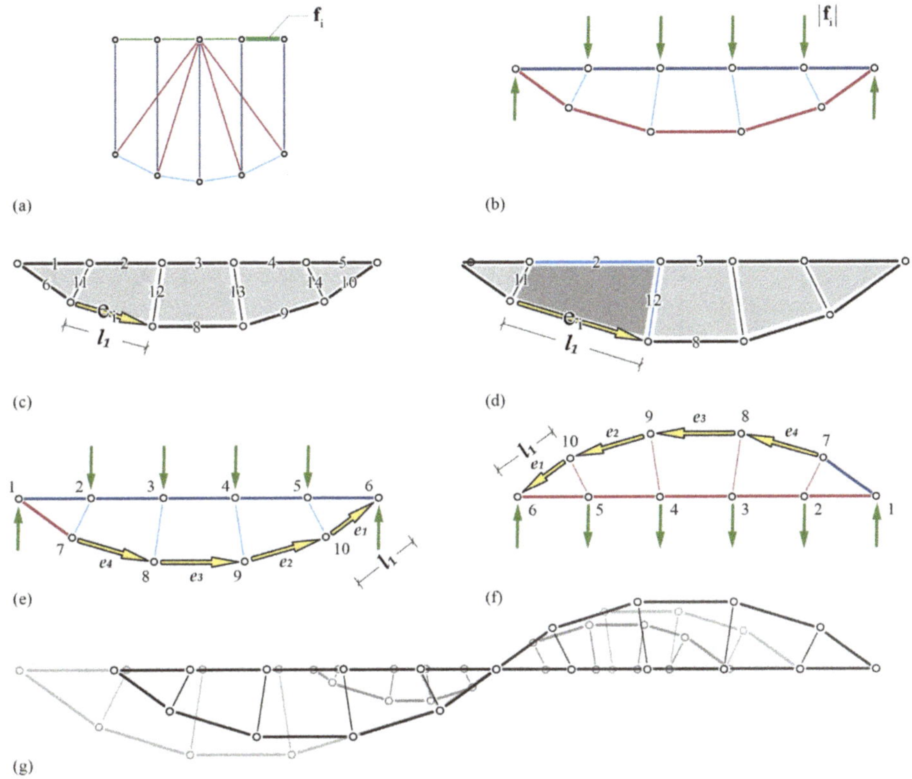

Figure 6.39 (a) Force diagram; (b) form diagram; (c) 14 edges are counted in the form diagram; (d) changing the edge length of e_i without changing the angles of the edges; (e), (f) two primary design schemes based on the length and the direction of the edge *ei*; and (g) various configurations of the form diagram made by changing the edge length from positive to negative.

tensile members on the bottom by changing the independent edge lengths from positive to negative. Pay attention to the changes in the location of the nodes labeled from 1 to 10 as the length of l_i changes from positive to negative. Note that the positive direction is a convention based on the initial construction of the form. As shown in Figure 6.39g, these two structural forms have rotational symmetry. Indeed, the initial form with all nodes and members flips over node 6 to make a new form diagram.

Transforming the Height

As noted in Example 6.6, the GDoF could not change the structure's height. This, in particular, is related to the triangular corners of the form that prevent changing the height or transformation of the structure in the *y*- direction. To overcome this limitation in design, we can remove the nodes at both ends of the structure and update the force diagram accordingly.

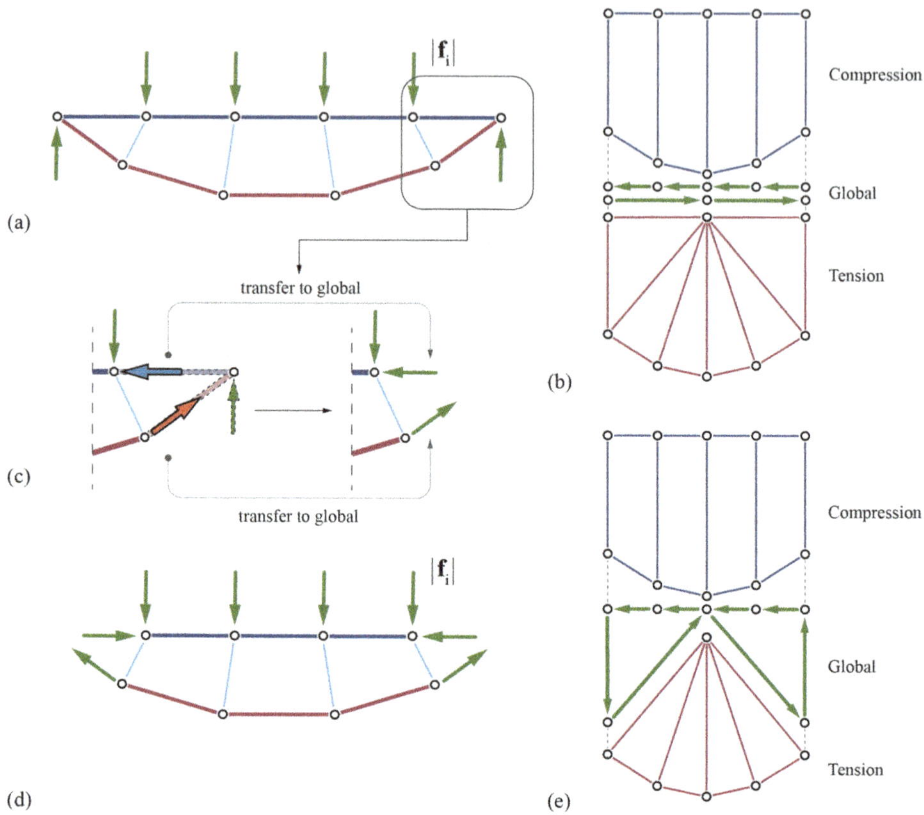

Figure 6.40 (a) Initial form diagram; (b) its reciprocal force diagram; (c) removing the corner nodes and adjusting the global force; (d) and (e) updated form and force diagram.

Figure 6.40a,b shows the form and force diagram of Example 6.6 with exploded layers of the force diagram. These layers include cells reciprocal to the compression nodes, tension nodes, and global forces. Removing the corner vertex changes the force diagram into the force diagram of Figure 6.40e. It adds two reaction forces to the system: a tension force in the bottom and a compression force on the top. These new reactions were initially the internal forces in the members connected to the corner node. This adjustment in the form and force diagram adds a new independent edge in the form that can transform the structure in the *y*- direction.

Figure 6.41a.1 shows the new form diagram with positive edge lengths for independent edges. Figure 6.41a.2 shows its counterpart with negative lengths for all its independent edges. Let us adjust the edge length l_1 in both configurations. By changing the length of l_1 from positive to negative in the first configuration, we can generate multiple form diagrams, as shown in Figure 6.41b.1, c.1, and d.1. This change transforms the structure in the *y*- direction. The same approach can be used for the form of Figure 6.41a.2, where the length of l_1 changes from negative to positive. Note that the rest of the independent edges preserve their edge lengths. The result is also multiple funicular forms with different internal forces in their members.

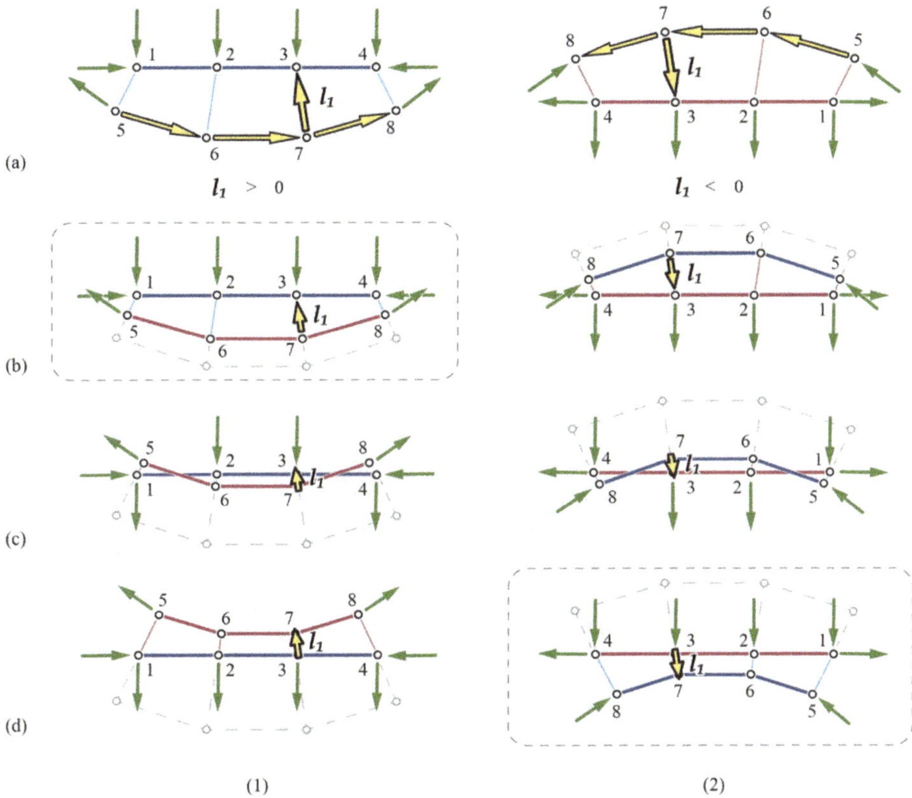

Figure 6.41 (a).1–(d).1 GDoF when all independent edges have positive length and l_1 changes from positive to negative; (a).2–(d).2 all independent edges have a negative length and l_1 changes from negative to positive.

Generating Undulating Funicular Beam

Multiple funicular forms were generated in Example 6.6 by manipulating the GDoF of the form diagram. An interesting idea in design emerged as a result of this exercise. What if we select from among these forms and combine them to get a more sophisticated form diagram?

Now that we can access multiple funicular systems with similar boundary forces and applied loads, we can combine them for design purposes. For instance, the funicular solution of Figure 6.41b.1 and the solution of Figure 6.41d.2 can be chosen to design an undulating beam system for a building structure. The illustrations of Figures 6.42 and 6.43 show the juxtaposition of these two systems in an undulating, multi-span beam structure.

6.6.4 Bidirectional Transformation of a 3D Network

A similar technique can generate various equilibrated structures designed by polyheral graphic statics. Consider the structural form of Figure 6.44a. This form is the 3D

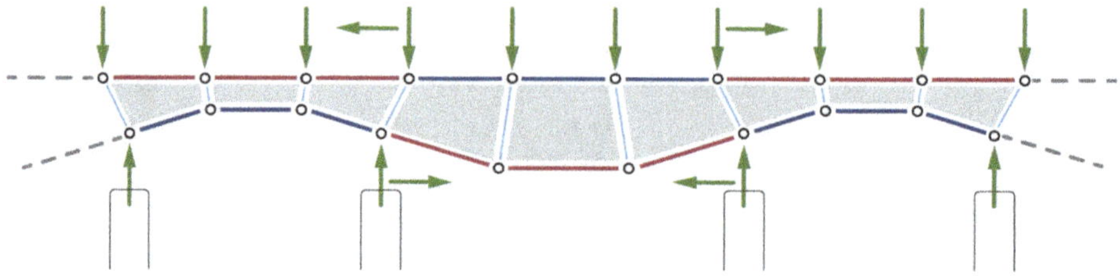

Figure 6.42 Column location for the continuous beam system.

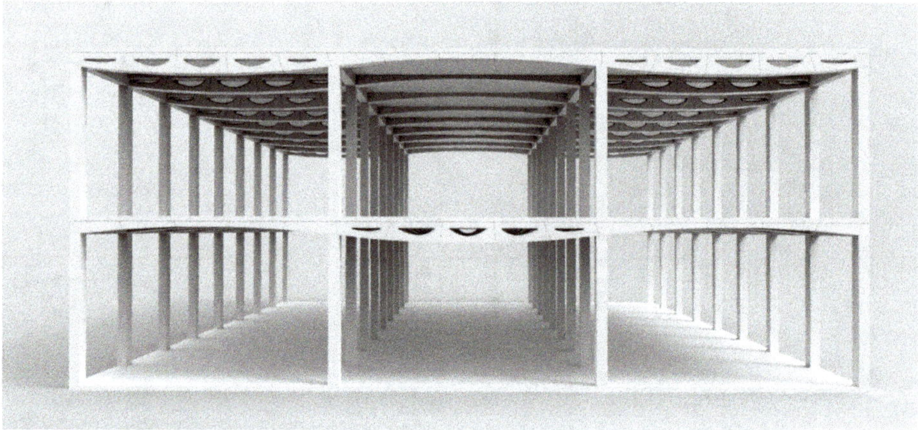

Figure 6.43 Undulating beam structure with potential porous materialization for all the modules.

equivalent of the previous 2D example. The form has multiple GDoFs whose exact number can be calculated using the algebraic methods of Chapter 7.

As shown, the initial structural form has a concave profile on the top and a convex profile on the bottom. The concave profile carries compressive internal forces, while the concave profile includes tensile members. Changing the length of the lateral independent edges from positive to negative gives another funicular form with a convex top and concave bottom with mixed types of forces, as shown in Figure 6.44b.

In the next step, one might change the length of the longitudinal independent edges from positive to negative, which gives the structure of Figure 6.44c with both concave profiles on the top and bottom where the location of the tension and compression members is flipped compared to the initial form.

Continuing the transformation by changing the lateral independent edges from negative to positive results in the case of Figure 6.44d with a convex top and concave bottom with mixed forces. Turning the longitudinal independent edges to a positive value will return us to the initial form.

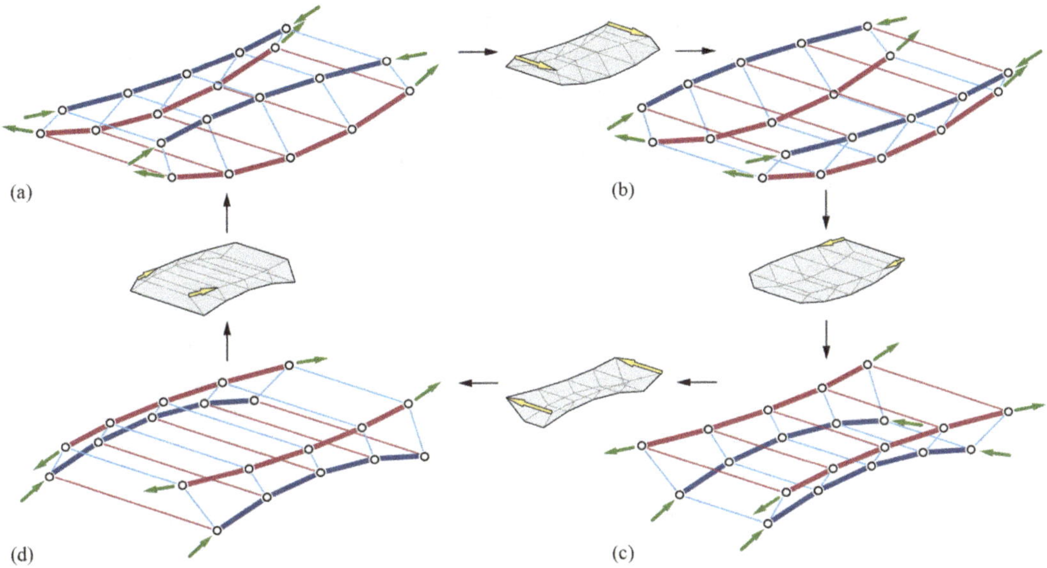

Figure 6.44 (a) The initial form diagram with a compression concave top and tension convex bottom profile; (b) changing the length of the lateral independent edges gives a funicular solution with a mixed force convex profile on the top and concave bottom; (c) changing the length of the longitudinal independent edges gives a system with tension concave profiles on top and compression concave in the bottom; and (d) changing the lateral independent edges gives a mixed convex top and mixed concave bottom.

6.7 Summary of the Chapter

The chapter expanded on the definition, meaning, and application of GDoF of the form and force diagrams in polyhedral graphic statics. The number of independent edges in the network is the GDoF of a network of closed polygons/polyhedrons. An independent edge is an edge whose length can be independently chosen in constructing a network. These edges then determine the lengths of the rest of the edges in the network.

The geometric degree of freedom has different applications in form and force diagrams. The GDoF of the force diagram in 2D and 3D allows exploring the static degrees of indeterminacies in the funicular form; the GDoF of the global force polygon/polyhedron corresponds to the external degrees of static indeterminacies in the funicular form.

Controlling the GDoF of a force diagram allows for removing reaction forces or applied loads in the boundary condition of a funicular form. This control can redistribute the internal forces in the structure's members, for example, from compression to tension, without changing the geometry of the funicular form and lead to exciting design options.

Like the force diagram, the form diagram also has GDoF. Using GDoF to manipulate the form diagram does change the form's configuration but does not change the force magnitude and the equilibrium of forces in the system. Meanwhile, it opens the door to various funicular solutions ranging from compression-only to compression-and-tension combined systems.

Identifying the exact number of GDoFs for both form and force diagrams in 3D is not a trivial task. It requires an algebraic formulation of the problem, which will be provided in Chapter 7.

7 Algebraic Formulation for Polyhedral Graphic Statics

Masoud Akbarzadeh, Marton Hablicsek, and Yao Lu

7.1 Introduction

As can be seen from Chapters 3 to 6, there are multiple approaches to constructing a form diagram given a force, or vice versa, based on reciprocal relationships. Some common implementations of two-dimensional graphics statics (2DGS) and polyhedral graphic statics (PGS) often rely on procedural techniques or iterative algorithms that modify local node locations until the reciprocal criteria are met. These implementations are closely linked to geometric concepts, making them straightforward and relatively easy to comprehend. However, despite their advantages, they have certain limitations in terms of capability and performance. Procedural techniques[9] can be tedious and labor-intensive, and they cannot be generalized to a wide range of inputs. On the other hand, nodal operation-based iterative algorithms[107,137] require an input angle tolerance as the criterion to terminate the iterations, meaning that the output will always have some level of deviation. The smaller the tolerance is, the longer the runtime will be, making it computationally intensive for complex geometries. Additionally, it cannot construct reciprocal diagrams with nonconvex polyhedrons that represent tension compression combined systems, which limits the flexibility of exploring the full solution space.

Algebraic formulations offer a novel approach to implementing both 2DGS and PGS. These formulations distill the fundamental mathematical essence of the reciprocal relationship and express it as algebraic equations, where the dual diagram is represented as a variable. Each solution to these equations corresponds to one potential dual diagram of the primal. Although this approach requires a deeper comprehension of the reciprocal relationship, it provides a tool that is more potent than traditional methods. By utilizing modern computational packages, the solution space for these equations can be efficiently solved and explored. This offers a robust foundation for the algebraic framework, yielding improved speed, flexibility, and precision. Furthermore, it facilitates more interactive explorations of the relationships between form and force, beneficial for education, research, and practice.[8,103]

7.2 Algebraic Construction of 2D Reciprocal Diagrams

The graph data structure can be employed to represent and interconnect form and force diagrams, finding extensive applications in various algebraic formulations. A graph consists of a set of vertices connected by edges. To provide a more accurate representation of the reciprocal diagrams in 2DGS, faces are introduced on top of the graph data structure. These faces are defined by the small polygonal cells enclosed by the edges.

The topological and geometric relationships between the form and force diagrams can be mathematically described using algebraically constrained linear equations. These equations are based on the closed edges of each face in the form and force diagrams. Their numerical solutions enable the direct analysis and immediate manipulation of truss structures in 2D.[58,138,139]

In the convention established in Chapters 1 to 6, the force diagram and its components are always labeled by a dagger (†) sign. This convention is slightly different in this chapter. We call the starting diagram *primal*, Γ, and the dagger (†) sign is used to denote the reciprocal diagram *dual*, $\Gamma^\dagger$. In general, the algebraic construction takes a known primal diagram and tries to construct its dual diagram. Both primal and dual diagrams can be either a force or a form diagram. The numbers of vertices, edges, and faces of the primal diagram are denoted by v, e, and f, respectively, and the ones of the dual diagram are super-scripted by a (†) symbol. The processes presented in this section are inspired by Mele and Block.[58] We invite interested readers to visit their paper "Algebraic Graph Statics" and also its extension as the "Bi-directional algebraic graphic statics" paper by Alic and Âkesson.[138]

7.2.1 Topological Relationships between 2D Reciprocal Diagrams

To help better describe the topological properties of the elements, the terms *external* (denoted with subscript $_{ext}$) and *internal* (denoted with subscript $_{int}$) can be introduced. As described earlier, a major topological difference between form and force diagrams in 2D is that force diagrams contain only closed faces, while form diagrams have open elements (Figure 7.1). In a form graph, the applied forces are represented as edges with only one vertex, which shows the locations of the forces, meaning that the other end is open. Such open edges are defined as external edges e_{ext}, the faces connected to the open edges are defined as external faces f_{ext}, and all other elements are internal. Those properties are different for a force graph, where all vertices and edges on the GFP are external, denoted as v_{ext} and e_{ext}, while others are internal. The form graph is the topological dual of the force graph both globally and locally, which is defined by:

- The number of vertices v of the primal graph Γ equals the number of faces $f^\dagger$ in the dual graph $\Gamma^\dagger$; the vertex v_i and face $f_i^\dagger$ are the dual of each other.
- The number of edges e in the primal graph Γ equals the number of edges $e^\dagger$ in the dual graph $\Gamma^\dagger$; the edges e_i and $e_i^\dagger$ are dual of each other.
- The number of faces f in the primal graph Γ equals the number of vertices $v^\dagger$ in the dual graph $\Gamma^\dagger$; the face f_i and vertex $v_i^\dagger$ are the dual of each other.
- If an element is internal/external, its dual is also internal/external.

Remark 7.1. In our current description, we have deliberately omitted external vertices in the form diagram and external faces in the force diagram to simplify explanations and enhance clarity. That also means all vertices in a form diagram are internal, and all faces in a force diagram are internal. However, alternative descriptions do exist where such external elements are included. In those descriptions, the force edges are no longer open, resulting in the presence of tail vertices that are defined as external vertices. Similarly, in the force diagram, the infinite area beyond the outermost boundary of the graph is considered an external face. Nevertheless, this definition introduces a problem in terms of duality. It creates one external face in the force diagram, but its dual representation in the form diagram consists of multiple external vertices. Consequently, the number of elements in the primal and dual graphs is no longer equal, which violates the definition of the dual relationship. While this approach may be useful in computational implementations, it can be challenging for beginners to understand due to the imbalance it introduces between the primal and dual elements.

Remark 7.2. It is important to note that all the examples presented in this section start from convex faces. While the reciprocal relationship and the algebraic formulation theoretically apply to nonconvex or even self-intersecting faces; however, it's a lot easier to start from convex ones and achieve nonconvex configurations through subsequent manipulations because identifying and establishing the correct topological data structure for such cases is challenging in implementation.

7.2.2 Geometric Relationship between 2D Reciprocal Diagrams

In some literature of graphical methods,[21,41,42,58] it is assumed that the corresponding edges of the diagrams, e_i and $e_i^\dagger$, are always parallel to each other. We are following a different convention by Rankine[5] and Maxwell[37] where the edges of the form and force diagrams are perpendicular to each other. These two conventions are **equivalent** under a rotation by 90°.

The topological and geometric relationships between the form and force diagrams are illustrated in Figure 7.1. The primal force diagram $\Gamma^\dagger$ (Figure 7.1a) has 7 vertices (named with $A - G$), 12 edges, and 6 faces (named with 1–6). Four boundary edges indicate external forces. In the corresponding form diagram shown in Figure 7.1b, there are 6 vertices (named with 1–6), 12 edges, and 7 faces (named with $A - G$). Note that the faces made by external forces are open on one side, and so are the external force edges. The external members are illustrated in green. Moreover, the corresponding edges of the two diagrams are perpendicular to each other.

Topological reciprocity dictates that each vertex in one diagram matches a face in its reciprocal diagram with the same name. For example, the vertex v_B in the force diagram corresponds to the face $f_B^\dagger$ in the force diagram, and the vertex $v_3^\dagger$ corresponds to the face f_3 in the force diagram. As discussed in Chapters 1 to 6, a face in the force diagram is also called a nodal force polygon (NFP). Therefore, the edge vectors of face f_3 correspond to the force vectors applied to vertex $v_3^\dagger$. The edge vectors of the NFP, connected tip-to-tail, form a loop that sums to a zero vector. This means that vertex $v_3^\dagger$ is under force equilibrium, as shown in Figure 7.1c,d. For clarity on the loops, each internal edge in the force diagram Γ can be viewed as two half-edge

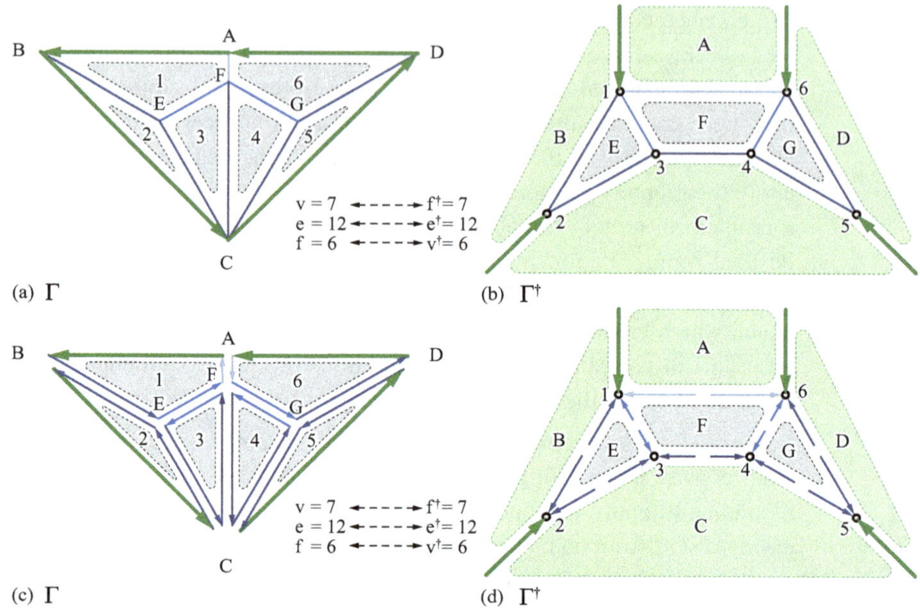

Figure 7.1 (a) A force diagram as the primal graph Γ. (b) A form diagram as the dual graph $\Gamma^{\dagger}$. (c) The internal edges of a force diagram can be viewed as two half-edges. (d) The half-edges correspond to the forces on the two ends of the respective form edge.

vectors with the same magnitude but opposite directions, belonging to different associated NFPs on two sides. Those two half-edges correspond to the forces at the ends of the respective form edge in $\Gamma^{\dagger}$, which also have the same magnitude but opposite directions. Note that the loop directions are consistent across all force polygons, either all clockwise or all counterclockwise, and they are determined according to the loop direction of the global force polygon (GFP) colored in green. Here, we define the NFP as positive when edge loops are counterclockwise, and negative otherwise. As can be seen in Figure 7.1c,d, positive NFPs assume that the edges in the form diagram create compression to the vertices. Similarly, negative NFPs assume tensions applied to the vertices. Whether the NFPs are positive or negative is important as it will be used to determine the tension-compression state of the form diagram to be generated.

The linear equations will be constructed based on those topological and geometrical relationships.

7.2.3 Choice of the Primal

Given one diagram, its dual diagrams can be found based on the topological and geometrical relationships. This process works for both form and force diagrams as input. If a force diagram is primal, then the process is called *form finding*, and if a form diagram is primal, then the force diagram will be constructed as its dual so the internal force distribution can be obtained, working as *structural analysis*. With a focus on form finding, this chapter always uses force diagrams as input.

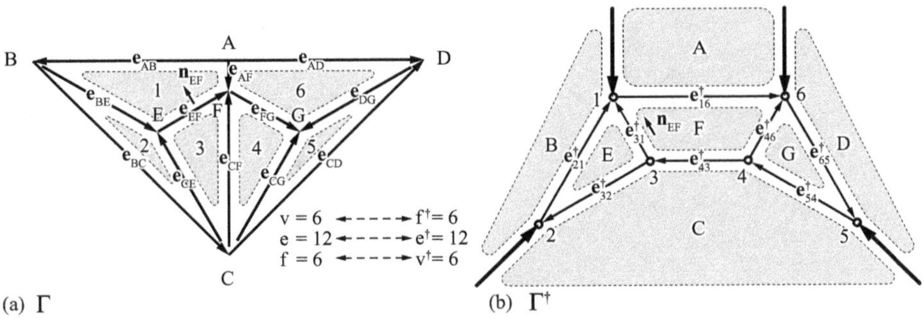

Figure 7.2 (a) and (b) The directed edges of the force diagram and its corresponding directed edge in the form graph.

7.2.4 Closing Equations in 2D

Based on the input primal diagram, a series of equations can be developed to describe the geometric and topological relationship of the primal and dual diagrams, where the unknown dual is represented by a vector variable $\mathbf{q}$, whose entries are the lengths of the internal dual edges. To better clarify this process, we first start with a simple motivating example.

A Motivating Example

Given the input primal diagram of Figure 7.2a, although the dual diagram is unknown at this stage, its topology and edge angles can already be determined based on the primal diagram, as illustrated in Figure 7.2b. First, the directions of all edges in both diagrams are defined. In the primal diagram, the directions of the edges are defined to always go from the end vertex with an earlier letter to that with a later letter. Although the initial edge direction does not matter and can be defined in any other way, we think this convention is easy to follow. Then, an edge vector connecting v_i to v_j is denoted by $\mathbf{e}_{ij}$ and can be described as follows:

$$\mathbf{e}_{ij} = \hat{\mathbf{e}}_{ij} \cdot q_{ij}.$$

Here $\hat{\mathbf{e}}_{ij}$ is the unit vector and q_{ij} is the signed length of the edge $\mathbf{e}_{ij}$ in the primal graph. A similar term can be used to describe an edge in the dual graph. Note that the edges of the primal and dual are perpendicular to each other, here we define the directions of the dual edges as in the primal edge directions rotated counterclockwise by 90°, which are also the normal directions of the primal edges, denoted as $\mathbf{n}_{ij}$. These normal directions are also associated with the starting and ending vertices of the dual edges. For example, if edge $\mathbf{e}_{EF}$ has a normal vector $\mathbf{n}_{EF}$ pointing from face f_3 to face f_1, then its corresponding edge in the dual diagram will be pointing from vertex $v_3^\dagger$ to vertex $v_1^\dagger$, with $\mathbf{n}_{EF}$ as its direction. Hence, the dual diagram edges can be described by the following equation

$$\mathbf{e}_{kl}^\dagger = \mathbf{n}_{ij} \cdot q_{kl}^\dagger, \tag{7.1}$$

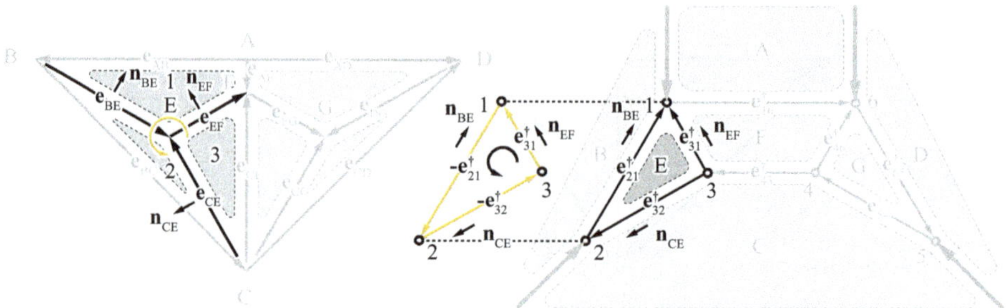

(a) Γ (b) $\Gamma^{\dagger}$

Figure 7.3 (a) and (b) The three edges e_{BE}, e_{CE}, and e_{EF} and their common vertex v_E in the primal diagram correspond to the reciprocal edges $e^{\dagger}_{21}$, $e^{\dagger}_{31}$, and $e^{\dagger}_{32}$ in the form diagram, which are connected into a loop and enclose the face $f^{\dagger}_{E}$.

where k and l denote the vertex indices of the dual graph, edge e_{ij} and edge $e^{\dagger}_{kl}$ are dual of each other, and $q^{\dagger}_{kl}$ is the signed length of the edge $e^{\dagger}_{kl}$ to be solved, which are assembled in the aforementioned vector variable $\mathbf{q}$. Note that the edges with an open end do not have lengths to solve, so they are not considered when writing equations.

Since every vertex in the primal is reciprocal to a polygonal face in the dual, the edge compatibility in the dual diagram can be expressed with some equations based on the fact that all edge vectors of each closed face sum up to a zero vector. Equations are not needed for the open faces. The closed faces in the dual diagram correspond to internal vertices in the primal diagram, and the open faces in the dual correspond to the external vertices. Take vertex v_E for example (Figure 7.3), the three connected edges e_{BE}, e_{CE}, and e_{EF} correspond to the three dual edges represented as vectors $\mathbf{e}^{\dagger}_{21}$, $\mathbf{e}^{\dagger}_{32}$, and $\mathbf{e}^{\dagger}_{31}$, respectively, and they are the boundary edges of the face $f^{\dagger}_{E}$ in the dual diagram. As a convention, the vertices are looped through in counterclockwise order to get the cycle of closed edge vectors, which can be written as

$$\mathbf{e}^{\dagger}_{12} + \mathbf{e}^{\dagger}_{23} + \mathbf{e}^{\dagger}_{31} = \mathbf{0},$$

where

$$\mathbf{e}^{\dagger}_{12} = -\mathbf{e}^{\dagger}_{21},$$

and

$$\mathbf{e}^{\dagger}_{23} = -\mathbf{e}^{\dagger}_{32}.$$

Combining this with Equation 7.1, we obtain a vector equation

$$-\mathbf{n}_{BE} \cdot q^{\dagger}_{21} - \mathbf{n}_{CE} \cdot q^{\dagger}_{32} + \mathbf{n}_{EF} \cdot q^{\dagger}_{31} = \mathbf{0}. \tag{7.2}$$

Since these normals are the edge vectors of the primal rotated counter-clockwise by 90°, we can get

$$\mathbf{n}_{BE} = \hat{\mathbf{e}}^{rot}_{BE},$$

$$\mathbf{n}_{CE} = \hat{\mathbf{e}}^{rot}_{CE},$$

$$\mathbf{n}_{EF} = \hat{\mathbf{e}}_{EF}^{rot}.$$

Combining the above with Equation 7.2, the following equation can be obtained

$$-\hat{\mathbf{e}}_{BE}^{rot} \cdot q_{21}^{\dagger} - \hat{\mathbf{e}}_{CE}^{rot} \cdot q_{32}^{\dagger} + \hat{\mathbf{e}}_{EF}^{rot} \cdot q_{31}^{\dagger} = \mathbf{0},$$

where $\hat{\mathbf{e}}_{ij}^{rot}$ can be easily computed from the primal diagram. For the 2D diagram, two equations can be separated from the above equation, one for the x- dimension and the other for the y- dimension.

We want to reiterate that the closing equations for primal and dual should be formulated around the *inner vertices* of the primal. This is because only such vertices have dual faces that are closed. External vertices correspond to external faces that are not closed; hence, there is no need for closing equations for them. If the primal is a form diagram, then all vertices are internal. If the primal is a force diagram, the internal vertices are those vertices that are not on the GFP.

In this example, the input primal diagram is a force diagram, so there are three internal vertices. For the dual diagram to be solved, we would have six equations in total to describe the closed faces. We call all these equations the system of *closing equations*.

7.2.5 Algebraic Formulation

The system of closing equations can be generalized and assembled in matrix form. There are two key pieces of information in each equation from the equation system: the signed lengths of the boundary edges of a dual face, and the x- or y- components of the unit directional vector of these edges.

For the signed edge lengths of the dual diagram to be solved, they are all assembled in a vector $\mathbf{q}$. The dimension of $\mathbf{q}$ is $[e \times 1]$ if the dual is a force diagram; otherwise, the dimension will be $[e_{int} \times 1]$, where e_{int} indicates the number of internal edges.

The x- or y- components of the unit directional vector of all dual edges can be assembled into two vectors. Since they are equivalent to the normal directions of the primal edges, they are denoted as $\mathbf{n}_x$ and $\mathbf{n}_y$, and their dimensions are the same as $\mathbf{q}$.

The face-to-edge connectivity in the dual diagram is equivalent to the vertex-to-edge connectivity in the primal graph, which can be described with a matrix $\mathbf{C}$ for the primal graph Γ to organize all equations. Again, if Γ is a form diagram, $\mathbf{C}$ is a $[v \times e]$ matrix, otherwise $[v_{int} \times e_{int}]$. In the case of this example, its dimension is $[3 \times 8]$.

The entries of the matrix are given as follows:

$$\mathbf{C}_{ij} = \begin{cases} -1 & \text{if vertex } i \text{ is the head of edge } j \\ 1 & \text{if vertex } i \text{ is the tail of edge } j \\ 0 & \text{vertex } i \text{ and edge } j \text{ are not associated} \end{cases}.$$

The i-th row $\mathbf{C}_i$ of this matrix works like a mask that filters out all edges connected to the i-th vertex in the primal diagram, or all boundary edges associated with the i-th face in the dual diagram. For the dual diagram, the entries of this connectivity matrix mean:

$$\mathbf{C}^{\dagger}_{ij} = \begin{cases} -1 & \text{if edge } j \text{ is a clockwise edge of face } i \\ 1 & \text{if edge } j \text{ is a counterclockwise edge of face } i\,. \\ 0 & \text{face } i \text{ and edge } j \text{ are not associated} \end{cases}$$

For the dual graph $\Gamma^{\dagger}$ to be constructed, we use the vectors $\mathbf{u}^{\dagger}$ and $\mathbf{v}^{\dagger}$ to denote the directed lengths along the x- and y- dimensions, respectively. The loop of closed edge vectors in each face $f_i^{\dagger}$ can be mathematically described as

$$\mathbf{C}_i\mathbf{u}^{\dagger} = 0, \quad \mathbf{C}_i\mathbf{v}^{\dagger} = 0. \tag{7.3}$$

Vectors $\mathbf{u}^{\dagger}$ and $\mathbf{v}^{\dagger}$ can be represented as

$$\mathbf{u}^{\dagger} = \mathbf{N}_x\mathbf{q}, \quad \mathbf{v}^{\dagger} = \mathbf{N}_y\mathbf{q}, \tag{7.4}$$

where $\mathbf{N}_x$ and $\mathbf{N}_y$ are the diagonal matrices whose diagonal entries are given by $\mathbf{n}_x$ and $\mathbf{n}_y$.

Combining the Equations 7.3 and 7.4 together, we obtain the matrix equations

$$\mathbf{C}_i\mathbf{N}_x\mathbf{q} = 0,$$
$$\mathbf{C}_i\mathbf{N}_y\mathbf{q} = 0,$$

and they have to be satisfied for every closed face of the graph $\Gamma^{\dagger}$. By putting these equations together, we get a matrix equation

$$\mathbf{A}\mathbf{q} = \mathbf{0}, \tag{7.5}$$

where $\mathbf{A}$ is a $[2v_{int} \times e_{int}]$ matrix, called the **closing matrix** of Γ described as

$$\mathbf{A} = \left(\frac{\mathbf{C}\mathbf{N}_x}{\mathbf{C}\mathbf{N}_y}\right).$$

Solutions to the closing equation $\mathbf{A}\mathbf{q} = \mathbf{0}$ provide all possible sets of edge lengths of the dual diagram $\Gamma^{\dagger}$, which can be used to construct the geometry of the dual diagram. This process will be shown later in Section 7.2.8. Each solution of this equation corresponds to a different dual diagram $\Gamma^{\dagger}$.

Remark 7.3. For those who are familiar with the concept of *force density* introduced by Schek,[48] the vector of edge lengths, $\mathbf{q}$, is related to it. Indeed, the force density of an edge in the form is defined by the ratios between the corresponding force diagram edge length and the form diagram edge length.

Example for the Closing Matrix

We use a simple example shown in Figure 7.4 to demonstrate how the closing matrix can be computed. In this example, the primal diagram Γ has seven vertices (labeled with $A, B, \ldots, G$). The locations of the three inner vertices are as follows:

$$\text{vertex } E\colon (-1, \sqrt{3}), \text{vertex } F\colon (0, 4\sqrt{3}/3), \text{vertex } G\colon (1, \sqrt{3}).$$

The dual diagram $\Gamma^{\dagger}$ has six vertices (as Γ has six faces), denoted by 1 to 6, whose locations are unknown at this stage. Furthermore, each diagram contains 12 edges. The direction of each edge in the primal diagram is already defined, which points from the

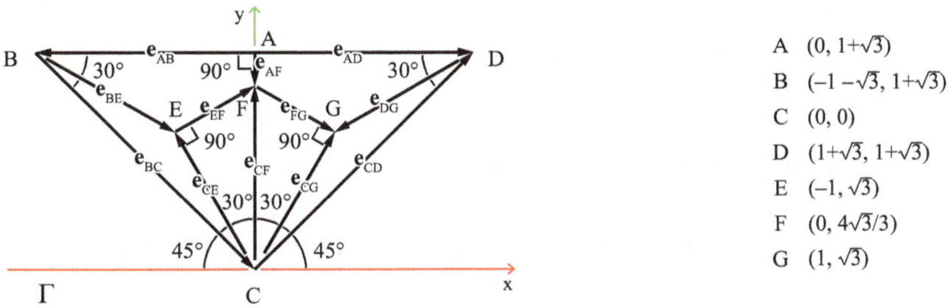

Figure 7.4 The edge angles and vertex coordinates of the starting force diagram Γ.

vertex of an earlier letter to the vertex of a later letter in our convention. Out of the 12 directed edges of $\Gamma^{\dagger}$, there are only eight associated with closed faces that we care about, and they are $\mathbf{e}^{\dagger}_{16}$, $\mathbf{e}^{\dagger}_{21}$, $\mathbf{e}^{\dagger}_{31}$, $\mathbf{e}^{\dagger}_{32}$, $\mathbf{e}^{\dagger}_{43}$, $\mathbf{e}^{\dagger}_{46}$, $\mathbf{e}^{\dagger}_{54}$, and $\mathbf{e}^{\dagger}_{65}$; we denote the edge lengths of these edges by $q^{\dagger}_{16}$, $q^{\dagger}_{21}$, $q^{\dagger}_{31}$, $q^{\dagger}_{32}$, $q^{\dagger}_{43}$, $q^{\dagger}_{46}$, $q^{\dagger}_{54}$, and $q^{\dagger}_{65}$, respectively.

Example 7.1 Compute the closing matrix for the example of Figure 7.4.

Construction steps:

1 Compute the direction of the normals by rotating the edge vectors of the primal diagram, namely

$$\mathbf{n}_{AF} = \hat{\mathbf{e}}^{rot}_{AF} = \hat{\mathbf{e}}^{\dagger}_{16} = (1, 0),$$

$$\mathbf{n}_{BE} = \hat{\mathbf{e}}^{rot}_{BE} = \hat{\mathbf{e}}^{\dagger}_{21} = \left(\frac{1}{2}, \frac{\sqrt{3}}{2}\right),$$

$$\mathbf{n}_{CE} = \hat{\mathbf{e}}^{rot}_{CE} = \hat{\mathbf{e}}^{\dagger}_{32} = \left(-\frac{\sqrt{3}}{2}, -\frac{1}{2}\right),$$

$$\mathbf{n}_{CF} = \hat{\mathbf{e}}^{rot}_{CF} = \hat{\mathbf{e}}^{\dagger}_{43} = (-1, 0),$$

$$\mathbf{n}_{CG} = \hat{\mathbf{e}}^{rot}_{CG} = \hat{\mathbf{e}}^{\dagger}_{54} = \left(-\frac{\sqrt{3}}{2}, \frac{1}{2}\right),$$

$$\mathbf{n}_{DG} = \hat{\mathbf{e}}^{rot}_{DG} = \hat{\mathbf{e}}^{\dagger}_{65} = \left(\frac{1}{2}, -\frac{\sqrt{3}}{2}\right),$$

$$\mathbf{n}_{EF} = \hat{\mathbf{e}}^{rot}_{EF} = \hat{\mathbf{e}}^{\dagger}_{31} = \left(-\frac{1}{2}, \frac{\sqrt{3}}{2}\right),$$

$$\mathbf{n}_{FG} = \hat{\mathbf{e}}^{rot}_{FG} = \hat{\mathbf{e}}^{\dagger}_{46} = \left(\frac{1}{2}, \frac{\sqrt{3}}{2}\right).$$

2 Write down the closing equation for the closed loop given by the face $f_E^\dagger$ in the dual diagram in terms of the edge normals

$$-\mathbf{n}_{BE}q_{21} - \mathbf{n}_{CE}q_{32} + \mathbf{n}_{EF}q_{31} = \mathbf{0}.$$

Substitute the edge normals in the equation above to obtain

$$-\left(\frac{1}{2}, \frac{\sqrt{3}}{2}\right)q_{BE} - \left(-\frac{\sqrt{3}}{2}, -\frac{1}{2}\right)q_{CE} + \left(-\frac{1}{2}, \frac{\sqrt{3}}{2}\right)q_{EF} = (0,0).$$

3 Similarly, write down the closing equation for the closed loop given by the face $f_F^\dagger$:

$$-\mathbf{n}_{EF}q_{31} - \mathbf{n}_{CF}q_{43} + \mathbf{n}_{FG}q_{46} - \mathbf{n}_{AF}q_{16} = \mathbf{0}.$$

Substitute again the edge normals in the equation to obtain

$$-\left(-\frac{1}{2}, \frac{\sqrt{3}}{2}\right)q_{31} - (-1,0)q_{43} + \left(\frac{1}{2}, \frac{\sqrt{3}}{2}\right)q_{46} - (1,0)q_{16} = (0,0).$$

4 Write down the closing equation for the closed loop given by the face $f_G^\dagger$:

$$-\mathbf{n}_{FG}q_{46} - \mathbf{n}_{CG}q_{54} - \mathbf{n}_{DG}q_{65} = \mathbf{0},$$

which yields the equation

$$-\left(\frac{1}{2}, \frac{\sqrt{3}}{2}\right)q_{46} - \left(-\frac{\sqrt{3}}{2}, \frac{1}{2}\right)q_{54} - \left(\frac{1}{2}, -\frac{\sqrt{3}}{2}\right)q_{65} = (0,0).$$

5 Stack the three vector equations into a matrix equation

$$\begin{pmatrix} 0 & -\frac{1}{2} & \frac{\sqrt{3}}{2} & 0 & 0 & 0 & -\frac{1}{2} & 0 \\ 0 & -\frac{\sqrt{3}}{2} & \frac{1}{2} & 0 & 0 & 0 & \frac{\sqrt{3}}{2} & 0 \\ -1 & 0 & 0 & 1 & 0 & 0 & \frac{1}{2} & \frac{1}{2} \\ 0 & 0 & 0 & 0 & 0 & 0 & -\frac{\sqrt{3}}{2} & \frac{\sqrt{3}}{2} \\ 0 & 0 & 0 & 0 & \frac{\sqrt{3}}{2} & -\frac{1}{2} & 0 & -\frac{1}{2} \\ 0 & 0 & 0 & 0 & -\frac{1}{2} & \frac{\sqrt{3}}{2} & 0 & -\frac{\sqrt{3}}{2} \end{pmatrix} \begin{pmatrix} q_{16} \\ q_{21} \\ q_{32} \\ q_{43} \\ q_{54} \\ q_{65} \\ q_{31} \\ q_{46} \end{pmatrix} = \begin{pmatrix} 0 \\ 0 \\ 0 \\ 0 \\ 0 \\ 0 \end{pmatrix}.$$

□

The matrix above is the *closing matrix*, and the corresponding equation system is the *closing equations*.

7.2.6 Optional Constraints to the Closing Matrix and Nonhomogeneous Linear Equation System

In practice, it is frequently necessary to consider specific lengths of the edges or specific locations for external forces or supports. In such cases, we need to constrain solutions to the given boundary conditions. The algebraic formulation can achieve constraints of edge lengths and vertex locations with additional constraint equations.

Optional Constraint Equations for Edge Lengths

Edge constraints set target lengths for the edges in the dual diagram, and each constrained edge requires one additional constraint equation. For example, if the i-th edge $e_i^\dagger$ is constrained to a target length l, its constraint equation may be written as

$$\mathbf{b}_i^T \mathbf{q} = l, \tag{7.6}$$

where $\mathbf{b}_i$ is a column vector with the same dimension as $\mathbf{q}$. Its entries are all zero (0) except at the index of $e_i^\dagger$ where it's one (1), which works as a mask to filter out the edge to be constrained. For a total number of b constrained edges, all b edge constraint equations may be assembled in the matrix form as

$$\mathbf{Bq} = \mathbf{l} \tag{7.7}$$

where the rows of the edge constraint matrix $\mathbf{B}$ are the row vectors $\mathbf{b}_i^T$, and $\mathbf{l}$ is a $[b \times 1]$ column vector with the entries being the target lengths.

Optional Constraint Equations for Vertex Locations

Since the lengths of the dual diagram edges are the variables we aim to solve for, the constraints imposed on the vertices must be expressed as equations involving these edge lengths. If there is only one constrained vertex for the dual diagram, it can be achieved simply by a translation from the original vertex location to the target location, and it will have no impact on the edge lengths. Vertex constraints will only have impacts on the edge lengths when the number of constrained vertices d is greater than one because the edges connecting those constrained vertices must have compatible lengths in order to fit in between.

The vertex constraint equations are written based on constrained vertex pairs. As a first step, we specify a constrained vertex $v_0^\dagger$ to be the starting vertex. Then, we use the Breadth-First Search (BFS) algorithm to find paths of edges that go from this initial vertex to all other constrained vertices $v_i^\dagger$. We can denote this path of edges as $\mathbf{e}_{i,1}^\dagger, \mathbf{e}_{i,2}^\dagger, \ldots, \mathbf{e}_{i,j}^\dagger$. Since the edges connect the vertices $v_0^\dagger$ and $v_i^\dagger$, we obtain a vector equation

$$\mathbf{e}_{i,1}^\dagger + \mathbf{e}_{i,2}^\dagger + \ldots + \mathbf{e}_{i,j}^\dagger = v_i^\dagger - v_0^\dagger = (y_i^\dagger, x_i^\dagger)^T - (y_0^\dagger, x_0^\dagger)^T, \tag{7.8}$$

where we consider the right-hand side as the vector between the vertices $v_0^\dagger$ and $v_i^\dagger$. Note that all vector segments in the path of edges should follow the direction from $v_0^\dagger$ to $v_i^\dagger$. Therefore, when using the previously defined edge normals $\mathbf{n}_{i,j}$ to represent edge direction vectors $\hat{\mathbf{e}}_{i,1}^\dagger, \ldots, \hat{\mathbf{e}}_{i,j}^\dagger$, we need to check whether they follow or are opposite to the path direction. Specifically, each term $\mathbf{e}_{i,j}^\dagger$ in Equation 7.8 is defined as:

$$\mathbf{e}_{ij}^\dagger = \begin{cases} -\mathbf{n}_{i,j} q_{i,j} & \text{if } \mathbf{n}_{i,j} \text{ is opposite to path direction} \\ \mathbf{n}_{i,j} q_{i,j} & \text{if } \mathbf{n}_{i,j} \text{ follows path direction} \end{cases}.$$

where $q_{i,1}, \ldots, q_{i,j}$ are the lengths of these edges. Equation 7.8 can then be rewritten as

$$\pm \mathbf{n}_{i,1} q_{i,1} \pm \mathbf{n}_{i,2} q_{i,2} \pm \ldots \pm \mathbf{n}_{i,j} q_{i,j} = v_i^\dagger - v_0^\dagger.$$

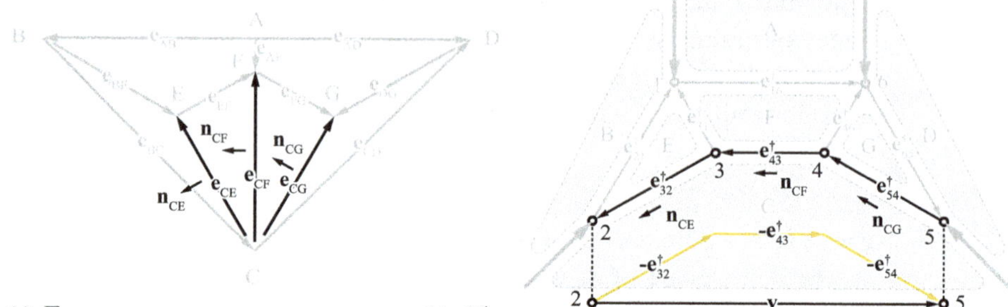

(a) Γ (b) Γ†

Figure 7.5 An example with vertices $v_2^\dagger$ and $v_5^\dagger$ constrained to points. The vector connecting the two constrained locations can be expressed with a path of edge vectors from the form diagram.

For example, in Figure 7.5, the vertices $v_2^\dagger$ and $v_5^\dagger$ are constrained to two points, and the vector going from $v_2^\dagger$ to $v_5^\dagger$ is denoted as $\mathbf{v}$. The connecting path going from $v_2^\dagger$ to $v_5^\dagger$ contains edges $e_{32}^\dagger$, $e_{43}^\dagger$, and $e_{54}^\dagger$. Since all three edge vectors are opposite to the path direction, we have

$$-\mathbf{e}_{32}^\dagger - \mathbf{e}_{43}^\dagger - \mathbf{e}_{54}^\dagger = \mathbf{v}.$$

For the x-coordinates, we obtain the constraint equation

$$\mathbf{l}_x^T \cdot \mathbf{q} = x_i^\dagger - x_0^\dagger, \tag{7.9}$$

where $\mathbf{l}_x$ is the vector of x-coordinates of the direction vectors $\pm\mathbf{n}_{i,1}, \ldots, \pm\mathbf{n}_{i,j}$; $\mathbf{q}$ is the vector of edge lengths. Similarly, we obtain equations for the y-coordinates

$$\mathbf{l}_y^T \cdot \mathbf{q} = y_i^\dagger - y_0^\dagger \tag{7.10}$$

where $\mathbf{l}_y$ is the vector of y-coordinates of the direction vectors $\pm\mathbf{n}_{i,1}, \ldots, \pm\mathbf{n}_{i,j}$. Those equations are able to constrain the vertices to the desired locations. In total, for d constrained vertices, we can have equations based on the $d-1$ pairs of constraint vertices, so there will be $2(d-1)$ constraint equations. Those equations can be stacked into a matrix equation as

$$\mathbf{Dq} = \mathbf{d}, \tag{7.11}$$

where $\mathbf{D}$ is the matrix obtained by stacking the row vectors $\mathbf{l}_x^T$ and $\mathbf{l}_y^T$; and $\mathbf{d}$ is the vector whose components are the $x_i^\dagger - x_0^\dagger$ and $y_i^\dagger - y_0^\dagger$ for all vertices $v_i^\dagger$.

7.2.7 Constrained Closing Equations and Solving Approaches

The closing equations, Equation 7.5, and all constraint equations (Equations 7.7 and 7.11) have to be satisfied simultaneously by the edge lengths of the form diagram. Stacking them together, we obtain a nonhomogeneous linear equation system

$$\mathbf{Mq} = \mathbf{t}, \tag{7.12}$$

where the matrix **M** is obtained by vertically stacking the matrices **A**, **B**, and **D**. The vector **t** is obtained by vertically stacking the vectors **0**, **b**, and **d**.

Before solving for the solutions, we can use the rank r of the closing matrix **M** to quickly evaluate the solution space of the dual diagram. Specifically, the difference between the number of unknown variables e_{int} (or e if the primal is a form diagram) and the rank r, written as $e_{int} - r$, shows the dimension of the solution space. Geometrically, it can be interpreted as the geometric degrees of freedom (GDoF) in the dual diagram, or, in other words, the number of independent edges that can have their edge lengths freely adjusted. For instance, the matrix from Example 7.1 has a rank of 6, and there are eight unknown internal edge lengths, so the GDoF of the solution is $8 - 6 = 2$, indicating that we can assign lengths to 2 independent edges in order to get one exact solution from the solution space.

Similarly, for a dual diagram whose GDoF is 0, there are no independent edges. Therefore, only one solution can be obtained for the system of equations. However, it's important to note that the situation is a little different when there are constraints. The system of equations may be overdetermined and no solution can be calculated, in which case the closing matrix has a full rank and the GDoF is also zero. This is often due to the incompatible constraints imposed on the dual diagram. This overconstraining issue cannot be detected solely from the rank of the closing matrix, and it will be further discussed in Section 7.2.8.

Presented later are several methods we can use to solve Equation 7.12, each with different advantages and disadvantages.

Moore–Penrose Inverse Method

The Moore–Penrose inverse (MPI) method is suitable for obtaining all solutions to a linear equation system.[140,141] The MPI, also called the pseudoinverse, of **M** is denoted by $\mathbf{M}^+$, which satisfies the following equations:

$$\mathbf{M}\mathbf{M}^+\mathbf{M} = \mathbf{M}, \quad \mathbf{M}^+\mathbf{M}\mathbf{M}^+ = \mathbf{M}^+.$$

Assume that the vector **t** is of the form **Mq** for some vector **q**. Multiplying the first equality by **q**, we have

$$\mathbf{M}\mathbf{M}^+\mathbf{M}\mathbf{q} = \mathbf{M}\mathbf{q}.$$

As a consequence, we obtain

$$\mathbf{M}\mathbf{M}^+\mathbf{t} = \mathbf{t}. \tag{7.13}$$

Therefore, if a solution to Equation 7.12 exists, then Equation 7.13 has to be satisfied. If Equation 7.13 doesn't hold, this usually means the dual diagram may be overconstrained. This happens when the vertex and edge constraints are incompatible with the closed edge vectors or competing with each other. The system of equations allows the overconstraining problems to be explicitly and effectively detected by checking whether Equation 7.13 holds.

If Equation 7.13 holds, from the first equality, any vector **q** of the format

$$\mathbf{q} = \mathbf{M}^+\mathbf{t} + (\mathbf{I} - \mathbf{M}^+\mathbf{M})\xi \tag{7.14}$$

solves the linear equation system Equation 7.12. Here, the matrix **I** denotes the $[e_{int} \times e_{int}]$ (or $[e \times e]$ when the primal is a form diagram) identity matrix and ξ is any $[e_{int} \times 1]$ (or $[e \times 1]$ when the primal is a form diagram) column vector. All solutions of Equation 7.12 can in fact be expressed as the form of Equation 7.14. As a result, users can freely choose the entries of the vector ξ, and each entry serves as the weight for the corresponding edge length. Equation 7.14 then yields a solution that is closest to the vector ξ (with the least square distance). For instance, assigning one (1) to all entries of ξ generates a solution with well-distributed edge lengths in the dual diagram, which can be leveraged from the standpoint of design. This method allows for global control over the edge lengths of the network by assigning entries of ξ.

For instance, linking back to the closing equations without the constraints shown in Example 7.1, the MPI of the closing matrix is

$$\mathbf{M}^{+} = \begin{pmatrix} -0.1515 & 0.1575 & -0.4848 & 0 & -0.1515 & -0.1575 \\ 0.394 & -1.1023 & 0.0606 & -1.1548 & -0.6061 & -0.6299 \\ 1.2072 & -0.4546 & 0.0525 & -1.0001 & -0.5249 & -0.5455 \\ 0.1515 & -0.1575 & 0.4848 & 0 & 0.1515 & 0.1575 \\ -0.5249 & 0.5455 & 0.0525 & 1.0001 & 1.2072 & 0.4546 \\ -0.6061 & 0.6299 & 0.0606 & 1.1548 & 0.394 & 1.1023 \\ -0.303 & 0.3149 & 0.0303 & -0.5774 & -0.303 & -0.3149 \\ -0.303 & 0.3149 & 0.0303 & 0.5774 & -0.303 & -0.3149 \end{pmatrix}.$$

Since **t** is a zero vector, the solution **q** becomes

$$\mathbf{q} = (\mathbf{I} - \mathbf{M}^{+}\mathbf{M})\xi.$$

Substitute $\mathbf{I} - \mathbf{M}^{+}\mathbf{M}$, and we can derive

$$\mathbf{q} = \begin{pmatrix} 0.515 & 0.061 & 0.052 & 0.485 & 0.052 & 0.061 & 0.03 & 0.03 \\ 0.061 & 0.242 & 0.21 & -0.061 & 0.21 & 0.242 & 0.121 & 0.121 \\ 0.052 & 0.21 & 0.182 & -0.052 & 0.182 & 0.21 & 0.105 & 0.105 \\ 0.485 & -0.061 & -0.052 & 0.515 & -0.052 & -0.061 & -0.03 & -0.03 \\ 0.052 & 0.21 & 0.182 & -0.052 & 0.182 & 0.21 & 0.105 & 0.105 \\ 0.061 & 0.242 & 0.21 & -0.061 & 0.21 & 0.242 & 0.121 & 0.121 \\ 0.03 & 0.121 & 0.105 & -0.03 & 0.105 & 0.121 & 0.061 & 0.061 \\ 0.03 & 0.121 & 0.105 & -0.03 & 0.105 & 0.121 & 0.061 & 0.061 \end{pmatrix} \xi.$$

Let all entries of ξ be 1, then we can calculate **q** as

$$\mathbf{q}^{T} = (1.286, 1.146, 0.994, 0.714, 0.994, 1.146, 0.574, 0.574),$$

which represents a set of well-distributed edge lengths (Figure 7.6a). A different ξ will yield a different solution. For example, replace the first entry of it with 2, which represents the weight of the edge $e_{16}^{\dagger}$, and we can get

$$\mathbf{q}^{T} = (1.801, 1.207, 1.046, 1.199, 1.046, 1.207, 0.604, 0.604).$$

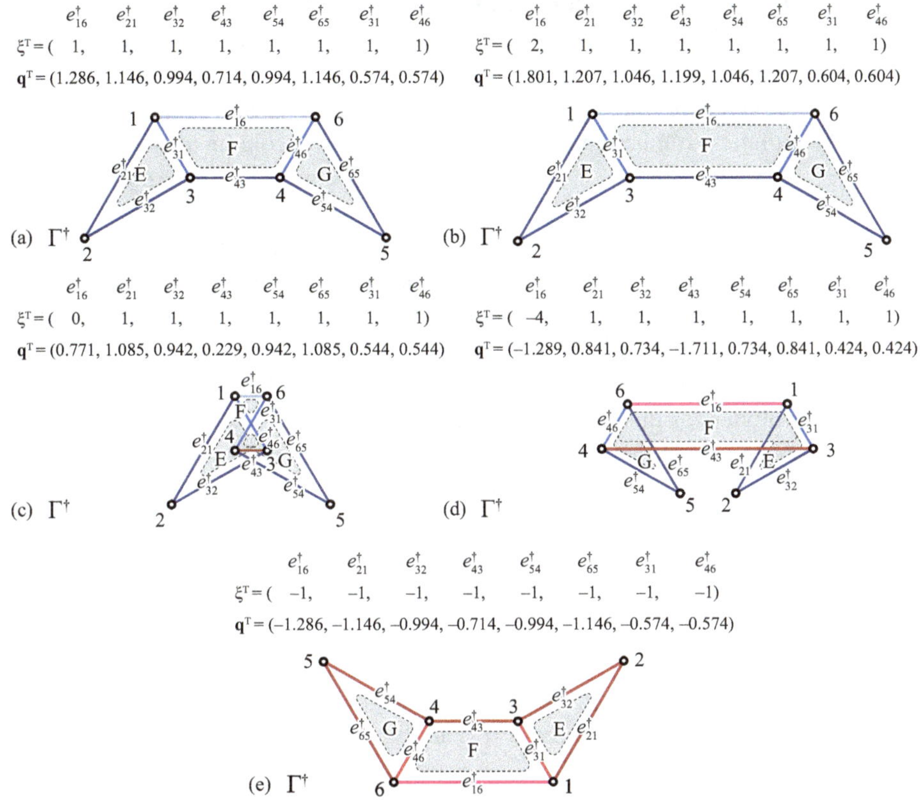

Figure 7.6 (a) to (e) Five dual diagrams generated with different input parameters ξ.

The length of edge $e^\dagger_{16}$ is increased and other edge lengths are also affected, as illustrated in Figure 7.6b. When the first entry is set to 0, the solution becomes

$$\mathbf{q}^T = (0.771, 1.085, 0.942, 0.229, 0.942, 1.085, 0.544, 0.544),$$

showing that edge $e^\dagger_{16}$ gets shorter, and edge $e^\dagger_{43}$ is affected and turned into negative length (Figure 7.6c). As a consequence, the face $f^\dagger_F$ is no longer convex and is turned into self-intersecting. When the first entry is set to -4, the solution becomes

$$\mathbf{q}^T = (-1.289, 0.841, 0.734, -1.711, 0.734, 0.841, 0.424, 0.424).$$

In this case, both edge $e^\dagger_{16}$ and edge $e^\dagger_{43}$ are turned into negative lengths (Figure 7.6d). When all entries are set to -1, the solution becomes

$$\mathbf{q}^T = (-1.286, -1.146, -0.994, -0.714, -0.994, -1.146, -0.574, -0.574),$$

and all edges are turned into negative (Figure 7.6e). One important observation from Example 7.1 is that, even though edge weights influence their lengths, a positive weight does not necessarily imply a positive length, and vice versa.

Reduced Row-Echelon Form Approach

The reduced row-echelon form (RREF) $\mathbf{A}^{rref}$ of the matrix $\mathbf{A}$ identifies the independent edges of the dual. Indeed, the independent edges correspond to those columns of $\mathbf{A}^{rref}$ where there is no pivot. The values of these independent edges corresponding to these columns can be represented by a $[(e-r) \times 1]$ column vector ζ. The geometry of the dual diagram (the lengths of its edges) is completely determined by the components of the ζ.

We formulate the computation of the remaining r edges from the $e-r$ independent edges as follows. We assume that the pivots of the matrix $\mathbf{A}^{rref}$ are in the main diagonal. This can be achieved by reordering the edges in a way that the independent edges have the biggest indices. Then we exclude all the zero rows of the matrix to obtain a $[r \times e]$ matrix $\mathbf{A}^{rref}_{r\times e}$. Because the pivots are in the main diagonal of $\mathbf{A}^{rref}_{r\times e}$, the first r columns of this matrix form the $[r \times r]$ identity matrix, $\mathbf{I}$. We denote the rest of the matrix by $\mathbf{R}$ that is a $[r \times (e-r)]$ matrix formed by the last $e-r$ columns. In other words, we can decompose $\mathbf{A}^{rref}_{r\times e}$ as

$$\mathbf{A}^{rref}_{r\times e} = (\mathbf{I}|\mathbf{R}). \tag{7.15}$$

Since the RREF has the same solutions as the original matrix, we have that

$$\mathbf{A}^{rref}_{r\times e}\mathbf{q} = \mathbf{0}. \tag{7.16}$$

We separate the $[e \times 1]$ column vector $\mathbf{q}$ into two pieces: (a) the vector consisting of the first r coordinates of $\mathbf{q}$ is denoted by $\mathbf{q}_r$, and (b) the $[(e-r) \times 1]$ column vector corresponding to the last $e-r$ entries is denoted by ζ. Using these notations, Equation 7.15 becomes

$$\mathbf{I}\mathbf{q}_r + \mathbf{R}\zeta = \mathbf{0}.$$

Therefore, the vector ζ determines $\mathbf{q}_r$:

$$\mathbf{q}_r = -\mathbf{R}\zeta.$$

We note that the vector ζ corresponds to the values of the independent edges. Thus, the lengths of the independent edges completely determine the geometry of the dual.

This method has limitations. Even though any values can be chosen for the independent edges, there is not much control over the rest of the edges' lengths. For instance, if all independent edges have positive lengths, then the rest of the edges may have negative values. In the case of form finding, this means that there is no guarantee that the resulting geometry will be a compression/tension-only system. To address this limitation, we suggest a third method which is an optimization approach to solve the closing equations (Equation 7.5).

Linear Programming Approach

In the constrained closing matrix, we can specify exact edge length or vertex location constraints. However, sometimes we want to use domains as constraints instead of exact numbers, for example, by setting the edge lengths to a specific range. Also, having a universal condition for the edges, such as making compression-only forms, is a bit challenging with the MPI method when the diagrams get complex because positive weights in ξ do not always indicate positive edge lengths. More importantly, in design practices, we would like to find solutions based on certain optimization criteria to help with design decisions. These limitations can be effectively addressed using linear programming.

For instance, in compression-only form-finding examples, the input is a convex network of polygons representing a force diagram and the dual is a funicular form that might have a large number of GDoF. In that case, we can specify optimization criteria to ensure that all edges of the dual will get positive lengths resulting in a compression-only network that also consists of convex-only polygons. In order to construct solutions with all positive edge lengths, we use the following linear optimization setup to find a dual diagram:

$$\text{Solve} \quad \begin{cases} \mathbf{Mq} = \mathbf{t} \\ \mathbf{q} \geq \mathbf{1} \\ \min(\mathbf{q} \cdot \lambda) \end{cases} . \qquad (7.17)$$

Here $\mathbf{1}$ is the $[e \times 1]$ column vector whose all entries are one (1). λ is any $[e \times 1]$ vector that can be chosen by the user. The solution of this linear programming setup is a solution vector $\mathbf{q}$ of the closing Equation 7.12 so that the entries of $\mathbf{q}$ are at least one (1) and, finally, it tries to find a solution that minimizes the objective function

$$\mathbf{q} \cdot \lambda = \sum_{i=1}^{e} q_i \lambda_i$$

for the user-defined vector λ. For instance, if $\lambda = \mathbf{1}$, then the objective function is the sum of the coordinates of $\mathbf{q}$, so the optimization approach minimizes the total lengths of the edges of the dual diagram. This approach will find a solution unless (Equation 7.12) the conditions cannot be satisfied simultaneously.

7.2.8 Construction of the Dual

After getting a solution for $\mathbf{q}$ that contains signed lengths, the vertex positions can be calculated through methods such as the algebraic approach and graph-search approach[135] in order to construct the dual diagram. Once again, the BFS algorithm is used. It traverses from a vertex $v_i^\dagger$ that has known coordinates, then gradually visits the adjacent vertices and calculates their positions based on the lengths of the connecting edges until all vertices are visited (Figure 7.29). If the form diagram has constrained vertices, the starting vertex $v_i^\dagger$ can be chosen from them. Otherwise, if no vertex is constrained, meaning that it can be constructed anywhere in the 3D space, a starting vertex needs to be specified and its position needs to be assigned.

For example, vertex $v_2^\dagger$ of the dual graph $\Gamma^\dagger$ of Figure 7.7a is the starting vertex with either a constrained or assigned location. Then, the BFS algorithm traverses every

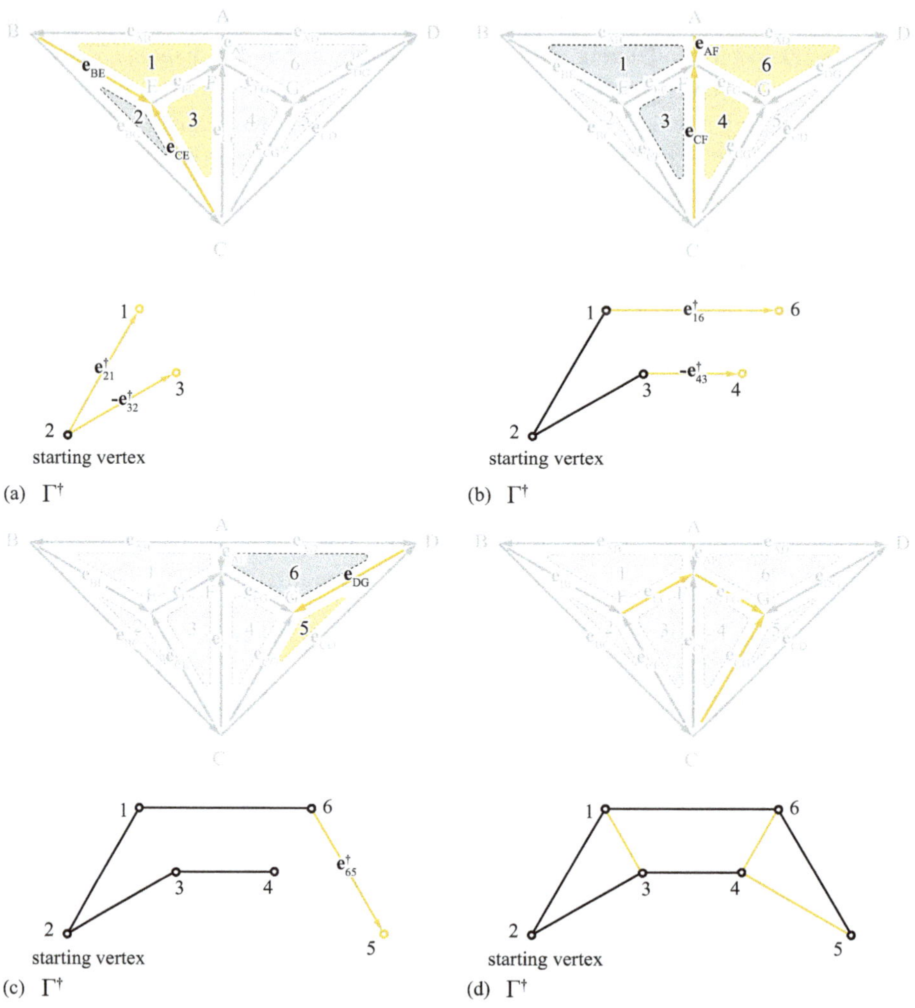

Figure 7.7 (a) to (d) Constructing the form diagram given a solution **q**.

vertex in the network from the starting vertex $v_2^\dagger$ that has a known position. Each time we reach a vertex with an unknown location, say vertex $v_1^\dagger$, we can have the edge vector $\mathbf{e}_{21}^\dagger$ connecting it with the known vertex $v_2^\dagger$. Since its direction $\hat{\mathbf{e}}_{21}^\dagger$ and the length of this edge q_{21} is known, we can easily compute the location of $v_1^\dagger$. This repeats until all vertex locations are found (Figure 7.7a–c). Then the remaining edges of $\Gamma^\dagger$ can be added using the topological connectivity described in Section 7.2.4, thus finishing the construction of the dual (Figure 7.7d).

Finally, the tension-compression states of the $e^\dagger$ edges are determined. When the input NFPs are positive, an edge with a positive length is under compression, while one with a negative length is under tension. Conversely, when the NFPs are negative, the interpretation of the edge lengths is reversed.

7.3 Form-Finding Applications of 2D Algebraic Graphic Statics

In this section, we discuss various applications of algebraic formulations for 2D form finding. These applications include the construction of networks that are exclusively compression- or tension-based, constraining the funicular solution to specific boundary conditions, and systems that integrate both compression and tension, such as re-entrant networks. The latter is particularly relevant in the design of auxetic materials.

7.3.1 Tension-Only and Compression-Only Form Finding

Since edge lengths are directly related to the tension and compression states, they can be used to help with tension-only and compression-only form finding. For an input force diagram with only positive NFPs, the resulting solution usually only contains positive edge lengths when the input parameter ξ of the MPI method has identical positive entries (with exceptions when the input force diagram is complex, in which case we can use linear programming instead to enforce tension- or compression-only). Figure 7.8a shows an input primal force diagram with 14 faces, and a compression-only form of a double-layer funicular bridge can be found using the MPI method with one (1) in all entries of ξ. Alternatively, if we use a negative one (−1) for all entries, the form of the bridge will become tension-only (Figure 7.8b).

7.3.2 Form Finding with Boundary Conditions

Using the same input force diagram, Figure 7.9 illustrates a series of double-layer funicular bridge forms with their ends constrained according to imaginary boundary conditions. In Figure 7.9a,b, vertices $v_1^\dagger$ and $v_8^\dagger$ are constrained to the same height and with a distance of 10. Changing the entries of ξ affects the edge lengths but always

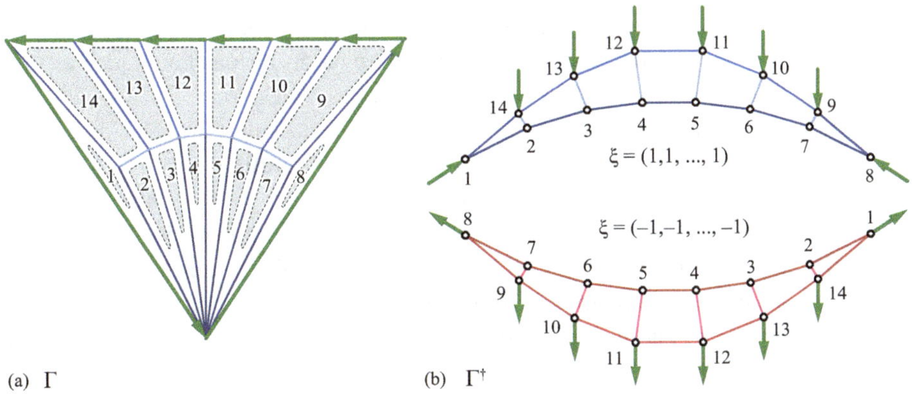

Figure 7.8 (a) An input primal force diagram Γ with 14 faces. (b) A compression-only and a tension-only form diagrams are generated with different ξ.

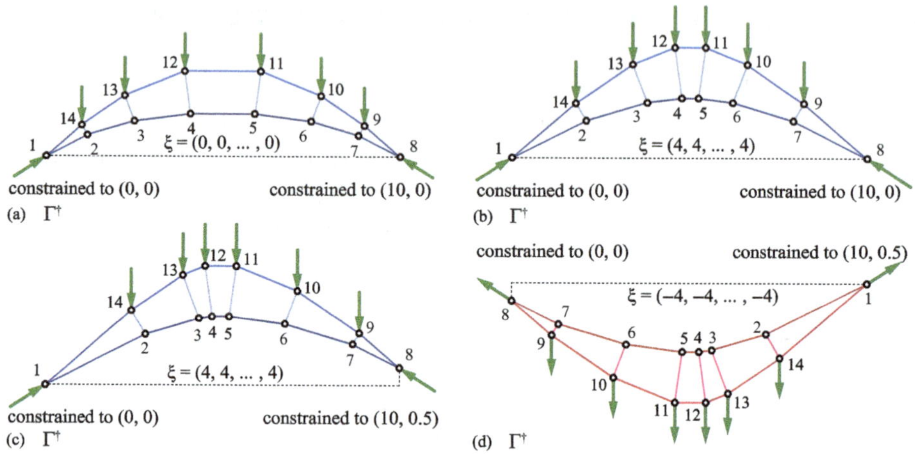

Figure 7.9 A series of double-layer funicular bridge forms with their ends constrained according to imaginary boundary conditions. (a) Vertices $v_1^\dagger$ and $v_8^\dagger$ are constrained to the same height, and all entries of ξ are zero. (b) Altering the entries of ξ to four yields a different form, but leaving vertices $v_1^\dagger$ and $v_8^\dagger$ unchanged. (c) For other boundary conditions, different forms can be generated simply by moving the constraint points. (d) The tension and compression states can be easily modified by switching the constraint locations of the two ends.

leaves the location of the two vertices fixed. Figure 7.9c,d shows asymmetrical funicular forms for other boundary conditions where the ends are constrained to different heights.

7.3.3 2D Auxetic Materials

One possible application of the algebraic formulation for 2D graphic statics is to leverage the GDoF and find geometries with auxetic behaviors, which are defined as having a negative Poisson's ratio. For instance, such materials expand perpendicular to the direction of an applied tensile force and contract when compressed.* This unique behavior is due to their particular internal concave geometrical configuration (Figure 7.10), which has found its way to be applied to various fields such as sensors, medical instruments, sportswear, aerospace devices, and so on.

The auxetic geometries can be generated using the GDoF in a two-step process (Figure 7.11). First, we identify a group of parallel edges in a form diagram with a direction **d** on a given convex tessellation (Figure 7.11a). Then, we convert the convex tessellation into a nonconvex system simply by reducing the corresponding weights in the input parameter ξ, which will flip the parallel edges as shown in Figure 7.11c,d. When forces are applied to the direction **d**, the geometry system will exhibit auxetic behavior. The geometric reasoning behind it is that the group of parallel edges can

* For further reading on this topic, please visit the references.[142–149]

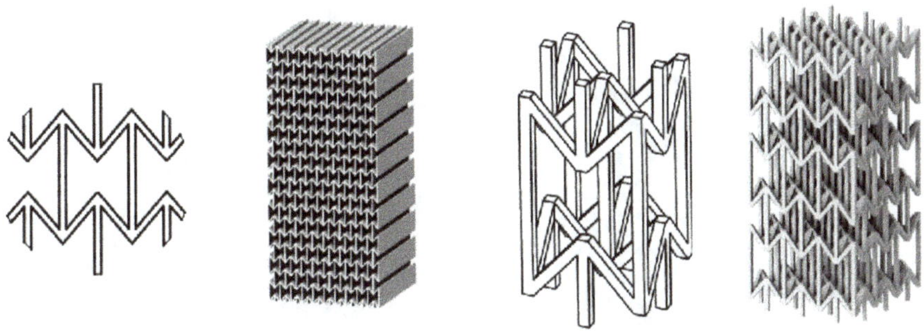

Figure 7.10 (a) A 2D re-entrant structure based on honeycomb geometry; (b) a 3D structure made of similar geometry.

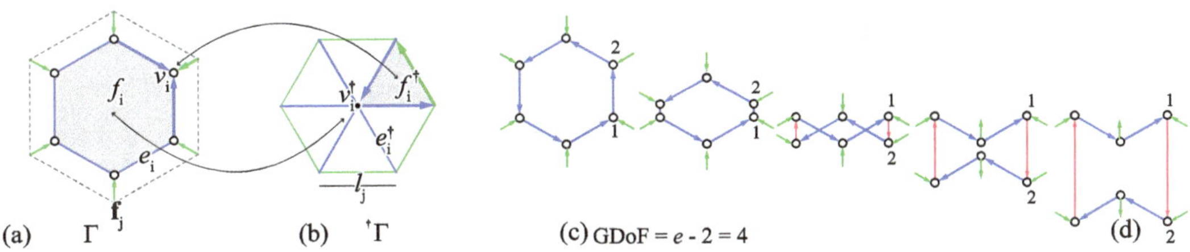

Figure 7.11 The form diagram (a) and the force diagram; (b) and their dual reciprocal elements; (c) Initial form diagram with the GDoF of 4 where the lengths of two parallel edges have been changed from positive value to negative value (d).

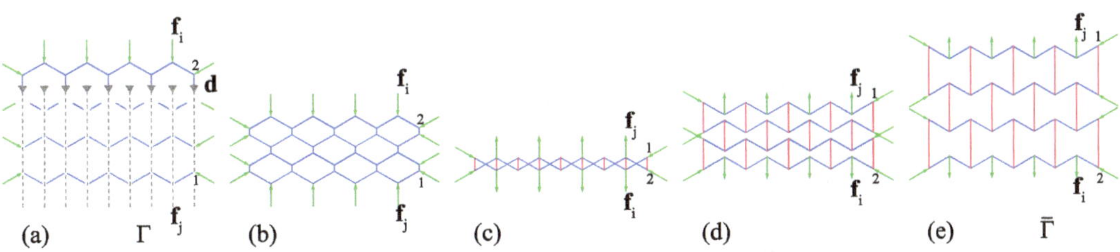

Figure 7.12 (a) The hexagonal tessellation with an overlaying TAD structure; (b) to (e) the transformation of the hexagonal tesselation into an auxetic configuration.

all be flipped to the other side simultaneously and therefore create nonconvex cells without creating self-intersections (Figure 7.12).

Figure 7.13a–e shows additional examples of convex tessellations and their related nonconvex auxetic geometry, together with the simulations of their auxetic behaviors. In addition to the most studied re-entrant auxetic geometry from the literature shown in Figure 7.13a1, our proposed method can work with more complex starting geometries and turn them into nonconvex configurations. For instance, Figure 7.13b1 consists

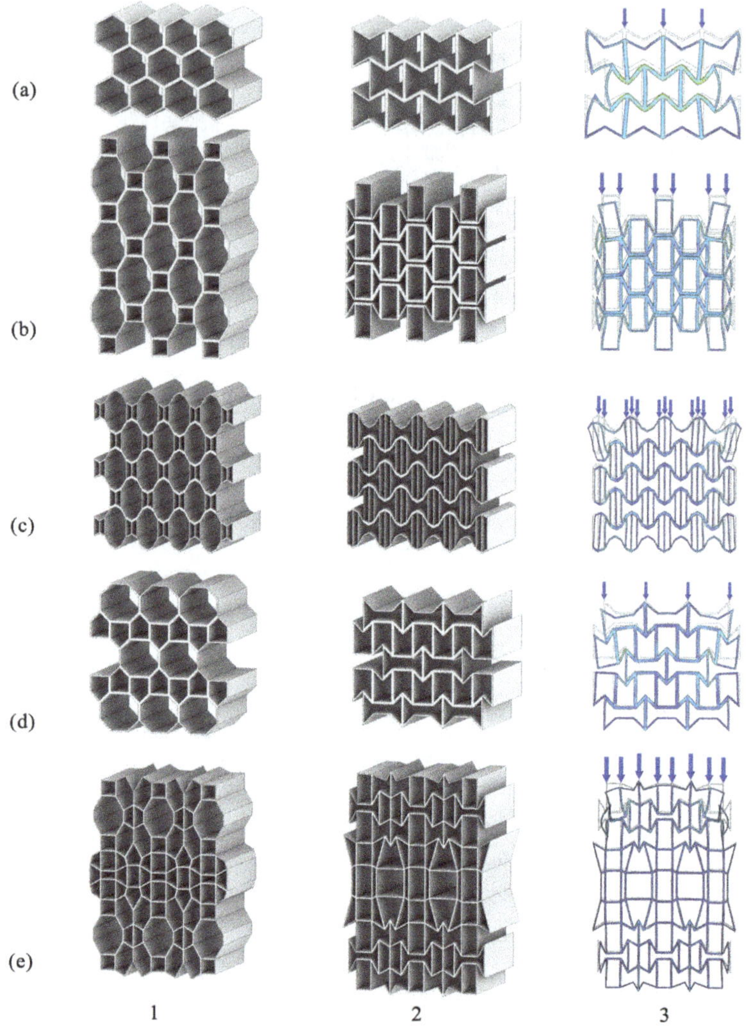

Figure 7.13 The diagrams in column (1) are examples of convex tessellations, which are transformed into auxetic systems in column (2); a preliminary finite element (FE) analysis in column (3) shows their auxetic behavior as the systems contract on its sides if compressive forces are applied on the top edges.

of squares and octagons; Figure 7.13c1 is the subdivided version of the sample of Figure 7.13b1 where the force diagram is subdivided into smaller internal polygons;[130] and Figure 7.13d1 is a network consisting of octagons and pentagons, while Figure 7.13e1 is a network consisting of quadrilaterals, pentagons, hexagons, and octagons. The simulations of their auxetic behaviors are illustrated in Figure 7.13, Column 3.

For a detailed discussion on the construction of auxetic systems, we refer the reader to Hablicsek and Akbarzadeh.[150]

7.4 Algebraic Construction of Polyhedral Reciprocal Diagrams

Sections 7.1 to 7.3 have provided an algebraic formulation for 2D graphic statics. The methodology explained for 2D construction can serve as a foundation for the 3D version of these diagrams. The algebraic formulation of the 3D situation has a similar theoretical framework to the 2D situation in terms of the concept of closed loops of edge vectors, the edge length and vertex location constraints, the approaches for solving the equation systems, and so on. This section is based on two recent publications of the authors.[135,151]

In the context of 3DGS using polyhedral reciprocal diagrams, both form and force diagrams consist of polyhedral cells, planar faces, straight edges, and vertices. Each edge in the reciprocal diagrams shares an identical vertex with its adjacent edges. As described by Akbarzadeh[9] and Hablicsek, Akbarzadeh, and Guo,[135] the edges and faces in the reciprocal diagrams have specific directions. The face normal must be consistent with its edge directions following the right-hand rule, the face orientations must be consistent within each cell, and the cell orientations must be consistent within each diagram. Therefore, each pair of adjacent faces shares two versions of an identical edge that have opposite directions. Similarly, each pair of adjacent cells shares two versions of an identical face that have opposite normal directions.

7.4.1 A Precise Definition of Form and Force Polyhedrons

Before we provide any algebraic formulation for the reciprocal construction of polyhedral diagrams, we would like to revisit and elucidate some of the key properties of the form and force diagrams and their differences that have been discussed on many occasions in Chapters 1 to 6. A more precise summary of these properties can be quite beneficial for the reader to thoroughly comprehend the properties of these diagrams before learning about their algebraic relationships.

Form Polyhedrons

As described in Chapters 1 to 6, in the context of polyhedral graphic statics, a form diagram consists of a group of polyhedrons with both open and closed cells. The open cells that are usually on the exterior of the system are connected to the externally applied forces and reactions of the supports.[102] Figure 7.14a,b shows a polyhedral joint with an externally applied force $\mathbf{f}_k$. This joint can be represented as a group of four open polyhedral cells, as shown in Figure 7.14b. These open cells include open faces that are shared by the applied loads or reaction forces.

Generally, a group of polyhedral cells with no open cell may represent a self-stressed system of forces with no externally applied loads.[17] Replacing the dashed edges in the form diagram of Figure 7.14b with additional members will turn the form into a self-stressed system.[152] In the context of this book, the form diagram always receives externally applied loads, and thus always includes open polyhedral cells.[75]

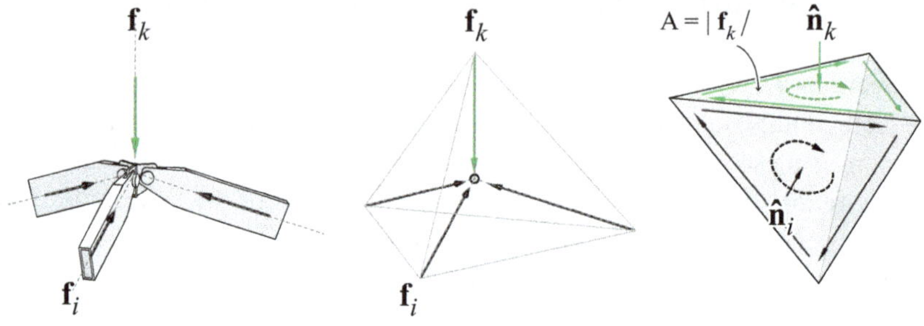

Figure 7.14 (a) A 3D structural joint with an applied force and internal forces in its members; (b) the form diagram/bar-node representation of the same joint in the context of 3DGS; and (c) the force diagram/polyhedron representing the equilibrium of the same node in 3DGS.

Force Polyhedrons

The force diagram, on the other hand, consists of closed convex and/or nonconvex polyhedral cells. Usually, in a group of polyhedral cells, there is a bigger polyhedron that is constructed by the external faces of the individual cells that are exposed to the outside. This polyhedron is referred to as the external polyhedron and the rest are called the internal cells.

Global and Nodal Force Polyhedrons

There is one cell in the group of polyhedrons that represents the global equilibrium of the system. The faces of this polyhedron are reciprocal to the external loads in the form diagram. Moreover, the closed cell volume shows that the sum of the external forces in the form diagram is zero. This polyhedron is referred to as the global force polyhedron (GFP). The rest of the cells represent the nodal equilibrium, and their faces are reciprocal and perpendicular to the internal members of the form. These cells are referred to as the nodal force polyhedrons (NFP).

The GFP can theoretically be any cell from a group of polyhedral cells. If the global force cell is the external polyhedron in the system, then the system can describe the equilibrium of compression/tension-only system of forces if all the NFPs are convex and all the edge lengths of the form diagram are positive. The global force polyhedron can also be any other cell in the group of cells.[9,153]

Remark 7.4. The GFP is not reciprocal to any vertex in the form diagram. Thus, its representative cell will not be included in the forthcoming topological description. However, its edges and faces are included in the topological relationships between the form and force diagrams.

The Orientation of a Face in a Polyhedral Cell

Each face in a closed polyhedron consists of a closed polygon of edge vectors with a consistent direction attached to each other tip to tail. This direction defines the unit

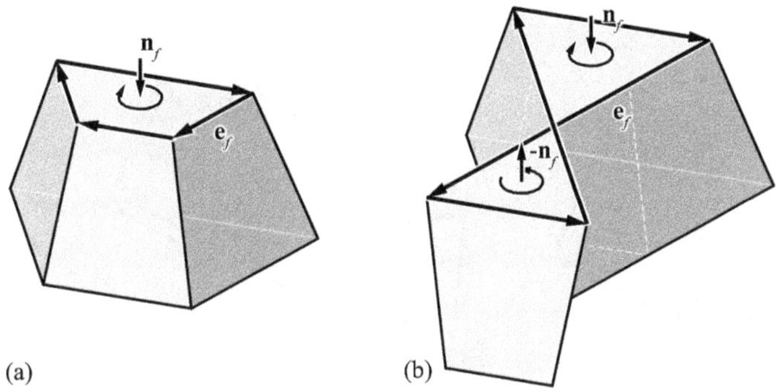

Figure 7.15 The orientation of a face and its edge vectors for (a) a convex face and (b) a self-intersecting face.

normal vector of the face using the right-hand rule, as illustrated in Figure 7.15a. Note that the geometry of the face can be either convex, nonconvex, or self-intersecting, but the directions of the edge vectors are consistently connected tip-to-tail in all cases. The normal direction of a face flips for self-intersecting parts of a face as discussed in many examples in Chapters 1 to 6, for instance, as shown in Figure 7.15b.

The Orientation of a Single Cell

Each cell in the force diagram consists of polygonal faces that meet at the edges of the cell. The orientation of the faces for a single with nonself-intersecting cell is consistent, meaning that all faces should have normal directions to the outside or inside of the cell. This phenomenon can be explained using Stokes' theorem.[89] For a single nonself-intersecting polyhedron with i faces, each having consistent unit normal vectors $\mathbf{n}_i^\dagger$ and face areas A_i, we have

$$\sum_{f_i^\dagger} \mathbf{n}_i^\dagger \cdot A_i = \mathbf{0}.$$

Following that theorem, every two adjacent faces share an equal but opposite edge vector, as shown in Figure 7.16a. Thus, the orientation of a single face in a cell can define the orientation of the rest of the faces of the cell requiring that the shared edge of the two adjacent faces have opposite directions. The same principle can be extended for a self-intersecting cell in Figure 7.16b. For a self-intersecting face, the normal directions will be flipped for certain regions. Still, the edge vectors are connected tip-to-tail and the normal directions of each region can be calculated using the right-hand rule, as shown in Figure 7.16b. The areas for the regions with flipped normal are considered negative, and the total area of the self-intersecting face is calculated by summing up the signed areas of all regions.

The Orientation of the Cells in the Force Diagram

Each nonintersecting cell in the force diagram has a direction of normal vectors pointing either toward the inside or outside of the cell. In the context of this chapter, we can

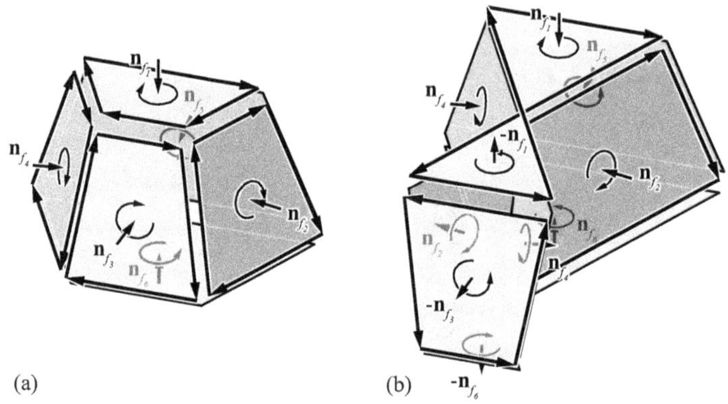

Figure 7.16 (a) A convex cell with consistent orientation of faces and with inward pointing normal vectors; and (b) a cell with a self-intersecting face with inward and outward pointing normal vectors.

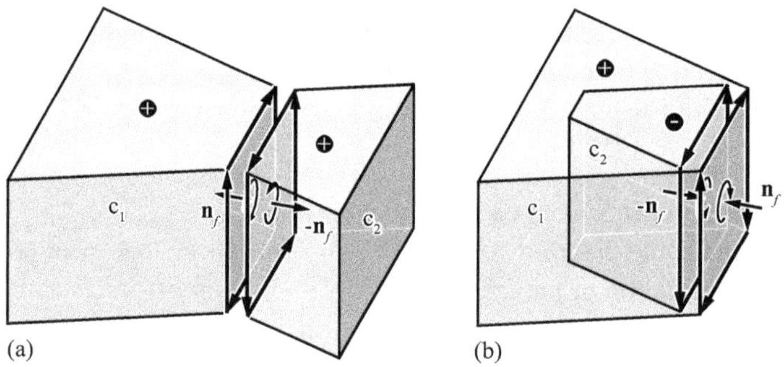

Figure 7.17 (a) Two positive cells sharing an identical face with opposite normal directions; (b) two overlapping cells, one positive and one negative, sharing an identical face with opposite normal directions.

call a cell with inward normals a *positive* cell that can represent a compression-only node. Also, a cell with outside normals can be referred to as a negative cell that may represent a tension-only node. In the case of a self-intersecting face, the part of the cell that has normals facing outside can be negative and the part with inward normals would be positive. Each pair of adjacent cells in the force diagram shares an identical face with equal area and opposite normal direction, as shown in Figure 7.17. Thus, the orientation of one cell in a group of polyhedral cells is sufficient to determine the direction of the rest of the cells in the force diagram.

For the force diagram, the orientation of the GFP matches the direction of the externally applied loads and reaction forces. As discussed, GFP is not counted as a cell in the force diagram, but its orientation defines the orientation of the rest of the cells in the system. In fact, the orientation of the faces of GFP can be viewed as being transferred to the faces shared by the adjacent NFP cells. The orientation of the rest of the NFPs in the group can be defined accordingly.

For instance, consider the force diagram of the truss of Figure 7.18a. The geometry of the force diagram in this example consists of five tetrahedrons including the

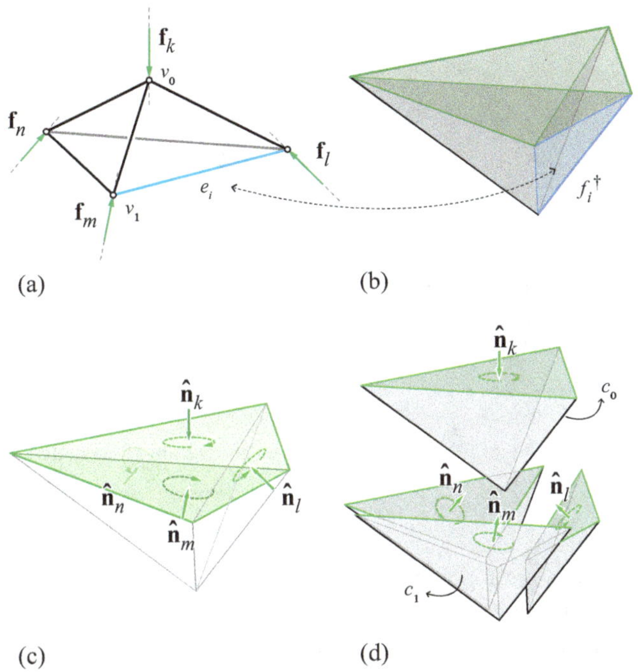

Figure 7.18 (a) A polyhedral truss with an applied load and reaction forces at the support and (b) its corresponding force diagram consisting of 10 faces and 5 polyhedral cells; (c) the global force polyhedron (GFP) with the direction of its faces toward inside of the cell; and (d) the faces of GFP constructs nodal force polyhedrons (NFP) whose orientations are inherited from the faces of the GFP (three cells toward outside (e.g., c_1) and one toward inside (c_0).

GFP (Figure 7.18b). The orientation of the GFP is determined by the direction of the externally applied loads and the reaction forces at the supports (Figure 7.18c). The force diagram is constructed by removing the GFP cell while transferring its face to the adjacent NFPs shown by green faces in Figure 7.18d.

The orientation of the NFPs will be determined by the orientation of the GFP; each face of the NFP that is shared with the GFP will have the same orientation as the GFP face, whereas the face shared by another NFP will always have an opposite normal direction (Figure 7.18b).

7.4.2 Topological Relationship between Form and Force Diagrams

We can use the example of Figure 7.19 to explain the topological relationship between reciprocal polyhedral diagrams. Similar to the convention presented in the 2D scenario, the starting diagram is called *primal*, Γ, and the reciprocal diagram is called *dual*, $\Gamma^\dagger$ (Figure 7.19a, b). The numbers of vertices, edges, faces, and cells of the primal are denoted as v, e, f, and c, respectively, and the ones of the dual are super-scripted with a dagger (†) symbol.

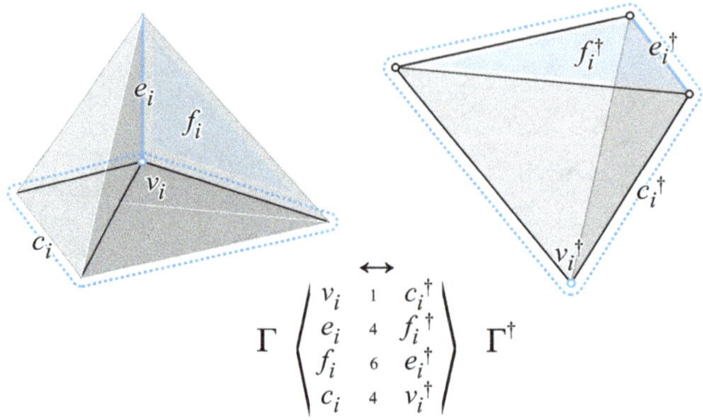

(a) (b)

Figure 7.19 (a) The *primal* diagram Γ and (b) its reciprocal diagram $\Gamma^\dagger$ as called *dual* and their corresponding components.

These two diagrams are topologically reciprocal, that is, the vertices v_i, edges e_j, faces f_k, and cells c_l of the primal topological map to the cells $c_i^\dagger$, faces $f_j^\dagger$, edges $e_k^\dagger$, and vertices $v_l^\dagger$, respectively, of the dual diagram.[75] Therefore, the numbers of reciprocal elements in both diagrams are the same. For instance, the number of vertices, v, of the primal is equal to the number of cells $c^\dagger$ in the dual, and so on. Adjacency is another key characteristic of reciprocity, meaning that if two elements in the primal diagram are adjacent, then their corresponding elements in the dual diagram are also adjacent. Moreover, each edge e_i of the primal is perpendicular to its corresponding face $f_i^\dagger$ in the dual.

Notably, the external faces of a force diagram correspond to the external edges that represent the applied load and reaction forces in the form diagram. Those external faces and edges are usually colored green. Unlike a force diagram where all elements are closed, a form diagram contains open elements. For example, vertices only exist at one end of the external edges in a form diagram, so such edges are open-ended. Also, all faces and cells containing open edges are also considered open. The closed faces in a form diagram are defined as internal, and they correspond to the internal edges of the force diagram. The edges of the internal faces are also defined as internal, and they correspond to the internal faces of the force diagram. All internal faces of the force diagram are shaded in blue, with the darkness indicating their respective areas. The darker the color, the larger the corresponding area. A similar color code is applied to the form diagram, where edges are colored blue for compression and red for tension. The darkness and thickness of each edge correspond to the area of its respective face in the force diagram.

The algebraic reciprocal construction with constraints starts from an input primal diagram. It aims to represent the dual diagram using a system of linear equations. Although both user-created form and force diagrams can be used as input as shown by Hablicsek et al.,[135] this chapter focuses on the form-finding aspect of it and always starts from force diagrams, so Γ and $\Gamma^{\dagger}$ always denote force and form diagrams, respectively. Moreover, starting from a form diagram, meaning that an analysis of its internal force distribution is wanted, is relatively difficult in the context of 3DGS as opposed to 2DGS because the forms with polyhedral features are not easy to construct directly.

7.4.3 Connectivity Matrices

To formulate the relationship between the primal and the dual diagrams, the association between the elements of each diagram is first described algebraically using multiple *connectivity* matrices.

Edge–Vertex

Let us consider the primal and dual diagrams of Figure 7.19a,b: the primal diagram includes arbitrarily indexed vertices and edges pointing from a vertex with a smaller number to a vertex with a bigger number (Figure 7.20). For the primal diagram, the connectivity matrix between the vertices and edges is a $[e \times v]$ matrix that is shown by $\mathbf{C}_{e \times v}$, described as

$$C_{e_i,v_j} = \begin{cases} +1 & \text{if vertex } v_j \text{ is the head of edge } e_i \\ -1 & \text{if vertex } v_j \text{ is the tail of edge } e_i \\ 0 & \text{otherwise.} \end{cases}$$

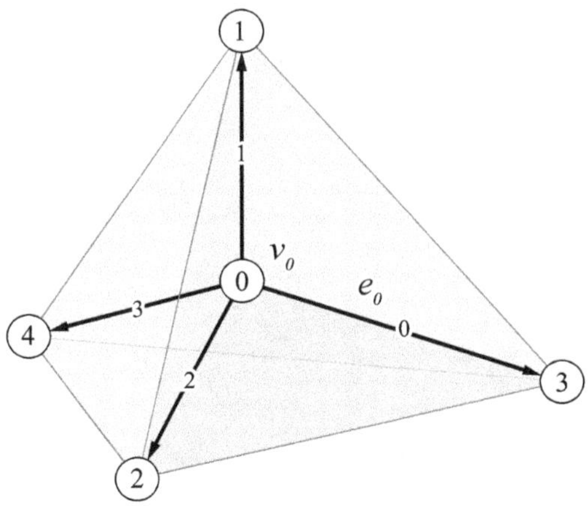

$$\mathbf{C}_{e \times v} = \begin{matrix} \\ 0 \\ 1 \\ 2 \\ 3 \end{matrix} \begin{matrix} \begin{matrix} 0 & 1 & 2 & 3 & 4 \end{matrix} \\ \begin{pmatrix} -1 & 0 & 0 & 1 & 0 \\ -1 & 1 & 0 & 0 & 0 \\ -1 & 0 & 1 & 0 & 0 \\ -1 & 0 & 0 & 0 & 1 \end{pmatrix} \end{matrix}$$

Figure 7.20 The primal diagram and the connectivity matrix given by its edges and vertices.

Since the edges and vertices of the primal correspond to faces and cells of the dual, matrix $\mathbf{C}_{e\times v}$ is equal to $\mathbf{C}_{f^{\dagger}\times c^{\dagger}}$ that represents the connectivity of the faces and cells of the dual diagram.

Edge–Face

The connectivity between edges and vertices, $\mathbf{C}_{e\times v}$, does not describe the topology of the primal completely, and further connectivity matrices are required to describe the topological relationships among other components of both primal and dual diagrams. Each face f_i of the primal has a unit normal vector $\hat{\mathbf{n}}_i$ where the direction of the normal may be chosen arbitrarily (Figure 7.21). This direction defines the orientation of the edges e_i on that face using the right-hand rule.

Therefore, for each edge e_i on the face f_i, there are two directions: one given by matrix $\mathbf{C}_{e\times v}$ and other defined by the direction of the unit normal of the face $\hat{\mathbf{n}}_i$ (Figure 7.21). Thus, for the edges and their connected faces in the primal complex, the edge-face connectivity matrix $\mathbf{C}_{e\times f}$ is a $[e \times f]$ matrix defined as

$$C_{e_i f_j} = \begin{cases} +1 & \text{if edge } e_i \text{ is an edge of face } f_j \\ -1 & \text{if the opposite of edge } e_i \text{ is an edge of face } f_j \\ 0 & \text{otherwise.} \end{cases}$$

Note that matrix $\mathbf{C}_{e\times f}$ can also describe the connectivity between the faces $f^{\dagger}$ and edges $e^{\dagger}$ of the dual complex and thus equals matrix $\mathbf{C}_{f^{\dagger}\times e^{\dagger}}$.

Face–Cell Connectivity Matrix

To complete the topological definition of the primal complex, the connectivity between the faces and cells of the primal should be described by an incidence matrix $\mathbf{C}_{f\times c}$. The

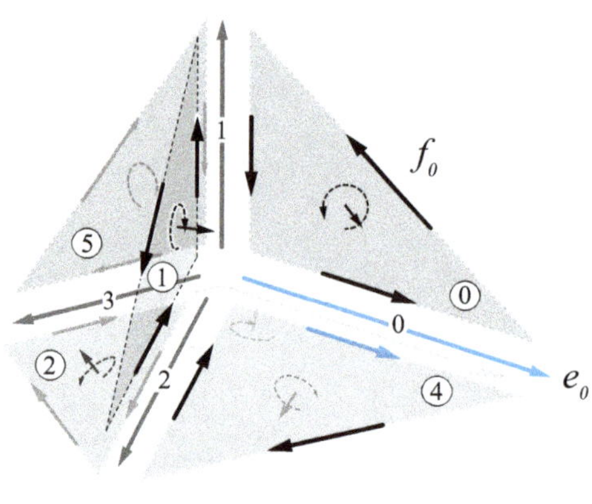

$$\mathbf{C}_{e\times f} = \begin{array}{c} \\ 0 \\ 1 \\ 2 \\ 3 \end{array} \begin{array}{c} \begin{array}{cccccc} 0 & 1 & 2 & 3 & 4 & 5 \end{array} \\ \begin{pmatrix} 1 & 0 & 0 & 1 & 1 & 0 \\ -1 & 1 & 0 & 0 & 0 & 1 \\ 0 & -1 & -1 & 0 & -1 & 0 \\ 0 & 0 & 1 & -1 & 0 & -1 \end{pmatrix} \end{array}$$

Figure 7.21 The connectivity of the faces and edges in the primal and its related matrix.

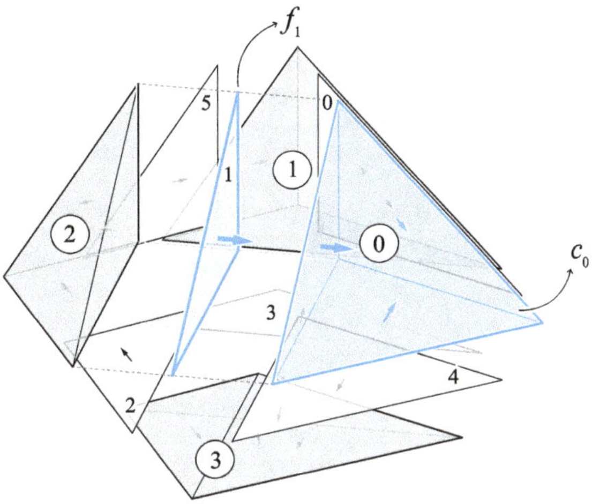

$$\mathbf{C}_{f\times c} = \begin{array}{c} \\ 0\\1\\2\\3\\4\\5 \end{array}\begin{array}{c} \begin{array}{cccc} 0 & 1 & 2 & 3 \end{array} \\ \begin{pmatrix} 1 & -1 & 0 & 0 \\ 1 & 0 & -1 & 0 \\ 0 & 0 & 1 & -1 \\ 0 & 1 & 0 & -1 \\ -1 & 0 & 0 & 1 \\ 0 & -1 & 1 & 0 \end{pmatrix} \end{array}$$

Figure 7.22 The connectivity of faces and cells of the primal and its incidence matrix.

orientation of each face f_i in the primal was chosen arbitrarily. However, the orientation of the cells is predefined as discussed in Section 7.4.1. We check the orientation of the face f_i with the orientation of the cell c_j. Hence, the incidence matrix for the faces and cells can be defined as (Figure 7.22):

$$C_{f_i,c_j} = \begin{cases} +1 & \text{if face } f_i \text{ has the same orientation as of } c_j \\ -1 & \text{if face } f_i \text{ has the opposite orientation as of } c_j \\ 0 & \text{otherwise.} \end{cases}$$

Since faces f and cells c of the primal correspond to $e^\dagger$ and $v^\dagger$ of the dual, the matrix $\mathbf{C}_{f\times c}$ can describe the topological relationship between the unknown edges and vertices of the dual complex and therefore is equal to an $[e^\dagger \times v^\dagger]$ matrix $\mathbf{C}_{e^\dagger \times v^\dagger}$ (Figure 7.23). Note that the direction of the edges of the dual is a result of the face-cell connectivity matrix. For instance, the edges are not necessarily directed from smaller indices to bigger indices.

7.4.4 3D Closing Equations for One Face

Linear equations can be developed to represent the closed edge vectors associated with each internal face in the form diagram. Let e_{int} and f_{int} denote the number of internal edges and faces, respectively, in the force diagram; let $e^\dagger_{int}$ and $f^\dagger_{int}$ denote the number of

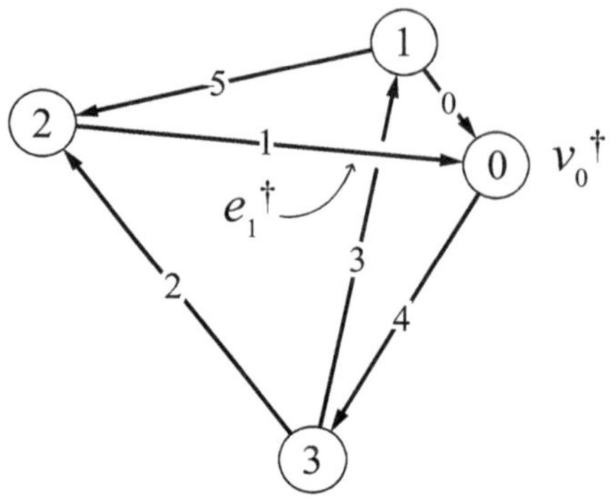

Figure 7.23 The edge-vertex connectivity of the dual diagram is the same as the face-cell connectivity of the primal of Figure 7.22.

internal edges and internal faces, respectively, in the form diagram. For each internal face $f_i^\dagger$ in the form diagram to be constructed, we can write an equation that shows its closed edges based on the edge lengths. By choosing a normal vector for each face $f_i^\dagger$, we can obtain a consistent edge orientation. We denote the unit direction vector $\hat{\mathbf{e}}_j^\dagger$ corresponding to the edge vectors $\mathbf{e}_j^\dagger$. Since each face provides a closed loop of edges, the sum of the edge vectors has to be a zero vector. Thus, we obtain a vector equation based on every edge length $q_j^\dagger$ of $\mathbf{e}_j^\dagger$ as

$$\sum_j \hat{\mathbf{e}}_j^\dagger q_j^\dagger = \mathbf{0}, \tag{7.18}$$

where the sum runs over the edges e_j of the face f_i. We remark that the edge lengths q_j can take both *negative and positive values*; changing the length of an edge to its negative means that we change the direction of the edge to the opposite direction without changing the direction of the unit vector.

For instance, in Figure 7.23, the top face provides a vector equation

$$\mathbf{e}_1^\dagger - \mathbf{e}_0^\dagger + \mathbf{e}_5^\dagger = \hat{\mathbf{e}}_1^\dagger q_1^\dagger - \hat{\mathbf{e}}_0^\dagger q_0^\dagger + \hat{\mathbf{e}}_5^\dagger q_5^\dagger = \mathbf{0}.$$

The direction vectors, $\hat{\mathbf{e}}_i$, either follow or are opposite to the loop direction defined by the chosen face normal $\mathbf{n}_i$. This sign depends on the face-cell connectivity matrix of the primal diagram.

Given these direction vectors, each face $f_i^\dagger$ of the polyhedron provides three equations: one equation for each of the x-, y-, and z- dimensions. Thus, Equation 7.18 can be described by a $[3 \times e_{int}^\dagger]$ matrix, $\mathbf{A}_{f_i^\dagger}$

$$\mathbf{A}_{f_i^\dagger}\mathbf{q} = \mathbf{0}, \tag{7.19}$$

where $\mathbf{q}$ denotes the signed lengths of all $e_{int}^\dagger$ internal edges assembled in a vector. This equation describes the geometry of the face $f_i^\dagger$.

7.4.5 The Closing Matrix

All the closing equations for the $f^{\dagger}_{int}$ internal faces in the dual diagram can be written in a matrix form.

It first starts from the coordinate difference vectors **u**, **v**, and **w**, which can be described as

$$\mathbf{u} = \mathbf{C}_{e \times v}\mathbf{x} \quad \mathbf{v} = \mathbf{C}_{e \times v}\mathbf{y} \quad \mathbf{w} = \mathbf{C}_{e \times v}\mathbf{z}. \tag{7.20}$$

Here, the vectors **x**, **y**, and **z** are the x-, y-, and z- coordinates of the vertices of the primal diagram, which are known. $\mathbf{C}_{e \times v}$ is the connectivity matrix for the edges and vertices of the primal. Similarly, the vectors $\mathbf{u}^{\dagger}$, $\mathbf{v}^{\dagger}$, and $\mathbf{w}^{\dagger}$, defined as the coordinate difference vectors of the vertices of the dual diagram, can describe the edges of the dual as

$$\mathbf{u}^{\dagger} = \mathbf{C}_{e^{\dagger} \times v^{\dagger}}\mathbf{x}^{\dagger} \quad \mathbf{v}^{\dagger} = \mathbf{C}_{e^{\dagger} \times v^{\dagger}}\mathbf{y}^{\dagger} \quad \mathbf{w}^{\dagger} = \mathbf{C}_{e^{\dagger} \times v^{\dagger}}\mathbf{z}^{\dagger}.$$

Here, the vectors $\mathbf{x}^{\dagger}$, $\mathbf{y}^{\dagger}$, and $\mathbf{z}^{\dagger}$ are the x-, y-, and z- coordinates of the vertices of the dual diagram. Since vertices and edges of the dual correspond to the cells and faces of the primal, we can write:

$$\mathbf{u}^{\dagger} = \mathbf{C}_{f \times c}\mathbf{x}^{\dagger} \quad \mathbf{v}^{\dagger} = \mathbf{C}_{f \times c}\mathbf{y}^{\dagger} \quad \mathbf{w}^{\dagger} = \mathbf{C}_{f \times c}\mathbf{z}^{\dagger}, \tag{7.21}$$

where $\mathbf{C}_{f \times c}$ is the connectivity matrix for the faces and cells of the primal.

Since the edge vectors of every face $f_i^{\dagger}$ should sum up to zero, each of the x-, y-, and z- dimensions also sums up to zero, written as

$$\mathbf{C}_{e_{int} \times f_{int}}\mathbf{u}^{\dagger} = \mathbf{0} \quad \mathbf{C}_{e_{int} \times f_{int}}\mathbf{v}^{\dagger} = \mathbf{0} \quad \mathbf{C}_{e_{int} \times f_{int}}\mathbf{w}^{\dagger} = \mathbf{0}. \tag{7.22}$$

Since the dual edges $e_i^{\dagger}$ and the corresponding normal vectors of the faces in the primal $\mathbf{n}_i$ are parallel, we can use those known normal vectors to represent the edge vectors of the dual diagram. Specifically, around each internal edge e_i and its adjacent faces f_{i-k} in the primal, the sum of the normal vector of the faces $\mathbf{n}_i$ multiplied by the length $q_i^{\dagger}$ of the corresponding edge $e_i^{\dagger}$ in the dual diagram should be the zero vector (Figure 7.24).

Let $\mathbf{N}_x$, $\mathbf{N}_y$, and $\mathbf{N}_z$ be the $[f_{int} \times f_{int}]$ diagonal matrices whose diagonal entries are the x-, y-, and z- coordinates (respectively) of the chosen unit normal vectors of the faces of the primal. Furthermore, let **q** denote the vector of the edge lengths of the dual diagram. Then, $\mathbf{u}^{\dagger}$, $\mathbf{v}^{\dagger}$, and $\mathbf{w}^{\dagger}$ can be written as

$$\mathbf{u}^{\dagger} = \mathbf{N}_x\mathbf{q} \quad \mathbf{v}^{\dagger} = \mathbf{N}_y\mathbf{q} \quad \mathbf{w}^{\dagger} = \mathbf{N}_z\mathbf{q}. \tag{7.23}$$

Combining Equations 7.22 and 7.23 results in

$$\mathbf{C}_{e_{int} \times f_{int}}\mathbf{N}_x\mathbf{q} = \mathbf{0} \quad \mathbf{C}_{e_{int} \times f_{int}}\mathbf{N}_y\mathbf{q} = \mathbf{0} \quad \mathbf{C}_{e_{int} \times f_{int}}\mathbf{N}_z\mathbf{q} = \mathbf{0}. \tag{7.24}$$

We call the $[3e_{int} \times f_{int}]$ matrix

$$\mathbf{A} = \begin{pmatrix} \mathbf{C}_{e_{int} \times f_{int}}\mathbf{N}_x \\ \hline \mathbf{C}_{e_{int} \times f_{int}}\mathbf{N}_y \\ \hline \mathbf{C}_{e_{int} \times f_{int}}\mathbf{N}_z \end{pmatrix}$$

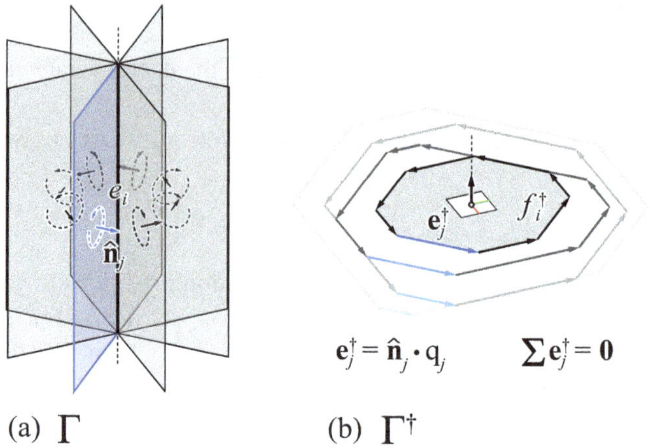

Figure 7.24 Going around each edge of the primal with its attached faces (a) provides the direction of the edge vectors of the corresponding face (b) in the reciprocal diagram where the sum of the edge vectors must be zero.

the *closing* matrix. Any solution of the equation system

$$\mathbf{Aq} = \mathbf{0} \tag{7.25}$$

gives us a set of edge lengths of the dual, and this can be used to construct the dual diagram. We call Equation 7.25 the *closing equations*.

7.4.6 An Alternative Closing Matrix with Reduced Dimension

The closing matrix $\mathbf{A}$ can be constructed in another way such that its size can be reduced to $[2e_{int} \times f_{int}]$. This optional step could significantly reduce the calculation time when solving the equation systems in subsequent steps.

In the previous formulation shown by Equation 7.22, each closed face in the form diagram is enforced by three equations that correspond to the x-, y-, and z- dimensions. This is redundant due to the planarity of all faces. Under a local coordinate system aligned with the face, this can be enforced by only two equations that correspond to the local x- dimension and local y- dimension (Figure 7.25).

Let $\mathbf{n}_x$, $\mathbf{n}_y$, and $\mathbf{n}_z$ be the $[e^\dagger_{int} \times 1]$ (equivalent to $[f_{int} \times 1]$) vector whose entries are the x-, y-, and z- coordinates (respectively) of the unit directional vectors of the $e^\dagger_{int}$ internal edges in the form diagram (equivalent to the unit normal vectors of the f_{int} internal faces in the force diagram). For each face $f_i^\dagger$ in the form diagram, let $\mathbf{P}_i$ be a $[f_{int} \times f_{int}]$ (or $[e^\dagger_{int} \times e^\dagger_{int}]$) diagonal matrix that describes the edge-face connectivity for a face $f_i^\dagger$, and its diagonal entries are defined as

$$\mathbf{P}_{j,j} = \begin{cases} +1 & \text{if edge } e_j^\dagger \text{ is an edge of face } f_i^\dagger \\ -1 & \text{if the opposite of edge } e_j^\dagger \text{ is an edge of face } f_i^\dagger \\ 0 & \text{otherwise.} \end{cases}$$

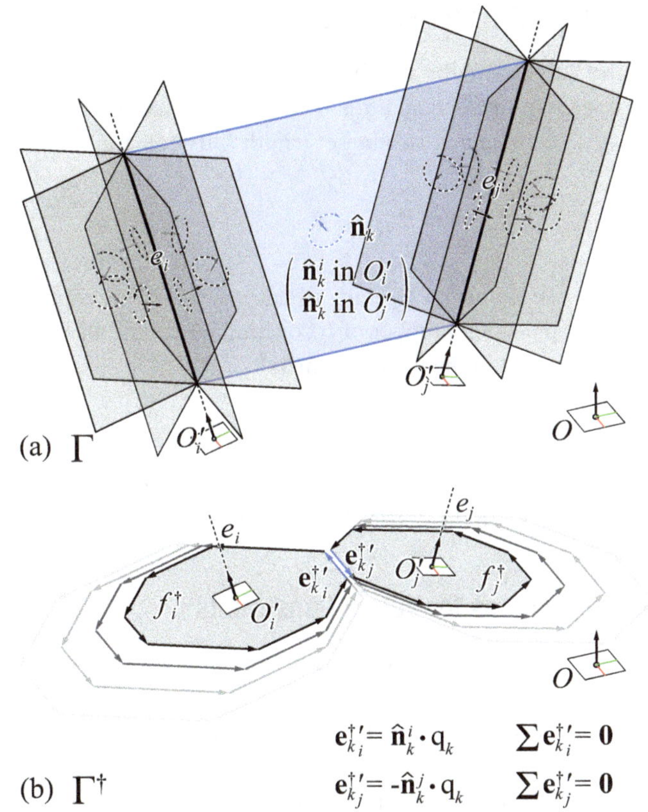

Figure 7.25 Closing equations are written based on the local coordinate system aligned with each face $f_i^\dagger$. (a) Two internal edges e_i, e_j, and all connected faces f_k in a force diagram. (b) The corresponding internal face $f_i^\dagger, f_i^\dagger$, and the surrounding edges $e_k^\dagger$ in the form diagram.

Then, a local coordinate system of $f_i^\dagger$ can be constructed by taking its center as the local origin and its normal vector as the local z- direction (Figure 7.25). This is followed by transforming the unit directional vectors of the $e_{int}^\dagger$ edges in the form diagram to the local coordinate system. Let $\mathbf{n}'_{xi}$ and $\mathbf{n}'_{y_i}$ be the $[e_{int}^\dagger \times 1]$ vectors whose entries are the x- and y- coordinates (respectively) of the unit directional vectors transformed to the local coordinate system of $f_i^\dagger$. The two closing equations can be written as

$$(\mathbf{P}_i \mathbf{n}'_{xi})^T = 0 \quad (\mathbf{P}_i \mathbf{n}'_{y_i})^T = 0.$$

Therefore, for all $f^\dagger$ faces, the simplified $[2f_{int}^\dagger \times e_{int}^\dagger]$ (or $[2e_{int} \times f_{int}]$) closing matrix $\mathbf{A}$ can be rewritten as

$$\mathbf{A} = \begin{pmatrix} \vdots \\ (\mathbf{P}_i \mathbf{n}'_{xi})^T \\ (\mathbf{P}_i \mathbf{n}'_{y_i})^T \\ (\mathbf{P}_{i+1} \mathbf{n}'_{x(i+1)})^T \\ (\mathbf{P}_{i+1} \mathbf{n}'_{y(i+1)})^T \\ \vdots \end{pmatrix}.$$

7.4.7 Additional Linear Equations for Edge Constraints

Edge constraints set target lengths for the internal edges in the form diagram, and each constrained edge requires one additional constraint equation. For example, if the i-th internal edge $e_i^\dagger$ is constrained to a target length l, its constraint equation may be written as

$$\mathbf{b}_i^T \mathbf{q} = l,$$

where $\mathbf{b}_i$ is the $[f_{int} \times 1]$ column vector with all entries zero (0) except at the index of $e_i^\dagger$ where it's one (1). For a total number of b constrained edges, all b edge constraint equations may be assembled in the matrix form as

$$\mathbf{Bq} = \mathbf{l},$$

where the rows of the $[b \times f_{int}]$ edge constraint matrix $\mathbf{B}$ are the row vectors $\mathbf{b}_i^T$, and $\mathbf{l}$ is a $[b \times 1]$ column vector with the entries being the target lengths.

7.4.8 Additional Linear Equations for Vertex Constraints

Vertex constraints set target locations for the vertices, and they can be points, lines, or planes. A target point fixes all three coordinates of the constrained vertex, while target lines and target planes allow the vertex to move on them. Nonlinear geometries like curves and curved surfaces are not covered in this study as these geometric constraints are difficult to represent through a linear equation system.

If there is only one constrained vertex for the form diagram, it can be achieved simply by a translation originating from the original vertex location to the target location, and it doesn't affect the lengths of the edges. Vertex constraints have influences on the edge lengths only if the number of constrained vertices d is greater than one, and, in such cases, additional vertex constraint equations are needed.

Points as Vertex Constraints

From all constraint types, the point constraint is the simplest and is hereby used to start the description. The vertex constraint equations are developed based on paired constrained vertices. For a pair of vertices $v_i^\dagger$ and $v_j^\dagger$ that are constrained to points $\mathbf{p}_i$ and $\mathbf{p}_j$, a path of edges that connects them can be found using BFS algorithm, then the equations can be written since the edges on the path must have compatible lengths in order to fit in between the two vertices (Figure 7.26).

Let $\mathbf{v}$ be the vector going from $\mathbf{p}_i$ to $\mathbf{p}_j$, and let u, v, and w be the x-, y-, and z-components of $\mathbf{v}$, respectively. Let $\mathbf{Q}$ be a $[f_{int} \times f_{int}]$ (or $[e_{int}^\dagger \times e_{int}^\dagger]$) diagonal matrix, and its diagonal entries are defined as

$$\mathbf{Q}_{k,k} = \begin{cases} 0 & \text{if edge } e_k^\dagger \text{ not on the path} \\ +1 & e_k^\dagger \text{ on the path, follows direction from } v_i^\dagger \text{ to } v_j^\dagger \\ -1 & e_k^\dagger \text{ on the path, follows direction from } v_j^\dagger \text{ to } v_i^\dagger. \end{cases}$$

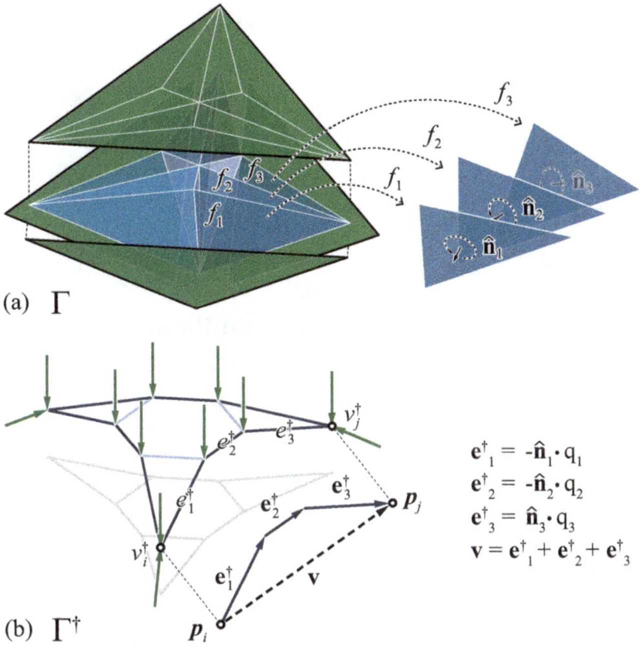

Figure 7.26 The path of edges connecting a pair of constrained vertices $v_i^\dagger$ and $v_j^\dagger$ need to have compatible lengths to fit in between.

Then the three equations for this pair of constrained vertices can be written as

$$(\mathbf{Qn}_x)^T \mathbf{q} = u \quad \left(\mathbf{Qn}_y\right)^T \mathbf{q} = v \quad (\mathbf{Qn}_z)^T \mathbf{q} = w. \tag{7.26}$$

Note that there could be multiple paths connecting $v_i^\dagger$ and $v_j^\dagger$, however, the result does not depend on the choice of path that is used.

When the number of constrained vertices d is greater than 2, all relative positions can be extensively described by $d-1$ pairs of constrained vertices. For the i-th vertex pair, the three constraint equations can be written as

$$(\mathbf{Q}_i\mathbf{n}_x)^T \mathbf{q} = u_i \quad \left(\mathbf{Q}_i\mathbf{n}_y\right)^T \mathbf{q} = v_i \quad (\mathbf{Q}_i\mathbf{n}_z)^T \mathbf{q} = w_i,$$

and the $3 \times (d-1)$ constraint equations can be assembled in matrix form as

$$\mathbf{Dq} = \mathbf{d},$$

where $\mathbf{D}$ is the $[3(d-1) \times f_{int}]$ vertex constraint matrix, written as

$$\mathbf{D} = \begin{pmatrix} \vdots \\ (\mathbf{Q}_i\mathbf{n}_x)^T \\ (\mathbf{Q}_i\mathbf{n}_y)^T \\ (\mathbf{Q}_i\mathbf{n}_z)^T \\ \vdots \end{pmatrix},$$

and **d** is the $[3(d-1)\times 1]$ column vector written as

$$\mathbf{d} = \begin{pmatrix} \vdots \\ u_i \\ v_i \\ w_i \\ \vdots \end{pmatrix}.$$

Mixed Constraint Types of Points, Lines, and Planes

Unlike point constraints where vertex locations are completely fixed, a line constraint or plane constraint releases the restriction in one or two dimensions and therefore allows it to move. The vertex constraint equations for mixed constraint types are also based on constrained vertex pairs $v_i^\dagger$ and $v_j^\dagger$.

There is an assumption in this chapter that one vertex of each pair, say $v_i^\dagger$, needs to be constrained to a point, while the constraint of $v_j^\dagger$ can be any type. That also means that for d constrained vertices, at least one vertex of them needs to be constrained to a point. This is because when none of the vertices has a fixed position, the difference between their positions will be difficult to represent. Although this can be resolved by setting up a reference point for one of the vertices and introducing its coordinates as three extra variables, they still need to be manually chosen after solving the equation system. Therefore, it's convenient to have at least one vertex fixed from the beginning to simplify the problem.

Figure 7.27 shows form diagram examples with line constant and plane constraint, all derived from the same force diagram illustrated in Figure 7.26a. Since target lines and planes may have 3D orientations, a local coordinate system O' that is aligned with the line or plane can help reduce the number of equations needed and thus simplify the problem. If $v_j^\dagger$ is constrained to a line, then the local coordinate system can be constructed with the unit directional vector of the line as the local z-axis and any point on the line as the origin. In such case, the x- and y- differences of $v_i^\dagger$ and $v_j^\dagger$ in the local coordinate system are constant (Figure 7.27a). If it's constrained to a plane, then the unit normal direction of the plane will be used as the local z-axis, and the origin can be any point on the plane. In this case, the z- difference of $v_i^\dagger$ and $v_j^\dagger$ in the local coordinate system is constant (Figure 7.27b). All points and vectors changed to this local coordinate system are super-scripted with a prime symbol (′). Let $\hat{\mathbf{x}}'$, $\hat{\mathbf{y}}'$, and $\hat{\mathbf{z}}'$ denote the unit directional vectors of the axes of the local coordinate system.

In the coordinate system, the paths of edges connecting $v_i^\dagger$ and $v_j^\dagger$ must still have compatible lengths. In order to have the equations in the local system, the target location $\mathbf{p}_i$ and the unit direction vectors of all edges $e_i^\dagger$ need to perform a basis change. In the local coordinate system, let v' be the vector going from $\mathbf{p}_i$ to any point $\mathbf{p}_j$ on the constraint geometry. Here the origin of the local coordinate system is used as $\mathbf{p}_j$ for the sake of clarity; let u', v', and w' be the x-, y-, and z-components of v', respectively; let n'_x, n'_y, n'_z be the $[e_{int}^\dagger \times 1]$ (equivalent to $[f_{int} \times 1]$) vectors whose entries are the x-, y-, and z- coordinates (respectively) of the transformed unit directional vectors of the $e_{int}^\dagger$

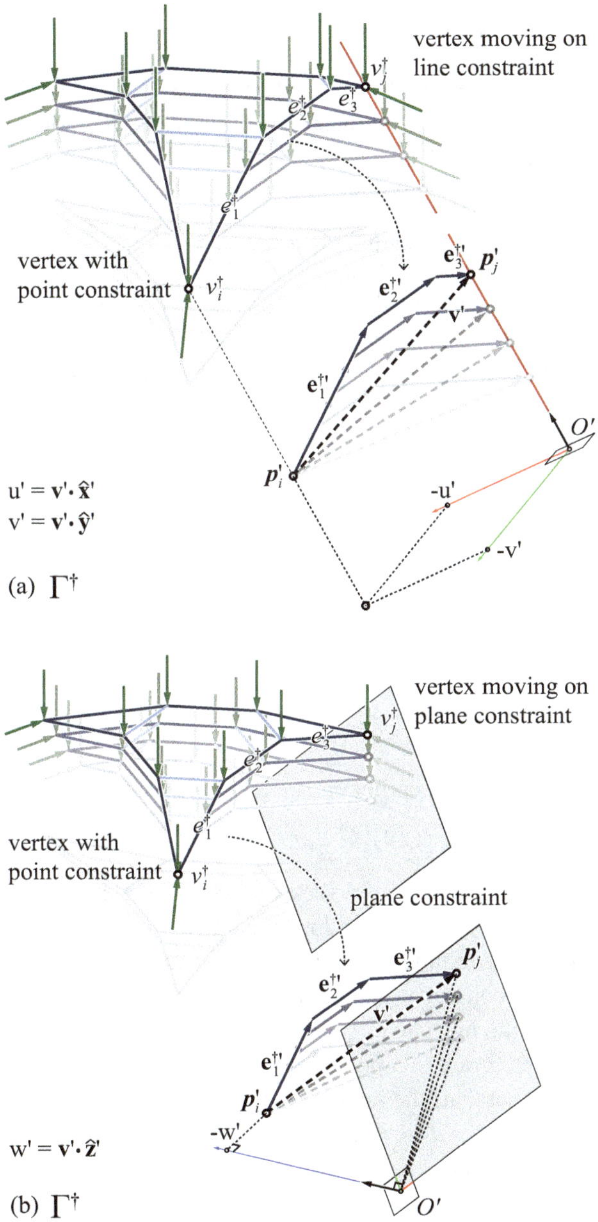

Figure 7.27 (a) When $v_j^\dagger$ is constrained to a line, the *x*- and *y*- differences of $v_i^\dagger$ and $v_j^\dagger$ in the local coordinate system are constant. (b) When $v_j^\dagger$ is constrained to a plane, the *z*- difference of $v_i^\dagger$ and $v_j^\dagger$ in the local coordinate system is constant.

internal edges in the form diagram (equivalent to the normal vectors of the f_{int} internal faces in the force diagram). If $v_j^\dagger$ is constrained to a line, then only two equations are enough to describe the constraints and they are written as

$$(\mathbf{Qn}_x')^T \mathbf{q} = \mathrm{u}' \quad \left(\mathbf{Qn}_y'\right)^T \mathbf{q} = \mathrm{v}'.$$

This is because only the local x- and y- coordinates of $v_j^\dagger$ are constrained, and the local z- coordinate is free to change. If $v_j^\dagger$ is constrained to a plane, then only one equation is needed, and it can be described as

$$\left(\mathbf{Qn}_z'\right)^T \mathbf{q} = \mathrm{w}'.$$

Similarly, this is because only the local z- coordinate of $v_j^\dagger$ is constrained, and the local x- and y- coordinates are free to change.

There is an alternative approach for developing the constraint equations for a mixed type of constraint geometries without going through the change of coordinate system. Let $v_i^\dagger$ be a constrained vertex, and assume that $v_j^\dagger$ is bounded to a plane p given by the equation

$$\mathbf{n}_p\mathbf{p}_j = d_p, \tag{7.27}$$

where $\mathbf{n}_p$ denotes a normal vector of the plane and $\mathbf{p}_j$ denotes the coordinate vector of the vertex $v_j^\dagger$.

Subtracting $\mathbf{n}_p\mathbf{p}_i$ from both sides of Equation 7.27, the following equation can be obtained:

$$\mathbf{n}_p(\mathbf{p}_j - \mathbf{p}_i) = d_p - \mathbf{n}_p\mathbf{p}_i.$$

The left-hand side of the equation earlier can be rewritten using Equation 8.10

$$\mathbf{n}_p\,(\mathbf{QN})^T\,\mathbf{q} = d_p - \mathbf{n}_p\mathbf{p}_i,$$

where the matrix $\mathbf{Q}$ describes a path from vertex $v_i^\dagger$ to $v_j^\dagger$ and $\mathbf{N}$ is the $[e_{int}^\dagger \times 3]$ matrix obtained by stacking the vectors $\mathbf{n}_x$, $\mathbf{n}_y$, and $\mathbf{n}_z$ column-wise.

Therefore, a constraint that bounds a vertex to a plane provides one additional linear equation in $\mathbf{q}$. Similarly, a constraint that bounds a vertex to a line provides two additional linear equations as the line can be represented as the intersection of two planes; a constraint that bounds a vertex to a point provides three additional linear equations as the point can be represented as the intersection of three planes.

To summarize the earlier, for d ($d > 2$) vertices constrained to p points, l lines, and n planes where d equals the sum of p, l, and n, a total number of $3(p-1) + 2l + n$ constraint equations are needed to describe the constraints. Accordingly, the shape of the vertex constraint matrix $\mathbf{D}$ becomes $[(3(p-1) + 2l + n) \times f_{int}]$, and that of the column vector $\mathbf{d}$ becomes $[(3(p-1) + 2l + n) \times 1]$.

The construction of the vertex constraint matrix $\mathbf{D}$ needs to find a connecting path for each constrained vertex pair $v_i^\dagger$ and $v_j^\dagger$ in the form diagram. The BFS algorithm can be used for this purpose. It starts by using one constrained vertex as the origin to perform a BFS to traverse the network of the form diagram. During the BFS, a search tree can be constructed with a constrained vertex being the root node, other vertices being its descendant, and links being the connecting edges. The constrained vertex pairs can be obtained from this search tree. For all other constrained vertices,

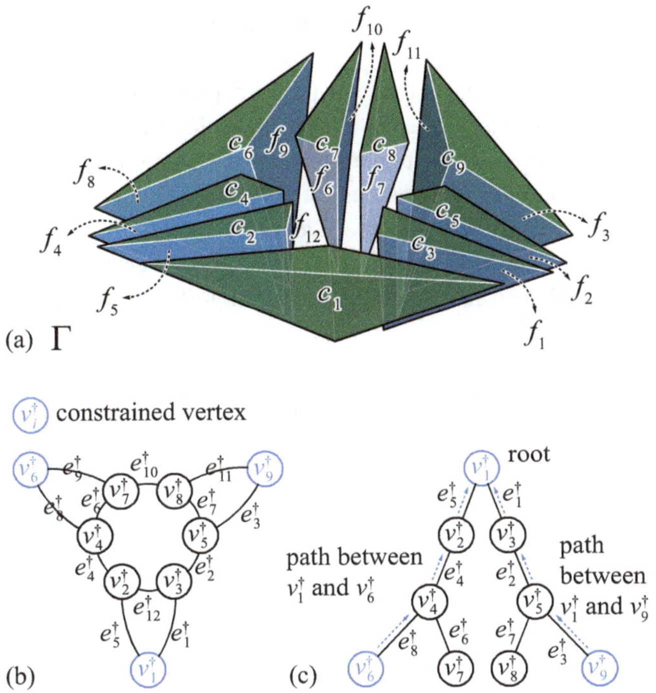

Figure 7.28 (a) Exploded force diagram showing individual cells. (b) The adjacency graph of the cells. (c) The BFS tree.

their connecting paths with the origin vertex can be obtained by traversing the tree up until the root is reached (Figure 7.28). Once the paths are found, the vertex constraint matrix **D** can be written.

7.4.9 Constrained Linear Closing Equations

All closing equations, edge constraint equations, and vertex constraint equations are assembled as a system of nonhomogeneous constrained linear closing equations written as follows:

$$\mathbf{Mq} = \mathbf{t}. \tag{7.28}$$

Here the $[(2e_{int}+b+3(p-1)+2l+n)\times f_{int}]$ matrix **M** is named the constrained closing matrix, obtained by vertically stacking the closing matrix **A**, the edge constraint matrix **B**, and the vertex constraint matrix **D**:

$$\mathbf{M} = \begin{pmatrix} \mathbf{A} \\ \mathbf{B} \\ \mathbf{D} \end{pmatrix}. \tag{7.29}$$

The $[(2e_{int}+b+3(p-1)+2l+n)\times 1]$ column vector **t** is obtained by vertically stacking the $[2e_{int} \times 1]$ zero vector, the vector **l**, and the vector **d**. Each solution of **q** represents a set of edge lengths of the form diagram that satisfy all closing requirements and constraints.

The form diagram may have overconstraining problems when constraints are incompatible with the planarity or competing with each other. In such cases, the constrained closing equations are overdetermined, and Equation 7.28 has no solution. However, the least-square solution defines the closest solution to satisfying all closing and constraint requirements such that the Euclidean norm is minimized. Some point constraints can be replaced with line constraints or plane constraints to help mitigate overconstraining problems because they allow for more flexibility in vertex locations.

7.4.10 The Geometric Degrees of Freedom of the Form Diagram

When the constrained closing equations are underdetermined, the solution may not be unique. In this case, the set of solutions forms an affine subspace inside $\mathbb{R}^{f_{int}}$ meaning that all solutions lie on a linear subspace translated by some vector.

The dimension of the linear subspace indicates the aforementioned GDoF of the form diagram. In other words, the GDoF is defined by the number of edges m that need to have their lengths assigned before an exact solution can be determined. This GDoF can be calculated according to the relationship between the numbers of independent equations and unknowns, where the rank r of the constrained closing matrix $\mathbf{M}$ is used:

$$m = f_{int} - r. \tag{7.30}$$

If a solution to Equation 7.28 exists, the GDoF is always greater than or equal to zero. Zero GDoF indicates that there is a unique solution for $\mathbf{q}$ and the form diagram is determinate, whereas a positive GDoF means there are infinitely many solutions.

7.4.11 Algebraic Representation of the Force Diagram

A system of linear equations can also be constructed to represent the input force diagram itself based on the closed coplanar edge vectors of its own f faces rather than the internal faces of its dual diagram such that the force diagram can be designed and modified. In this case, $\hat{\mathbf{n}}_i$ will be the unit direction vector of the edge e_i in the force diagram, the dimension of the closing matrix $\mathbf{A}$ becomes $[2f \times e]$, and the dimension of the constrained closing matrix $\mathbf{M}$ becomes $[(2f + b + 3(p-1) + 2l + n) \times e]$. The details are similar to those described earlier and hence omitted. The solutions of this system of equations show all force diagrams that have the same topology as well as edge and face orientations, and they represent different force distributions for the corresponding form diagram. This can be helpful because the force diagram can be adjusted without breaking the reciprocity between form and force diagrams. Related examples are presented in Section 7.6.5.

7.4.12 Solving the Constrained Closing Matrix

As in Section 7.2.7, there are various methods to solve the closing equations for the 2D scenario. Here, we use the same methods for three dimensions.

Moore–Penrose Inverse Method

The constrained closing matrix $\mathbf{M}$ is usually not invertible. All solutions of the non-homogeneous constrained linear closing equation system (Equation 7.28) may be calculated using the MPI method.[140,141] The MPI of the constrained matrix $\mathbf{M}$ is denoted $\mathbf{M}^+$. A solution to Equation 7.28 exists only when the following identity holds

$$\mathbf{M}\mathbf{M}^+\mathbf{t} = \mathbf{t}. \tag{7.31}$$

Moreover, all solutions of $\mathbf{q}$ are given by

$$\mathbf{q} = \mathbf{M}^+\mathbf{t} + [\mathbf{I} - \mathbf{M}^+\mathbf{M}]\xi, \tag{7.32}$$

where $\mathbf{I}$ is the $[f_{int} \times f_{int}]$ identity matrix and ξ is a user-specified $[f_{int} \times 1]$ column vector input. For instance, assigning 1 to all components gives us a dual solution with well-distributed edge lengths. When $\mathbf{M}$ has full column rank, which means $\mathbf{I}-\mathbf{M}^+\mathbf{M}$ is a zero matrix, the solution $\mathbf{q} = \mathbf{M}^+\mathbf{t}$ is unique. Otherwise, the solution $\mathbf{q}$ depends on the parameter ξ that can be chosen freely. In general, the i-th entry of ξ loosely controls the signed length of edge $e_i^\dagger$ when it's not constrained. In fact, if ξ is a solution of Equation 7.28, then we have

$$\mathbf{q} = \mathbf{M}^+\mathbf{t} + [\mathbf{I} - \mathbf{M}^+\mathbf{M}]\xi = \mathbf{M}^+\mathbf{t} + \xi - \mathbf{M}^+\mathbf{t} = \xi, \tag{7.33}$$

in other words, the solution given by the MPI method coincides with ξ. On the contrary, if edge $e_i^\dagger$ has its length constrained, the i-th entry of ξ will have no effect on the solution.

The most computationally intensive task is computing $\mathbf{M}^+$. However, once computed and stored, assigning ξ and getting $\mathbf{q}$ is extremely fast, allowing for the efficient generation of a wide range of solutions. The accumulation of numerical errors may lead to an incorrect result during the computation. This can be resolved by cutting off small singular values when calculating $\mathbf{M}^+$.

Reduced Row-Echelon Form Approach

Since the dimension of the solutions of the closing equation system is $f_{int}-r$, therefore, we have exactly $f_{int} - r$ independent equations in the closing matrix. This means that we can specify the length of $f_{int} - r$ edges of the dual and the rest of the edges will be determined accordingly. Simply put, a user can interact with $f_{int} - r$ independent edges to manipulate the geometry of the dual. This number is equal to the GDoF of the dual.

The independent edges of the dual diagram can be identified from the RREF $\mathbf{M}^{rref}$ of the matrix $\mathbf{M}$. Indeed, the rank of $\mathbf{M}$ equals the number of pivots in $\mathbf{M}^{rref}$; in turn, the independent edges correspond to those columns of $\mathbf{M}^{rref}$ where there is no pivot. The coordinates corresponding to these columns can be represented by a $[(f_{int} - r) \times 1]$ column vector ζ. Since the lengths of independent edges can be freely chosen, any choice of ζ will determine the geometry of the dual.

To address the approach mathematically, we reorder the columns of the $\mathbf{M}^{rref}$ matrix so that the pivots are in the matrix's main diagonal. We delete the rows of the matrix, which are all zero, to obtain a $[r \times f_{int}]$ matrix $\mathbf{M}^{rref}_{r\times f_{int}}$. The first r columns of this matrix form the $[r \times r]$ identity matrix, $\mathbf{I}$. We denote the $[r \times (f_{int} - r)]$ matrix formed by the last $f_{int} - r$ columns by $\mathbf{E}$, so that

$$\mathbf{M}^{rref}_{r\times f_{int}} = (\mathbf{I}|\mathbf{E}). \tag{7.34}$$

We can use $\mathbf{M}^{rref}_{r\times f_{int}}\mathbf{q} = \mathbf{0}$ as the new closing matrix in Equation 7.25 as

$$\mathbf{M}^{rref}_{r\times f_{int}}\mathbf{q} = \mathbf{0}. \tag{7.35}$$

The solutions to Equation 7.35 are the same as the solutions to Equation 7.28, except that the coordinates of the solution vector are reordered as the last $f_{int} - r$ coordinates corresponding to the independent edges.

We denote the $[r \times 1]$ column matrix corresponding to the first r coordinates of $\mathbf{q}$ by $\mathbf{q}_r$ and the $[(f_{int} - r) \times 1]$ column matrix corresponding to the last $f_{int} - r$ columns by ζ. Using these notations, Equation 7.34 becomes

$$\mathbf{I}\mathbf{q}_r + \mathbf{E}\zeta = \mathbf{0}.$$

Therefore, the vector ζ, which corresponds to the length of the independent edges, determines the rest of the solution vector, indeed,

$$\mathbf{q}_r = -\mathbf{E}\zeta.$$

As a result, the user can choose the lengths of the independent edges to manipulate the geometry of the dual.

Although any (positive/negative) values can be chosen for the independent edges, there is no guarantee that if all the independent edges have positive values, the rest of the edges will also be positive and the resulting geometry will be a compression/tension-only system (edges with positive lengths). To address this limitation, we suggest using a linear programming approach to solve the closing matrix.

Linear Programming Approach

Just like the 2D scenario mentioned earlier, in 3D, we can use domains as constraints, like specifying edge length ranges. However, enforcing universal conditions for edges, especially in complex diagrams with the MPI method, can be challenging. Additionally, we often seek solutions based on specific optimization criteria in design practices, which linear programming can effectively address.

We can use the following linear optimization setup to find a dual diagram with all positive edge lengths:

$$\text{Solve} \quad \begin{cases} \mathbf{Mq} = \mathbf{t} \\ \mathbf{q} \geq \mathbf{1} \\ \min(\mathbf{q}\cdot\lambda) \end{cases} \tag{7.36}$$

where $\mathbf{1}$ is the $[f_{int} \times 1]$ vector whose all coordinates are one (1) and λ is a vector that can be chosen by the user. The solution of this linear programming setup is a solution vector $\mathbf{q}$ of the closing Equation 7.28, whose coordinates are at least one (1) minimizing the objective function

$$\mathbf{q} \cdot \lambda = \sum_{i=1}^{f_{int}} q_i \lambda_i.$$

Different λ vectors yield different objective functions. For instance, the objective function given by $\lambda = \mathbf{1}$ is the sum of the lengths of the edges of the dual. Hence, the solution of Equation 7.36 is a solution that minimizes the total edge lengths of the dual given that all edges are of length at least one (1). A similar approach can be used to generate structural solutions in 3D with the minimum load paths that can significantly reduce the use of materials in the structure as suggested by Beghini et al.[55]

7.4.13 Construction of the Dual in 3D

After getting a solution for $\mathbf{q}$ which contains signed lengths for all $e^{\dagger}_{int}$ edges, the vertex positions can be calculated through methods such as the algebraic approach and graph-search approach for the construction of the form diagram. Once again, the BFS algorithm is used. Just like the process presented in 2D, it traverses from a vertex $v^{\dagger}_{i}$ that has known coordinates, then gradually visits the adjacent vertices and calculates their positions based on the lengths of the connecting edges until all vertices are visited (Figure 7.29). If the form diagram has constrained vertices, the starting vertex $v^{\dagger}_{i}$ can be chosen from them. Otherwise, if no vertex is constrained, meaning that it can be constructed anywhere in the 3D space, a starting vertex needs to be specified and its position needs to be assigned.

7.4.14 Detection of Overconstraining Problems

As mentioned in Section 7.4.9, the form diagram may be overconstrained if the vertex and edge constraints are incompatible with the planarity or competing with each other. The system of equations allows the overconstraining problems to be explicitly detected using the MPI method.

When a form diagram is overconstrained, Equation 7.31 doesn't hold, meaning that the equation system is overdetermined and there will be no solution. This serves as a criterion to assess whether a form diagram is overconstrained. However, when ξ is a zero vector, Equation 7.32 gives a minimum norm solution (namely $\mathbf{M}^{+}\mathbf{t}$) that is the closest to satisfying all constraints, meaning that

$$||\mathbf{Mq} - \mathbf{t}|| \geq ||\mathbf{MM}^{+}\mathbf{t} - \mathbf{t}|| \tag{7.37}$$

holds for any possible vector $\mathbf{q}$ where $||.||$ denotes the Euclidean norm. Although this minimum norm solution of edge lengths can be used to construct a form diagram using the process described in Section 7.4.13, the resulting form diagram is usually incorrect due to nonplanarity and edge angle deviations. Figure 7.30a shows the minimum norm solution of an overconstrained form diagram generated from the same force diagram as Figure 7.28a. Vertices $v^{\dagger}_{1}$, $v^{\dagger}_{6}$, and $v^{\dagger}_{9}$ are constrained to points $\mathbf{p}_1$, $\mathbf{p}_2$, and $\mathbf{p}_3$, respectively, where the three vertex constraints cannot be simultaneously satisfied. The nonzero angle deviations are displayed on each respective edge.

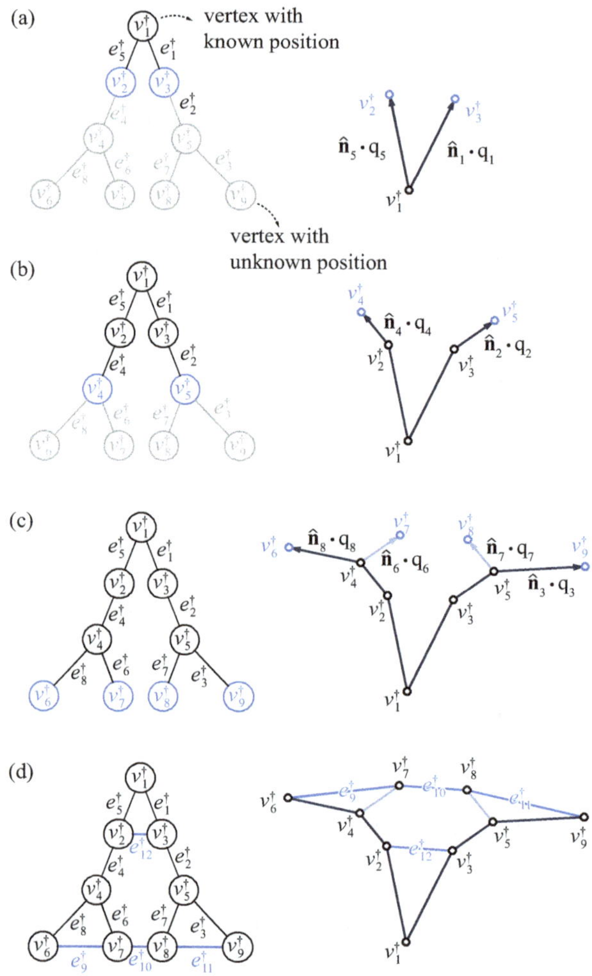

Figure 7.29 The construction of the form diagram once a solution $\mathbf{q}$ is obtained.

The problem of overconstraining may be mitigated or resolved by using more forgiving constraint types such as lines and planes. Figure 7.30b shows the minimum norm solution of the form diagram with point constraint $\mathbf{p}_3$ replaced with a line constraint. This form is less overconstrained as can be seen from the reduced angle deviation. Figure 7.30c shows that the overconstraining issue is resolved by replacing the point constraint $\mathbf{p}_3$ with a plane constraint.

7.4.15 Identify Tensile and Compressive Members

The state of tension or compression of each edge in the form diagram is still undetermined at this moment. To find that, we need to compare *the topological and geometric directions* of the edge $e^{\dagger}_{j,k}$ of the dual, which has vertices $v^{\dagger}_j$ and $v^{\dagger}_k$.

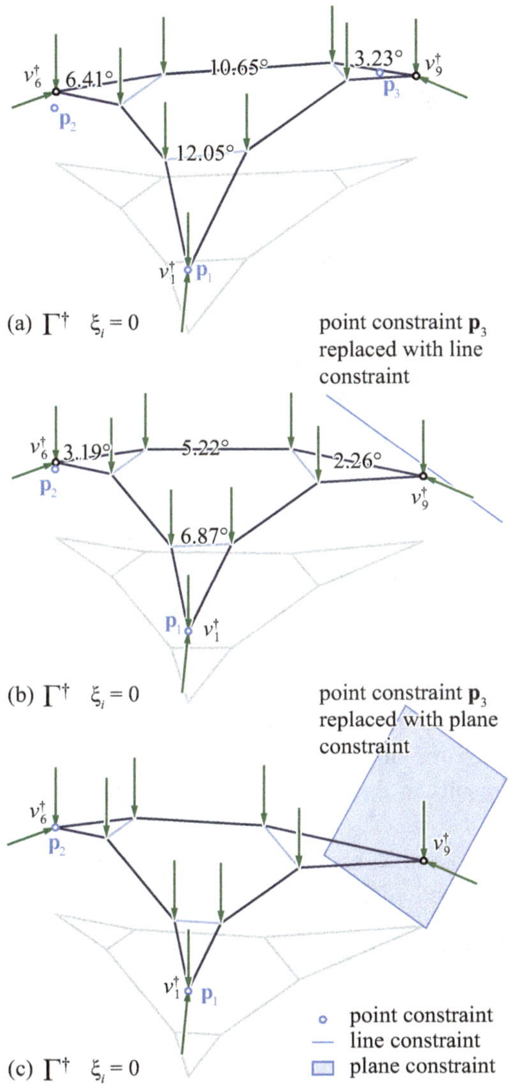

Figure 7.30 The problem of overconstraining can be mitigated or resolved by using more forgiving constraint types such as lines and planes. Overconstrained form diagrams usually have angle deviations on some edges as shown in (a) and (b). (a) The minimum norm solution of an overconstrained form diagram with vertices $v_1^\dagger$, $v_6^\dagger$, and $v_9^\dagger$ constrained to points $\mathbf{p}_1$, $\mathbf{p}_2$, and $\mathbf{p}_3$, respectively. The positions of $v_6^\dagger$ and $v_9^\dagger$ deviate from the target points. (b) The angle deviations are reduced after replacing point constraint $\mathbf{p}_3$ with a line constraint. (c) The overconstraining issue is resolved after replacing point constraint $\mathbf{p}_3$ with a plane constraint.

The topological order of the vertices of the dual is given by the connectivity matrix $\mathbf{C}_{f \times c}$ of the primal. That is, the order of vertices $v_j^\dagger$ and $v_k^\dagger$ is given by the order of c_j and c_k of the primal and we define this as the topological direction. The topological direction of the edge $e_{j,k}^\dagger$ is defined by the direction of the reciprocal normal $\mathbf{n}_{j,k}$ of the face f_i pointing from the cell c_j to c_k. Moreover, the geometric direction of the edge vector $\mathbf{e}^{\dagger}_{j,k}$ is calculated from the geometry of the dual diagram and is defined as the vector starting from $v^{\dagger}_j$ and ending at $v^{\dagger}_k$ in the dual. We can compare the topological and geometric directions by considering their dot product

$$\psi_{e_i} = \mathbf{e}^{\dagger}_{\mathrm{j,k}} \cdot \mathbf{n}_{\mathrm{i}}, \tag{7.38}$$

where ψ is the dot product of the two directions. According to the following definition, we can find the type of internal force in each member:

$$\text{if GFP}\begin{cases} \text{negative (inward)} \begin{cases} \text{if } \psi_{e_{j,k}} > 0 : e^{\dagger}_i \text{ is compressive} \\ \text{if } \psi_{e_{j,k}} < 0 : e^{\dagger}_i \text{ is tensile} \end{cases} \\ \text{positive (outward)} \begin{cases} \text{if } \psi_{e_{j,k}} > 0 : e^{\dagger}_i \text{ is tensile} \\ \text{if } \psi_{e_{j,k}} < 0 : e^{\dagger}_i \text{ is compressive} \end{cases} \end{cases}$$

For instance, if the GFP is the external cell and its direction is inward, then the directions of all the NFPs are consistent and inward. In such cases, simply the sign of $q^{\dagger}_i$ can define whether the member is in tension or compression. If $q^{\dagger}_i$ corresponding to the length of the edge $e^{\dagger}_i$ is positive, then the edge $e^{\dagger}_i$ is a compressive member, and if it is negative, it will be a tensile member. Likewise, if every entry $q^{\dagger}_i$ in the solution $\mathbf{q}$ is positive, then the dual is a compression-only system, and if all are negative, all the edges are tensile depending on the direction of the GFP (Figure 7.31a,b). Choosing any other cell as the GFP results in a form with combined tensile and compressive forces (Figure 7.32).

In general, there are four possible scenarios in the form finding of polyhedral graphic statics using algebraic methods:

- Compression/tension only form finding where (1) the force diagram consists of convex NFPs, (2) the GFP is the external polyhedron encompassing all the internal cells, and (3) all the edge lengths of the form diagram are positive.

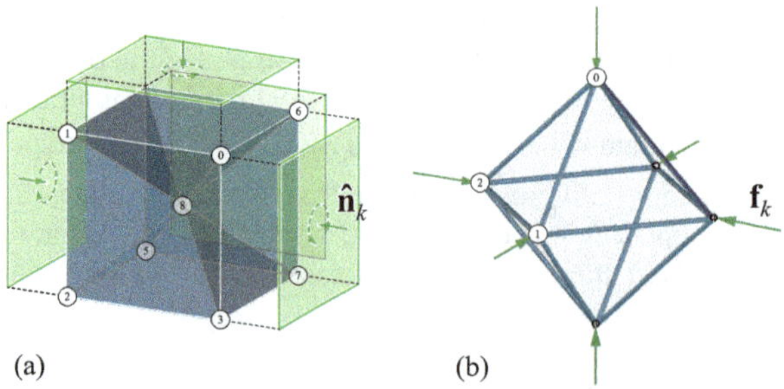

(a) (b)

Figure 7.31 (a) A group of polyhedral cells as the primal where GFP is the external cell; (b) the dual complex representing the form diagram resulted from algebraic approach; (c) the same polyhedrons as the form diagram where the vertices of the external polyhedron define the externally applied loads; and (d) its reciprocal force diagram.

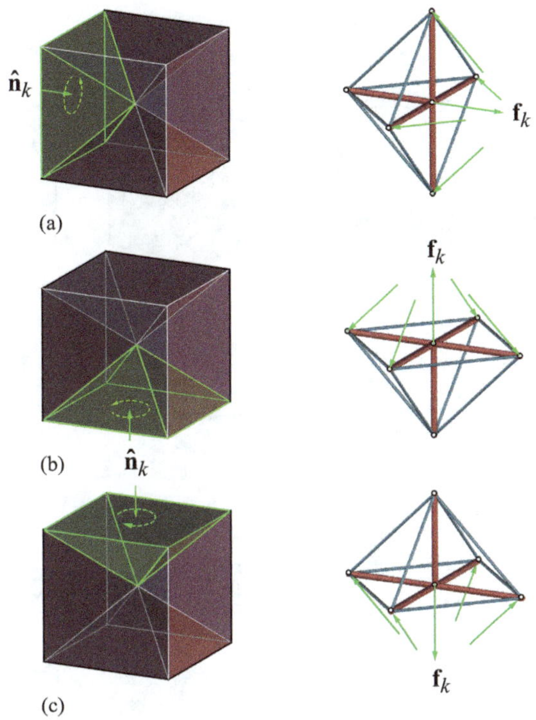

Figure 7.32 Choosing a different GFP results in compression and tension combined systems.

- Compression-and-tension combined form finding where (1) the force diagram consists of convex cells, (2) GFP is the external polyhedron, and (3) the edge lengths of the dual diagram can be manipulated to have both positive and negative values.
- Compression-and-tension combined form finding where (1) the GFP is not the external force polyhedron, and (2) the NFPs might overlap with each other.
- Compression-and-tension combined form finding by constraining the areas of the faces of the force diagram.

7.4.16 Implementation, Computation Complexity, and Comparison with the Iterative Approach

The algebraic procedures described earlier are implemented in the 3D modeling environment of Rhino3D® and Grasshopper3D® as a plug-in named PolyFrame 2[154] that is readily available for researchers and designers to explore. The handy data management framework of the reciprocal diagrams is developed by Nejur and Akbarzadeh[155] based on a winged-edge data structure (WED) proposed by Kremer et al.[156] Matrix computations are accomplished using Numpy.NET, a .NET binding for the scientific computing library Numpy.[157]

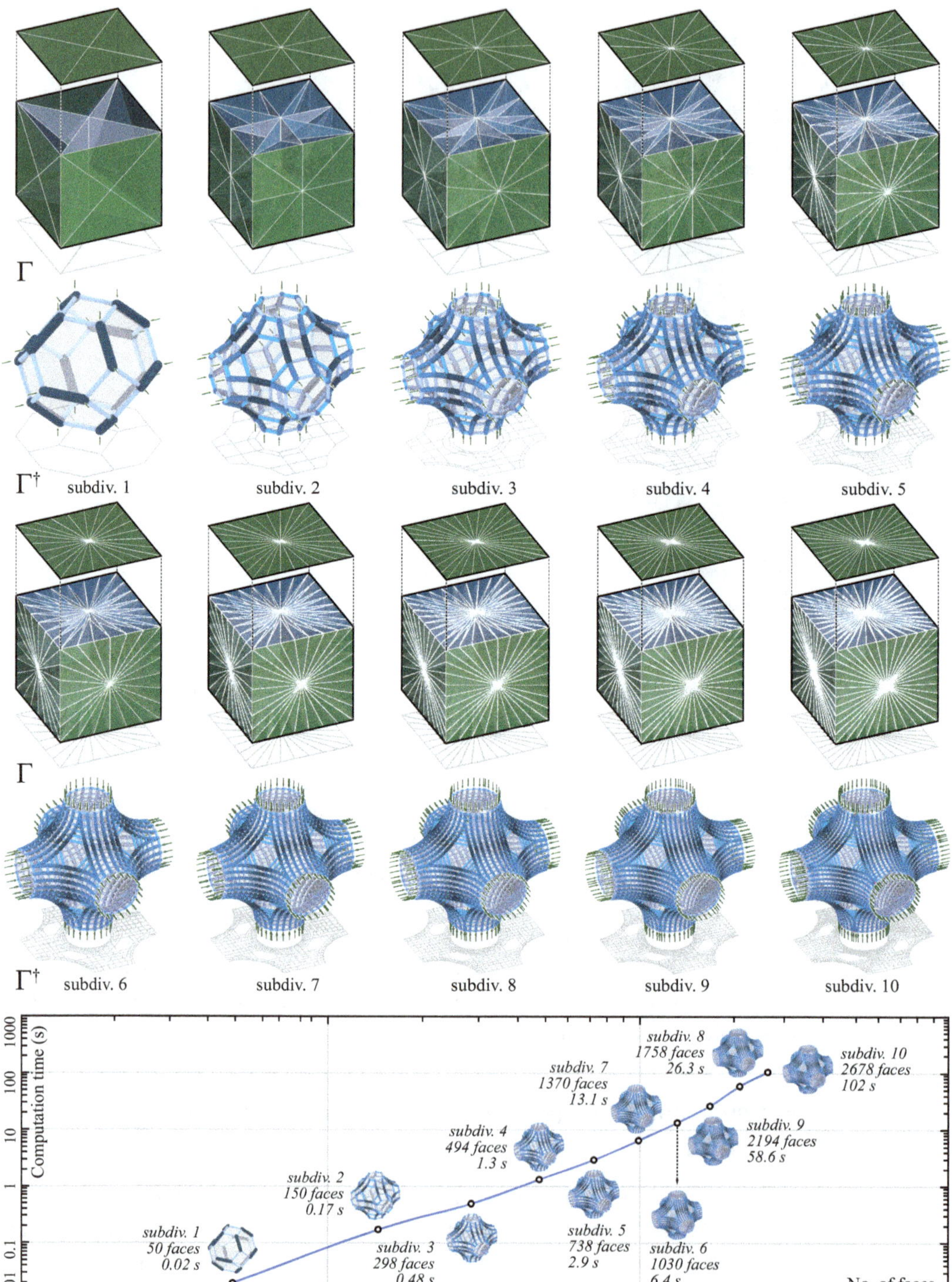

Figure 7.33 Runtime of 3DGS approximations of Schwarz P minimal surface in different resolutions.

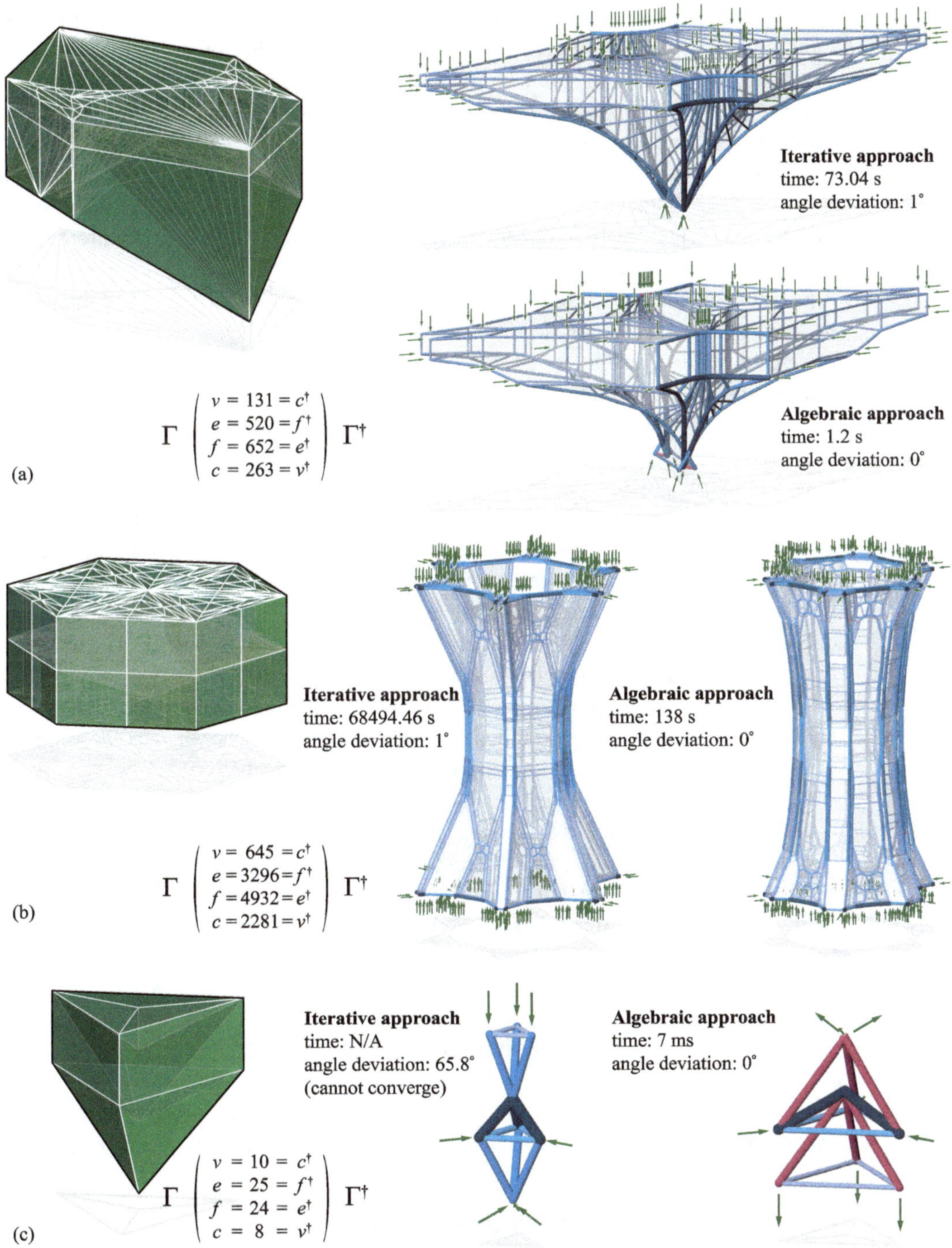

Figure 7.34 A comparative performance study between the algebraic approach and the iterative approach.

In general, it takes a group of surfaces, which represents a force diagram, as input and creates the algebraic representation of either the force diagram itself or its corresponding form diagram, up to the user's choice. The algebraic representation encompasses all possible solutions of edge lengths $\mathbf{q}$ within the solution space. Through a second input of ξ, a specific solution can be selected from the solution space. After getting a solution $\mathbf{q}$ for the edge lengths, the corresponding form diagram or a new force diagram can be constructed. The examples in this section utilize only the algebraic representation of form diagrams for the sake of clarity and simplicity.

A performance test is presented to show the computation complexity of the algebraic implementation (Figure 7.33). The tested models are 3DGS approximations of Schwarz P minimal surface in different resolutions.[84] Despite the exponential growth in runtime, the algorithm can efficiently process complex geometries within a relatively short timeframe.

Several benchmark models are also tested to compare the algebraic approach with the iterative approach described by Nejur and Akbarzadeh[137] (Figure 7.34). The algebraic approach outperforms the iterative approach in terms of both precision and speed. The iterative approach needs a maximum allowed deviation angle for perpendicularity as the stopping criteria, and the smaller the angle, the longer it will take. In most cases, the result cannot reach zero angle deviation. For the algebraic approach, the results without overconstraining are precise and have almost zero angle deviation due to the nature of close-form solutions. For over-constraining problems, the algebraic method can explicitly tell whether a form or force diagram is overconstrained, whereas the iterative method cannot. In terms of speed, different constraint conditions are less likely to affect the runtime of the algebraic approach, whereas the runtime of the iterative approach heavily depends on how different the goal is from the initial state. In all benchmark models, the algebraic approach is approximately tens to hundreds of times faster than the iterative approach that uses 1 degree as the maximum allowed deviation. Besides, the algebraic method works with diagrams with concave polyhedrons, which is difficult to achieve using the iterative method. All tests mentioned earlier were performed single-threaded on a laptop with a Xeon E3-1505M CPU and 32 GB of RAM.

7.5 Applications of Algebraic Polyhedral Graphic Statics

This section outlines the ways of using algebraic polyhedral graphic statics to find various structural forms based on the MPI method, supplemented by several examples.

7.5.1 Compression-Only or Tension-Only Form Finding

The force diagram shown in Figure 7.35a has convex cells and 377 faces that result in a form diagram with 133 GDoF. The GFP of this force diagram is the exterior cell that wraps the external faces of the NFPs illustrated by green faces in the figure. The closing equations around the internal edges of the primal are solved using the MPI method. In this method, positive weights in ξ for the edges of the dual diagram usually result in positive edge lengths (with exceptions when input diagrams are complex). In this case, when the directions of the faces of GFP are inward and all edge lengths have positive values, the result will be a compression-only shell (Figure 7.35b). If we change the GFP to negative, meaning that its faces are outward, then with all positive edge lengths we will have a tension-only system.

7.5.2 Compression-tension Combined Form-Finding Using GDoF

Tension-compression combined forms can also be easily created by leveraging the GDoF of the form diagram. Negative weights in ξ tend to reduce the lengths of edges and flip the directions into negative. So, by having mixed positive and negative weights, the result tends to be tension-compression combined.

Shell Structures with Both Tension and Compression Members

Figure 7.35c,d shows two different form diagrams generated from the same force diagram (Figure 7.35a) with both tension and compression members. In $\Gamma_2^\dagger$ (Figure 7.35c), all the parallel edges e_{i-n} have positive weights, while the undulating edges e_{j-m} receive negative weights. Assigning positive values to the undulating edges and negative values to the parallel ones generates $\Gamma_3^\dagger$ (Figure 7.35d) with the opposite direction of the forces in the boundary conditions. The most interesting feature of using this method in design is that a designer can have direct control over designing synclastic structural forms such as a compression-only shell as well as anticlastic structural forms with both tension and compression members for the same force equilibrium. Moreover, all the faces of these structural configurations are *planar* due to the planarity constraints of the reciprocal polyhedrons, which significantly facilitates their fabrication process.

Spatial Structures with Both Tension and Compression Members

In more complex spatial configurations, manipulating or adjusting the GDoF can lead to more geometrical variations. Figure 7.36 shows a primal force diagram with 66 faces and convex polyhedral cells, which results in a spatial form with 9 GDoF. Note that in this case, the direction of normals of GFP points outwards the faces, thus reciprocal to a tension-only structural form as its dual. Figure 7.36c–f shows various spatial configurations with both tension and compression forces by assigning both positive and negative values to the edge members of $\Gamma^\dagger$ with all negative edge lengths. For instance, in Figure 7.36c, edges e_5 and e_7 have received positive values while other edges have received negative values.

Note that this spatial configuration does not include an internal cell, and visiting a system with such a property might open the possibility of designing counterintuitive equilibrium states in the 3D space. Figure 7.37 shows multiple spatial configurations with both tension and compression members derived from the same force diagram. $\Gamma_{1-3}^\dagger$ show the change of weights for the tagged members and the resulting geometrical configurations. One interesting feature of these solutions is the formation of

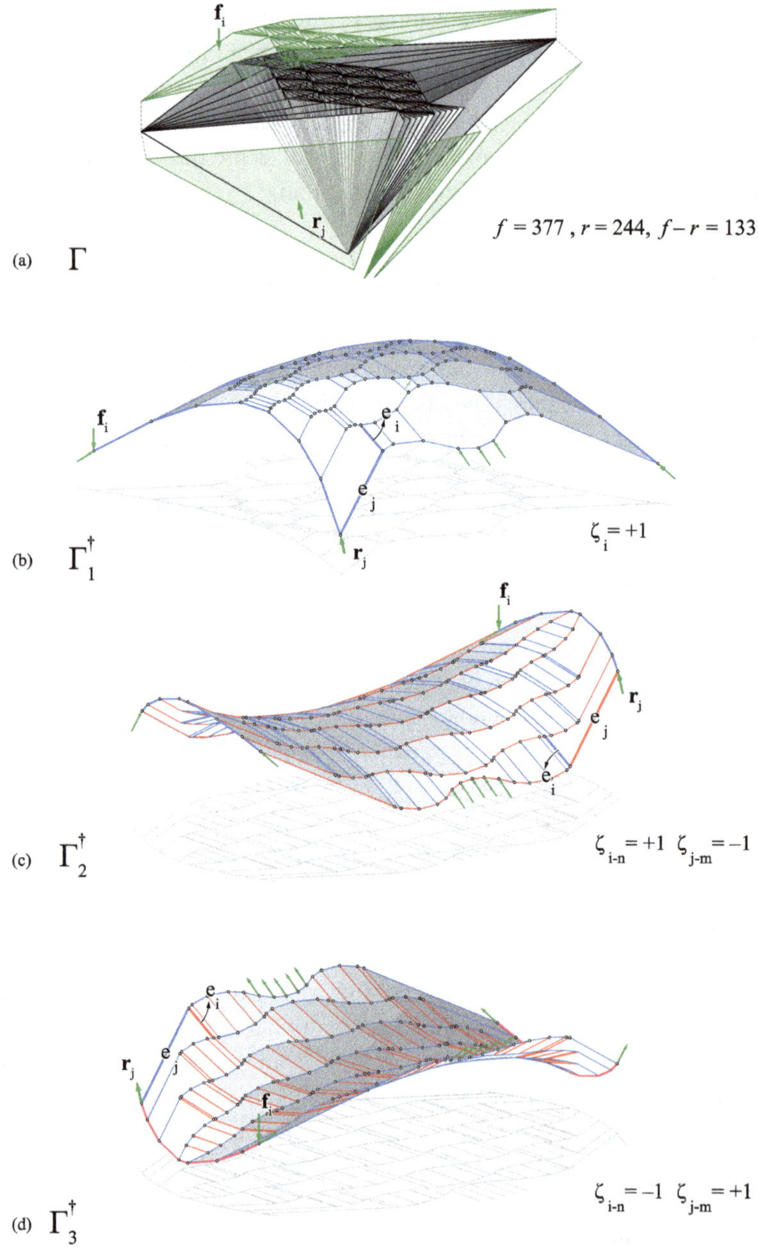

Figure 7.35 (a) A force diagram consisting of convex polyhedral cells with 377 faces resulting in (b) a (synclastic) compression-only form with 133 degrees of freedom; (c) and (d) two different (anticlastic) shells with both tension and compression members by assigning both negative and positive values to the edges of the form (a).

nonconvex polyhedral cells in the resulting forms. For instance, note the middle convex cell in the structural form of $\Gamma_1^\dagger$. This convex cell will turn into a nonconvex cell with two concave faces in $\Gamma_2^\dagger$ and a nonconvex cell with four concave faces in $\Gamma_3^\dagger$.

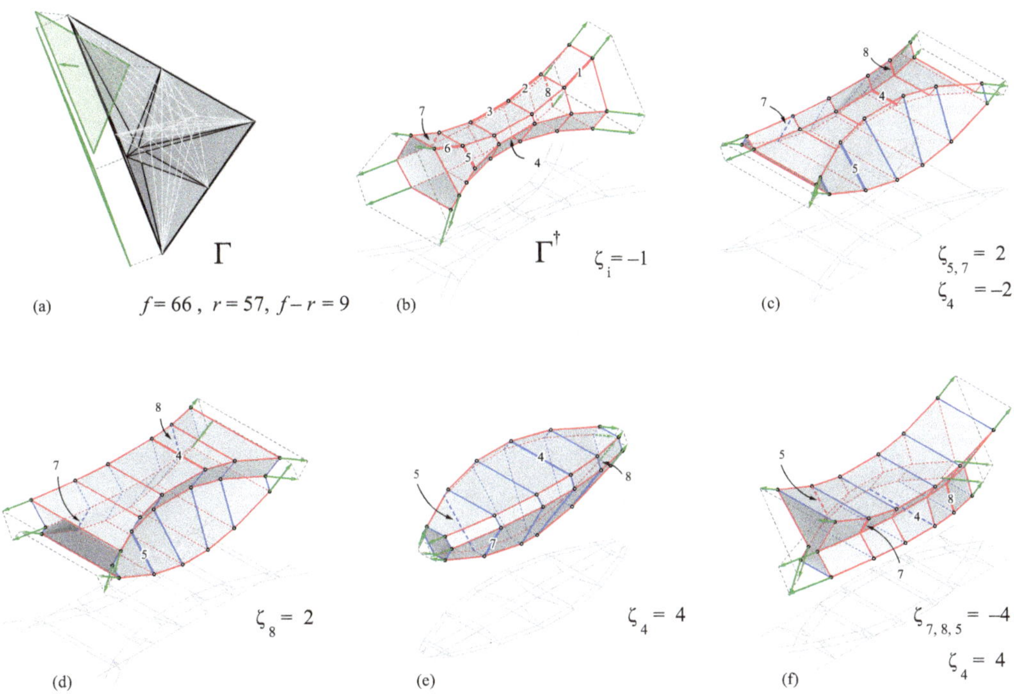

Figure 7.36 (a) A force diagram with 66 faces and convex polyhedral cells; (b) a tension-only reciprocal structural form with GDoF 9; (c) to (f) various spatial configurations with both compression and tension members for the same force diagram.

7.5.3 Compression-Tension Combined Form Finding by Assigning Different GFPs

Figure 7.32 in Section 7.4.15 shows multiple form-finding examples by choosing different GFPs from the same polyhedral system shown in Figure 7.31a. The input polyhedral system consists of five convex cells including four pyramids and a cube. Choosing any pyramid as GFP and assigning either a positive or negative direction to it results in a different dual diagram. Note that in every resulting dual diagram, one vertex disappears, which is the intersection of the externally applied loads. This vertex corresponds to the GFP cell that is eliminated from the force diagram. Furthermore, in all cases, the dual diagram consists of compression and tension members. Note that in these examples, the interior spaces of the convex NFP cells overlap, for instance, the pyramids occupy the space bounded by the cube.

7.5.4 Manipulation of Form Diagrams Using Constraints

Generating and manipulating form diagrams by altering the entries of ξ can sometimes be nonintuitive and may not yield the desired geometries. Utilizing edge and vertex constraints offers a more intuitive approach to control and adjust them.

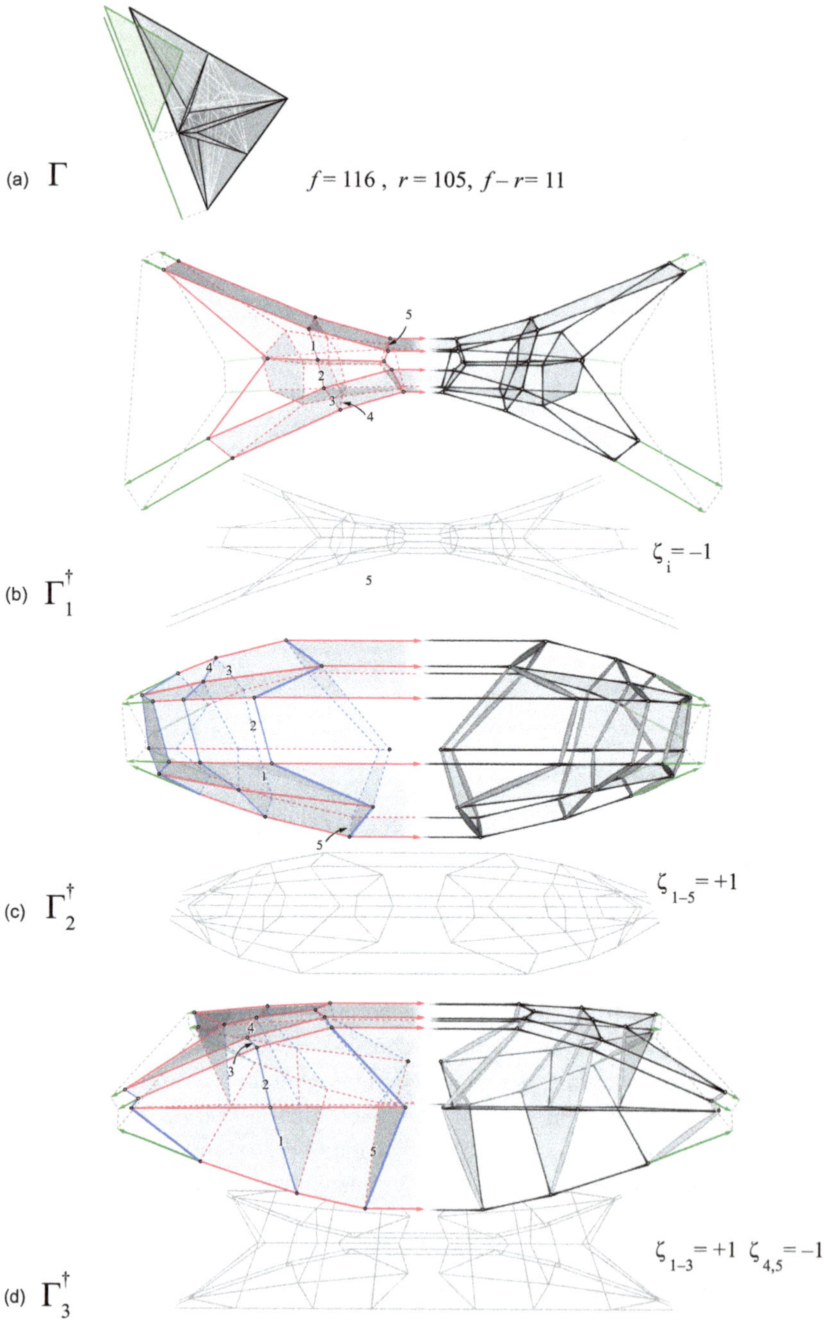

Figure 7.37 (a) A force diagram including 116 faces and (b) a tension-only spatial structural form with 11 GDoF; (c) change in the lengths value of certain edges results in spatial structural forms with both tension and compression members with planar faces as shown in (c) and (d).

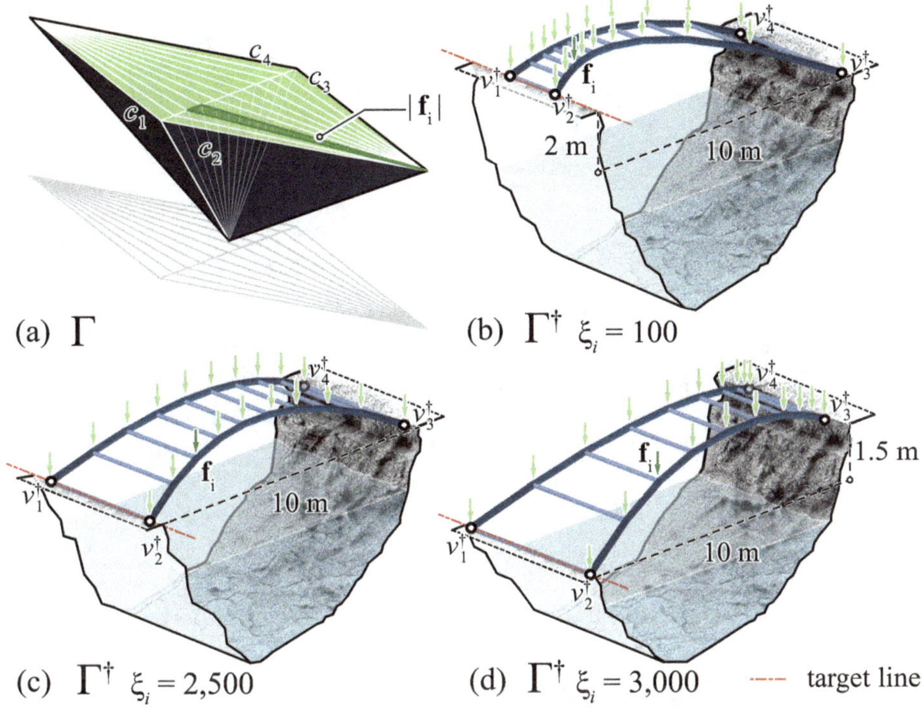

Figure 7.38 Simple bridge designs with constraints set up in accordance with different site conditions.

Form Finding Based on Site Conditions

Having edge and vertex constraints is convenient for accommodating design adjustments based on real-world site conditions. Figure 7.38 shows a simple design scenario where a bridge is needed above a 10 m-wide river. In order to design a funicular bridge that fits between the banks, constraint lines and points can be used as target geometries to constrain the four corners of the bridge. For different site conditions, the bridge geometry can be easily adjusted by setting new target positions for the constrained vertices.

Intuitive Exploration of the Solution Space

Due to the GDoF, it is possible to generate an infinitude of drastically different form diagrams from the same force diagram. This solution space of the form diagram can be explored by specifying a different input parameter ξ. Although this is theoretically doable, modifying ξ can be tedious and counterintuitive. This is because ξ is a $[f_{int} \times 1]$ vector, and its dimension can be large for complex form diagrams. Moreover, it's difficult to directly perceive the relationship between ξ and the resulting form diagram. Edge and vertex constraints can provide a different approach for intuitively exploring the solution space. For example, Figure 7.39 illustrates different bridge geometries designed for the same site. Those different geometries have different forms and states of tension and compression. This is achieved by precisely modifying the height and

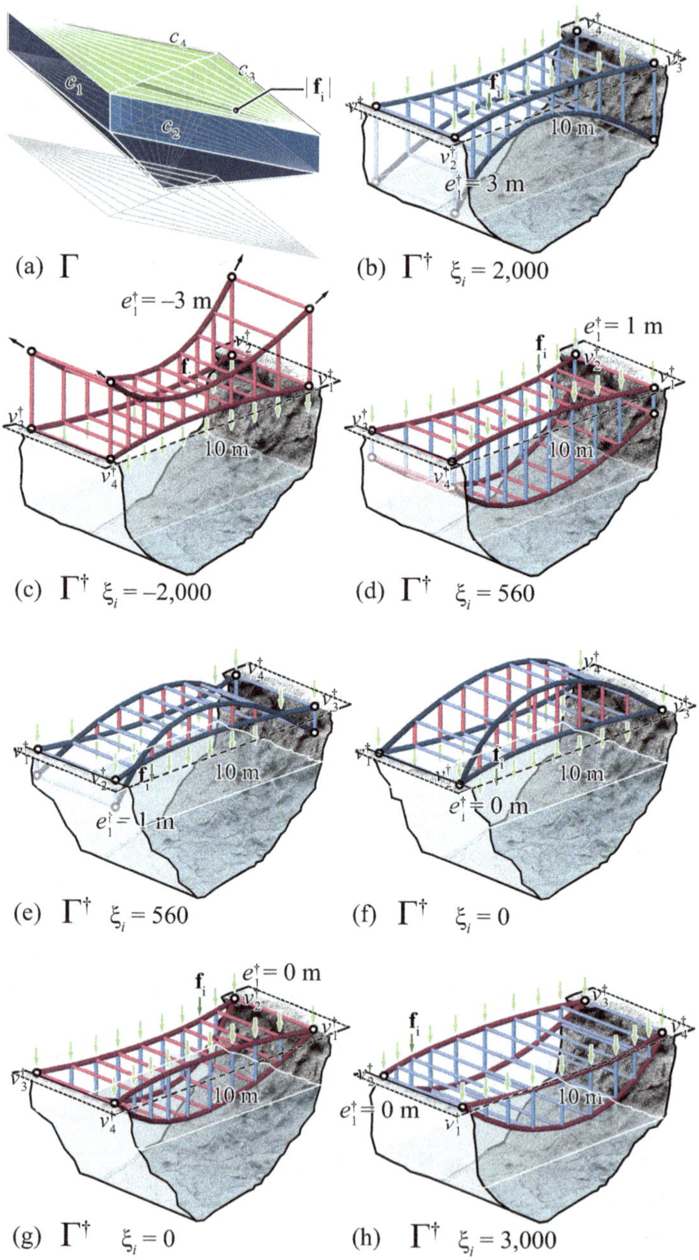

Figure 7.39 A family of bridge geometries with different geometries and states of tension and compression. (a) The shared force diagram. (b) A compression-only geometry. (c) A tension-only geometry. (d) to (h) Tension compression combined geometries.

end positions of the bridge. Figure 7.40 shows a force diagram that is shared by a family of funicular shells depicted in Figure 7.41, whose four corner vertices $v_1^\dagger$, $v_2^\dagger$, $v_3^\dagger$, and $v_4^\dagger$ are constrained to different points. Changing their target point locations results

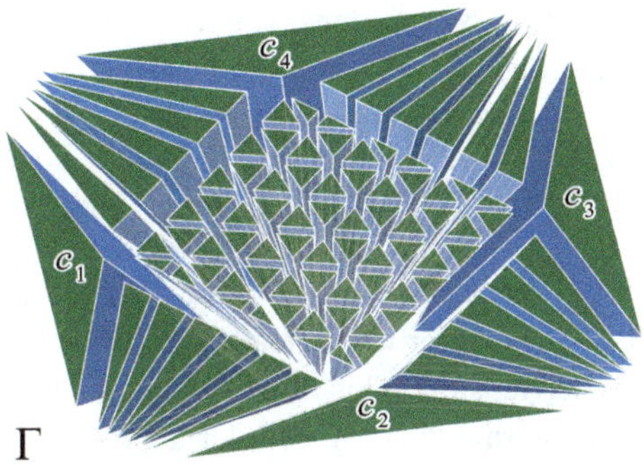

Figure 7.40 The exploded force diagram shared by the family of funicular shells.

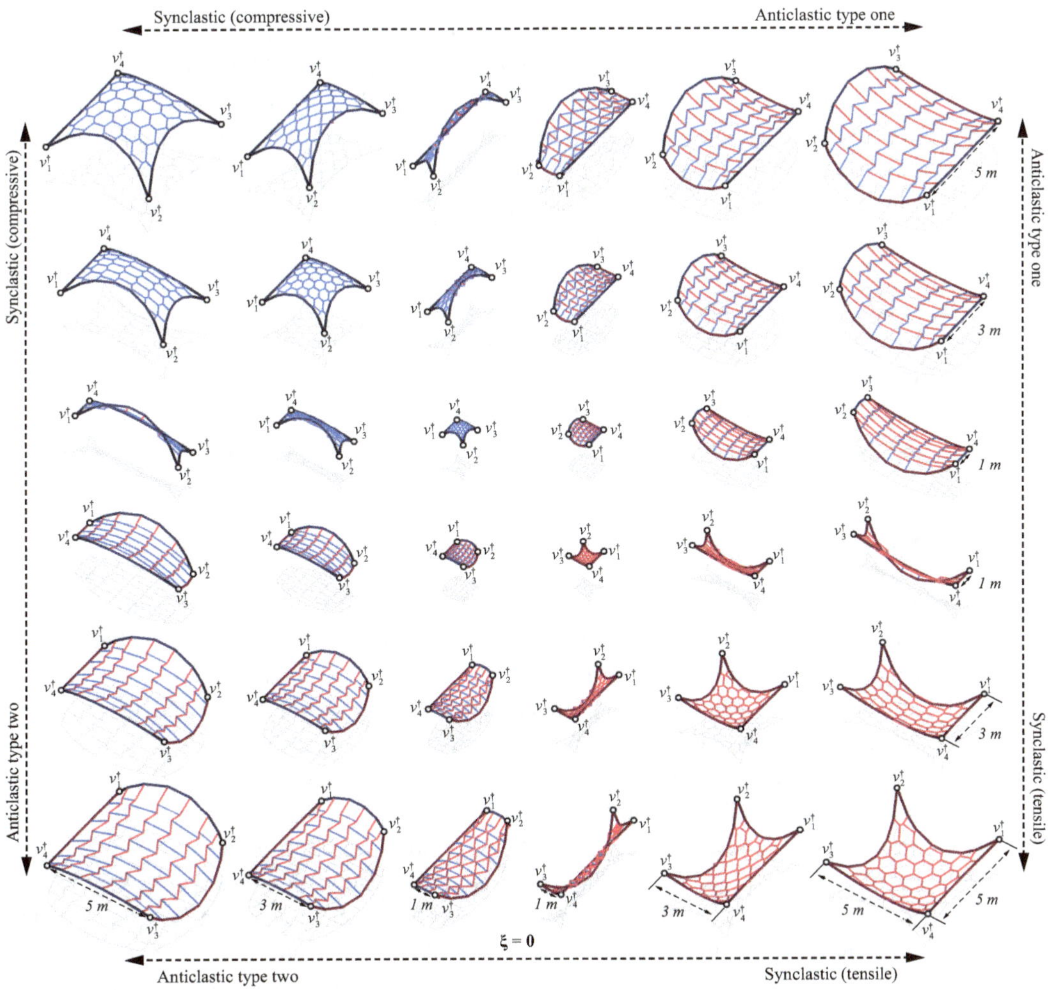

Figure 7.41 A family of funicular shells generated based on the same force diagram, and their four corner vertices $v_1^\dagger$, $v_2^\dagger$, $v_3^\dagger$, and $v_4^\dagger$ are constrained to different points.

in an array of forms, ranging from compression or tension only to tension-compression combined, from synclastic to anticlastic.

7.5.5 3D Auxetic Structures

Auxetic metamaterials show unusual reaction behaviors under uni-axial compression or tension forces. They contract perpendicular to the load direction under an applied compressive force and expand under tension. This behavior, namely negative Poisson's ratio, heavily depends on the internal concave cellular geometries according to Reid et al.[158,159] Section 7.3.3 has shown that the algebraic formulation of graphic statics is a suitable tool for designing 2D auxetic metamaterials with concave geometric arrangements. This may also be extended to three dimensions. Algebraic 3DGS is capable of creating those 3D geometries with concave cells in two different ways. First, it can start from a convex configuration and modify the strut lengths based on the GDoF to turn certain cells into a concave configuration (Figure 7.42a–e). Alternatively, the concave configuration can be generated directly from a force diagram with concave cells (Figure 7.42f–j). The auxetic behaviors of the two structures are then verified using the finite element method (FEM). The specimens have a footprint of 50 mm by 50 mm, with all edges materialized as 1.2 mm-diameter struts using a linear elastic material. The bottom vertices are fixed in z- direction, and a total vertical force of 10 kN is distributed to all top vertices. The simulation results are shown in Figure 7.42e,j, which indicates a negative Poisson's ratio. These approaches have the potential to significantly accelerate the development and exploration of various auxetic metamaterials. It provides a powerful tool for the science and engineering communities to further investigate and develop new auxetic metamaterials and explore their potential applications in areas such as biomedicine, aerospace, and advanced manufacturing.

7.5.6 Form Finding for Various Loading Scenarios

A force diagram is only able to represent the equilibrium of a single loading scenario, and the corresponding form diagram will only be in equilibrium under that specific situation. In practice, the initial force diagram usually represents the combination of structural self-weight and the primary static loads. However, dynamic loads must also be taken into consideration if their magnitudes are comparable to the dead loads. Figure 7.43 shows a 3D bridge design example that accounts for vertical live loads by using multiple force diagrams to indicate different live load situations. All force diagrams have the same topology but different areas on the top faces. In Figure 7.43a, each top face of Γ_1 represents the tributary dead load of the structure with an area of $|\mathbf{f}|$, while Γ_2 to Γ_8 have several enlarged top faces. Each enlarged face has an additional area of $|2\mathbf{f}|$, indicating the live load at the corresponding vertex. For each force diagram, a corresponding form diagram is generated and constrained to the same span, load position, and height (Figure 7.43b). By overlaying those form diagrams, a new

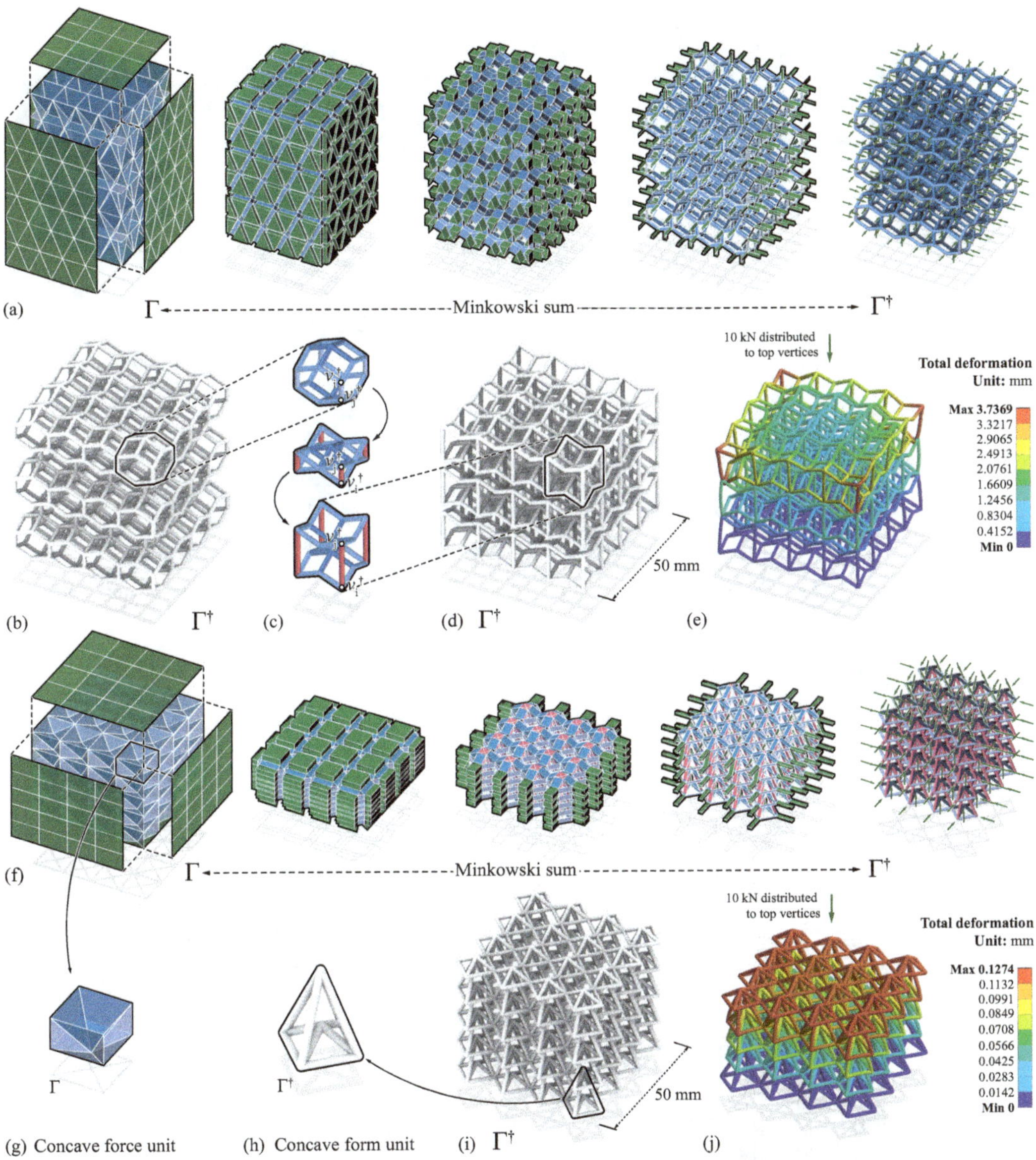

Figure 7.42 Two ways of generating structures with auxetic behavior using algebraic 3DGS. (a) to (e) A form diagram is turned from a convex configuration to a concave configuration through edge length manipulation. (f) to (j) A form diagram with concave configuration directly generated from a force diagram with concave cells. (a) and (f) Use Minkowski sum[10] to show the transition between force and form diagrams and to illustrate the reciprocity. (e) and (j) show analysis results using FEM that indicate negative Poisson's ratios.

bridge geometry can be obtained, which contains load paths for all live load cases (Figure 7.43c). While this new bridge geometry may alter the static load represented by Γ_1, this preliminary exploration provides insights for further design developments.

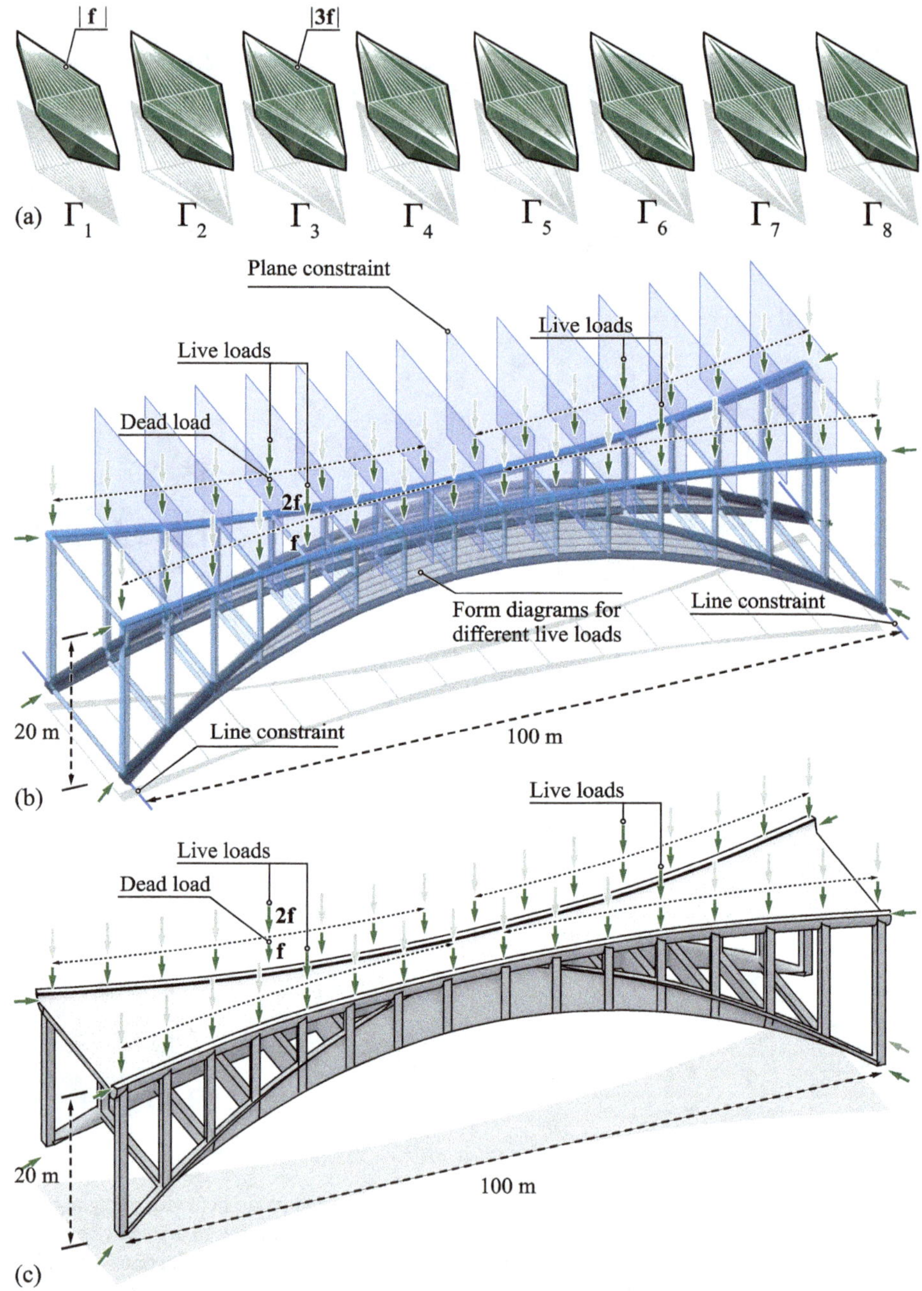

Figure 7.43 A 3D bridge design example that accounts for vertical live loads by using multiple force diagrams to indicate different situations of live loads. (a) Eight force diagrams that show different live load situations. Γ_1 represents the dead loads of the structure, each top face has an area of $|\mathbf{f}|$ indicating the tributary dead load on the top vertex. Γ_2 to Γ_8 have several enlarged top faces, representing vertices with applied live loads. (b) Form diagrams are generated with the same span, load positions, and height for all force diagrams. (c) A final bridge geometry obtained by overlapping those form diagrams.

7.6 Constrain Face Areas

We have shown that both edge lengths and vertex locations can be constrained using linear equations. Based on two recent publications of the authors,[151,160] this section is going to provide a way to mathematically control the areas of faces by controlling the length of their edge vectors. This will allow for advanced form finding and analysis applications using polyhedral graphic statics, such as constraining face sizes of the form diagram for sheet-based construction or exploring different force diagrams without changing the form. To start with, we relate the area of a single face to the lengths of its edge vectors.

7.6.1 A Quadratic Formulation for the Area of a Face

The area of a face f_i is expressed with a quadratic equation, Equation 7.42, based on the edge vectors of the face. This follows the convention described by Akbarzadeh and Hablicsek and O'Rourke.[160,161]

Consider a face f_i with k vertices: $v_0, v_1, \dots, v_{k-1}$. We denote the edges connecting the vertices v_i and v_{i+1} by e_i. Let $\mathbf{n}$ be a chosen unit normal vector of the face f_i. Using the right-hand rule, the normal $\mathbf{n}$ provides the direction of the edges. We denote the directional edges by vectors $\mathbf{e}_0, \mathbf{e}_1, \dots, \mathbf{e}_{k-1}$. For the sake of simplicity, we are going to use a cyclic order for the edges in the following explanations, meaning e_k and e_{k+1} will also denote the edges e_0, e_1, and so on.

The area of the face f_i is computed with the following process:

- We divide the face f_i into k triangles; each triangle is given by one of the e_i and the geometric center of vertices, O.
- We compute the area of each triangle by computing its height H_i, which is the distance from O to the edge e_i (Figure 7.44b).

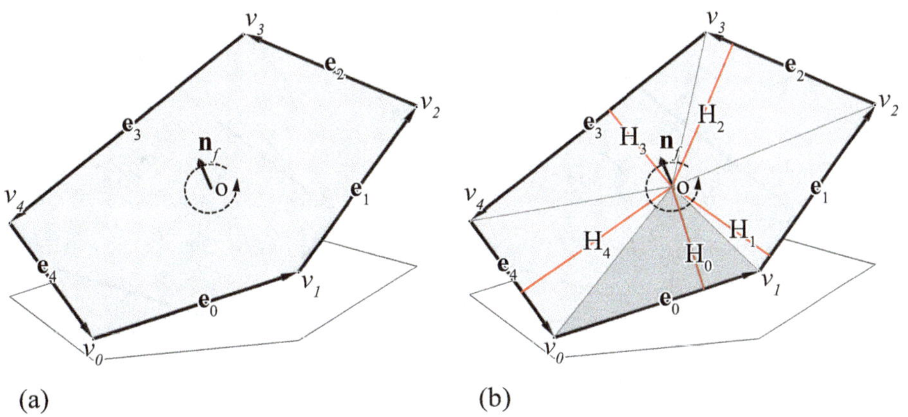

Figure 7.44 (a) Vertices and edge vectors of the face f with a normal direction $\mathbf{n}_f$; and (b) dividing the face into triangles with a base $\mathbf{e}_i$ and the height H_i from the centroid of the vertices.

Average Height

The height H_i, that is, the perpendicular distance of O from the edge e_i, is the average of the signed projected distances $h_{i,j}$ of the vertices v_0, v_1, ..., v_{k-1} from e_i (Figure 7.45d)

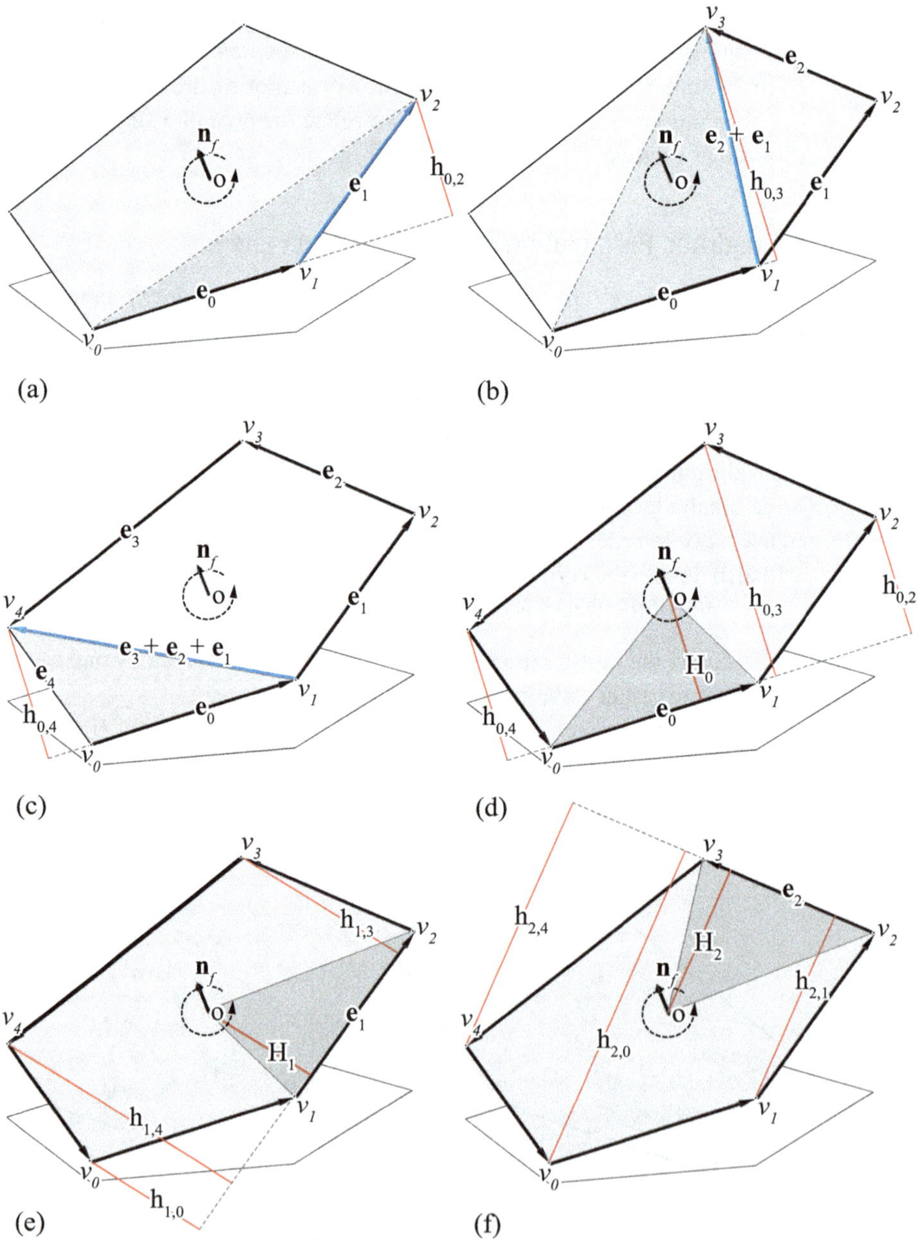

Figure 7.45 Finding the average height H_i for each edge $\mathbf{e}_i$ by constructing triangles starting from (a) $\mathbf{e}_i$ and $\mathbf{e}_{i+1}$, and (b) $\mathbf{e}_i$ and $\mathbf{e}_{i+1} + \mathbf{e}_{i+2}$ until we find all heights of the vertices in (c) and (d); and repeating the same process for other edges to find all H_is in (e) and (f).

$$H_i = \frac{1}{k}\sum_{j=1}^{k} h_{i,j}.$$

The signed distances, $h_{i,j}$, can be computed from the areas of the triangles given by the edge e_i and vertex v_j. Explicitly, we compute the area of the triangle given by the edge e_i and the vertex v_j in two ways:

- Using $h_{i,j}$, we have that the signed area of the triangle is

$$\frac{1}{2}|\mathbf{e}_i|h_{i,j}\mathbf{n}.$$

- Using the cross product, we know that the signed area of the triangle is

$$\frac{1}{2}\mathbf{e}_i \times \left(\sum_{l=i}^{j-1} \mathbf{e}_l\right).$$

Comparing the two approaches, we obtain[160]

$$h_{i,j} = \sum_{l=i}^{j-1} \eta_{i,l}|\mathbf{e}_l| \tag{7.39}$$

where $\eta_{i,l}$ is the unique scalar satisfying

$$\eta_{i,l}\mathbf{n} = \frac{\mathbf{e}_i \times \mathbf{e}_l}{|\mathbf{e}_i| \cdot |\mathbf{e}_l|}.$$

The scalars $\eta_{i,l}$ only depend on the angle between the directional edges $\mathbf{e}_i$ and $\mathbf{e}_j$. Using Equation 7.39, we can compute the average heights H_i.

$$H_i = \frac{1}{k}\sum_{j=2}^{k-1} h_{i,i+j} = \frac{1}{k}\sum_{j=2}^{k-1}\sum_{1\le m\le j-1} \eta_{i,i+m}|\mathbf{e}_{i+m}| = \tag{7.40}$$

$$= \frac{1}{k}\sum_{j=1}^{k-2}(k-j-1)\eta_{i,i+j}|\mathbf{e}_{i+j}| \tag{7.41}$$

As a consequence, we obtain a quadratic formula for the area of the face f in the edge lengths of the edges of f

$$A_f = \frac{1}{2}\sum_{i=0}^{k-1}|\mathbf{e}_i|H_i = \frac{1}{2k}\sum_{i=0}^{k-1}\sum_{j=0}^{k-2}(k-j-1)\eta_{i,i+j}|\mathbf{e}_i||\mathbf{e}_{i+j}|. \tag{7.42}$$

Example 7.2 Consider the pentagon of Figure 7.44. The coordinates of the vertices of the pentagon are the following: $v_0 = (0,0,0)$, $v_1 = (6.79,-18,-9)$, $v_2 = (35.3,-21.5,-1)$, $v_3 = (55.7,-8.9,14.5)$, and finally $v_4 = (36.3,8.9,18.7)$. We compute the area of the face A_f using the method explained in Section 7.6.1.

Construction steps:

1 First, choose a unit normal vector for the face. This can be done by taking the cross product of the vectors $\mathbf{e}_0 = (6.79,-18,-9)$ and $\mathbf{e}_1 = (28.51,-3.5,8)$, connecting the points v_0 and v_1, and v_1 and v_2, respectively:

$$\mathbf{n}{:} = \frac{\mathbf{e}_0 \times \mathbf{e}_1}{|\mathbf{e}_0| \cdot |\mathbf{e}_1|} = (-0.29, -0.51, 0.81).$$

2 The choice of this unit normal vector yields directions on the edges:

$$\mathbf{e}_0 = (6.79, -18, -9); \quad \mathbf{e}_1 = (28.51, -3.5, 8); \quad \mathbf{e}_2 = (20.4, 12.6, 15.5);$$

$$\mathbf{e}_3 = (-19.4, 17.8, 3.2); \quad \mathbf{e}_4 = (-36.3, -8.9, -18.7).$$

3 Compute the scalars $\eta_{i,j}$. For instance, the equation

$$\eta_{0,2}\mathbf{n} = \frac{\mathbf{e}_0 \times \mathbf{e}_1}{|\mathbf{e}_0| \cdot |\mathbf{e}_1|} = (-0.28, -0.49, 0.77)$$

shows that $\eta_{0,2} = 0.957$.

4 After the computation of all the $\eta_{i,j}$, compute the quadratic equation for the area A_f of the face

$$\begin{aligned} A_f &= 2.872|\mathbf{e}_1||\mathbf{e}_2| + 1.85|\mathbf{e}_1||\mathbf{e}_3| - 0.506|\mathbf{e}_1||\mathbf{e}_4| + 1.893|\mathbf{e}_2||\mathbf{e}_3| + \\ &+ 1.358|\mathbf{e}_2||\mathbf{e}_4| - 0.38|\mathbf{e}_2||\mathbf{e}_5| - 0.925|\mathbf{e}_3||\mathbf{e}_1| + 2.97|\mathbf{e}_3||\mathbf{e}_4| + \\ &+ 0.577|\mathbf{e}_3||\mathbf{e}_5| + 1.012|\mathbf{e}_4||\mathbf{e}_1| - 0.679|\mathbf{e}_4||\mathbf{e}_2| + 2.722|\mathbf{e}_4||\mathbf{e}_5| + \\ &+ 2.986|\mathbf{e}_5||\mathbf{e}_1| + 0.761|\mathbf{e}_5||\mathbf{e}_2| - 0.288|\mathbf{e}_5||\mathbf{e}_3|. \end{aligned} \tag{7.43}$$

□

Quadratic Form

Note that Equation 7.42 is a *quadratic form*, meaning that it is polynomial with terms of degree two. Given a quadratic form in n variables, its coefficients can be arranged into an $[n \times n]$ matrix.

In our case, we can formulate the right-hand side of Equation 7.42 in a quadratic form given by a matrix $\mathbf{M}_f$, whose entries are given by the coefficients:

$$\mathbf{M}_{f,ij} = \begin{cases} (k - j + i - 1)\eta_{i,j} & \text{if } j > i \\ (i - j - 1)\eta_{i,j} & \text{if } j < i\,. \\ 0 & \text{if } i = j \end{cases} \tag{7.44}$$

Indeed, we can rewrite Equation 7.42 as a quadratic form

$$A_f = \frac{1}{2k}\mathbf{q}^T\mathbf{M}_f\mathbf{q}, \tag{7.45}$$

where $\mathbf{q}$ is the column vector of the edge lengths

$$\mathbf{q} = (|\mathbf{e}_1|, |\mathbf{e}_2|, \dots, |\mathbf{e}_k|)^T.$$

In Example 7.3, construct the matrix Mf for a simple example.

Example 7.3 Calculate the matrix $\mathbf{M}_f$ for the example of the pentagon of Figure 7.44.

Construction steps:

1 Compute the area equation, Equation 7.43 as in Example 7.2.

2 Assemble the coefficients of Equation 7.43 into a [5 × 5] matrix

$$\mathbf{M}_f = \begin{pmatrix} 0 & 2.872 & 1.85 & -0.506 & 0 \\ 0 & 0 & 1.893 & 1.358 & -0.38 \\ -0.925 & 0 & 0 & 2.97 & 0.577 \\ 1.012 & -0.679 & 0 & 0 & 2.722 \\ 2.986 & 0.761 & -0.288 & 0 & 0 \end{pmatrix}$$

Here, the order of the edges is given as e_0, e_1, e_2, e_3, and e_4.

We remark also that most entries of the matrix are positive; however, we can see that $\mathbf{M}_{f,14}$ has to be negative, since the vectors $\mathbf{e}_0, \mathbf{e}_3$, and $\mathbf{n}$ have negative orientation with respect to the local coordinate of the surface (Figure 7.47).

□

Symmetric Matrix

Usually, the matrix $\mathbf{M}_f$ is not a symmetric matrix. However, the computations may become simpler if the matrix is symmetric. Indeed, the matrix $\mathbf{M}_f$ can be turned into a symmetric matrix. Since

$$\left(\mathbf{q}^T\mathbf{M}_f\mathbf{q}\right)^T = \mathbf{q}^T\mathbf{M}_f^T\mathbf{q}$$

the quadratic form given as

$$\frac{1}{2k}\mathbf{q}^T\mathbf{M}_f^T\mathbf{q}$$

also computes the area of the face.

As a consequence,

$$A_f = \frac{1}{2k}\mathbf{q}^T\frac{\mathbf{M}_f + \mathbf{M}_f^T}{2}\mathbf{q},$$

and in this case, the corresponding matrix

$$\frac{\mathbf{M}_f + \mathbf{M}_f^T}{2} \tag{7.46}$$

is symmetric. For the sake of computational simplicity, from now on, we assume that the matrix $\mathbf{M}_f$ appearing in Equation 7.42 is always symmetric.

7.6.2 Constraining One Face to a Target Area

Once we have the quadratic formulation of the face area, we can constrain a given face f_i to a target area. A face f_i illustrated in Figure 7.44a is used as an example. Initially, without any constraint, we have five unknowns, which are the edge lengths of the five edges $e_{0\text{–}4}$. There are three closing equations based on Equation 7.19, in $x-$, $y-$, and $z-$ around the face f_i, and one of them is redundant.[136] As a consequence, the dimension of the possible solutions, that is, the GDoF of the face, satisfying the closing equations is

$$e - 2 = 3.$$

Instead of solving the quadratic area equation, Equation 7.45, for three unknowns, we constrain two of them and solve the quadratic equation for only one unknown. Using this technique, we can significantly reduce the complexity of finding a solution for this quadratic equation. This provides additional design possibilities for the user, as we either allow the user to define up to two edge lengths out of three or we use the existing edge lengths for two edges and compute the area based on the last unknown edge.

In general, our goal is to simplify solving the quadratic equation of the face by solving it for only one edge length.

Computing $GDoF_f$ of a Face

The dimension of the possible solutions for the geometry of the face is called the Geometric Degrees of Freedom ($GDoF_f$). In fact, $GDoF_f$ describes the dimension of the family of polygons with edges parallel to the edges of the initial face, which is always equal to:

$$e - 2.$$

The GDoF is also equal to the number of independent edges in each face. In fact, the lengths of the independent edges can define the lengths of the rest of the edges and the geometry of the face.[162] The independent edges can be found using the RREF method, see Section 7.4.12.

Specifying the edge lengths for the $e-2$ independent edges yields a unique solution for the geometry of the face. It's very likely this solution will not have the desired target area. However, we can specify $e-3$ independent edges and then use the area equation to solve for the last one.

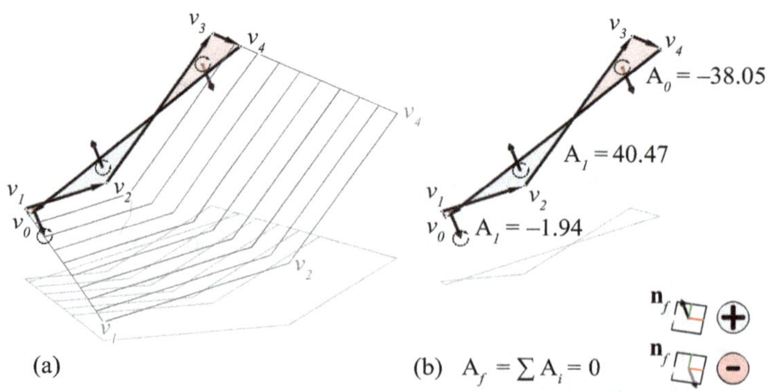

Figure 7.46 (a) Multiple steps to compute the new face geometry with the preassigned area (for visualization purposes only); and (b) the computed face geometry with zero area.

Constrained Geometric Degrees of Freedom ($CGDoF_f$)

The user-defined edge lengths provide linear equations for the edge lengths that are in general nonhomogeneous. As a consequence, the dimension of the solution space for possible geometries of the face may decrease significantly.

The user may overdetermine the system, for instance, by assigning too many edge lengths. To avoid this problem, we compute the dimension of the constrained solution space, called the constrained geometric degrees of freedom (CGDoF), using RREF.

The result of this method classifies the edges into the following classes:

- the fixed edges, e_{fix}: the edges chosen by the user with predefined edge lengths (these edges are always dependent edges of the equation system);
- the nonfixed dependent edges, e_{nfd}: the dependent edges which are not predefined by the user, and
- the independent edges, e_{ind}.

To solve the quadratic equation for the area, we reduce the number of independent edges e_{ind} to one by assigning the existing edge length for all independent edges except one. The last remaining independent edge is called the critical independent edge, e_{ci}, the length of which we find using the quadratic equation. This method will be described in detail in Section 7.6.3.

Defining the Constrained Equations for a Face

In Section 7.6.1, we expressed the area of a face polygon based on a quadratic form of the edge lengths. Now, we develop linear, nonhomogeneous constrained equations describing the geometry of the face with preassigned lengths for certain edges of the face.

We can write the edge $\mathbf{e}_i$ with a predefined length q_i as a constraint vector equation in the following way:

$$\mathbf{l}_i^T \mathbf{q} = q_i, \tag{7.47}$$

where $\mathbf{l}_i$ is the $[e \times 1]$ column vector whose entries are all zero (0) except at the index of e_i where it is one (1).

Similarly, multiple constraints, that is, other fixed edge lengths, can be written as a matrix equation

$$\mathbf{L}_f \mathbf{q} = \mathbf{l}_f, \tag{7.48}$$

where the rows of $\mathbf{L}_f$ are the row vectors $\mathbf{l}_i^T$ and $\mathbf{l}_f$ is the vector whose entries are the q_i, the predefined edge lengths.

Together with the closing equations of 7.19, we obtain all the linear equations describing the linear constraints, which results in the constraint equation system

$$\mathbf{R}_f \mathbf{q} = \mathbf{r}_f, \tag{7.49}$$

where the matrix $\mathbf{R}_f$ is obtained by stacking the matrices $\mathbf{A}_f$ and $\mathbf{L}_f$

$$\mathbf{R}_f = \begin{pmatrix} \mathbf{A}_f \\ \mathbf{L}_f \end{pmatrix},$$

and the vector $\mathbf{r}_f$ is obtained as stacking the zero vector and the vector $\mathbf{l}_f$ together

$$\mathbf{r}_f = \begin{pmatrix} \mathbf{0} \\ \mathbf{l}_f \end{pmatrix}$$

We call the matrix $\mathbf{R}_f$ the constraint matrix and the vector $\mathbf{r}_f$ the constraint vector.

Analyzing the Constraint Equation System

The constraint equation system, Equation 7.49, is, in general, a nonhomogeneous, linear equation system. The solution space of this equation system is often not a linear subspace but rather the empty set or an affine subspace of the possible solution space $\mathbf{R}^e$. Here, e denotes the number of edges. The dimension of this affine subspace is the constrained geometric degrees of freedom of the face ($CGDoF_f$). The $CGDoF_f$ is the geometric degrees of the face after applying the edge constraints by the user.

It is also possible to have no solutions for Equation 7.49. In this case, we say that the $CGDoF_f$ is $-\infty$. If there exists a solution to Equation 7.49, then the $CGDoF_f$ is a non-negative integer, which is the dimension of the affine subspace formed by the solutions. When the $CGDoF_f$ is zero, the constraint equations have a unique solution. If the $CGDoF_f$ is positive; then there are many significantly different solutions to the constraint equations.

In order to compute the $CGDoF_f$, we use the reduced row-echelon form ($\mathbf{RREF}_f$) of the matrix obtained from the constraint matrix, $\mathbf{R}_f$, and the constraint vector $\mathbf{r}_f$:

$$\left(\mathbf{R}_f \mid \mathbf{r}_f\right).$$

The $CGDoF_f$ can be easily computed from this RREF, but the following two possibilities might occur:

- If there exists a row of $\mathbf{RREF}_f$, so that the last entry is one, but all other entries are zero:

$$\left(0 \ \ldots \ 0 \mid 1 \right)$$

  then the constraint equation, Equation 7.49, has no common solution. In this case, the $CGDoF_f$ is $-\infty$, and the user needs to modify the constraints and/or release some of the constrained edges from their input.
- Otherwise, we have at least one solution. In this case, the $CGDoF_f$ equals $e - r$, where e is the number of edges of the face and the number of columns of the constraint matrix $\mathbf{R}_f$, and r is the rank of the $\mathbf{RREF}_f$. This rank equals the number of pivots, and as a consequence, in this case, it also equals the rank of $\mathbf{R}_f$.

Solving the Area Equation

The main idea of manipulating the edge lengths of the face to obtain the required area is to solve the area equation, Equation 7.45, by reducing the number of unknown edges into a single unknown edge length. We reduce the unknowns by finding all the independent edges of the constraint Equation 7.49 and assigning either the current values or user-defined values to them.

From now on, we assume that there exists a solution to Equation 7.49. The columns in $\mathbf{RREF}_f$ corresponding to the pivots are called the pivotal columns. The non-pivotal columns correspond to the so-called independent edges, whose lengths can be manipulated freely. Once the values for the independent edges are set (possibly by the user), there is a unique solution to Equation 7.49.

The edges corresponding to the pivotal columns are the ones that depend on the independent ones. These edge lengths will be updated so that Equation 7.49 is satisfied. Our method for solving area Equation 7.45 is to set as many values of the lengths of the independent edges as possible. In this case, it is one less than the number of independent edges: $e - r - 1$. The length of the last independent edge, q_{ci}, is the length of the *critical independent* edge, e_{ci}. This length will be treated as a variable for which we will solve Equation 7.45.

To solve Equation 7.45, we organize the edges according to the form of $\mathbf{RREF}_f$ into multiple vectors:

1. The vector corresponding to the critical edge is defined as an $[e \times 1]$ column vector $\mathbf{q}_{ci}$:

$$\mathbf{q}_{ci,i} = \begin{cases} q_{ci} & \text{if i is the index of the } e_{ci} \\ 0 & \text{otherwise} \end{cases}$$

   Here q_{ci}, the edge length of the e_{ci} edge, is unknown and we will solve the area equation for q_{ci}.
2. The vector of the edge lengths of the *noncritical independent* edges, e_{nci}, is defined as $\mathbf{q}_{nci}$, which is an $[e \times 1]$ column vector:

$$\mathbf{q}_{nci,i} = \begin{cases} q_i & \text{if i is the index of an } e_{nci} \text{ edge} \\ 0 & \text{otherwise} \end{cases}$$

   where q_i are the current edge lengths of the e_{nci} edges.
3. Similarly, the vector of the edge lengths of the fixed/predefined edges, e_{fix}, is defined by $\mathbf{q}_{fix}$ as

$$\mathbf{q}_{fix,i} = \begin{cases} q_i & \text{if i is the index of a user-selected edge} \\ 0 & \text{otherwise} \end{cases}$$

   where q_i denotes the length of the user-selected edge. The indices of fixed edges are indices of some of the pivotal columns. These q_i are fixed in the beginning of the problem and won't be updated.
4. Finally, the vector of edge lengths of the *nonfixed dependent* edges, e_{nfd}, is defined by vector $\mathbf{q}_{nfd}$ as

$$\mathbf{q}_{nfd,i} = \begin{cases} q_i & \text{if i is the index of an } e_{ndf} \\ 0 & \text{otherwise} \end{cases}$$

   The edge lengths q_i of the e_{nfd} edges can be computed from the lengths of the edges corresponding to e_{ci}, e_{nci}, and e_{fix}. The edge lengths will be updated in order to satisfy the area equation.

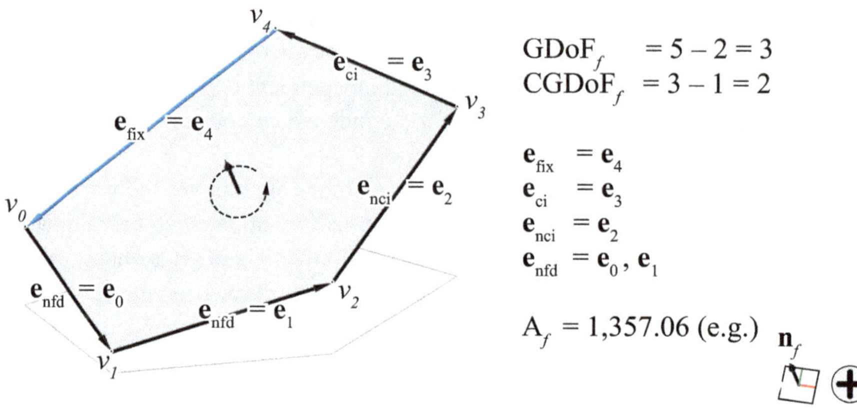

Figure 7.47 A sample face showing the edge vectors, its normal direction and the choices of e_{fix}, e_{cr}, e_{nci}, and e_{nfd}.

After setting up the notations, we begin to solve the area Equation 7.45.

Since any edge is either e_{ci}, e_{nci}, e_{fix}, or e_{nfd}, we have equality

$$\mathbf{q} = \mathbf{q}_{ci} + \mathbf{q}_{nci} + \mathbf{q}_{fix} + \mathbf{q}_{nfd}. \tag{7.50}$$

Moreover, the lengths of the e_{nfd} depend linearly on the lengths of e_{ci}, e_{nci}, and e_{fix}, hence there exist an $[e \times e]$ square matrix, $\mathbf{H}$ and vectors $\mathbf{h}$, $\mathbf{g}$ so that:

$$\mathbf{q}_{nfd} = \mathbf{H}\mathbf{q}_{nci} + \mathbf{h}q_{ci} + \mathbf{g} \tag{7.51}$$

The matrix $\mathbf{H}$ and the vectors $\mathbf{h}$ and $\mathbf{g}$ can be computed from the RREF form easily as follows.

The matrix $\mathbf{H}$ is a matrix whose entries are mostly 0 except for the entries corresponding to the columns of e_{nci} and to the rows of the e_{nfd}, where the value has the opposite sign than the corresponding value in the $\mathbf{RREF}_f$ matrix of $\left(\mathbf{R}_f \mid \mathbf{r}_f\right)$.

The vector $\mathbf{h}$ is a vector whose entries are mostly zero (0) except at the entries corresponding to the column of the e_{ci} edge and to rows of the dependent edges, where the value has the opposite sign than the corresponding value in the $\mathbf{RREF}_f$ matrix of $\left(\mathbf{R}_f \mid \mathbf{r}_f\right)$.

The vector $\mathbf{g}$ is the contribution coming from the fixed edges. This is a vector whose entries are mostly zero (0), except for entries corresponding to the indices of the e_{nfd} edges, when the entry is the last entry of the corresponding row of the $\mathbf{RREF}_f$ matrix $\left(\mathbf{R}_f \mid \mathbf{r}_f\right)$.

We simplify Equation 7.50 slightly. Consider the $[e \times e]$ square matrix $\mathbf{Id}_{nci}$, which is the identity restricted to the e_{nci} edges and 0 elsewhere. Since,

$$\mathbf{Id}_{nci}\mathbf{q}_{nci} = \mathbf{q}_{nci}$$

we have that

$$\mathbf{q}_{nfd} + \mathbf{q}_{nci} = \mathbf{H}'\mathbf{q}_{nci} + \mathbf{h}q_{ci} + \mathbf{g}$$

where $\mathbf{H}'$ is the matrix $\mathbf{H} + \mathbf{Id}_{nci}$.

Thus, by Equation 7.50, we have

$$\mathbf{q} = \mathbf{q}_{ci} + \mathbf{H}'\mathbf{q}_{nci} + \mathbf{q}_{fix} + \mathbf{h}q_{ci} + \mathbf{g}.$$

Let us denote $\mathbf{h}'$ by the vector obtained by adding a 1 to the vector $\mathbf{h}$ at the index of the critical edge, that is, $\mathbf{h}'q_{ci} = \mathbf{q}_{ci} + \mathbf{h}q_{ci}$. Then, we have

$$\mathbf{q} = \mathbf{H}'\mathbf{q}_{nci} + \mathbf{q}_{fix} + \mathbf{h}'q_{ci} + \mathbf{g} \tag{7.52}$$

Now, we can solve the area Equation 7.45 by plugging in the right-hand side of Equation 7.52 into $\mathbf{q}$: the quantity

$$\frac{1}{2k}\left(\mathbf{H}'\mathbf{q}_{nci} + \mathbf{q}_{fix} + \mathbf{h}'q_{ci} + \mathbf{g}\right)^T \mathbf{M}_f \left(\mathbf{H}'\mathbf{q}_{nci} + \mathbf{q}_{fix} + \mathbf{h}'q_{ci} + \mathbf{g}\right) \tag{7.53}$$

computes the area of the face, A_f. Rearranging the terms, we obtain a quadratic equation for q_{ci}:

$$aq_{ci}^2 + bq_{ci} + c = 0 \tag{7.54}$$

where

$$a = \mathbf{h}'^T\mathbf{M}_f\mathbf{h}'$$

and

$$b = 2\mathbf{H}'^T\mathbf{M}_f\left(\mathbf{H}'\mathbf{q}_{nci} + \mathbf{q}_{fix} + \mathbf{g}\right)$$

and

$$c = \left(\mathbf{H}'\mathbf{q}_{nci} + \mathbf{q}_{fix} + \mathbf{g}\right)^T \mathbf{M}_f \left(\mathbf{H}'\mathbf{q}_{nci} + \mathbf{q}_{fix} + \mathbf{g}\right) - 2kA_f$$

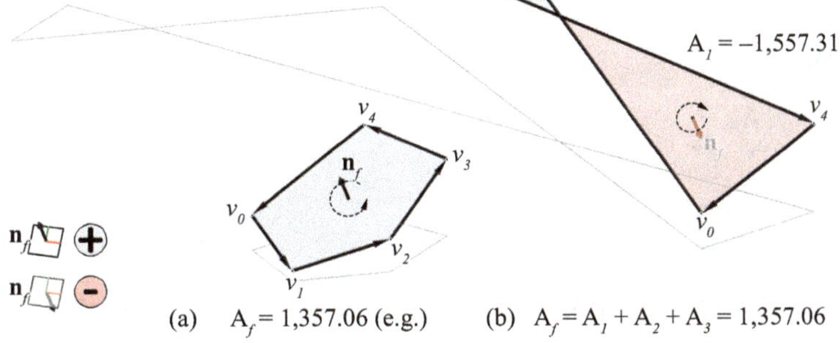

Figure 7.48 There are at least two polygons with significantly different geometries that represent the same area.

We can solve this quadratic equation (using the quadratic formula) to obtain possibly two solutions for q_{ci}:

$$q_{ci} = \frac{-b \pm \sqrt{b^2 - 4ac}}{2a}$$

Number of Solutions

Depending on the target area, A_f, we might have different numbers of solutions for Equation 7.54. It is possible to have no solution, a unique solution, or two significantly different solutions (see Figure 7.48).

Depending on the sign of A, a large positive or a small negative prescribed area ensures that we have multiple (two) solutions.

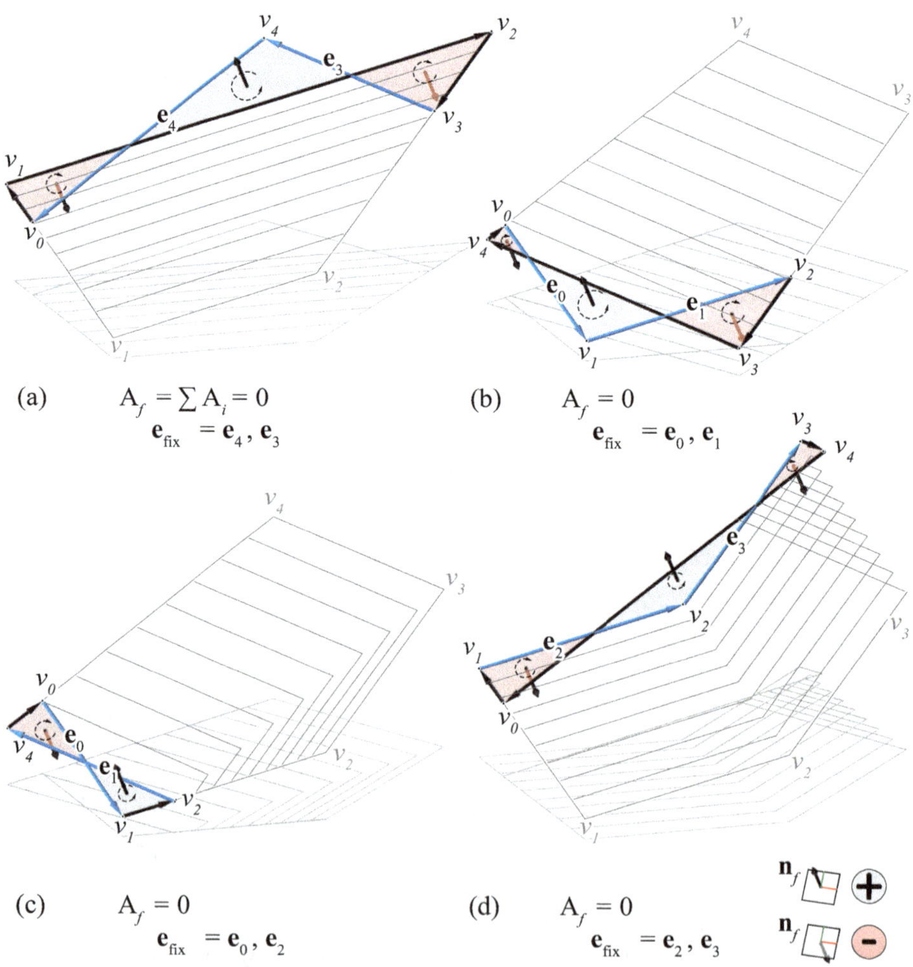

Figure 7.49 (a) to (d) Various zero-area computations for a starting face with the area A_f and different chosen fixed edges.

Updating the Edges of the Face

Once we compute the length q_{ci} of the e_{ci} edge, we can update the lengths of the dependent edges using Equation 7.51. Now, all the lengths of the edges are computed. The face corresponding to the edge lengths has the required area and the edges of the face satisfy the constraint equation system, Equation 7.49, while only the lengths of the e_{nfd} edges and the length of the e_{ci} were manipulated (Figure 7.49).

An Example

In the Example of 7.4, the geometry used in Example 7.2 is revisited here to give a case study for the process described earlier.

Example 7.4 Let us revisit the geometry of Example 7.2. Calculate its geometry while keeping the same target area, and select e_3 and e_4 as the fixed edges.

Construction steps:

1 Denote the unknowns (the edge lengths of the pentagon) by $q_0, q_1, \dots, q_4$.

2 First, consider the closing equations for this face. There are two such equations, one for the x- coordinate and one for the y- coordinate (the equation for the z-coordinate is redundant[136]):

$$\begin{aligned} 0.319q_0 + 0.956q_1 + 0.714q_2 - 0.713q_3 + 0.867q_4 &= 0 \\ -0.847q_0 - 0.117q_1 + 0.445q_2 + 0.663q_3 - 0.215q_4 &= 0. \end{aligned} \tag{7.55}$$

3 The constraints that e_3 and e_4 are fixed edges yield another two equations

$$q_3 = 58.24 \quad \text{and} \quad q_4 = 41.78, \tag{7.56}$$

where 58.24 and 41.78 are the original edge lengths of e_3 and e_4, respectively.

4 Simplify the closing equations, Equation 7.55, using the constraint equations, Equation 7.56

$$\begin{aligned} 0.319q_0 + 0.956q_1 + 0.714q_2 &= 5.3 \\ -0.847q_0 - 0.117q_1 + 0.445q_2 &= -29.63. \end{aligned} \tag{7.57}$$

Using the RREF, the equations can be written as

$$\begin{aligned} q_0 - 0.659q_2 &= 24.69 \\ q_1 + 0.996q_2 - &= 74.25 \end{aligned} \tag{7.58}$$

in other words,

$$\begin{aligned} q_0 &= 0.659q_2 + 24.69 \\ q_1 &= -0.996q_2 + 74.25 \end{aligned} \tag{7.59}$$

5 Plug the equations, Equations 7.59 and 7.56

$$q_0 = 0.659q_2 + 24.69, \quad q_1 = -0.996q_2 + 74.25, \quad q_3 = 58.24 \quad q_4 = 41.78$$

into the area equation, Equation 7.43 with

$$A_f = 0$$

obtaining the quadratic equation

$$1.796q_2^2 + 390.646q_2 + 1898.751 = 0.$$

This equation has two solutions, namely

$$q_2 = -4.974 \text{ and } 212.535.$$

As a consequence, there are two significantly different solutions for the geometry of this face. The updated (self-intersecting) face corresponding to $q_2 = -4.974$ is shown in Figure 7.46.

□

7.6.3 Constraining Multiple Faces

When multiple internal/external faces need to be constrained to different target areas, the task becomes more complex as the faces are interrelated. This section provides a multi-step, inductive process to tackle this problem. In each step, we compute the geometry of a single-face f_i with the assigned area as described in Section 7.6.2. Then, we update the geometry of the whole system with the new edge lengths using the MPI method. It repeats this process for the next constrained face until there is no face left to change.

Sequentially Prescribing Area for Each Constrained Face

First, we specify the fixed edges from the face f_i with prescribed edge lengths and use the RREF method to solve the quadratic area Equation 7.45 described in Section 7.6.2. Then, to preserve the area of the next face, we need to maintain the edge lengths that preserve the already constrained faces. So, a list e_{fix} is used to keep track of all fixed edges in the polyhedral system. This list will be populated when we sequentially constrain the areas. When all faces are done, the list e_{fix} will have a series of edge lengths that satisfy all target areas.

Solving the Nonhomogeneous Equation System

Next, the list of edge lengths is used together with the closing equations (Equation 7.25) to solve for the lengths of all other edges:

$$\mathbf{Rq} = \mathbf{r}, \tag{7.60}$$

where the constraint matrix

$$\mathbf{R} = \begin{pmatrix} \mathbf{A} \\ \mathbf{L} \end{pmatrix}$$

is built from the closing matrix **A** (see Equation 7.25). The constraint equations

$$\mathbf{Lq} = \mathbf{l}$$

come from the fixed edges (see Equation 7.48). Similarly, the vector

$$\mathbf{r} = \begin{pmatrix} \mathbf{0} \\ \mathbf{l} \end{pmatrix}$$

is obtained from the edge vector of the fixed edges **l**.

We may go back to Section 7.4.12 to review the calculation details. If

$$\mathbf{RR}^{+}\mathbf{r} = \mathbf{r}$$

holds, then Equation 7.60 has solutions, and each solution **q** is in the format

$$\mathbf{q} = \mathbf{R}^{+}\mathbf{r} + (\mathbf{I} - \mathbf{R}^{+}\mathbf{R})\xi, \tag{7.61}$$

where ξ can be any vector with the same dimension of **q**.

7.6.4 Form-Finding Examples by Controlling Force Diagram Areas

Being able to constrain face area in design is quite useful because the form-finding process can proceed in multiple stages. It may start with a compression-only form-finding stage where a group of convex cells describes a force diagram from which a compression-only system can be constructed. In the next stage of design, particular boundary forces in the force diagram are eliminated in the system by making their face areas zero using the GDoF of the force diagram. Then, the magnitudes of the internal forces are updated in the form diagram without changing the geometry of the form.

Figures 7.50, 7.51, 7.52, and 7.53 illustrate several examples that use this approach. In the examples shown in Figure 7.50, the intention is to highlight certain properties of the constrained polyhedral computation. In all the examples, the chosen face is highlighted by an orange shade and the constrained edges are highlighted by blue color.

For instance, in Figure 7.50a, the target area for the chosen face is zero and an edge has been chosen that does not belong to the selected face. Figure 7.50b shows an animated drawing for clarification purposes. As shown in Figure 7.50c, the selected face turns into a self-intersecting face. In the next example, in Figure 7.50d, the top face is set to be zero with the same constrained edge. As a result of this computation, the face collapses to a line due to its rectangular geometry. Also, the normal of the side faces, $\mathbf{n}_{f_0}$, flips, which means that the magnitude of the internal force in the corresponding form will change (Figure 7.50h,i). In the example of Figure 7.50g, one of the vertical side faces with a constrained top edge is chosen and the target area is also set to zero. As illustrated in Figures 7.50h and 7.50i, the rest of the vertical faces will disappear after the polyhedral computation. This method can certainly be applied to more than one face in the polyhedral system. In Figure 7.50j, one external and one internal face are chosen with a zero area target, and as illustrated in Figure 7.50k–o, the computation process proceeds sequentially by first computing the geometry of the face f_0 in

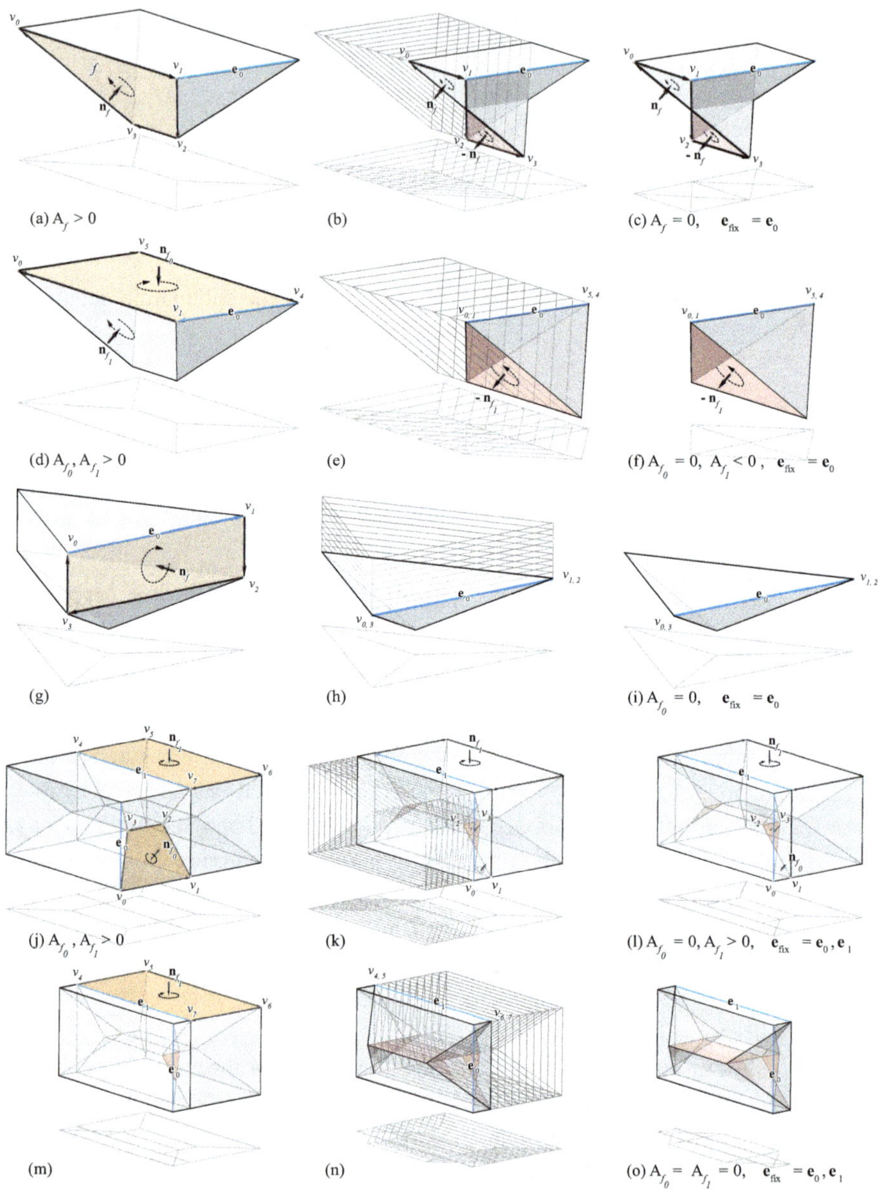

Figure 7.50 (a), (d), (g), (j), and (m) Multiple polyhedral geometries with selected faces (orange) and user-assigned fixed edges; (b), (e), (h), (k), and (n) the face area computation and visualization in multiple steps (for visualization purposes only); and (c), (f), (i), (l), and (o) the resulting polyhedral geometries with zero areas for the selected faces.

Figure 7.50l and then recomputing the polyhedral geometry to solve for the new face f_1. In this process, multiple other faces will also turn into a self-intersecting face, as shown in Figure 7.50l. In the final geometry, the face f_1 will collapse to an edge e_1 (Figure 7.50o).

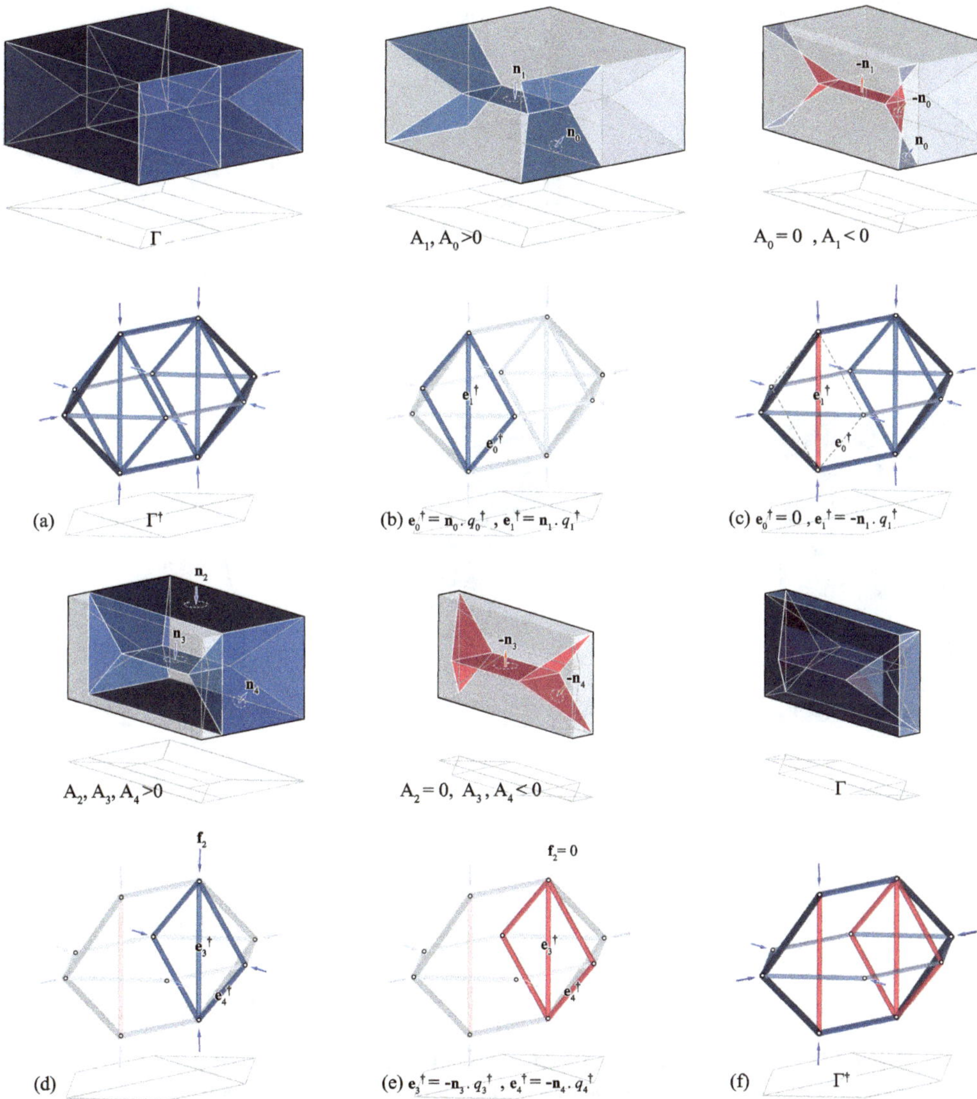

Figure 7.51 (a) Initial force polyhedron of the Figure 7.50j (top) and its reciprocal form (bottom) as a compression-only system; (b) highlighted internal faces of the polyhedron before transformation (top) and the corresponding members in the form (bottom); (c) zero area faces and their updated normal directions and the resulting zero-force members in the form; (d), (e), and (f) highlighted faces in the second step of the transformation and the updated form with new internal force distribution.

In most of these examples, the target areas were intentionally set to be zero to highlight its resulting reciprocal structural form. Figure 7.51a shows the same polyhedron of Figure 7.50j. In this figure, certain faces in the force diagram and their corresponding edges in the form diagram are highlighted to show the effect of changing the areas on the magnitude of the internal forces. Starting from the force polyhedron of

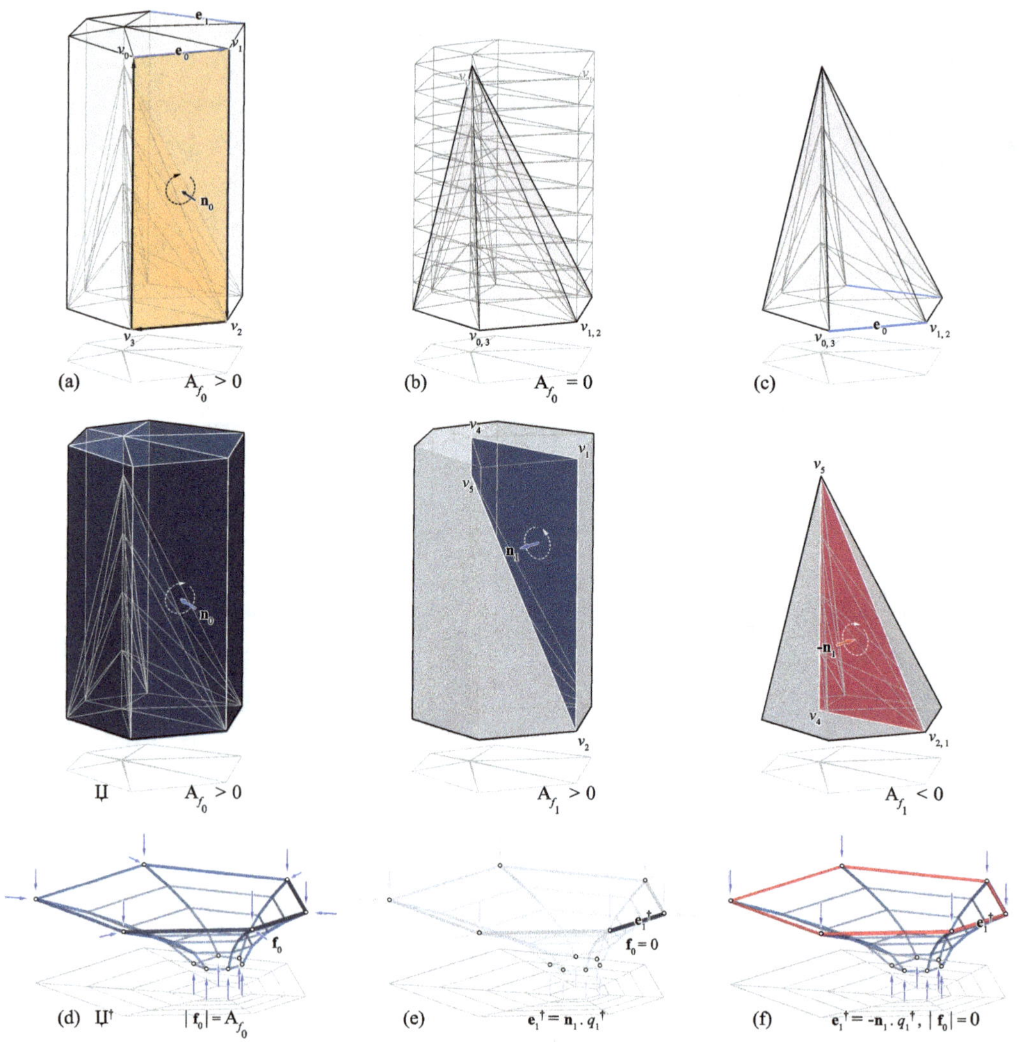

Figure 7.52 (a) A force polyhedron and a selected face with zero area target, and two fixed edges on the top; (b) the transformation animation (for visualization purposes only); (c) the resulting force polyhedron where the selected face and all other side faces collapse to a line; (d), (e), and (f) the force and form diagrams and their transformation after changing the areas.

Figure 7.51a (top) and its compression-only form (bottom), faces f_1 and f_0 are emphasized in Figure 7.51b (top) and their corresponding compression-only edges in the form. The result of the zero area computation results in face f_0 to turn into a self-intersecting face together with other similar faces attached to the edges of the face f_1. Besides, face f_1 is flipped as well as the direction of its normal vector. Note that as a result of this transformation, the internal force in the corresponding member of the face f_0 is decreased to zero. This is a fascinating effect in the equilibrium of polyhedral frames, as it describes the internal equilibrium of forces in a polyhedral system

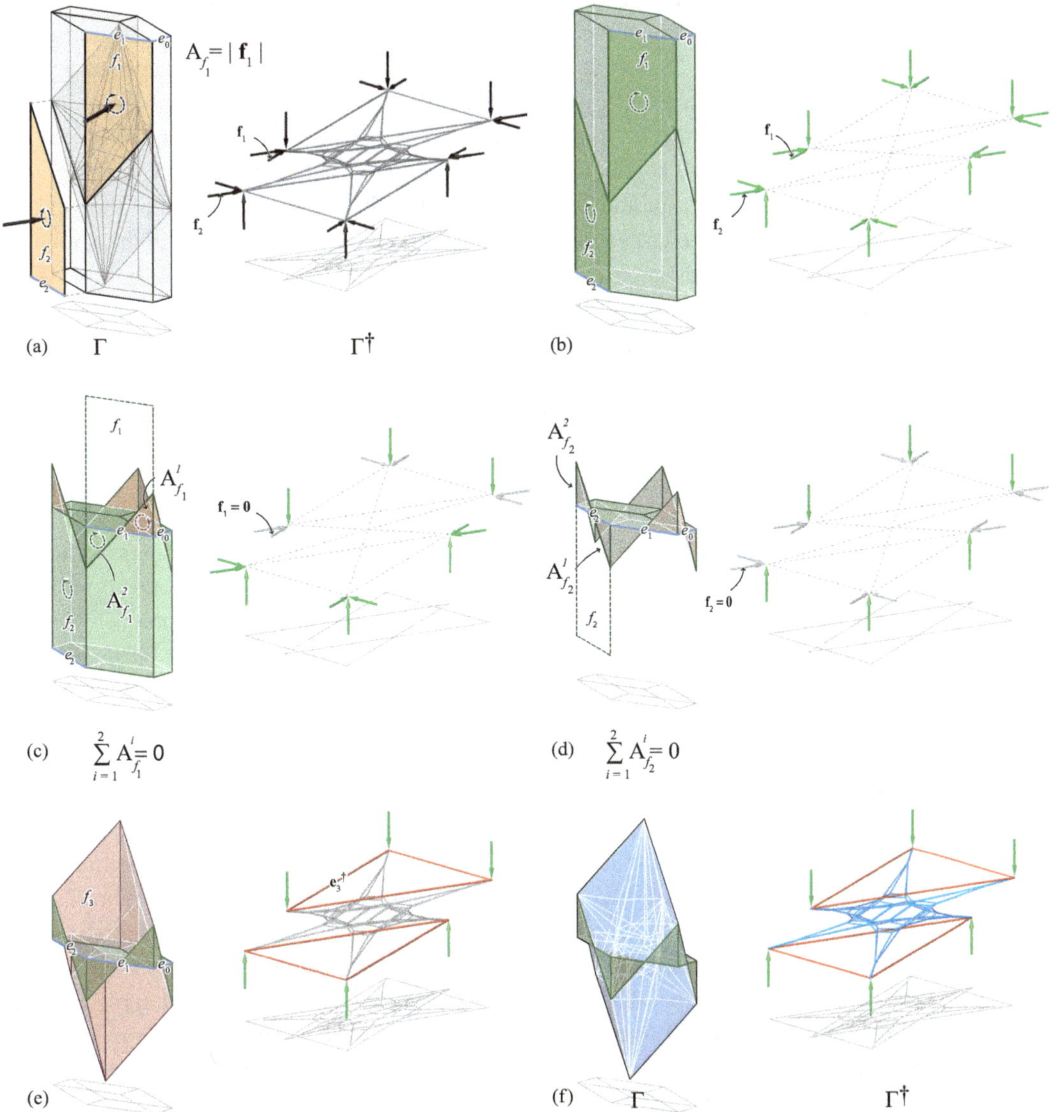

Figure 7.53 (a) The force Γ and form $\Gamma^{\dagger}$ diagrams of a compression-only funicular polyhedral structural form; (b) changing the areas of the chosen faces of the force polyhedron to zero; (c) the first step of the computation; (d) the second step of the computation; (e) the tensile members; and (f) the entire updated polyhedron and its corresponding structural form.

where the edges or the members can be removed from the system without disturbing the internal and external equilibrium. The zero-force edges have been previously observed in some polyhedral reciprocal diagrams by D'Acunto et al.,[53] but there was no method to compute them, particularly in self-intersecting faces as described in the methodology section. Also, the change in the direction of the normal of the face f_1 results in a reversal of the internal force in the edge e_1 of the form diagram.

Figure 7.51d emphasizes the faces f_2, f_3, and f_4 and the faces that will be affected as a result of the second step of the computation. As shown in Figure 7.51e, the normal of the faces f_3 and f_4 will invert the direction of the internal forces in the form diagram. This transformation also removes the applied load in the system as the area of the face f_2 is zero. In another example, the zero-area faces are used to completely remove the external horizontal forces in the system. Figure 7.52 shows a force diagram and its corresponding form in another transformation. In this example, one of the vertical faces is chosen and the target area is set to zero. Note that as animated in the drawings of Figure 7.52b, all the side faces collapse into a line and the top and bottom faces of the polyhedron become coplanar. This transformation results in the disappearance of all the horizontal applied loads in the system, as shown in Figure 7.52f. The most interesting geometric outcome of this process is the transformation of the internal face f_1. Face f_1 changes its direction and so does its corresponding edge at the boundary of the form diagram. The resulting structure shows a funnel-shaped compression-only structure with tensile members on the top.

Figure 7.53 illustrates another example of using this approach in the structural form finding of a funicular spatial structural form with both tensile and compressive forces in equilibrium. Figure 7.53a shows the force diagram consisting of groups of polyhedral cells with planar faces. These faces establish groups of convex polyhedral cells as a force diagram Γ that corresponds to a compression-only form $\Gamma^\dagger$.[75] The designer chooses f_1 and f_2 of the global (external) force polyhedron and assigns zero for their target areas (Figure 7.53b). Figure 7.53c shows the first step of the computation where the area of the face f_1 becomes zero and consequently its corresponding force in the form diagram $\Gamma^\dagger$ disappears.

Figure 7.53d shows the second step of the computation that makes the area of the face f_2 zero. Together with faces f_1 and f_2 all surrounding faces will also become zero, and consequently, all the horizontal forces in the system will disappear (Figure 7.53d). Note that in all steps of the computation, the rest of the faces of the polyhedral system will be updated to match the coordinates of the newly created zero-area faces. As a result, the faces corresponding to the top and bottom members flip, which reverses the direction of the internal force into tension. The resulting structural form is a funicular form with both tensile and compressive forces with no horizontal reactions at the support (Figure 7.53f).

Figure 7.54 shows the built structure with tension members on the top and compression members on the bottom.

7.6.5 An Alternative Approach of Controlling Face Areas of The Force Diagram

The quadratic formulation has a drawback related to its performance. Solving quadratic equations is more computationally expensive compared to linear equations. Additionally, when dealing with multiple constrained faces, this formulation solves the constraints sequentially. The solution for the previous target area determines the edge lengths for the surrounding edges, which serve as a precondition for the next

Figure 7.54 The built structure was designed using the methods of this section as part of the *Spatial Efficiency* Exhibition at the Center for Architecture and Design in Philadelphia in January 2020.

area constraint. Consequently, there is a possibility of overconstraining when the target areas are incompatible with each other.

There is a more efficient alternative to controlling face areas only by assigning edge length constraints. Although manually assigning edge lengths can be challenging for constrained faces with numerous edges, it can still be very useful when the number of edges is low. Compared to the approach of constraining face areas through quadratic formulations, changing face areas only through edge lengths is more intuitive and computationally efficient. The main limitation is that faces with more edges require more edge length constraints to determine their area. Figure 7.55 illustrates examples from two built design projects (Hedracrete and Saltatur) where the lateral external forces were difficult to provide in the actual construction. In Figure 7.55b where the external side faces of the force diagram are rectangles, their areas can be reduced to zero by constraining the length of edge e_1 to zero. In the more complex example shown in Figure 7.55e where the external side faces are trapezoids, their areas are turned to zero through self-intersection. The area calculation for self-intersecting faces has been exhaustively explained. In the new equilibrium states of both examples, some edges are turned into tension, which can be materialized and treated differently from the compression members.

A force diagram may have all its external face areas reduced to zero while keeping nonzero internal face areas (Figure 7.56). In this case, the corresponding form diagram represents a self-stressed structure because the edges have forces without any external load. This can be used to detect the static indeterminacy in a form diagram since statistically determinate structures cannot have any state of self-stress. However, it cannot be concluded that a form diagram is statically determinate if the internal faces cannot keep a nonzero area when constraining external face areas to zero. This is due to the polyhedral nature of the force diagram, which only shows a subset of all possible equilibrium states.

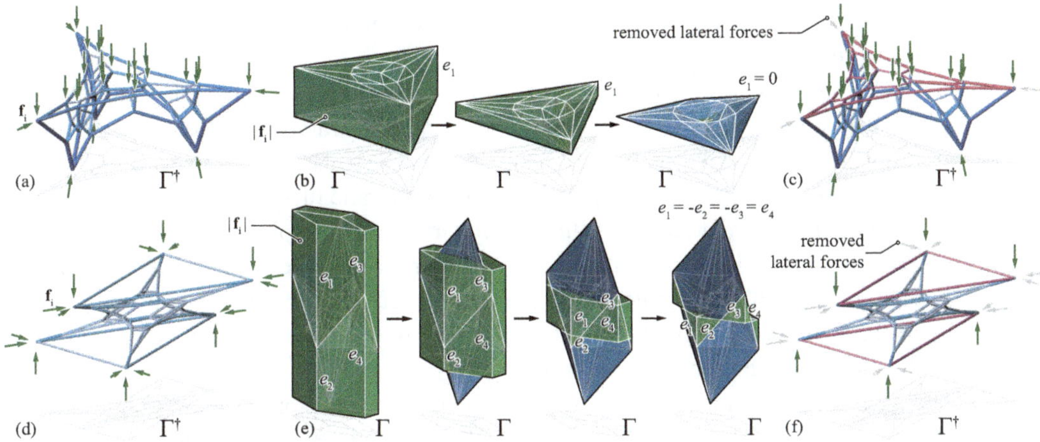

Figure 7.55 Two built design projects are used as examples to show the removal of lateral external forces. Those forces are removed by constraining the corresponding faces to zero area in the force diagram.

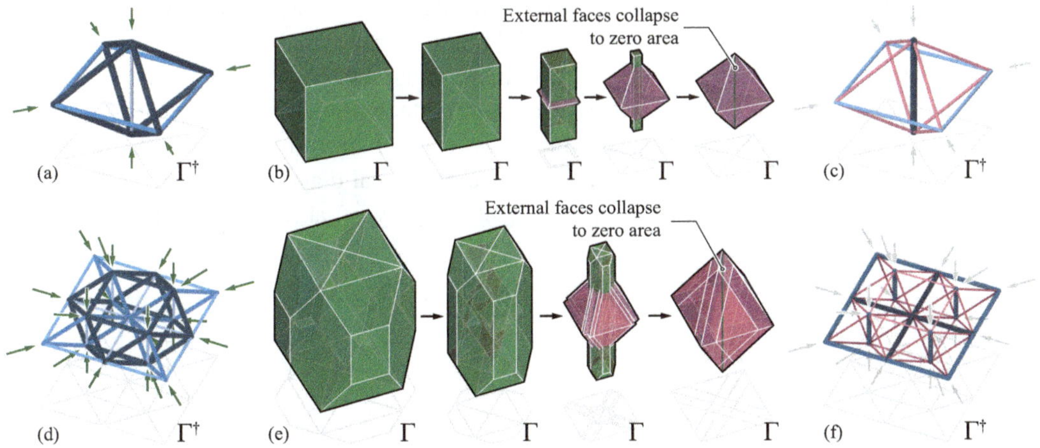

Figure 7.56 Some force diagrams may allow all external faces constrained to zero area while keeping internal face areas, and the corresponding form diagrams will represent a self-stressed structure.

7.7 Summary and Further Readings

This chapter introduces algebraic formulations as an innovative approach to 2DGS and 3DGS. These formulations capture the essence of the reciprocal relationship, representing the dual diagram algebraically. While requiring deeper insight into reciprocity, this method surpasses traditional approaches in its potency. With the aid of modern computational tools, we can efficiently navigate the solution space of these equations. The resultant algebraic framework promises increased efficiency, flexibility, and precision, enhancing our understanding of the relationship between form and force. Such advancements prove invaluable not only for academic pursuits but also for practical applications.

The method of algebraic graphic statics comes with certain limitations. Primarily, it's suited for linear static issues and linear constraints. Handling more intricate problems, such as the previously mentioned constraints on face areas, might not be straightforward. Such challenges may require quadratic functions, increasing the computational effort and cost for solutions. In terms of the implementation, although the data structure allows self-intersecting faces and cells to exist through later manipulation, the initial construction of the data structure cannot accept self-intersecting surfaces, that is, all cells in the starting primal diagram need to be convex; otherwise, the topology cannot be detected and defined properly.

We invite the interested reader to visit other related topics under graphic statics including vector-based 3D graphic statics[41,52,163] and approaches using projections of polyhedral systems to higher dimensions and back by Konstantatou and McRobie[164] and Baranya.[165] Furthermore Lee et al.,[18] proposed a method called *disjointed force polyhedra* where the equilibrium of the system was computed by constructing a single convex polyhedron for each node using extended gaussian image algorithm and then areas of the adjacent faces have matching areas but are allowed to have different shapes.[11,18] In McRobie,[10] a methodology was introduced for the computation and visualization of the Maxwell (in two dimensions) and Rankine (in three dimensions) diagrams that are reciprocal to a truss subjected to load. The formulation of these reciprocals is grounded in the Minkowski summation of the polyhedral stress functions inherent to both the original and the reciprocal diagrams. This approach aligns seamlessly with 3D graphic statics that utilize polyhedral reciprocal diagrams.

8 Broader Applications of Polyhedral Graphic Statics

Masoud Akbarzadeh, Yefan Zhi, Georgios Athanasopoulos, Marina Konstantatou, Maximilian E. Ororbia, Salma Mozaffari, Mostafa Akbari, Hao Zheng, Hossein Mofatteh, Hamid Akbarzadeh, Armin Mirabolghasemi, Zakaria Hsain, James Pikul, Yao Lu, Shu Yang, Kun-Hao Yu, Hua Chai, Mohammas Bolhassani, Paria Yavartanoo, Allan McRobie, Joseph Robert Yost, and Jens Schneider

8.1 Introduction

In Chapters 1 to 7, we presented various topics on the geometric design optimization of funicular structural form using the methods of polyhedral graphic statics. In this chapter, we will show the broader application of polyhedral graphic statics in other fields of science and briefly introduce research directions and topics that go beyond the polyhedral limitations of this method.

The explicit relationship between the form of the structure and the force equilibrium makes this method a novel and intuitive design approach that combines design and engineering that could be combined with other fields of science where this explicit relationship is reasonably demanded. In this chapter, we will visit a series of research projects using polyhedral graphic statics in the initial research steps. It established a rigorous starting point from which these investigations can depart into very interesting undiscovered research directions in the future.

8.2 Analyzing Natural Structures and Feeding Machine Learning Models

Engineers, scientists, and designers have long been inspired by the morphological and mechanical properties of natural/biological structures and used them in the design of the next generation of advanced materials and structures on various scales.[166,167] The dragonfly wing is an example of a high-performance, lightweight structure, providing a repository of design inspirations.[168–171]

Valuable research has been conducted on the three-dimensional (3D) morphology of the dragonfly wing, including the ridges and valleys of the wing.[172–175] Nevertheless, we deem that the two-dimensional (2D) pattern of the wing still possesses exciting properties that are yet to be studied and exploited in the design of architected materials or structures.

Interestingly, the 2D reciprocal diagrams of graphic static that Maxwell proposed[17] have been used to analyze natural structures. For instance, they were used to describe the equilibrium of forces in the spider web, consisting of convex polygons. It is justified that the spider web can be created from a polyhedron projection on a 2D plane.[106] The convex-dominant pattern of the dragonfly has never been studied using his proposed reciprocity, which inspired us to analyze it.

8.2.1 The Force Diagram of the Wing

The structural geometry of the dragonfly wing consists of veins connected by surface patches. The projected geometry of the dragonfly veins or the skeleton of the wing on a 2D plane may represent a structural network made of bars and nodes from which Maxwell's reciprocal diagrams can be extracted (Figure 8.1a–c). Figure 8.1a shows the structural network of the wing, and Figure 8.1b shows its dual reciprocal diagram with their related components highlighted in Figure 8.1c. Based on Maxwell's reciprocity, the reciprocal dual diagram for the internal network of the wing may represent a possible equilibrium condition for the internal forces in the wing's structure as follows: each node of the form diagram is reciprocal to a closed polygon of force in the dual diagram. That is, if the internal force in each edge connected to a node matches

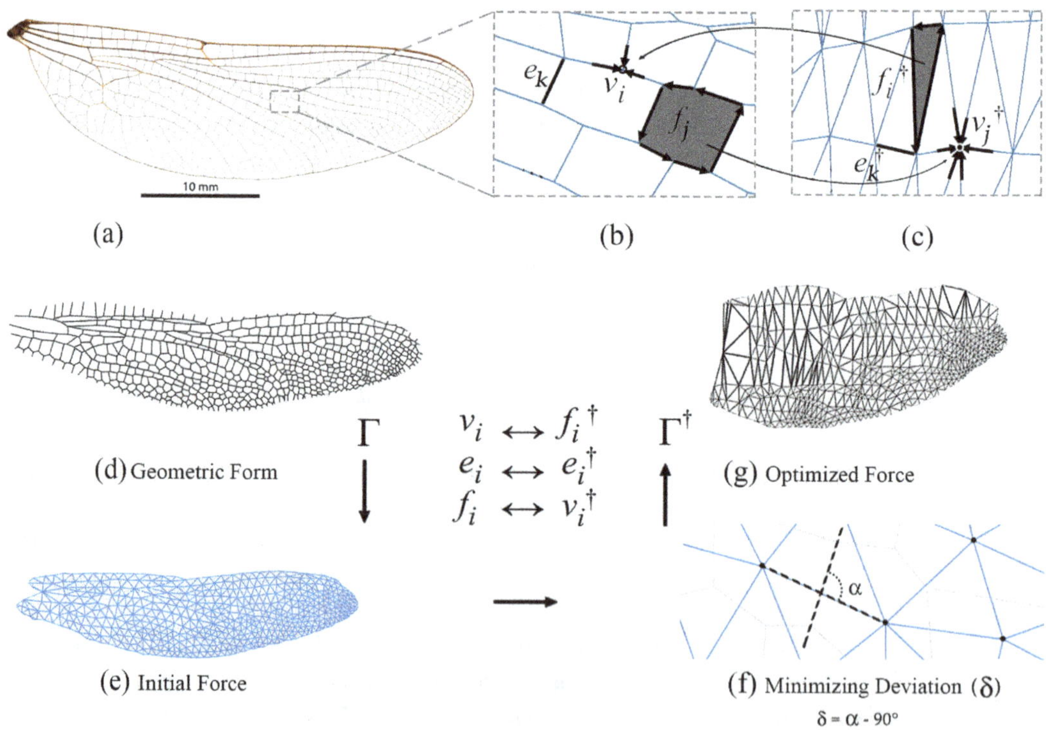

Figure 8.1 (a) The structural network of the dragonfly wing; (b) form diagram of the structure including vertices (v), edges (e), and faces (f); (c) the dual reciprocal force diagram with its vertexes ($v^\dagger$), edges ($e^\dagger$), and faces ($f^\dagger$); and a computational flowchart for the construction of the dual from the main network of the wing; (d) the input form graph; (e) the topology of the force; (f) iterative solutions; and (g) the final force diagram. Adapted with permission.

the length of its corresponding edge in the force polygon, then the system will be in equilibrium.

The equation system for the algebraic representation of the dual diagram is indeterminate and can have infinite possible solutions. To solve this equation system, we divide the wing's geometry into two parts by separating the internal network from the boundary edges. This means that the system of equations to solve for the dual can also be divided into equations related to the main network and the equations of the boundary edges and their nodes. The key observation is that the internal network of the dragonfly wing has a unique dual/reciprocal diagram up to scaling (Figure 8.1d–g).

The force equilibrium in the projected network of the wing structure, excluding the boundary edges, mainly consists of triangles representing a geometrically determinate network. Based on this dual diagram, we can conclude that the internal structural skeleton of the wing is a statically determinate system (Figure 8.1g).

8.2.2 Reconstructing the Wing's Main Network from Its Force Diagram

Once the force diagram is constructed for the original network, we can use it to reconstruct a new network of the wing. The dual relationship between these graphs allows us to indefinitely build one from the other using the iterative method of Nejur and Akbarzadeh.[137] Note that although the force diagram is geometrically determinate, its reciprocal dual, that is, the network of the wing, has geometric degrees of freedom (GDoF) larger than one. In fact, for the same force diagram, there are infinite reciprocal form diagrams that can be constructed with different edge lengths, all representing a statically determined network in equilibrium. Besides, the magnitude of the forces in the members stays intact, as only one force diagram represents the equilibrium. Therefore, to get the closest possible solution to the dragonfly wing, we need the original network's edge lengths and use them to reconstruct a statically determinate network similar to the wing.

8.2.3 Matching the Thickness of the Real Members

Visualizing the internal forces as diameters in the wing members provides a thickened wing network, as illustrated in Figure 8.2a. We assume a circular section for the generated network to compare with the real wing since the outer shape of the cross-section of the actual members is circular.[176] Note that the cross-section of the real members is not uniform,[177] and for this research, we only compare the outer diameter of the real and the generated members.

Figure 8.2d shows an example of the comparison of the generated dragonfly wing and the real dragonfly wing by measuring the radius of each member (Figure 8.2d). The structural thickness is predicted with an accuracy of 92.5%. A surprising observation is that the network with member diameters defined by the geometric equilibrium of forces matches the 2D projection of the real structure of the wing with its member

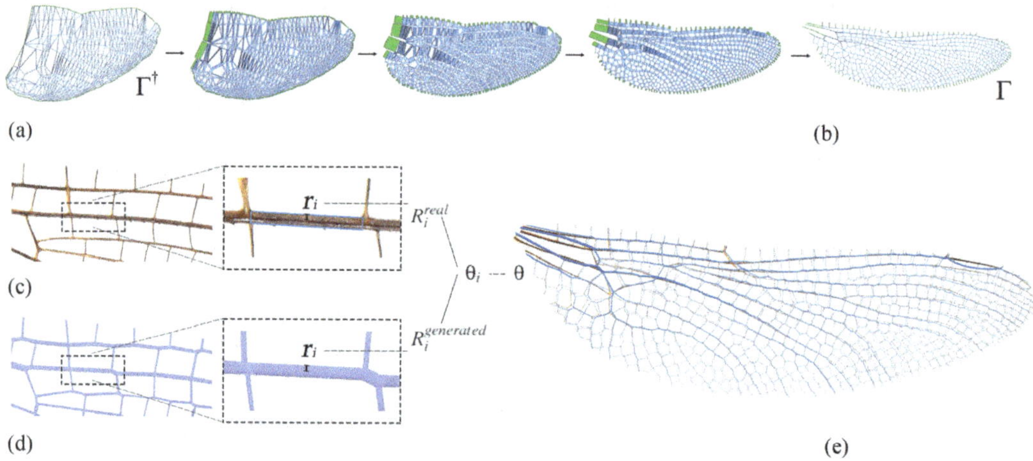

Figure 8.2 (a) Combining force and form diagrams of the wing using the Minkowski sum; (b) comparing the structural thickness of (c) the real wing (brown) and (d) the reconstructed form (blue); r: measured radius; R: normalized radius; θ: accuracy; $\theta = 1 - |R^{real} - R^{generated}|$; (e) superimposition of the reconstructed and the real wing. Adapted with permission.

diameters with an acceptable tolerance range (Figure 8.2d,e). The sizing can be visualized by combining form and force diagrams into a single diagram using Minkowski sum,[53] as illustrated in Figure 8.2a.

This observation has four major outcomes: (1) the internal network of the wing has a unique dual reciprocal diagram that can show a determinate equilibrium of forces in the network; (2) the location of the primary and secondary members of the real wing matches the reconstructed structural network made by graphic statics; (3) the primary members in the form of the wing corresponds to a particular region in the force diagram including force triangles with larger edge lengths; this information is quite useful in the regeneration of the wing and the location of potential primary and secondary members; and (4) the determinate solution for the dual graph of the main network can be used to find a dual diagram for the boundary members by matching the diameters of the natural wing and the static force equilibrium in the nodes (Figure 8.2).

8.2.4 The In-Plane Equilibrium Disclaimer

Undoubtedly, the mentioned observation does neither imply that the entire structure of the cellular core of the wing is in static equilibrium nor does it say that the static equilibrium is sufficient to describe the behavior of the cellular core of the wing or its loading conditions. We should acknowledge that the biological and environmental loading conditions, including the sophisticated loading patterns that led to the formation of the wing's geometry over the years, are quite complex, as they have already been investigated.[178,179] Nevertheless, our observation provides researchers with additional insight that the morphological configuration of the veins may suggest an in-plane, static equilibrium of forces as a possible force equilibrium in the system. Furthermore, the geometric representation of forces can be used to regenerate lightweight structural design solutions for various boundaries inspired by the wing's geometry.

8.2.5 Force in the Boundary Members

We have shown that the internal network of the wing with the member sizes proportional to the edge lengths of the reciprocal diagram matches the members in the real wing. We can use these results as deductive reasoning to find the thickness of the external members and complete a determinate dual diagram. As illustrated in Figure 8.3c,d, we introduce a set of external loads at the periphery of the wing called *virtual forces*. We find these loads such that if applied to the boundaries, they would produce the same internal forces proportional to the diameters of the members in the real wing. To find the magnitude of such forces, we measure the diameter of the actual boundary members and solve for an external loading that can create such internal forces in those members proportional to their diameter. Then, a separate force diagram is constructed for the boundary members and combined with the determinate force diagram of the internal network of the wing (Figure 8.3a–f). The resulting combined dual diagram is determinate and can be used to describe the geometry of the entire wing. Note that the entire wing geometry represents an indeterminate system, and our dual diagram can present an equilibrium condition among others where the internal forces match each member's diameters.

The determinate force diagram of the wing with its convex polygons includes information about the primary and secondary veins and their diameters. Thus, it will be used as valuable data to learn from and apply its design logic to similar cases. These initial results lay the basis for the second phase of this research, which is related to the work's generative use of machine-learning algorithms with no information on the initial topology of the network.

8.2.6 Generating Structural Networks Using Machine Learning Models

We further use machine learning methods to generate the structural networks by inputting the form and force diagrams of the dragonfly wing as the training dataset for the generation process. We identify the main regions in the force diagram, including triangles with similar edge lengths and areas (Figure 8.4). These regions correspond to the primary and secondary veins of the wing. We then reconstruct these regions using machine learning and triangulate those regions to get force polygons that complete the force diagram. Subsequently, a vector-based machine learning model is used to construct the structural network of the wing as a dual for the regenerated force diagram. This machine-learning model predicts edge lengths for the structural network close to the wing's actual geometry. In this process, we use the force diagram to ensure that the resulting structural network generated by machine learning is in static equilibrium. For further information about the details, please visit the article by Zheng et al.[85]

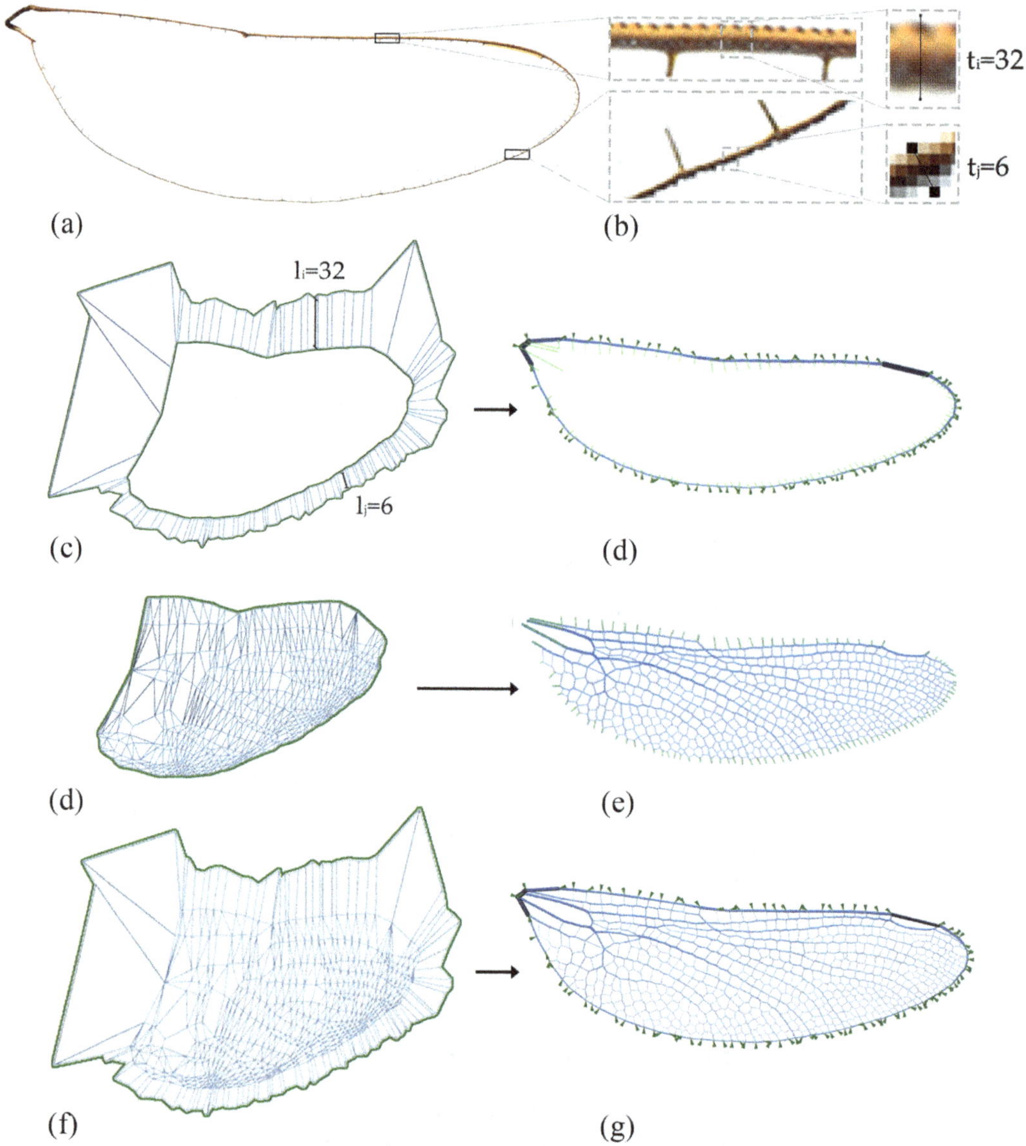

Figure 8.3 (a) and (b) Measuring the pixel width and mapping to (c) an indeterminate external force diagram; (d) reconstruction of the boundary edges; (d) and (e) the determinate force and form diagrams of the internal network; and (f) and (g) combining the indeterminate force of the boundaries and the determinate force of the internal members to complete the geometry of the wing with the boundary edges.

8.2.7 Designing an Airplane Wing Using the Proposed Method

The novelty of this work is to replicate the structural patterns of some natural species using reciprocal diagrams of graphic statics and machine learning methods. The out-of-plane performance of the dragonfly wing in nature motivated us to test

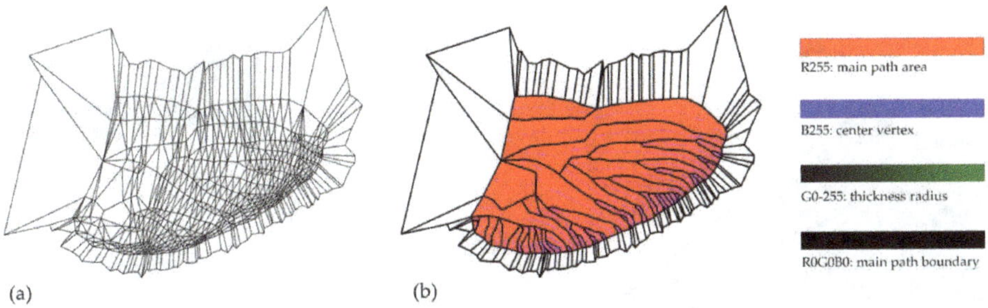

Figure 8.4 Main path combining the vertex information: (a) original force diagram; and (b) force diagram with main paths and pixel values. Adapted with permission.

this research's real-world application in designing structures for airplane wings as a cantilever structure.

Airplane wings are among the structures that require to be stiff and lightweight simultaneously. Even a small improvement in achieving a higher stiffness-to-weight ratio leads to a considerable reduction in material and fuel consumption in the aerospace sector. For a long time, airplane wings have been fabricated with straight ribs. Advances in manufacturing have enabled the fabrication of advanced materials and structures with complex architecture from nanoscale macroscale.* However, obtaining a very stiff design for the airplane wing is a crucial challenge.

The loading scenarios of the dragonfly wing and the airplane wing have some common similarities; both withstand the lifting force from beneath. Therefore, features from the dragonfly wing can be transformed to design an airplane wing. Besides, we learn from the dragonfly that the design needs a larger external force in the front than smaller external forces in the back. The generated force diagram matches this principle that the internal forces in the front are larger than the ones in the back. Also, the structural form contains similar features, such as the main paths of the dragonfly wing.

In the current study, we utilized the developed machine learning-assisted design methodology to introduce dragonfly-inspired structures as the core of the airfoil wing, assuming that the dragonfly's wings generate a great lift force while being lightweight and stiff.[179,189] We can input the human-defined boundary with the trained models and ask the machine-learning model to generate the structural form within the given boundary. Figure 8.5 shows one of the applications of designing an airplane wing.

Specifically, we investigate dragonfly-inspired designs on the 2D extruded airframe of the Boeing 777 wing at the 1:120 scale. We could observe the improvement of the structural efficiency of the wings from the perspective of stiffness and weight. In this regard, the out-of-plane stiffness of the wing concerning point load and distributed air pressure has been studied. Numerical and experimental results depict dragonfly design can improve out-of-plane stiffness by 25%, leading to lighter designs. In the next paragraph, we will provide detailed explanations of the methodology and findings of this research.

* For further information on this topic, please visit.[84,171,180–188]

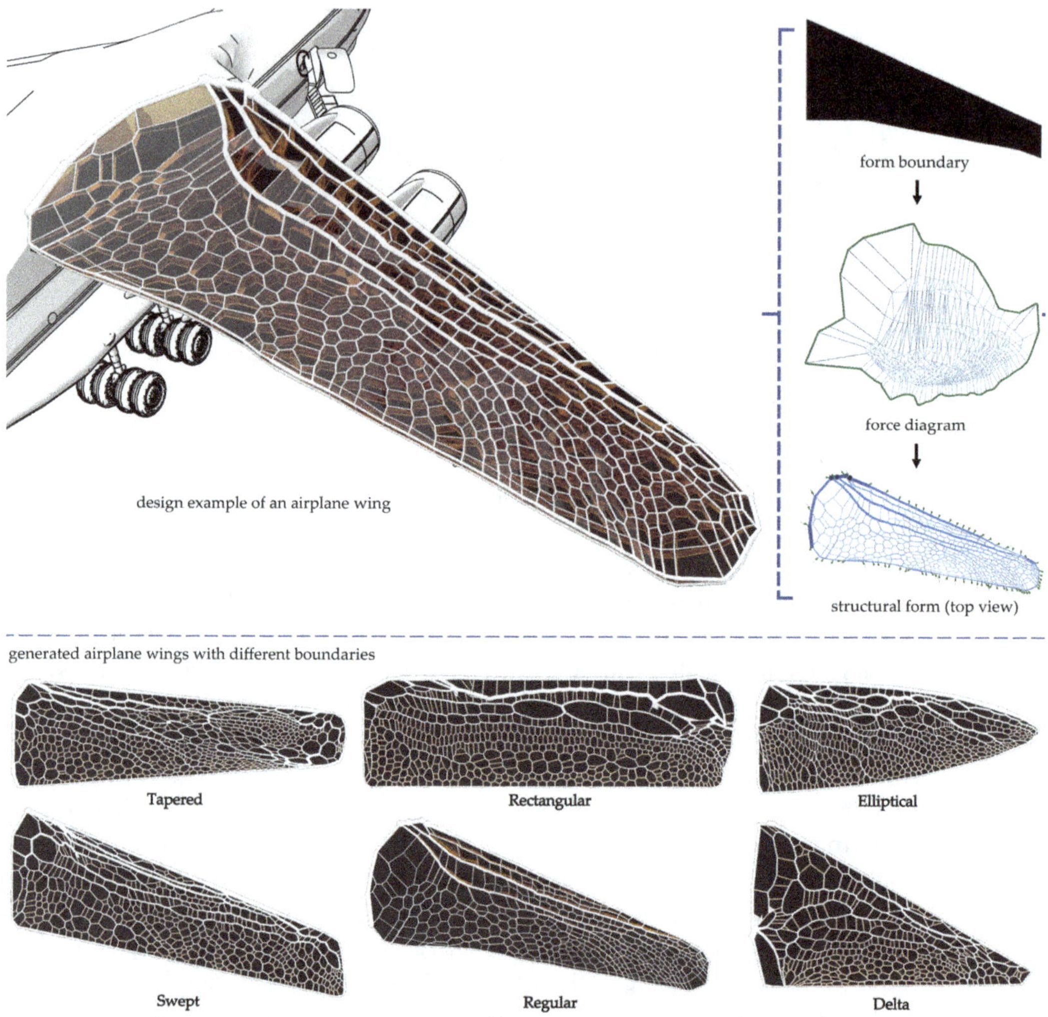

Figure 8.5 A broader application of our method in generating airplane wings. Adapted with permission.

We generated four dragonfly structures with alternative subdivision densities correlated to their architectural features (i.e., number of constitutive struts), namely V_0, V_1, V_2, and V_3 designs (Figure 8.6). Besides, we have considered a minimum thickness limitation in our design to reflect the constraints of additive manufacturing technologies for the 3D printing of small features. Adopting 0.6 mm as the minimum strut thickness, manufacturable by common fused deposition modeling (FDM), we have also designed four samples (i.e., V_0^L, V_1^L, V_2^L, and V_3^L) with manufacturing constraints. Samples tested under point loads have a thickness and weight of 7.8 mm and 30 g, and those tested in the wind tunnel have a thickness of 2.32 and 3.85 mm and a weight of 9 g and 14.8 g, respectively. All samples are designed with the same volume fraction of 0.38.

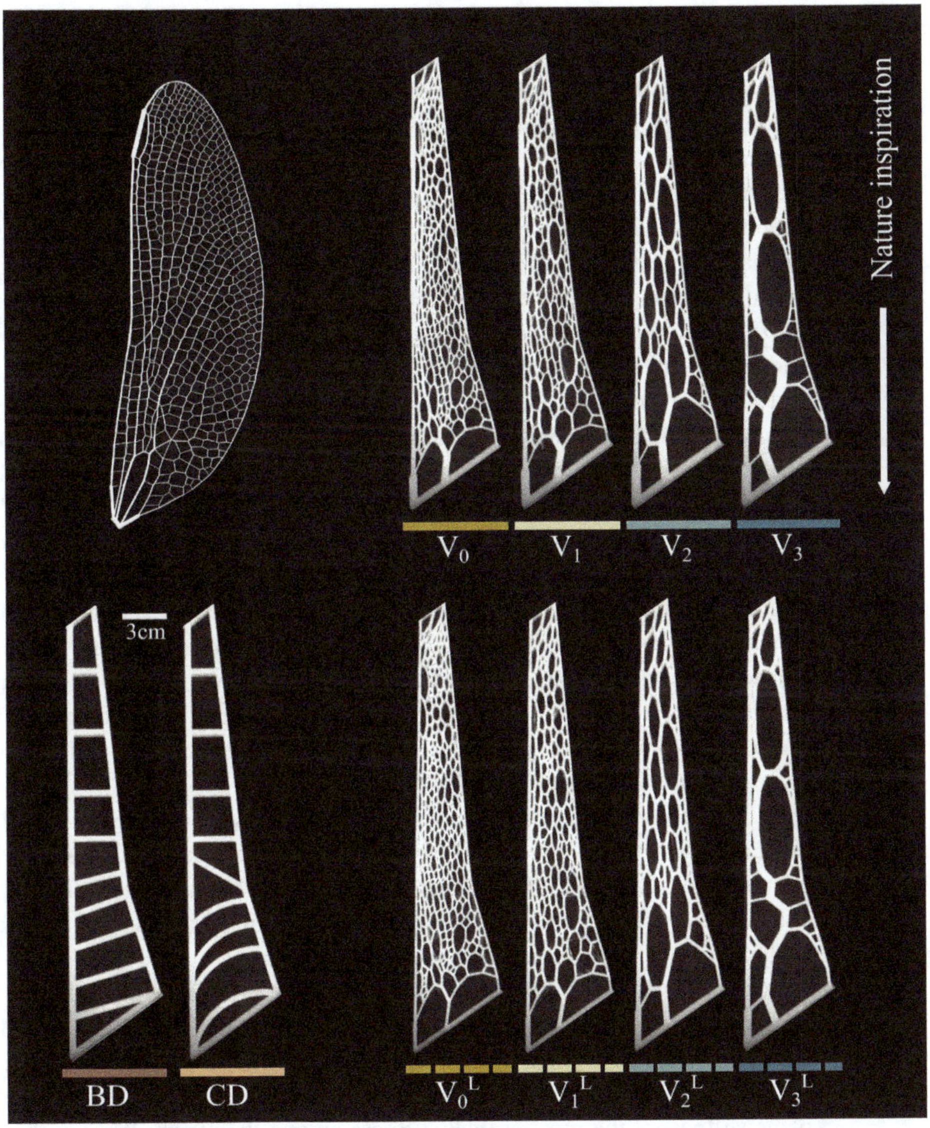

Figure 8.6 3D-printed samples for conventional designs (BD and CD) and dragonfly-inspired designs V_0 to V_3 and V_0^L to V_3^L. Adapted with permission.

Mechanical properties studied here for the wing structure are out-of-plane, torsional, and in-plane stiffness and maximum strength. Due to the importance of out-of-plane stiffness, it has been evaluated under both points and a uniformly distributed load (wind pressure). Torsional and in-plane stiffness have been measured under point load. The maximum strengths of the 3D-printed samples, made out of polylactic acid (PLA), are evaluated under out-of-plane two-sided clamped deformation and tail rotation.

Initially, the developed bio-inspired designs are studied under a simplified load case (i.e., point load) to better understand the effect of geometrical constraints, minimum beam width, and subdivision density on the wing designs' structural performance of the dragonfly-inspired cellular core.

For the baseline comparison, two wing core patterns have been chosen. The first traditional design with straight ribs has conventionally been used in designs of airplanes, and the curved rib design has recently been suggested to improve the performance of the wing structures.[190] This study compares all results to the wing with the traditional pattern (straight ribs) as a baseline. The experimental results for structural stiffness evaluated under a point load condition indicate that dragonfly-inspired designs can improve out-of-plane and torsional stiffness of the wing compared to the basic design (BD), especially for V_2^L and V_3^L (Figure 8.6). However, the BD performs very well under in-plane loading conditions. It is found that reducing the number of features and adopting the minimum thickness limitations improve the strength of the structure, which may emanate from manufacturing defects for samples with high subdivision density and very thin struts. In addition to improving the strength of V_2^L and V_3^L designs compared to BD, the ultimate structural failure section occurs at the tip of dragonfly designs. It is worth mentioning that airplane wings in service are under a distributed load and are clamped to the airplane fuselage, leading to maximum shear force and moment at the fixed supports. Therefore, failure at the tail found in the dragonfly-inspired designs is an advantage compared to the failure at the front of the basic and curved rib designs. Finite element (FE) studies on the bio-inspired designs also reveal a similar trend for the designed wings' out-of-plane, torsional, and in-plane stiffness.

To further justify the applicability of the dragonfly-inspired cellular core of the wing designs, we have evaluated the performance of 3D-printed samples under constant air pressure in the wind tunnel (Figure 8.7). As the designed samples are 2D extruded and do not have an airfoil, the out-of-plane distributed loading is the most effective measure for evaluating their performance as airplane wings. Figure 8.6 shows that dragonfly-inspired design without manufacturing limitation considerations can perform significantly better than the BD, for example, up to 25% improvement in out-of-plane stiffness for V_2 design compared to the BD. In our study, the V_2 design demonstrates the highest out-of-plane stiffness under pressure. Since the lift force on airfoils is significantly higher than the drag force (the lift-to-drag ratio is commonly in the range of 10–20[191]) and the smallest dimension of the wing is its height, out-of-plane stiffness is the most critical stiffness among the others. It is worth mentioning that the previously reported curved rib design with a higher thickness ($t = 3.85$ mm) also outperforms the BD in out-of-plane stiffness. In addition to the stiffness, the V_2 design with lower height has higher natural frequencies, which implies less risk of resonance under dynamic loads or higher flapping frequencies for designing flying microrobots.[192–195]

The developed biomimicry approach presented in this study can be applied to designing the cellular core of alternative lightweight wings to improve their structural performance. The airfoil shape that sandwiches the bio-inspired cellular cores is assumed to be unchanged since the wing's aerodynamic performance relies on the airfoil's geometry,[196–199] the dragonfly-inspired reticulated core has a negligible effect on the aerodynamic performance.

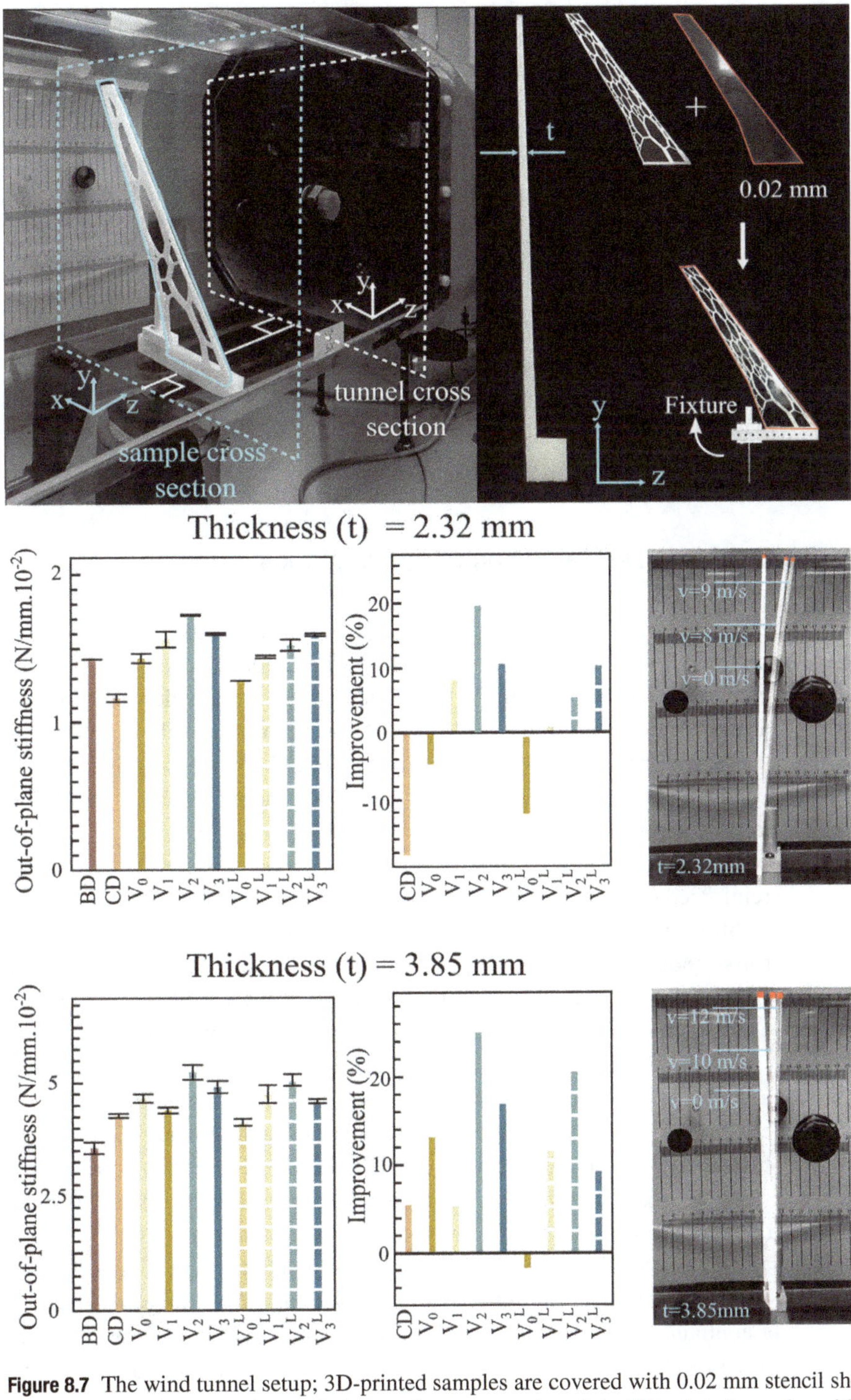

Figure 8.7 The wind tunnel setup; 3D-printed samples are covered with 0.02 mm stencil sheets for conducting the test and out-of-plane stiffness properties of the 3D-printed samples with 2.32 mm thickness and 3.85 mm thickness under uniform air pressure. Adapted with permission.

The numerical and experimental results suggest that biomimicry of the dragonfly pattern for designing the core of airplane wings can have advantages in terms of mechanical performance. In this respect, a dragonfly possesses long and relatively rigid wings that, through a high rate of flapping, bear high pressure.[179,200–202] Through veins and stems, dragonfly wings form very stiff structures; mimicking such structures can be advantageous for long and cantilevered structures such as commercial airplane wings. Besides biomimicry, the design methodology based on graphic statics assumes that all struts are under the same axial stress. Although this approach neglects bending moments in the struts, it provides a well-connected pattern with proper thicknesses in the skeleton of the wing. The adopted methodology also offers flexibility in defining subdivision density for creating new cellular structures. This parameter can control the stiffness of cellular structures.* As we know from experiments, the generated wings with smaller subdivision density usually show higher stiffness since the distribution of the structural weight tends to be concentrated more on the main structural members.

8.3 Strut and Tie Method and Graphic Statics

2D graphic statics can be applied to generate *strut-and-tie models* (STMs), also known as *truss models* for the plastic design of reinforced concrete. A STM consists of compressive struts and tensile ties representing the force flow or a load transfer mechanism inside a continuum. The model has been intuitively developed to understand the behavior of cracked reinforced concrete in the plastic region, where concrete only contributes to the compression and steel is activated in tension.[203,204] The concept has later found justification in the *lower bound theorem of plasticity*,[205–208] where in combination with constant stress fields, an appropriate tool for the design and detailing of reinforced concrete[209,210] was provided.

Strut-and-tie models are discrete representations of constant stress fields. While truss models are suitable for calculating the amount and location of reinforcements, stress fields provide a better estimate of the strut locations and help detail nodes and check the concrete behavior. In contrast to the intuitive nature of truss models, stress fields are developed as the direct application of plasticity theory.[207,208] The stress field approach is a simplified design tool to visualize the magnitude of forces in truss models and the required concrete dimensions to carry compression. A combination of truss models with stress fields is typically used in engineering practice to provide safe solutions since it complements the equilibrium and yield criteria of the lower bound theorem of plasticity.[205,206,211,212]

The strut-and-tie method also has a tremendous educational advantage over the FEA methods as it simplifies the understanding of bending and shear transfer inside a continuous element. Thus, it aids in a simplified calculation of the required reinforcement quantity and location. Beyond the educational value, the method is used

* See the studies on the mechanical performance of the subdivided polyhedral systems by Akbari et al.[84]

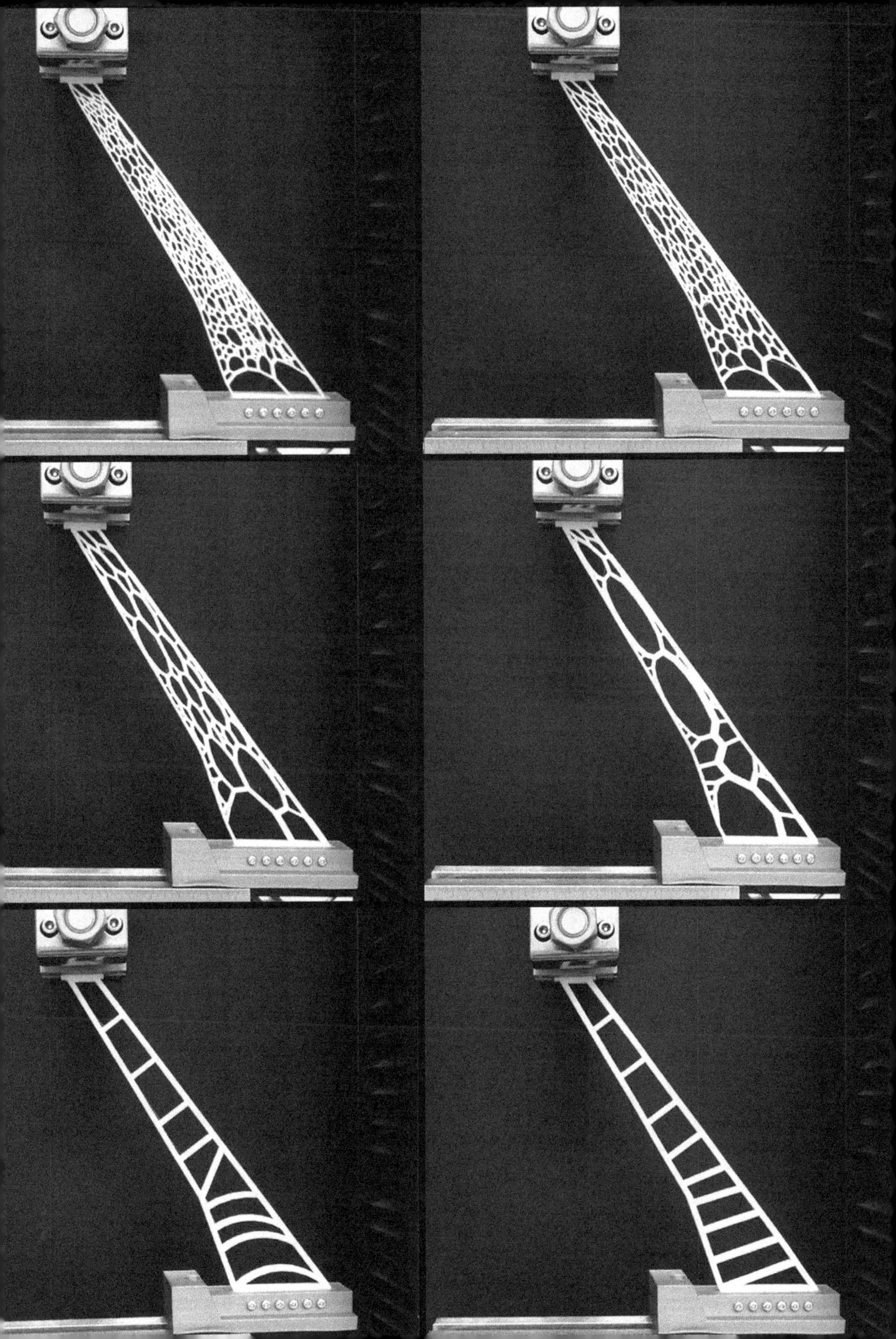

in practice and introduced in concrete design guidelines for structural design and analysis.[213–217]

Over the past three decades, several guidelines and tools have been introduced to help create STMs and stress fields, and graphic statics is among those methods.[16,206,218] Finding an in-equilibrium truss with graphic statics has the advantage of assuring the nodal and external equilibrium and controlling the magnitude of forces in truss members by visualization of the forces in the force diagram. Graphic static has also been combined with other methods in applying the strut-and-tie method; examples include its combination with a load path minimization algorithm for finding optimum truss models[219] and with the geometrical Minkowski sum operation to generate constant stress fields.[53]

8.3.1 An Integrated Algorithm

Although applying graphic statics in strut-and-tie modeling provides the opportunity to create various in-equilibrium load path scenarios with the known magnitude of forces, it has the drawback of requiring an input geometry or an initial truss model usually achieved through trial-and-error. To automate the generation of the initial truss model, we can introduce a combined algorithm of a *ground truss optimization* algorithm with algebraic graphic statics.* The novelty of the integrated algorithm is that an initial geometry is not required, and a force diagram will also be generated, which is used to create constant stress fields using the Minkowski sum. Also, the produced form and force diagrams can be used further to apply modifications and improve the models.

The novel integrated algorithm uses the similarities of the equilibrium equations in algebraic graphic statics and the plastic formulation of layout optimization[220,221] to generate the solutions. The linear force–stress relationship of this formulation allows the objective function to either aim for volume or load path minimization, which theoretically yields the same output.[221] The summation of each truss member's internal force multiplied by the length of the member is called *load path*.[55,219] Suppose the cross-section of each truss member is the size proportional to its internal force using the force diagram. This number may represent the total volume of the construction material needed to build the structure. Therefore, minimizing a structure's *load path* can lead to material and cost minimization. The load path minimization for *layout* and *shape optimization* for a truss using algebraic formulation have been addressed in the literature, and we encourage the readers to visit the works of Beghini et al. and Alic and Persson.[55,219]

The starting geometry for layout optimization is a ground truss. The ground truss is defined in two common ways: first-order and full ground truss (e.g., Figure 8.8a,b). For a ground truss with e_{gt} members (edges), v_{gt} nodes (vertices), and $\mathbf{n}_r$ support reactions (i.e., the number of restricted degrees of freedom), the plastic formulation is given as:

* Using the ground truss optimization, also known as *layout or truss topology optimization*, to generate STMs is not new, and it was first introduced by Ali and White.[70]

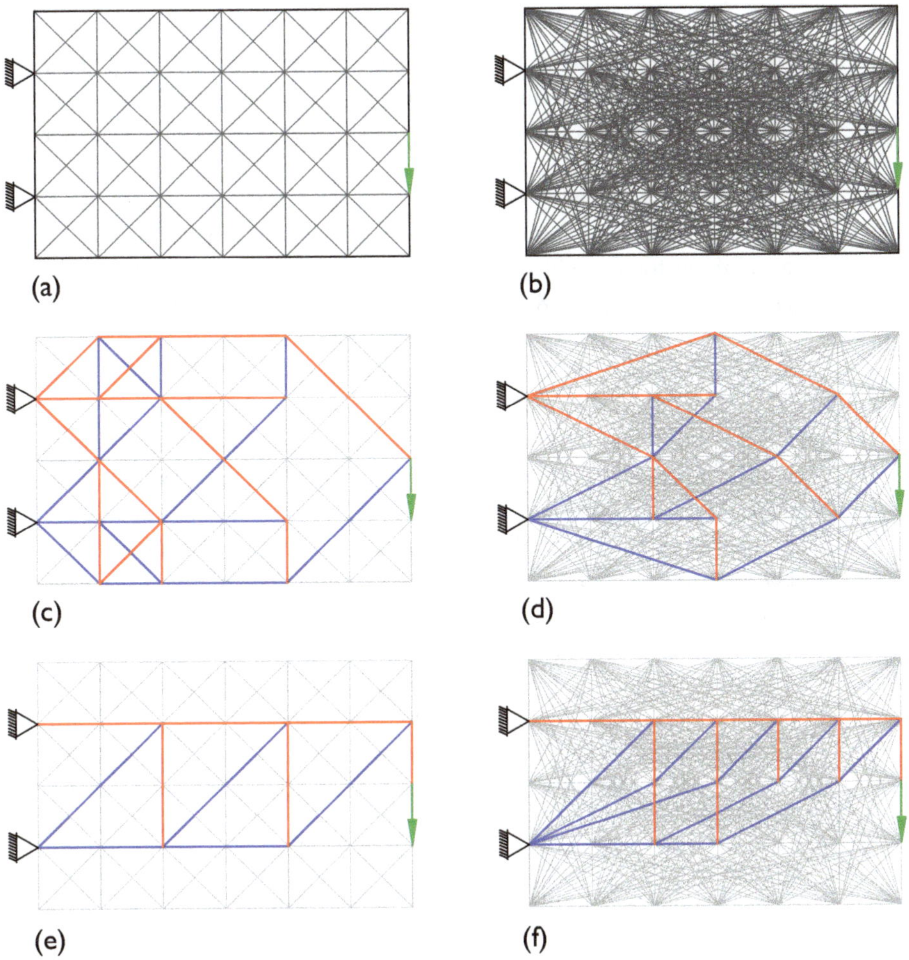

Figure 8.8 (a) First-order ground truss; (b) full ground truss; (c) and (d) optimum layouts for (a) and (b); and (e) and (f) optimum layouts for (a) and (b) with the additional constraint of Equation 8.1e. Adapted with permission.

$$\min_{\mathbf{a},\mathbf{f}_{gt}} \quad V = \mathbf{l}^T\mathbf{a} \tag{8.1a}$$

$$\text{s.t.} \quad \mathbf{A}_{opt}\mathbf{f}_{gt} = \mathbf{f} \tag{8.1b}$$

$$-\sigma_C\mathbf{a} \leq \mathbf{f}_{gt} \leq \sigma_T\mathbf{a} \tag{8.1c}$$

$$\mathbf{a} \geq \mathbf{0} \tag{8.1d}$$

$$\mathbf{f}_{gt,diag} \leq \mathbf{0} \tag{8.1e}$$

where V is a scalar for a total volume of the structure, $\mathbf{l}^\mathbf{T}$ $[1 \times e_{gt}]$ and $\mathbf{a}$ $[e_{gt} \times 1]$ are the vectors of length and cross-sectional area of the ground truss members. The objective in Equation 8.1a is to minimize the total volume subjected to the following constraints: (1) Equation 8.1b is the equilibrium constraint with $\mathbf{A}_{opt}$ the $[(2v_{gt} - n_r) \times e_{gt}]$

equilibrium matrix, $\mathbf{f}_{gt}$ the $[e_{gt} \times 1]$ member forces vector, and $\mathbf{f}$ the external loads' vector. (2) Equation 8.1c is the stress constraint with σ_C and σ_T as factored yield stresses in compression (concrete) and tension (steel). (3) Equation 8.1d limits the areas to nonnegative values. (4) Equation 8.1e constrains the force magnitudes in the diagonal members, $\mathbf{f}_{gt,diag}$, to nonpositive values (i.e., defined as either compressive or zero forces).

The optimization outputs are the ground truss member forces, $\mathbf{f}_{gt}$, and cross-sectional areas, **a**. Members with zero or minimal cross-sectional areas, that is, zero or minimal forces, will be removed, leading to an optimized truss as a subset of the ground truss. Figure 8.8c,d shows the results of the original layout optimization plastic formulation, that is, using the constraints in Equation 8.1b to 8.1d, and Figure 8.8e,f shows the solutions with all constraints. The additional constraint in Equation 8.1e makes the reinforcement layouts more practical with only vertical and horizontal alignments. Although the solutions with the full ground truss input in Figure 8.8d,f include intersecting tension and compression members, which is undesirable for a STM, the results of Figure 8.8 also illustrate how the optimization solutions depend on the ground truss definition. To avoid overlapping tension and compression members and to have a model with practical reinforcement layouts, we decided to use the first-order ground truss and the constraints (Equation 8.1b to 8.1e).

The equilibrium matrix $\mathbf{A}_{opt}$ in Equation 8.1b is constructed similarly to matrix $\mathbf{A}$ in Equation 7.5, with the difference that the rows related to support reaction edges and the columns corresponding to applied loads and support edges are removed. That is why the applied load vector $\mathbf{f}$ shows up on the right side of the equation. The optimization solution provides vector $\mathbf{f}_{gt}$, then, the support reactions can be calculated as:

$$\mathbf{f}_r = \bar{\mathbf{A}}_{opt}\mathbf{f}_{gt}, \tag{8.2}$$

where the nonzero elements of $\mathbf{f}_r$ $[2v_{gt} \times 1]$ are the applied loads and the support reactions, and $\bar{\mathbf{A}}_{opt}$ is a $[2v_{gt} \times e_{gt}]$ modified $\mathbf{A}_{opt}$ matrix, where the columns corresponding to applied loads and support edges are removed. After the real force magnitudes for all the edges are known, they can be translated into edge lengths of the force diagram $\mathbf{q}$ through a fixed scale. Once $\mathbf{q}$ is known, the force diagram can be constructed using the method explained in Section 7.2.8, as shown in Figure 8.9a–c.

The availability of the form and force diagrams allows the creation of discontinuous stress fields with *hydrostatic nodes* (i.e., the geometry of each node is obtained using the constant factored yield compressive strength σ_C)[205] using the Minkowski sum operation (Figure 8.9d).[222,223]

With the help of an optimization algorithm, we produce a truss or load path inside a boundary volume accounting for the boundary conditions. Also, regardless of the input or resulting truss's static degree of (in)determinacy, the optimization solution provides the force magnitudes for the optimized truss without the concern of picking the appropriate input edge lengths for truss analysis. The input forces are always the applied loads assigned through **f**.

Figure 8.10(1)–(7), a–d shows the application of the integrated algorithm along with the implementation of Minkowski sum to a simply supported beam, a stepped beam,

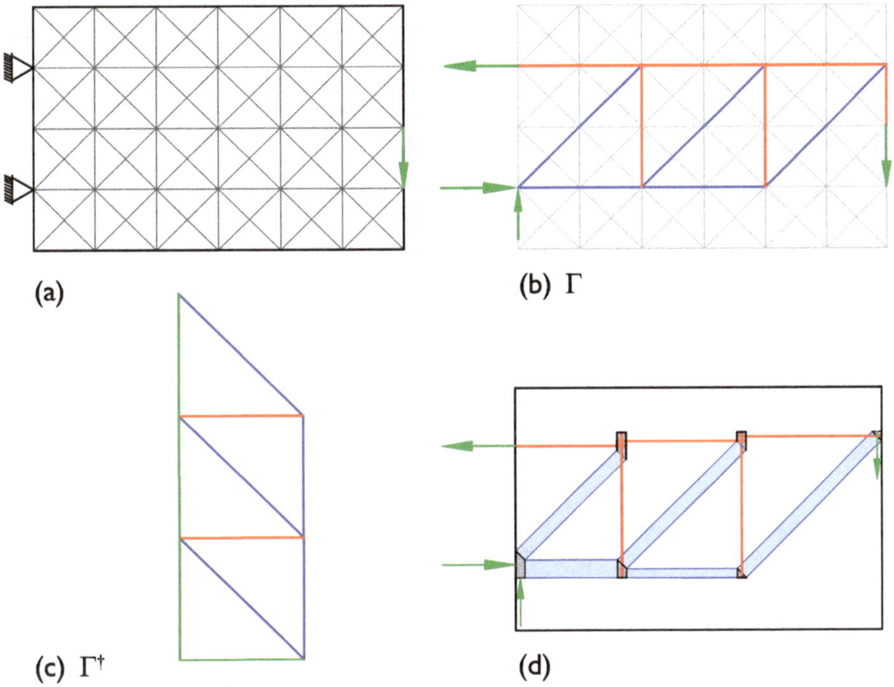

Figure 8.9 (a) Input boundary conditions and ground truss; (b) output form diagram (STM); (c) output force diagram; and (d) output constant stress fields (as a convention by structural engineers, the center lines of the tension fields are shown to illustrate the position of the reinforcing bars). Adapted with permission.

a frame joint, a dapped-end beam, a hammerhead pier, a wall with an opening, and a corbel column. Besides initiating a practical STM (for designing reinforced concrete), the method enables the visualization of forces through the force diagram and illustrates the concrete element's sufficiency to carry compression through constant stress fields.

8.3.2 Modifications with Graphic Statics

Proper procedures can be performed on the force diagram with the help of graphic statics to modify the original solutions from the integrated algorithm. These procedures ensure static equilibrium by securing the closeness of the nodal and global polygons in the force diagram. The new strut-and-tie solutions can be reconstructed after the modifications of the force diagram. Figure 8.11 shows a sample of the step-wise transformations that can be implemented to generate more elaborate STM solutions and their force diagrams while maintaining static equilibrium.

The step-wise subdivision of the form and force diagrams in Figure 8.11(1)–(4) illustrates the change from a point load in (a) to a distributed load in (e) for the STM of the simply supported beam in 8.10(1b). The truss models in Figure 8.11(1) and (3) are carrying the loads directly to the supports; the difference relies on the concrete's load transfer behavior, where in Figure 8.11(1), an arching funicular and in Figure 8.11(3),

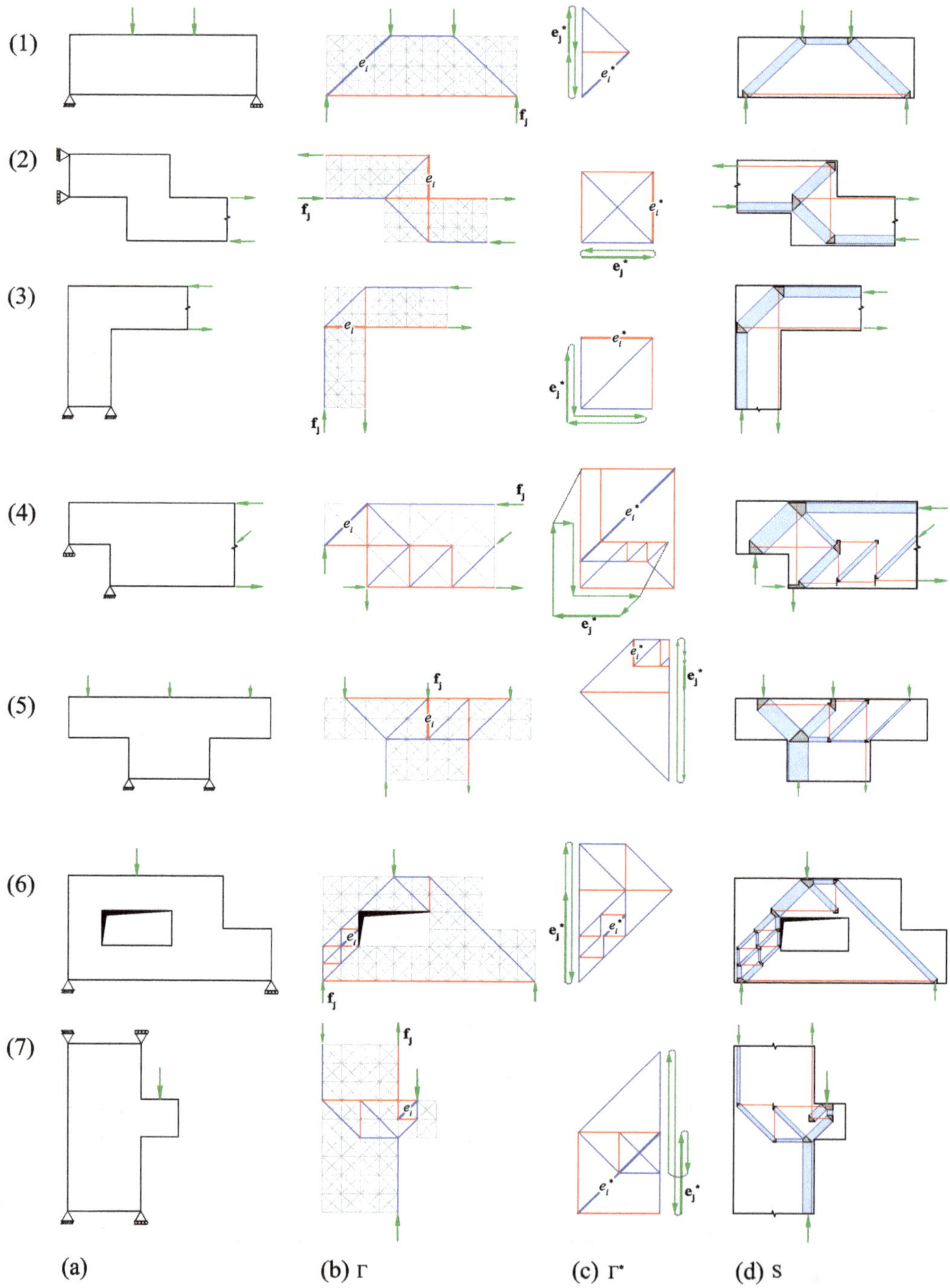

Figure 8.10 (a) Input design domain and boundary conditions; (b) output form diagram (STM); (c) output force diagram; and (d) output stress fields. Adapted with permission.

a fan-shaped action occurs. Whether a fan or an arch mechanism occurs in a certain member depends, among other things, on the slenderness, the reinforcement ratio, and the loading history. In an elastic solution, the stiffest mechanisms will be formed.[224] Considering that the funicular behavior uses a shorter load path and assuming that the model with the least load path has less deformation and therefore is stiffer, the STM in Figure 8.11(1e) compared to Figure 8.11(3e) is likelier to happen in the elastic range.

The other exciting modification is to transfer the load indirectly to the supports by adding stirrups, as in Figure 8.11(5) and (6), where a combined funicular and fan-shaped behavior occurs. For example, in Figure 8.11(5e), a portion of the external load transfers with the fan-shaped struts to the horizontal tie and then passes through the stirrups to the arching struts on the top. Eventually, it transfers along the other portion of the external load through the arch to the supports. For a detailed discussion on the integrated algorithm and applying the modifications using algebraic formulations, we refer the reader to study the works of Mozaffari et al.[222,223]

8.3.3 Materializing Using Multi-Material Additive Manufacturing

Members of truss structures predominantly experience axial loads as tensile or compressive forces acting along their longitudinal axes. To ensure stability and load-bearing capacity, the materials employed in these members must exhibit adequate resistance to deformation, specifically elongation or shortening, under the influence of the applied forces. The truss structures are often made of discrete members for ease of construction. Advances in additive manufacturing have led to some breakthroughs in constructing truss systems by depositing multiple materials on a continuous toolpath.[225]

Utilizing fiber-reinforced materials with fibers aligned in the longitudinal direction can enhance the tensile strength and stiffness of the members, making them well-suited for withstanding tensile loads. Conversely, nonfiber-reinforced materials demonstrate fine compressive strength when oriented along their short axial direction, making them more appropriate for resisting compressive forces in truss members.

The holistic fabrication of a truss structure through multi-material additive manufacturing necessitates several critical steps. First, tensile and compressive members within the structure must be identified. Subsequently, the local segment orientation is determined based on the stress distribution. Utilizing the stress condition, a toolpath is designed, and appropriate materials are assigned for each member. The material data is then integrated into the toolpath and converted into G-code, guiding the 3D printer during manufacturing. The 3D printer must be able to process and deposit multiple materials simultaneously, ensuring accurate and efficient fabrication of the truss structure. We use the classic Pratt truss as a case study to demonstrate our approach.

8.3.4 Translating Stress Fields into Continuous Toolpath Design

The mechanical property of a parallel print is similar to that of wood lumbers: the longitudinal direction takes more tensile forces, while the tangential direction takes

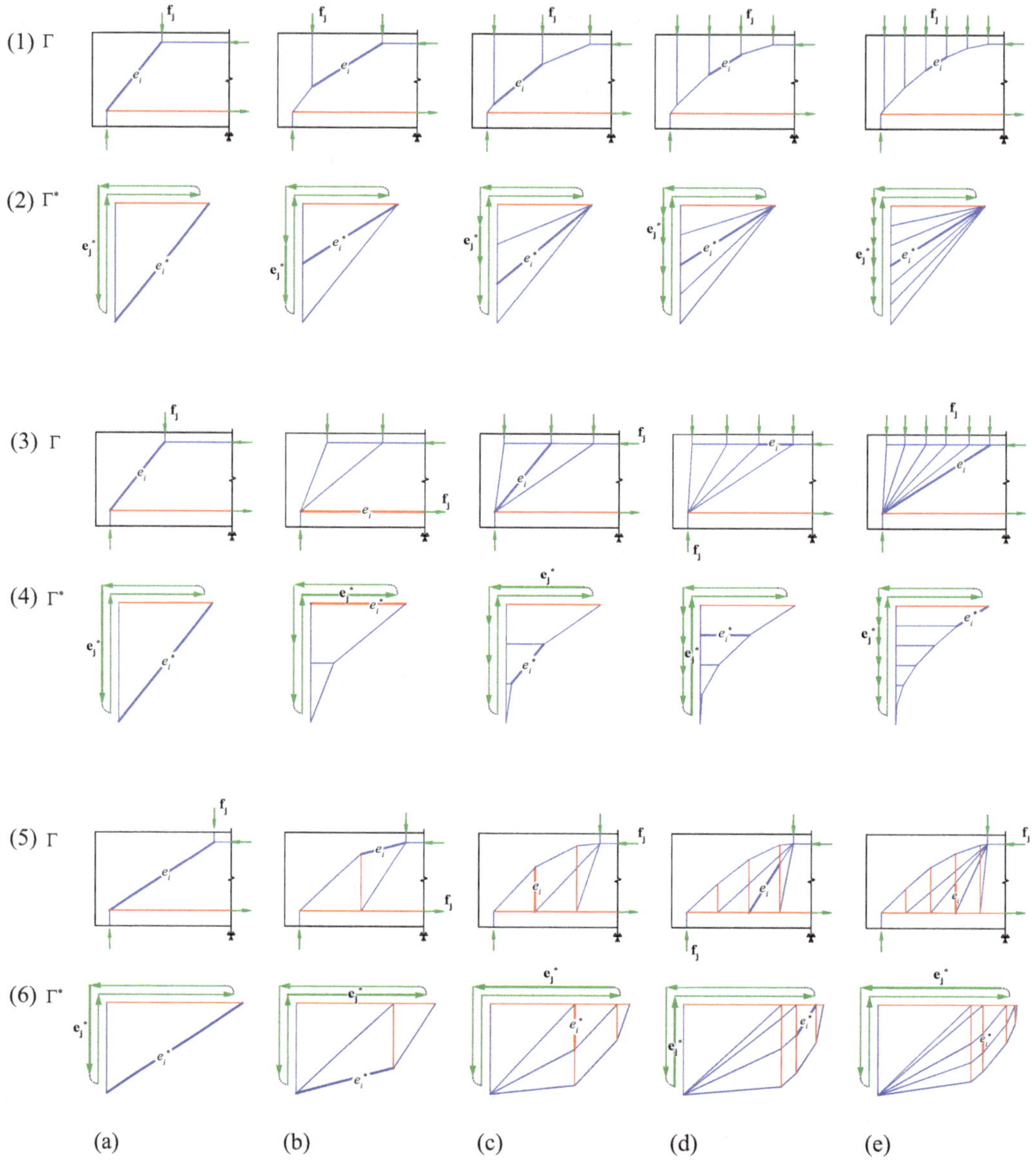

Figure 8.11 Step-wise subdivision of external load through arching/funicular action(1a)–(1e) and fan-shaped behavior (3a)–(3e); (5a)–(5e) step-wise addition of stirrups (combination of the funicular and fan-shaped behavior); (2a)–(2e), (4a)–(4e), and (6a)–(6e) step-wise modifications of their corresponding force diagrams. Adapted with permission.

more compression. To utilize such knowledge, the local orientation of the toolpaths is determined by the type of stress (Figure 8.12a,b): tension or compression. When a member of the truss takes tensile forces, its toolpaths are parallel to the alignment of the member. Whereas, if it takes compressive forces, its toolpaths are parallel to its

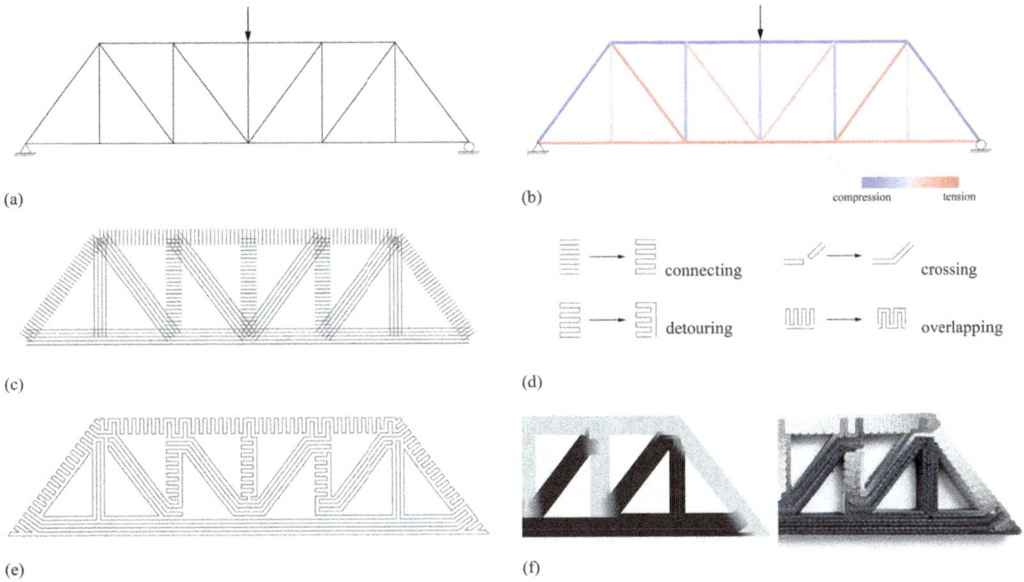

Figure 8.12 Multi-material printed truss toolpath generation process: (a) input curves; (b) axial force conditions solved by Karamba3D; (c) raw curves according to the axial forces; (d) postprocessing methods to form a continuous toolpath; (e) the final toolpath; and (f) left: multi-material visualization of the print, right: multi-material printing

cross-section. Figure 8.12c shows this step of generating raw toolpaths as discontinuous lines.

A continuous toolpath for 3D printing significantly saves production time and avoids underfilling and overfilling to enhance the bound.[226] Postprocessing (Figure 8.12d) is introduced to make the toolpath continuous. The curves are connected, primarily in a zig-zag fashion, forming several loops. The loops are then merged by crossing and detouring. Finally, we modify the toolpaths at the joints to create overlapping for better bounding.

Figure 8.12e shows the continuous planar toolpath. Multi-material information is then embedded to print the tensile part using carbon fiber filament and the compressive part using regular PLA filament. Figure 8.12f shows the visualization of the material-coded toolpath with the printed object.

To validate the improvement in the mechanical performance of the components fabricated using our proposed method, we prepared two truss samples (Figure 8.13) to conduct a three-point bending test. Each sample was printed with a layer height of 0.8 mm and an extrusion bandwidth of 2 mm. The dimensions of all samples were 250 mm in length, 65 mm in height, and 16 mm in thickness.

Calculated from toolpath length, the multi-material experimental specimen contains 52.5% of the toolpath that withstands tensional stress, while another 47.5% of the toolpath bears compression stress. The multi-material experimental specimen is 166 g, while the single-material control group specimen is 179 g.

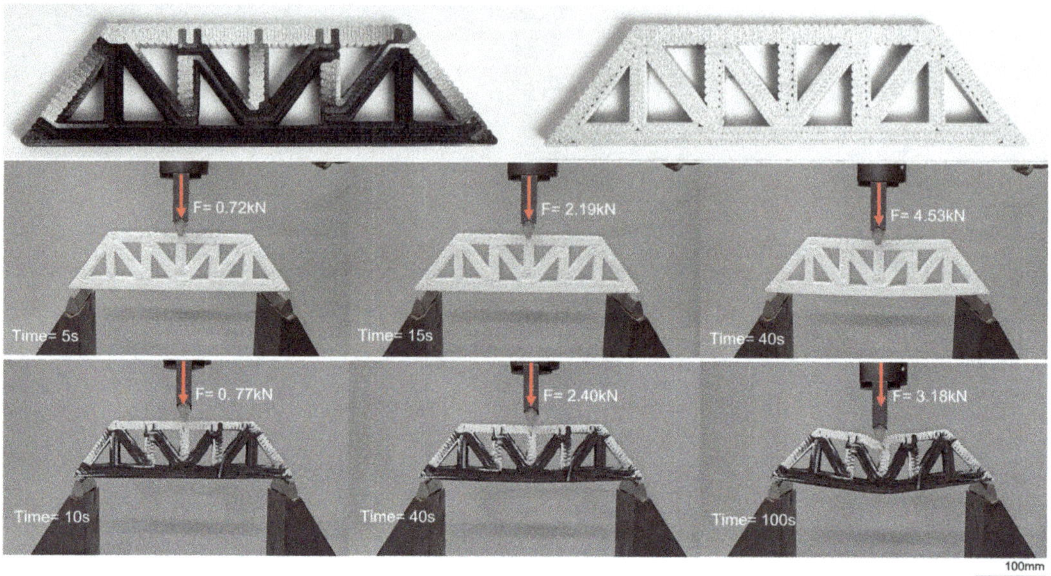

Figure 8.13 Top: Two samples for the three-point bending test. Left: multi-material experimental specimen printed with HATCHBOX white PLA and OVERTURE Carbon Fiber reinforced PLA. Right: specimen printed with HATCHBOX white PLA only. Bottom: a three-point bending test for examining structures' fracture behavior.

The three-point bending test showed that the single-material control specimen yielded under a load of 0.83 kN (Figure 8.13). In contrast, the multi-material experimental specimen reaches the yield point at 1.16 kN, and one of the compressive members begins to buckle, causing overall plastic deformation in the truss. As expected, the multi-material experimental specimen exhibits a higher stiffness than the single-material specimen. For further discussion on this topic, we invite the reader to visit the article by Teng et al.[225]

8.4 From Cellular to Shellular Systems and Their Applications in Architected Materials

The effective properties of low-density cellular solids heavily rely on their underlying architecture; thus, a variety of explicit and implicit techniques have been proposed for designing cellular geometries.[183,227–229,229] However, most of these techniques fail to present a correlation between architecture, internal forces, and effective mechanical/structural properties. Polyhedral graphic statics can serve as an alternative design strategy based on the static equilibrium of forces, the equilibrium of polyhedral frames, and the reciprocity of form and force. This approach presents a geometric relationship among the truss system architecture, topological dual, and equilibrium of forces based on polyhedral graphic statics. Consequently, we can make periodic strut-based cellular architectures and manipulate them to construct shell-based

(shellular) cellular solids with distinctive mechanical properties. Validated by experimental tests on 3D-printed cellular polyhedral materials, we can systematically design cellular mechanical metamaterials by alternative strut materialization schemes and rational addition of bracing struts to demonstrate properties ranging from bending- to stretching-dominated ones, realize meta fluidic behavior in architected solids, or create novel hybrid shellulars.

8.4.1 Architecture of Cellular Materials

Cellular solids can be defined as an assembly of random or periodic unit cells comprised of straight/curved struts, flat/curved plates, and/or solid bulks that cannot be considered struts or plates packed together to fill a predefined space. Their growing range of utilization spans numerous applications, such as in lightweight load-bearing structures, energy absorbers, tissue scaffolds, heat exchangers, and batteries, to name a few.[230,231] Developing unit cells that can be tesselated in 2D or 3D arrays is a common approach for designing periodic cellular solids.[232] Based on the geometrical configurations of the unit cell, 3D cellular materials are typically classified as strut-, polyhedron-, and shell-based or shellular architectures,[118,183,233,234] even though a unit cell can be a hybrid of these classes.

The multifunctional properties of cellular materials are mainly governed by the architectural and topological features of their nano/micro/mesostructure and the force flow within their elements rather than merely depending on the properties of their constituent materials.[228,230,235] Designing the architecture of cellular solids extends the property space of known materials, showing how their properties, for example, density, Young's modulus, and yield stress, correlate with each other.[236] As an example, by designing the architecture of nickel micro-lattices, researchers have extended the property space of nickel for densities, elastic modulus, and strengths.[230,237] Hence, to systematically enhance the material property space, a design methodology must be used to precisely control the geometry, topology, and force flow within the system.

8.4.2 Stretching versus Bending Dominated Cellular Solids

The mechanical behavior of cellular materials under uni-axial tension spans from bending- to stretching-dominated; within the former, members mostly undergo bending and transversal loading and sustain little to no axial loads, while the opposite holds true for the latter where most of the microstructure experiences axial tension/compression; this results in much higher stiffness/strength-to-weight ratios for stretching-dominated cellular solids.[118,238,239] Under this description, a cellular solid such as the simple cubic truss can act as stretching-dominated along one direction while being bending-dominated along others; nevertheless, the accepted definition generally includes any arbitrary tensile/compressive loading along any direction, rendering the simple cubic truss as a bending-dominated lattice. Higher stiffness and strength of stretching-dominated cellular solids might suggest that they are more weight efficient; however, the structural efficiency is not limited to the concept of the maximum load per weight and instead is based on the specific performance requirements,[240] which might include other functionalities such as mechanical unfeelability, acoustic, and auxetic properties that do not necessarily correspond to higher stiffness or strength. Consequently, stretching-dominated cellular materials are used as load-bearing structures, while bending-dominated cellular solids find potential applications in energy absorption and vibration insulation.[241]

It has been shown that at small relative densities $\bar{\rho}$ (defined as the measured density ρ of a mono-material cellular solid divided by that of its constituent material ρ_s), mechanical properties such as effective moduli of stretching-dominated strut-based cellular materials correlate linearly with $\bar{\rho}$, whereas in the bending-dominated type, they decrease more rapidly as $\bar{\rho}$ lessens. For instance, the elastic modulus of 3D open-cell bending-dominated foams generally correlates with ρ^2.[238,242] As a result, the relationship between $\bar{\rho}$ and effective Young's modulus $\bar{E}$ is often used as an indicator of the degree of stretching dominancy of a cellular material; the closer to linear correlation, the more stretching-dominated.

8.4.3 Designing Low-density Cellular Materials by Subdividing a Polyhedral Force Diagram

The methods of polyhedral graphic statics can be applied in designing a force diagram, reciprocal to the geometry of the periodic unit cell. This approach provides a new route for controlling the topology of cellular architectures. Various subdivisions of a force diagram result in alternative topologically different structural forms for a given boundary condition, as shown extensively in Chapters 1 to 7. Moreover, we can design periodic cellular and shellular unit cells using polyhedral graphic statics.

Increasing the number of subdivisions in the force diagram results in a form diagram with shorter edges. In Figure 8.14, a tetrahedron as a global force polyhedron is reciprocal to a node with four applied loads, two upwards and two downwards, in the form diagram. In this example, the force diagram has two skew edges, $l_1^\dagger$, red, and $l_2^\dagger$, black, that can be subdivided recursively to make smaller tetrahedrons. In this process, the skew edges $l_1^\dagger$ and $l_2^\dagger$ are divided into an equal number of segments such that each segment of the former will make a tetrahedron with its associated segment in the latter (Figure 8.14b).[120] This particular type of subdivision is named *anticlastic subdivision*, and the form diagram resulting from this subdivision is a polyhedral surface with an anticlastic curvature. This topic was covered extensively in Chapter 4. In this type of subdivision, the nodal cells are tetrahedrons, resulting in reciprocal, valency-four vertices in the form diagram (Figure 8.14c,d).

The Role of Labyrinths in Design

As shown in Figure 8.14d, the curvature axes in an anticlastic surface, l_1 and l_2, are parallel to the subdivision axes in the reciprocal force diagrams, $l_1^\dagger$ and $l_2^\dagger$. These axes are referred to as *labyrinths* in the shellular material design.

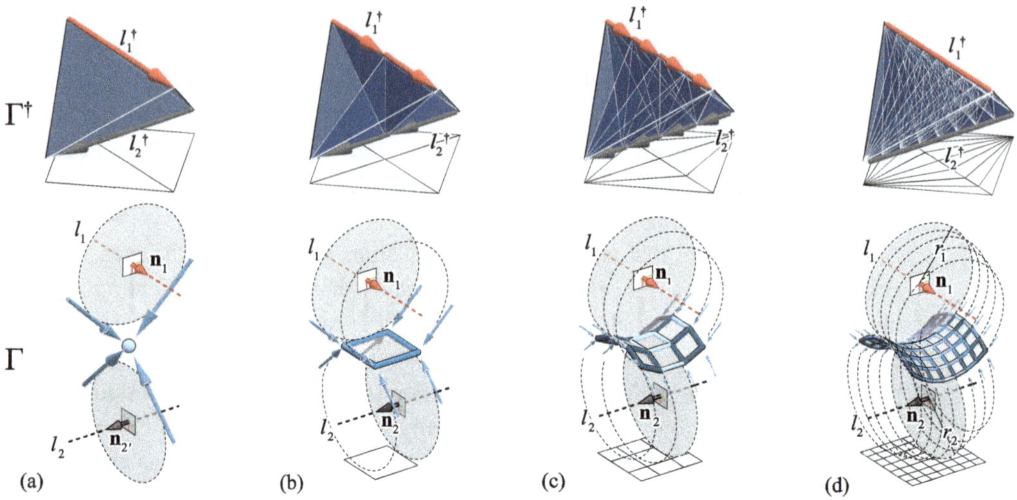

Figure 8.14 (a) to (d) The process of subdividing the labyrinths and generating a shellular anticlastic frame. Adapted with permission.

An anticlastic surface geometry of Figure 8.14d divides the 3D space into two subspaces separated by the shell geometry. These subspaces can also be identified by their labyrinths representing the connectivity graph of these subspaces in complex shellular geometries. The labyrinths in anticlastic geometries are two intertwined graphs whose edges are always in a *skew* position with each other. The two labyrinths interpenetrate each other while the anticlastic surface is in between (Figure 8.14d). The geometry of the surface changes when the angle of these labyrinths changes.[119] Interestingly, the labyrinths in the form and force diagrams are related.

Translating a Strut-based Cellular to a Shellular Unit Cell

A cuboid force diagram can be subdivided into anticlastic geometry and aggregated in three directions to make a shellular material. Consider the cube of Figure 8.15; the cube is decomposed into nonoverlapping tetrahedrons. One can select two skew edges from each of these tetrahedrons to create two intertwined labyrinth graphs that are separated. This process continues until all the cells are parsed, and their labyrinth edges are determined. Finding skew edges results in three different pairs of labyrinth edges in each tetrahedron and thus three geometrically different labyrinth graphs (see Chapter 4). If the labyrinths are subdivided into tetrahedrons in the force diagram, it gives shellular geometry in the form diagram.

Approximating Well-known Minimal Surfaces Using PGS

Figure 8.16 shows subdividing a cube to make a reciprocal Schwartz P unit cell. Well-known minimal surfaces include two sets of labyrinths that can be used to approximate their geometry. By changing the level of the anticlastic subdivision, one can control the resolution of the minimal surface from a strut-based cellular to a shellular unit cell.

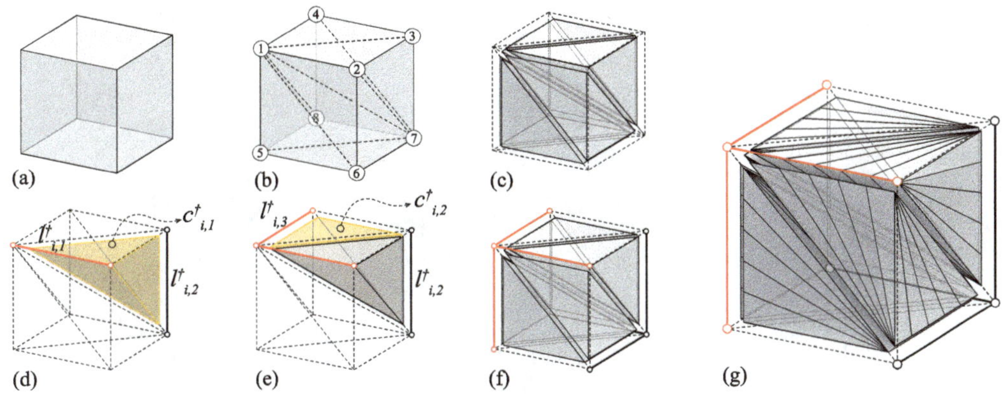

Figure 8.15 (a) to (g) The process of starting from a single cube and tetrahedralizing the cube, identifying the skew edges for each tetrahedron sequentially and subdividing the skew edges. Adapted with permission.

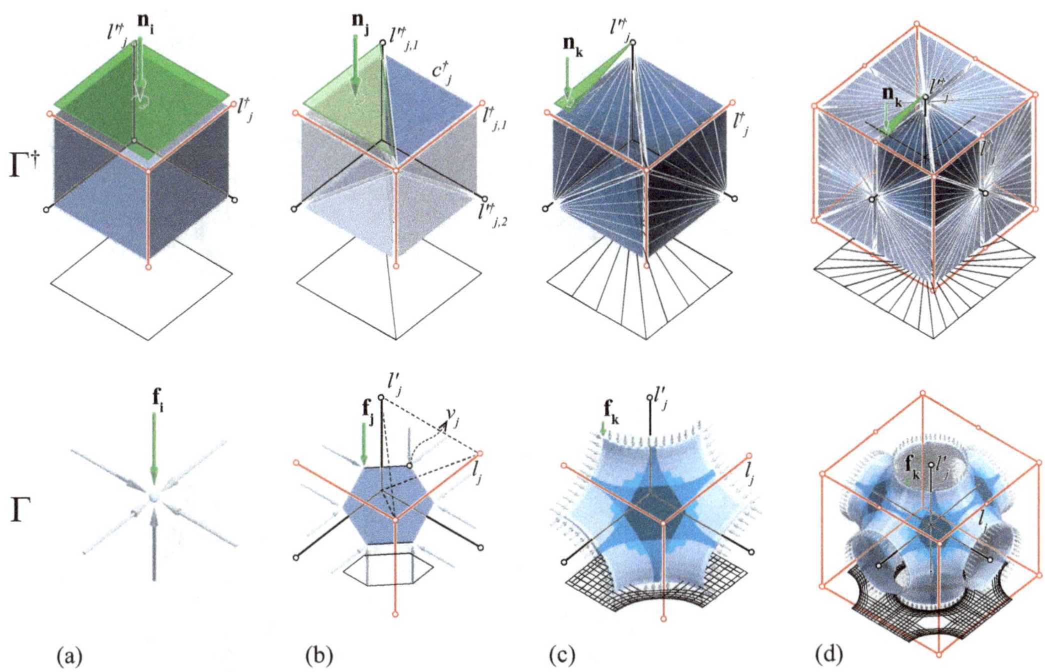

Figure 8.16 (a) to (d) Subdividing a cube to get the approximate Schwartz P unit cell.

Figures 8.17 and 8.18 illustrate different examples of shellular unit cells designed based on this method. In each example, the tetrahedralization of the one-eighth section of a force diagram and the force and form diagram of a shellular funicular unit cell have been shown. Each design involves mirroring and aggregating cuboid designs to establish force diagrams reciprocal to alternative shellular unit cells for material design.

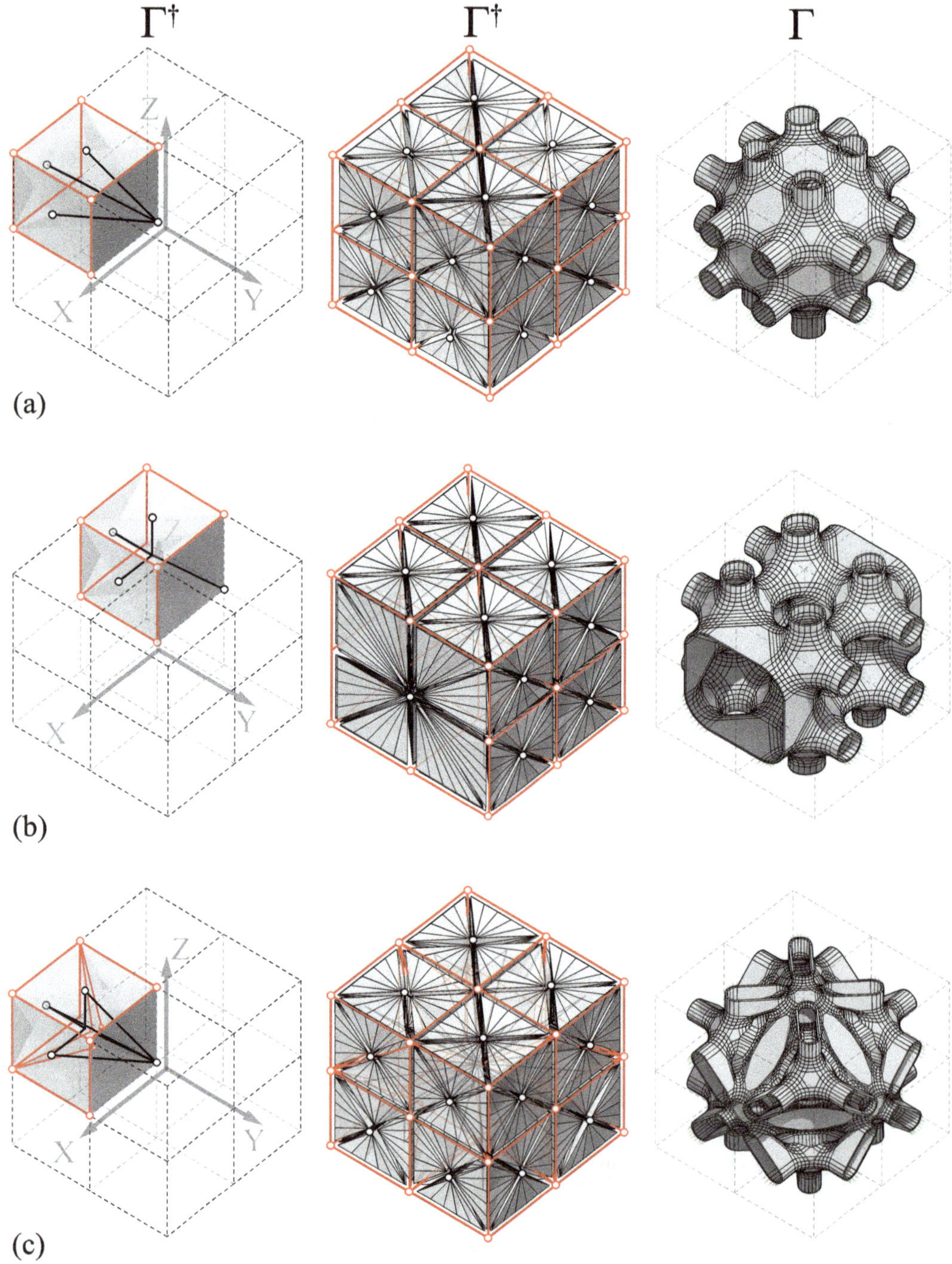

Figure 8.17 (a) to (c) Multiple *Schwarz P*-based hybrid unit cell designs. Adapted with permission.

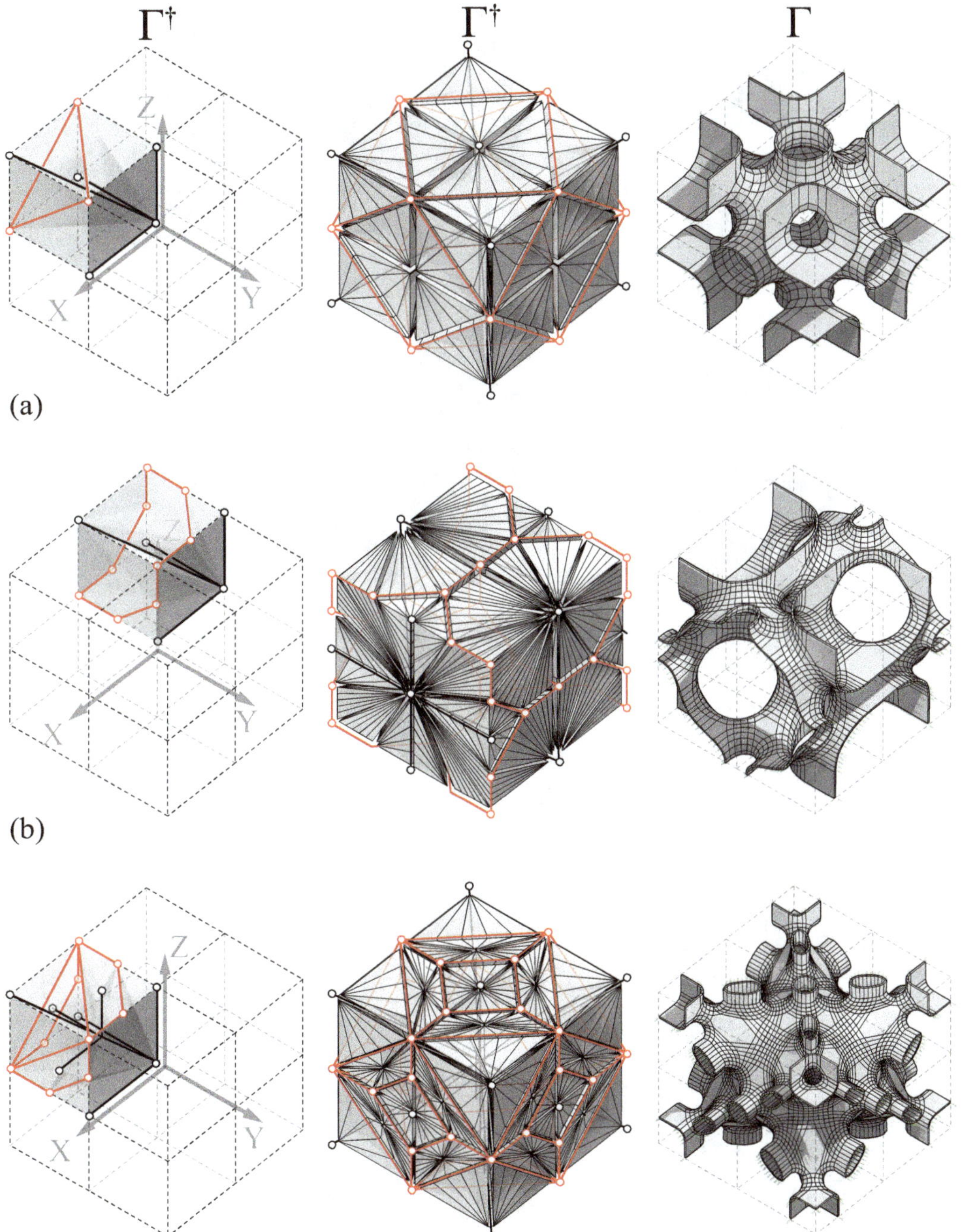

Figure 8.18 (a) to (c) Different examples of Schwartz *P* and Schwartz *W* hybrid unit cell designs. Adapted with permission.

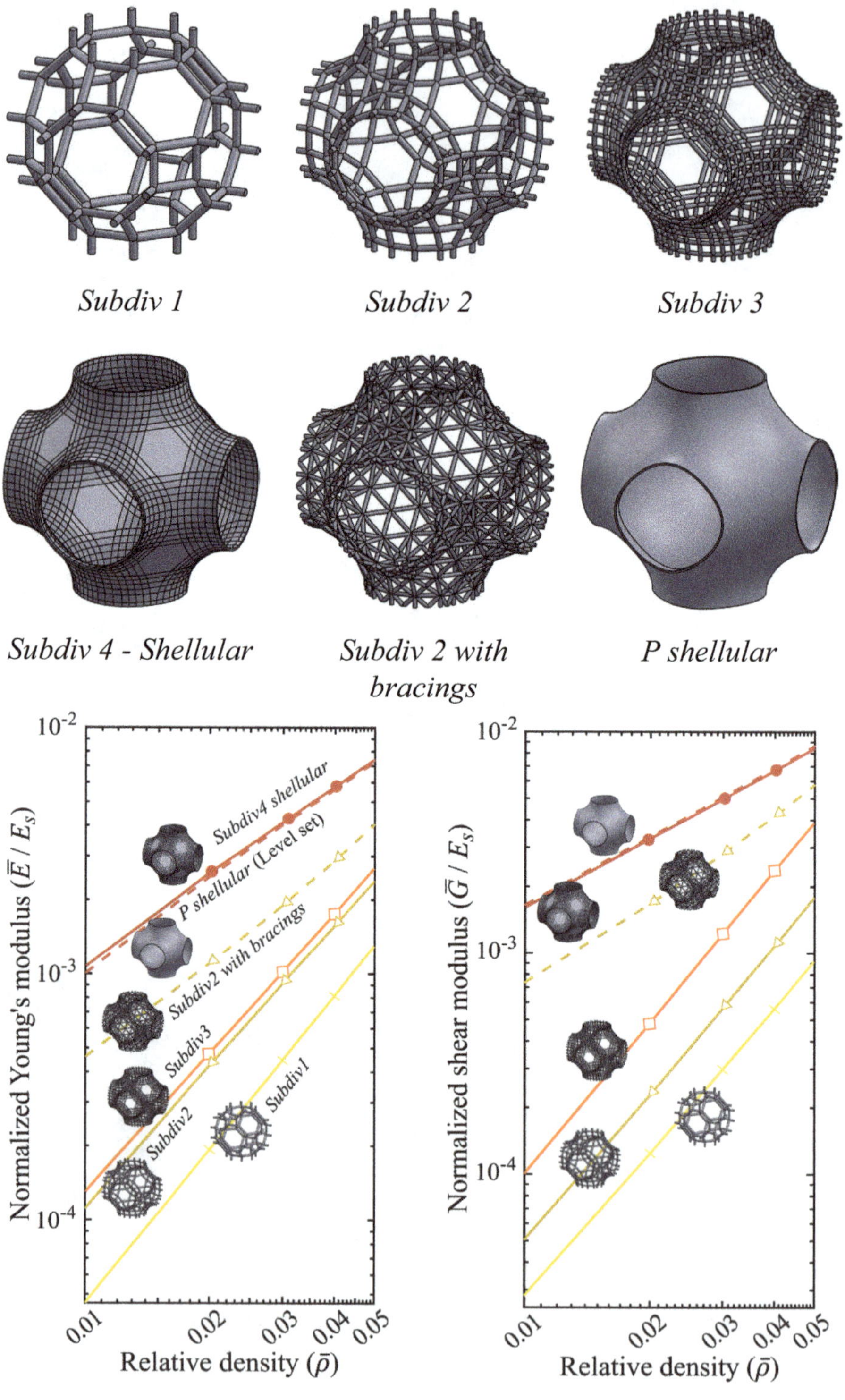

Figure 8.19 Effective mechanical properties of a family of cellular materials resembling *Schwarz P*. (a) Young's modulus, (b) Shear modulus.

8.4.4 Architected Cellular Materials Mimicking *Schwarz P*

We constructed a family of truss-like cellular and shellular architectures resembling *Schwarz P* using the subdivision techniques discussed earlier. We compared their normalized effective Young's and shear moduli ($\bar{E}/E_s$ and $\bar{G}/E_s$) and Poisson's ratios ($\bar{\nu}$) in Figure 8.19, determined by conducting a multiscale homogenization technique.[243–245]

As shown, the mechanical properties of the designed shellular based on *subdivision 4* match those of the level-set-based *P shellular*. It is also expected that increasing the subdivision number shall decrease the differences between the two since the subdivision scheme here acts as a structural form refinement that controls the smoothness of the shell. Although the same trend can be observed for the elastic moduli of the truss-like cellular counterparts and the cellular materials get stiffer by increasing the subdivision number from 1 to 3, their effective mechanical properties cannot reach those of the *P shellular* as the selected subdivision scheme cannot cover the whole surface.

Furthermore, struts are always parallel to one of the coordinate planes using this subdivision scheme; hence, even with several subdivisions, each small polygonal part of the cell does not behave as a 2D isotropic material, and unlike *P shellular*, the local mechanical response of the mesh-like network is anisotropic throughout the cell. It is possible to reduce the in-plane mechanical anisotropy of a polygon by introducing diagonal members, referred to here as bracings. Similarly, bracings can also be added to the discussed empty hexagons. It is worth mentioning that these bracings are not part of the original idealized design methodology. These bracings are being introduced to showcase how reducing the number of inextensional mechanisms and increasing the number of states of self-stress of the originally designed ideal truss can stiffen the cellular material counterpart and shift its properties toward those of the *P shellular*.

Considering that introducing the bracings invalidates some of the original design considerations, including the equal stresses in the struts that are used to determine the diameter ratios of the constitutive struts, bracings are added while all struts and bracings have equal diameters as presented in Figure 8.19a,b. Compared to the original *Subdiv2*, *Subdiv2 with bracings* is considerably stiffer and its effective E, G, and ν are significantly closer to the *P shellular*.

While Young's and shear moduli of the three designed strut-based materials (*Subdiv1*, *2* and *3*) are generally orders of magnitude smaller than those of the *P shellular*, their bulk moduli K are relatively closer to the *P shellular*, which originates from the assumed triaxial loading as their design boundary conditions. In this regard, the idealized truss architectures are all mechanisms; hence, small $\bar{E}$ and $\bar{G}$; however, they are statically determinate under a periodic triaxial loading, which results in high $\bar{K}$ values. Comparing *P shellular* with the truss-like architectures without bracings at $\bar{\rho} = 0.01$, $\bar{E}$ of the shellular cell is more than ten times greater; however, its $\bar{K}$ is only about two times greater. This good performance of the three subdivision cells under triaxial loading can be correlated to their Poisson's ratios that are close to 0.5. Using the relation between K, E, and ν of the orthotropic materials with cubic rotational symmetries that are given by $\bar{K} = \bar{E}/[3(1 - 2\bar{\nu})]$, a Poisson's ratio close to 0.5 creates a very small denominator, which significantly amplifies the right-hand-side division and consequently results in a high bulk modulus even when Young's modulus is small.

8.4.5 Concluding Remarks

As discussed, polyhedral graphic statics provides a novel systematic approach for the topological design of lightweight architected strut- or shell-based cellular materials. The proposed methodology provides a synergistic correlation between the internal force distribution, geometry, and external boundary conditions, paving the way toward rational designs of periodic or nonperiodic funicular truss architectures. The successive distribution of the internal forces through numerous struts with progressively shorter lengths can transform a bending-dominated strut-based funicular structure into a shellular architecture with significantly improved mechanical properties. Furthermore, the relationships between the underlying labyrinths of the form and force diagrams of the shellular structures are used to devise unprecedented shell-based architectures, either by generating new labyrinths or by deforming and/or combining existing ones.

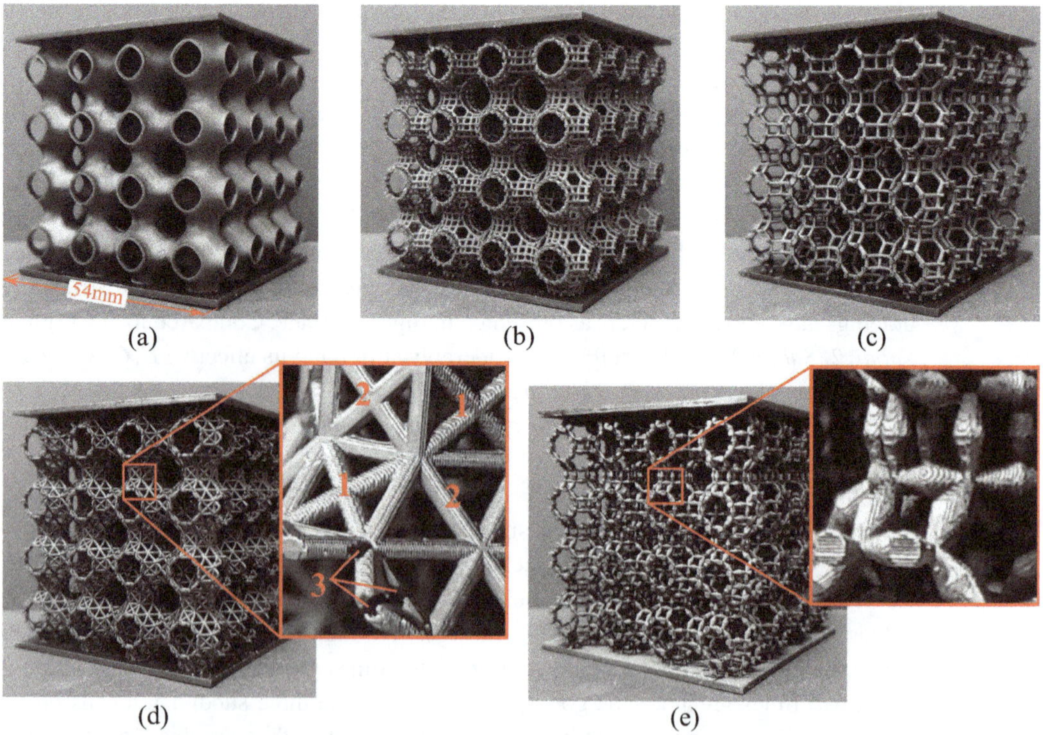

Figure 8.20 3D-printed cellular samples: (a) *P Shellular*, (b) *Subdiv1*, (c) *Subdiv2*, (d) *Subdiv1 with bracing*, and (e) *Subdiv1 with double cone struts*. Adapted with permission.

While one can choose any arbitrary criterion to control the shape and cross-section of the struts in a funicular architecture, cylindrical struts based on constant stress or diameter and double-cone struts are adopted to materialize a selected set of designed periodic cells. Additionally, introducing bracing members to the sub-architecture of a funicular cell is discussed as an alternative intermediate step in transforming truss cells to shellulars. The wide range of attained effective mechanical properties of the selected cellular structures with relative densities lower than 0.05 demonstrates the robustness of the proposed design methodology and materialization schemes in developing application-specific lightweight cellular materials spanning from soft and bending-dominated to stiff and stretching-dominated architected materials for structural applications. The developed methodology can serve as a framework for developing novel shellulars with hybrid topologies that present tuned mechanical properties with potential applications in the engineering of tissue scaffolds or for designing materials with exotic meta-fluid behavior. Ultimately, the numerically obtained mechanical properties are compared with experimental tests on 3D-printed cellular samples (Figure 8.20a–e). The observed agreement between the two further endorses the presented design and materialization techniques and highlights their applications in the design of cellular and shellular funicular materials.

8.5 Designing Repairable and Self-healing Metallic Structures

8.5.1 A Paradigm Shift

The current paradigm for designing load-bearing structures is heavily focused on preventing or delaying crack growth and fracture. Designers and engineers conduct probabilistic and computational analyses and empirical investigations to consider the microstructure of structural materials, assess the distribution of defects and inclusions, account for possible loading scenarios, and predict useful life and the likelihood of structural failure. Additionally, safety factors are used to guard against brittle or fatigue failure, resulting in structures with greater volume and mass than strictly necessary.[246,247]

Shifting the engineering design paradigm from one focused on failure to one focused on continuous repair and healing would result in materials and structures that not only experience damage and still sustainably recover their original properties and functions but also adjust their physical properties in response to user input or environmental stimulus to, for instance, avoid future damage. Most importantly, this is achieved without having to decommission parts, manufacture replacements, or even, if warranted, employ any human or robotic intervention. Polyhedral graphic statics can accelerate this paradigm shift by designing structures with optimized geometry, mechanical behavior, and failure characteristics. Coupled with metal additive manufacturing and a recently developed repair technique termed "electrochemical healing,"[248,249] polyhedral graphic statics could enable highly complex cellular/shellular funicular metallic structures that can be effectively healed at room temperature and

restored to their original mechanical strength and function with minimal material and energy input.

8.5.2 Healing Metals using Electrochemistry

While the development of self-healing polymers has seen significant success through a variety of physical and chemical mechanisms,[250–252] metals have proven more difficult to repair or endow with self-healing capacity.[253] Since metallic bonds are stronger than the supramolecular bonds (e.g., hydrogen bonds) that hold polymer networks, a higher energy input is required to induce transport or phase transition in metals. At ambient conditions (~20°C, 1 bar), metal atoms in a solid metallic phase possess very slow transport rates, with their diffusion coefficients ranging from 10^{-45} to $10^{-35} m^2/s$.[254] Thus, high-temperature techniques such as welding and brazing have long been used to repair metals. However, the increasing ubiquity of digital manufacturing and "unweldable" alloys, as well as the integration of metals with polymers and electronics, call for radically different repair approaches.

One such approach is electrodeposition (or electrochemical plating) (Figure 8.21a). This technique employs an electric potential gradient (i.e., voltage) that is either spontaneous or caused by an applied electric field to induce ion transport between two electrodes (a cathode and an anode) immersed in a liquid or solid electrolyte. On the cathode surface, electrons combine with nearby ions, which are reduced to metal atoms that agglomerate on the surface to form a solid metal deposit. For example, nickel ions, Ni^{2+}, are reduced and become nickel atoms as follows: $Ni^{2+}(aq) + 2e^{-} \rightarrow Ni(s)$. The cathode reaction may involve a complicated multi-step reduction process or undesirable side reactions (e.g., hydrogen evolution). An oxidation reaction occurs at the anode, where ions are released in the electrolyte and electrons are released into the

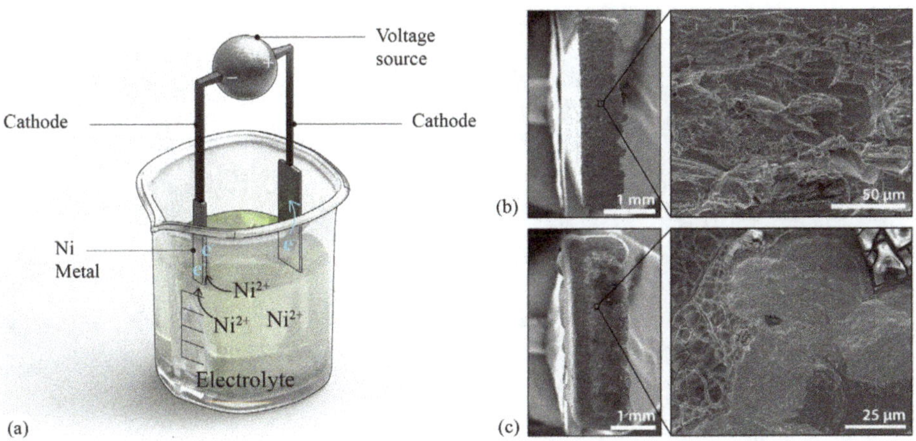

Figure 8.21 (a) Illustration of an electrochemical cell for nickel electrodeposition and SEM image of a cross-section of half-dogbone Ni 200 sample before (b) and after healing (c). Reproduced under a Creative Commons License.[255]

anode. During nickel electrodeposition, if a nickel anode is present, the oxidation reaction is the reverse of the reduction reaction shown earlier, and there is no net change in the concentration of nickel ions in the electrolyte (Figure 8.21b,c).

Employing electrodeposition to repair metals is attractive because metal atoms are transported as ions dissolved in a solvent, as opposed to their reduced metallic form, which allows them to migrate under the influence of an electric field and increases their diffusion coefficients by about 30 orders of magnitude. The diffusivity of nickel, for example, increases from about $2.10^{-41} m^2/s$ for Ni atoms in solid metal to about $1.10^{-9} m^2/s$ for Ni^{2+} ions in water. Thus, an electrochemical healing approach can overcome the lagging kinetics of metals without resorting to high-temperature heat.

In a recent work,[249] we showed that room-temperature area-selective nickel electrodeposition can enable 100% recovery of tensile strength after full rupture in a variety of metallic alloys (a 99% nickel alloy, a low-carbon steel, and two "unweldable" aluminum alloys). We achieved this via a mechanics-based theoretical model that guided the tailoring of geometric and electrochemical process parameters to optimize the recovery of tensile strength with minimum energy and material expenditure (Figure 8.22a,b). Prior work on the electrochemical joining of metal parts required customized multi-step processes with acid pre-treatments and plating in different electrolyte chemistries to achieve high adhesion of plated nickel to different alloys,[256–259] our approach overcame this limitation. It enabled reliable restoration of strength in various alloys using only one well-studied, widely used, and relatively inexpensive aqueous electrolyte: nickel sulfamate.

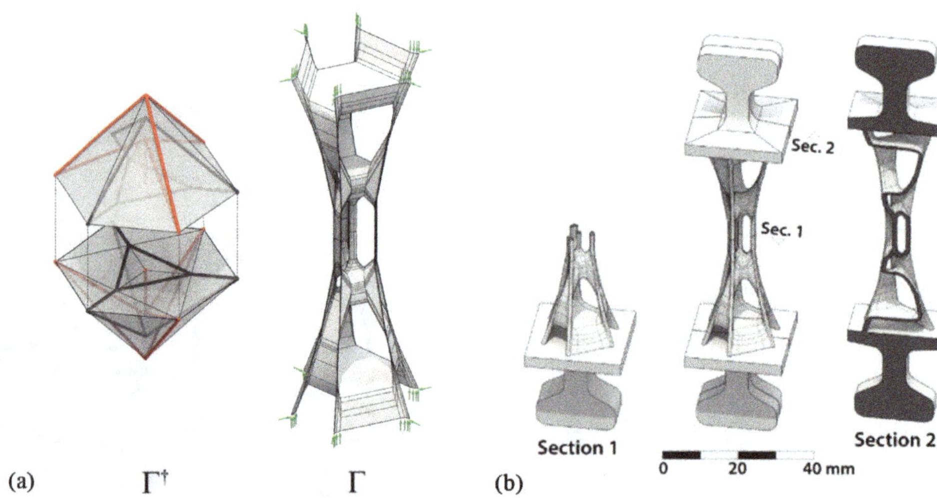

Figure 8.22 The shellular structure illustrated through (a) force and form diagrams and (b) the tensile testing connections and the cross-section and the longitudinal section of the sample. Adapted with permission.[249]

8.5.3 Healing a 3D-Printed Metallic Shellular Structure Designed by Polyhedral Graphic Statics

In addition to effectively repairing a variety of alloys, electrochemical healing enabled tensile strength recovery in 3D-printed metal structures with complex geometries, which would be impossible to repair using conventional welding. Using polyhedral graphic statics as a design method and selective laser melting of an aluminum-silicon-magnesium alloy ($AlSi_{10}Mg_{0.35}$) for 3D printing, we fabricated a funicular shellular structure that carries internal forces axially with no transverse stress and is designed to fail at its central section, which includes three outer struts and a middle strut that is difficult to access, as shown in Figure 8.23. This structure is a type of funicular polyhedral frame designed using the technique of topological subdivision, which generates two-manifold anticlastic surface-based structures with cross-sections normal to the direction of the internal force flow under the external loadings. Owing to its geometric complexity and the inaccessibility of the middle strut, this structure would not be effectively repaired by conventional welding. It would, thus, need to be reprinted or replaced.

The structure was conformally coated with poly(chloro-p-xylylene), also known as Parylene C, a vapor-deposited polymer with excellent barrier properties, chemical stability, and dielectric strength.[260] Using a sacrificial mask, the central four-strut section was left uncoated to ensure nickel was electrodeposited only in this central section during healing. We loaded the structure in tension until the four struts fractured, healed it, then fractured it in tension again to assess the recovery of mechanical properties

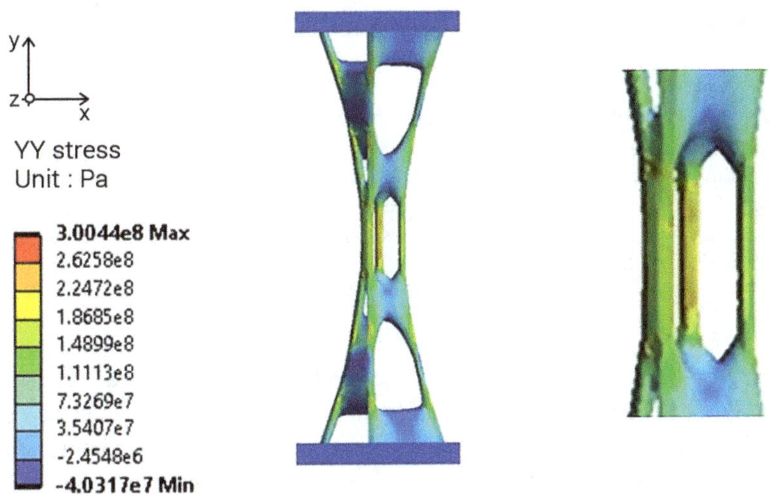

(a) (b)

Figure 8.23 (a) YY-stress distribution in the structure due to tensile loading along the Y-axis as calculated by finite element analysis and (b) the zooming detail of the central column. Adapted with permission.[249]

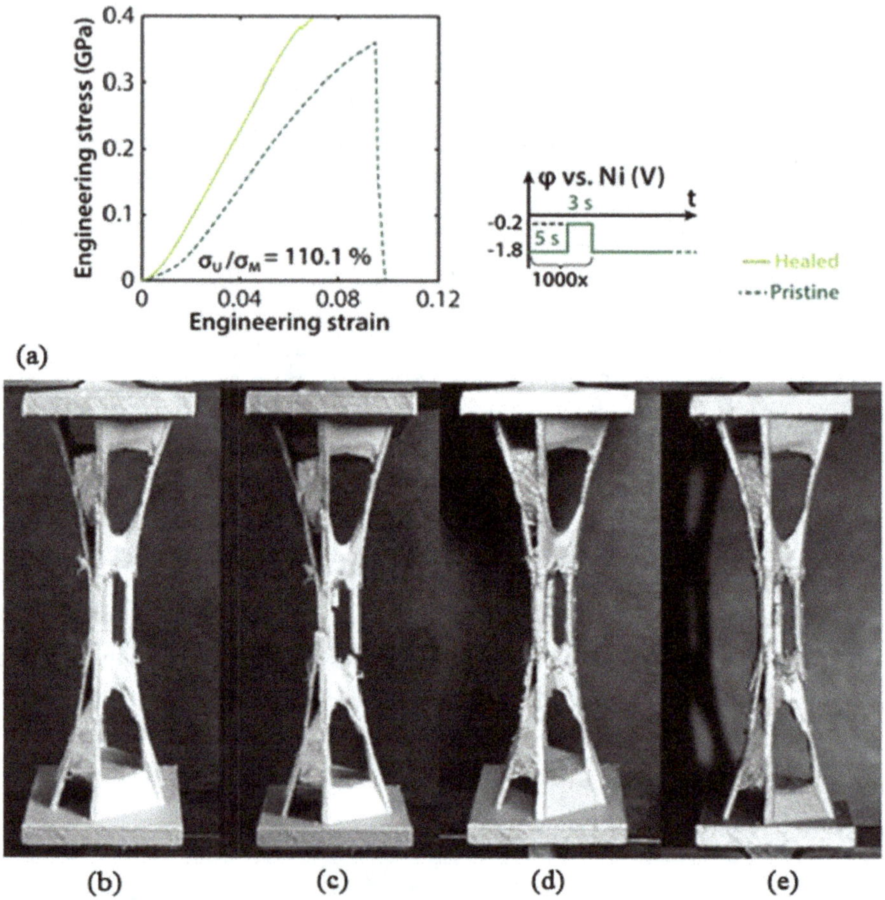

Figure 8.24 (a) Electrochemical healing using a process that combined pulsed and potentiostatic plating resulted in full restoration of tensile strength, as shown in a comparison of stress-strain data in the pristine state and after healing. Photographs of the 3D-printed metallic shellular structure (b) in its pristine form, (c) after tensile fracture, (d) after healing, and (e) after a second fracture. Notice the electrodeposited nickel on the four-strut section of the structure in (d) and (e). Adapted with permission.[249]

(Figure 8.24a–d). We immersed the fractured structure in a nickel sulfamate electrolyte to begin healing. In contrast to the healing process that we used to heal the alloys mentioned earlier (constant voltage or potentiostatic plating at −1.8V vs. *Ni*), we used an enhanced process in which we apply 1,000 cycles of pulsed plating (one cycle consists of plating at −1.8V vs. *Ni* for 5 seconds followed by −0.2V vs. Ni for 3 seconds) before proceeding with potentiostatic plating at −1.8V vs. *Ni*. We have shown in an earlier work that pulsed plating mitigates diffusion-limited electrodeposition, which improves nickel adhesion to the healed metals and prevents void formation at the fracture gaps.[261] Applying this enhanced healing process (pulsed + potentiostatic) to deliver a total charge input of 1,200 mAh enabled the recovery of 110.1% of the pristine tensile strength (Figure 8.24e). By maximizing nickel density at the fracture gap, this enhanced process can be attractive in cases where there are constraints either on energy (e.g., a battery-powered robot) or on the metal area available for plating.

This proposed room-temperature electrochemical healing approach could potentially enable a universal and scalable approach to repair fractured 3D-printed metal parts. Electrochemical healing is also a promising approach to enable exciting technologies in electrically controlled autonomous and preemptive repair, as well as mitigate the high environmental and energetic costs associated with the current approach of recycling and reprinting fractured parts. Notably, the effective repair of metal parts that extends their operational lifetimes could reduce demand for metal mining and refining, thus potentially mitigating the carbon emissions intensity of structural metals by up to 400 kg of CO_2 per ton of metal.[262]

8.5.4 Toward a Healing-Focused Engineering Design Paradigm

Polyhedral graphic statics may become an advantageous design tool within a new engineering design paradigm focused on continuous healing, rather than failure prevention. Coupled with ongoing advancements in metal 3D printing, this design tool can pave the way for highly complex shellular funicular metallic structures that provide enhanced physical properties (i.e., strength-to-weight ratio, fracture toughness, thermal transport rate), reduce the rates of resource usage and waste, and improve the aesthetics of the built environment.

More importantly, the design of these structures can be readily tailored to facilitate healing when damage or fracture occurs. Since these structures are designed from a reciprocal compression/tension-only anticlastic shell with transverse and bending stresses eliminated by design, the structure can be designed with a priori knowledge of the loading and deformation behavior and of the location of crack initiation. Thus, the material required for healing (e.g., electrolyte, nickel anode) can be pre-positioned at or swiftly transported to the location of crack initiation, which reduces the energy and material mass needed for effective healing.

Prior work[248] on healing cellular nickel has shown that the application of a conformal passivating polymer coating, with a failure strain close to but lower than the failure strain of the underlying metal, can enable fracture sensing and healing initiation without the need for visual inspection. This would be particularly beneficial in complex structures with high tortuosity or concealed members. In practical terms, the fracture of the polymer coating under tensile loading would expose the underlying metal to the electrolyte, which would lead to a change in open-circuit voltage, a change that could be used as a signal to initiate, with or without human intervention, the electrochemical healing process. Depending on the deformation behavior of the coating, healing can be initiated after a metal fracture or, interestingly, after plastic deformation to pre-emptively strengthen local stress concentrations.[248] Designing metal structures and parts with the expressed intention of facilitating pre-emptive healing would be a critical building block of the new healing-focused engineering design paradigm.

8.6 Multilayer Sheet-Based Lightweight Funicular Structures

The space frames designed using the methods of polyhedral graphic statics have certain shortcomings. When the actual loads are different from the design loads, the nodes of the 3D frameworks may undergo considerable bending moments, and the structure's safety relies heavily on the nodal bending resistance. Moreover, the unique geometries of bars and nodes usually lead to high costs during fabrication and assembly when it comes to complex irregular space frames. It's worth noting that the forms found by polyhedral graphic statics have intrinsic planarity that can be harnessed to design sheet-based structure systems, mitigating the mentioned challenges of the space frames. Sheet-based systems have advantages over space frames because of their material accessibility, processibility, low cost, and applicability to large scales.[121] Sheet materials can be easily processed by various fabrication techniques, such as laser cutting, computer numerically controlled (CNC) milling, CNC bending, waterjet cutting, and so on. In terms of structural performance, a sheet-based system provides more stability. It is less vulnerable to various loading scenarios because the forces can be transferred and distributed across the faces.

Sheet-based structures designed using polyhedral graphic statics are advantageous because sheet elements constrain the nodal kinematic degrees of freedom and provide more load paths. However, not all sheet elements in the funicular polyhedral systems are necessary as long as the load paths are preserved and the system does not have kinematic degrees of freedom. Removing redundant sheets can reduce the material cost, and the structural performance can be further improved.

In this research, we use matrix analysis methods developed by Pellegrino and Calladine[263] to calculate funicular networks' kinematic degrees of freedom. Pellegrino and Calladine[263] formulated an algorithm that evaluates the framework's performance rapidly by determining the rank of the kinematic matrix and the bases of its four linear algebraic vector subspaces. Specifically, it offers complete details of any states of inextensional deformation a framework may possess. For the face and hinge systems, matrix analysis is also used in the folding simulation of rigid origami, where the loops of bars simulate rigid faces and the connections between faces act as hinges. The idea of representing triangulated origami as a pin-jointed framework was first proposed by Schenk and Guest.[264] Filipov et al.[265] improved this method with new triangulation schemes for quadrilateral facets. Zhang et al.[266] further generalized the triangulation schemes for any n-gons.

The design principle is inspired by trusses, where the beam members are connected to be geometrically restrained. Therefore, when a truss is loaded, the forces are mostly transferred through the axial directions of the beam members without needing much nodal bending resistance.

Similarly, this geometric restraining effect is desired for a sheet-based system such that its load-bearing capacity does not rely much on the edge bending resistance.

In more technical terms, this "restraining" effect can be described as zero kinematic degrees of freedom, meaning the structure has no mechanism. The matrix analysis method is incorporated for kinematic analyses to determine whether a structure is kinematically determinate or indeterminate. The matrix analysis method has been used to analyze pin-jointed inextensional frameworks.

A truss model is adapted for the analysis of rigid face and frictionless hinge systems because a rigid face can be simulated by a cluster of kinematically determinate pin-jointed bars. The performance of this face-hinge system is a good indicator of the performance of an actual structure constructed with rigidly connected sheet materials. Zero kinematic degrees of freedom implies a structure with more stability and better performance. This adapted matrix analysis approach is then incorporated into a computational pipeline to help find an efficient set of faces adequate for constructing a kinematically determinate face-hinge structure.

This section is organized into three parts. First, the base geometry is generated using graphic statics. Next, the matrix analysis method is adapted for the kinematic analysis of sheet-based structures. This is then incorporated into a computational workflow that helps determine an efficient group of faces adequate to keep the kinematic stability and load paths.

8.6.1 Polyhedral Form Finding

In this section, a single-layer funicular shell is used as an example for the explanation and demonstration of the design principles (Figure 8.25). As mentioned earlier, the intrinsic planarity allows the form to be delivered as a faceted shell in addition to a space frame (Figure 8.25d). The goal is to find an efficient set of faces that does not have any mechanisms while maintaining the structural form, that is, keeping all the load paths. Not all faces in the original faceted geometry are needed to achieve the kinematic determinacy, hence a computational pipeline is devised to help find a reduced number of faces that satisfy the requirements.

8.6.2 Matrix Analysis and Adaptation

Before diving into the details of the computational pipeline, the matrix analysis method for the pin-jointed inextensional framework and its adapted method for face-and-hinge structures are explained first as they are the basis of the computational pipeline. The kinematic analysis of a pin-jointed inextensional framework starts from the assembly of the kinematic matrix following the steps formulated by Pellegrino and Calladine.[263] The form can be depicted with three characteristics: v vertices connected

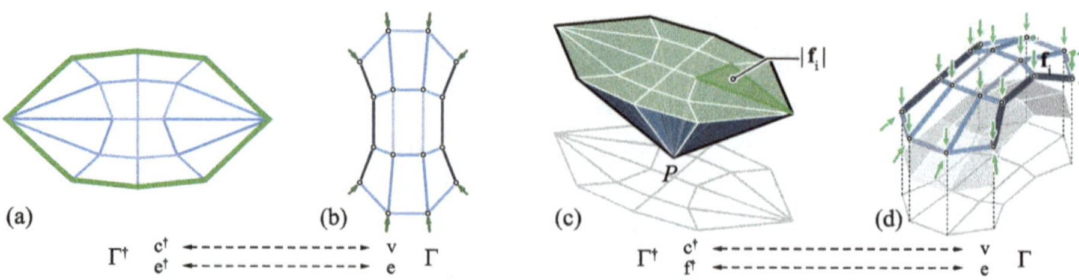

Figure 8.25 Form-finding of a single-layer funicular shell.

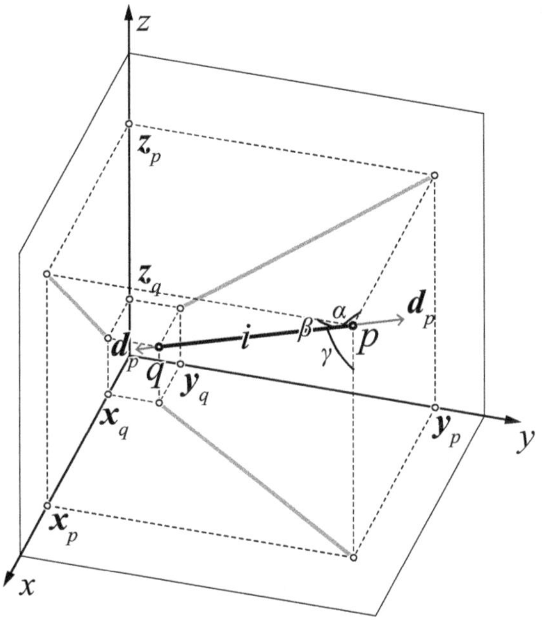

Figure 8.26 An example edge and the geometric attributes describing nodal displacements and edge elongation.

by e edges and constrained by k kinematic constraints (each defined as one constrained degree of freedom in X, Y, or Z directions) to a rigid foundation.

There are also two sets of kinematic variables to be considered: the elongation di for each edge i, and the displacements d_{jx}, d_{jy}, and d_{jz} along X, Y, and Z axes in 3D Euclidean space for each vertex j. Their relationship (illustrated in Figure 8.26) can be written as

$$\begin{aligned}\delta_i \times l_i = {} & (x_p - x_q)d_{px} + (y_p - y_q)d_{py} + (z_p - z_q)d_{pz} \\ & - (x_p - x_q)d_{qx} - (y_p - y_q)d_{qy} - (z_p - z_q)d_{qz}\end{aligned} \tag{8.3}$$

where l_i is the length of edge i and p and q are the endpoints of edge i. Assemble all equations for the e edges in matrix form as

$$\mathbf{\Delta} = \mathbf{A} \cdot \boldsymbol{d} \tag{8.4}$$

Here Δ is the vector of e elongation coefficients, each defined as $\delta_i \times l_i$

$$\mathbf{\Delta} = \begin{pmatrix} \delta_1 \times l_1 \\ \vdots \\ \delta_i \times l_i \\ \vdots \\ \delta_e \times l_e \end{pmatrix} \tag{8.5}$$

A is the e by $3v - k$ kinematic matrix, written as

$$\mathbf{A}^T = \begin{pmatrix} \vdots & \vdots & \vdots \\ \cdots & x_p - x_q & \cdots \\ \cdots & y_p - y_q & \cdots \\ \cdots & z_p - z_q & \cdots \\ \vdots & \vdots & \vdots \\ \cdots & x_q - x_p & \cdots \\ \cdots & y_q - y_p & \cdots \\ \cdots & z_q - z_p & \cdots \\ \vdots & \vdots & \vdots \end{pmatrix} \tag{8.6}$$

$\boldsymbol{d}$ is the vector of $3v - k$ displacements, written as

$$\boldsymbol{d} = \begin{pmatrix} \vdots \\ d_{px} \\ d_{py} \\ d_{pz} \\ \vdots \\ d_{qx} \\ d_{qy} \\ d_{qz} \\ \vdots \end{pmatrix} \tag{8.7}$$

The kinematic indeterminacy m, meaning the number of independent mechanisms, can then be determined by the relationship between the numbers of equations and unknowns, where an important concept of rank r_A comes into play:

$$m = 3v - k - r_A \tag{8.8}$$

It's important to note that, as stated by Pellegrino, the kinematic indeterminacy here may include the rigid body motion of the framework. In other words, when no vertex is constrained to a foundation, a framework will have a kinematic indeterminacy of greater than or equal to 6, including three translational and three rotational rigid body degrees of freedom. In the scope of this paper, the rigid body motions are named external kinematic indeterminacy m_{ex}, and the mechanisms at the vertices are named internal kinematic indeterminacy m_{in}. They satisfy the equation

$$m = m_{ex} + m_{in} \tag{8.9}$$

When detecting the internal mechanisms of the structure, the external indeterminacy m_{ex} should be excluded. The calculation of m_{ex} is based on the k kinematic constraints to the rigid foundation and is omitted here. All examples in this paper are set up with adequate kinematic constraints to the rigid foundation such that m_{ex} is zero. For the example geometry, there are 24 bars, 12 unconstrained vertices, and 4 other vertices set up as fully constrained (Figure 8.25d). The rank of the kinematic

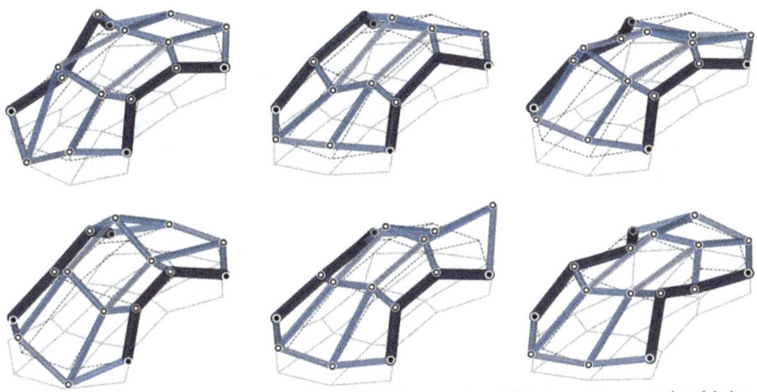

Figure 8.27 Exaggerated infinitesimal deformations of the framework.

matrix is calculated to be 12, therefore the internal kinematic indeterminacy is 12 given that no rigid body motion is possible, indicating that the framework has 12 internal independent mechanisms.

The locations of the mechanisms can then be obtained by solving for the vertices that have potential displacements. Since the edges in the framework are assumed to be rigid, there is no elongation in any edge, hence Equation 8.5 can be replaced by

$$\mathbf{0} = \mathbf{A} \cdot \boldsymbol{d} \tag{8.10}$$

The potential displacements of the vertices can be obtained by solving for $\boldsymbol{d}$, which is equivalent to solving for the null space of **A**. With SciPy,[267] the 36 by 12 orthonormal basis of the null space can be computed through singular value decomposition (SVD). The linear combination of the 12 columns represents the possible infinitesimal displacements of the unconstrained vertices. If an unconstrained vertex is unmovable, meaning that it's geometrically restrained by other edges, its displacements will always be zero under any linear combinations. Some random displacement scenarios are exaggerated and visualized in Figure 8.27, and the locations of the mechanisms can be found. As a side note, each displacement scenario is interrelated to a set of external loads that cannot be balanced, the magnitudes and directions of those loads are not discussed here.

When the structure is built with rigid faces, some additional planar constraints are needed. Those constraints can be implemented using helper edges and vertices for the "stiffening" effect. The edges related to each face need to form a rigid body such that the resulting new pin-jointed framework performs like a face and hinge structure (Figure 8.28).

The mathematical relationship between the number of required helper edges and the number of polygonal face sides is then established following a method proposed by Zhang et al.[266] As exploited by,[264] a triangular framework can be directly used for the folding simulation of triangular origami without any helper edges and vertices. For

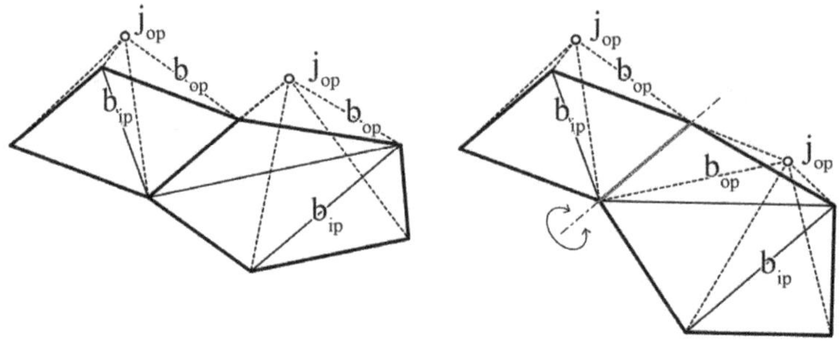

Figure 8.28 Pin-jointed framework performs like a face hinge structure.

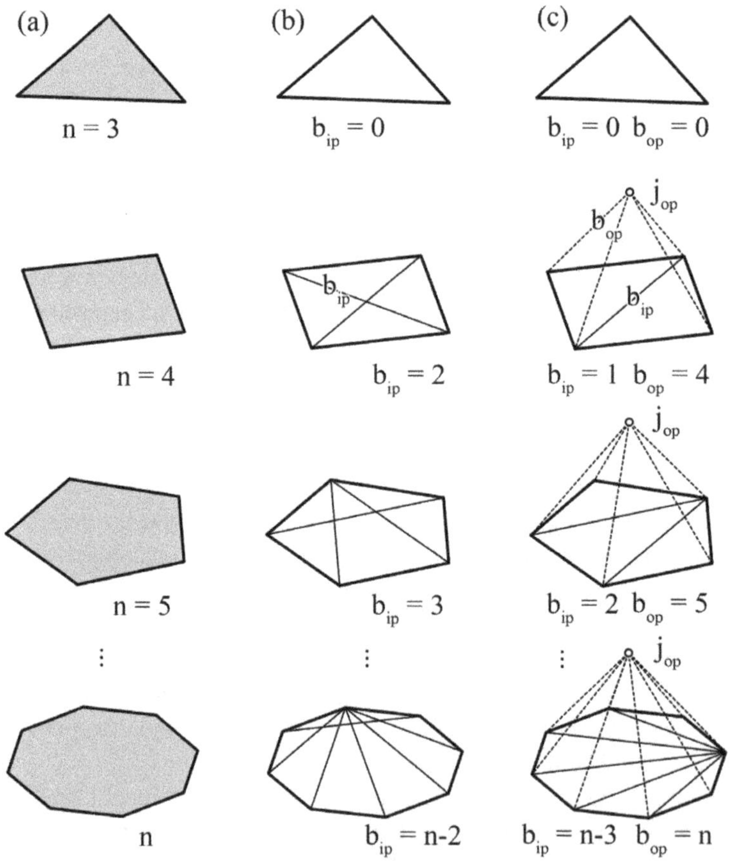

Figure 8.29 Helper edges and vertices are added for the simulation of rigid faces.

any side count that is larger than 3, helper edges and vertices are needed (Figure 8.29). Later, the kinematic indeterminacy and locations of mechanisms can be determined for the framework with certain rigid faces. As illustrated in Figure 8.30, the kinematic indeterminacy of the pin-jointed framework is suppressed with an increasing number

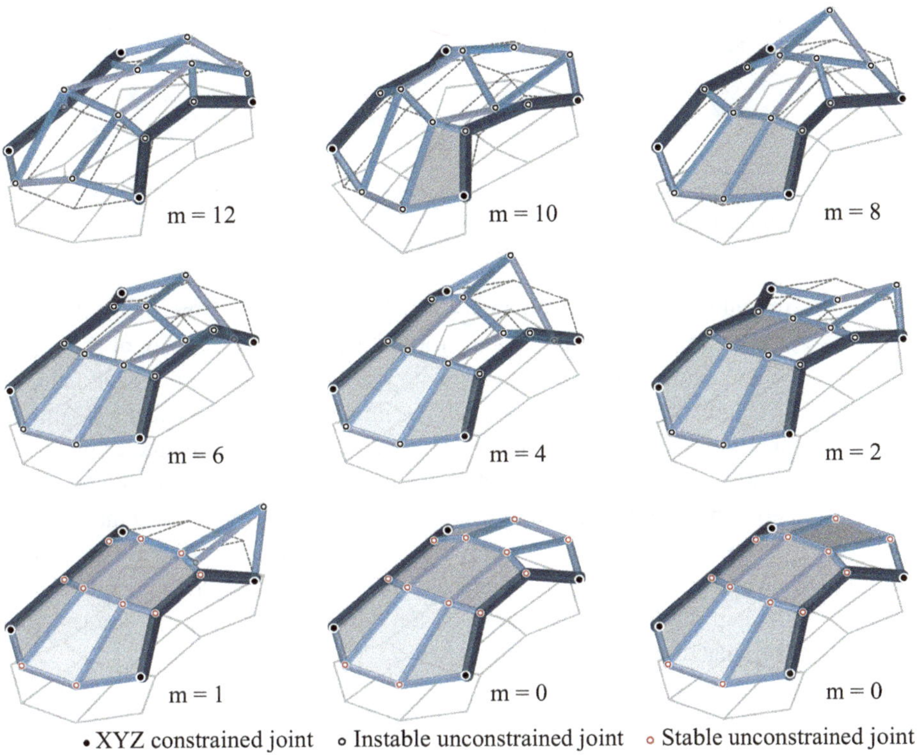

Figure 8.30 Reduce kinematic indeterminacy with more rigid faces.

of rigid faces, and the framework becomes kinematically stable after seven rigid faces are added.

8.6.3 Algorithmic Design

After establishing the analysis method that is compatible with rigid faces, an iterative algorithm is then devised to help determine the reduced group of faces that are sufficient to keep the kinematic stability and load paths. In each iteration, one rigid face is added to the framework, and it stops when the internal kinematic indeterminacy becomes zero. Figure 8.30 shows the decreased kinematic indeterminacy as more faces are added to the framework. Notably, the sequence of adding rigid faces significantly affects the result of this algorithm. For example, Figure 8.31a–d shows four cases of adding seven rigid faces, in which three become stable while one is still kinematically unstable. Besides, since the goal is to use only sheet materials to construct the structure, naked edges are not allowed in the result (Figure 8.31e). To obtain an efficient number of faces and, therefore, achieve higher structural efficiency, the sequence of adding rigid faces needs careful consideration.

The algorithm starts with the input of polyhedral geometry, including all vertices v_i, edges e_i, and faces f_i. The vertices and edges are used to construct the initial framework, and the faces are the candidates to be added. Next, the constraints are set, and

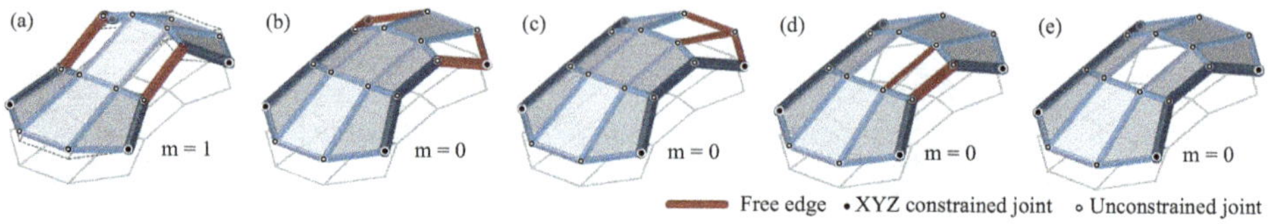

Figure 8.31 (a) to (d) Kinematic indeterminacies on different sets of seven faces, and (e) the solution without naked edges.

the initial m_{in} of the framework is calculated. There is no face stiffened at this point. To help determine the sequence for adding faces, the concept of priority is introduced, where a larger priority means a face will be added first. For each face, its priority *pri* is calculated based on its area a, the number of neighbor faces f_n, and the number of single-valence edges e_f it has. A larger area leads to a smaller priority because less area means less material and hence higher efficiency. More neighbor faces lead to a larger priority because it tends to reduce more degrees of kinematic indeterminacy. In the case of sheet-only systems, any face with an edge of single valence, meaning that the edge only belongs to one face candidate, has an infinite priority since it is required to keep the load paths. Based on the description earlier, the priority function can be formulated as

$$pri = \begin{cases} \infty & e_f > 0 \\ x(1-a) + yf_n & e_f = 0 \end{cases} \tag{8.11}$$

where a is the face area mapped to range 0–1; x and y are coefficients that can be adjusted to tune the weights of a and f_n. Also, this pipeline allows design decisions to be incorporated into the priority function.

If certain face candidates are required due to the functionality of the structure, its priority will be overwritten to positive infinite. Contrarily, if any face candidate is unwanted, its priority will be overwritten to zero. After that the priorities of all face candidates are calculated, and the faces are sorted with descending priority. Then, faces are iteratively added to the framework. In each iteration, the face with the highest priority is removed from the list of candidates and "stiffened." The stiffening is realized by adding helper vertices according to the rule described earlier in Figure 8.29. This is then followed by calculating the new m_{in} with all helper vertices and edges. If m_{in} stays the same compared to the last iteration, meaning that this newly added face doesn't help constrain the mechanisms, it will be "unstiffened" by removing the corresponding helper vertex and edges. This process repeats until m_{in} becomes zero. The final step is to add additional faces in order to eliminate the naked edges and vertex-to-vertex connections since the structure is designed to be built with only sheet materials and the original load. As a result, the output faces function as a kinematically stable face-hinge structure that maintains all primary load paths.

Here, the aforementioned method is used to design a bridge to attest to the proposed method. A comparative study is carried out using finite element analysis (FEA). A small-scale physical model is also made to explore the connection details between the sheets.

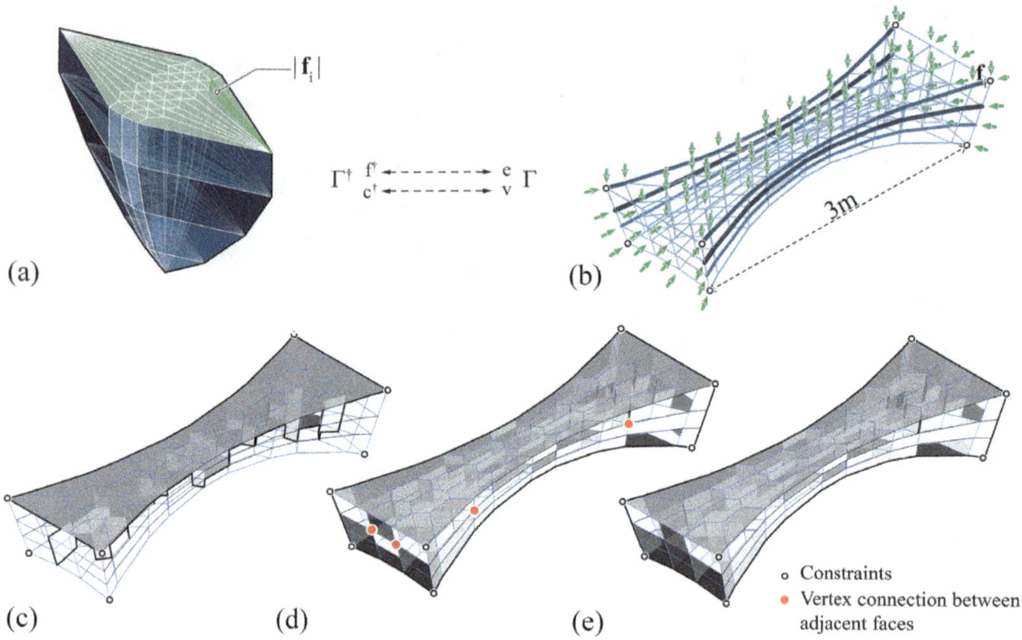

Figure 8.32 (a)–(e) Steps for generating the bridge.

Base Geometry and Generation Process

The form and force diagrams generated using polyhedral graphic statics are shown in Figure 8.32a,b. 10 kN is used as the total design load applied on the top of the structure, and the total span is set to 3 m. The face-adding process begins after having the base framework geometry. Eight vertices are first chosen as pin anchors to constrain the structure (Figure 8.32b). No faces are added at this point. Due to its functionality as a bridge, 31 top faces are determined as must-haves for people to walk on. Later, the iterative face-adding algorithm is invoked, which finds the additional faces with the least area that reduce the internal kinematic indeterminacy to zero, meaning there is no mechanism across the structure (Figure 8.32c). Next, a secondary iterative algorithm is needed to add the minimum set of faces for removing naked edges (Figure 8.32d). The resulting structure may have vertex-to-vertex connections between adjacent faces, as illustrated in Figure 8.32d, which causes problems for materialization. Therefore, one further action is taken to add additional faces that help eliminate those vertex-to-vertex connections. The final structure is shown in Figure 8.32e.

Comparative Numerical Study

To further understand the mechanical performance of the design, a comparative numerical study is carried out on both the sheet-based structure and space frame using the finite element method (FEM). Structural steel is used as the material, and the total material usage is controlled at 66.5 kg for both structures. The structures are simply supported on the vertices of two ends of the bridge, and they are simulated under two static loading scenarios: first under the design load of 10 kN distributed on the top

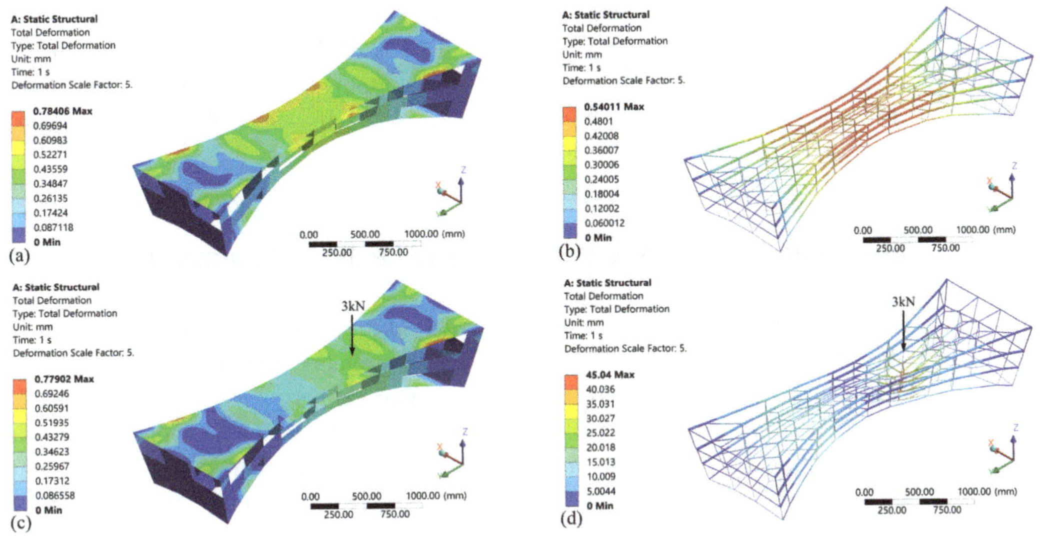

Figure 8.33 (a) to (d) Comparative study with FEA on the sheet-based and framework structures under two load cases

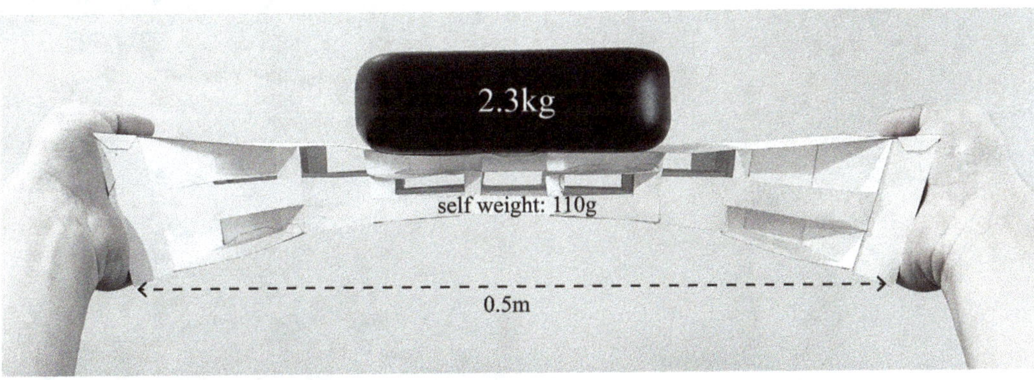

Figure 8.34 A simple physical load test on the small-scale model made of Bristol paper.

vertices (Figure 8.33a,b), then under a point load of 3 kN (Figure 8.33c,d). For the first loading scenario, both structures reported a max displacement below 0.8 mm, and the space frame slightly outperforms the sheet-based structure. For the second loading scenario, the max displacement of the sheet-based structure remains at a low level. However, the space frame reports that of more than 45 mm, indicating a high risk of failure. The results show that although the sheet-based structure performs slightly worse than the space frame under the design loads, it's potentially more versatile in taking a wider range of loads in real-world applications.

Small-Scale Physical Model

A 1:6 physical model spanning 0.5 m is made using Bristol paper (Figure 8.34). For a complex nonmanifold geometry like this, two techniques are used to facilitate the fabrication and assembly. First, the 283 faces are merged into 35 continuous strips and unrolled onto flat sheets to be laser-cut and assembled with fewer parts. Second, all edge connections are realized by small overlaps (tabs) bonded with glue. The model can take 2.3 kg of load with a span of 0.5 m and a self-weight of 110 g, manifesting minor deflections.

8.7 3D-printed Structures with Embedded Anticlastic Geometries

The following project demonstrates research on the structural design, optimization, fabrication, and assembly of a post-tensioned, prefabricated concrete canopy with embedded anticlastic surfaces (Figure 8.35). Our primary construction material for the building structure is low-emission, carbon-storing concrete, including cementitious nano/microparticles from lime and particular biominerals that can maintain the concrete's printability while adding the benefits of reducing embodied carbon.

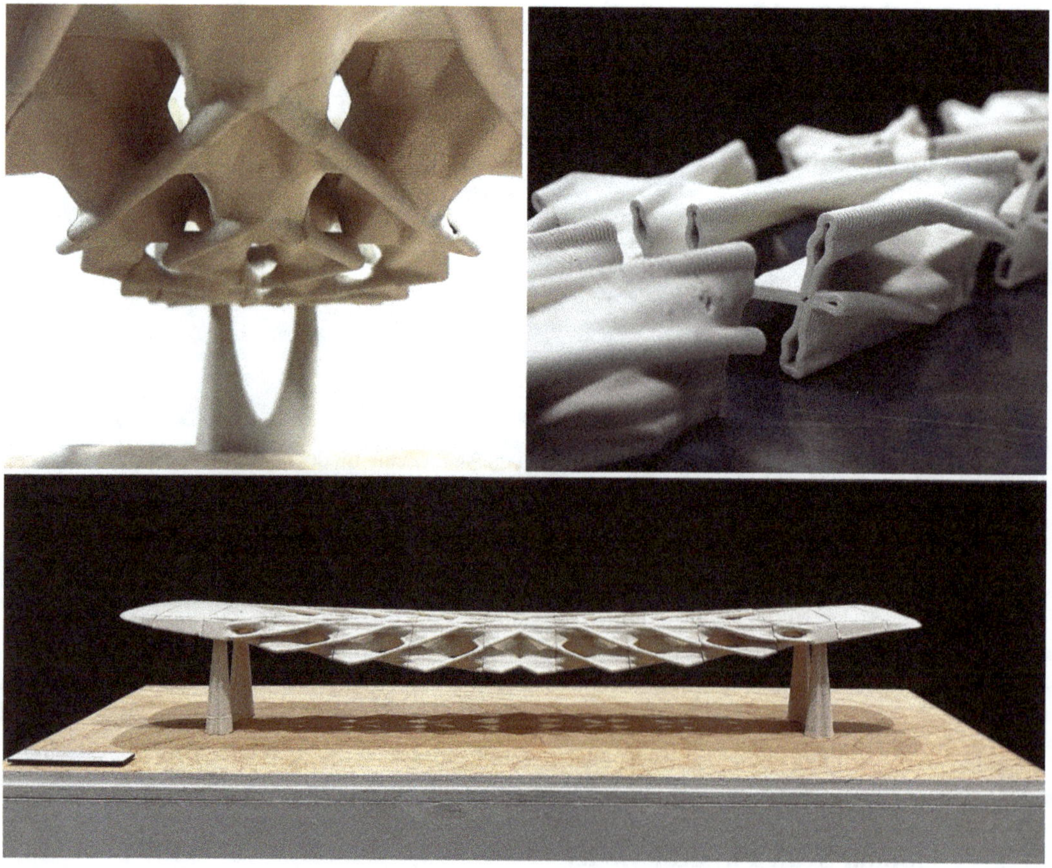

Figure 8.35 3D-printed, post-tensioned structure with embedded anticlastic surfaces. Clay-printed and steel wire-tensioned prototype produced for a feasibility study.

The structural form of this canopy carries the applied loads using both tension and compression members to extend the application of such versatile systems in constructing buildings with lower carbon emissions and minimized waste. In the past couple of years, discrete, compression-only structural systems have received much attention due to the minimization of the construction material, mass, and embodied energy.[268,269]

8.7.1 Structural Form Finding

While compression-only systems are excellent in minimizing mass, their application in multi-story buildings may require extensive external provisions to maintain compression-only load paths in their optimized structural form. We use an optimized structure with mixed compression and tensile members to demonstrate the application of polyhedral graphic statics in the design of prefabricated, 3D-printed concrete structures. The funicular form's structural design and optimization process were discussed in detail in Chapter 5.

In the context of 3D-printed concrete structures, maintaining constant tension forces is particularly beneficial for several reasons: (a) 3D-printed concrete materials exhibit anisotropic behavior, meaning their properties vary depending on the direction of the applied load. Ensuring constant tension forces allows for better control of material deposition and orientation, ultimately maximizing the strength and durability of the printed structure; (b) in the 3D-printing process, concrete is deposited layer-by-layer. Maintaining constant tension forces helps to ensure consistent load distribution across layers, reducing the risk of localized stress concentrations or delamination, which could compromise the structural integrity; (c) by maintaining constant tension forces, the design process can be streamlined to account for the constraints and requirements of the 3D printing technology. This simplification enables a more efficient fabrication process, reducing potential errors or inconsistencies during construction; and (d) in concrete structures, cracks may develop due to shrinkage or tensile stresses. Ensuring constant tension forces in the design helps distribute these stresses more evenly, mitigating the risk of crack formation and propagation and improving the structure's durability.

8.7.2 Embedding the TPMS Geometry

Figure 8.36 summarizes the structural design and materialization of the structure. After constructing the bar-node model of the structural form, the next step is to design the cross-sections for materialization and construction. We used triply periodic minimal surfaces (TPMS) to materialize the structure. Triply periodic minimal surface TPMS is a formula-based porous microstructure with zero mean curvature.[270,271] Owing to the surface's interconnectedness, the shape can be modified using the equations' designated parameters to control porosity and volume for better structural performance.[272] There are different types of TPMS based on trigonometric functions with specific geometric properties. As a result, complex topologies of microstructure can easily be generated. Figure 8.37 shows the computed TPMS shapes from the fundamental

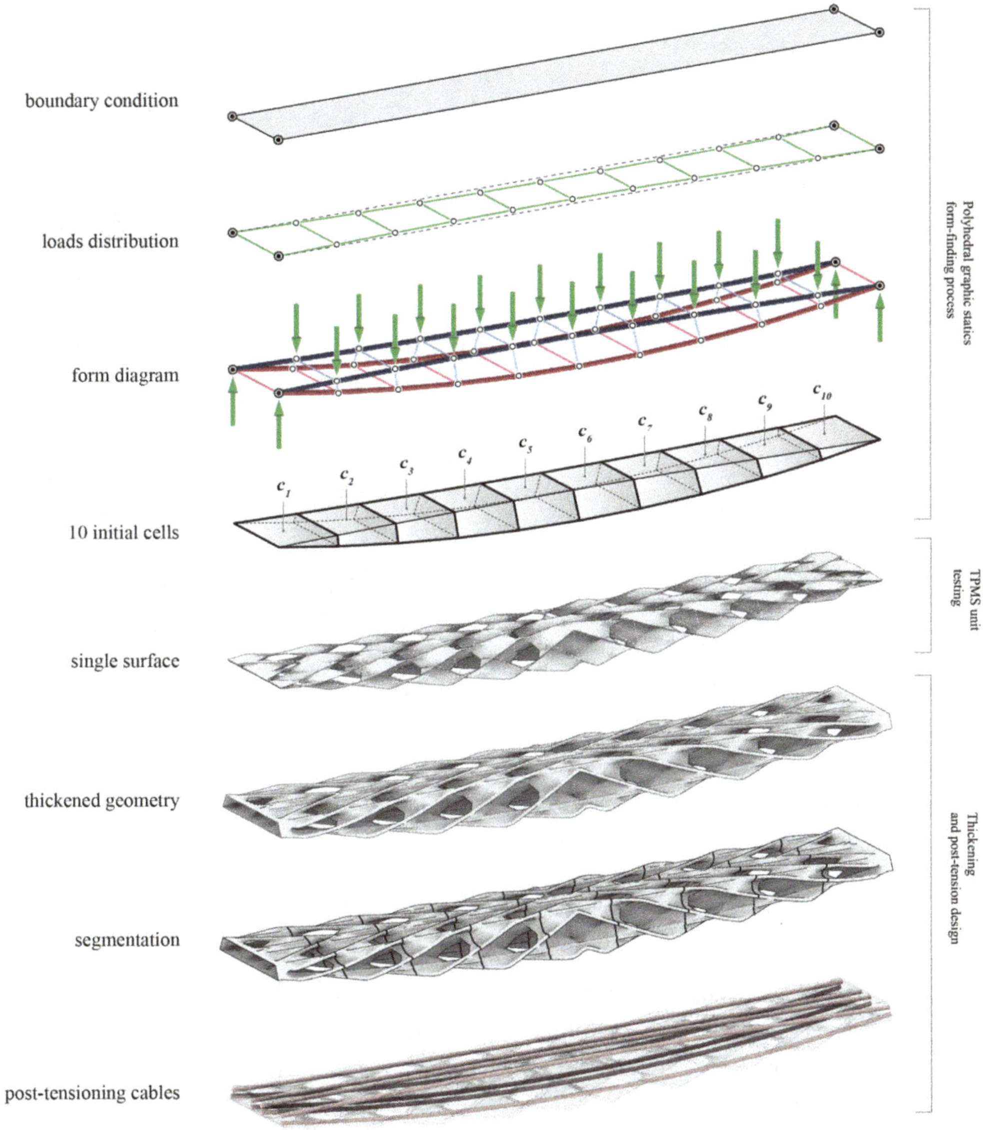

Figure 8.36 Different steps of structural design: including form finding, contorting TPMS geometry, adding thickness to the TPMS surface, slicing, and including the post-tensioning cables.

shapes such as primitive and gyroid to the more complex geometries such as Lidinoid and F-Koch.

TPMS geometries are usually represented by cubic boundaries, so they can be repeated easily in three directions. They have continuous surfaces with periodicity in three dimensions with maximized surface area. These surfaces are suitable for generating optimized structures, providing efficient load distribution, reduced material usage, and enhanced geometric rigidity.

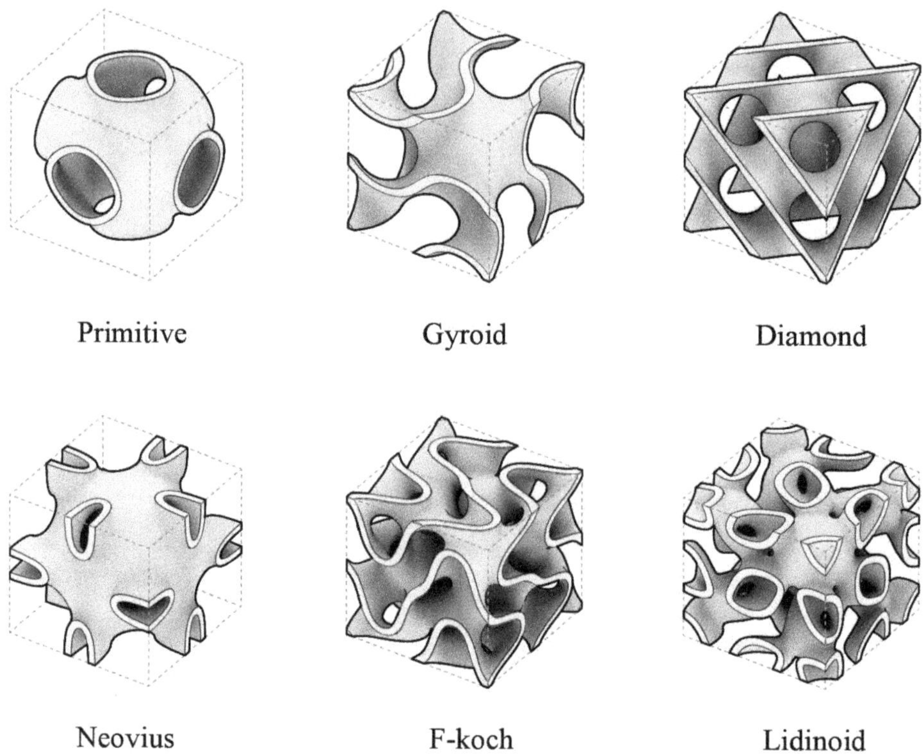

Figure 8.37 Different TPMS structures from primitive to high-complexity geometry: Schwarz P, gyroid, diamond, neovius, F-koch, lidinoid.

The geometry of TPMS structures can populate the cells of the form diagram. Based on the desired structural and aesthetic properties, a suitable TPMS unit, such as the Gyroid, Schwarz P, or Schwarz D, and diamond surfaces, can be chosen for this purpose and fit the cells in the form diagram by adjusting its scale, orientation, and position (Figure 8.36). Note that the thrust network designed by polyhedral graphic statics should be embedded within the TPMS geometry to ensure the best performance. Diamond TPMS showed better alignment with the geometry of the form diagram and was used in this design.

The polyhedral cells of the structure contort the diamond TPMS geometry to align with the principal stress directions. The TPMS units' anticlastic geometry enhances the structural form's geometric stiffness. Its maximized surface area can be used for carbon capturing, while minimizing the cross-section of the parts.[273] The anticlastic geometry of the TPMS models was further optimized for printability using robotic concrete printing. Signed distance function (SDF) modeling[131,274,275] combined the funicular form and the TPMS geometry into a smooth, unified model for fabrication and FE analysis.

8.7.3 Slicing and Overhang Analysis

Concrete 3D printing has become an increasingly popular method for constructing complex geometries quickly and efficiently. However, one of the primary challenges of printing is the limitation in printing overhangs. An overhang in a 3D-printed structure is any part that extends horizontally beyond its supporting base without any direct support below it. Fresh concrete mixtures are typically heavy and exhibit low tensile strength, which makes it difficult for the material to maintain its shape and structural integrity when subjected to the forces acting on overhangs. Concrete 3D printing typically involves depositing material in layers, each relying on the previous one for support. When printing overhangs, the lack of direct support underneath the printed material may cause it to sag, deform, or collapse under its weight before it can set and gain strength. Printed parts take time to cure and gain their final strength. During curing, overhangs are susceptible to deformation or collapse due to the material's low initial strength. Optimization algorithms were developed to effectively slice the geometry with nonparallel planes perpendicular to the direction of compression and tension force flow following the structural form finding[276] (Figure 8.38).

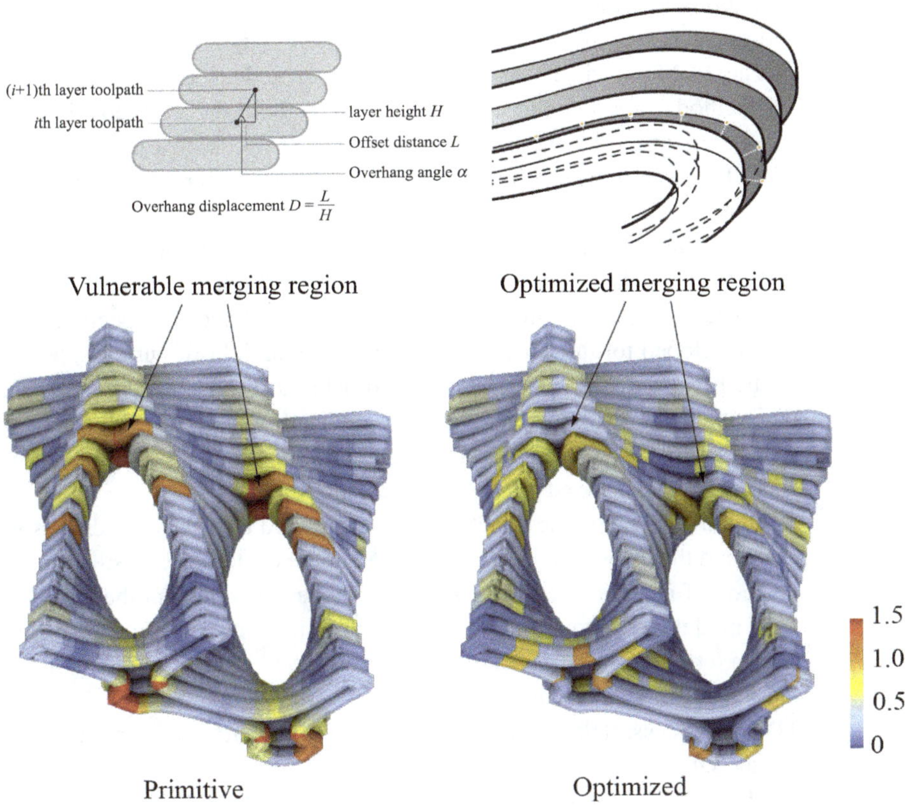

Figure 8.38 Overhang of the layers and its optimization to reduce the problem.

After the nonparallel slicing process, the geometry of the parts included multiple disconnected loops, and precise control of the flow and speed for high-quality printing was needed. The additional computational algorithm was developed for efficient noncontinuous printing to minimize the material dripping at each loop's start and end points and to avoid a potential collision of the extruder and the printed parts.

8.7.4 Structural Analysis

The behavior of a given structure subjected to specific loads can be assessed experimentally or numerically. Experimental testing is inherently more accurate, yet it can be expensive and has various limitations during design when the evaluation of multiple alternatives is necessary. Numerical approaches can be seen as conducting virtual experiments, wherein the structure is mathematically modeled and analyzed to evaluate the behavior of the structure. Careful consideration must be given when defining and selecting the material parameters that are assigned to the numerical model. These parameters should closely match the material's actual properties such that an accurate simulation can be conducted. Typically, the numerical model can be calibrated after conducting an experimental test. This calibration allows for the repetition of analyses across a vast range of cases, which is ideal for conducting parametric studies.

The FEM is widely recognized and accepted as one of the primary numerical methods in structural analysis. In this method, the geometry is discretized into small elements, with the behavior of each element defined by a mathematical function. To learn more about FEM, interested readers are directed to the following references.[277–279]

To evaluate the behavior of the prefabricated, post-tensioned concrete structure with embedded anticlastic surfaces and compare it with other commonly constructed spanning structures, a preliminary FE analysis was conducted. Specifically, a FE model was developed for the structure with embedded anticlastic surfaces, herein referred to as the proposed structure, and analyzed using a displacement control analysis. Four other incumbent structures were also evaluated (B–E in Figure 8.39). All structures span the same distance (10.6 meters) and have the same total volume of both concrete and reinforcement used for the proposed structure.

To fairly compare the structures, all considered boundary conditions were kept the same and the displacement was measured at the center of each structure. The number, position of reinforcement, and volume of concrete were kept the same. All structures were modeled as a continuum in Abaqus™ using the same number of elements. However, a tetrahedral mesh was used in the proposed structure instead of the hexahedral mesh as generated for the incumbent structures, due to the complexity of embedded TPMS surfaces. The applied load and displacement of all structures are plotted in Figure 8.40.

The T-shape beam, labeled E, has the greatest stiffness compared to the other structures, up to 1.5 mm displacement. The rectangular beam, labeled D, has the second-strongest stiffness. Both rectangular structures, labeled B and C, show

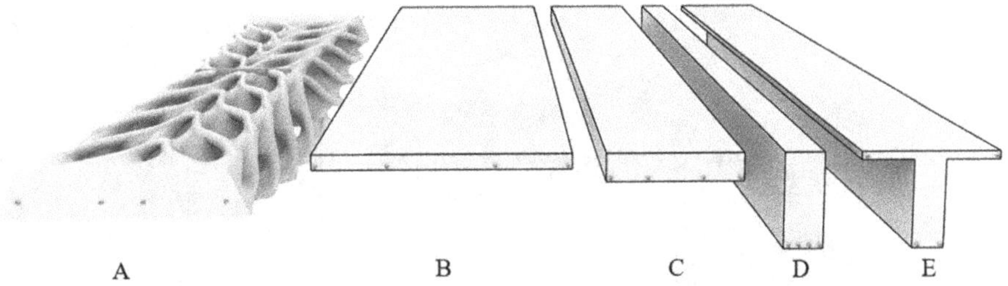

Label	Description	Dimensions
A	Proposed structure	10'600 mm length (all)
B	Beam with the same projected area	1858 x 114 mm (W x H)
C	Beam with the same walking path width	994 x 214 mm (W x H)
D	Beam with the same (maximum) depth	294 x 721 mm (W x H)
E	T-shape beam	994 x 60 / 231 x 661 mm (W x H)

Figure 8.39 Illustration of the different structure geometries considered as well as their dimensions. Location of post-tensioning steel cables is indicated by gray points.

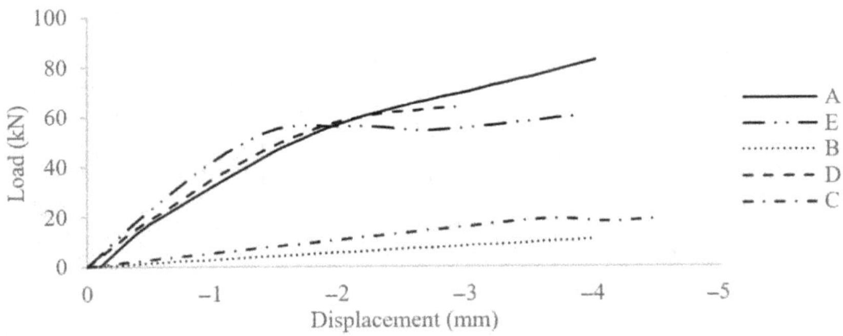

Figure 8.40 Load-displacement comparison between different geometries with the same volume.

significantly less stiffness. The proposed structure, labeled A, suffers a 14% reduction in strength in the early stage compared to structure E. However, after this point, the proposed structure starts to gain strength as the others weaken. By increasing the displacement, all the beams experience plastic deformation, while the proposed structure still shows elastic behavior, which is preferred in structural design. The observed elastic behavior is mainly due to the nature of compression-only by showing a higher stiffness.

8.7.5 3D-printed, Post-tensioned Structural Prototype

A noncarbon-absorbing prototype structure was fabricated and constructed to assess 3D-printing prefabrication, assembly procedures, and the structural system's general performance. The structure was designed as a modular system, consisting of nine individual segments. Each segment was prefabricated using a two-component (2k)

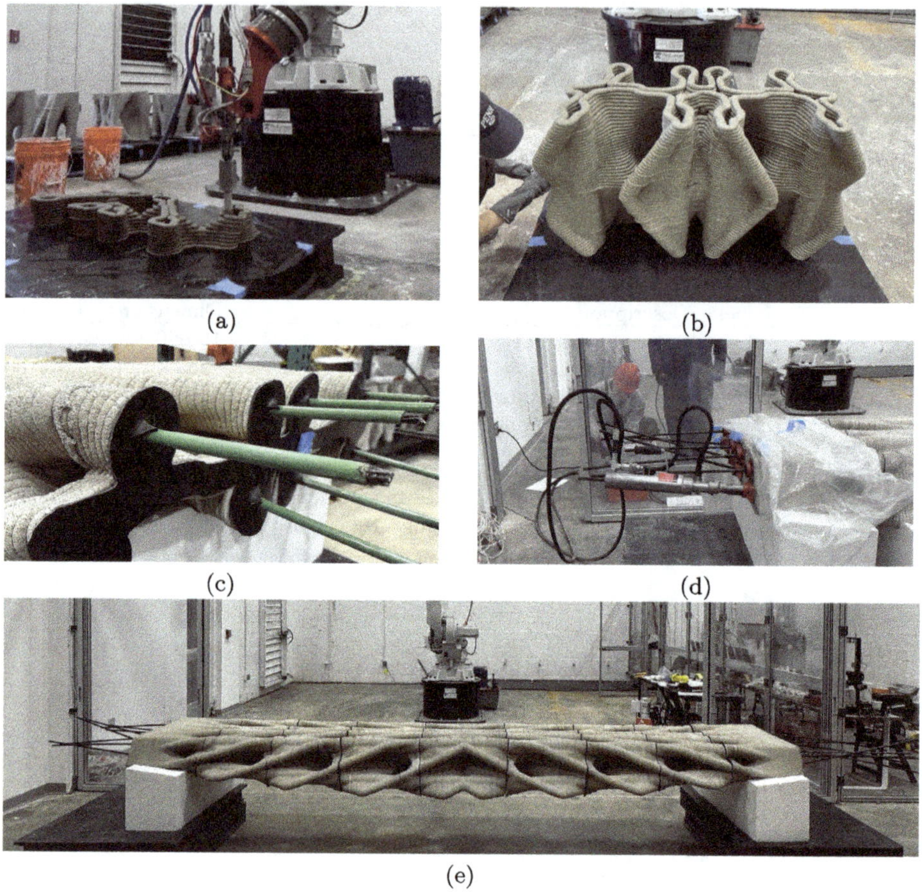

Figure 8.41 Fabrication and construction of a 5-m structure prototype with an embedded TPMS, diamond geometry. (a) 3D concrete printing of one of the nine segments using an ABB robotic printer with a concrete 2k printing end effector; (b) final printed middle segment; (c) steel cables being threaded through the structural segments; (d) post-tensioning of the steel cables, compressing the segments together, and forming the structure; and (e) the final 3D-printed, post-tensioned structure prototype spanning and supporting its self-weight.

concrete material and a robotic printer. Two-component concrete consists of two separate materials, typically a dry component mix of cement, aggregates, and additives and a liquid component of water, chemical additives, or other binding agents. The nine individual printed segments were post-tensioned together to form the structure in Figure 8.41.

Having tested the structure's ability to be post-tensioned and span while supporting its self-weight, its structural capability was further tested by conducting a sandbag test. Sandbags, each weighing ≈25 kg, were placed at the mid-span of the structure. In total, 40 bags were placed, stacked on top of each other, amounting to ≈1,000 kg. The structural prototype was capable of bearing this load without failing, and a small amount of deflection was visually observed, indicating the structure's good performance. The structural prototype supporting the sandbags is shown in Figure 8.42.

Figure 8.42 The post-tensioned structural prototype was load tested by placing sandbags, each weighing ≈25 kg, at around mid-span. A total of 40 bags were placed on the canopy amounting to ≈1,000 kg. The structure was able to support this load.

8.7.6 Carbon-absorbing 3D-Printed Recipe

Concrete, the world's second-most widely used material after water, is primarily composed of cement. Still, its production process is carbon-intensive, contributing to a significant portion of greenhouse gas emissions. About 7% of anthropogenic CO_2 emissions worldwide stem from cement production, exacerbating climate change and global warming.[280] The cement industry has embraced three primary strategies to mitigate its carbon footprint: (1) using low-emission secondary fuels, (2) carbon capture and storage, and (3) reducing clinker usage. Among these approaches, carbon capture and storage within concrete, where CO_2 is stored in a carbonate form, has emerged as a promising solution with notably low energy consumption requirements.[281] Extensive research has demonstrated significant advancements in carbonating concrete, from the mixing process to the carbonation curing of cast concrete and repurposing waste concrete aggregates.[282] The advent of 3D concrete printing presents a transformative opportunity in construction by reducing waste without the need for a scaffold, saving time and labor costs while potentially enabling intricate and efficient architectural designs. Leveraging 3D concrete printing, the development of 3D-printed carbon-absorbing concrete ushers in a new era of material innovation for carbon capture and storage. It will reduce material usage and energy consumption during fabrication while maximizing surface area for efficient carbon capture and absorption within the structure.

Several key parameters come into play to develop concrete that is both 3D printable and highly efficient at carbon absorption. The properties of the concrete mixture, such as its rheological properties, setting time, and carbonation efficiency, are crucial to the material design and selection. Desired rheological properties include shear-thinning

characteristics, where the mixture's viscosity decreases under high shear rates, facilitating pumping and extrusion. After extrusion, the concrete should immediately retain its shape. This will maintain the printed structures without significant geometry changes by balancing dynamic and static forces. Thixotropy testing quantifies the material's ability to regain viscosity after high shear rate extrusion, indicating its suitability for 3D printing. Additionally, the material must demonstrate shape retainability, resisting deformation from the self-weight of the structure and supporting the weight of additional layers stacked on top of it, referred to as buildability. Both properties can be characterized through static yield stress and plastic viscosity testing in rheology. These parameters act as benchmarks to evaluate if a mixture is appropriate for 3D concrete printing. Concrete suitable for 3D printing typically comprises cement, aggregates, water-reducing admixtures, viscosity-modifying admixtures, and water. To obtain the desired rheological properties for 3D printing, it requires careful formulation. Furthermore, the setting time of the concrete mixture plays a critical role. Rapid hydration can decrease workability, which affects the open time for 3D printing and overall printability.

Portland cement is made up of tricalcium silicate ($3CaO \cdot SiO_2$), dicalcium silicate ($2CaO \cdot SiO_2$), tricalcium aluminate ($3CaO \cdot Al2O_3$), and tetra-calcium aluminoferrite ($4CaO \cdot Al_2O_3Fe_2O_3$). Upon mixing with water, calcium silicate hydrate ($3CaO \cdot 2SiO_2 \cdot 3H_2O$, referred to as C-S-H) and calcium hydroxide ($Ca[OH]_2$, referred to as CH) are formed. Following the carbonation reactions between CO_2, C-S-H, and CH, cement is hardened, forming thermally stable, denser carbonated products, calcium carbonate ($CaCO_3$), and calcium silicate in the form of C3S and/or C2S that further strengthen the concrete in the following days. The effectiveness of carbon capture and storage, as well as the concrete's mechanical properties and durability, hinges on factors such as mineral particle size and morphology, chemical composition (e.g., Ca/Si ratio), carbonation time, pre-moist curing time, curing temperature, curing pressure, and environmental humidity. Pore size and porosity in concrete also significantly impact carbonation efficiency. While larger pores and higher porosity could increase the rate of CO_2 diffusion within the printed concrete, they could adversely lower the mechanical strength of the printed concrete. Careful balancing of different needs when developing carbon-absorbing concrete is necessary. As climate change continues, printed concrete should possess long-term durability against extreme conditions, such as acidic environments and freeze-thaw cycles. It remains a subject of study to determine the long-term capacity for CO_2 absorption and storage (e.g., ¿50 years) in relation to a building's lifetime and whether minerals can be regenerated for continuous CO_2 capture and storage.

Impact on US Construction Industry

The proposed technology will reduce the overall carbon emissions of concrete construction by optimizing the design, using carbon-absorbing concrete, reducing material waste, and eliminating the need for formwork. In turn, the savings will be propagated to all building weights in the lateral and foundation systems of the structure, further realizing dramatic savings compared to traditional processes. Contractors and design firms that incorporate these methodologies will gain strategic competitive advantages to provide these services at a saving to the owners and developers of construction projects. Such research will encourage other researchers and engineers to develop the relevant building codes, thus expanding the use as a mainstream approach. As the skill sets for building construction are broadened, the demand for these services will help to develop a more extensive network of designers, contractors, and suppliers to adopt this approach.

Impacts on Waste Reduction and Carbon Mission, and Low-income Housing
The prefabrication strategy yields enormous potential to economize the design of individual elements and the entire structural system. It results in faster erection times, which also reduces soft construction costs. The technology will reduce the disproportional amount of waste in landfills and the overall carbon emissions of concrete construction. By first optimizing the design, less material will be required overall, thus reducing emissions. Less material will be wasted in prefabrication methods compared to construction on a project site. The printed elements will also eliminate the need for formwork, a significant waste and inefficiency of concrete construction. Besides, the material to be disposed of after the design life will be reduced. Although the primary objective of this research was to provide an innovative design approach for a medium-sized commercial building, low-income housing, as well as immediate housing for postdisaster sites, could be the most promising byproducts of this research. The house geometry can be divided into multiple segments assembled on-site in a few days/hours.

8.8 Nonpolyhedral Structures

The reciprocal relationship between polyhedral form and force diagrams limits applications of 3D polyhedral graphic statics to structures with underlying topologies that are planar or proper cell decompositions, where the consolidated polyhedral force diagrams can be deconstructed into a disjoint union of convex polyhedrons. Polyhedral form diagrams (i.e., subdivided tree structures, faceted domes, polyhedral mesh or surface structures, crystalline aggregations, etc.) represent only a small subset of spatial structures in equilibrium. The framework for networks of disjointed force polyhedrons presented in this section extends applications of polyhedral graphic statics to nonpolyhedral structures. This section also presents a few examples to demonstrate how nonpolyhedral design applications can be explored. This section is an excerpt from Lee et al.[18]

8.8.1 Network of Disjointed Force Polyhedrons

In a consolidated polyhedral force diagram $\Gamma^{\perp}$, every constituent force polyhedron is constrained to have the same *contact face* geometries and areas as each of its

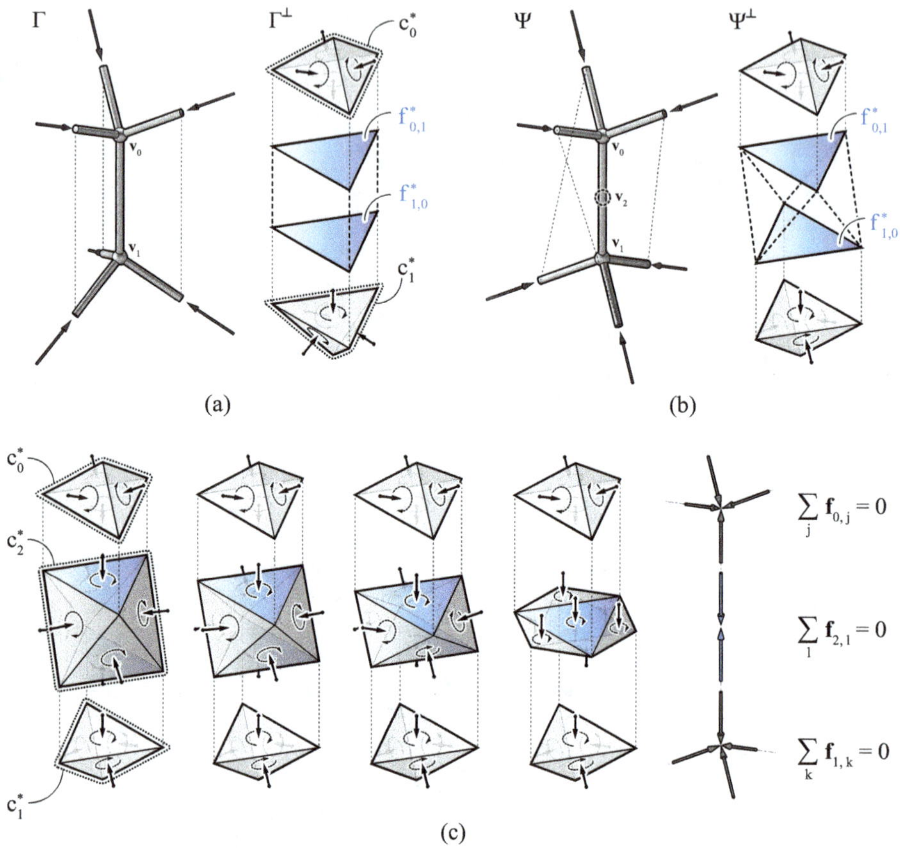

Figure 8.43 (a) Two nodes in equilibrium where the two polyhedrons have matching contact face geometries and areas; (b) two nodes in equilibrium where the two neighboring polyhedrons are disjointed with mismatching contact face geometries; and (c) fictitious node $\mathbf{v}_2$, and the corresponding polyhedron c_2^* linking the two cells with mismatching contact faces, in convex and collapsed states.

neighboring polyhedrons (Figure 8.43a). It is impossible to construct $\Gamma^\perp$ for equilibrated spatial structures that are nonpolyhedral (i.e., structures with overlapping members, nonplanar faces, etc.). Figure 8.43b shows a simple two-node, nonpolyhedral structure Ψ where the two corresponding force polyhedrons have mismatching contact faces. A fictitious node can be added in the middle of the bar connecting the two nodes, where the local equilibrium at that fictitious node can be represented with a collapsed force polyhedron c_2^* (Figure 8.43c). While the collapsed polyhedrons are necessary for graphically representing and describing the equilibrium of nonpolyhedral form diagram Ψ, they do not need to be visualized for practical purposes in an interactive design environment.

A nonpolyhedral form diagram Ψ and its corresponding $\Psi^\perp$ do not have a reciprocal relationship. In a $\Psi^\perp$, the contact faces between the constituent force polyhedrons and each of their neighboring polyhedrons have the same areas but can have different geometries. The intrinsic geometric constraints of $\Gamma^\perp$ (contact faces between any two adjacent polyhedrons that have the same geometries and areas, and the boundary faces

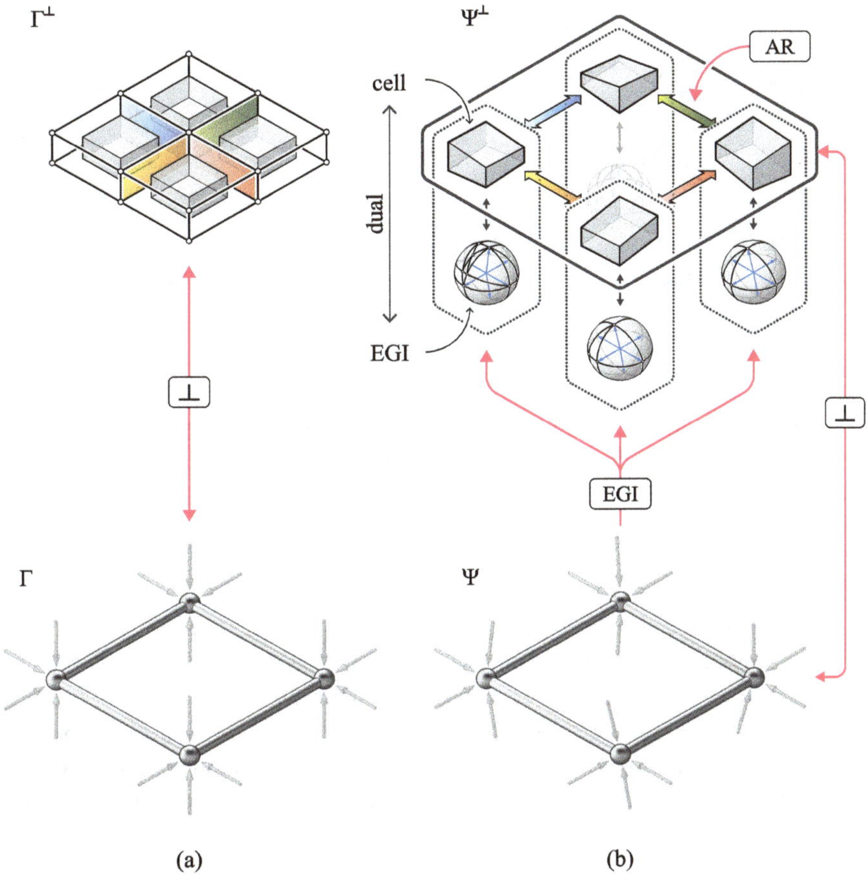

Figure 8.44 (a) Typical datastructure setup for reciprocal form (bottom) and force diagrams (top); and (b) a datastructure setup for a network of disjointed force polyhedrons.

that together form the global force polyhedron) are enforced algorithmically in a $\Psi^{\perp}$ (Figure 8.44a,b):

- Extended Gaussian image (EGI) algorithm ensures that the topology and geometry of the individual cells are correct for the given force distributions at the corresponding node of the structure.
- Arearization (AR) algorithm ensures that all pairs of adjacent cells have contact faces with equal areas.
- $\perp$ algorithm enforces the perpendicularity between Ψ and $\Psi^{\perp}$.

8.8.2 Nonpolyhedral Applications

Improved Visualization for 3D Graphic Statics

Some spatial structures have complicated $\Gamma^{\perp}$ with complex faces and cells, as well as overlapping elements that make the diagrams more difficult to read and understand visually. The self-stressed Jessen icosahedral tensegrity structure (Figure 8.45a)

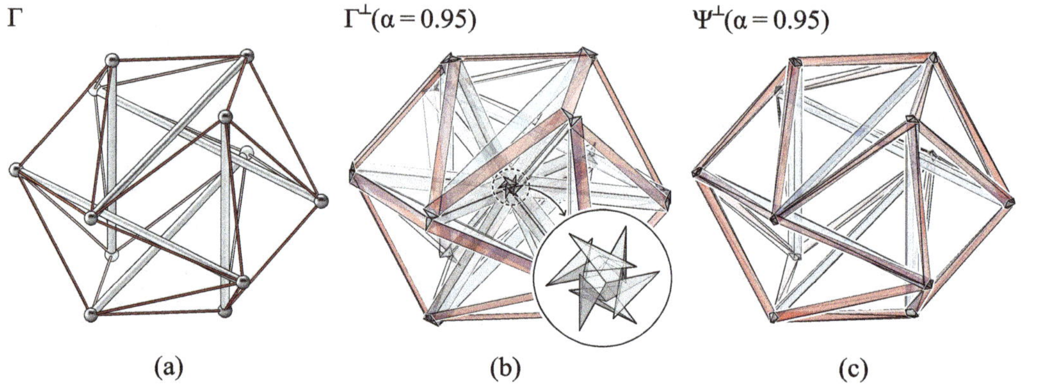

Figure 8.45 (a) Γ of a Jessen icosahedral tensegrity structure; (b) the $\Gamma^{\perp}(\alpha)$ using complex cells and "zero bars" (after[289]); and (c) $\Psi^{\perp}(\alpha)$ without the use of any complex cells, "zero-volume cells" or "zero bars."

is one such example, where the $\Gamma^{\perp}$ (zoomed-in diagram in Figure 8.45b) has no visual resemblance to Γ. The *unified diagram*, $\Gamma^{\perp}(\alpha)$, which represents both the geometry and internal forces of a structure in a single diagram, can improve the legibility of $\Gamma^{\perp}$[283] (Figure 8.45b). By parametrically modifying the scaling factor α, all cells of Γ are scaled relative to their corresponding nodes of Γ such that the distance between any pair of adjacent cells is $\alpha \cdot L$, where L is the length of the corresponding member of Γ. All of the complex faces of the central, "zero-volume cell" c_0^* (which in this case has the same boundary coordinates as $\Gamma^{\perp}$, Figure 8.45b) have areas of zero. While necessary for constructing the $\Gamma^{\perp}$, the "zero-volume cell" adds little benefit to a user who may prioritize quick visual feedback and interactivity with the diagram in a Computer-Aided Design (CAD) environment. In contrast, the unified diagram of the nonpolyhedral representation of the force polyhedra $\Psi^{\perp}(\alpha)$ (Figure 8.45c) is drawn with the same scaling factor as $\Gamma^{\perp}(\alpha)$ (Figure 8.45b), but without the "zero-volume cell" and radial "zero bars," and therefore reducing a significant amount of clutter and resulting in a much more visually discernable representation of equilibrium.

2D and 3D Combined Structures

Using prism-like cells with *virtual faces* (two faces of a prismic force polyhedron that are opposite and parallel and have the same areas that represent two external forces that cancel each other) to represent 2D nodes in equilibrium, $\Psi^{\perp}$s for structures with both 2D and 3D nodes can now be constructed. Figure 8.46 is a twisting arch bridge with 2D nodes along the arch (where the three members at a node are coplanar) and point loads applied to the deck. The combination of 2D and 3D nodes in the same structure means that the members that form a face of the form diagram are no longer necessarily planar and thereby allow the incorporation of twisted faces and features into the structure.

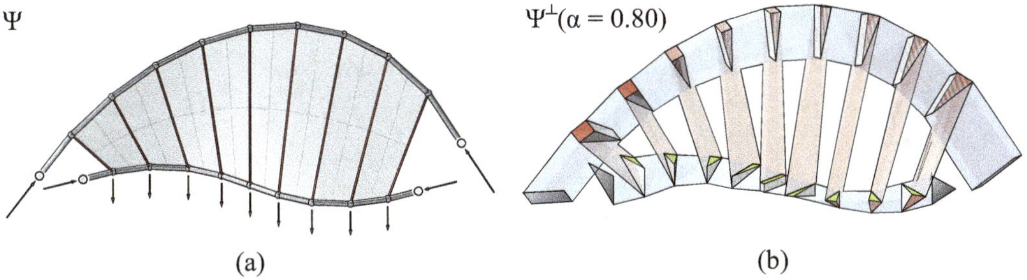

Figure 8.46 Twisting arch bridge with nonplanar faces and the corresponding $\Psi^{\perp}(\alpha)$.

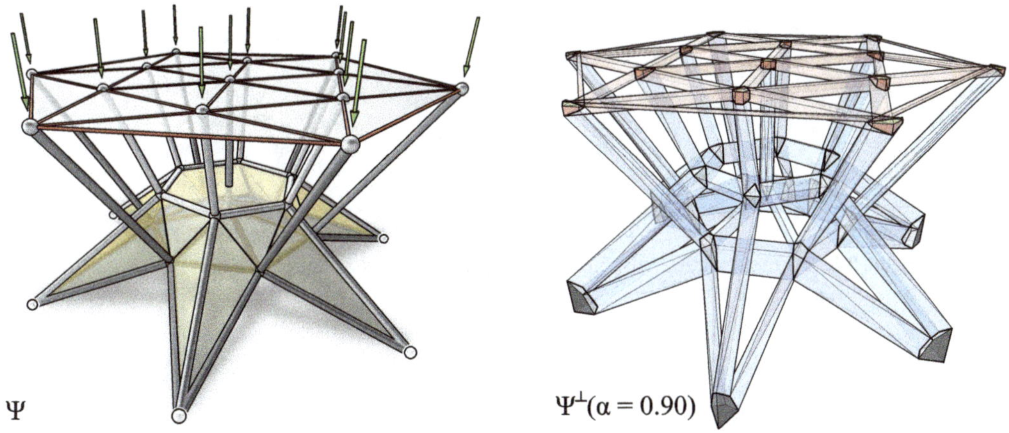

Figure 8.47 A layered and self-overlapping shell structure supporting a flat surface that is uniformly loaded.

Overlapping Structures

Because of the dual and reciprocal constraints between polyhedral form and force diagrams, $\Gamma^{\perp}$ cannot be constructed for structures that have crossing or overlapping edges. The example in Figure 8.47 is a layered shell structure with overlapping vertical support elements. Although this structure is in equilibrium, a $\Gamma^{\perp}$ cannot be constructed. However, a $\Psi^{\perp}$ with force polyhedrons that are individually in equilibrium can be constructed.

Nonpolyhedral Structural Typologies

The spatial tree structure shown in Figure 8.48 was generated by node-by-node transformations, which resulted in an equilibrated and yet nonpolyhedral form diagram. $\Psi^{\perp}$s allows exploration of equilibrated structures that do not have polyhedral geometries, and investigation of free-form designs that are more organic in their aesthetic is now possible. Furthermore, any force equilibrium of structures generated with other form-finding methods such as FDM and TNA can be translated into $\Psi^{\perp}$s.

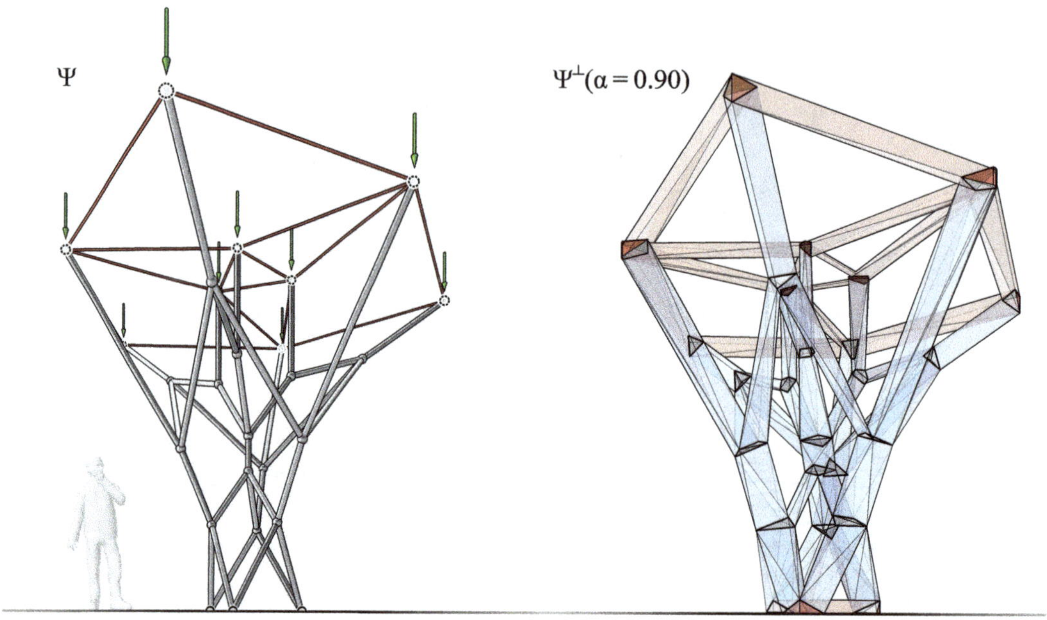

Figure 8.48 A tree structure with nonpolyhedral subspaces as a result of additive, node-by-node transformations of Ψ.

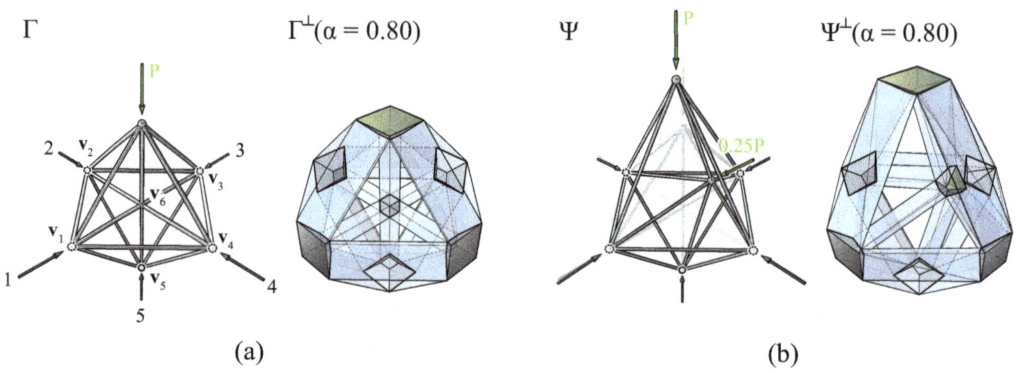

Figure 8.49 (a) An indeterminate truss with four horizontally restrained supports (1–4) and one pinned support (5), and the corresponding $\Gamma^{\perp}(\alpha)$; and (b) the same truss with a point load applied to the interior node, and the updated $\Psi^{\perp}(\alpha)$.

Placing Point Loads Anywhere

One of the main limitations of previous graphic statics applications is that the external loads must be applied at the periphery of the structure, meaning there cannot be any "inner leaves."[122] This is also true for polyhedral graphic statics, where any inner leaves or crossing members mean that a topological dual does not exist. With $\Psi^{\perp}$, point loads can be placed anywhere on either the interior or boundary nodes of the structure. Figure 8.49a is an indeterminate truss with four horizontally restrained supports (1–4) and one pinned support at the bottom (5). If a point load were to be applied to the inner, central node of the structure, a $\Gamma^{\perp}$ cannot be constructed as there does not exist a topological dual for such a configuration of edges. However, in Figure 8.49b, a point load of $0.25P$ is applied to the central node at an arbitrary angle, from which a new Ψ and $\Psi^{\perp}$ were found. The newly added freedom to place loads anywhere in the structure allows investigation of irregular loading scenarios and asymmetric loading conditions and potentially incorporates self-weight loads into the structure.

Tributary Area

While polyhedral subdivisions[130] and transformations[126] allow the generation of intricate spatial structures, the resulting distribution of applied loads on the structure does not end up representing realistic loading scenarios. Consider the spatial tree structure shown in Figure 8.50a, which is designed to support a triangle-shaped roof that weighs P. Suppose that the points on top and the bases of the structures are finalized by the design team and are fixed for the remainder of the design exploration. As shown in Figure 8.50a, the distribution of applied forces as a result of polyhedral transformations of the form diagram often does not correctly reflect the true tributary areas of the structure. With the top and base points fixed and the correct distribution of applied loads imposed, the new shape of the design is found (Figure 8.50b). A designer is typically concerned with rapid shape explorations during the early stages of design. As the design is gradually finalized and a specific topology of the structure is chosen, the designer can begin to adjust the design load case of the initial form diagram to a more correctly calibrated loading scenario to continue developing the design toward a more realistic version of the initial concept.

8.8.3 Future Work

The presented framework for networks of disjointed force polyhedrons broadens the range of applications of polyhedral graphic statics that are freed from any global geometric and topological constraints. The topology of the individual force polyhedrons can change independently from one another. However, the inherent dual and reciprocal relationship is broken and lost. Because there is no longer an explicit geometric constraint that governs the entire force diagram, sufficient constraints need to be imposed manually by the user to obtain stable solutions. While these constraints can be used to explore different equilibrium solutions, it may not be obvious to first-time users what the reasonable domains are for various parameters and constraints. In future investigations, a better means of informing and guiding the user regarding the control and enforcement of constraints will help address solutions that are over- or underconstrained.

8.9 Advanced Topics: Completeness, Bending, and Kinematics

Chapters 1 to 7 in this volume have demonstrated the power of polytopic graphic statics as a design tool for exploring innovative and efficient structural forms. Section 8.10 describes research underway to develop techniques capable of addressing several

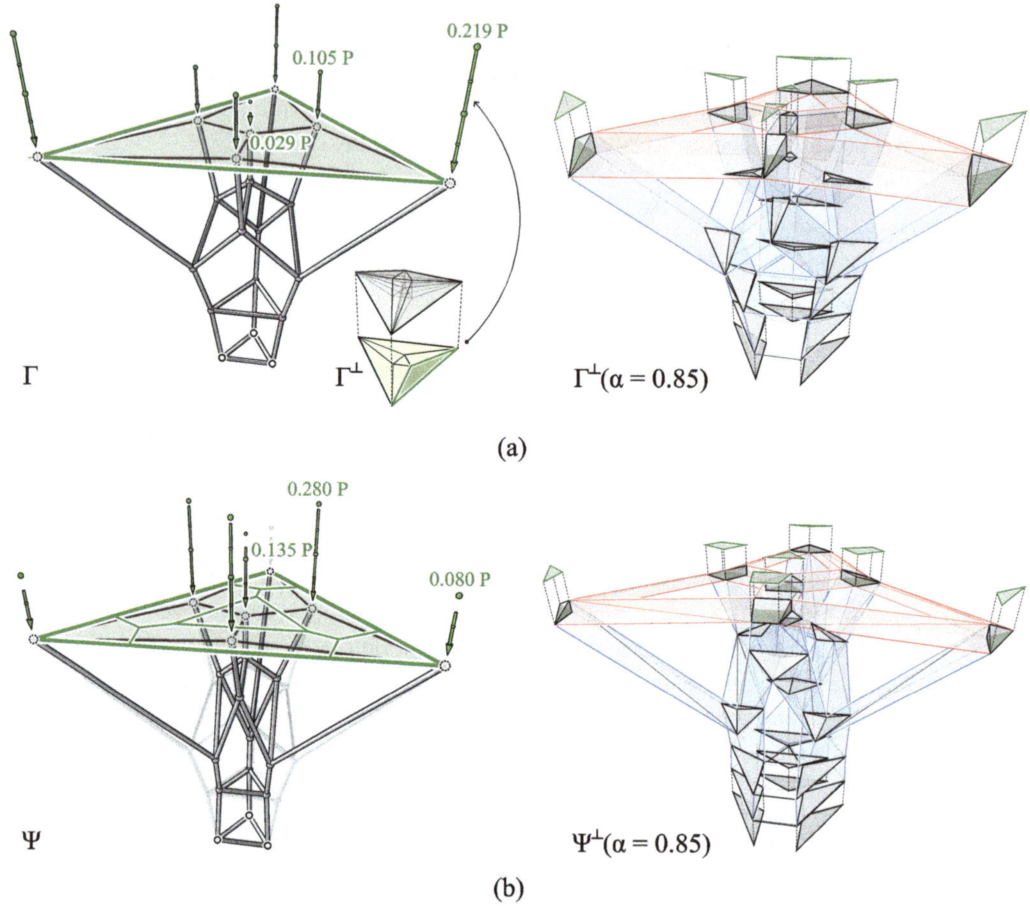

Figure 8.50 (a) A tree structure generated through polyhedral subdivisions and transformations of the form diagram, with a distribution of applied loads that does not reflect the correct tributary areas and (b) the same structure with the correct distribution of applied loads according to the actual tributary areas and the subsequently form-found, new tree structure.

limitations in the existing procedures. The three main problems discussed here are completeness, bending, and kinematics:

- *Completeness*. It is well known that trusses exist – even quite simple ones – whose internal forces cannot be readily represented in polytopic form. The problem is that the polygons at the end nodes of a bar may have equal areas but differing geometry, such that nodal polyhedra cannot be readily assembled into an overall polytope.
- *Bending*. A significant driver for the recent upsurge of interest in graphical methods has been the desire to create design methods appropriate for creating materially efficient structures. Part of that ansatz is the preference for structures that avoid bending under dominant load cases, with applied loads carried to the ground via axial compression and tension wherever possible. However, many structural forms – such as beams and slabs – necessarily involve bending. Even

structures designed to avoid bending may encounter it when the applied loads differ from those of the primary design load cases.

- *Kinematics*. The equilibrium of forces within a structure is, loosely speaking, only half of the structural analysis. The other half is the kinematics – the deflections and rotations of the members and joints. In linear elastic analysis, the compatibility matrix is the transpose of the equilibrium matrix. The polytopic methods described thus far are a powerful approach to equilibrium, and Section 8.9.1 provide an outline of how it may be possible to construct analogous graphical techniques for the compatibility part of the problem.

8.9.1 Completeness and Bending

These two issues will be discussed together. They are different issues, but the approaches to addressing them share much in common. The first step is to generalize the geometrical concepts of polygons, polyhedra, and polytopes. In particular, we relax the requirement that polygonal faces need to be flat. By allowing more generally warped polygons, a richer set of problems becomes amenable to analysis, of which the axial forces in trusses are a subset.

Progress in this direction arose with the paper of Williams and McRobie[284] that explained how discontinuous Airy stress functions could be used to represent bending moments in 2D graphic statics (see Figure 8.51). Prior to this, there was an implicit assumption that the Airy stress functions representing the bar forces in 2D trusses were continuous polyhedra. Faces of polyhedra were necessarily plane, and while there was a discontinuity of slope at the face edges, there was (traditionally) no discontinuity of value. The Williams–McRobie paper relaxed that. Stress functions were still linear over the faces of polyhedra but no longer had to match in value at the face edges. The paper then showed that the discontinuity in stress function value at the face edges provided a neat graphical representation of bending moments in the structural bars. Bar forces need no longer be aligned with the axes of the bars, and the misalignment in force direction could be used to represent shear forces.

The next step appeared in *The Geometry of Structural Equilibrium*[285] where the Williams–McRobie concept of representing bending moments and shears in 2D was

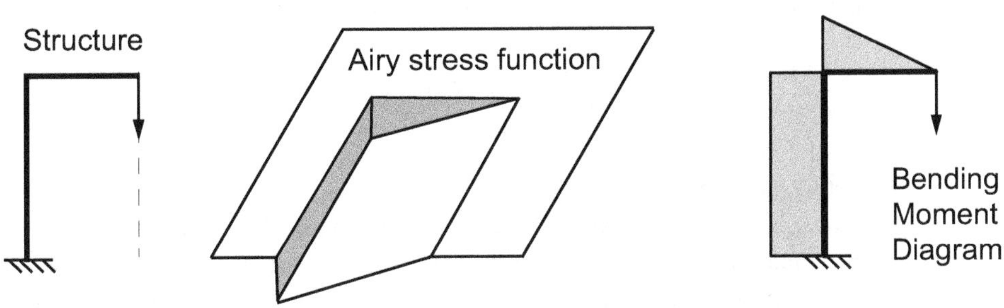

Figure 8.51 A simple cantilever frame and the associated discontinuous Airy stress function whose discontinuities correspond to the bending moment diagram (BMD).

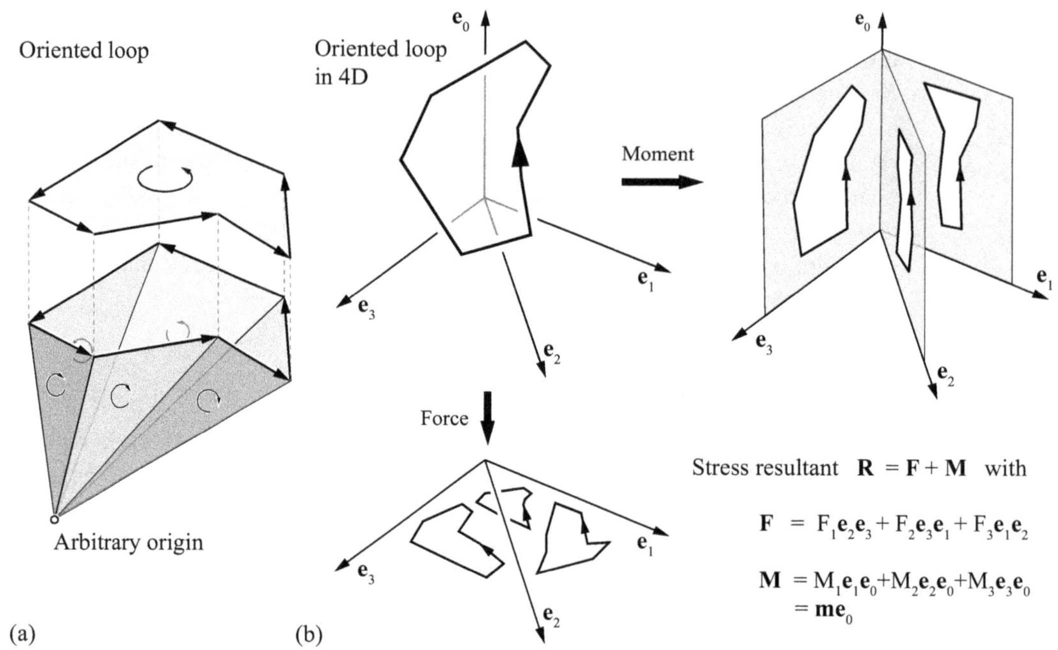

Figure 8.52 (a) The oriented area of any loop can be found by summing the oriented areas of the triangular sides formed by a coning of the loop. With oriented areas defined by bivectors using the wedge product, this construction works in any dimension. In 3D, this corresponds to the areas given by the magnitude of the vector cross products. (b) A loop in four dimensions has six bivector components, three of which can represent force, with the other three representing moment.

generalized to deal with 3D frame structures. The resulting description is essentially the Rankine-like polytopic stress function approach described in the rest of this volume but with moments included. Frame structures are still described as systems of line-like members connecting nodes, all arranged into assemblages of polyhedra, which may be recognized as polytopes. However, the faces of polyhedra need not be plane, and the lines representing bars need not be straight. Rather than representing the force in a bar by the perpendicular area of a flat polygon, the first step in generalization is to replace the flat polygons with more general closed space curves. For such general loops in 3D space, an oriented area can be readily defined by a version of the "shoe-lace formula."

$$A = \frac{1}{2}\sum_{i=1}^{n} \mathbf{r}_i \times \mathbf{r}_{i+1}$$

illustrated in Figure 8.52, where the summation is taken cyclically over the edge vectors surrounding the face.

Any loop can be "coned." That is, vectors can be drawn from an arbitrary origin to all points on the loop. This creates a pyramid-like object with triangular sides whose base is the loop, the warped polygon of interest. Since, by hydrostatics, the total oriented area of the surface of a closed object is zero, the oriented area of the warped polygon must be equal and opposite to the sum of the oriented areas of the triangular

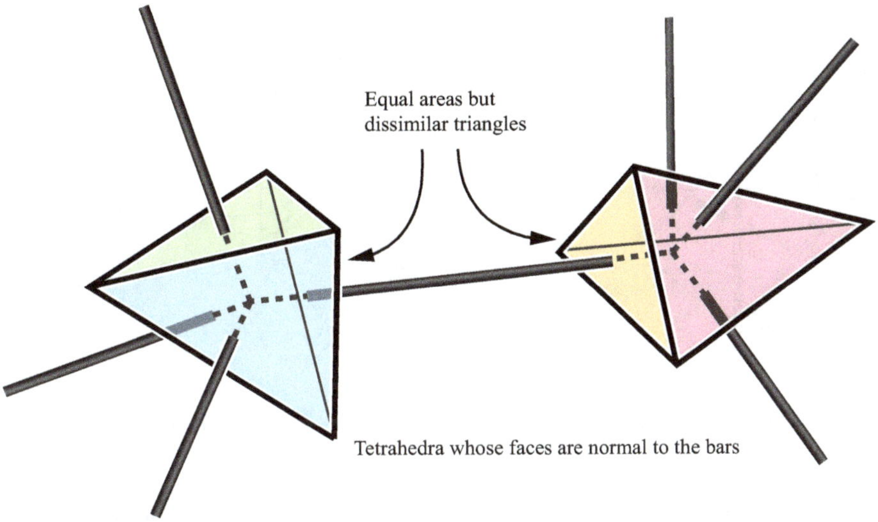

Figure 8.53 Rankine incompleteness. The areas of the triangles facing each other at either end of the central bar have equal areas, each equal to the tension in the bar. However, the two triangles have differing geometry, so the force polyhedra at the end nodes cannot be conjoined to create a dual Rankine polytope.

sides. The oriented area of each triangle is a vector of magnitude equal to the area of the triangle and in a direction normal to the face of the triangle. The sum of such vectors is also a vector that represents (up to sign) the oriented area of the more general warped polygon of interest. The negative sign can be removed if the triangle orientations match the vectors' orientations around the loop. If desired, the three components of the vector can be taken to give the $x-$, $y-$, and $z-$ components of the force that the loop represents.

In this generalized version of the Rankine construction, reciprocal to any point on a bar of the structure is a loop in the force space, whose oriented area defines a vector whose magnitude and direction represent the force acting within the bar at that point. This force vector need not be parallel to the axis of the bar and may thus represent both axial and shear forces, and there will thus be associated moments. However, before addressing moments, we consider some axial-only solutions that do not have a standard Rankine polytopic representation. The problem is known as *Rankine incompleteness*. Figure 8.53 gives an example. The nodes at the bar ends are each dual to cells, and the faces of the cells corresponding to the bar force must have equal area, but they do not match geometrically. The reciprocal polygons differ in shape. A state of self-stress can exist, but a simple Rankine dual cannot be constructed.

The paper "Rankine reciprocals with Zero Bars"[286] introduced the idea that there were many cases where, although it appeared that a truss possessed a state of self-stress that could not be represented by a polytopic dual, the self-stress could actually be represented by the inclusion of fictitious "Zero Bars." These bars have forces represented by polygons that cross themselves, having regions of positive and negative oriented area such that the total oriented area is zero.

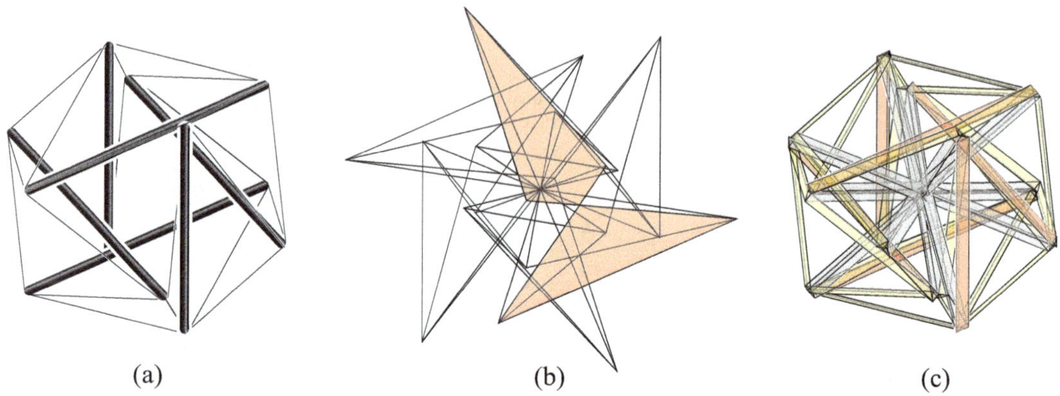

Figure 8.54 (a) a Jessen icosahedral tensegrity. This has only two cells (two co-located icosahedra); thus, the Rankine reciprocal should have only two points, which is insufficiently rich to represent a meaningful state of self-stress. However, if the icosahedron is coned with zero bars to a central node, a reciprocal can be defined. (b) the Rankine reciprocal, with one of the zero-area polygons of one such zero bar highlighted. (c) The Rankine–Minkowski diagram shows that the bars in the icosahedron carry a state of self-stress, with the radial bars carrying zero force.

Figure 8.54 shows a Jessen icosahedral tensegrity. It has a single state of self-stress, yet since the structure has only two cells (two coexistent icosahedra), a traditional Rankine reciprocal would consist of only two points, which does not have sufficient richness to describe the state of self-stress. By adding a node at the center of the icosahedron and connecting all the tensegrity nodes to that central node via zero bars, one of the icosahedral cells may be subdivided into twenty pyramidal cells. A stress function may then be constructed that is linear across each cell and continuous across cell boundaries, allowing a reciprocal structure to be constructed. Bars in the tensegrity are reciprocal to flat polygons perpendicular to the bars and have oriented areas corresponding to forces carried in the self-stressed state. Reciprocal to the Zero Bars radiating from the central node are flat polygons of zero area. These may be loosely described as fish-like, with the positive area of the head and body regions being matched by the negative area of the tail region. There is thus no radial force: all forces are carried out on the tensegrity itself.

If the faces of polyhedra are allowed to be nonflat polygons, then the solution space is expanded further, giving even more possibilities for finding and representing equilibrium solutions. If there is no state of self-stress with a purely axial set of forces for a given structure, then rather than returning a null result, a search algorithm may return an equilibrium solution with nonaxial forces and accompanying moments instead. Misalignment of forces and bars should thus not necessarily be interpreted as an error; instead, it may be a perfectly good model of a frame structure that possesses states of self-stress that include moments.

Indeed, if nonflat polygons are allowed, all possible states of self-stress of any 3D frame can be represented, including all states of self-stress that are purely axial. The incompleteness issue has thus been solved.

To represent moments, the Maxwell–Rankine stress function F is considered as providing an additional dimension, taking the analysis from the usual 3D space of $(x, y,$ and $z)$ into the four-dimensional (4D) space of $(F, x, y,$ and $z)$. The total stress resultant at any cross-section on a bar may again be given by the oriented area of a general (nonflat) loop. Still, in the 4D setting, the loop has six two-dimensional projections. The areas of the loop, when projected onto the pure space planes xy, yz, and zx, will be used to give the three components of the force, as described previously. The projections on the xF, yF, and zF planes will be used to give the three components of the moment at that cross-section. The two components of the bending moment can be so represented, along with the torsional component that is oriented along the bar. Since vector cross-products are not defined in 4D space, the required version of the shoelace formula to calculate the oriented area of a general polygon in 4D uses the wedge product of Grassmann calculus:

$$\mathbf{A} = \frac{1}{2}\sum_{i=1}^{n} \mathbf{r}_i \wedge \mathbf{r}_{i+1}$$

The result is a general bivector giving the oriented area for the loop, whose six components represent the stress resultant of force and moment at the bar cross-section under consideration (see Figure 8.52).

The simplest structure that can carry self-stress is a loop. We may imagine a loop that is cut at one section. Equal and opposite forces and moments are applied to the two exposed faces, which are then rejoined, locking in the stresses. This is the *form loop*. The state of self-stress throughout the loop is uniquely defined when the stress resultant $\mathbf{R} = \mathbf{F}+\mathbf{M}$ is defined at a single cross-section of the loop. That stress resultant could be specified in a number of ways, perhaps by giving a force vector and a moment vector. Here, though, we can choose to specify the stress resultant by means of a *force loop*. This is a loop in the reciprocal 4D stress space defined by an ordered set of n nodes. The coordinates of a reciprocal node are of the form $\mathbf{A}^K = (a_0, \mathbf{a})$, where a_0 is a scalar and $\mathbf{a}$ is a vector (a_1, a_2, a_3) in the 3D subspace reciprocal to x, y, and z. The six bivector projections of this force loop define the stress resultant at that cut face. The stress resultant at any other cross-section of the loop at a point spatially separated by $\mathbf{X} = (0, \mathbf{x})$ with $\mathbf{x} = (\Delta x, \Delta y, \Delta z)$ is then defined[285] via the Corsican sum $\mathbf{S} = \bigcup(\mathbf{X}.\mathbf{A}^K, \mathbf{x} + \mathbf{a}^K)$.

Within the 3D space, xyz, the Corsican sum is simply a translation of the force loop along the vector $\mathbf{x}$, and therefore it does not change the areas of the force loop projected onto the xy, yz, and zx planes. The force thus remains constant on different cut sections of the form loop, as required by force equilibrium.

The components in the fourth dimension do, however, change in such a way that moment equilibrium is satisfied. That is, if the stress resultant at a cross-section at the point P with coordinate $\mathbf{X}_P$ is $\mathbf{R}_P = \mathbf{F}_P + \boldsymbol{m}_P\mathbf{e}_0$, the stress resultant at a cross-section at the point Q with coordinates $\mathbf{X}_Q = \mathbf{X}_P + \mathbf{X}$ is $\mathbf{R}_Q = \mathbf{F}_Q + \boldsymbol{m}_Q\mathbf{e}_0$ with $\mathbf{F}_Q = \mathbf{F}_P$ and $\boldsymbol{m}_Q = \boldsymbol{m}_P + \boldsymbol{x} \times \boldsymbol{f}$. Here $\mathbf{f} = F_1\mathbf{e}_1 + F_2\mathbf{e}_2 + F_3\mathbf{e}_3$ is the force expressed as a vector, and $\boldsymbol{x} \times \boldsymbol{f}$ is the extra moment at the second point due to the additional lever arm offset $\boldsymbol{x}$.

The mathematical machinery for creating generalized Rankine-like reciprocal diagrams thus exists and is capable of representing not only all possible axial-only states of self-stress in a truss structure but all possible states of self-stress, including those with axial and shear forces and with bending moments and torsions. Bars need not be straight, graphs need not be planar, and the faces of polyhedra do not need to be flat.

In summary, by allowing polyhedra whose faces are nonflat polygons, a wider vocabulary of Rankine reciprocal diagrams becomes possible, allowing states of self-stress to be represented that have no traditional Rankine reciprocal. These may or may not include moments. It can be readily shown[287] that there is no incompleteness problem for such a description. Any state of self-stress in any framed structure may be so represented.

8.9.2 Kinematics

Similar to the way that polyhedral and polytopic stress functions can be used to represent the equilibrium of forces in truss and frame structures, it is possible to construct related geometric descriptions of the kinematics of the displacements of such structures. The essential ideas are spelled out in McRobie et al.[288]

In analyzing the equilibrium of pin-jointed trusses, a vector representing the externally applied nodal forces may be ascribed at every node. Similarly, in the analysis of compatibility, a displacement vector may be ascribed at every node to define the deformation of the truss. If all such displacement vectors are gathered into a single vector space with the same origin, they define a set of points, and two such points can be connected by a line if they correspond to nodes at the ends of a shared bar. We shall call this the displacement diagram.

Bar extensions are related to the relative displacements of adjacent nodes, and thus, the connecting lines in the displacement diagram give relative displacement vectors. Assuming displacements are small, the components of these that are oriented along the original bar direction give the bar extensions. Multiplying these bar extensions by the bar stiffness gives the bar forces, which must be the same as those given by the force diagram.

The displacement diagram is a 3D network of nodes connected by edges with the same topological connectivity as the form diagram of the original structure. It follows immediately that since the force diagram is the topological dual of the form diagram, it is also the topological dual of the displacement diagram. Many of the geometric ideas developed for graphic statics between form and force can thus also be developed for displacement and force. One interesting construction is to combine the force and displacement diagrams into a single unified diagram similar to the Maxwell–Minkowski and Rankine–Minkowski diagrams of 2D and 3D graphic statics. Where the equilibrium versions led to the visualization of load paths, the compatibility versions led to interesting visualizations of virtual work, which is one of the mainstays of structural analysis and design.

Displacement diagrams are not new. The earliest examples are by Williot,[289] with subsequent enhancements by Mohr.[290] The two-volume textbook by

Müller-Breslau[291,292] describes them in detail, as does the more recent book by Rao.[293] Displacement diagrams are mostly used in mechanical engineering, supplemented by acceleration diagrams created on the same principle: all nodal acceleration vectors are drawn from a single origin and connected as per the form diagram connectivity. Using such diagrams in design practice has gradually been replaced by computational techniques. However, even today, this graphical method is taught in the early years of most engineering curricula to give students a deep understanding of the fundamentals of kinematics.

8.9.3 Maxwell–Williot and Rankine–Williot Diagrams

Rather than pursuing Williot's original graphical methods, we follow a novel approach suggested by graphic statics. Here, a displacement diagram for a truss is constructed by first decomposing the force diagram into its constituent pieces and then sliding the pieces apart. The displacement diagram emerges in the gaps between the separating pieces. The general idea is shown in Figure 8.55 for the simplest 2D truss, the three-spoked triangular wheel (after the famous Figure 1 of Maxwell[17]). The form and force diagrams are shown at the top of the figure. Simple support boundary conditions have been ascribed to the form diagram to specify the kinematics. The central node is pinned, and the uppermost node is on a roller support that prevents horizontal motion. The line connecting the central and upper nodes is now interpreted not as a bar but as the line of action of an externally applied force F acting upwards on the upper node. The problem is determining the nodal displacements of the truss under the action of load F.

The procedure is demonstrated in Figure 8.55c. The central node N_4 is a pinned support, so we begin by taking the force polygon N^4 reciprocal to this, shown as the large shaded triangle. Node N_4 is connected to the lower two nodes N_2 and N_3 by flexible bars, which will deform under the loading. The lower two nodes will thus move. The force polygons N^2 and N^3 reciprocal to these two lower nodes are thus slid to displaced locations relative to the force polygon N^4 of the central node. The gap between the two relocated force polygons N^2 and N^3 will correspond to the extension or compression of the horizontal bar connecting the nodes N_2 and N_3, as per the central diagram of Figure 8.55c. Finally, the force polygon N^1 reciprocal to the upper node N_1 is slid to a position offset from the edges of the force polygons N^2 and N^3 of the two lower nodes N_2 and N_3 to which it is connected by bars. The vertical offset that has opened up between the main and upper force polygons N^1 and N^4 is the vertical displacement of the upper node N_1 under the action of the applied force F. Indeed, the small polygons that open at all nodes between the force polygons in the Maxwell-Williot diagram (Figure 8.55d) can be assembled to create the overall displacement diagram of all nodes (Figure 8.55e).

The only part of the procedure not yet defined is how far to slide the various force polygons. In general, only a component of the displacement vector will be known, the component parallel to the bar in question, which is the bar extension. Since the force in the bar is known from the force diagram, the bar extension can be calculated if the bar

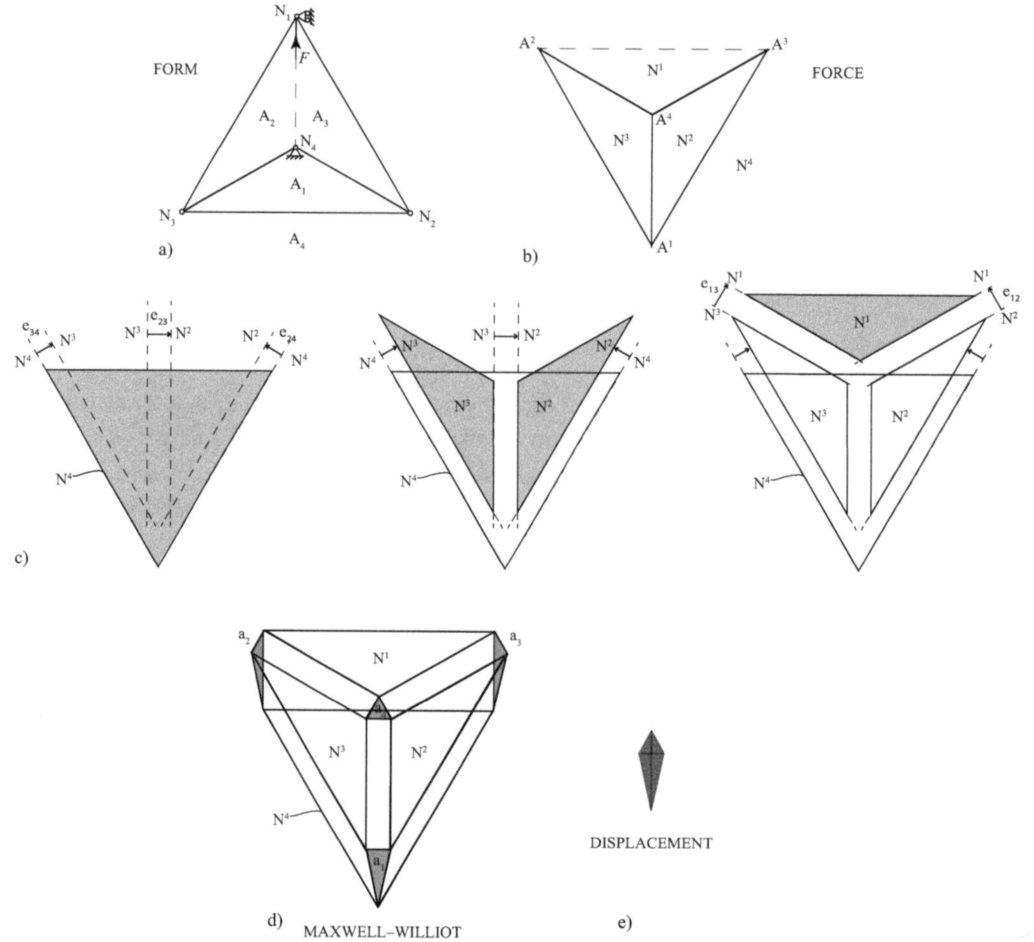

Figure 8.55 The construction of a displacement diagram via the Maxwell–Williot diagram. (a) and (b) Reciprocal form and force diagrams of a simple truss; (c) moving apart the pieces of the force diagram; (d) the Maxwell–Williot diagram; and (e) the displacement diagram created within the gaps of the Maxwell–Williot diagram.

stiffness is known. This gives the required displacement component. This is illustrated by the first two diagrams of Figure 8.55c. If the lower sloping bars compress axially by amounts e_{24} and e_{34}, then force polygons N^2 and N^3 are offset from the sides of the main force polygon N^4 by perpendicular distances of this amount. Their final location is determined by the extension e_{23} of the lower horizontal bar: the force polygons N^2 and N^3 may be slid relative to the sloping sides of the main force polygon N^4, while always maintaining the correct perpendicular offsets e_{24} and e_{34} until they reach a horizontal separation equal to e_{23}.

The upper force polygon can be readily located if the bar extensions e_{12} and e_{13} have been calculated because the perpendicular offsets by these amounts from the force polygons N^2 and N^3 dictate the only possible location for the final piece N^1.

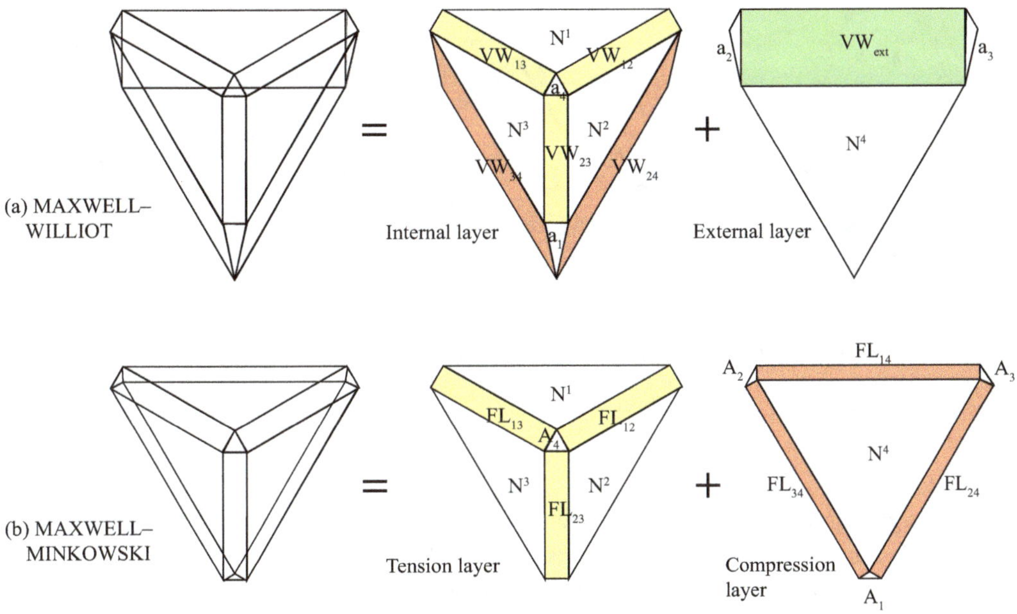

Figure 8.56 (a) The areas of the Maxwell–Williot diagram give a graphic representation of the equality between internal and external work. (b) Similarly, areas of the Maxwell–Minkowski diagram represent the equality between tension and compression load paths.

As well as creating the displacement diagram, the procedure has produced the Maxwell–Williot diagram. This unifying graphical representation combines the force and displacement diagrams with interspersed quadrilaterals. It is similar to the Maxwell–Minkowski diagrams, which combine form and force.[294,295] However, displacement vectors are not perpendicular to the nearby force vectors, such that the interspersed quadrilaterals are parallelograms rather than rectangles. The area of each quadrilateral is thus the product of a force vector with a displacement, which is the work done. This is similar to the visual portrayal of the Maxwell Load Path terms $T.L$ that emerge as the area of the rectangles in Maxwell–Minkowski diagrams. It is immediately evident that the Maxwell–Williot diagram of Figure 8.55d is a doubly-layered diagram and that by simple consideration of the diagram areas, the parallelograms that have opened up on one layer must have the same area as the parallelograms opened on the other layer (Figure 8.56). This area of equality is a visual representation of the equality of internal and external work.

Given that real work is the integral of stress times strain, with both rising from zero, the actual work done is equal to half the areas of the parallelograms. There is, however, no requirement that the displacement vectors were in any way caused by the force vectors. Bar forces could be held constant, and small displacement vectors could be applied. This corresponds to a graphical representation of virtual work with the full areas of the parallelograms representing the various contributions to virtual work. The Principle of virtual work, which "dots" an equilibrium set of forces with a compatible set of displacements and equates internal and external contributions, is

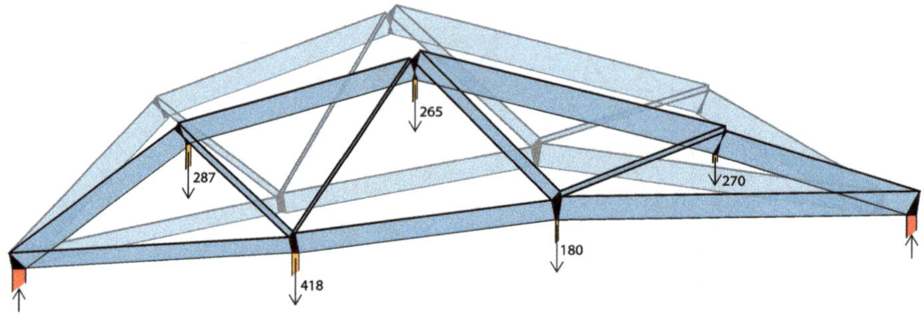

Figure 8.57 The deformed Maxwell–Williot diagram for a truss bridge. The elastic deformations were calculated via the Maxwell–Williot procedure and combined with the form diagram. See McRobie et al.[288]

arguably the most important principle in structural analysis, and the Maxwell–Williot diagram gives a vivid display of its inner workings.

Combining form, force, and displacement into a single diagram is even possible. Form and displacement have the same topology; thus, any linear combination of the two is reciprocal to the force diagram. The resulting combination is called the *deformed Maxwell–Williot diagram*. Figure 8.57 gives an example of a simple bridge truss.

The way that polygons of the force diagram were slid relative to one another to create the Maxwell–Williot diagram leads to a method for identifying mechanisms within 2D trusses. Since adjacent force polygons' pieces are moved apart by the extension of the bar that connects the corresponding nodes, it follows that if two polygons are slid along the adjoining edge without separating, the bar in question underwent no extension. If it is possible to slide all polygons in this manner, this clearly corresponds to a deformation with no bar extensions – that is, to a mechanism. Figure 8.58 gives an example, identifying two mechanisms in a simple bridge truss.

The foregoing description of graphic kinematics extends immediately from 2D Maxwell reciprocals to 3D Rankine reciprocals. Figure 8.59 is the 3D version of the three-spoked wheel example earlier. An outer tetrahedron has four internal spokes connecting the corners to a central node. The structure is on simple supports. These consist of a main pin support at the central node, with three roller supports at the outer nodes to prevent rotation. A vertical load is applied at the lower node. The line connecting the central and lower nodes is thus the line of action of this force rather than a bar. Figure 8.59a,b shows the form diagram and the reciprocal Rankine diagram with polyhedra whose polygonal face areas represent the forces. Figure 8.59c shows how the reciprocal force cells may be slid apart to create the vectors of a displacement diagram at the nodes.

Figure 8.60 shows the displacement diagram in more detail, and Figure 8.61 shows the resulting unified diagram of form and displacement combined with the forces.

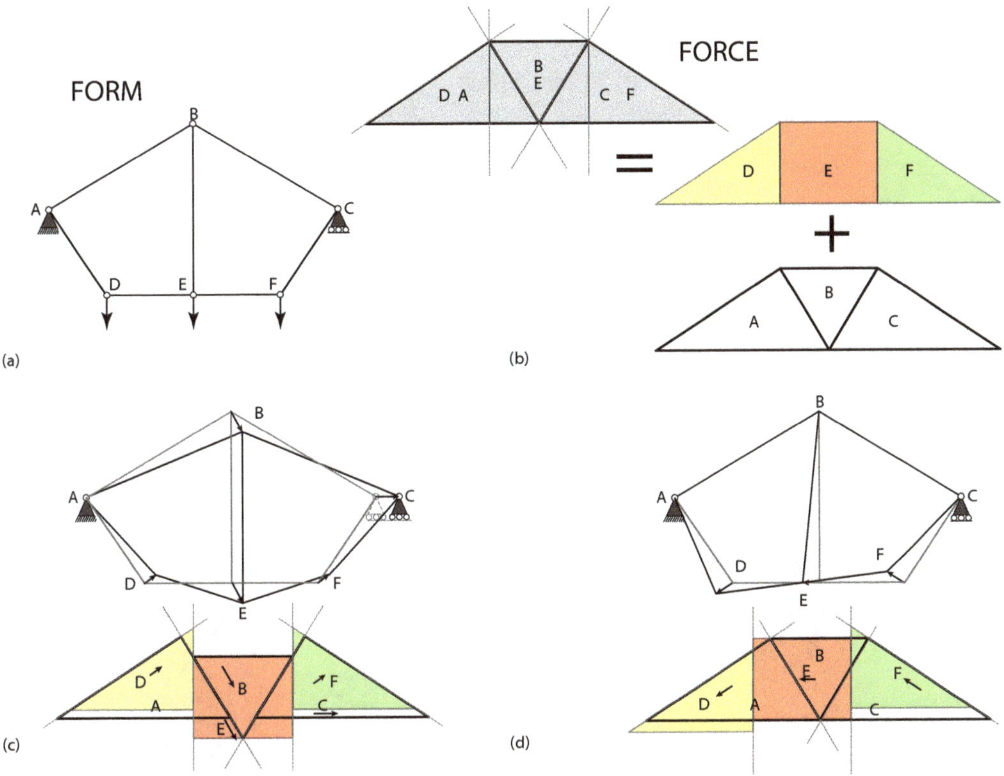

Figure 8.58 Graphic kinematics, using the Maxwell–Williot diagram to identify mechanisms in a simple bridge truss: (a) The form diagram; (b) the pieces of the doubly-layered reciprocal force diagram; and (c) and (d) show how sliding the force polygons without separation leads to the nodal displacement vectors of the two mechanisms.

This gives clear visual expression to the displaced shape and, via the triangular cross-sectional areas of the bars, to the associated bar forces.

The work associated with the deformation is now given by (half) the volumes of the various parallelopipeds that have opened up between the force cells. As in the 2D case, moving the force cells apart with compatible small displacements and obtaining a 3D graphical representation of the principle of virtual work would have been possible.

An example of mechanism analysis via sliding force cells along common boundaries is given in Figure 8.62. This is the Jessen icosahedron considered in Section 8.9.1. Using zero bars, a valid Rankine reciprocal diagram was created by connecting the outer icosahedron to a central node. The complicated geometry of the resulting Rankine reciprocal is shown in Figure 8.62b,c, with Figure 8.62d showing how the force cells may be slid relative to each other without separation. Each such motion of a force cell corresponds to a displacement of the original node to which the cell is reciprocal, and the mechanism that has been so identified is shown in Figure 8.62e.

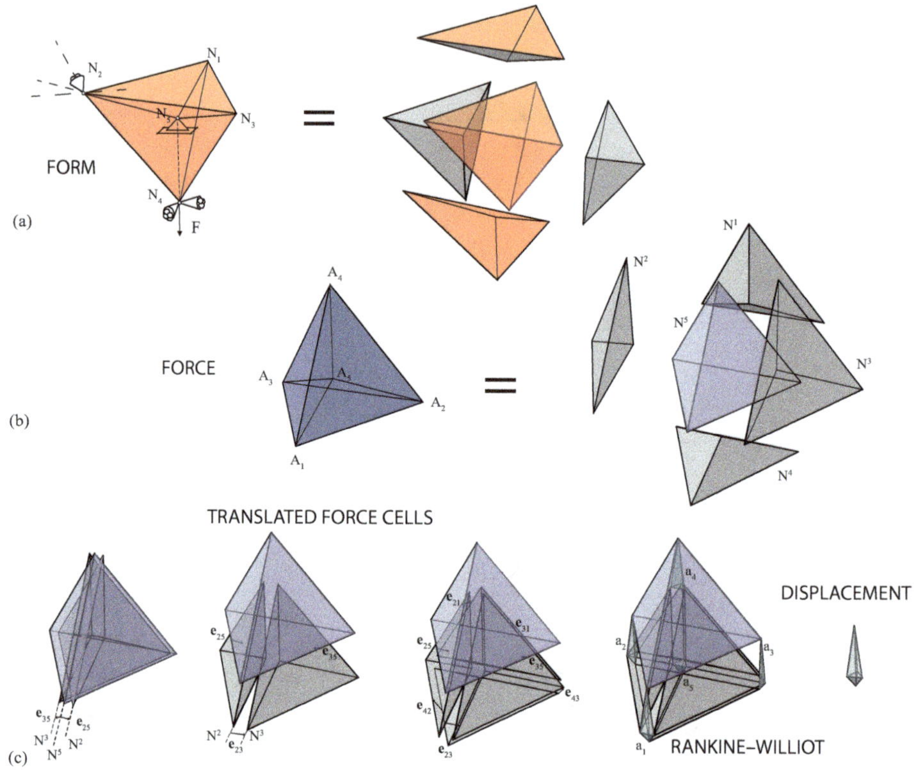

Figure 8.59 Graphic kinematics of a simple 3D structure. (a) and (b) The form diagram and its Rankine reciprocal. (c) Sliding apart the force cells to create the vectors of the displacement diagram.

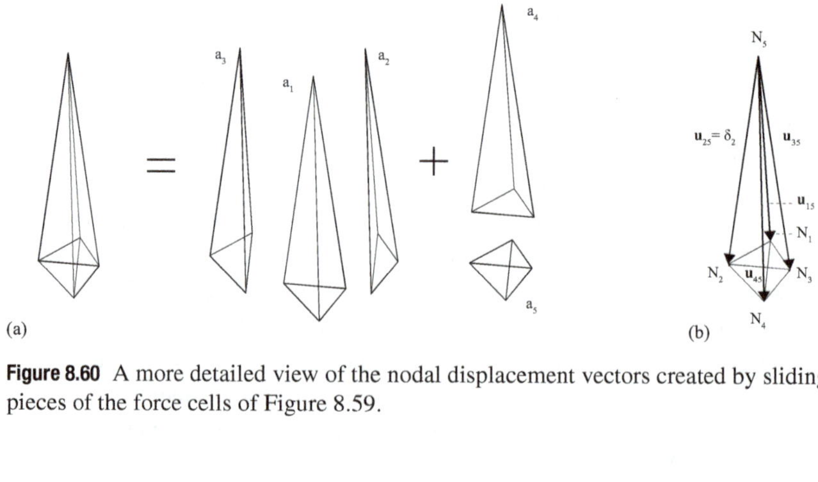

Figure 8.60 A more detailed view of the nodal displacement vectors created by sliding apart the pieces of the force cells of Figure 8.59.

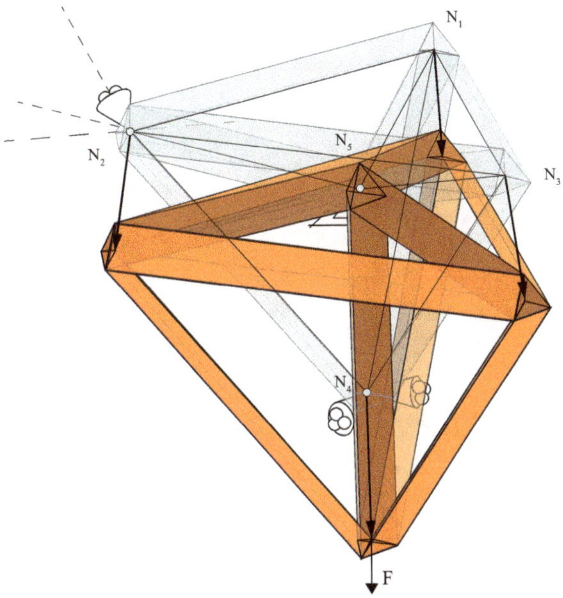

Figure 8.61 The deformed Rankine–Williot diagram for the structure of Figure 8.59.

8.10 Chapter Summary

In this chapter, we showed the broader application of polyhedral graphic statistics in other fields of science and briefly introduced research directions and topics that go beyond the polyhedral limitations of this method. Particularly, we show a research project in which graphical methods were used to analyze the structural pattern of a dragonfly wing. The result was then combined with machine learning methods to generate the structure of the wing of an airplane with enhanced out-of-plane performance. We also visited applications in the design of strut-and-tie structures for referenced concrete and its further application in designing multi-material structural components where the direction of the deposition of material is adjusted with respect to the internal force flow to maximize mechanical performance. The application of polyhedral graphic statics was shown in the design of cellular solids and briefly discussed how particular subdividing of the force diagram can control the stress distribution in the system and the overall behavior of the structure from a bending dominant to a stretching dominant system. We also showed the application of the structures designed using polyhedral graphic statics in self-healing structural components and 3D-printed structural systems with maximized surface area and minimized mass. Another important topic was the extension of the methods of polyhedral graphic statics to nonpolyhedral systems using disjointed force polyhedra. In the end, advanced topics related to completeness, being, and kinematics in polyhedral graphic statics were discussed, which opened the door to many further research directions in this field.

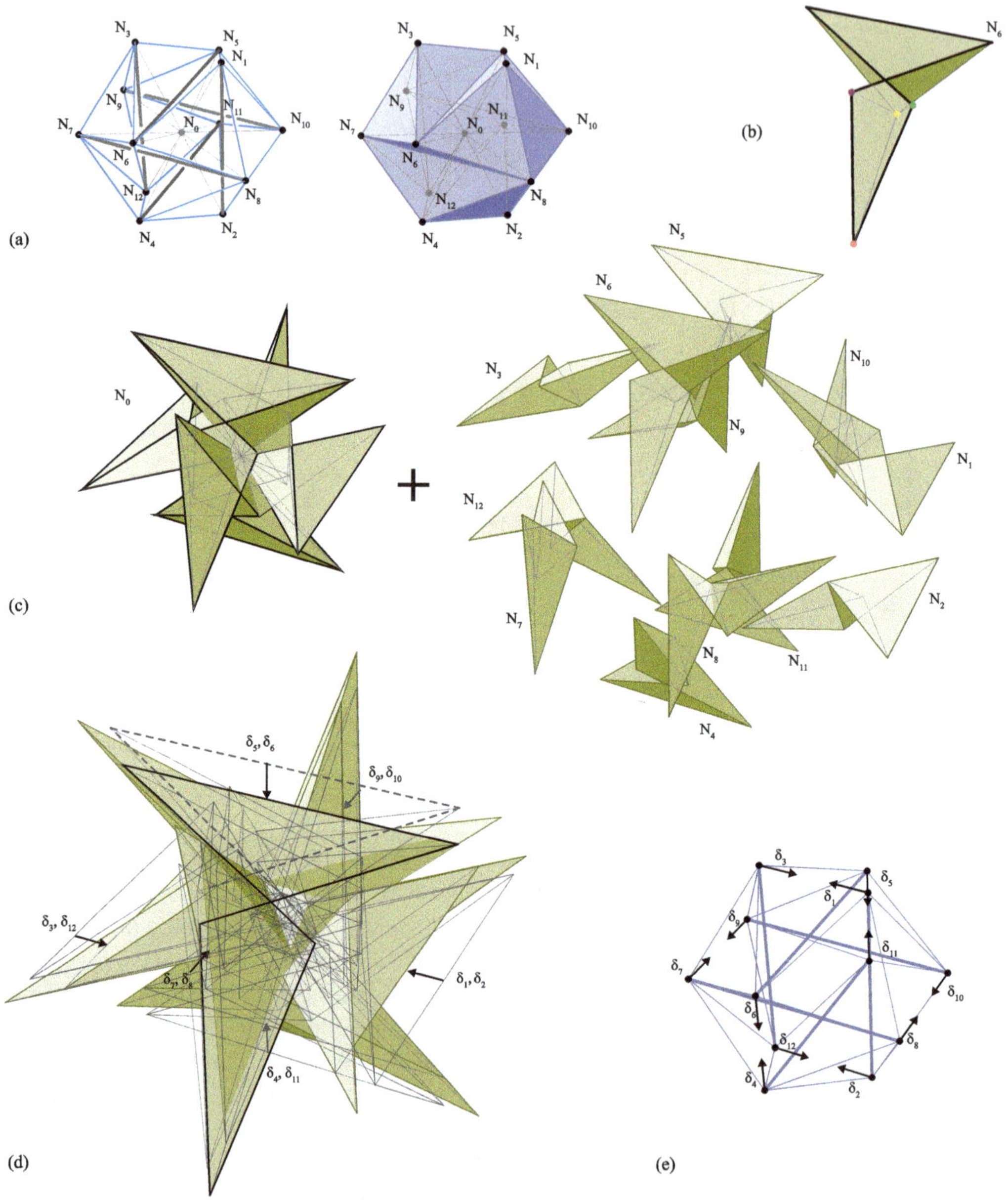

Figure 8.62 Mechanism analysis of the Jessen icosahedral tensegrity by sliding force cells along common boundaries. (a) The form diagram and its coming; (b) a representative reciprocal force polygon. The face dual to the zero bar of the coning is visible; (c) the Rankine reciprocal; (d) sliding force cells along common faces; and (e) the mechanism identified.

References

1. G. Hegemier and W. Prager, "On Michell trusses," *International Journal of Mechanical Sciences*, vol. 11, no. 2, pp. 209–215, 1969.
2. A. G. M. Michell, "Lviii: The limits of economy of material in frame-structures," *The London, Edinburgh, and Dublin Philosophical Magazine and Journal of Science*, vol. 8, no. 47, pp. 589–597, 1904.
3. W. F. Baker, L. L. Beghini, A. Mazurek, J. Carrion, and A. Beghini, "Maxwell's reciprocal diagrams and discrete michell frames," *Structural and Multidisciplinary Optimization*, vol. 48, pp. 267–277, 2013.
4. S. Timoshenko, *History of Strength of Materials: With a Brief Account of the History of Theory of Elasticity and Theory of Structures*. Lowell, MA: Courier Corporation, 1983.
5. M. Rankine, "Principle of the equilibrium of polyhedral frames," *Philosophical Magazine*, vol. 27, no. 180, p. 92, 1864.
6. R. Cruise, "The 300m Eiffel Tower," *Construction History*, vol. 35, no. 2, pp. 43–66, 2020.
7. L. K. Eaton, "Hardy Cross and the 'moment distribution method'," *Nexus Network Journal*, vol. 3, no. 2, pp. 15–24, 2001.
8. P. Block, *Thrust Network Analysis: Exploring Three-Dimensional Equilibrium*. PhD thesis, MIT, Cambridge, MA, 2009.
9. M. Akbarzadeh, *3D Graphic Statics Using Polyhedral Reciprocal Diagrams*. PhD thesis, ETH Zürich, Zürich, Switzerland, October 2016.
10. A. McRobie, "Maxwell and Rankine reciprocal diagrams via Minkowski sums for two-dimensional and three-dimensional trusses under load," *International Journal of Space Structures*, vol. 31, pp. 203–216, August 2016.
11. J. Lee, *Computational Design Framework for 3D Graphic Statics*. PhD thesis, Department of Architecture, ETH Zurich, Zurich, 2018.
12. P. Block, T. Van Mele, and M. Rippmann, "Geometry of forces: Exploring the solution space of structural design," *GAM*, vol. 12, pp. 28–37, April 2016. Special issue S. Peters and A. Trummer (eds.) – "Structural Affairs" Faculty of Architecture, Graz University of Technology – BIRKHÄUSER, 2016.
13. P. Block, T. Van Mele, M. Rippmann, F. Ranaudo, C. Calvo Barentin, and N. Paulson, "Redefining structural art: Strategies, necessities and opportunities," *The Structural Engineer*, vol. 98, pp. 66–72, January 2020.
14. Bhooshan, Shajay, A. Dell'Endice, F. Ranaudo, T. Van Mele, and P. Block, "Unreinforced concrete masonry for circular construction," *Architectural Intelligence*, vol. 3, no. 1, p. 7, 2024.
15. T. Van Mele, M. Rippmann, L. Lachauer, and P. Block, "Geometry-based understanding of structures," *Journal of the International Association of Shell and Spatial Structures*, vol. 53, no. 2, pp. 285–295, 2012.

16. E. Allen and W. Zalewski, *Form and Forces: Designing Efficient, Expressive Structures*. New York, NY: John Wiley & Sons, 2009.
17. J. C. Maxwell, "On reciprocal figures and diagrams of forces," *Philosophical Magazine and Journal Series*, vol. 4, no. 27, pp. 250–261, 1864.
18. J. Lee, T. Van Mele, and P. Block, "Disjointed force polyhedra," *Computer-Aided Design*, vol. 99, pp. 11–28, June 2018.
19. M. Konstantatou, *Geometry-Based Structural Analysis and Design via Discrete Stress Functions*. PhD thesis, University of Cambridge, 2020.
20. A. McRobie, M. Konstantatou, G. Athanasopoulos, and L. Hannigan, "Graphic kinematics, visual virtual work and elastographics," *Royal Society Open Science*, vol. 4, no. 5, p. 170202, 2017.
21. W. S. Wolfe, *Graphical Analysis: A Text Book on Graphic Statics*. New York, NY: McGraw-Hill Book Company, Inc., 1921.
22. F. Neumeyer, *Friedrich Gilly. Essays zur Architektur 1796–1799*. Berlin: Ernst & Sohn, 1997.
23. Die Architektur der Spongiosa. (Zehnter Beitrag zur Mechanik des menschlichen Knochgerüstes) (pp. 615–628, 1 lith. Tafl.). *Archives of Anatomy and Physiology*, 1867/5. – Leipzig, Veit et Comp., 1867, 8°, pp. 547–658, 3 gefalt. lith. Taf., orig. Broschur. First Edition! "In 1867 Georg Hermann von Meyer (1815–1892) published a paper on the architecture of spongy bone which led, indirectly, to Julius Wolff's historic studies of the relation between form and function." – David LeYey, *The History of Orthopaedics*, p. 299 [Attributes: First Edition; Soft Cover]
24. R. Hooke, *A Description of Helioscopes, and Some Other Instruments*. London: John Martyn, 1675.
25. P. Block, M. DeJong, and J. Ochsendorf, "As hangs the flexible line: Equilibrium of masonry arches," *Nexus Network Journal*, vol. 8, no. 2, pp. 13–24, 2006.
26. F. Otto and B. Rasch, *Finding Form: Towards an Architecture of the Minimal*. Edition Axel Menges, 1996.
27. D. P. Billington, *The Art of Structural Design: A Swiss Legacy*. New Haven, CT: Yale University Press, 2003.
28. G. R. Collins, "Antonio Gaudi: Structure and form," *Perspecta 8: The Yale Architectural Journal*, pp. 63–90, 1963.
29. B. Addis, "History of using models to inform the design and construction of structures," in *Proceedings of IASS Symposium 2004, Shell and Spatial Structures: From Models to Realization* (Montpelier, France), 2004.
30. S. Stevin, *Der Beghinselen der Weeghconst van Christoffel Platijn*. Amsterdam: Françoys van Raphelinghen, 1586.
31. P. Varignon, *Nouvelle mecanique ou statique*. Paris: at Claude Jombert, 1725.
32. J. V. Poncelet, *Traitè des propriètès projectives des figures*. Paris: Wiley, 1822.
33. K. Chatzis, "La réception de la statique graphique en france durant le dernier tiers du xixe siècle," *Revue d'histoire des mathématiques*, vol. 10, 2004.
34. K. E. Kurrer, *The History of the Theory of Structures: From Arch Analysis to Computational Mechanics*. Berlin: Ernst & Sohn, 2008.
35. G. Lamè and E. Clapeyron, "Mèmoire sur la construction des polygones funiculaires," *Journal du génie civil*, vol. I, 1828.
36. M. Rankine, *A Manual of Applied Mechanics*. London: R. Griffin, 1858.
37. J. C. Maxwell, "On reciprocal figures, frames and diagrams of forces," *Transactions of the Royal Society of Edinburgh*, vol. 26, no. 1, pp. 1–40, 1870.

38. A. F. Möbius, "Kann von zwei dreiseitigen Pyramiden eine jede in Bezug auf die andere um- und eingeschrieben zugleich heiβen?" *Journal für die reine und angewandte Mathematik*, vol. 1828, no. 3, pp. 273–278, 1828.
39. K. Culmann, *Die Graphische Statik*. Zürich: Verlag Meyer & Zeller, 1864.
40. E. Scholz, *Symmetrie, Gruppe, Dualität. Zur Beziehung zwischen theoretischer Mathematik und Anwendungen in Kristallographie und Baustatik des 19. Jahrhunderts*. Basel: Birkhäuser, 1989.
41. L. Cremona, *Graphical Statics: Two Treatises on the Graphical Calculus and Reciprocal Figures in Graphical Statics*. Oxford: Clarendon Press, 1890.
42. R. H. Bow, *Economics of Construction in Relation to Framed Structures*. London: Spon, 1873.
43. W. Zalewski and E. Allen, *Shaping Structures: Statics*. New York: Wiley, 1998.
44. A. Kilian and J. Ochsendorf, "Particle-spring systems for structural form finding," *Journal of the International Association for Shell and Spatial Structures*, vol. 46, no. 2, pp. 77–85, 2005.
45. A. Kilian, "Cadenary tool v.1 [computer software]." www.designexplorer.net/project pages/cadenary.html, 2004 (Accessed March 1, 2009).
46. D. Piker, "Kangaroo: Form finding with computational physics," *Architectural Design*, vol. 83, no. 2, pp. 136–7, 2013.
47. K. Linkwitz and H. J. Schek, "Einige Bemerkung von vorsgepannten Seilnetzkonstruktionen," *Ingenieur-Archiv*, vol. 40, pp. 145–158, 1971.
48. H. J. Schek, "The force density method for form finding and computation of general networks," *Computer Methods in Applied Mechanics and Engineering*, vol. 3, no. 1, pp. 115–134, 1974.
49. M. R. Barnes, "Form finding and analysis of tension structures by dynamic relaxation," *International Journal of Space Structures*, vol. 14, no. 2, pp. 89–104, 1999.
50. A. Föppl, *Das Fachwerk im Raume*. Leipzig: Verlag von B.G. Teubner, 1892.
51. M. J. Schrems and T. Kotnik, "Statically motivated form-finding based on extended graphical statics (EGS)," in *Open Systems: Proceedings of the 18th International Conference on Computer-Aided Architectural Design Research in Asia (CAADRIA 2013)* (R. Stouffs, P. Janssen, S. Roudavski, and B. Tuner, eds.), National University of Singapore, 2013.
52. P. D'acunto, J. P. Jasienski, P. O. Ohlbrock, C. Fivet, J. Schwartz, and D. Zastavni, "Vector-based 3D graphic statics: A framework for the design of spatial structures based on the relation between form and forces," *International Journal of Solids and Structures*, vol. 167, pp. 58–70, 2019.
53. A. McRobie, "Maxwell and Rankine reciprocal diagrams via Minkowski sums for 2D and 3D trusses under load," *International Journal of Space Structures*, vol. 31, pp. 115–134, 2016.
54. J. Heyman, *The Stone Skeleton: Structural Engineering of Masonry Architecture*. Cambridge: Cambridge University Press, 1997.
55. L. L. Beghini, J. Carrion, B. A., A. Mazurek, and W. F. Baker, "Structural optimization using graphic statics," *Structural and Multidisciplinary Optimization*, vol. 49, no. 3, pp. 351–366, 2013.
56. C. Fivet and D. Zastavni, "Constraint-based graphic statics: New paradigms of computer-aided structural equilibrium design," *Journal of the International Association of Shell and Spatial Structures*, vol. 54, no. 4, pp. 271–280, 2013.

57. W. F. Baker, L. L. Beghini, A. Mazurek, J. Carrion, and A. Beghini, "Maxwell's reciprocal diagrams and discrete michell frames," *Structural Multidisciplinary Optimization*, vol. 48, pp. 267–277, 2013.
58. T. V. Mele and P. Block, "Algebraic graph statics," *Computer-Aided Design*, vol. 53, pp. 104–116, 2014.
59. S. Adriaenssens, P. Block, D. Veenendaal, and C. Williams, eds., *Shell Structures for Architecture: Form Finding and Optimization*. London: Taylor & Francis – Routledge, 2014.
60. E. Morsch, "Der eisenbetonbique, seine theorie undanwendung," *ReinforcedConcrete, Theory and Application), Verlag Konrad Wittwer, Stuttgart*, 1912.
61. T. J. Hughes, *The Finite Element Method: Linear Static and Dynamic Finite Element Analysis*. Lowell, MA: Courier Corporation, 2012.
62. J. Schlaich, K. Schäfer, and M. Jennewein, "Toward a consistent design of structural concrete," *PCI Journal*, vol. 32, no. 3, pp. 74–150, 1987.
63. A. Muttoni, M. Fernández Ruiz, and F. Niketic, "Design versus assessment of concrete structures using stress fields and strut-and-tie models," *ACI Structural Journal*, vol. 112, no. S49, pp. 605–616, 2015.
64. C. Meléndez, P. F. Miguel, and L. Pallarés, "A simplified approach for the ultimate limit state analysis of three-dimensional reinforced concrete elements," *Engineering Structures*, vol. 123, pp. 330–340, 2016.
65. G. Meschke, B. Pichler, and J. G. Rots (eds.), *Computational Modelling of Concrete Structures: Proceedings of the Conference on Computational Modelling of Concrete and Concrete Structures (EURO-C 2018), February 26–March 1, 2018, Bad Hofgastein, Austria*. London: CRC Press, 2018.
66. M. P. Bendsøe, *Topology Optimization*. Berlin, Germany: Springer, 2009.
67. M. P. Bendsøe and O. Sigmund, "Material interpolation schemes in topology optimization," *Archive of Applied Mechanics*, vol. 69, no. 9–10, pp. 635–654, 1999.
68. H. A. Eschenauer and N. Olhoff, "Topology optimization of continuum structures: A review," *Applied Mechanics Reviews*, vol. 54, no. 4, pp. 331–390, 2001.
69. Q. Q. Liang, Y. M. Xie, and G. P. Steven, "Topology optimization of strut-and-tie models in reinforced concrete structures using an evolutionary procedure," *ACI, Structural Journal*, vol. 97, no. 2, pp. 322–332, 2000.
70. M. A. Ali and R. N. White, "Automatic generation of truss model for optimal design of reinforced concrete structures," *Structural Journal*, vol. 98, no. 4, pp. 431–442, 2001.
71. H. G. Kwak and S. H. Noh, "Determination of strut-and-tie models using evolutionary structural optimization," *Engineering Structures*, vol. 28, no. 10, pp. 1440–1449, 2006.
72. M. Bruggi, "Generating strut-and-tie patterns for reinforced concrete structures using topology optimization," *Computers and Structures*, vol. 87, no. 23–24, pp. 1483–1495, 2009.
73. M. Victoria, O. M. Querin, and P. Martí, "Generation of strut-and-tie models by topology design using different material properties in tension and compression," *Structural and Multidisciplinary Optimization*, vol. 44, no. 2, pp. 247–258, 2011.
74. Y. Yang, C. D. Moen, and J. K. Guest, "Three-dimensional force flow paths and reinforcement design in concrete via stress-dependent truss-continuum topology optimization," *Journal of Engineering Mechanics*, vol. 141, no. 1, p. 04014106, 2014.
75. M. Akbarzadeh, T. Van Mele, and P. Block, "On the equilibrium of funicular polyhedral frames and convex polyhedral force diagrams," *Computer-Aided Design*, vol. 63, pp. 118–128, 2015.

76. M. Bolhassani, A. Tabatabaei Ghomi, A. Nejur, and M. Akbarzadeh, "Structural behavior of a cast-in-place funicular polyhedral concrete: Applied 3D graphic statics," in *Proceedings of the IASS Symposium 2018, Creativity in Structural Design*, vol. 2018 (MIT, Boston, MA), pp. 1–8, International Association for Shell and Spatial Structures (IASS), July 16–20, 2018.
77. M. Akbarzadeh, M. Mahnia, R. Taherian, and A. H. Tabrizi, "Prefab, concrete polyhedral frame: Materializing 3D graphic statics," in *Proceedings of the IASS Annual Symposium 2017, "Interfaces: Architecture. Engineering. Science"* (M. G. Annette Bögle, ed.), (Hamburg, Germany), September 2017.
78. M. Bolhassani, M. Akbarzadeh, M. Mahnia, and R. Taherian, "On structural behavior of the first funicular polyhedral frame designed by 3D graphic statics," *Journal of Building Engineering*, vol. 14, pp. 56–68 2017.
79. M. Akbarzadeh, M. Bolhassani, A. Nejur, J. R. Yost, C. Byrnes, J. Schneider, U. Knaack, and C. Borg Costanzi, "The design of an ultra-transparent funicular glass structure," in *Structures Congress 2019*, (Orlando, FL), April 24–27, 2019.
80. J. R. Yost, M. Bolhassani, P. A. Chhadeh, L. Ryan, J. Schneider, and M. Akbarzadeh, "Mechanical performance of polyhedral hollow glass units under compression," *Engineering Structures*, vol. 254, p. 113730, 2022.
81. Y. Lu, A. Seyedahmadian, P. A. Chhadeh, M. Cregan, M. Bolhassani, J. Schneider, J. R. Yost, G. Brennan, and M. Akbarzadeh, "Funicular glass bridge prototype: Design optimization, fabrication, and assembly challenges," *Glass Structures and Engineering*, vol. 7, pp. 319–330, June 2022.
82. A. Tabatabaei Ghomi, M. Bolhassani, A. Nejur, and M. Akbarzadeh, "Effect of subdivision of force diagrams on the local buckling, load-path and material use of founded forms," in *Proceedings of the IASS Symposium 2018, Creativity in Structural Design* (MIT, Boston, MA), July 16–20, 2018.
83. M. Akbari, M. Bolhassani, and M. Akbarzadeh, "From polyhedral to minimal surface funicular spatial structures," in *Proceedings of IASS Symposium 2019 and Structural Membranes 2019, FORM and FORCE* (Barcelona, Spain), October 7–10, 2019.
84. M. Akbari, A. Mirabolghasemi, M. Bolhassani, A. Akbarzadeh, and M. Akbarzadeh, "Strut-based cellular to shellular funicular materials," *Advanced Functional Materials*, vol. 32, no. 14, p. 2109725, 2022.
85. H. Zheng, H. Mofatteh, M. Hablicsek, A. Akbarzadeh, and M. Akbarzadeh, "Dragonfly-inspired wing design enabled by machine learning and Maxwell's reciprocal diagrams," *Advanced Science*, vol. 10, no. 18, p. 2207635, 2023.
86. Y. Lu, T. Alsalem, and M. Akbarzadeh, "A method for designing multi-layer sheet-based lightweight funicular structures," *Journal of the International Association for Shell and Spatial Structures*, vol. 63, p. 252–262, December 2022.
87. D. S. Richeson, *Euler's Gem: The Polyhedron Formula and the Birth of Topology*. Princeton, NJ: Princeton University Press, 2010.
88. E. Kreyszig, *Advanced Engineering Mathematics*, 10th edition, Wiley, 2011.
89. G. G. Stokes, *Mathematical and Physical Papers*. Cambridge: Cambridge University Press, 1905.
90. E. D. Demaine and J. O'Rourke, "Reconstruction of polyhedra," in *Geometric Folding Algorithms: Linkages, Origami, Polyhedra*, pp. 339–357. Cambridge: Cambridge University Press, 2007.

91. C. B. Barber, D. P. Dobkin, and H. Huhdanpaa, "The quickhull algorithm for convex hulls," *ACM Transactions on Mathematical Software (TOMS)*, vol. 22, no. 4, pp. 469–483, 1996.
92. B. K. P. Horn, "Extended Gaussian images," *Proceedings of the IEEE*, vol. 72, no. 2, pp. 1671–1686, 1984.
93. J. Roach and J. Wright, "Spherical dual images: A 3D representation method for solid objects that combines dual space and gaussian spheres," in *Proceedings of the 1986 IEEE International Conference on Robotics and Automation*, vol. 3, pp. 1087–1092, April 1986.
94. S. Moni, "A closed-form solution for the reconstruction of a convex polyhedron from its extended gaussian image," in *[1990] Proceedings. 10th International Conference on Pattern Recognition*, vol. 1, pp. 223–226, June 1990.
95. National Gallery of Art Catalog of an exhibition held at the National Gallery of Art, Mar. 29–July 12, 1998; San Francisco Museum of Art, Sep. 4–Dec. 1, 1998.
96. S. Huerta, "Structural design in the work of gaudi," *Architectural Science Review*, vol. 49, no. 4, pp. 324–339, 2006.
97. G. Poleni, *Memorie Istoriche della gran cupola del tempio vaticano*. No. 33 in Padova: Nella Stamperia del Seminario, 1748.
98. I. A. Karnovsky, *Theory of Arched Structures: Strength, Stability, Vibration*. Berlin, Germany: Springer Science & Business Media, 2011 Dec 29.
99. C. Fivet and D. Zastavni, "Robert maillart's key methods from the Salginatobel Bridge design process (1928)," *Journal of the International Association for Shell and Spatial Structures*, vol. 53, no. 1, pp. 39–47, 2012.
100. D. P. Billington, "Deck-stiffened arch bridges of robert maillart," *Journal of the Structural Division*, vol. 99, no. 7, pp. 1527–1539, 1973.
101. M. Ros, "Versuche und erfahrungen an ausgeführten eisenbeton-bauwerken in der schweiz 1924–1937. 1," *Ergänzung*, vol. 1939, pp. 1941–1942, 1938.
102. M. Akbarzadeh, T. Van Mele, and P. Block, "Compression-only form finding through finite subdivision of the external force polygon," in *Proceedings of the IASS-SLTE Symposium 2014* (O. J.B. and R. Tarczewski, eds.), (Brasilia, Brazil), International Association of Shell and Spatial Structures, p. 1, 2014.
103. M. Rippmann, L. Lachauer, and P. Block, "Interactive vault design," *International Journal of Space Structures*, vol. 27, no. 4, pp. 219–230, 2012.
104. B. G. Baumgart, "A polyhedron representation for computer vision," in *Proceedings of the May 19–22, 1975, National Computer Conference and Exposition 1975, May 19*, pp. 589–596, ACM Digital Library.
105. A. Nejur and M. Akbarzadeh, "Constrained manipulation of polyhedral systems," in *Proceedings of the IASS Symposium 2018, Creativity in Structural Design* (MIT, Boston, MA), July 16–20, 2018.
106. W. Whiteley, P. F. Ash, E. Bolker, and H. Crapo, "Convex polyhedra, dirichlet tessellations, and spider webs," in *Shaping Space: Exploring Polyhedra in Nature, Art, and the Geometrical Imagination* (M. Senechal, ed.), (New York, NY), pp. 231–251, Springer, 2013.
107. M. Akbarzadeh and A. Nejur, "PolyFrame: Structural form finding tool using 3D graphic statics." www.food4rhino.com/app/polyframe, 2017–2019.
108. E. Bompiani *et al.*, "DJ Struick, lectures on classical differential geometry," *Bulletin of the American Mathematical Society*, vol. 57, no. 2, pp. 154–155, 1951.
109. J. Ochsendorf and M. Freeman, *Guastavino Vaulting: The Art of Structural Tile*. Princeton NJ: Princeton Architectural Press, 2013.

110. D. Waite and P. Gioia. "United States patents held by the Rafael Guastavinos, father and son," *APT Bulletin: The Journal of Preservation Technology*, vol. 30, no. 4, pp. 59–156, 1999.
111. M. Mofid, R. Alizadegan, B. Ashori, and A. Yavari, "Computing the kern of a general cross-section," *Advances in Engineering Software*, vol. 32, no. 12, pp. 949–955, 2001.
112. H. B. Wilson and L. H. Turcotte, "Determining the kern for a compression member of general cross-section," *Advances in Engineering Software*, vol. 17, no. 2, pp. 113–123, 1993.
113. S. Anderson, E. Dieste, and S. Hochuli, *Eladio Dieste: Innovation in structural art.* Princeton, NJ: Princeton Architectural Press, 2004.
114. K. Van Balen and E. Verstrynge (eds), *Structural Analysis of Historical Constructions: Anamnesis, Diagnosis, Therapy, Controls: Proceedings of the 10th International Conference on Structural Analysis of Historical Constructions* (SAHC, Leuven, Belgium, 13–15 September 2016). CRC Press, 2016 Nov 3.
115. P. Cassinello, "The evolution of concrete shells; innovations by Ildefonso Sánchez del Río," *Informes de la Construcción*, vol. 65, no. 530, pp. 147–154, 2013.
116. C. V. Ortíz, D. L. Chalarca, M. G. Restrepo, and N. M. Arango. *Arquitectura moderna en Medellín 1947–1970.* Medellín: Universidad Nacional de Colombia. 2010.
117. F. Woodman, *The Architectural History of King's College Chapel: And Its Place in the Development of Late Gothic Architecture in England and France*. London: Routledge, 2023.
118. S. C. Han, J. W. Lee, and K. Kang, "A new type of low density material: Shellular," *Advanced Materials*, vol. 27, no. 37, pp. 5506–5511, 2015.
119. W. Fischer and E. Koch, "Genera of minimal balance surfaces," *Acta Crystallographica Section A*, vol. 45, no. 10, pp. 726–732, 1989.
120. M. Akbari, A. Mirabolghasemi, H. Akbarzadeh, and M. Akbarzadeh, "Geometry-based structural form-finding to design architected cellular solids," in *ACM SYMPOSIUM ON COMPUTATIONAL FABRICATION, SCF'20*, (Virtual Event, USA), Association for Computing Machinery, ACM, November 5–6 2020.
121. M. Akbari, Y. Lu, and M. Akbarzadeh, "From design to the fabrication of shellular funicular structures," in *Proceedings of the Association for Computer-Aided Design in Architecture (ACADIA)*, 2021.
122. T. Van Mele and P. Block, "Algebraic graph statics," *Computer-Aided Design*, vol. 53, pp. 104–116, 2014.
123. A. McRobie, W. Baker, T. Mitchell, and M. Konstantatou, "Mechanisms and states of self-stress of planar trusses using graphic statics, part II: Applications and extensions," *International Journal of Space Structures*, vol. 31, no. 2–4, pp. 102–111, 2016.
124. A. Marmier and K. E. Evans, "Flexibility in MOFs: Do scalar and group-theoretical counting rules work?," *Dalton Transactions*, vol. 45, no. 10, pp. 4360–4369, 2016.
125. J. Lee, C. Fivet, and C. Mueller, "Modelling with forces: Grammar-based graphic statics for diverse architectural structures," in *Modelling Behaviour: Design Modelling Symposium 2015* (pp. 491–504). Springer International Publishing, 2015.
126. J. Lee, T. V. Meled, and P. Block, "Form-finding explorations through geometric transformations and modifications of force polyhedrons," in *Proceedings of IASS Annual Symposia*, vol. 2016, pp. 1–10, International Association for Shell and Spatial Structures (IASS), 2016.

127. H. N. Timo, A. Masoud, and G. Per, "Addressing buckling of compression members using subdivision of force diagrams," in *Proceedings of the IASS Annual Symposium 2017, Interfaces: Architecture. Engineering. Science*, IASS, 2017.
128. J. Conzett and M. Mostafavi, "Structure as space: Engineering and architecture in the works of Jürg Conzett and his partners," (Mohsen Mostafavi, ed.) (Text Accompanying Photos of Jürg Conzett and Bruno Reichlin), Architectural Association, 2006.
129. D. Zastavni, "The structural design of Maillart's Chiasso Shed (1924): A graphic procedure," *Structural Engineering International*, vol. 18, no. 3, pp. 247–252, 2008.
130. M. Akbarzadeh, T. Van Mele, and P. Block, "Spatial compression-only form finding through subdivision of external force polyhedron," in *Proceedings of the International Association for Shell and Spatial Structures (IASS) Symposium* (Amsterdam), p. 1, International Association of Shell and Spatial Structures, August 2015.
131. M. Bernhard, M. Hansmeyer, and B. Dillenburger, "Volumetric modelling for 3D printed architecture," in *AAG – Advances in Architectural Geometry* (L. Hesselgren, A. Kilian, O. Sorkine Hornung, S. Malek, K.-G. Olsson, and C. J. K. Williams, eds.), (Göteborg, Sweden), pp. 392–415, Klein Publishing GmbH, 2018.
132. W. E. Lorensen and H. E. Cline, "Marching cubes: A high resolution 3D surface construction algorithm," *Proceedings of the 14th Annual Conference on Computer Graphics and Interactive Techniques – SIGGRAPH '87*, vol. 21, no. 4, pp. 163–169, 1987.
133. M. Akbarzadeh, M. Mahnia, R. Taherian, and A. H. Tabrizi, "Prefab, concrete polyhedral frame: Materializing 3D graphic statics," in *Proceedings of IASS Annual Symposia*, vol. 2017, pp. 1–10, International Association for Shell and Spatial Structures (IASS), 2017.
134. E. Simiu, *Design of Buildings for Wind: A Guide for ASCE 7-10 Standard Users and Designers of Special Structures*. New York, NY: John Wiley & Sons, 2011.
135. M. Hablicsek, M. Akbarzadeh, and Y. Guo, "Algebraic 3D graphic statics: Reciprocal constructions," *Computer-Aided Design*, vol. 108, pp. 30–41, 2019.
136. M. Akbarzadeh, M. Hablicsek, and Y. Guo, "Developing algebraic constraints for reciprocal polyhedral diagrams of 3D graphic statics," in *Proceedings of the IASS Symposium 2018, Creativity in Structural Design* (MIT, Boston, MA), July 16–20, 2018.
137. A. Nejur and M. Akbarzadeh, "Polyframe, efficient computation for 3D graphic statics," *Computer-Aided Design*, vol. 134, p. 103003, 2021.
138. V. Alic and D. Âkesson, "Bi-directional algebraic graphic statics," *Computer-Aided Design*, vol. 93, pp. 26–37, 2017.
139. R. Maia Avelino, J. Lee, T. Van Mele, and P. Block, "An interactive implementation of algebraic graphic statics for geometry-based teaching and design of structures," in *Proceedings of the International fib Symposium on the Conceptual Design of Structures 2021* (C. Fivet, P. D'Acunto, M. Fernández Ruiz, P. O. Ohlbrock, eds.), (Attisholz Areal, Switzerland), pp. 447–454, September 2021.
140. E. H. Moore, "On the reciprocal of the general algebraic matrix," *Bulletin of the American Mathematical Society*, vol. 26, no. 9, pp. 385–396, 1920.
141. R. Penrose, "A generalized inverse for matrices," *Mathematical Proceedings of the Cambridge Philosophical Society*, vol. 51, no. 3, pp. 406–413, 1955.
142. J. N. Grima, R. Gatt, and P.-S. Farrugia, "On the properties of auxetic meta-tetrachiral structures," *Physica Status Solidi (b)*, vol. 245, no. 3, pp. 511–520, 2008.
143. M. Schenk and S. D. Guest, "Geometry of miura-folded metamaterials," *Proceedings of the National Academy of Sciences*, vol. 110, no. 9, pp. 3276–3281, 2013.

144. Y. Cho, J.-H. Shin, A. Costa, T. A. Kim, V. Kunin, J. Li, S. Y. Lee, S. Yang, H. N. Han, I.-S. Choi, and D. J. Srolovitz, "Engineering the shape and structure of materials by fractal cut," *Proceedings of the National Academy of Sciences*, vol. 111, no. 49, pp. 17390–17395, 2014.
145. R. Gatt, L. Mizzi, J. I. Azzopardi, K. M. Azzopardi, D. Attard, A. Casha, J. Briffa, and J. N. Grima, "Hierarchical auxetic mechanical metamaterials," *Scientific Reports*, vol. 5, no. 1, p. 8395, 2015.
146. T. A. M. Hewage, K. L. Alderson, A. Alderson, and F. Scarpa, "Double-negative mechanical metamaterials displaying simultaneous negative stiffness and negative poisson's ratio properties," *Advanced Materials*, vol. 28, no. 46, pp. 10323–10332, 2016.
147. R. M. Neville, F. Scarpa, and A. Pirrera, "Shape morphing Kirigami mechanical metamaterials," *Scientific Reports*, vol. 6, no. 1, p. 31067, 2016.
148. W. Wu, D. Qi, H. Liao, G. Qian, L. Geng, Y. Niu, and J. Liang, "Deformation mechanism of innovative 3D chiral metamaterials," *Scientific Reports*, vol. 8, no. 1, p. 12575, 2018.
149. P. Koudelka, O. Jiroušek, T. Fíla, and T. Doktor, "Compressive properties of auxetic structures produced with direct 3D printing," *Materiali in Tehnologije*, vol. 50, no. 3, pp. 311–317, 2016.
150. M. Hablicsek and M. Akbarzadeh, "Designing the geometry of auxetic materials using graphic statics," *Proceedings of the IASS Annual Symposium 2020/21 and the 7th International Conference on Spatial Structure*, 2021.
151. Y. Lu, M. Hablicsek, and M. Akbarzadeh, "Algebraic 3D graphic statics with edge and vertex constraints: A comprehensive approach to extend the solution space for polyhedral form-finding," *Computer-Aided Design*, vol. 166, p. 103620, January 2024.
152. C. R. Calladine, "Buckminster fuller's 'tensegrity' structures and Clerk Maxwell's rules for the construction of stiff frames," *International Journal of Solids and Structures*, vol. 14, no. 2, pp. 161–172, 1978.
153. J. Lee, T. Van Mele, and P. Block, "Form-finding explorations through geometric manipulations of force polyhedrons," in *Proceedings of the International Association for Shell and Spatial Structures (IASS) Symposium 2016* (Tokyo, Japan), September 2016.
154. PSL, "PolyFrame 2." www.food4rhino.com/app/polyframe-2, May 2023.
155. A. Nejur and M. Akbarzadeh, "Polyframe, efficient computation for 3D graphic statics," *Computer-Aided Design*, vol. 134, p. 103003, May 2021.
156. M. Kremer, D. Bommes, and L. Kobbelt, "Openvolumemesh – A versatile index-based data structure for 3D polytopal complexes," in *Proceedings of the 21st International Meshing Roundtable* (X. Jiao and J.-C. Weill, eds.), (Berlin, Heidelberg), pp. 531–548, Springer, 2013.
157. C. R. Harris, K. J. Millman, S. J. van der Walt, R. Gommers, P. Virtanen, D. Cournapeau, E. Wieser, J. Taylor, S. Berg, N. J. Smith, R. Kern, M. Picus, S. Hoyer, M. H. van Kerkwijk, M. Brett, A. Haldane, J. F. del Río, M. Wiebe, P. Peterson, P. Gérard-Marchant, K. Sheppard, T. Reddy, W. Weckesser, H. Abbasi, C. Gohlke, and T. E. Oliphant, "Array programming with NumPy," *Nature*, vol. 585, pp. 357–362, September 2020.
158. D. R. Reid, N. Pashine, J. M. Wozniak, H. M. Jaeger, A. J. Liu, S. R. Nagel, and J. J. de Pablo, "Auxetic metamaterials from disordered networks," *Proceedings of the National Academy of Sciences*, vol. 115, pp. E1384–E1390, February 2018.
159. D. R. Reid, N. Pashine, A. S. Bowen, S. R. Nagel, and J. J. D. Pablo, "Ideal isotropic auxetic networks from random networks," *Soft Matter*, vol. 15, no. 40, pp. 8084–8091, 2019.

160. M. Akbarzadeh and M. Hablicsek, "Algebraic 3D graphic statics: Constrained areas," *Computer-Aided Design*, vol. 141, p. 103068, 2021.
161. J. O'Rourke, "Convex polyhedra realizing given face areas," *CoRR*, abs/1101.0823, 2011.
162. M. Akbarzadeh and M. Hablicsek, "Geometric degrees of freedom and non-conventional spatial structural forms," in *Impact: Design with All Senses* (Berlin, Germany), September 23–25, 2020.
163. P. O. Ohlbrock, P. D'acunto, J. P. Jasienski, and C. Fivet, "Vector-based 3D graphic statics (part iii): Designing with combinatorial equilibrium modelling," in *Proceedings of IASS Annual Symposia 2016 Sep 26* (Vol. 2016, No. 15, pp. 1–10). International Association for Shell and Spatial Structures (IASS).
164. M. Konstantatou and A. McRobie, "3D graphic statics: Geometric construction of global equilibrium," in *Proceedings of the International Association for Shell and Spatial Structures (IASS) Symposium* (Tokyo, Japan), International Association of Shell and Spatial Structures, 2015.
165. T. Baranyai, "Analytical graphic statics," *International Journal of Space Structures*, vol. 36, no. 2, pp. 117–126, 2021.
166. J. Cai, H. Chen, Y. Li, and A. Akbarzadeh, "Lessons from nature for carbon-based nanoarchitected metamaterials," *Small Science*, vol. 2, no. 12, p. 2200039, 2022.
167. S. E. Naleway, M. M. Porter, J. McKittrick, and M. A. Meyers, "Structural design elements in biological materials: Application to bioinspiration," *Advanced Materials*, vol. 27, no. 37, pp. 5455–5476, 2015.
168. C. M. Bhadra, V. K. Truong, V. T. H. Pham, M. A. Kobaisi, G. Seniutinas, J. Y. Wang, S. Juodkazis, R. J. Crawford, and E. P. Ivanova, "Antibacterial titanium nano-patterned arrays inspired by dragonfly wings," *Scientific Reports*, vol. 5, p. 16817, 2015. https://doi.org/10.1038/srep16817.
169. P. N. Sivasankaran, T. A. Ward, E. Salami, R. Viyapuri, C. J. Fearday, and M. R. Johan, "An experimental study of elastic properties of dragonfly-like flapping wings for use in biomimetic micro air vehicles (BMAVs)," *Chinese Journal of Aeronautics*, vol. 30, no. 2, pp. 726–737, 2017. https://doi.org/10.1016/j.cja.2017.02.011.
170. A. Khaheshi, S. Gorb, and H. Rajabi, "Triple stiffness: A bioinspired strategy to combine load-bearing, durability, and impact-resistance," *Advanced Science*, vol. 8, no. 11, p. 2004338, 2021.
171. C. Göttler, K. Elflein, R. Siegwart, and M. Sitti, "Spider origami: Folding principle of jumping spider leg joints for bioinspired fluidic actuators," *Advanced Science*, vol. 8, no. 5, p. 2003890, 2021.
172. S. Donoughe, J. D. Crall, R. A. Merz, and S. A. Combes, "Resilin in dragonfly and damselfly wings and its implications for wing flexibility," *Journal of Morphology*, vol. 272, no. 12, pp. 1409–1421, 2011. https://doi.org/10.1002/jmor.10992.
173. K. Tsuyuki, S. Sudo, and J. Tani, "Morphology of insect wings and airflow produced by flapping insects," *Journal of Intelligent Material Systems and Structures*, vol. 17, pp. 743–751, 2006.
174. H. Rajabi, N. Ghoroubi, M. Malaki, A. Darvizeh, and S. Gorb, "Basal complex and basal venation of odonata wings: Structural diversity and potential role in the wing deformation," *PloS one*, vol. 11, no. 8, p. e0160610, 2016.
175. R. Wootton, "The geometry and mechanics of insect wing deformations in flight: A modelling approach," *Insects*, vol. 11, no. 7, p. 446, 2020.

176. X.-J. Li, Z.-H. Zhang, Y.-H. Liang, L.-Q. Ren, M. Jie, and Z.-G. Yang, "Antifatigue properties of dragonfly pantala flavescens wings," *Microscopy Research and Technique*, vol. 77, no. 5, pp. 356–362, 2014.
177. H. Rajabi, A. Shafiei, A. Darvizeh, J.-H. Dirks, E. Appel, and S. N. Gorb, "Effect of microstructure on the mechanical and damping behaviour of dragonfly wing veins," *Royal Society Open Science*, vol. 3, no. 2, p. 160006, 2016.
178. J. Sun and B. Bhushan, "The structure and mechanical properties of dragonfly wings and their role on flyability," *Comptes Rendus Mécanique*, vol. 340, no. 1–2, pp. 3–17, 2012.
179. S. Jongerius and D. Lentink, "Structural analysis of a dragonfly wing," *Experimental Mechanics*, vol. 50, no. 9, pp. 1323–1334, 2010.
180. A. Vyatskikh, S. Delalande, A. Kudo, X. Zhang, C. M. Portela, and J. R. Greer, "Additive manufacturing of 3D nano-architected metals," *Nature Communications*, vol. 9, no. 1, pp. 1–8, 2018.
181. X. Zhang, A. Vyatskikh, H. Gao, J. R. Greer, and X. Li, "Lightweight, flaw-tolerant, and ultrastrong nanoarchitected carbon," *Proceedings of the National Academy of Sciences*, vol. 116, no. 14, pp. 6665–6672, 2019.
182. H. Y. Sarvestani, A. Akbarzadeh, A. Mirbolghasemi, and K. Hermenean, "3D printed meta-sandwich structures: Failure mechanism, energy absorption and multi-hit capability," *Materials and Design*, vol. 160, pp. 179–193, 2018.
183. J. Shi, H. Mofatteh, A. Mirabolghasemi, G. Desharnais, and A. Akbarzadeh, "Programmable multistable perforated shellular," *Advanced Materials*, vol. 33, no. 42, p. 2102423, 2021.
184. M. C. Fernandes, J. Aizenberg, J. C. Weaver, and K. Bertoldi, "Mechanically robust lattices inspired by deep-sea glass sponges," *Nature Materials*, vol. 20, no. 2, pp. 237–241, 2021.
185. D. Gu, X. Shi, R. Poprawe, D. L. Bourell, R. Setchi, and J. Zhu, "Material-structure-performance integrated laser-metal additive manufacturing," *Science*, vol. 372, no. 6545, p. eabg1487, 2021.
186. J. Fan, L. Zhang, S. Wei, Z. Zhang, S.-K. Choi, B. Song, and Y. Shi, "A review of additive manufacturing of metamaterials and developing trends," *Materials Today*, 2021.
187. H. Mofatteh, B. Shahryari, A. Mirabolghasemi, A. Seyedkanani, R. Shirzadkhani, G. Desharnais, and A. Akbarzadeh, "Programming multistable metamaterials to discover latent functionalities," *Advanced Science*, vol. 9, no. 33, p. 2202883, 2022.
188. A. Khaheshi and H. Rajabi, "Mechanical intelligence (MI): A bioinspired concept for transforming engineering design," *Advanced Science*, vol. 9, no. 32, p. 2203783, 2022.
189. J. Xu, T. Liu, Y. Zhang, Y. Zhang, K. Wu, C. Lei, Q. Fu, and J. Fu, "Dragonfly wing-inspired architecture makes a stiff yet tough healable material," *Matter*, vol. 4, no. 7, pp. 2474–2489, 2021.
190. N. Aage, E. Andreassen, B. S. Lazarov, and O. Sigmund, "Giga-voxel computational morphogenesis for structural design," *Nature*, vol. 550, no. 7674, pp. 84–86, 2017.
191. R. Martinez-Val, E. Perez, and J. F. Palacin, "Historical evolution of air transport productivity and efficiency," in *43rd AIAA Aerospace Sciences Meeting and Exhibit*, 2005.
192. J.-S. Chen, J.-Y. Chen, and Y.-F. Chou, "On the natural frequencies and mode shapes of dragonfly wings," *Journal of Sound and Vibration*, vol. 313, no. 3, pp. 643–654, 2008. https://doi.org/10.1016/j.jsv.2007.11.056.

193. J. Wakeling and C. Ellington, "Dragonfly flight. II. Velocities, accelerations and kinematics of flapping flight," *The Journal of Experimental Biology*, vol. 200, no. 3, pp. 557–582, 1997.
194. X. Liu, C. Hefler, W. Shyy, and H. Qiu, "The importance of flapping kinematic parameters in the facilitation of the different flight modes of dragonflies," *Journal of Bionic Engineering*, vol. 18, no. 2, pp. 419–427, 2021.
195. T. S. Clawson, S. Ferrari, S. B. Fuller, and R. J. Wood, "Spiking neural network (SNN) control of a flapping insect-scale robot," in *2016 IEEE 55th Conference on Decision and Control (CDC)*, pp. 3381–3388, IEEE, 2016.
196. A. Ribeiro, A. Awruch, and H. Gomes, "An airfoil optimization technique for wind turbines," *Applied Mathematical Modelling*, vol. 36, no. 10, pp. 4898–4907, 2012.
197. J. Chen, Q. Wang, X. Pang, S. Li, and X. Guo, "Improvement of airfoil design using smooth curvature technique," *Renewable energy*, vol. 51, pp. 426–435, 2013.
198. P. A. Henne and R. D. Gregg III, "New airfoil design concept," *Journal of Aircraft*, vol. 28, no. 5, pp. 300–311, 1991.
199. H. Wen, S. Sang, C. Qiu, X. Du, X. Zhu, and Q. Shi, "A new optimization method of wind turbine airfoil performance based on Bessel equation and gabp artificial neural network," *Energy*, vol. 187, p. 116106, 2019.
200. E. Appel, L. Heepe, C.-P. Lin, and S. N. Gorb, "Ultrastructure of dragonfly wing veins: composite structure of fibrous material supplemented by resilin," *Journal of Anatomy*, vol. 227, no. 4, pp. 561–582, 2015.
201. H. Rajabi, M. Moghadami, and A. Darvizeh, "Investigation of microstructure, natural frequencies and vibration modes of dragonfly wing," *Journal of Bionic Engineering*, vol. 8, no. 2, pp. 165–173, 2011.
202. T. Deubel, S. Wanke, C. Weber, and F. Wedekind, "Modelling and manufacturing of a dragonfly wing as basis for bionic research," *International Design Conference – Design 2006*, 2006.
203. W. Ritter, "The Hennebique construction method," *Die Bauweise Hennebique, Schweizerische Bauzeitung*, vol. 33, no. 7, pp. 41–61, 1899.
204. E. Mörsch, *Der Eisenbetonbau, seine Theorie und Anwendung*. Stuttgart: Verlag von Konrad Witter, 5 ed., 1922.
205. J. Schlaich and K. Schäfer, "Design and detailing of structural concrete using strut-and-tie model," *The Structural Engineer*, vol. 69, pp. 113–125, March 19, 1991.
206. A. Muttoni, J. Schwartz, and B. Thürlimann, *Design of Concrete Structures with Stress Fields*. Berlin, Germany: Springer Science & Business Media, 1996.
207. D. Drucker, "On structural concrete and the theorems of limit analysis," *International Association for Bridge and Structural Engineering (IABSE)-Reports*, vol. 21, pp. 45–59, 1961.
208. M. Nielsen, M. Braestrup, B. Jensen, and F. Bach, "Concrete plasticity, beam shear–shear in joints–punching shear," *Special Publication, Danish Society for Structural Science and Engineering*, pp. 1–129, 1978.
209. P. Marti, "Basic tools of reinforced concrete beam design," *ACI Journal*, vol. 82, no. 4, pp. 46–56, 1985.
210. V. Sigrist, M. Alvarez, and W. Kaufmann, "Shear and flexure in structural concrete beams," in *Comité Euro-International du Béton (CEB) Bulletin d'information No. 223, "Ultimate limit state design models", A state-of-art report*, 1995.

211. K. H. Reineck, M. S. Lourenço, J. F. Almeida, and S. A. Haugerud, "Paper 16: Gaining experience with strut-and-tie models for the design of concrete structures," *Design Examples for Strut-and-tie Models: Technical Report No. 61*, p. 197, 2011.
212. K. Reineck and L. Novak, *Further Examples for the Design of Structural Concrete with Strut-and-Tie Models* (No Title), 2010.
213. AASHTO-LRFD, *Bridge Design Specifications (7th ed.)*. American Association of State Highway and Transportation Officials, 2014.
214. Standard AA, *Building code requirements for structural concrete (ACI 318-11)*. American Concrete Institute, 2011 Aug.
215. CSA, *Canadian Standards Association: Design of Concrete Structures*, 2004.
216. EuroCode, *Eurocode 2: Design of Concrete Structures – Part 1-1: General Rules and rules for buildings (EN 1992-1-1:2004)*. European Committee for Standardization, Brussels, 2004.
217. CEB-FIP, *fib Model Code*. International Federation for Structural Concrete (fib), Lausanne, Switzerland, 2010.
218. C. Fivet and D. Zastavni, "A fully geometric approach for interactive constraint-based structural equilibrium design," *Computer-Aided Design*, vol. 61, pp. 42–57, 2015.
219. V. Alic and K. Persson, "Generating initial reinforcement layouts using graphic statics," in *Proceedings of the IASS Symposium, Creativity in Structural Design* (Caitlin Mueller, Sigrid Adriaenssens, eds.), (MIT, Boston), July 16–20, 2018.
220. U. Kirsch, *Structural Optimization: Fundamentals and Applications*. Berlin, Heidelberg: Springer Science and Business Media, 2012.
221. T. Zegard and G. Paulino, "Grand – Ground structure based topology optimization for arbitrary 2D domains using matlab," *Structural and Multidisciplinary Optimization*, vol. 50, no. 5, pp. 861–882, 2014.
222. S. Mozaffari, M. Akbarzadeh, and T. Vogel, "Graphic statics in a continuum: Strut-and-tie models for reinforced concrete," *Computers and Structures*, vol. 240, p. 106335, 2020.
223. S. Mozaffari, *Computational Strut-and-Tie Modeling: Explorations of Algebraic Graphic Statics and Layout Optimization*. PhD thesis, ETH Zurich, 2021.
224. W. Kaufmann, "Advanced structural concrete lecture notes," 2019. [Accessed January 30, 2020].
225. T. Teng, Y. Zhi, K.-H. Yu, S. Yang, and M. Akbarzadeh, "Continuous multi-filament 3D printing for tension-compression structure components," in *Proceedings of IASS 2023 Symposium Integration of Design and Fabrication* (Melbourne, Australia), July 10–14, 2023.
226. M. Bi, L. Xia, P. Tran, Z. Li, Q. Wan, L. Wang, W. Shen, G. Ma, and Y. Xie, "Continuous contour-zigzag hybrid toolpath for large format additive manufacturing," *Additive Manufacturing*, vol. 55, p. 102822, 2022.
227. A. Mirabolghasemi, A. Akbarzadeh, D. Rodrigue, and D. Therriault, "Thermal conductivity of architected cellular metamaterials," *Acta Materialia*, vol. 174, pp. 61–80, 2019.
228. J. Shi and A. Akbarzadeh, "Architected cellular piezoelectric metamaterials: Thermo-electro-mechanical properties," *Acta Materialia*, vol. 163, pp. 91–121, 2019.
229. H. Chen, J. Shi, and A. Akbarzadeh, "Curved architected triboelectric metamaterials: Auxeticity-enabled enhanced figure-of-merit," *Advanced Functional Materials*, vol. 33, no. 49, p. 2306022, 2023.

230. T. A. Schaedler and W. B. Carter, “Architected cellular materials,” *Annual Review of Materials Research*, vol. 46, pp. 187–210, 2016.

231. S. S. Hoseini, A. Seyedkanani, G. Najafi, A. P. Sasmito, and A. Akbarzadeh, “Multi-scale architected porous materials for renewable energy conversion and storage,” *Energy Storage Materials*, vol. 59, p. 102768, 2023.

232. M. Munsch, “Laser additive manufacturing of customized prosthetics and implants for biomedical applications,” in *Laser additive manufacturing* (pp. 399–420). Woodhead Publishing, 2017.

233. L. J. Gibson and M. F. Ashby, *Cellular Solids: Structure and Properties*. Cambridge, UK: Cambridge University Press, 1999.

234. G. Savio, S. Rosso, R. Meneghello, and G. Concheri, “Geometric modeling of cellular materials for additive manufacturing in biomedical field: A review,” *Applied Bionics and Biomechanics*, vol. 2018, p. 1654782, 2018.

235. J. R. Greer and V. S. Deshpande, “Three-dimensional architected materials and structures: Design, fabrication, and mechanical behavior,” *MRS Bulletin*, vol. 44, no. 10, pp. 750–757, 2019.

236. “Granta Design Ltd, CES selector.” Computer software, 2005.

237. T. A. Schaedler, A. J. Jacobsen, A. Torrents, A. E. Sorensen, J. Lian, J. R. Greer, L. Valdevit, and W. B. Carter, “Ultralight metallic microlattices,” *Science*, vol. 334, no. 6058, pp. 962–965, 2011.

238. V. Deshpande, M. Ashby, and N. Fleck, “Foam topology: Bending versus stretching dominated architectures,” *Acta Materialia*, vol. 49, no. 6, pp. 1035–1040, 2001.

239. S. C. Han and K. Kang, “Another stretching-dominated micro-architectured material, shellular,” *Materials Today*, vol. 31, pp. 31–38, 2019.

240. J. Hong, T. Li, H. Zheng, and Y. Ma, “Applications of structural efficiency assessment method on structural-mechanical characteristics integrated design in aero-engines,” *Chinese Journal of Aeronautics*, vol. 33, no. 4, pp. 1260–1271, 2020.

241. M. G. Lee, J. W. Lee, S. C. Han, and K. Kang, “Mechanical analyses of “shellular”, an ultralow-density material,” *Acta Materialia*, vol. 103, pp. 595–607, 2016.

242. M. F. Ashby, “The properties of foams and lattices,” *Philosophical Transactions of the Royal Society A: Mathematical, Physical and Engineering Sciences*, vol. 364, no. 1838, pp. 15–30, 2006.

243. H. Niknam, A. Akbarzadeh, D. Rodrigue, and D. Therriault, “Architected multi-directional functionally graded cellular plates,” *Materials and Design*, vol. 148, pp. 188–202, 2018.

244. H. Niknam and A. Akbarzadeh, “Thermo-mechanical bending of architected functionally graded cellular beams,” *Composites Part B: Engineering*, vol. 174, p. 107060, 2019.

245. M. Eynbeygui, J. Arghavani, A. Akbarzadeh, and R. Naghdabadi, “Anisotropic elastic-plastic behavior of architected pyramidal lattice materials,” *Acta Materialia*, vol. 183, pp. 118–136, 2020.

246. M. Kutz, ed., *Mechanical Engineers' Handbook*. Hoboken, NJ: Wiley, 3rd ed., 2006.

247. J. N. Reddy, *An Introduction to Nonlinear Finite Element Analysis: With Applications to Heat Transfer, Fluid Mechanics, and Solid Mechanics*. Oxford, UK: Oxford University Press, 2014.

248. Z. Hsain and J. H. Pikul, “Low-energy room-temperature healing of cellular metals,” *Advanced Functional Materials*, vol. 29, no. 43, p. 1905631, 2019.

249. Z. Hsain, M. Akbari, A. Prasanna, Z. Jiang, M. Akbarzadeh, and J. H. Pikul, "Electrochemical healing of fractured metals," *Advanced Materials*, vol. 35, no. 24, pp. 1–10, 2023.
250. B. Blaiszik, S. Kramer, S. Olugebefola, J. Moore, N. Sottos, and S. White, "Self-healing polymers and polymer composites," *Annual Review of Materials Research*, vol. 40, pp. 179–211, 2010.
251. F. Herbst, D. Döhler, P. Michael, and W. H. Binder, "Self-healing polymers via supramolecular forces," *Macromolecular Rapid Communications*, vol. 34, no. 3, pp. 203–220, 2013.
252. S. R. White, N. R. Sottos, P. H. Geubelle, J. S. Moore, M. R. Kessler, S. R. Sriram, E. N. Brown, and S. Viswanathan, "Autonomic healing of polymer composites," *Nature*, vol. 409, no. 6822, pp. 794–797, 2001.
253. N. van Dijk and S. van der Zwaag, "Self-healing phenomena in metals," *Advanced Materials Interfaces*, vol. 5, no. 17, pp. 1–13, 2018.
254. M. Inman and L. Barr, "Diffusion of antimony in copper single crystals," *Acta Metallurgica*, vol. 8, no. 2, pp. 112–116, 1960.
255. Z. Hsain, *Room-Temperature Electrochemical Healing of Structural Metals*. PhD thesis, University of Pennsylvania, 2022.
256. J. Dini and H. Johnson, "Joining by electroplating," *Metals Engineering Quarterly*, vol. 14, no. 1, pp. 6–12, 1974.
257. J. Dini and H. Johnson, "Preparation of metals for plating: Ring shear tests," *Surface Technology*, vol. 5, no. 5, pp. 405–417, 1977.
258. J. W. Dini and H. R. Johnson, "Plating on some difficult-to-plate metals and alloys," Tech. Rep., Sandia Laboratories, Livermore, CA, 1980.
259. H. R. Johnson and J. W. Dini, "Beryllium windows joined by electroplating," *Review of Scientific Instruments*, vol. 46, no. 1, pp. 109–110, 1975.
260. W. F. Beach, "Xylylene polymers," 2004.
261. Z. Hsain, Z. Jiang, and J. H. Pikul, "Enabling effective electrochemical healing of structural steel," *Multifunctional Materials*, vol. 4, no. 2, p. 024004, 2021.
262. K. Daehn, R. Basuhi, J. Gregory, M. Berlinger, V. Somjit, and E. A. Olivetti, "Innovations to decarbonize materials industries," *Nature Reviews Materials*, vol. 7, no. 4, pp. 275–294, 2022.
263. S. Pellegrino, and C. R. Calladine, "Matrix analysis of statically and kinematically indeterminate frameworks," *International Journal of Solids and Structures*, vol. 22, no. 4, pp. 409–428, 1986.
264. M. Schenk, S. D. Guest, et al., "Origami folding: A structural engineering approach," *Origami*, vol. 5, pp. 291–304, 2011.
265. E. T. Filipov, K. Liu, T. Tachi, M. Schenk, and G. H. Paulino, "Bar and hinge models for scalable analysis of origami," *International Journal of Solids and Structures*, vol. 124, pp. 26–45, 2017. www.sciencedirect.com/science/article/pii/S0020768317302408
266. T. Zhang, K. Kawaguchi, and M. Wu, "A folding analysis method for origami based on the frame with kinematic indeterminacy," *International Journal of Mechanical Sciences*, vol. 146–147, pp. 234–248, 2018. www.sciencedirect.com/science/article/pii/S0020740318315376
267. P. Virtanen, et al., "SciPy 1.0: Fundamental algorithms for scientific computing in Python," *Nature Methods*, vol. 17, pp. 261–272, 2020.
268. D. L. López, D. Veenendaal, M. Akbarzadeh, and P. Block, "Prototype of an ultra-thin, concrete vaulted floor system," in *Proceedings of IASS Annual Symposia 2014* (Vol.

2014, No. 15, pp. 1–8). International Association for Shell and Spatial Structures (IASS), Brasilia, Brazil.

269. W. Hawkins, J. Orr, T. Ibell, and P. Shepherd, "A design methodology to reduce the embodied carbon of concrete buildings using thin-shell floors," *Engineering Structures*, vol. 207, p. 110195, 2020.
270. H. A. Schwarz, *Gesammelte Mathematische Abhandlungen*. New York: Chelsea Publishing Company, 1933.
271. E. R. Neovius, "Bestimmung zweier spezieller periodischer minimal flachen," *Akad. Abhandlungen*, Helsinki: J.C. Frenkel & Sohn, 1883.
272. H. Karcher and K. Polthier, "Construction of triply periodic minimal surfaces," *Philosophical Transactions of the Royal Society of London. Series A: Mathematical, Physical and Engineering Sciences*, vol. 354, no. 1715, pp. 2077–2104, 1996.
273. Y. Wu, J. Fang, C. Wu, C. Li, G. Sun, and Q. Li, "Additively manufactured materials and structures: A state-of-the-art review on their mechanical characteristics and energy absorption," *International Journal of Mechanical Sciences*, vol. 246, p. 108102, 2023.
274. J. F. Blinn, "A generalization of algebraic surface drawing," *ACM Transactions on Graphics*, vol. 1, no. 3, pp. 235–256, 1982.
275. M. Bernhard, M. Bolhassani, and M. Akbarzadeh, "Performative porosity – Adaptive infills for concrete parts," in *Proceedings of the IASS Annual Symposium 2020/21* (Surrey, UK), p. 1, 2021.
276. Y. Zhi, H. Chai, T. Teng, and M. Akbarzadeh, "Local optimization of self-supporting shell structures in 3D printing: A skeleton method," in *Proceedings of IASS 2023 Symposium Integration of Design and Fabrication* (Melbourne, Australia), July 10–14, 2023.
277. K.-J. Bathe, *Finite Element Procedures*. Hoboken, NJ: Prentice Hall Pearson Education, 2006.
278. M. Bolhassani, A. A. Hamid, A. C. Lau, and F. Moon, "Simplified micro modeling of partially grouted masonry assemblages," *Construction and Building Materials*, vol. 83, pp. 159–173, 2015.
279. F. Yavartanoo and T. H.-K. Kang, "Dry-stack masonry wall modeling using finite-element method," *Journal of Structural Engineering*, vol. 148, no. 11, p. 04022176, 2022.
280. T. Gao, L. Shen, M. Shen, F. Chen, L. Liu, and L. Gao, "Analysis on differences of carbon dioxide emission from cement production and their major determinants," *Journal of Cleaner Production*, vol. 103, pp. 160–170, 2015.
281. E. I. Koytsoumpa, C. Bergins, and E. Kakaras, "The CO_2 economy: Review of CO_2 capture and reuse technologies," *The Journal of Supercritical Fluids*, vol. 132, pp. 3–16, 2018.
282. S.-C. Kou, B.-J. Zhan, and C.-S. Poon, "Use of a CO_2 curing step to improve the properties of concrete prepared with recycled aggregates," *Cement and Concrete Composites*, vol. 45, pp. 22–28, 2014.
283. G. Zanni and G. R. Pennock, "A unified graphical approach to the static analysis of axially loaded structures," *Mechanism and Machine Theory*, vol. 44, no. 12, pp. 2187–2203, 2009.
284. C. Williams and A. McRobie, "Graphic statics using discontinuous airy stress functions," *International Journal of Space Structures*, vol. 31, no. 2–4, pp. 121–134, 2016.
285. A. McRobie, "The geometry of structural equilibrium," *Royal Society Open Science*, vol. 4, no. 160759, 2017.
286. A. McRobie, "Rankine reciprocals with Zero Bars," *ResearchGate*, 2017.
287. A. McRobie, "A complete graphic statics for self-stressed 3D frames," in *Proceedings of the IASS Symposium 2018, Creativity in Structural Design* (C. Mueller and

S. Adriaenssens, eds.) (MIT, Boston, MA), International Association of Shell and Spatial Structures, July 2016.

288. A. McRobie, M. Konstantatou, G. Athanasopoulos, and L. Hannigan, "Graphic kinematics, visual virtual work and elastographics," *Royal Society Open Science*, vol. 4, no. 170202, 2017.

289. V.-J. Williot, *Notions pratiques sur la statique graphique*. Paris: E. Lacroix, 1878.

290. O. C. Mohr, "Über Geschwindigkeitspläne und Beschleunigungspläne kinematischer Kette," *Der Civilingenieur*, vol. 33, no. 631, pp. 879–893, 1887.

291. H. F. B. Müller-Breslau, *Die graphische Statik der Baukonstruktionen, Vol. 1*. Stuttgart, Germany: Alfred Kröner, 1905.

292. H. F. B. Müller-Breslau, *Die graphische Statik der Baukonstruktionen, Vol. 2*. Leipzig, Germany: Baumgärtner, 1892.

293. D. S. P. Rao, *Graphical Methods in Structural Analysis*. Hyderabad, India: Universities Press, 1997.

294. A. McRobie, "Maxwell and Rankine reciprocal diagrams via Minkowski sums for 2D and 3D trusses under load," *International Journal of Space Structures*, vol. 31, p. 203–216, 2016.

295. G. Zanni and G. Pennock, "A unified graphical approach to the static analysis of axially loaded structures," *Mechanism and Machine Theory*, vol. 44, pp. 2187–2203, 2009.

Index

For EU product safety concerns, contact us at Calle de José Abascal, 56–1°,
28003 Madrid, Spain or eugpsr@cambridge.org.

www.ingramcontent.com/pod-product-compliance
Lightning Source LLC
LaVergne TN
LVHW062200200926
848484LV00006B/1289

* 9 7 8 1 1 0 8 4 9 4 7 3 1 *